T0185443

Lecture Notes in Computer Science 11817

More information about this series at http://www.springer.com/series/7409

Weiwei Ni · Xin Wang ·
Wei Song · Yukun Li (Eds.)

Web Information Systems and Applications

16th International Conference, WISA 2019
Qingdao, China, September 20–22, 2019
Proceedings

 Springer

Editors
Weiwei Ni (iD)
Southeast University
Nanjing, China

Xin Wang (iD)
Tianjin University
Tianjin, China

Wei Song (iD)
Wuhan University
Wuhan, China

Yukun Li (iD)
Tianjin University of Technology
Tianjin, China

ISSN 0302-9743 ISSN 1611-3349 (electronic)
Lecture Notes in Computer Science
ISBN 978-3-030-30951-0 ISBN 978-3-030-30952-7 (eBook)
https://doi.org/10.1007/978-3-030-30952-7

LNCS Sublibrary: SL3 – Information Systems and Applications, incl. Internet/Web, and HCI

This Springer imprint is published by the registered company Springer Nature Switzerland AG
The registered company address is: Gewerbestrasse 11, 6330 Cham, Switzerland

Preface

It is our great pleasure to present the proceedings of the 16th Web Information Systems and Applications Conference (WISA 2019). WISA 2019 was organized by the China Computer Federation Technical Committee on Information Systems (CCF TCIS) and Qingdao University. WISA 2019 provided a premium forum for researchers, professionals, practitioners, and officers closely related to information systems and applications, to discuss the theme of future intelligent information systems with big data, focusing on difficult and critical issues, and promoting innovative technology for new application areas of information systems.

WISA 2019 was held in Qingdao, Shandong, China, during September 20–22, 2019. The theme of WISA 2019 was "Intelligent Information Systems," which focused on intelligent cities, government information systems, intelligent medical care, and information system security, and emphasized the technology used to solve the difficult and critical problems in data sharing, data governance, knowledge graph, and block chains.

This year we received 154 submissions, each of which was assigned to at least three Program Committee (PC) members to review. The peer-review process was double-blind. The thoughtful discussions on each paper by the PC resulted in the selection of 39 full research papers (an acceptance rate of 25.32%) and 33 short papers. The program of WISA 2019 included keynote speeches and topic-specific invited talks by famous experts in various areas of intelligent information systems to share their cutting edge technologies and views about the academic and industrial hotspots. The other events included industrial forums, CCF TCIS salon, and PhD forum.

We are grateful to the general chairs Prof. Ruixuan Li (Huazhong University of Science and Technology) and Prof. Fengjing Shao (Qingdao University), all PC members, and external reviewers who contributed their time and expertise to the paper reviewing process. We would like to thank all the members of the Organizing Committee, and many volunteers, for their great support in the conference organization. Especially, we would also like to thank publication chairs Prof. Wei Song (Wuhan University) and Prof. Yukun Li (Tianjin University of Technology) for their efforts in the publication of these proceedings. Many thanks to all the authors who submitted their papers to the conference.

August 2019

Weiwei Ni
Xin Wang

Organization

Steering Committee

Baowen Xu	Nanjing University, China
Ge Yu	Northeastern University, China
Xiaofeng Meng	Renmin University of China, China
Yong Qi	Xi'an Jiaotong University, China
Chunxiao Xing	Tinghua University, China
Ruixuan Li	Huazhong University of Science and Technology, China
Lizhen Xu	Southeast University, China

General Chairs

Ruixuan Li	Huazhong University of Science and Technology, China
Fengjing Shao	Qingdao university, China

Program Committee Chairs

Weiwei Ni	Southeast University, China
Xin Wang	Tianjin University, China

Local Chairs

Zhenkuan Pan	Qingdao University, China
Jianbo Li	Qingdao University, China

Publicity Chair

Lizhen Xu	Southeast University, China

Publication Chairs

Wei Song	Wuhan University, China
Yukun Li	Tianjin University of Technology, China

Website Chairs

Weifeng Zhang	Nanjing University of Post and Telecommunications, China
Jing Li	Qingdao University, China

Program Committee

Zhifeng Bao	RMIT University, Australia
Yu Cao	The University of Massachusetts Lowell, USA
Lemen Chao	Renmin University of China, China
Lin Chen	Nanjing University, China
Yanhui Ding	Shandong Normal University, China
Wenzhong Guo	Fuzhou University, China
Qinming He	Zhejiang University, China
Tieke He	Nanjing University, China
Shujuan Jiang	China University of Mining and Technology, China
Weijin Jiang	Xiangtan University, China
Wang-Chien Lee	Penn State University, USA
Bin Li	Yangzhou University, China
Chunying Li	Guangdong Polytechnic Normal University, China
Feifei Li	University of Utah, USA
Jianxin Li	Deakin University, Australia
Qingzhong Li	Shandong University, China
Ruixuan Li	Huazhong University of Science and Technology, China
Yukun Li	Tianjin University of Technology, China
Zhenxing Li	Agilecentury, China
Xuemin Lin	University of New South Wales, Australia
Chen Liu	North China University of Technology, China
Genggeng Liu	Fuzhou University, China
Huan Liu	Arizona State University, USA
Qing Liu	Renmin University of China, China
Youzhong Ma	Luoyang Normal University, China
Weiyi Meng	SUNY Binghamton University, USA
Weiwei Ni	Southeast University, China
Baoning Niu	Taiyuan University of Technology, China
Zhiyong Peng	Wuhan University, China
Yong Qi	Xi'an Jiaotong University, China
Weiguang Qu	Nanjing Normal University, China
Jiadong Ren	Yanshan University, China
Yonggong Ren	Liaoning Normal University, China
Derong Shen	Northeast University, China
Baoyan Song	Liaoning University, China
Wei Song	Wuhan University, China
Haojun Sun	Shantou University, China
Yong Tang	South China Normal University, China
Xiaoguang Wang	Virginia Tech University, USA
Xin Wang	Tianjin University, China
Xingce Wang	Beijing Normal University, China
Shengli Wu	Jiangsu University, China
Feng Xia	Dalian University of Technology, China

Contents

Machine Learning and Data Mining

Online Anomaly Detection via Incremental Tensor Decomposition 3
  Ming Gao, Yunbing Zong, Rui Li, Chengzhi An, Qiang Duan,
  Qingshan Yin, and Xiangcheng Wang

Intelligent Trader Model Based on Deep Reinforcement Learning 15
  Daoqi Han, Junyao Zhang, Yuhang Zhou, Qing Liu, and Nan Yang

BiRNN-DKT: Transfer Bi-directional LSTM RNN
for Knowledge Tracing 22
  Bin Xu, Sheng Yan, and Dan Yang

Semi-supervised Learning to Rank with Uncertain Data 28
  Xin Zhang, ZhongQi Zhao, ChengHou Liu, Chen Zhang, and Zhi Cheng

Signed Network Embedding Based on Noise Contrastive Estimation
and Deep Learning 40
  Xingong Chang, Wenqiang Shi, and Fei Zhang

Under Water Object Detection Based on Convolution Neural Network 47
  Shaoqiong Huang, Mengxing Huang, Yu Zhang, and Menglong Li

Face Photo-Sketch Synthesis Based on Conditional
Adversarial Networks 59
  Xiaoqi Xie, Huarong Xu, and Lifen Weng

Research and Implementation of Vehicle Tracking Algorithm Based
on Multi-Feature Fusion. 72
  Heng Guan, Huimin Liu, Minghe Yu, and Zhibin Zhao

An Enterprise Competitiveness Assessment Method Based
on Ensemble Learning. 79
  Yaomin Chang, Yuzheng Li, Chuan Chen, Bin Cao, and Zhenxing Li

A Method for Duplicate Record Detection Using Deep Learning. 85
  Qing Gu, Yongquan Dong, Yang Hu, and Yali Liu

Case Facts Analysis Method Based on Deep Learning. 92
  Zihuan Xu, Tieke He, Hao Lian, Jiabing Wan, and Hui Wang

The Air Quality Prediction Based on a Convolutional LSTM Network   98
  Canyang Guo, Wenzhong Guo, Chi-Hua Chen, Xin Wang,
  and Genggeng Liu

Cloud Computing and Big Data

Core Solution Computing Algorithm of Web Data Exchange   113
  Yuhang Ji, Gui Li, Zhengyu Li, Ziyang Han, and Keyan Cao

A Novel Adaptive Tuning Mechanism for Kafka-Based
Ordering Service   119
  Li Xu, Xuan Ma, and Lizhen Xu

Author Name Disambiguation in Heterogeneous Academic Networks   126
  Xiao Ma, Ranran Wang, and Yin Zhang

Is Bigger Data Better for Defect Prediction: Examining the Impact
of Data Size on Supervised and Unsupervised Defect Prediction   138
  Xinyue Liu and Yanhui Li

Using Behavior Data to Predict the Internet Addiction
of College Students   151
  Wei Peng, Xinlei Zhang, and Xin Li

Detection of Entity-Description Conflict on Duplicated Data Based
on Merkle-Tree for IIoT   163
  Yan Wang, Hui Zeng, Bingqing Yang, Zhu Zhu, and Baoyan Song

Anti-money Laundering (AML) Research: A System for Identification
and Multi-classification   169
  Yixuan Feng, Chao Li, Yun Wang, Jian Wang, Guigang Zhang,
  Chunxiao Xing, Zhenxing Li, and Zengshen Lian

Research of Query Verification Algorithm on Body Sensing Data
in Cloud Computing Environment   176
  Yanfang Gao, Lan Yao, and Jinyue Yu

Analysis of Macro Factors of Welfare Lottery Marketing Based
on Big Data   189
  Cheng Li, Hua Shao, Tiancheng Zhang, and Ge Yu

Information Retrieval

An Inflection Point Based Clustering Method for Sequence Data   201
  Ying Fan, Yilin Shi, Kai Kang, and Qingbin Xing

Efficient Large-Scale Multi-graph Similarity Search Using MapReduce   213
Jun Pang, Minghe Yu, and Yu Gu

A Subgraph Query Method Based on Adjacent Node Features
on Large-Scale Label Graphs   226
Xiaohuan Shan, Jingjiao Ma, Jianye Gao, Zixuan Xu, and Baoyan Song

Semantic Web Service Discovery Based on LDA Clustering   239
Heng Zhao, Jing Chen, and Lei Xu

Research and Implementation of Anti-occlusion Algorithm for Vehicle
Detection in Video Data.   251
Yongqi Wu, Zhichao Zhou, Lan Yao, Minghe Yu, and Yongming Yan

Temporal Dependency Mining from Multi-sensor Event Sequences
for Predictive Maintenance   257
Weiwei Cao, Chen Liu, and Yanbo Han

An Anomaly Pattern Detection Method for Sensor Data.   270
Han Li, Bin Yu, and Ting Zhao

Natural Language Processing

A Sequence-to-Sequence Text Summarization Model with Topic Based
Attention Mechanism.   285
Heng-Xi Pan, Hai Liu, and Yong Tang

Sentiment Tendency Analysis of NPC&CPPCC in German News   298
Ye Liang, Lili Xu, and Tianhao Huang

Ensemble Methods for Word Embedding Model Based on Judicial Text   309
Chunyu Xia, Tieke He, Jiabing Wan, and Hui Wang

Multilingual Short Text Classification Based on LDA
and BiLSTM-CNN Neural Network   319
Meng Xian-yan, Cui Rong-yi, Zhao Ya-hui, and Zhang Zhenguo

Data Privacy and Security

Random Sequence Coding Based Privacy Preserving Nearest Neighbor
Query Method   327
Yunfeng Zou, Shiyuan Song, Chao Xu, and Haiqi Luo

Multi-keyword Search Based on Attribute Encryption   340
Xueyan Liu, Tingting Lu, Xiaomei He, and Xiaotao Yang

Private Trajectory Data Publication for Trajectory Classification   347
    Huaijie Zhu, Xiaochun Yang, Bin Wang, Leixia Wang,
    and Wang-Chien Lee

Adaptive Authorization Access Method for Medical Cloud Data Based
on Attribute Encryption   361
    Yu Wu, Nanzhou Lin, Wei Song, Yuan Shen, Xiandi Yang, Juntao Zhang,
    and Yan Sun

Dummy-Based Trajectory Privacy Protection Against Exposure
Location Attacks   368
    Xiangyu Liu, Jinmei Chen, Xiufeng Xia, Chuanyu Zong, Rui Zhu,
    and Jiajia Li

Privacy Protection Workflow Publishing Under Differential Privacy   382
    Ning Wu, Jiaqiang Liu, Yunfeng Zou, Chao Xu, and Weiwei Ni

Knowledge Graphs and Social Networks

A Cross-Network User Identification Model Based
on Two-Phase Expansion   397
    Yue Kou, Xiang Li, Shuo Feng, Derong Shen, and Tiezheng Nie

Link Prediction Based on Node Embedding and Personalized
Time Interval in Temporal Multi-relational Network   404
    Yuxin Liu, Derong Shen, Yue Kou, and Tiezheng Nie

A Unified Relational Storage Scheme for RDF and Property Graphs   418
    Ran Zhang, Pengkai Liu, Xiefan Guo, Sizhuo Li, and Xin Wang

Construction Research and Application of Poverty Alleviation
Knowledge Graph   430
    Hongyan Yun, Ying He, Li Lin, Zhenkuan Pan, and Xiuhua Zhang

Graph Data Retrieval Algorithm for Knowledge Fragmentation   443
    Wang Jingbin and Lin Jing

A Method of Link Prediction Using Meta Path and Attribute Information . . .   449
    Zhang Yu, Li Feng, Gao Kening, and Yu Ge

TransFG: A Fine-Grained Model for Knowledge Graph Embedding   455
    Yaowei Yu, Zhuoming Xu, Yan Lv, and Jian Li

Adjustable Location Privacy-Preserving Nearest Neighbor Query Method   467
    Linfeng Xie, Zhigang Feng, Cong Ji, and Yongjin Zhu

Semi-supervised Sentiment Classification Method Based on Weibo
Social Relationship   480
   Wei Liu and Mingxin Zhang

Organization and Query Optimization of Large-Scale Product Knowledge . . .   492
   You Li, Taoyi Huang, Hao Song, and Yuming Lin

Drug Abuse Detection via Broad Learning   499
   Chao Kong, Jianye Liu, Hao Li, Ying Liu, Haibei Zhu, and Tao Liu

Semi-supervised Meta-path-based Algorithm for Community Detection
in Heterogeneous Information Networks.   506
   Limin Chen, Yan Zhang, and Liu Yang

CLMed: A Cross-lingual Knowledge Graph Framework
for Cardiovascular Diseases   512
   Ming Sheng, Han Zhang, Yong Zhang, Chao Li, Chunxiao Xing,
   Jingwen Wang, Yuyao Shao, and Fei Gao

How to Empower Disease Diagnosis in a Medical Education System
Using Knowledge Graph   518
   Samuel Ansong, Kalkidan F. Eteffa, Chao Li, Ming Sheng, Yong Zhang,
   and Chunxiao Xing

Blockchain

Blockchain Retrieval Model Based on Elastic Bloom Filter   527
   Xuan Ma, Li Xu, and Lizhen Xu

An Empirical Analysis of Supply Chain BPM Model Based
on Blockchain and IoT Integrated System   539
   Ruixue Zhao

A Trusted System Framework for Electronic Records Management
Based on Blockchain.   548
   Sixin Xue, Xu Zhao, Xin Li, Guigang Zhang, and Chunxiao Xing

A Blockchain Based Secure E-Commerce Transaction System   560
   Yun Zhang, Xiaohua Li, Jili Fan, Tiezheng Nie, and Ge Yu

Query Processing

Data-Driven Power Quality Disturbance Sources Identification Method   569
   Qi Li, Jun Fang, and Jia Sheng

Research and Implementation of a Skyline Query Method for Hidden
Web Database   575
   Zhengyu Li, Gui Li, Ziyang Han, Ping Sun, and Keyan Cao

Research on Real-Time Express Pick-Up Scheduling Based on ArcGIS
and Weighted kNN Algorithm   582
   Yi Ying, Kai Ren, and Yajun Liu

Similarity Histogram Estimation Based Top-k Similarity Join Algorithm
on High-Dimensional Data.   589
   Youzhong Ma, Ruiling Zhang, and Yongxin Zhang

Hybrid Indexes by Exploring Traditional B-Tree and Linear Regression.   601
   Wenwen Qu, Xiaoling Wang, Jingdong Li, and Xin Li

Crowdsourced Indoor Localization for Diverse Devices
with RSSI Sequences.   614
   Jing Sun, Xiaochun Yang, and Bin Wang

Grading Programs Based on Hybrid Analysis   626
   Zhikai Wang and Lei Xu

Transferring Java Comments Based on Program Static Analysis   638
   Binger Li, Feifei Li, Xinlin Huang, Xincheng He, and Lei Xu

A Monitoring Mechanism for Electric Heaters Based on Edge Computing . . .   644
   Jing Wang, Zihao Wang, and Ling Zhao

GRAMI-Based Multi-target Delay Monitoring Node Selection Algorithm
in SDN   650
   Zhi-Qi Wang, Jian-Tao Zhou, and Lin Liu

Research on Fuzzy Adaptive PID Fuzzy Rule Optimization Based
on Improved Discrete Bat Algorithm.   662
   Xuewu Du, Mingxin Zhang, and Guangtao Sha

Recommendations

A Tag-Based Group Recommendation Algorithm   677
   Wenkai Ma, Gui Li, Zhengyu Li, Ziyang Han, and Keyan Cao

A Graph Kernel Based Item Similarity Measure
for Top-N Recommendation   684
   Wei Xu, Zhuoming Xu, and Bo Zhao

Mining Core Contributors in Open-Source Projects   690
   Xiaojin Liu, Jiayang Bai, Lanfeng Liu, Hongrong Ouyang, Hang Zhou,
   and Lei Xu

A QueryRating-Based Statistical Model for Predicting Concurrent
Query Response Time   704
    Zefeng Pei, Baoning Niu, Jinwen Zhang, and Muhammad Amjad

Application of Patient Similarity in Smart Health: A Case Study
in Medical Education.   714
    Kalkidan Fekadu Eteffa, Samuel Ansong, Chao Li, Ming Sheng,
    Yong Zhang, and Chunxiao Xing

Author Index   721

Machine Learning and Data Mining

Online Anomaly Detection via Incremental Tensor Decomposition

Ming Gao[1], Yunbing Zong[2], Rui Li[3], Chengzhi An[3], Qiang Duan[3],
Qingshan Yin[3], and Xiangcheng Wang[2(✉)]

[1] Baidu Knows Business Department, Baidu, Beijing, China
[2] Department of Data and Enterprise Intelligence Products,
Inspur Group, Jinan, China
wangxiangcheng@inspur.com
[3] Inspur AI Research Institute, Inspur Group, Science and Technology Park,
No. 1036, Langchao Road, Jinan Hi-Tech Development Zone,
Jinan, People's Republic of China

Abstract. Anomaly detection, though, is a common and intensively studied data mining problem in many applications, its online (incremental) algorithm is yet to be proposed and investigated, especially with respect to the use of tensor technique. As online (incremental) learning is becoming increasingly more important, we propose a novel online anomaly detection algorithm using incremental tensor decomposition. The online approach keeps updating the model while new data arrive, in contrast to the conventional approach that requests all data to rebuild the model. In addition, the online algorithm can not only track the trend in time evolving data, but also requests less memory since only the new data is necessary for model updating. The experimental results show that the presented algorithm has sound discriminative power that is essential for anomaly detection. In addition, the number of anomalies can be flexibly adjusted by the parameters in the algorithm, which is necessary in some real-world scenarios. The effects of these parameters are also consistent using two experimental datasets.

Keywords: Tensor decomposition · Online machine learning · Anomaly detection · Incremental · Streaming

1 Introduction

Anomaly detection [3] is a classic problem existing in a variety of areas, such as fraud detection in finance. It aims at finding patterns that behave differently from the rest of the main population. Anomaly detection and outlier detection are often used interchangeably, and it does not necessarily requires labelled data. In fact, unsupervised anomaly detection is of great practical interest, since

---

M. Gao, Y. Zong and R. Li—Both authors contributed equally to this study.

© Springer Nature Switzerland AG 2019
W. Ni et al. (Eds.): WISA 2019, LNCS 11817, pp. 3–14, 2019.
https://doi.org/10.1007/978-3-030-30952-7_1

labelled data are scare to obtain in many real-world applications. We reveal several challenges in anomaly detection: (a): The definition of normality is a hard task due to the absence of labelled data; (b): The noise and trend in data lend the problem even more difficult; (c): Abnormal patterns may evolve with time.

To date, time evolving data are becoming more prevalent. At the same time, different problems, such as changing patterns, arise due to the streaming/non-stationary data. As a result, online machine learning or similarly incremental machine learning draws much attention in recent years [4]. In contrast to the conventional machine learning approach, incremental machine learning keeps updating the model with respect to the underlying characteristics of the incoming data. The classic approach trains the model using the entire data, which may result in two drawbacks, namely neglecting the time-evolving phenomenon and costing too much time. An early example of incremental machine learning is the very fast decision tree (VFDT) proposed by Domingos [5], modifying the decision tree using Hoeffding's inequality as a statistics to handle the streaming data. Massive online analysis (MOA) [2] is particularly developed for analysis of incremental data with abundant methods devised for classification, regression etc.

Commonly, machine learning and data mining algorithms deal with matrix. In some scenarios, tensor may be more suitable to represent the data, which is an extension to matrix. In the movie recommendation application, a matrix can be formed by users and items, and a tensor can be shaped by adding time as the third dimension. The relational learning via tensor approach proposed by Nickel [13] stresses that the tensor factorization can capture the essential information via the decomposed latent factors. With all information reside in a tensor, the interrelationship between different dimensions can be better captured by tensor compared to matrix. Hence, decomposing the tensor uncovers useful patterns which may not be obtained by matrix factorization.

Having mentioned the usefulness of anomaly detection, online machine learning and tensor decomposition, we propose a new anomaly detection algorithm using incremental tensor decomposition, and demonstrate the results using two datasets. This work is a contribution to the anomaly detection community by introducing the tensor decomposition in the manner of incremental learning. The unsupervised algorithm is suitable for applications where the data are multivariate and vary with time, such as data collected by sensors used in IoT scenarios.

The paper is organized as follows: The introduction is followed by related work. The third section presents the preliminary knowledge on tensor, and the incremental tensor decomposition is introduced subsequently. The experimental results on two datasets are then shown in a later section, along with results discussion. Finally, conclusions are drawn in the last section.

2   Related Work

Recently, tensor factorization has gain popularity due to its ability in finding interesting patterns in data. A survey [14] offers an overview of a range of applications based on tensor decomposition. They show that the tensor decomposition is able to extract hidden correlations and meaningful structure in the data mining field. Furthermore, a few open challenges in tensor decomposition are highlighted for future research direction. Kolda and Bader [9] study several methods of higher-order tensor decompositions and suggest relevant applications as well as available software. CANDECOMP/PARAFAC (CP) [8] and Tucker decompositions [17] are the primary focused methods. Moreover, they mention that a flourish of research focuses on more efficient and better methods and a range of applications is developed. A similar work [16] introduces an incremental tensor analysis (ITA) that efficiently computes a summary for high-order data and finds latent correlations. They propose three related methods that are dynamic, streaming and window-based, showing significant gain in performance. Another work conducted by Sun et al. [15] further shows experiments on anomaly detection and multi-way latent semantic indexing using two real-world large datasets. The result shows that the dynamic tensor analysis and streaming tensor analysis are effective and efficient. They can find interesting pattern, such as outliers. Most recently, Gujral et al. [7] introduce a sample-based batch incremental tensor decomposition algorithm (SamBasTen). This algorithm can update the existing decomposition without recomputing the entire decomposition by summarizing the existing tensor and incoming updates. Then, it updates the reduced summary space. This method can be regarded as an approach to decomposing the tensor incrementally, rather than an approach devised for anomaly detection.

A research work [6] suggests an online self-organizing map (SOM) to model the large number of behavior patterns in the crowed. The proposed online learning method has been proved to efficiently reduce the false alarms while still keep the ability to detect most of the anomalies. In 2011, Li et al. [12] presents a tensor-based incremental learning algorithm for anomaly detection in background modeling. Compared to the other vector-based methods, this algorithm is a robust tensor subspace learning algorithm that has ability to fit the variation of appearance model via adaptively updating the tensor subspace. Their experimental results show the robustness of tensor subspace-based incremental learning for anomaly detection. An unsupervised incremental learning method [18] was developed to detect maritime traffic patterns. It can automatically derive knowledge of maritime traffic in an unsupervised way without any Automatic Identification System (AIS) description. In 2014, Rikard Laxhammar and Göran Falkma [10] propose an improved online learning and sequential anomaly detection algorithm based on their previous works. The new algorithm is called Sequential Hausdorff Nearest-Neighbor Conformal Anomaly Detector (SHNN-CAD) that is a kind of parameter-light anomaly detection algorithm. It has good performance on unsupervised online learning and sequential anomaly detection in trajectories. The biggest advantage of SHNN-CAD is the ability to achieve competitive classification performance by utilizing minimum parameter tuning. A review [19] introduces the challenges, current techniques and applications with respect to the online aggregation.

3   Methods

Prior to introducing the proposed method, we first briefly reveal the related knowledge as background information.

3.1   Background Knowledge

Tensor is a natural extension to matrix, with N distinct dimensions (also known as orders, modes, ways). The data used in this study can be denoted as an order-3 tensor $T \in \mathbb{R}^{N \times P \times Q}$, where N, P, Q are the dimensions. n, p, q can take on the specific value in the N, P and Q respectively. An order-3 tensor T can be factorized (decomposed) into three components bases (factor matrices A, B, C) with a pre-defined rank R. The CANDECOMP/PARAFAC (CP) [8] decomposition[1] is employed in this work. In contrast to the tensor decomposition, principal component analysis (PCA) is a classic unsupervised dimensionality reduction technique, being applied to numerous applications such as face recognition. PCA assumes that the low-dimensional manifold is an affine space, transforming the data into another linear space. The original data $X \in \mathbb{R}^{N \times M}$ can be reduced to new data $X \in \mathbb{R}^{N \times R}$ $(R \ll M)$ by the projection matrix $U \in \mathbb{R}^{R \times M}$. Figure 1 (a) shows the basic concepts relevant to tensor, while sub-figures (b) and (c) demonstrate how the tensor data are constructed in our experiments.

$$T \approx \sum_{r=1}^{R} \lambda_r \mathbf{a}_r \circ \mathbf{b}_r \circ \mathbf{c}_r. \tag{1}$$

Equation 1 suggests that the tensor T can be computed by the outer product ("$\circ$") of $\mathbf{a}_r \in \mathbb{R}^N$, $\mathbf{b}_r \in \mathbb{R}^P$ and $\mathbf{c}_r \in \mathbb{R}^Q$. Given the rank R (dimension after decomposition), the factor matrices are in the form of $A = [\mathbf{a}_1 \ \mathbf{a}_2 \cdots \mathbf{a}_R] \in \mathbb{R}^{N \times R}$, $B = [\mathbf{b}_1 \ \mathbf{b}_2 \cdots \mathbf{b}_R] \in \mathbb{R}^{P \times R}$ and $C = [\mathbf{c}_1 \ \mathbf{c}_2 \cdots \mathbf{c}_R] \in \mathbb{R}^{Q \times R}$ respectively. Generally, the A, B, C are normalized by a scalar λ_r, representing some latent structure in the data such that they can be used to perform data mining task [14].

3.2   Incremental Tensor Decomposition

According to the work proposed by Kolda [9], the CP decomposition tries to approximate the tensor T with R components, i.e.,

$$\min_{\hat{T}} \|T - \hat{T}\| \quad \text{with} \quad \hat{T} = \sum_{r=1}^{R} \lambda_r \mathbf{a}_r \circ \mathbf{b}_r \circ \mathbf{c}_r. \tag{2}$$

The solution of Eq. 2 can be reduced to a least-squares problem by fixing all but one matrix using the alternating least squares (ALS) approach. As suggested in the work [9], $T_1^{A,BC}$ is roughly equal to $A(C \odot B)^T$. $T_1^{A,BC} \in \mathbb{R}^{N \times PQ}$ represents that the data T_1 is unfolded by concatenating the second and third

---

[1] Tensor decomposition and tensor factorization (TF) are often used interchangeably.

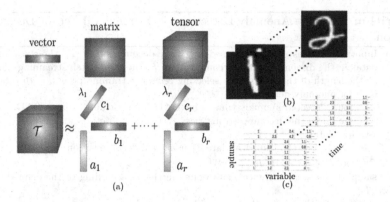

Fig. 1. (a): Illustration of vector, matrix, tensor and CP decomposition. (b): The construction of tensor data using MNIST dataset. (c): The construction of tensor data using the real-world (in-house) dataset.

mode (order) into a long matrix, while keeping the first mode intact[2]. Similarly, $\mathcal{T}_1^{B,AC} \in \mathbb{R}^{P \times NQ}$ is achieved by unfolding the first and third mode while keeping the second mode unchanged. According to the Eq. 2, the optimal solution is then boiled down to $\hat{A} = \mathcal{T}_1^{A,BC}[(C \odot B)^T]^+$. One of the core parts in our algorithm is to keep updating the matrix $\hat{A}$ (similarly $\hat{B}$ and $\hat{C}$) to perform online anomaly detection. To simplify the computation, we give the detailed following proof showing that $\hat{A}$ can be calculated straightforwardly[3].

$$\hat{A} = \mathcal{T}_1^{A,BC}[(C \odot B)^T]^+ \tag{3}$$

$$= \mathcal{T}_1^{A,BC}[((C^TC) \circ (B^TB)) * (C \odot B)^+]^+ \tag{4}$$

$$\text{due to } (C \odot B)^T(C \odot B) = (C^TC) \circ (B^TB), \text{cf. [11]} \tag{5}$$

$$\text{hence } (C \odot B)^T = ((C^TC) \circ (B^TB)) * (C \odot B)^+ \tag{6}$$

$$= \mathcal{T}_1^{A,BC}(C \odot B)^+((C^TC) \circ (B^TB)) \tag{7}$$

The detailed algorithm description and a graphical illustration are presented in Algorithm 1 and Fig. 2 respectively. It is essential to explain some key parameters in the algorithm before elaborating the approach.

1. S: mini-batch, a chunk of streaming data that are supposed to be processed.
2. T: reference interval length, every T data points are selected as reference (non-anomaly) points for similarity calculation.

---

[2] We use A, B, C on the superscripts to represent the respective modes in the tensor $\mathcal{T}$.
[3] "$\odot$" denotes the Khatri–Rao product, "$\circ$" denotes the vector outer product, "$*$" denotes the Hadamard product that is the elementwise matrix product, the superscript "$+$" denotes pseudo inverse.

Algorithm 1. Online Anomaly Detection via Incremental Tensor Decomposition

---

Data: Input: T_1: the historical data used for tensor decomposition, T_2: the streaming data. R: the dimension after decomposition. S: the mini-batch streaming data. T: the length of the interval for selecting reference training points. δ: threshold used for anomaly detection.

Result: Output: detected anomaly time point OAD_t of a certain instance OAD_i.

1 The first phase infers the decomposed matrices A, B, C, and detects all anomaly points using historical data.

2 $A, B, C \leftarrow$ decomposes the historical data T_1 using CP decomposition.

3 $\hat{A} \leftarrow T_1^{A,BC} \times (C \odot B) * (B^T B * C^T C)^+$.

4 $T_1^T \leftarrow$ sampled training (reference) data points in the T_1 according to the defined interval T.

5 **for** $i = 1 : |T_1|$, "$| \bullet |$" denotes the cardinality of a set **do**

6   **for** $j = 1 : |T_1^T|$ **do**

7      $\text{similarity}(i, j) = \sqrt{\sum(\hat{A}(i, :) - \hat{A}(T_1^T(j), :))^2}$

8   **end**

9 **end**

10 $d \in \mathbb{R}^{N \times 1} \leftarrow \min(\text{similarity}(i, :))$, the smallest distance of each point to sampled data

11 $d_{\max} \leftarrow \max(\max(\text{similarity}(i, :)))$, the greatest distance used for distance normalization

12 $d \leftarrow \frac{d}{d_{\max}}$

13 anomalies $\leftarrow$ for all the $d > \delta$

14 The second phase detects anomalies on the incoming streaming data.

15 **while** T_2 is not empty **do**

16    $i = i + 1$, new instances in T_2 go to T_1 for following similarity computation

17   **if** $(i - 1 \mod T) = 0$ **then**

18   append index i to the T_1^T

19   **end**

20    $\hat{A}(i, :) \leftarrow T_2^{A,BC} \times (C \odot B) * (B^T B * C^T C)^+$.

21   **for** $i = 1 : |T_1|$ **do**

22   **for** $j = 1 : |T_1^T|$ **do**

23        $\text{similarity}(i, j) = \sqrt{\sum(\hat{A}(i, :) - \hat{A}(T_1^T(j), :))^2}$

24   **end**

25   **end**

26    $d_{\text{score}} \leftarrow \min(\text{similarity}(i, :))$

27    $d_{\max}^s \leftarrow \max(\max(\text{similarity}(i, :)))$, the s in $d_{\max}^s$ denotes the distance in the streaming data

28    $d_{\text{score}} \leftarrow \frac{d_{\text{score}}}{d_{\max}^s}$

29 **end**

30 **for** $j = 1 : |S|$ **do**

31    $distance = \sqrt{\frac{\sum(T_2(i,j,:) - T_2(i-1,j,:))^2}{\sum T_2(i,j,:)}}$

32   **if** $distance > \delta$ **then**

33   collect anomaly time point and the instance as results

34   **end**

35 **end**

36 $T_1 \leftarrow T_1 + T_2$

37 $\hat{A} \leftarrow T_1^{A,BC} \times (C \odot B) * (B^T B * C^T C)^+$

38 $\hat{B} \leftarrow T_1^{B,AC} \times (C \odot A) * (A^T A * C^T C)^+$.

39 $\hat{C} \leftarrow T_1^{C,AB} \times (B \odot A) * (A^T A * B^T B)^+$.

40 Go to the while loop to continue the process.

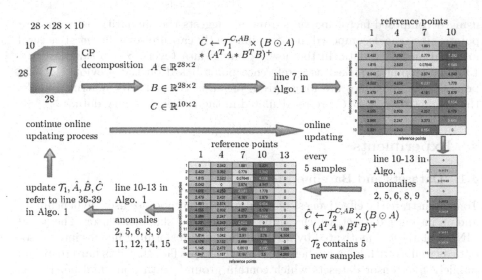

Fig. 2. An illustrative workflow of proposed online anomaly detection algorithm.

3. δ: threshold, a data point is abnormal if the similarity is greater than the threshold.
4. R: dimension after CP decomposition.
5. $\mathcal{D}_1$: the historical data (or base data) used for CP decomposition.
6. $\mathcal{D}_2$: the streaming data to be processed.

We take the MNIST dataset (cf. Sect. 4.1) as a running example (cf. Fig. 2), setting parameters $S = 5$, $T = 3$, $\delta = 0.1$, $R = 2$, with 10 examples (images) in $\mathcal{T}_1$, i.e., $\mathcal{T}_1 \in \mathbb{R}^{28 \times 28 \times 10}$. We format the tensor data by adding images on the third dimension, with an aim to detect images that are not "1". Therefore, the matrix $\hat{C}$ should be used and updated to compute the abnormal level (dissimilarity). Alternatively, we would then compute the matrix $\hat{A}$ if we form the tensor as $\mathcal{T}_1 \in \mathbb{R}^{10 \times 28 \times 28}$. After CP decomposition, we obtain the decomposed matrices A and B that can further approximate the $\hat{C}$. It is worthy mentioning that C and $\hat{C}$ are numerically very close. In the subsequent procedures, $\hat{C}$ is constantly updated so that we also apply $\hat{C}$ in the very beginning. $\mathcal{T}_1 \in \mathbb{R}^{28 \times 28 \times 10}$ is transformed into a matrix $\mathcal{T}_1^{C,AB} \in \mathbb{R}^{10 \times 784}$ along the first and second modes ($28 \times 28 = 784$), while the third mode is remained. Within the first ten samples in $\mathcal{T}_1$, sample points $1, 4, 7, 10$ are chosen as the reference (training, normal) points because the reference interval length $T = 3$. Thus, the distance between each data point to the reference point can be readily calculated by the Euclidean distance based on $\hat{A}$, cf. line 7 in Algorithm 1. Taking the smallest distance values for each line in the similarity matrix (line 10 in Algorithm 1), we receive a vector that can be directly used for anomaly judgement using the δ. The anomaly point is determined if its minimum distance to all the reference points is greater than the pre-defined threshold, which suggests that it is dissimilar to normal points and hence should be noted as anomaly. Calculation of the Euclidean distance using decomposed factor matrices directly suggests the similarity between data points, since the decomposed matrices inherently capture some latent structural information in the data. In the newly coming data (every $S = 5$ samples), the $13th$ data point is picked as a reference point based on the previous 10 base points. New anomalies are found given an updated similarity matrix. Finally, the key matrices $\hat{A}, \hat{B}, \hat{C}$ are recalculated in the presence of new data.

4   Experiments

4.1   Datasets and Baseline Methods

To demonstrate the effectiveness of the proposed method, we use one benchmark and an in-house dataset, which can be formed in the shape of tensor. The use of MNIST dataset is motivated by its feasibility in transforming images into tensor data as well as with known labels. Due to the simple fact that it is hard to obtain publicly free tensor datasets which contain ground truth abnormal labels, we, therefore, employ the in-house dataset to demonstrate the experimental results.

Introduction to MNIST Dataset: The MNIST dataset is composed of handwritten digits, containing 60,000 training images with each of dimension 28×28. It has been intensively used for image classification task, in particular for deep learning. The image data can be represented as a two-dimensional matrix, and a tensor is formed when we stack the images on top of each other. In the experimental setting, we regard digit "1" as the normal class with 6742 instances, and all the rest as the abnormal class. As we know the true labels of the data, we therefore employ the F1-Score $= 2((precision \times recall)/(precision + recall))$ to compare the results.

Introduction to In-house Dataset: Different from the MNIST data, the in-house data come from a real-world scenario that contains five same type equipments running in long period of time. Ten variables represent the equipments' running condition. Therefore, time (2039 recordings), equipments (five) and variables (ten) constitute the tensor data of size $2039 \times 5 \times 10$ as depicted in Fig. 1 (c). The anomaly can first be detected by the time. Further, we are able to know which equipment is running abnormally by calculating the distance (using variables) to the neighbouring time points. However, the dataset is in shortage of ground truth anomaly labels, although there may be some abnormal states in the data. Hence, it is a completely unsupervised anomaly detection task.

Introduction to Compared Methods: We compare the methods with different updating batch sizes, namely Online Anomaly Detection (OAD) with mini-batch of size 10, i.e., OAD-10 and so on. Regarding the MNIST, we apply the well-known k-means clustering as a baseline comparison, since both k-means and the presented approach are unsupervised algorithms. We used the Matlab tensor toolbox [1] implementation in the experiments. To the best of our knowledge, similar algorithms using tensor for online anomaly detection are very rare. Thus, k-means is the only chosen compared method.

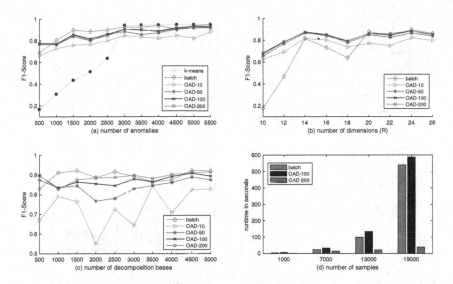

Fig. 3. MNIST dataset results. (a): The F1-Score affected by number of anomalies, $R = 20$, $T = 20$, $\delta = 0.1$. (b): The F1-Score affected by the number of dimensions (R), $T = 20$, $\delta = 0.1$, T_1 contains 5000 samples as initial base, T_2 contains 6742 anomalies and 1742 normal points. (c): The F1-Score affected by the size of initial decomposition bases, $T = 20$, $\delta = 0.1$, decomposition base T_1 and T_2 contain 13484 samples as a whole. (d): Runtime comparison.

4.2 Experimental Results

In Fig. 3 (a), the number of anomalies is randomly sampled from all digits except "1", i.e., the rest numbers are treated as abnormal ones. As the number of anomalies increases, the k-means ($k = 2$) performs gradually better which means that k-means has trouble in separating abnormal points when the two clusters are too imbalanced. The batch method, without online updating, seems to behave the best. We see better performance when the size of mini-batches increases. Figure 3 (b) reveals that we may obtain poor results if the R is set to a low value, especially for the batch method. Figure 3 (c) illustrates that the performance tend to be poor when the mini-batch size is small. In terms of the running time, the time drops dramatically when the mini-batch reaches 200. In fact, the smaller the mini-batch the longer running time.

Regarding the in-house real-world dataset, we investigate the effects of several parameters to offer an insight into parameters. From Fig. 4 (a), we see that smaller mini-batch finds more anomalies than greater ones. In Fig. 4 (b), we observe more anomalies as the interval becomes larger, since fewer data points are selected as reference (normal) points in larger interval. In Fig. 4 (c), the goodness of fit (definition is referred to [7]) is a measure of how well the decomposed model explains the underlying data, which were tested using the remaining data points after excluding the base data. Figure 4 (d) clearly shows that smaller mini-batches cost more time.

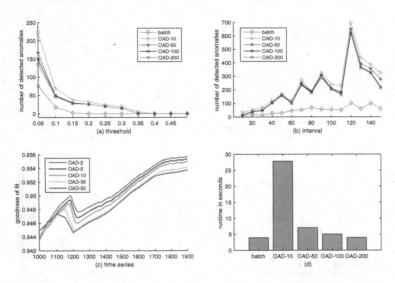

Fig. 4. In-house dataset results, $R = 5$, T_1 contains 1000 samples as initial base. (a): The number of anomalies affected by the thresholds, $T = 30$. (b): The number of anomalies affected by the interval length (c): The goodness of fit varies with the time series. (d): Runtime comparison.

In Fig. 5, we can conclude three trends, (1): The number of anomalies declines as we increase the decomposition bases. (2): The number of anomalies is inverse proportional to the value of R. (3): Small mini-bathes detect more anomalies, which is consistent with the observation in MNIST data.

4.3 Discussion of Experimental Results

From above experiments, we see that the mini-batch size S and reference interval length T play essential role in finding the number of anomalies. The introduction of T may lead to some biased results, because the selected reference points (depends on value T) can be actually anomalies whereas they are regarded as normal training data for model updating. As we frequently update the model, more anomalies can be detected. A larger S also means less computational time. The number of decomposition base (historical data for initial model) is not as critical as S and T. The parameter threshold δ is certainly an important factor influencing the number of anomalies, which can be tuned according to domain specific knowledge or by a training dataset.

From the algorithmic point of view, the algorithm can still be further studied in following aspects. A forgetting factor can be introduced to serve as a memory mechanism to remember or forget the historical data. In addition, we may consider excluding the anomaly points prior to updating the model, which would yield different results and may be beneficial in some scenarios. Furthermore, we may use more benchmark datasets to validate the approach. Regarding the

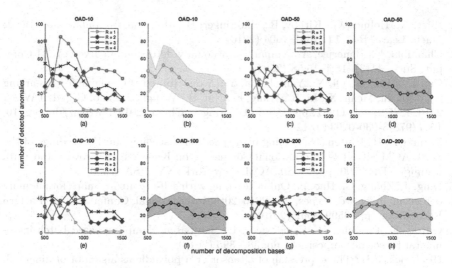

Fig. 5. The number of detected anomalies affected by the size of initial decomposition bases using various parameter settings. The green line in subplot (b) is the average value calculated from the four lines in subplot (a), and the shaded area represents the standard deviation of these lines. (Color figure online)

algorithm, extra efforts need to be paid to study the higher order (mode) incremental tensor learning, which is beyond the scope of this work.

5 Conclusion

The present study proposed a new online anomaly detection algorithm using incremental tensor decomposition. The updating process is achieved by refreshing the factor matrices, and the anomalies are detected by computing a distance based on the information captured in these matrices. Since anomaly detection is often a hard task by itself due to the fact of lacking true anomaly labels, we therefore used MNIST dataset to show the discriminative effectiveness. The other real-world in-house dataset demonstrates the effects of various parameters influencing the results. Our proposed algorithm can be a suitable approach for applications that involve large amount of data, because online algorithm does not request large memory space, whereas tensor decomposition on a large matrix can be computationally very expensive. The online algorithm can be a considered candidate for applications in which the data is hard to fit in memory and change with time, because the updating algorithm and the tunable parameters allow the model to track the changes.

References

1. Brett, W.B., Tamara, G.K., et al.: MATLAB tensor toolbox version 2.6, February 2015

2. Bifet, A., Holmes, G., Kirkby, R., Pfahringer, B.: Moa: massive online analysis. J. Mach. Learn. Res. **11**, 1601–1604 (2010)
3. Chandola, V., Banerjee, A., Kumar, V.: Anomaly detection: a survey. ACM Comput. Surv. **41**(3), 15:1–15:58 (2009)
4. Cai, D., Hou, D., Qi, Y., Yan, J., Lu, Y.: A distributed rule engine for streaming big data. In: Meng, X., Li, R., Wang, K., Niu, B., Wang, X., Zhao, G. (eds.) WISA 2018. LNCS, vol. 11242, pp. 123–130. Springer, Cham (2018). https://doi.org/10.1007/978-3-030-02934-0_12
5. Domingos, P., Hulten, G.: Mining high-speed data streams. In: Proceedings of the Sixth ACM SIGKDD International Conference on Knowledge Discovery and Data Mining, KDD 2000, pp. 71–80. ACM, New York, NY, USA (2000)
6. Feng, J., Zhang, C., Hao, P.: Online learning with self-organizing maps for anomaly detection in crowd scenes. In: 2010 20th International Conference on Pattern Recognition, pp. 3599–3602 (2010)
7. Gujral, E., Pasricha, R., Papalexakis, E.: Sambaten: Sampling-based batch incremental tensor decomposition, pp. 387–395 (2018)
8. Hitchcock, F.L.: The expression of a tensor or a polyadic as a sum of products. J. Math. Phys. **6**(1), 164–189 (1927)
9. Kolda, T.G., Bader, B.W.: Tensor decompositions and applications. SIAM Rev. **51**(3), 455–500 (2009)
10. Laxhammar, R., Falkman, G.: Online learning and sequential anomaly detection in trajectories. IEEE Trans. Pattern Anal. Mach. Intell. **36**(6), 1158–1173 (2014)
11. Lev-ari, H.: Efficient solution of linear matrix equations with application to multistatic antenna array processing. Commun. Inf. Syst. **5**, 123–130 (2005)
12. Li, J., Han, G., Wen, J., Gao, X.: Robust tensor subspace learning for anomaly detection. Int. J. Mach. Learn. Cybern. **2**(2), 89–98 (2011)
13. Nickel, M., Tresp, V.: Tensor factorization for multi-relational learning. In: Blockeel, H., Kersting, K., Nijssen, S., Železný, F. (eds.) ECML PKDD 2013. LNCS (LNAI), vol. 8190, pp. 617–621. Springer, Heidelberg (2013). https://doi.org/10.1007/978-3-642-40994-3_40
14. Papalexakis, E.E., Faloutsos, C., Sidiropoulos, N.D.: Tensors for data mining and data fusion: models, applications, and scalable algorithms. ACM Trans. Intell. Syst. Technol. **8**(2), 16:1–16:44 (2016)
15. Sun, J., Tao, D., Faloutsos, C.: Beyond streams and graphs: Dynamic tensor analysis. In: Proceedings of the 12th ACM SIGKDD International Conference on Knowledge Discovery and Data Mining, KDD 2006, pp. 374–383. ACM, New York, NY, USA (2006)
16. Sun, J., Tao, D., Papadimitriou, S., Yu, P.S., Faloutsos, C.: Incremental tensor analysis: theory and applications. ACM Trans. Knowl. Discov. Data **2**(3), 11:1–11:37 (2008)
17. Tucker, L.R.: Some mathematical notes on three-mode factor analysis. JPsychometrika **31**(3), 279–311 (1966)
18. Vespe, M., Visentini, I., Bryan, K., Braca, P.: Unsupervised learning of maritime traffic patterns for anomaly detection. In: 9th IET Data Fusion & Target Tracking Conference (DF & TT 2012): Algorithms & Applications, p. 14. IET (2012)
19. Li, Y., Wen, Y., Yuan, X.: Online aggregation: a review. In: Meng, X., Li, R., Wang, K., Niu, B., Wang, X., Zhao, G. (eds.) WISA 2018. LNCS, vol. 11242, pp. 103–114. Springer, Cham (2018). https://doi.org/10.1007/978-3-030-02934-0_10

Intelligent Trader Model Based on Deep Reinforcement Learning

Daoqi Han, Junyao Zhang, Yuhang Zhou, Qing Liu[✉],
and Nan Yang

School of Information, Renmin University of China, Beijing 100872, China
qliu@ruc.edu.cn

Abstract. The stock market has the characteristics of changing rapidly, having many interference factors, and yielding insufficient period data. Stock trading is a game process under incomplete information, and the single-objective supervised learning model is difficult to deal with such serialization decision problems. Reinforcement learning is one of the effective ways to solve these problems. This paper proposes an ISTG model (Intelligent Stock Trader and Gym) based on deep reinforcement learning, which integrates historical data, technical indicators, macroeconomic indicators, and other data types. The model describes evaluation criteria and control strategies. It processes long-period data, implements a replay model that can incrementally expand data and features, automatically calculate reward labels, constantly train intelligent traders, and moreover, proposes a method of directly calculating the single-step deterministic action values by price. Upon testing 1479 stocks with more than ten years' data in the China stock market, ISTG's overall revenue reaches 13%, which is better than overall - 7% of the buy-and-hold strategy.

Keywords: Deep reinforcement learning · DQN ·
One-step deterministic action value · Quantization strategy

1 Introduction

Deep reinforcement learning model is known for solving process decision problems and achieving remarkable results in fields such as bio-simulation, machine translation, industrial manufacturing, automatic driving control, optimization and scheduling, gaming [1]. The model has also been studied for stock forecasting and operational strategies in the financial sector [2–4].

In the quantitative operation of financial markets, in order to conduct a comprehensive assessment, an operational strategy was formed by researchers using product value, numerous indicators, and surrounding environmental factors, building such strategy can often be hindered by the three following conditions:

- The product has insufficient information and cannot be accurately valued.
- The effect is worse if only based on a few indicators.
- According to summarized indicators and fixed operational strategies, it is not possible to adapt to environmental changes dynamically.

W. Ni et al. (Eds.): WISA 2019, LNCS 11817, pp. 15–21, 2019.
https://doi.org/10.1007/978-3-030-30952-7_2

A stock trader based on deep reinforcement learning technology also faces the above problems. In view of this, we extend the DQN algorithm [5, 6] to implement the Intelligent Stock Trader and Gym model (ISTG). The model can find investment opportunities more frequently and accurately, conduct end-to-end practices, and optimize operational strategies. Moreover, the trader can automatically adjust to environmental changes.

2 Related Work

Financial markets have uncertainties and timing characteristics due to the interaction of a large number of complex factors. The quantitative investment [7] is built upon rigorous models, effectively capturing opportunities, and automatic execution. Therefore, it is urgent to study the suitable intelligent decision model for investment.

DeepMind [5] first proposed DQN that combines the CNN model and Q-learning. This solves the problem that traditional Q-learning is challenging to process high-dimensional data. Then double DQN [8] uses two Q networks, one responsible for selecting actions, the other for calculations, and periodically updating the computing network to overcome the over-optimization of Q-learning. In [9], Zhao proposed a multi-objective classification Mao-CNN model, which performs better than traditional CNN.

Deep reinforcement learning has been applied in the financial pairing, high-frequency trading, and portfolio. Moody et al. [10] proposed a combination of Recurrent Reinforcement Learning (RLL) and Q-learning learning algorithms to train trading systems. However, the training data is a single index product with a more extended period and monthly market. Deng et al. [11] constructed the DRL model and used machine learning techniques in parameter initialization, feature learning, and denoising. Qi Yue et al. [4] applied the deterministic strategy gradient method DDPG to portfolio management, and dynamically adjusted the weight of assets in the portfolio with the optimal, but there is no reasonable method for selecting a portfolio.

3 ISTG Model

3.1 DQN Methodology

The main goal of the ISTG model is to conduct the best action strategy according to the historical market in a particular market, carry out the trading operation, and maximize the ultimate profit of the specified period.

Referring to the classic DQN method, this paper uses the CNN network to learn and output action value. The Q-learning method interacts with the environment continuously to obtain training data with reward labels. It then establishes memory queues for storing millions of frames, and randomly samples small batches for model training.

3.2 Strategy

The theoretical basis of reinforcement learning is Markov decision process (MDP). The MDP model is a quintuple $<S, P, A, R, \gamma>$, which includes: finite state set S, state transition probability P, finite action set A, return function R, and discount factor γ after calculating future return discount. The cumulative return G_t is defined as:

$$G_t = R_{t+1} + \gamma R_{t+2} + \cdots = \sum_{k=0}^{\infty} \gamma^k R_{t+k+1} \tag{1}$$

The goal of reinforcement learning is to find the optimal strategy π. The strategy π is the distribution of action a for the given state s:

$$\pi(a|s) = P[A_t = a \mid S_t = s] \tag{2}$$

Trading strategies are: 1 controlling the number of single purchases, 2 controlling the risk position, 3 controlling the range of the ups and downs, 4 controlling the stop loss, and 5 taking profit. Intelligent traders should select high-quality stocks, buy and sell at the right time, maximize the portfolio while minimizing the operations.

The system's workflow (shown in Fig. 1) has been designed to implement the optimal solution abiding to the principle mentioned above. The Original data module collects and processes related data including intelligent trader status, environmental status, sequence status. The Agent module forms a multi-day time window input matrix.

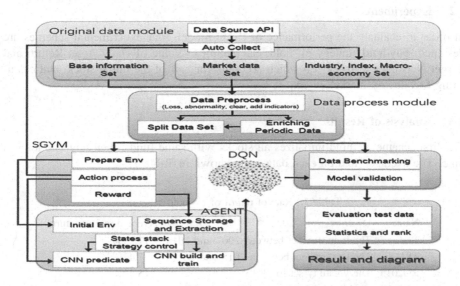

Fig. 1. Intelligent trader workflow

3.3 Features

In order to establish an SYGM environment in Fig. 1, this paper divides stock information into four categories, with 37 features:

- Pre-operation status of an intelligent agent (last operation, last return, current funds, stock balance, total value, yield)
- Status of stock market (date, opening price, highest price, closing price, lowest price, trading volume, turnover, code)
- Status of market analysis indicator (up and down, moving average, MACD, RSI, ROC, KDJ, BOLL, OBV);
- Status of macroeconomic indicators (M1 growth with the same period last year, M2 growth with the same period last year, the demand deposit rate, regular 3-month deposit rate, regular 6-month deposit rate, and US dollar exchange rate)

4 Experiment

4.1 Data Preparation

The market data includes 1479 stocks from 2007 to 2018 in China, the Shanghai Composite Index, and macroeconomic data. The Data Preprocessing module performs zero-value filling on the missing fields. It normalizes field data to [0,1], and codes 0-n class labels for date and stock code fields. Then, the results are divided into four subsets: 1 the train set TN1 between 2007 and 2014, 2 the test set TS1 between 2015 and 2017, 3 the volatility set RG2015 and 4 the overall decline set RB2018.

4.2 Experiments

In order to evaluate the performance of ISTG model, four experimental schemes are designed: buy-hold strategy ev_hold, calculating Q-Value of Daily Asset Return and Target Network ev_tq, ev_tqh which accounts for half of the initial stock, and calculating Q-value by market ev_mq.

4.3 Analysis of Results

The first scheme ev_hold initializes all stocks with same funds, then buy one lot at a time. The average yield on each data set is shown in Table 1:

Table 1. Rate of return of ev_hold scheme

Name	Describe	Average rate of return
TN1	The train data set between 2007 and 2014	77.29%
TS1	The test data set between 2015 and 2017	−7.42%
RG2015	The volatility set for 2015	15.42%
RB2018	The decline data set for 2018	−18.07%

The second scheme ev_tq initializes all stocks with the same sufficient funds. As the number of rounds is continuously increased, the return value is gradually stabilized. After 5,000 rounds, the return on assets becomes stable. The rate of return in the learning phase can reach a maximum of 5000%, a minimum of −24%, and an average of 22%.

The critical evaluation indicators of the ev_tq scheme test are shown in Table 2. Compared with the ev_hold scheme, the average of the yield and the Sharpe ratio is high.

Table 2. Comparison of indicators between the ev_tq and ev_hold schemes on the test data set

Indicators	Scheme	Mean	Max	Min	Std
Return rate	ev_tq	**13.73%**	**487.34%**	**−71.22%**	57.37%
	ev_hold	−7.42%	328.99%	−76.10%	45.52%
Sharp ratio	ev_tq	**−0.057%**	1.14%	−2.84%	0.312%
	ev_hold	−0.165%	1.176%	−2.84%	0.277%
Max withdraw	ev_tq	−61%	0%	**−88%**	**11%**
	ev_hold	13.73%	487.34%	−71.22%	57.37%

On the test data set, the process of the total yield of the ev_hold and ev_tq scheme is recorded and analyzed. By calculating the daily asset mean and standard deviation of 1479 stocks, the trend of total return rate in 3 years is shown in Fig. 2(a), ev_tq exceeds ev_hold in each period.

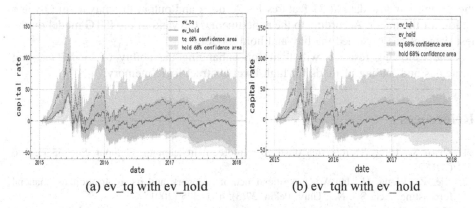

(a) ev_tq with ev_hold   (b) ev_tqh with ev_hold

Fig. 2. Total return rate on capital

As ev_tq scheme initializes all holding funds, the action of model learning is biased towards buying. Figure 2(a) shows that the effect of the scheme is the same as that of buying and holding. We therefore design third ev_tqh scheme, which tries to initialize half of the stock. Both buying and selling actions can be profitable at the beginning. After four rounds of training, the model can learn to reduce frequent operations and to make trading operations more balanced. The yield on the training set is up to 7000%, exceeding the effect of all initialization into funds. On the test data set, the effect of ev_tqh and ev_hold is shown in Fig. 2(b), and more stable than ev_tq.

For the four experimental schemes: ev_hold, ev_tq, ev_tqh and ev_mq, the statistical analysis of the total return rate (capital rate), the maximum withdraw rate (max withdraw rate), the results (in Fig. 3) showed that ev_tqh works best.

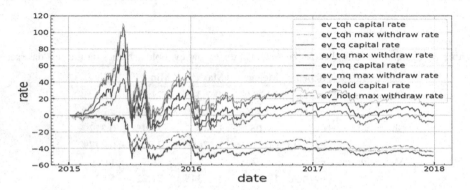

Fig. 3. Total return rate and withdrawal rate of four schemes

5   Conclusion

The ISTG model proposed in this paper uses DQN technology and selects the 12-year data of the China stock market for experiments. The CNN deep network is used to learn the sliding window data of 37 features for 20 days and output the maximum Q value action for the trader. According to the experiments, the yield of the ISTG model on the test data set has exceeded the buy-and-hold model.

The following research will focus on enhancing the ability of model learning strategies and finding methods for high-level abstract logic memory.

References

1. Zhao, X., Ding, S.: Research on deep reinforcement learning. Comput. Sci. **45**(7), 1–6 (2018)
2. Xie, Z., Wong, H., Ip, W.C.: Artificial neuron network and its application to financial forecasting. Acta Sci. Nat. Univ. Pekin. **37**(3), 421–425 (2001)
3. Man, Q.: An empirical research on the investment strategy of stock market based on deep reinforcement learning model. Guangdong University of Finance & Economics (2017)
4. Qi, Y., Huang, S.: Portfolio management based on DDPG algorithm of deep reinforcement learning. Comput. Mod. **2018**(05), 93–99 (2018)
5. Mnih, V., Kavukcuoglu, K., Silver, D., et al.: Playing Atari with deep reinforcement learning. arXiv preprint: arXiv:1312.5602 (2013)
6. Mnih, V., Kavukcuoglu, K., Silver, D., et al.: Human-level control through deep reinforcement learning. Nature **518**(7540), 529 (2015)

7. Gao, J.: The role of artificial intelligence in financial transactions and its future development direction. Electron. Technol. Softw. Eng. **2017**(18), 253 (2017)
8. Hasselt, H., Guez, A., Silver, D.: Deep reinforcement learning with double Q-learning. In: Proceedings of the Thirtieth AAAI Conference on Artificial Intelligence, pp. 2094–2100. AAAI Press (2016)
9. Zhao, H., Xia, S., Zhao, J., Zhu, D., Yao, R., Niu, Q.: Pareto-based many-objective convolutional neural networks. In: Meng, X., Li, R., Wang, K., Niu, B., Wang, X., Zhao, G. (eds.) WISA 2018. LNCS, vol. 11242, pp. 3–14. Springer, Cham (2018). https://doi.org/10.1007/978-3-030-02934-0_1
10. John Moody, M.S.: Reinforcement learning for trading systems and portfolios. In: Advances in Neural Information Processing Systems, vol. 17(5–6), pp. 917–923 (1998)
11. Deng, Y., Bao, F., Kong, Y., et al.: Deep direct reinforcement learning for financial signal representation and trading. IEEE Trans. Neural Netw. Learn. Syst. **28**(3), 653–664 (2017)

BiRNN-DKT: Transfer Bi-directional LSTM RNN for Knowledge Tracing

Bin Xu, Sheng Yan$^{(\boxtimes)}$, and Dan Yang

Northeastern University, NO. 3-11, Wenhua Road, Heping District, Shenyang, China
xubin@mail.neu.edu.cn, yansheng1117@foxmail.com, yangdan@ise.neu.edu.cn

Abstract. In recent years, online education is transforming from mobile education to intelligent education, and the rapid development of machine learning breathe into intelligent education with powerful energy. Deep Knowledge Tracing (DKT) is a state of the art method for modeling students' abilities which is changing by time. It can accurately predict students' mastery of knowledge or skill as well as their future performance. In this paper, we study the structure of the DKT model and proposed a new deep knowledge tracing model based on Bidirectional Recurrent Neural Network (BiRNN-DKT). We have also optimized the incorporating of data preprocessing and external features to improve model performance. Experiments show that the model can not only predict students' performance by capturing their history performance, but also get more accurate learning status simulation by integrating past and future context sequence information into the model multiple knowledge concepts. Compared with the traditional model, the proposed BiRNN-DKT gets an improvement in predicting students' knowledge ability and performance, and has great advantages in measuring the level of knowledge acquired by students.

Keywords: Knowledge tracing · Deep learning ·
Intelligent education · Data processing ·
Bidirectional Recurrent Neural Network · Student modeling

1 Introduction

Adaptive Learning (AL) is a concept that is now more talked about in the field of education technology. Its core problem is to improve students' learning efficiency through personalized learning path planning. Knowledge tracing is a time-based modeling of students' knowledge so that we can accurately predict students' mastery of knowledge points and the next performance of students. Early knowledge tracking models rely on first-order Markov models, such as Bayesian Knowledge Tracking (BKT) [2]. Deep Learning (DL) is an emerging method of machine learning research. The development of deep learning also provides a new method

---

This work was supported by the National Natural Science Foundation of China under Grant no. U1811261, and the National Natural Science Foundation of China under Grant no.51607029.

© Springer Nature Switzerland AG 2019
W. Ni et al. (Eds.): WISA 2019, LNCS 11817, pp. 22–27, 2019.
https://doi.org/10.1007/978-3-030-30952-7_3

for the development of adaptive learning. Deep Knowledge Tracking (DKT) [5] is a model that uses deep learning algorithms to evaluate student abilities using Recurrent Neural Networks (RNN). Some studies have shown that the DKT model has made great progress compared to the traditional Bayesian knowledge tracking model and has been proven to be used for course optimization [5,9,10]. For the data that crosses over a long time interval, LSTM-RNN is more suitable than base RNN [8].

The purpose of our research is to improve the effectiveness of the DKT model under the same data set, and introduce a deep knowledge tracking model based on the Bidirectional Recurrent Neural Network (BiRNN) to model the learning sequence and evaluate the students' ability at various moments.

2   Related Work

In order to build an accurate student model, we need to understand the process of student learning. The knowledge tracking task can be summarized as: the observation sequence of a student's performance on a particular learning task $x_0, x_1, x_2, ..., x_t$, predicting their next performance x_{t+1}.

2.1   Recurrent Neural Networks

RNN are mainly used to process and predict sequence data. In a Fully Connected Neural Network or Convolutional Neural Networks, the network results are from the input layer to the hidden layer to the output layer. The layers are fully connected or partially Connected, but the nodes between each layer are unconnected [1]. The source of the Recurrent Neural Networks is to characterize the relationship between the current output of a sequence and the previous information. From the network results, the RNN will remember the previous information and use the previous information to influence the subsequent output. RNN suffer from the now famous problems of vanishing and exploding gradients, which are inherent to deep networks.

2.2   Bidirectional Recurrent Neural Networks

BiRNN connects two hidden layers of opposite directions to the same output. With this form of generative deep learning, the output layer can get information from past (backwards) and future (forward) states simultaneously [3]. Those two states' output are not connected to inputs of the opposite direction states. By using two time directions, input information from the past and future of the current time frame can be used unlike standard RNN which requires the delays for including future information.

2.3   Deep Knowledge Tracing

DKT is a Seq2Seq model, and the structure of the model is LSTM [5]. The knowledge tracking problem can be described as: Given a student's observation sequence $x_0, x_1, x_2, \ldots$ to predict the next performance x_{t+1}, usually $x_t = \{q_t, a_t\}$, where q_t represents the question component of the answer (eg corresponding knowledge point), a_t represents whether the corresponding answer is correct, usually $a_t = \{0,1\}$. The DKT model is based on the LSTM model to model knowledge tracking tasks.

3 Methodology

In the time series, students do not have continuous consistency in the degree of mastery of knowledge points, and the fluctuations are large. It can be concluded that a student's mastery of knowledge in the sequence is not necessarily complete. Here, we will use the Bidirectional Recurrent Neural Network model to integrate past and future contextual sequence information for prediction in order to achieve a better simulation of student learning.

3.1 Model

The batch input $X_t \in R^{n*d}$ (the number of samples is n, the number of inputs is d) for a given time step t and the hidden layer activation function are $tanh$. In the architecture of the Bidirectional Recurrent Neural Network, it is assumed that the time-step forward hidden state is $H_t \in R^{n*h}$ (the number of positive hidden cells is h), and the reverse hidden state is $H'_t \in R^{n*h}$ (The number of reversed hidden cells is h). As shown in Fig. 1.

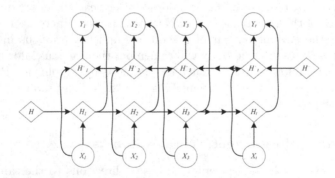

Fig. 1. Time expansion sequence of Bidirectional Recurrent Neural Network knowledge tracing model.

We can calculate the forward hidden state and the reverse hidden state separately :

$$H_t = tanh(X_t W_{xh}^{(f)} + H_t - W_{hh}^{(f)} + b_h^{(f)}) \tag{1}$$

$$H'_t = tanh(X_t W_{xh}^{(b)} + H'_t - W_{hh}^{(b)} + b_h^{(b)}) \tag{2}$$

The weight of the model is given by $W_{xh}^{(f)} \in R^{d*h}$, $W_{hh}^{(f)} \in R^{h*h}$, $W_{xh}^{(b)} \in R^{d*h}$, $W_{hh}^{(b)} \in R^{h*h}$. The model's Biases is $b_h^{(f)} \in R^{1*h}$, $b_h^{(b)} \in R^{1*h}$. Then we join the hidden states H_t and H_t' in both directions to get the hidden state H and input it to the output layer. The output layer calculates the output $O_t \in R^{n*q}$(the number of outputs is q):

$$Y_t = sigmoid(HW_{hq} + B_q) \tag{3}$$

The Weight $W_{hq} \in R^{2h*q}$ and the Biases $B_q \in R^{1*q}$ are the model parameters of the output layer. The number of hidden units in different directions can also be different.

The input data is x_t encoded by one-hot. If the model input data involves M knowledge components (such as knowledge points), each question has two results 0, 1 (corresponding to error and correct respectively), then the model input length is $2M$. For example, suppose the knowledge component of a topic is i. If the answer is correct, the $(i+1)^{th}$ bit of the corresponding input is 1 and the remaining positions are 0; if the answer is wrong, the ith bit is 1 and the remaining positions are 0. The output of the model is y_t, and the length is M, which corresponds to the degree of mastery of each knowledge component (corresponding to the correctness of the questions corresponding to the knowledge component).

4   Experiment

4.1   Datasets

The problem with the Skill Builder problem set is based on specific skills, and the question can have multiple skill tags. Students must answer three questions correctly to complete the assignment. If the student uses a guide ("Prompt" or "Resolve this issue as a step"), the issue will be flagged as incorrect.

We observed that the proportion of a few abnormal student data in the raw data was too high, and there was about 20% duplicate data in the ASSISTments data set. And when the problem involves multiple knowledge points, the same problem exists in the case of reusing data. So we cleaned the data and got a new data as input to our model. The statistics of the datasets are shown in Table 1.

Table 1. The statistics of the datasets.

Dataset	Records	Students	Exercise tags
Before data processing	401757	4217	124
After data processing	328292	4217	124

In real life, students have a lot of behavioral characteristics in the process of answering questions. This information can be of great help in assessing student abilities and predicting the correctness of the next question. Therefore, we added overlap_time (time of student overlap time), hint_count (number of student attempts for this question), and first_action (type of first action: try or ask for a prompt) in the model.

4.2 Implementation of Model

We decided to build the model based on Python and using the Keras API. The focus of Keras' development is to support rapid experimentation [7]. Ability to turn your ideas into experimental results with minimal delay. We implemented the model using the loss function used in the original DKT algorithm. Our hidden layer contains 200 hidden nodes. We use the Adam algorithm to optimize the stochastic objective function to speed up our training process. We set the batch size of each epoch to 100. We set Dropout to 0.4 to avoid over-fitting of the model and use 20% of the data set as a test dataset to verify model performance [4].

4.3 Result

We trained the original DKT model and the knowledge tracking model using Bidirectional Recurrent Neural Network on the preprocessed and post-processed datasets. We used Area Under Curve (AUC) and accuracy as evaluation criteria to evaluate the model. The higher the AUC, the better the effect of the model fitting student ability. The higher the accuracy, the better the global accuracy of the model. We will organize the prediction results obtained from the experiment into Tables 2 and 3. By observing the experimental results, we can see that the model has a better effect after adding the Bidirectional Recurrent Neural Network, and it has a good performance in predicting the correctness of the next answer. And after the data cleaning and the addition of external features, the effect of the model has been significantly improved. Therefore, it can be concluded that the improved model can better simulate the student's learning state and better predict the students' understanding of the knowledge points.

Table 2. Model prediction results without introducing external features.

Datasets	DKT		BiRNN-DKT	
	AUC	ACC	AUC	ACC
Before data processing	0.81	0.78	0.82	0.84
After data processing	0.82	0.79	0.84	0.90

Table 3. Model prediction results with introducing external features.

Datasets	DKT		BiRNN-DKT	
	AUC	ACC	AUC	ACC
Before data processing	0.78	0.79	0.81	0.87
After data processing	0.80	0.81	0.85	0.93

5   Conclusion and Future Work

In this paper, we have improved the original DKT model, and the new model has added the Bidirectional Recurrent Neural Network to the original. The new model can be used to model past and future students' learning of knowledge points. It effectively reflects the level of knowledge of students after reviewing and previewing, and makes the model's prediction of student performance better. The experimental data shows that compared with the traditional model, the deep knowledge tracking model based on BiRNN has a good performance in predicting students' problems, and has certain advantages in measuring students' level. Moreover, after preprocessing the data and adding external features, the performance of the model is improved.

In the future, we will explore how to improve the model to solve the problem of long sequences. We will validate the validity of the model on more data, compare and discuss the impact of more external features on the model to improve model performance. At the same time, the problem is extended to more general exercises, and knowledge tracking will be conducted in a wider range of fields and more perspectives to help students more effectively.

References

1. Williams, R.J., Zipser, D.: A learning algorithm for continually running fully recurrent neural networks. Neural Comput. **1**(2), 270–280 (1989)
2. Corbett, A.T., Anderson, J.R.: Knowledge tracing: modeling the acquisition of procedural knowledge. User Model. User-Adap. Inter. **4**(4), 253–278 (1994)
3. Schuster, M., Paliwal, K.: Bidirectional recurrent neural networks. IEEE Trans. Sig. Process. **45**(11), 2673–2681 (1997)
4. Srivastava, N., Hinton, G., Krizhevsky, A., Sutskever, I., Salakhutdinov, R.: Dropout: a simple way to prevent neural networks from overfitting. J. Mach. Learn. Res. **15**(1), 1929–1958 (2014)
5. Piech, C., et al.: Deep knowledge tracing. In: Advances in Neural Information Processing Systems, pp. 505–513 (2015)
6. ASSISTments Data (2015). https://sites.google.com/site/assistmentsdata/home/assistment-2009-2010-data/skill-builder-data-2009-2010. Accessed 07 Mar 2016
7. Ketkar, N.: Introduction to Keras. Deep Learning with Python, pp. 95–109. Apress, Berkeley (2017). https://doi.org/10.1007/978-1-4842-2766-4_7
8. Zheng, H., Shi, D.: Using a LSTM-RNN based deep learning framework for ICU mortality prediction. In: Meng, X., Li, R., Wang, K., Niu, B., Wang, X., Zhao, G. (eds.) WISA 2018. LNCS, vol. 11242, pp. 60–67. Springer, Cham (2018). https://doi.org/10.1007/978-3-030-02934-0_6
9. Wang, L., Sy, A., Liu, L., et al.: Deep knowledge tracing on programming exercises. In: Proceedings of the Fourth (2017) ACM Conference on Learning@ Scale, pp. 201–204. ACM (2017)
10. Zhang, L., Xiong, X., Zhao, S., et al.: Incorporating rich features into deep knowledge tracing. In: Proceedings of the Fourth (2017) ACM Conference on Learning@ Scale, pp. 169–172. ACM (2017)

Semi-supervised Learning to Rank with Uncertain Data

Xin Zhang[1,2(✉)], ZhongQi Zhao[2], ChengHou Liu[3], Chen Zhang[1,2], and Zhi Cheng[1,2]

[1] Anhui Provincial Engineering Laboratory of Big Data Technology Application for Urban Infrastructure Technology,
Department of Computer Science and Technology, Hefei University, Hefei, China
zxscce@163.com, zhangchen0304@163.com, cz_ganen108@126.com
[2] Suzhou Vocational University, Suzhou, China
zzqi@jssvc.edu.cn
[3] Changan Automobile R&D Center, Hefei, China
chenghouliu@163.com

Abstract. Although, semi-supervised learning with a small amount of labeled data can be utilized to improve the effectiveness of learning to rank in information retrieval, the pseudo labels created by semi-supervised learning may not reliable. The uncertain data nearby the boundaries of relevant and irrelevant documents for a given query has a significant impact on the effectiveness of learning to rank. Therefore, how to utilize the uncertain data to bring benefit for semi-supervised learning to rank is an excellent challenge. In this paper, we propose a semi-supervised learning to rank algorithm, that builds a query-quality predictor by utilizing uncertain data. Specially, this approach selects the training queries following the empirical observation that the relevant documents of high quality training queries are highly coherent. This approach learns from the uncertain data to predict the retrieval performance gain of a given training query by making use of query features. Then the pseudo labels for learning to rank are aggregated iteratively by semi-supervised learning with the selected queries. Experimental results on the standard LETOR dataset show that our proposed approaches outperform the strong baselines.

Keywords: Uncertain data · Semi-supervised learning · Effectiveness

1 Introduction

Learning to rank, a family of algorithms that utilizes training data, is one of the most popular weighting schemes in information retrieval [1]. Training data in

Supported by the Natural Science Foundation of Hefei University (18ZR07ZDA, 19ZR04ZDA), National Nature Science Foundation of China (Grant No. 61806068), the natural science research key project of Anhui university (Grant No. KJ2018A0556), the grant of Natural Science Foundation of Hefei University (Grant No. 16-17RC19,18-19RC27).

© Springer Nature Switzerland AG 2019
W. Ni et al. (Eds.): WISA 2019, LNCS 11817, pp. 28–39, 2019.
https://doi.org/10.1007/978-3-030-30952-7_4

learning to rank is composed of a set of feature vectors and a relevance label that specifies whether the document is relevant in response to the user input query. Due to the fact that the creation of such relevance labels usually involves exhaustive efforts and cost in the judgments, the effectiveness of learning to rank hints by lacking labeled data [2]. To deal with the problem of lacking training data in learning to rank, semi-supervised learning methods that utilize large amount of unlabeled data are applied. It is often supposed that the documents in the top-ranked positions in the initial retrieval results are highly relevant to the given query, while the ones in the bottom-ranked positions are irrelevant to the given query. However, it is not the case in many occasions which lead to poor retrieval effectiveness for semi-supervised learning to rank. For example, on MQ2008, a subject of LETOR 4.0 [7], only 13.32% of the top-5 retrieved documents are relevant for the given queries. If the traditional semi-supervised learning to rank is applied on MQ2008, the retrieval effectiveness may hurt. On the other hand, the uncertain data scattering around the boundaries between relevant and irrelevant documents for a given query is critical for improving the retrieval effectiveness of learning to rank. Therefore, it is necessary to consider the most uncertain data in semi-supervised learning to improve the quality of labeled data in order to guarantee the retrieval effectiveness of learning to rank algorithms.

In this paper, we propose to make using of uncertain data in semi-supervised learning algorithms to filter the training queries for learning to rank. Instead of using all the training queries in traditional semi-supervised learning, only the good-quality queries are used in our proposed semi-supervised learning to rank. Empirically, the relevant documents of high quality training queries are highly coherent. Thus, in this paper we attempt to learn a query-quality predictor from the existing few labels and uncertain data. More specifically, the uncertain data close to the class boundaries is added in semi-supervised learning to construct a query-quality predictor by building a relationship between the query features and the retrieval information gains. During this process, the training data with the best learning effectiveness is chosen to construct the predictor for estimating the query quality on a per-query basis.

The major contributions of the paper are two-fold. First, we propose to build a query-quality predictor to estimate the quality of the training queries with the most uncertain data from the normalized query commitment features and the information gains. Only high-quality queries are leveraged to aggregate pseudo examples in our semi-supervised learning to rank. Second, this paper proposes a workable approach that effectively simulates the training data without necessarily involving human labels to obtain effective retrieval performance. Experimental results on the standard dataset, namely LETOR, show that our method outperforms the strong baselines.

The rest of the paper is organized as follows. Section 2 presents the research on learning to rank, semi-supervised learning in information retrieval and some learning research that utilizes unlabeled data. Section 3 introduces the uncertain-data based on semi-supervised learning to rank method in detail. Section 4 introduces the experimental settings and explains the strong baselines employed in the evaluation. Besides, the experimental results are discussed in this section. Finally, Sect. 5 concludes the work and suggests future directions.

2 Related Work

2.1 Learning to Rank

Learning to rank is a family of learning methods and algorithms for ranking [1]. The typical input setting in learning to rank is a triplet, namely training data, which includes the query, the feature vectors indicating the query-document relationships and the relevance of the documents to the given query. In learning to rank, a ranking function is learned based on this training data, and then applied to the test data. There are three types of learning to rank algorithms: the pointwise, pairwise, and listwise approaches [1]. Extensive experimental results have demonstrated that the pairwise and listwise approaches usually outperform the pointwise methods [1, 8, 9]. The representations of input and loss functions are the main differences among these approaches. For example, the object pairs are input in the pairwise learning process; while the document lists rather than the document pairs are input in the listwise approach. And, the pairwise learning aims to minimizing the errors of dis-classification of the document pairs. While in the listwise approach, the objective is the optimization of the ranking list using any metric between the probability distributions.

Great improvement brought by incorporating various sources of features in learning to rank for information retrieval has been demonstrated both in the research and industry domains [3–6]. Recently, deep learning has been widely used in information retrieval [35]. However, the retrieval performance of learning to rank may be affected by the lack of labeled data. In applications such as building systems for domain-specific areas and learning personalized rankers for individuals or enterprises, there often exists a common phenomenon of lacking high-quality labeled data. In contrast, large amount of unlabeled data, i.e. the query document pairs without explicitly given relevance degrees, can be easily fetched and utilized in semi-supervised learning for learning to rank.

2.2 Semi-supervised Learning

The idea of semi-supervised learning is to learn both from small number of labeled data and large amount of unlabeled data [10, 11]. Thus, it is often adopted to generate labels for facilitating training data for learning to rank. Traditionally, semi-supervised learning is applied in classification tasks in information retrieval. For instance, an efficient pairwise ranking model constructed both from positive and unlabeled data is proposed by [12]. To improve the effectiveness of pseudo relevance feedback, an adaptive co-training method is also proposed to classify good feedback documents from bad ones [13].

There are also some research about how to improve the effectiveness of information retrieval by applying semi-supervised learning in learning to rank. The benefits brought by incorporating unlabeled data in document ranking is proved by Usunier [14]. Duh and Kirchhoff propose to create better features for the specific test queries from the test data [16]. In this research, the traditional information retrieval model and semi-supervised learning model are combined to tag the unlabeled data by Li [17]. Szummer and Yilmaz introduce a preference regularzier favoring the similar items to capture the mainfold structure of the unlabeled data in semi-supervised learning [2]. [15] proposes to improve the discriminantability of classier by applying semi-supervised learning from unlabeled data. [18] proposes to predict defaults in social lending by applying an ensemble semi-supervised learning method.

2.3   Learning with Uncertain Data

A semi-supervised batch mode for visual concept recognition is proposed by means of exploiting the whole active pool to evaluate the uncertainty of the data [33]. [34] proposes piecewise linear kernel-based support vector clustering as a new approach tailored to data-driven robust optimization by converting the distributional geometry of massive uncertain data to a compact convex uncertainty set. Hong et. al propose a flexible and efficient method to balance between time complexity and accuracy for noisy and uncertain data [19]. Rhee et al. propose a combined learning paradigm to leverage the strong points of semi-supervised learning and the most uncertain data [20]. Some research validates that, if variances over the inputs-outputs are known, it is possible to compute the expected covariance matrix of the data in order to obtain a posterior distribution over functions [21]. Liang et al. classify uncertain data streams by utilizing a aviaries of the concept-adapting very fast decision tree algorithm [22]. Zhu et al. propose to deal with the uncertain data by applying an inductive logic programming framework to modify the performance evaluation criteria [23].

3   Framework of Semi-supervised Learning to Rank with Uncertain Data

The framework of our proposed method, which learns a classifier with uncertain data to select the high quality queries is illustrated in Fig. 1. These selected queries are utilized by semi-supervised learning to generate the pseudo labels iteratively for learning to rank. In the following sections, we introduce our proposed semi-supervised learning method with uncertain data in detail.

3.1   Extraction of Query Features

How to represent the query features that indicate the query quality is an excellent challenge. Although, there are some methods, such as clarity [24], Weighted Information Gain [25] and Normalized Query Commitment [26], we decide to choose normalized query commitment. It is mainly due to the fact that it is more easily to compute the normalized query commitment on large standard dataset for learning to rank. Assuming that the query-shift is correlated with the high deviation of retrieval scores in the ranking list, normalized query commitment estimates the potential amount of query drift in the top-ranked documents to predict the given query performance.

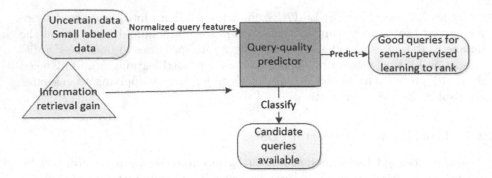

Fig. 1. Framework of the proposed method.

In this paper, we exploit 25 query performance features to represent the query-document relationship, shown as follows.

⋆ LMIR.ABS-based normalized query commitment query features. The normalized query commitment score is computed on the basis of LMIR.ABS, and it includes the different domains of the given document, namely body, anchor, title, URL and the whole document.

⋆ LMIR.DIR-based normalized query commitment query features. The normalized query commitment score is computed on the basis of LMIR.DIR. Different domains of the given document, such as body, anchor, title, URL and the whole document are incorporated.

⋆ LMIR.JM-based normalized query commitment query features. The normalized query commitment score is computed on the basis of LMIR.JM, and it includes the different domains of the given document, such as body, anchor, title, URL and the whole document.

⋆ The normalized query commitment score is computed on the basis of TF*IDF, and it includes the different domains of the given document, such as body, anchor, title, URL and the whole document.

⋆ Normalized query commitment query features on the basis of BM25. The normalized query commitment score is computed on the basis of BM25, and it includes the domains of the body, anchor, title, URL and the whole document.

The number of documents in the result-list D_q^k is set to 5 rather than 100 as in [26], since the initial retrieval list offered in LETOR 4.0 is treated as a substitution of a whole corpus.

3.2 Construction of Query Quality Predictor on the Basis of Uncertain Data

In this paper, the query quality is measured by the benefit, namely the retrieval performance gain brought by the given query. We aims to selecting high-quality queries to improve the performance of learning to rank effectively through the construction of the query-quality predictor. In detail, the retrieval performance gain is the Δ MAP, namely the map difference between the traditional retrieval model and the learning to rank model. During the training process, each query available is represented by the exploited features as introduced in Sect. 3.1. The next step is to build a relationship between the normalized query commitment features and the information retrieval gain for each given query by applying a classifying algorithm. Thus, a query-quality predictor is learned from the normalized query commitment features. Besides, we take the most uncertain data into consideration during the process of building the query-quality predictor. The query-quality predictor not only learns from the small number of labeled examples, but also from the most uncertain ones. We attempt to sample the examples by adopting the uncertainty strategy, a commonly used data labeling strategy in active learning [27]. Thus, the instances close to the class boundary are considered as the uncertain ones. In other words, we firstly learn a model by leveraging the most confident examples and then greedily add uncertain examples randomly to the original training data to re-train the ranking model. In the proposed method, the uncertain examples sampling is an enumeration process, and it can be infeasible if the dataset is very large. Therefore, in our method, the process of random sampling is conducted 15–20 times, which appears to have near optimal results in our experiments. The training data with the best learning effectiveness is chosen to construct the classifier for estimating the query quality.

Traditionally, all the queries available are taken as input for learning to rank, just ignoring the quality of the queries. However, in our method we can estimate the quality of a candidate query by applying the learned predictor, in order to determine whether a given query should be included in the training set for the semi-supervised learning. The candidate queries are categorized as two classes, the good ones and the bad ones. The good query, namely the behaved-well query, is able to bring great retrieval performance gain; and the bad one may not bring performance gain or just a little gain. Thus, in the following step, the queries that can bring benefit in retrieval performance are chosen for semi-supervised learning to aggregate the pseudo labels for learning to rank.

Once a query is selected as the training query for semi-supervised learning, it can be utilized in transduction to iteratively aggregate pseudo labels for learning to rank. In semi-supervised learning, we adopt similar approach as [13] did, namely the high top-ranked documents of the retrieval results obtained from traditional model are treated as the initial relevant training data; and the very bottom-ranked documents are extracted as the initial irrelevant examples. The algorithm iteratively generates the pseudo examples from the remaining unlabeled data to construct the pseudo training set for the learning to rank algorithms.

3.3 Classification Algorithm

There exists various kinds of classification algorithms for distinguishing the good-quality queries from the bad ones, such as Naive Bayes, Logistic Regression and the relatively sophisticated support vector machines. The effectiveness of these classification algorithms has been proven in previous literature. In this paper,

Logistic Regression and Naive Bayes are chosen as the classifiers to learn from the uncertain data for our proposed method. Next, we briefly introduce logistic regression and Naive Bayes.

Logistic Regression (LR) is a probabilistic statistical classification model for binomial regression [29]. It is commonly used to describe data and to explain the relationship between one dependent binary variable and one or more independent variables. Logistic regression not only indicates where the boundary between the classes is, but also indicates that the class probabilities depend on distance from the boundary. Suggest that there exists a vector X = X1, X2, ..., Xn. Xi can be discrete or continuous variables. Logistic regression aims to learning the probability of $P(Y|X)$ from the distribution of X and Y.

$$P(Y = 1|X) = \frac{1}{1 + exp(w_0 + \sum_{i=1}^{n} w_i X_i)} \tag{1}$$

and

$$P(Y = 0|X) = \frac{exp(w_0 + \sum_{i=1}^{n} w_i X_i)}{1 + exp(w_0 + \sum_{i=1}^{n} w_i X_i)} \tag{2}$$

where w_i is the weight for each vector. In order to minimize the mis-classification rate, Y is set to be 1 when $p \geq 0.5$, and Y is set to be 0 when $p < 0.5$. This means logistic regression is a linear classifier.

Naive Bayes (NB) is a probabilistic learning algorithm with the assumption of strong class condition independence, namely the presence of a particular feature of a class is unrelated to the presence of any other features [28]. The major advantage of NB is its reduced complexity because of the hypothesis mentioned above. Another advantage of NB is that only a small amount of labeled data is required by applying NB to estimate the means and variances of the variables for classification. Previous research shows that NB classifiers have worked quite well in many complex real world situations.

4   Experiments and Results

4.1   Dataset

Experiments are conducted on the standard learning to rank dataset LETOR [7]. LETOR, released by Microsoft Research Asia, is a benchmark collection for learning to rank in information retrieval. LETOR is widely used in information retrieval which was constructed based on multiple data corpora and query sets. The data was partitioned into five folds for cross validation, and standard evaluation tools were provided. In addition, the ranking performances of several state-of-the-art ranking methods were also provided, which can serve as baselines for newly developed methods. In LETOR, two document corpora, namely the "Gov"corpus and the OHSUMED corpus, are collected, due to the fact that the two corpora are publicly available and widely used in previous learning to rank research. LETOR can be used in dierent kinds of researches. In our experiments, the queries without relevant documents are deleted. Specifically, the experiments are conducted on LETOR 2.0 and 4.0 respectively. LETOR 2.0 [31] contains the topic distillation task of TREC 2003 (TD2003) and the topic distillation task of TREC 2004 (TD2004). For LETOR 4.0, "MQ2007" and "MQ2008" stand for Million Query track of TREC 2007 and TREC 2008, respectively. MAP (mean average precision) and M_nDCG (mean normalized discounted cumulative gain) are used to evaluate the retrieval effectiveness of our proposed approach on LETOR. Note that a ⋆ in all the following tables indicates a statistically significant difference between the baseline and our proposed method. The significance test is conducted using T test at the 0.05 level. Our choice of T test is mainly due to the fact that it is commonly used as a standard statistical test for IR experimentation on LETOR 4.0.

4.2   Evaluation Design

The aim of our experiments is to evaluate the effectiveness of the proposed approach for semi-supervised learning to rank. Good queries are selected by the established query-quality predictor. And then they are used to generate the pseudo examples iteratively by the proposed approaches for learning to rank to produce the list of returned documents for the given test queries. In particular, two baselines are adopted as follows. –First, we compare the proposed method, namely the semi-supervised learning to rank with uncertain data (USTL), to the previous transductive learning approach (TL) which extracts the pseudo training examples from all the queries available [30]. In our experiments, the TL baseline extracts the initial negative examples from the very bottom-ranked documents, as this has the best effectiveness on LETOR 4.0 according to the results. –Second, we comapre the method to OptPPC, as proposed in [32]. In [32], Geng et al. propose to select an optimal subset of the original training data by maximizing the "pairwise preference consistency". Experimental results on LETOR 2.0 demonstrate that OptPPC achieves the best retrieval effectiveness when different granularity noise is added into the training data.

4.3   Experimental Results

Firstly, we evaluate the proposed method to TL. In this level of experiments, we examine the effectiveness of our proposed approach USTL and TL with no labeled data available, as shown in Table 1. Two types of classification strategies, namely NB and LR are adopted, as shown in this table. When taking NB as the learning algorithm, we can see that the experimental results are comparable in most cases, except that a significant improvement of 1.69% in M_nDCG is observed on MQ2008. However, significant improvements in MAP and M_nDCG are observed both on MQ2007 and MQ2008 while LR is applied to classify the candidate queries. Specially, on MQ2007 MAP improves 1.35% significantly by applying our proposed method when compared to TL and a notable improvement of 1.48% in M_nDCG is observed on this standard dataset. Although, no improvement in MAP is obtained on MQ2008 while applying LR algorithm, M_nDCG is improved notably with an enhance of 1.03%. From Table 1, we can see that when

Table 1. Comparison of USTL with TL on LETOR 4.0.

MQ2007			
		NB	LR
	TL	USTL	
MAP	0.5186	0.5172,−0.27%	**0.5256,+1.35%**
M_nDCG	0.5532	0.5579,+0.85%	**0.5614,+1.48%**
MQ2008			
		NB	LR
MAP	0.6533	0.6586,+0.81%	0.6533,+0.00
M_nDCG	0.6632	**0.6744,+1.69%**	**0.6700,+1.03%**

LR is used to build the relationship between MAP and the queries features, our proposed method performs well on both MQ2007 and MQ2008, while NB only perform well on MQ2008 with M_nDCG as the evaluation measure. We suggest the quantity of the training data has an important impact on the effectiveness of NB due to the fact that the features are independent of each other in this classifying algorithm. Besides, we suggest that the method for optimizing the parameters of NB, namely maximum likelihood estimation, also plays a role in the retrieval effectiveness of USTL.

Table 2. Comparison of the proposed USTL method to OptPPC.

TD2003							NB	LR
		Noise					NB	LR
	Noise rate	0.25	0.30			Noise rate	0.2638	0.2754
OptPPC	NDCG@5	**0.2630**	0.2320	STL		NDCG@5	0.2578	0.2489
	NDCG@10	**0.2680**	0.2350			NDCG@10	0.2480	0.2278
TD2004							NB	LR
		Noise					NB	LR
	Noise rate	0.20	0.25			Noise rate	0.2276	0.2146
OptPPC	NDCG@5	0.3020	0.2990	STL		NDCG@5	0.3385	**0.3495**
	NDCG@10	0.2780	0.2750			NDCG@10	0.3452	**0.3678**

Table 2 shows the experimental results, which are obtained on LETOR 2.0 by applying USTL and OptPPC, respectively. On TD2003, considering that the noise rates by applying USTL are close to 0.25 and 0.30, results obtained by OptPPC with the two noise levels are taken as the baselines. From Table 2, we can see that on TD2003, no advantages by applying USTL are shown. On TD2004, the noise rates in the pseudo examples by applying NB and LR are 0.2438 and 0.2276, respectively. Thus, we compare our proposed method to

OptPPC with 0.20 and 0.25 noise levels. The experimental results of USTL outperform OptPPC on TD2004 obviously. It is suggested that relatively low effectiveness of USTL on TD2003 is due to the fact that only 14% of all the top-5 documents on the entire dataset are relevant to the queries. Therefore, the pseudo labels generated by USTL are somehow low-quality which may hurt the retrieval effectiveness of USTL.

5   Conclusions and Future Work

In this paper, a semi-supervised learning to rank approach with uncertain data is proposed. Specifically, we first build a classifier with uncertain data to select the high quality queries by predicting the retrieval performance gain brought by each given query. These selected queries are utilized by semi-supervised learning to generate the pseudo labels iteratively for learning to rank. To examine the retrieval effectiveness of our proposed method, experiments are conducted on the standard collection LETOR. Results show that, in most cases statistically significant improvement is able to be observed over the baselines by applying the proposed approach. Besides, our study indicates the possibility of simulating the training data without involving human labels for the given new queries.

In future, we plan to investigate the retrieval effectiveness of the semi-supervised learning to rank method with uncertain data in different domains,such as in diversified search. In this task, diverse retrieval documents are preferred. Therefore, different evaluation of the query-quality may be presented in diversified search. Besides, to validate the retrieval effectiveness of our proposed method, more benchmark algorithms are needed to compare with the proposed USTL method. Also, the optimization of the proposed algorithm is needed to be considered in our future work.

Acknowledgements. This work is supported in part by the Natural Science Foundation of Hefei University (18ZR07ZDA,19ZR04ZDA), National Nature Science Foundation of China (Grant No. 61806068), the natural science research key project of Anhui university (Grant No. KJ2018A0556), the grant of Natural Science Foundation of Hefei University (Grant No. 16-17RC19,18-19RC27).

References

1. Liu, T.: Learning to rank for information retrieval. Found. Trends Inf. Retrieval **3**, 225–331 (2011)
2. Szummer, M., Yilmaz, E.: Semi-supervised learning to rank with preference regularization. In: Proceedings of the 20th ACM Conference on Conference on Information and Knowledge Management, CIKM 2011, pp. 269–278 (2011)
3. van den Akker, B., Markov, I., de Rijken, M.: ViTOR: learning to rank webpages based on visual features. In: The Web Conference (2019)
4. Zoghi, M., Tunys, T., Ghavamzadeh, M., Kveton, B., Szepesvari, C., Wen, Z.: Online learning to rank in stochastic click models. In: Proceedings of the 20th ACM Conference on Conference on Proceedings of the 34th International Conference on Machine Learning, pp. 4199–4208 (2017)

5. Severyn, A., Moschitti, A.: Learning to rank short text pairs with convolutional deep neural networks. In: Proceedings of the 38th International ACM SIGIR Conference (2015)
6. Wang, B., Klabjan, D.: An attention-based deep net for learning to rank (2017). arXiv preprint arXiv
7. Qin, T., Liu, T.: Introducing LETOR 4.0 datasets. Technical Report Microsoft Research Asia (2013)
8. Ganjisaffar, Y., Caruana, R., Lope, C.: Bagging gradient-boosted trees for high precision, low variance ranking models. In: Proceedings of the 34th International ACM SIGIR Conference on Research and Development in Information Retrieval, SIGIR 2011, pp. 85–94. ACM, New York, NY, USA (2011)
9. Joachims, T.: Optimizing search engines using clickthrough data. In: Proceedings of the Eighth ACM SIGKDD International Conference on Knowledge Discovery and Data Mining, KDD 2002, pp. 133–142. ACM, New York, NY, USA (2002)
10. Hu, H., Sha, C., Wang, X., Zhou, A.: A unified framework for semi-supervised Pu learning. World Wide Web 17(4), 493–510 (2014)
11. Chapelle, O., Schlkopf, B., Zien, A.: Semi-Supervised Learning, 1st edn. MIT Press, Cambridge (2010)
12. Sellamanickam, S., Garg, P., Selvaraj, S.K.: A pairwise ranking based approach to learning with positive and unlabeled examples. In: Proceedings of the 20th ACM International Conference on Information and Knowledge Management, CIKM 2011, pp. 663–672. ACM, New York, NY, USA (2011)
13. Huang, J.X., Miao, J., He, B.: High performance query expansion using adaptive co-training. Inf. Process. Manage. 49(2), 441–453 (2013). https://doi.org/10.1016/j.ipm.2012.08.002
14. Usunier, N., Truong, V., Amini, M.R., Gallinari, P., Curie, M.: Ranking with unlabeled data: a first study. In: Proceedings of NIPS Workshop (2005)
15. Zhang, L., Ma, B., He, J., Li, G., Huang, Q., Tian, Q.: Adaptively unified semi-supervised learning for cross-modal retrieval. In: Proceedings of the Twenty-Sixth International Joint Conference on Articial Intelligence(IJCAI-17) (2017)
16. Duh, K., Kirchhoff, K.: Learning to rank with partially-labeled data. In: Proceedings of the 31st Annual International ACM SIGIR Conference on Research and Development in Information Retrieval, SIGIR 2008, pp. 251–258. ACM, New York, NY, USA (2008)
17. Li, M., Li, H., Zhou, Z.H.: Semi-supervised document retrieval. Inf. Process. Manage. 45, 341–355 (2009)
18. Kim, A., Cho, S.-B.: An ensemble semi-supervised learning method for predicting defaults in social lending. Eng. Appl. Artif. Intell. 81, 193–199 (2019)
19. Hong, T.P., Tseng, S.S.: A generalized version space learning algorithm for noisy and uncertain data. IEEE Trans. Knowl. Data Eng. 9(2), 336–340 (1997)
20. Rhee, P.K., Erdenee, E., Kyun, S.D., Ahmed, M.U., Jin, S.: Active and semi-supervised learning for object detection with imperfect data. Cogn. Syst. Res. 45, 109–123 (2017)
21. Dallaire, P., Besse, C., Chaib-draa, B.: An approximate inference with Gaussian process to latent functions from uncertain data. Neurocomputing 74, 1945–1955 (2011)
22. Liang, C., Zhang, Y., Shi, P., Hu, Z.: Information sciences learning very fast decision tree from uncertain data streams with positive and unlabeled samples. Inf. Sci. 213, 50–67 (2012)
23. Zhu, M., Gao, Z., Qi, G., Ji, Q.: DLP learning from uncertain data. Tsinghua Sci. Technol. 15, 650–656 (2010)

24. Cronen-Townsend, S., Zhou, Y., Croft, W.B.: Predicting query performance. In: Proceedings of the 25th Annual International ACM SIGIR Conference on Research and Development in Information Retrieval, SIGIR 2002, pp. 299–306. ACM, New York, NY, USA (2002)

25. Zhou, Y., Croft, W.B.: Query performance prediction in web search environments. In: Proceedings of the 30th Annual International ACM SIGIR Conference on Research and Development in Information Retrieval, SIGIR 2007, pp. 543–550. ACM, New York, NY, USA (2007)

26. Shtok, A., Kurland, O., Carmel, D., Raiber, F., Markovits, G.: Predicting query performance by query-drift estimation. ACM Trans. Inf. Syst. **30**(2), 11:1–11:35 (2012)

27. Reitmaie, T., Calma, A., Sick, B.: Transductive active learning –a new semi-supervised learning approach based on iteratively refined generative models to capture structure in data. Inf. Sci. **293**, 275–298 (2015)

28. Webb, G.I., Boughton, J.R., Wang, Z.: Not so naive bayes: aggregating one-dependence estimators. Mach. Learn. **58**(1), 5–24 (2005)

29. Palei, S.K., Das, S.K.: Logistic regression model for prediction of roof fall risks in bord and pillar workings in coal mines: an approach. Saf. Sci. **47**(1), 88–96 (2009)

30. Zhang, X., He, B., Luo, T.: Transductive learning for real-time twitter search. In: The International Conference on Weblogs and Social Media (ICWSM), pp. 611–614 (2012)

31. Liu, T., Xu, J., Qin, T., Xiong, W., Li, H.: LETOR: benchmark dataset for research on learning to rank for information retrieval. In: SIGIR 2007 Workshop on Learning to Rank for Information Retrieval (2007)

32. Geng, X., Qin, T., Liu, T., Cheng, X., Li, H.: Selecting optimal training data for learning to rank. Inf. Process. Manage. **47**(5), 730–741 (2011)

33. Yang, Y., Ma, Z., Nie, F., Chang, X., Hauptmann, A.G.: Multi-class active learning by uncertainty sampling with diversity maximization. Int. J. Comput. Vis. **113**, 113–127 (2015)

34. Shang, C., Huang, X., You, F.: Data-driven robust optimization based on kernel learning. Comput. Chem. Eng. **106**, 464–479 (2017)

35. Liu, J., Cui, R., Zhao, Y.: Multilingual short text classification via convolutional neural network. In: Meng, X., Li, R., Wang, K., Niu, B., Wang, X., Zhao, G. (eds.) WISA 2018. LNCS, vol. 11242, pp. 27–38. Springer, Cham (2018). https://doi.org/10.1007/978-3-030-02934-0_3

Signed Network Embedding Based on Noise Contrastive Estimation and Deep Learning

Xingong Chang, Wenqiang Shi[(✉)], and Fei Zhang

School of Information, Shanxi University of Finance and Economics,
Taiyuan, China
842171357@qq.com

Abstract. Network embedding is a representation learning method to learn low-dimensional vectors for vertices of a given network, aiming to capture and preserve the network structure. Signed networks are a kind of networks with both positive and negative edges, which have been widely used in real life. Presently, the mainstream signed network embedding algorithms mainly focus on the difference between positive and negative edges, but ignore the role of empty edges. Considering the sparsity of signed networks, a signed network embedding algorithm NDSE is proposed based on noise contrastive estimation model and deep learning framework. The algorithm emphasizes the role of empty edges and optimizes a carefully designed objective function that preserves both local and global network structures. Empirical experiments prove the effectiveness of the NDSE on three real data sets and one signed network task.

Keywords: Network embedding · Signed networks · Empty edges · Link prediction

1 Introduction

The purpose of network embedding is to learn the low-dimensional vector representations for vertices in the network while preserving the network structure. Given a signed network $G = (V, E)$, where V is the set of vertices, and E is the set of edges, the problem of signed network embedding aims to represent each vertex $v \in V$ into a low-dimensional space R^d, i.e., learning a function $f: V \rightarrow R^d$, where $d \ll |V|$. In the space R^d, the structural characteristics of G are preserved.

During embedding, if x_i and x_j are the embedding vectors of vertices v_i and v_j, many researchers [1, 2] think that x_i and x_j should be more similar if there is a positive edge between v_i and v_j. On the contrary, they should be dissimilar if the edge between v_i and v_j is negative. But a more common scenario is that there is no edge between v_i and v_j, i.e. the edge between them is empty, because of the sparsity of networks in real life. In this case, should x_i and x_j be similar or dissimilar?

In this paper, we investigate the effect of empty edges in the signed network embedding. We propose the NDSE (Noise contrastive estimation and Deep learning based Signed network Embedding). The major contributions of this paper are as follows:

W. Ni et al. (Eds.): WISA 2019, LNCS 11817, pp. 40–46, 2019.
https://doi.org/10.1007/978-3-030-30952-7_5

(1) Design an objective function that reflects the relationship between empty, positive and negative edges;
(2) Propose a deep learning algorithm NDSE based on the noise contrastive estimation model for signed network embedding, which learns low-dimensional vector representations for vertices by optimizing the objective function; and
(3) Conduct experiments on three signed networks to demonstrate the effectiveness of the proposed algorithm NDSE.

The rest of this paper is organized as follows. Section 2 summarizes the related work. Section 3 introduces the NDSE in details. Section 4 presents the experimental results. Finally we conclude in Sect. 5.

2 Related Work

The existing signed network embedding algorithms can be categorized into two categories: (1) spectral analysis based embedding and (2) deep learning based embedding. SL [3], SNS [4] and BNS [4] fall into the first category, they define different Laplacian matrices and calculate the top-d eigenvectors of them, which are combined as the embedding vectors for vertices. SiNE [1] and DNE-SBP [2] belong to the latter. SiNE has three striking features: the structural balance theory is used to design its objective function, an ANN framework is proposed to optimize the objective function, a virtual node is introduced to deal with the imbalance problem caused by the lack of negative edges. Also inspired by the structural balance theory, DNE-SBP employs a semi-supervised stacked auto-encoder to reconstruct the adjacency matrix of a given signed network and uses the pairwise constraints to make the positively connected nodes much closer than the negatively connected nodes in the embedding space.

SiNE is most relevant to our work. However, it doesn't consider the effect of empty edges and the introduction of the virtual node may be lack of respect for reality.

3 Signed Network Embedding Based on Noise Contrastive Estimation and Deep Learning

Let $G = (V, E)$ be a signed network where $V = \{v_i\}_{i=1}^{n}$ is a set of n vertices and $E = \{e_{ij}\}_{i,j=1}^{n}$ is a set of edges. Particularly, any edge e_{ij} can be 1, -1 or 0 where $e_{ij} = 1$ denotes a positive edge between v_i and v_j, $e_{ij} = -1$ denotes a negative edge, and $e_{ij} = 0$ denotes an empty edge.

3.1 Objective Function

Noise contrastive estimation (NCE) [5] is a general parameter estimation method that follows these two steps: (1) A two-class training dataset P is generated. For each positive example, several negative examples (noise) are sampled, All (positive, negative) pairs constitute P. (2) Model parameters can be estimated as a binary classification problem to maximize conditional log-likelihood of P.

If $e_{ij} = 1$ and $e_{ik} = -1$, according to the extended structural balance theory [6], v_i is likely to be more similar to the user with a positive edge, i.e. v_j, than a user with a negative edge, i.e. v_k, which can be mathematically modeled as:

$$f(x_i, x_j) \geq f(x_i, x_k) + \delta_0 \tag{1}$$

where x_i denotes the d-dimensional representation of vertex v_i, $f(x_i, x_j)$ a measure of the similarity between x_i and x_j. δ_0 is a threshold used to regulate the difference between two similarities. A large δ_0 will push v_i, v_j more close and v_i, v_k more far away.

If v_i is a fully positive vertex, i.e., it has no negative adjacent edges. For each $e_{ij} = 1$, we sample an empty edge $e_{ik} = 0$, where v_k is as close to v_i as possible, because the farther away the vertex is, the less impact it has. Then, we have

$$f(x_i, x_j) \geq f(x_i, x_k) + \delta_1 \tag{2}$$

Similarly, If v_i is a fully negative vertex, it has no positive adjacent edges. For each $e_{ij} = -1$, We sample an empty edge $e_{ik} = 0$ near v_i's neighborhood. Then we have

$$f(x_i, x_k) \geq f(x_i, x_j) + \delta_2 \tag{3}$$

Parameters δ_0, δ_1 and δ_2 are thresholds used to regulate similarities. Intuitively, $\delta_1 < \delta_0$, $\delta_2 < \delta_0$. By sampling we can construct 3 training data sets:

$$P_0 = \{(v_i, v_j, v_k) | e_{ij} = 1, e_{ik} = -1, v_i, v_j, v_k \in V\}$$

$$P_1 = \{(v_i, v_j, v_k) | e_{ij} = 1, e_{ik} = 0, v_i \text{ is fully positive}, v_i, v_j, v_k \in V\}$$

$$P_2 = \{(v_i, v_j, v_k) | e_{ij} = -1, e_{ik} = 0, v_i \text{ is fully negative}, v_i, v_j, v_k \in V\}$$

Based on NCE and Eqs. (1), (2) and (3), the objective function for signed network embedding guided by the extended structural balance theory can be written as:

$$
\begin{aligned}
\min_{X,\theta} \frac{1}{C} [& \sum\nolimits_{(x_i, x_j, x_k) \in P_0} max(0, f(x_i, x_k) + \delta_0 - f(x_i, x_j)) \\
+ & \sum\nolimits_{(x_i, x_j, x_k) \in P_1} max(0, f(x_i, x_k) + \delta_1 - f(x_i, x_j)) \\
+ & \sum\nolimits_{(x_i, x_j, x_k) \in P_2} max(0, f(x_i, x_j) + \delta_2 - f(x_i, x_k))] + \alpha \left(\Re(\theta) + ||X||_F^2 \right)
\end{aligned}
\tag{4}
$$

where $C = |P_0| + |P_1| + |P_2|$, $X = \{x_1, x_2, \ldots, x_n\}$ is the low-dimensional representation of vertices, and θ is a set of parameters to define the function f. $R(\theta)$ is the regularizer to avoid overfitting and α is a parameter to control the contribution of the regularizers.

3.2 Architecture of NDSE

With the objective function given above, the task now is to find a function f that is able to give good similarity measure and learn good representations of vertices. Due to the powerful function approximation ability of deep learning [7], we use the same deep learning framework of SiNE, which defines f with θ and optimizes the objective function in Eq. (4). The framework is a 2 hidden layer neural network with triplets from P_0, P_1 and P_2 as input, *tanh* or *ReLU* as inner or output layer activation functions respectively. See reference [1] for details.

3.3 Training NDSE

The training algorithm for NDSE is summarized in Algorithm 1. From line 1 to line 9, we prepare the mini-bath training triplets. In line 10, we initialize the parameters of the deep network and the low-dimensional representations and we train the deep network from line 11 to line 18. The time complexity of NDSE is $\mathcal{O}(tC(2dd_1 + d_1d_2))$, where t is the number of epochs; $C = |P_0| + |P_1| + |P_2|$, d is the dimension of embedding; d_i is the number of nodes in the i-th layer of the deep network.

Algorithm 1 NDSE

Input: Signed network $G = (V, E)$, d, δ_0, δ_1, δ_2, α

Output: vector representation of vertices X

1. Initialize $P_0 = \emptyset$, $P_1 = \emptyset$, $P_2 = \emptyset$
2. for i=1: n do
3.   if v_i has both positive and negative edges
4.   for each pair $(e_{ij} = 1, e_{ik} = -1)$, put (v_i, v_j, v_k) in P_0
5.   if v_i is fully positive
6.   for each $e_{ij} = 1$, sample $e_{ik} = 0$, put (v_i, v_j, v_k) in P_1
7.   if v_i is fully negative
8.   for each $e_{ij} = -1$, sample $e_{ik} = 0$, put (v_i, v_j, v_k) in P_2
9. prepare min-batch from P_0, P_1, P_2
10. initialize the parameters of the deep network of NDSE
11. repeat
12.   for each mini-batch do
13.   forward propagation
14.   backpropagation
15.   update network parameters
16.   update related X
17. until convergence
18. return X

---

4 Experimental Results

In this section, we conduct experiments to evaluate the effectiveness of NDSE and factors that could affect the performance of NDSE. The experiments are conducted on three real-world signed network datasets, i.e., Epinions [8], Slashdot [8], and Wikipedia [8]. In the experiment, we set d as 20, $(\delta_0, \delta_1, \delta_2) = (1, 0.3, 0.7)$.

4.1 Link Prediction in Signed Network

For each dataset, we randomly select 80% edges as training set and the remaining 20% as test set. We use the training set to learn the signed network embedding. With the learned signed network embedding, we train a logistic regression classifier on training dataset, and then use the classifier to predict link. We use AUC to assess the performance. The random selection is carried out 5 times independently and the average AUC is reported in Table 1. We can easily see, NDSE performs best on all three datasets. Since DNE-SBP, SL, SNS and BNS are matrix based, it's difficult to implement on Epinions because of its huge size.

Table 1. AUC on Wikipedia, Slashdot and Epinions

Algorithm	Wikipedia (n = 7114, m = 99862)	Slashdot (n = 82062, m = 498532)	Epinions (n = 131513, m = 708507)
NDSE	**0.87**	**0.88**	**0.86**
SiNE	0.82	0.82	0.81
DNE-SBP	0.85	0.86	–
SL	0.60	0.85	–
SNS	0.70	0.77	–
BNS	0.65	0.82	–

In signed networks, negative edges are valuable [8]. So we also conduct experiments to predict negative edges. The results are shown in Table 2. It can be seen that the recall rates of NDSE are higher than those of SiNE. It shows that the predictive performance of the algorithm is improved by considering the effect of empty edges.

Table 2. Recall rates on negative edge prediction

Algorithm	Epinions	Slashdot	Wikipedia
NDSE	**0.83**	**0.84**	**0.83**
SiNE	0.78	0.82	0.81

4.2 Parameter Analysis

In this subsection, we investigate the impact of δ_0, δ_1, δ_2, d. Throughout the experiments for parameter sensitivity analysis, we randomly select 80% edges as training set and the remaining 20% as test set. The random selection is repeated 5 times and the average AUC and recall will be reported.

To investigate the impact of $(\delta_0, \delta_1, \delta_2)$, we set the dimension d to be 20 and let $(\delta_0, \delta_1, \delta_2)$ take (1, 0.1, 0.9), (1, 0.2, 0.8), (1, 0.3, 0.7), (1, 0.4, 0.6) in turn. Experimental results in Table 3 show the combination (1, 0.3, 0.7) is more preferable.

To investigate the sensitivity of NDSE on d, we fix $(\delta_0, \delta_1, \delta_2) = (1, 0.3, 0.7)$ and vary d as $\{20, 32, 64\}$. Experiment results in Table 4 show that a value of d around 20 gives relatively good performance. Apparently, small d leads to information lost and embeddings lack of representation capacity. When d is large, the embedding tends to overfit.

Table 3. The impact of different thresholds

$(\delta_0, \delta_1, \delta_2)$	(1, 0.1, 0.9)	(1, 0.2, 0.8)	(1, 0.3, 0.7)	(1, 0.4, 0.6)
Recall	0.81	0.80	**0.83**	0.81
AUC	0.85	0.85	**0.85**	0.85

Table 4. The impact of different dimensions

d	20	32	64
Recall	**0.83**	0.80	0.80
AUC	**0.85**	0.84	0.84

5   Conclusion

Most of the existing signed network embedding algorithms focus on the difference between positive and negative edges, while little work focuses on the role of empty edges. In this paper, we propose NDSE based on noise contrastive estimation model and deep learning framework. In particular, we propose an objective function which fully considers the effect of empty edges and use a deep learning framework to optimize this objective function. Via experiments on three signed networks, we demonstrate that the embedding learned by NDSE can significantly improve the link prediction performance compared to representative baseline methods.

There are several directions needing further investigation. First, we will investigate how the embedding benefits other signed network mining tasks, such as node classification, anomaly detection, etc. Second, for the directed signed networks, we will investigate how does status theory work. Finally, the existing network embedding methods are all about embedding a network into Euclidean space, perhaps embedding in non-Euclidean space is an interesting research direction.

References

1. Wang, S., Tang, J., Aggarwal, C., Chang, Y., Liu, H.: Signed network embedding in social media. In: Proceedings of the 2017 SIAM International Conference on Data Mining, Commonwealth of Pennsylvania, pp. 327–335. SIAM (2017)
2. Xiao, S., Fu-Lai, C.: Deep network embedding for graph representation learning in signed networks. IEEE Trans. Cybern., 1–8 (2018)

3. Kunegis, J., Schmidt, S., Lommatzsch, A., Lerner, J., De Luca, E.W., Albayrak, S.: Spectral analysis of signed graphs for clustering, prediction and visualization. In: Proceedings of the 2010 SIAM International Conference on Data Mining, PA, USA, pp. 559–570. SIAM (2010)
4. Zheng, Q., Skillicorn, D.B.: Spectral embedding of signed networks. In: Proceedings of the 2015 SIAM International Conference on Data Mining, PA, USA, pp. 55–63. SIAM (2015)
5. Gutmann, M., Hyvärinen, A.: Noise-contrastive estimation: a new estimation principle for unnormalized statistical models. In: Yee, W.T., Mike, T. (eds.) Proceedings of the Thirteenth International Conference on Artificial Intelligence and Statistics, Sardinia, Italy. PRML, vol. 9, pp. 297–304. Proceedings of Machine Learning Research (2010)
6. Cygan, M., Pilipczuk, M., Pilipczuk, M., Wojtaszczyk, J.O.: Sitting closer to friends than enemies, revisited. In: Rovan, B., Sassone, V., Widmayer, P. (eds.) MFCS 2012. LNCS, vol. 7464, pp. 296–307. Springer, Heidelberg (2012). https://doi.org/10.1007/978-3-642-32589-2_28
7. Tian, B., Xing, C.: Deep learning based temporal information extraction framework on Chinese electronic health records. In: Meng, X., Li, R., Wang, K., Niu, B., Wang, X., Zhao, G. (eds.) WISA 2018. LNCS, vol. 11242, pp. 203–214. Springer, Cham (2018). https://doi.org/10.1007/978-3-030-02934-0_19
8. Leskovec, J., Huttenlocher, D., Kleinberg, J.: Signed networks in social media. In: Proceedings of the 28th International Conference on Human Factors in Computing Systems, pp. 1361–1370. ACM, New York (2010)

Under Water Object Detection Based on Convolution Neural Network

Shaoqiong Huang[1,2], Mengxing Huang[1,3](✉), Yu Zhang[1,2], and Menglong Li[1,2]

[1] State Key Laboratory of Marine Resource Utilization in South China Sea,
Hainan University, Haikou 570228, China
huangmx09@163.com
[2] College of Computer and Cyberspace Security, Hainan University,
Haikou 570228, China
[3] College of Information and Communication Engineering, Hainan University,
Haikou 570228, China

Abstract. Traditional underwater image processing methods have low accuracy and slow recognition speed. Underwater multi-target recognition is an important part of multi-target tracking. To solve this problem, an underwater multi-target detection method based on multi-scale convolution neural network (MSC-CNN) is proposed. Firstly, according to the problem of low contrast caused by the absorption and attenuation of underwater light, using the attenuation theory to equalize the dark channel of image to reduce the interference noise of background light in order to enhance the feature signal of object. Secondly, using MSC-CNN to fuse the feature maps generated by different convolution layers in the form of down-sampling and 1*1 convolution to extract the feature. Using the idea of Faster R-CNN for reference, the region-of-interest network is used to select regions of interest and share convolution layers. Finally, Softmax Loss is used to calculate classification probability and non-maximum suppression to correct the border position. The accuracy and test time are taken as evaluation metrics, and the recognition effect of different methods for underwater image is analyzed by comparing with the existing top algorithms. The experimental results of underwater images show that this method has advantages under various evaluation metrics.

Keywords: Object detection · Convolution neural network · Multi-scale

1 Introduction

There are many obvious shortcomings in the existing ocean detection methods [1]: Global Positioning System (GPS) has high navigation accuracy on land, but it is no longer suitable for the ocean because of the absorption of electromagnetic waves by sea water. Inertial navigation system and sonar system usually have low accuracy and lack of intuition. Compared with the complex and high-cost technology, the exploration and development of oceans by computer vision technology has the advantages of low cost, simple operation, high resolution and

© Springer Nature Switzerland AG 2019
W. Ni et al. (Eds.): WISA 2019, LNCS 11817, pp. 47–58, 2019.
https://doi.org/10.1007/978-3-030-30952-7_6

convenient information acquisition. It can be used for the protection and monitoring of marine resources, the prediction and exploration of marine economy, etc. However, due to the attenuation of light in water, the underwater imaging effect is not good.

Deep convolutional neural networks have been widely used in many areas [2], such as target detection [3], speech recognition [4]. In the field of object detection, region-based convolutional neural network (R-CNN) has developed rapidly. Among them, Faster R-CNN [5] model won the first prize in the COCO detection contest in 2015, and achieved the highest target detection accuracy in PASCAL VOC [21].

In the field of underwater recognition, paper [6] proposes an image retrieval method based on salient regions and spatial pyramids. The most important task of this method is to find the region of interest (ROI) by saliency algorithm, which can reduce the impact of background, and then extract quality features to enhance accuracy, but it is inefficient and takes up a lot of memory. Because deep convolution neural network contains different convolution layers, it can help improve classification and extract high quality feature information. paper [7] uses deep neural network to classify large-scale data, and achieves better classification results, but it is not suitable for areas with small amount of data. Another method [8] summarizes the results of SIATMMLAB using advanced in-depth learning model. In the task of species identification, the authors use different frameworks to detect the boundary boxes containing foreground, and then fine-tune a pre-trained neural network for classification. Another convolution neural network method [9] aims to detect and recognize different kinds of fish from underwater images using Fast R-CNN features. However, the robustness of the improved Faster R-CNN method for multi-scale characteristics in reference [10] still needs to be improved.

To sum up, first of all, due to the influence of underwater image imaging effect, it is not effective to directly use the original image to identify the image, so this paper uses dark channel priori to preprocess the image. Secondly, the object scales in the underwater images are quite different. If the multi-scale problem in target detection can be solved, not only the detection performance of small targets will be improved, but also the overall performance will be improved. In order to alleviate the over-fitting problem, the sample image will be expanded by rotation, translation and flipping, and the data volume will be increased. Therefore, aiming at these characteristics, this paper proposes an underwater multi-target detection method based on multi-scale characteristics. The purpose of this paper is to solve the existing problems of underwater image recognition.

Therefore, the main contributions of this paper are as follows:

(1) The multi-scale convolution neural network is used to obtain the feature map of underwater target, which combines the convolution calculation results of multi-scale, and improves the recognition accuracy.
(2) The low image quality area of underwater target recognition is studied, and an improved dark channel preprocessing method is proposed to improve the accuracy of the results.

2   Underwater Image Preprocessing Method

The serious attenuation of light in water leads to poor underwater image quality. The imaging process of underwater image is similar to that of haze image. The attenuation and scattering of the medium will result in low contrast and low visibility of the image, and the two imaging models are similar. Therefore, in theory, the background scattering of underwater image can be removed by image defogging. In most regions, some pixels always have at least one color channel with very low values. For any input image, its dark channel [12] is represented as shown in Eqs. (1)–(6). In Eq. (1), Jc represents each channel of the color image and (x) represents a window centered on the pixel X; In Eq. (2), I(x) is the image to be restored, J(x) is the original image, A is the background light component, t(x) is the transmittance; For the case of underwater image processing, the known condition is I(x), and the target value J (x) is required to be solved. Eq. (2) is slightly transformed into Eq. (3); According to the dark channel prior, the transmittance can be obtained as shown in Eq. (4), wheredefaults to 0.95; According to the attenuation theory of light, most bands of light will be strongly absorbed and attenuated when they propagate underwater. Only blue-green light has the smallest attenuation coefficient and the strongest penetration ability in water, while red light has the weakest penetration ability. Therefore, the imaging effect in sea water is blue-green. To solve the above problems, background light A is set as Eq. (5); Background light A takes the highest intensity pixel value in the dark channel of G and B channels and the lowest intensity pixel value in R channels. Estimated transmittance t(x) is usually refined by soft matting or steering filter [6] to avoid block effect in restored images. The reconstructed underwater image can be obtained as in Eq. (6) by substituting the obtained transmittance t(x) with the background light value into Eq. (2).

$$J^{dark}(x) = \min_{y \in \Omega(x)} \left(\min_{c \in \{r,g,b\}} J^c(y) \right) \tag{1}$$

$$I(x) = J(x)t(x) + A(1 - t(x)) \tag{2}$$

$$\frac{I^c(x)}{A^c} = t(x)\frac{J^c(x)}{A^c} + 1 - t(x) \tag{3}$$

$$t(x) = 1 - \omega \min_{y \in \Omega(x)} \left[\min_c \frac{J^c(y)}{A^c} \right] \tag{4}$$

$$A^c = \max_{y \in \Omega(x)} \left[\min_{c1 \in \{G,B\}} J^{c1}(y) \right] + \left[\min_{c2 \in \{R\}} J^{c2}(y) \right] \tag{5}$$

$$J(x) = \frac{I(x) - A^c}{t(x)} + A^c \tag{6}$$

There is an important problem in underwater image recognition, that is, the image quality of underwater image is not good and the contrast is not high. If the original graph is directly used to train the model, it is easy to cause problems such as slow fitting speed and low accuracy of the model. In this paper, a dark channel processing strategy is added to the training stage to improve the contrast of the sample image. Experiments show that the dark channel strategy is helpful to improve the ability of underwater image target recognition.

3   Underwater Target Recognition Method

3.1   Convolutional Neural Networks

Convolutional Neural Networks (CNN) is a kind of feedforward neural networks including convolution computation. It is a deep learning method for image recognition based on multi-layer neural networks. CNN is an unsupervised feature learning algorithm with multi-layer network structure. Low-level network mainly extracts low-level features, while high-level network mainly cognizes shape or object. The main structure of CNN includes convolution layer, pool layer and full connection layer. Its network structure is shown in Fig. 1.

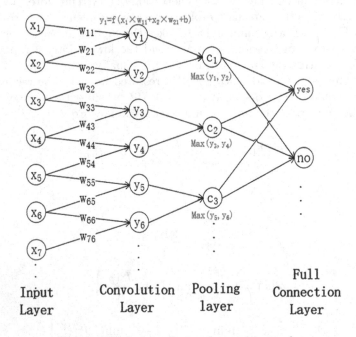

Fig. 1. Simple convolutional neural network.

Convolution layer is composed of several convolution units. The purpose of convolution operation is to extract different features of input. The first convolution layer may only extract some low-level features such as edges and lines. The higher-level network can extract more abstract and semantically rich features iteratively from the low-level features. For convolution layer i, its input is $x^{i-1} = \{x_1^{i-1}, x_2^{i-1}, x_3^{i-1} \cdots, x_m^{i-1}\}$, its corresponding convolution core is $k^i = \{k_1^i, k_2^i, k_3^i \cdots, k_n^i\}$, and its output is $x^i = \{x_1^i, x_2^i, x_3^i \cdots, x_n^i\}$, where $x_j^i = f\left(x_\lambda^{i-1} \times \sum k_v^i + b\right)$, b are biased values and f are activation functions.

The pooling layer is a network structure that performs down-sampling. Max pooling is the most common among many pooling functions. It divides the input image into several fixed regions and maximizes the output region for each sub-region. The reason why this mechanism can be effective is that the relative position of a good feature with other features is very important. In the process of downsampling, the pool layer can effectively control the amount of data, the number of parameters and the amount of calculation will also decrease, which to some extent also controls the over-fitting.

Full connection layer is a combination of feature maps obtained by convolution calculation in convolution layer. At this time, the feature map loses its high-dimensional structure in the full connection layer, and is expanded into vectors and transferred to the output layer through the activation function.

3.2 Object Detection in Faster R-CNN

CNN is an important milestone in the application of convolutional neural network (CNN) to object detection. Fast R-CNN and Faster R-CNN are improved versions of CNN. Among them, Faster R-CNN optimizes Fast R-CNN by using RegionProposal Network, and the detection accuracy and speed have been greatly improved. Faster R-CNN for object detection mainly consists of three steps: firstly, feature extraction of the image to generate a feature map; secondly, RegionProposal Network is used to extract candidate regions in the feature map, that is, to select possible object regions; then, the feature map corresponding to the candidate regions is classified and the boundary box regression is carried out. Although Faster R-CNN uses RegionProposal Network and improves detection speed on the basis of sharing convolution computation in area generation network and Fast R-CNN, it still does not solve the multi-scale problem. In this paper, the idea of Faster R-CNN is used for reference, and the concept of shared convolution of classification and detection network is used to further improve the recognition speed.

3.3 Multiscale Convolutional Neural Network

CNN has a pyramid feature structure, which includes the semantic level from low to high and the spatial characteristics from strong to weak. In the past, when using convolution neural network to extract features, only the result of the last convolution layer is often used, which makes the extracted features have strong semantic features but lose a large number of contour features. If the feature map can be used to fuse the low-level contour features with the high-level semantic features, the expression effect of the extracted features can be enhanced and the recognition accuracy can be improved. The feature pyramid structure [15] can construct a feature pyramid with high-level semantics. The feature pyramid structure is used to fuse multi-layer features to improve the feature extraction of convolution neural network. output feature mapping at multiple levels Proportionally. The characteristic pyramid structure is shown in Fig. 2.

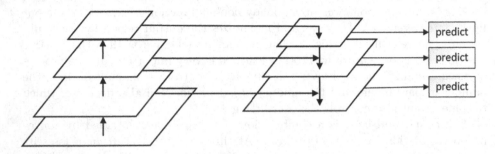

Fig. 2. Feature pyramid structure.

The network takes a single-scale image of any size as input. In the feature pyramid structure, it contains bottom-up path, top-down path and horizontal connection in a full convolution way.

Bottom-up path: In the feature pyramid network, the bottom-up path is the feedforward convolution operation of the backbone network. The scaling step of each stage is 2. The feature map of the convolution neural network is output proportionally. Top-down paths: In feature pyramid networks, top-down paths can obtain some higher resolution features by twice sampling high-level feature maps of pyramids with strong semantic information. Horizontal connection: By combining top-down features and bottom-up feature mapping, the final features will have richer semantics and more obvious contour information.

The MSC-CNN used in this paper includes five main convolution stages. The residual structures of RESNET are shown in Table 1. They are conv1, conv2_x, conv3_x, conv4_x and conv5_x, respectively. The output of these five stages is fused with multi-scale features, and the multi-scale output results are fed into the ROI POOLING layer, which improves the original single-scale convolution features into multi-scale convolution features and further enhances the robustness.

The conv1 uses a convolution core with a size of 7*7 and a step size of 2 to process the input image of 224*224 to get 112*112 dimension output. The conv2_x first process the input from conv1 with a maxpool, then use convolution cores with sizes of 1*1, 3*3 and 1*1, with channels of 64, 64 and 256 to get 56*56 dimension output.The conv3_x use convolution cores with sizes of 1*1,3*3 and 1*1, with channels of 128, 128 and 512 to get 28*28 dimension output. The conv4_x use convolution cores with sizes of 1*1, 3*3 and 1*1, with channels of 256, 256 and 1024 to get 14*14 dimension output. The conv5_x use convolution cores with sizes of 1*1, 3*3 and 1*1, with channels of 512, 512 and 2048 to get 7*7 dimension output.

Table 1. Convolutional structure of MSC-CNN.

Convolution stage	Input size	Output size
conv1	7*7 @ 64	112*112
conv2_x	3*3 max pool	56*56
	1*1 @ 64	
	3*3 @ 64	
	1*1 @ 256	
conv3_x	1*1 @ 128	28*28
	3*3 @ 128	
	1*1 @ 512	
conv4_x	1*1 @ 256	14*14
	3*3 @ 256	
	1*1 @ 1024	
conv5_x	1*1 @ 512	7*7
	3*3 @ 512	
	1*1 @ 2048	

The function of RegionProposal Network is to input an image and output a batch of rectangular candidate regions. The RegionProposal Network consists of convolution layer. Its output is divided into two ways, one is the probability fraction of the object and non-object, the other is the output of four parameters related to the boundary box, including the central coordinates x and y, the width W and the length h of the boundary box. Generally, the RegionProposal Network uses single-scale feature map to select three size of areas and three length-width ratios of anchor frames for a pixel (that is, nine anchor frames for each pixel), while the improved RegionProposal Networkk generates five feature maps of different scales on one graph, and selects anchor frames with fixed area and length-width ratio in each feature map, totally selecting 15 anchor frames. MSC-CNN is used to improve the multi-scale performance of the RegionProposal Network, the network which only uses single-scale feature map has multi-scale characteristics. Experiments show that this improvement improves the recognition accuracy.

In this work, we use MSC-CNN model to complete underwater image recognition. The main steps are as follows:

(1) Data enlargement and dark channel pretreatment.
(2) The pre-processed image is input into the backbone MSC-CNN, and the convoluted image is fused with multi-scale feature map.
(3) The convolution feature is divided into two paths, one way input into the RegionProposal Network to get the feature information of candidate box.

(4) Another way is sent to the pooling layer of the region of interest, which combines with the information from the region generation network to generate the location information of the candidate box and pass it to the next layer.
(5) After obtaining the feature map of the anchor frame, the classification probability of the candidate frame and the regression of the position coordinates are calculated by using soft Max and non-maximum suppression respectively.

4   Underwater Target Recognition Method

In order to train MSC-CNN, the data used in this paper come from the Internet. We collected and labeled the pictures of ten kinds of underwater fishes. After data enlargement, a multi-objective data set of underwater fishes containing 3000 pictures was created, all of which are real underwater images. In the experiment, the pictures are transformed into VOC2007 dataset format for training according to the need of the experiment. The validity of the proposed method is verified by testing the model with 500 test images.

4.1   The Experimental Data

The experimental results show that the dark channel priori algorithm used in this paper is helpful to enhance the color and visibility, highlight the color and contour of the target, and has a good effect on subsequent target recognition. Subsequent experiments show that the dark channel algorithm used in this paper is helpful to improve the accuracy of underwater image recognition.

(a) Original picture.   (b) Equilibrium Gray His- (c) Dark Channel Process.
                         togram.

Fig. 3. The processing effect of the algorithm (Picture 1).

Figures 3a and 4a are underwater images of . It can be seen from the original image that, due to the influence of artificial light source, the original image is brighter overall, but the contrast between target color and background color is not obvious, and the information of target contour is not prominent. If used directly, it is not conducive to underwater target recognition. Compared with the equalized gray histogram method, the algorithm proposed in this paper is

(a) Original picture.   (b) Equilibrium Gray His- (c) Dark Channel Process.
togram.

Fig. 4. The processing effect of the algorithm (Picture 2).

more effective. Figures 3c and 4c are the results of image restoration based on this algorithm. It can be seen that the restored image highlights the color and contour of the target fish ground more clearly, which will be helpful for target recognition.

4.2 Experimental Results of Fish Target Detection

The dataset in this paper contains 10 species of fish, including fish images of various postures. Experiments show that MSC-CNN has a good effect on underwater target recognition for underwater image pre-processed in dark channel. Figure 5 is the experimental result of the test sample of this algorithm. It can be seen that the algorithm adapts to multi-scale targets, and can recognize multi-target when part of the body of the target fish is covered. It can recognize many kinds of fish in a picture, and the recognition effect is good. The algorithm uses rectangular boxes of different colors to represent different kinds objects.

In this work, a lot of experiments have been done to obtain the best parameters. Firstly, the shared convolution layer in MSC-CNN is initialized by using ImageNet classified pre-training network model. Secondly, model training and testing are carried out under the framework of GTX 1080TI GPU and TensorFlow. Experiments show that when the basic learning rate is 0.0001, the momentum is 0.9, the batch is 1 and the number of iterations is 30K, the loss is minimum and the detection performance is the best. The comparison between MSC-CNN and several deep learning methods is shown in Table 2. Among them, MSC-CNN test time and accuracy on test set and verification set are superior to the comparison algorithm. By observing different iterations in Table 3, it is found that the network achieves 0.025 loss when the number of iterations is 30k, and the effect is the best.

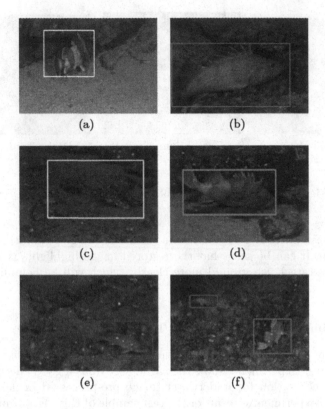

(a)   (b)

(c)   (d)

(e)   (f)

Fig. 5. MSC-CNN recognition results.

Table 2. Comparison of different methods.

Methods	Test Time(s)	Accuracy of training set/%	Verification set accuracy/%
MSC-CNN	**0.436**	**99.25**	**98.03**
CNN	0.452	98.19	96.54
Fast R-CNN	3.28	97.98	95.68
Faster R-CNN	0.471	97.65	96.31

Table 3. Comparison of the effect of different iterations.

Iteration times	MSC-CNN total loss	RPN total loss
10k	0.058	2.364
20k	0.044	4.851
30k	0.025	0.032
40k	0.024	0.033

5   Conclusion and Expectation

In order to solve the problem that the object recognition effect of underwater image is not ideal due to the refraction of light in complex underwater environment, this paper proposes an underwater multi-target detection method based on MSC-CNN, which adds a dark channel prior algorithm to the underwater image characteristics. The method enhances the contrast of the target image by improving the dark channel prior algorithm, classifies and recognizes the preprocessed image according to MSC-CNN, and finds the location and classification of the object in the image. The experimental results show that the method proposed in this paper can effectively remove background light interference, and has good effect and high accuracy for underwater image target detection. It is suitable for fish detection in uncontrolled environment. Because of the difficulty of underwater image collection, this paper does not use underwater images other than fish data images for training and detection. In the future work, the autonomous collection of image data should be combined with the underwater vehicle, and the application of the object recognition should be further combined with the underwater vehicle to further improve the recognition efficiency and real-time.

Acknowledgment. This research received financial support from the Key R&D Project of Hainan province (Grant No: ZDYD2019020), the National Key R&D Program of China (Grant No: 2018YFB1404401 and 2018YFB1404403), the National Natural Science Foundation of China (Grant No: 61662019 and 61862020), the Education Department of Hainan Province (Grant No: Hnky2019-22), the Higher Education Reform Key Project of Hainan province (Hnjg2017ZD-1) and Academician Workstation in "Hainan" Intelligent Healthcare Technologies.

References

1. He, B., Wang, G.: Seabed image stitching algorithm with different view points. Microcomput. Inf. **26**, 152–154 (2005)
2. Zhao, H., Xia, S., Zhao, J., Zhu, D., Yao, R., Niu, Q.: Pareto-based many-objective convolutional neural networks. In: Meng, X., Li, R., Wang, K., Niu, B., Wang, X., Zhao, G. (eds.) WISA 2018. LNCS, vol. 11242, pp. 3–14. Springer, Cham (2018). https://doi.org/10.1007/978-3-030-02934-0_1
3. Ancuti, C., Ancuti, C.O., Haber, T., et al.: Enhancing underwater images and videos by fusion. In: IEEE Conference on Computer Vision and Pattern Recognition, pp. 81–88. IEEE (2012)
4. Chiang, J.Y., Chen, Y.C.: Underwater image enhancement by wavelength compensation and dehazing. IEEE Trans. Image Proc. **21**(4), 1756–1769 (2012)
5. Ren, S., He, K., Girshick, R., et al.: Faster R-CNN: towards real-time object detection with region proposal networks. IEEE Trans. Pattern Anal. Mach. Intell. **39**(6), 1137–1149 (2017)
6. Zhang, M., Wu, J., Yu, H., et al.: A Novel Fish Image Retrieval Method Based on Saliency Spatial Pyramid. In: 14th International Symposium on Pervasive Systems, Exeter, United Kingdom, 2018, pp. 312–317 (2017)
7. Sarigül, M., Avci, M.: Comparison of different deep structures for fish classification. In: International Conference on CIT (2017)

8. Zhuang, P., Xing, L., Liu, Y., et al.: Marine animal detection and recognition with advanced deep learning models. In: Working Notes of CLEF (2017)
9. Li, X., Shang, M., Qin, H., Chen, L.: Fast accurate fish detection and recognition of underwater images with fast R-CNN. In: OCEANS 2015 MTS/IEEE Washington, Chicago. IEEE (2018)
10. Lin, W., He, Z.: Application of faster R-CNN model in vehicle detection. Comput. Appl. **38**(3), 666–670 (2018)
11. Garcia, R., Nicosevici, T., Cufi, X.: On the way to solve lighting problems in underwater imaging. In: Oceans, vol. 2, pp. 1018–1024. IEEE (2003)
12. He, K., Sun, J., Tang, X.: Single image haze removal using dark channel prior. In: IEEE Conference on Computer Vision & Pattern Recognition. IEEE (2009)
13. Yao, X., Wan, L., Huo, H., et al.: Aircraft target detection in high-resolution remote sensing images based on multi-structure convolutional neural network. Comput. Eng. **43**(1), 259–267 (2017)
14. He, K., Sun, J., Tang, X.: Guided image filtering. In: Daniilidis, K., Maragos, P., Paragios, N. (eds.) ECCV 2010. LNCS, vol. 6311, pp. 1–14. Springer, Heidelberg (2010). https://doi.org/10.1007/978-3-642-15549-9_1
15. Lin, T.Y., Dollar, P., Girshick, R., et al.: Feature pyramid networks for object detection, pp. 936–944 (2016)
16. He, K., Zhang, X., Ren, S., et al.: Deep residual learning for image recognition, pp. 770–778 (2015)
17. Lécun, Y., Bottou, L., Bengio, Y., et al.: Gradient-based learning applied to document recognition. Proc. IEEE **86**(11), 2278–2324 (1998)
18. Chuang, M.C., Hwang, J.N., Williams, K.: Supervised and unsupervised feature extraction methods for underwater fish species recognition. In: Computer Vision for Analysis of Underwater Imagery, pp. 33–40. IEEE (2014)
19. Krizhevsky, A., Sutskever, I., Hinton, G.E.: Imagenet classification with deep convolutional neural networks. In: Advances in Neural Information Processing Systems, pp. 1097–1105 (2012)
20. He, K., Gkioxari, G., Dollar, P., Girshick, R.: Mask r-cnn. arXiv:1703.06870 (2017)
21. Everingham, M., Van Gool, L., Williams, C.K.I., Winn, J., Zisserman, A.: The pascal visual object classes (VOC) challenge. IJCV **88**(2), 303–338 (2010)

Face Photo-Sketch Synthesis Based on Conditional Adversarial Networks

Xiaoqi Xie[1], Huarong Xu[2], and Lifen Weng[3]($\boxtimes$)

[1] Electrical Engineering, Xiamen University of Technology, Xiamen, China
xqixie@163.com
[2] Computer Science and Technology, Xiamen University of Technology, Xiamen, China
hrxu@xmut.edu.cn
[3] Academy of Design Arts, Xiamen University of Technology, Xiamen, China
2009990509@xmut.edu.cn

Abstract. Face Photo-Sketch synthesis is designed to generate an image contains rich personal information and details facial. It is widely used in law enforcement and life areas. Although some existing methods have achieved remarkable results, however, due to the gap between the synthetic and real image distribution, the synthetic image does not achieve the expected effect. To solve this problem, In this paper, we proposed conditional generation adversarial networks (CGAN) based on arts drawing steps constrain. We divide the process into two stages. In the first stage, we take the original face photos which concat noise z as input, generating stage 1 low resolution synthesize sketches. In the second stage take the stage 1 results and original face photo as inputs, yielding stage 2 high resolution synthesize sketches, which can express more natural textures and details. To add realism, we train our network using an adversarial loss. Experiments have shown that, Compared with previous methods, our results generate visually comfortable face sketches. And express more natural textures and details.

Keywords: Face Photo-Sketch synthesis ·
Arts drawing steps constrain ·
Conditional generation adversarial networks

1 Introduction

The face sketch has rich shadow texture and strong three-dimensional texture, which can vividly express the characteristics and personality of the character, so it is widely used in law implementation. For example, in the process of law enforcement, in most cases, the real face of the suspect cannot be obtained. At this time, the best alternative is usually based on sketches of eyewitness recalls which can help the police quickly narrow down and lock the suspect [5,11,13,18,25]. In terms of life crafts, passenger often keep a sketch by the painter in a tourist attraction or scenic spot as a souvenir. In addition, it has

Supported by Project of Fujian Science and Technology Department 2019I0036.

a wide range of applications, such as digital entertainment [9,14]. At present, face sketches are mostly drawn by painters or professional artists create them through painting tools. They need long-term practice to have this skill, and this kind of creative method is inefficient. So, the automatic generated face sketch technique is becoming the research hot in these application area.

There are some algorithms such as Sketch2Photo [4] and Photos-ketcher [1], require a large number of fine extraction features, and need complex processing as cutting images to make the image more realistic. In recent years, the convolutional neural network in deep learning is very popular and provides a powerful method for facial sketch synthesis [3,19,31]. Among them, the extended GAN network [8] application to the face sketch synthesis method is especially prominent. Image-to-Image Translation with Conditional Adversarial Networks [8] uses CGAN to solve face sketch synthesis. This network can not only learn the mapping relations between input image and output image, but also learn the loss function for training mapping relations. Unpaired Image-to-Image Translation Zhu et al. [35] proposed use Cycle-Consistent Adversarial Networks (CycleGAN) to solve where paired training data does not exist problem. The article introduces a cycle consistency loss to make $F(G(X)) \approx X$ (and vice versa). However, the meaning is to convert source domain images back into target domain images, and its distribution is still the same.

Although these methods have achieved remarkable results, due to the limitations of their structural consistency, accurate depicting face photos/sketches remains challenging. In addition, the synthesized image mostly yield blurred effects, leading to the synthesized image is unrealistic. In order to solve this problem, we proposed conditional generation adversarial networks (CGAN) based on arts drawing steps constrain.

2   Related Work

In generation adversarial network we must train two models at the same time. The generator mainly learns the real image distribution to make the image generated by itself more realistic, to fool the discriminator. The discriminator needs to make a true and false judgement on the received picture. The probability that the prediction of the fixed image is true is close to 0.5 [6]. Since then, many improvements and interesting applications have been proposed [7]. recently, the great success achieved by conditional Generative Network(cGAN) [15]. In a series of image-to-image translation tasks [10]. For example, CycleGAN [35] generate high quality sketches using multi-scale discriminators [26]. Wang et al. [27] recommend to first generate a sketch using the vanilla cGAN [26] and refine it using a postprocessing method called back projection. Jun Yu et al. [12] proposed to improve the CA-GAN after GAN to generate a realistic synthesized sketch. Di et al. combining a convolutional variational Autoencoder with cGAN for attribute-aware facial sketch synthesis [3]. Stacked networks have also made great progress in this area, such as image super-resolution reconstruction and generate high resolution image with photo-realistic details [12]. They have all achieved very good results. Affected by these success stories, we are interested in using CGAN to generate sketch photos, but found that the effect is not the best, so it is proposed based on arts drawing steps constrain CGAN to improve the effect. Our method divide into two stages and is different from the previous one when choosing generation and discriminator. We are Using Encoder-Decoder [2] as Generator, and for discriminator we use patchGAN [16].

Encoder-Decoder is a very common model framework for deep learning. Many previous solutions [10,29,30,34] to problems in this area have used an encoder-decoder network [17]. For example Unsupervised algorithm auto-encoding [2], application of image caption, the neural network machine translation NMT model all trained with encoder-decoder. Therefore, encoder-decoder not a model, but a type of framework. The Encoder and Decoder sections can be anything, as text, voice, image. Models can be CNN [32], RNN, LSTM, and more. So we choose Encoder-Decoder. It is an end-to-end learning algorithm.

GANs network structure has been shown in some experiments to be unsuitable for the field of images requiring high resolution and hight detail retention. Some people have designed PatGAN [16] based on this situation. The difference between PatchGAN is in the discriminator. The general GAN is a vector that only need to output a true or false which represents the evaluation of the whole image, but the PatvhGAN output is an $N \times N$ matrix, each element such as a(i,j), has only two choices of True or False(label is a matrix of $N \times N$, each element is True or False), and the result is often through the convolution layer to achieve. The discriminator makes a true and false discriminant for each patch, and average the results of all patches of a picture as the final discriminator output.

3   Methodology

3.1   Model

Figure 1 shows the model structure. We all know in order to express details and textures, the painter's painting is divided into many steps. In order to produce high resolution sketches like a painter, we divided our process into two stages based on arts drawing steps constrain. In the first stage, we take the original face photos which concat noise z as input, generating stage 1 low resolution synthesize sketches x_{g1}. In the second stage take the stage 1 results x_{g1} and original face photo as inputs, yielding stage 2 high resolution synthesize sketches x_{g2}, which can express more natural textures and details. G we choose encoder-decoder, D we use PatchGAN.

Generator(G1,G2). The G1 network take photos and noise as input, G1 structure as follows: Conv1(16)-Conv2(32)-Pooling2(32)-Conv3(64)-Poling 3(32)-Conv4(128)-Conv5(128)-Deconv6(128)-Deconv7(64)-conv8(32)-conv9(1). Conv(K) denotes K-channel convolutional layers, pooling is max-pooling. Convolution and polling layers are used to extract advanced feature. G2 network have two modules. The input one is G1 generated synthetic sketches and another input is photos. G2 structure same as G1.

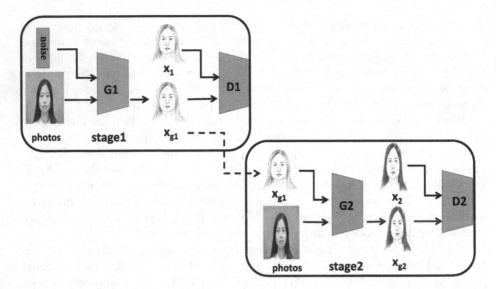

Fig. 1. Two-stage trainging networks. Stage 1 generates x_{1g}. Stage 2 concat photos and x_{1g} put into G2 generates sketch image.

Discriminator(D1,D2) using patch-Decoder discriminator D1, there are 16 patches in every iterative training with G1. PatchGAN discriminator preserving of high-frequency details. All the convolutional layers in D1 have a filter size of 3×3. D2 structure same as D1.

3.2 Data Set

We build a two stages face photo-sketch database. 188 faces provided by CUHK students database [24]. For each face, there are two stages sketches by our professional painter and a photo taken in a frontal pose, under normal lighting condition. In CHUK student database, we choose 88 faces for training and 100 faces for test.

3.3 Objective Function

The GANs there are two networks generators and discriminators. The goal of G is to generate a real images as much as possible to deceive the discriminant network D. The goal of D is to separate the G generated image from the real image. Conditional GAN is another variant where the generator is conditioned on additional variables such as category labels, partial data for image restoration, date from different modalities. If the conditional variables is a category label, it can be seen that CGAN is an improvement to turning a purely unsupervised GAN into a supervised model. The objective of a conditional GAN can be defined as:

$$L_{cGAN}(G, D) = E_{x,y}[log D(x, y)] + E_{x,z}[log(1 - D(x, G(x, z)))] \qquad (1)$$

where z, is noise as input, x, is a corresponding face image, y, the output images, are sampled from the true data distribution $P_{data}(x, y)$.

To add realism, we train our network using an adversarial loss, similar to Generative Adversarial Networks(GANs) [6], such that the synthesized sketches are indistinguishable from target photos. In order to measures the error between the synthesized sketch image and target sketch, we complement the adversarial loss with a regularization loss [8, 21, 23]. Stage 2 is built on stage 1 to generate realistic target sketches. where $x_g = (x, z)$ is generated by stage 1. Stage 2 not using noise, assuming that the noise has already been maintained in the stage 1 generated x_g, Our final objective is define as follow:

$$G^* = L_A + \lambda L_1 \tag{2}$$

where L_A is the adversarial loss, L_1 is the loss based on the L_1-norm between the synthesized image and the target. λ is weight. G is tries to minimized this loss to updates its parameters, against D is that tries to maximized loss is defined as

$$L_A = \min_G \max_D L_{cGAN} \tag{3}$$

The L_1 loss is present synthesized sketch image and the target sketch loss. x indicate target sketch, x_g indicate synthesized sketch.

$$L_1 = \| x - x_g \|_1 \tag{4}$$

3.4 Optimization

To optimize our networks, follow [8, 12] we alternate between one gradient descent step on D, then one step on G. All networks are trained using Adam solver. In our experiments, we use batch size 1 and momentum parameters $\beta = 0.9$. Follow the GAN approach [14], we model include two-stage, every stage has a generator and a discriminator. which are sequentially denoted by $G1$, $D1$, $G2$, $D2$. As follows Algorithm 1.

4 Experimental Results

4.1 Data Set

We build a two stages face photo-sketch database. 188 faces provided by CUHK students database [24]. For each face, there are two stages sketches by our professional painter and a photo taken in a frontal pose, under normal lighting condition. In CHUK student database, we choose 88 faces for training and 100 faces for test.

In Sect. 4.2, we use CUHK database train our model and do three experiments. first, experimental setting and evaluation of the proposed method are discussed in detail. Results are compared with previous state-of-the-art-methods. Second, we compare with who also use two-stage generation model. Third, we do internal comparisons, using the same structural model for one-stage and two-stage sketch generation comparisons. We then provide a quantitative comparison with the method mentioned in the first part in Sect. 4.2. Finally, we show a few examples of our model results in Fig. 5.

Alogrithm 1 Optimization procedure of our network

Input: Set of synthetic images x_{g1} ,x_{g2} ,and real images x_1,x_2,a face photo y,max number of steps(T);

Outoput:optimal G1,D1,G2,D2;

initial G1,D1,G2,D2;

for t=1,. . .,T **do**

1. select one training instance:

{Set of synthetic images x_{g1} ,x_{g2} ,and real images x_1,x_2,a face photos y}

2. Estimate the first stage sketch imag:

   x_{g1} =G1(y,z)

3. Estimate the first stage sketch imag:

   x_{g2} =G2(x_{g1} ,y)

4. Update D1:

   D1=argmin$_{D1}$L$_A$(G1,D1)

5. Update D2:

   D2=argmin$_{D2}$L$_A$(G2,D2)

6. Update G1:

   G1=argmax$_{G1}$L$_A$(G1,D1)+λL$_1$(x_1,x_{g1})

7. Update G2:

   G2=argmax$_{G2}$L$_A$(G2,D2)+λL$_1$(x_2,x_{g2})

end

4.2 Qualitative Results

In the first set of experiments, we perform face photo-sketch synthesis on the CUFS datasets and compared with existing advanced methods: MWF [33], MRF [28] and cGAN [8]. And we attempt to recover the image directly from attributes without going to the intermediate stage of sketch. The entire network in Fig. 2 is trained stage-by-stage using caffe.

In the second set of experiments. As show Fig. 3, we compare results with CA-GAN and SCA-GAN [12], which also uses a two-stage generation model.

In the third set of experiments, And we attempt to recover the image directly without going to the intermediate stage of sketch. As show in Fig. 4. And we show some our model results in Fig. 5.

4.3   Quantitative Results

In the CGNs, the discriminator and generator objective functions are usually used to measure how they do each other. But this does not measure the quality and diversity of the generated images. Usually, we use the IS (inception score) and FID (Fr'echet Inception distance) indicator to evaluate different and GAN model. In this work, we choose the Fr'echet Inception distance (FID) to evaluate

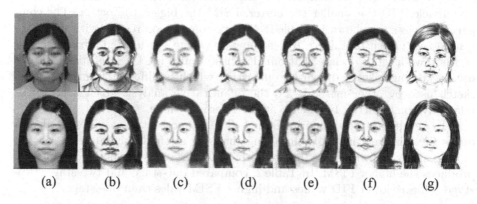

(a)   (b)   (c)   (d)   (e)   (f)   (g)

Fig. 2. Comparison of the sketch synthesis result using three different approach and our model: (a) photos; (b) sketches by artist from CUFS; (c) MWF; (d) MRF; (e) cGAN; (f) our model; (g) sketches by our artist

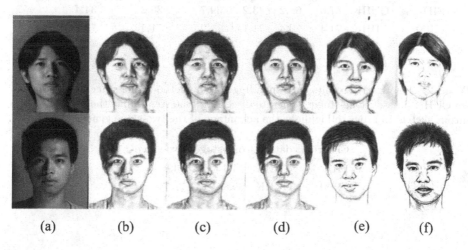

(a)   (b)   (c)   (d)   (e)   (f)

Fig. 3. (a) photos; (b) cGAN; (c) CA-GAN; (d) SCA-GAN; (e) our model; (f) sketches by our artist

the realism and variation of synthesized sketches [12,20]. FID values present distances between synthetic and real data distributions. Therefore, lower FID is better. We use all test samples to compute the FID. In FID we use inception network to extract the characteristics of the middle layer [12,22].

We also use the Feature Similarity index Metric (FSIM) [22] between the synthesized image and the corresponding ground truth image to objectively evaluate the quality of the synthesized image [12]. These two indicators have been compared in article [12], we compare the experimental data of our model based on this article. FID the smaller the better. FSIM the bigger the better. The comparison results are shown in Table 1. And we comparison results with one-stage synthesized sketch In Table 2, which show two-stage synthesized advantage.

Here we have to declare that our database is also a CUFS database, but we use a two-stage sketch drawn by our artist, which is different from the CUFS sketch used by other methods. So the quantitative analysis given here is also relatively comparative. In a sense, it has reference value.

The results are presented in Table 1. The proposed two-stage method produces the relatively low FID. Implying that our method generated images are more realistic than the ones generated by most previous methods. our method produces the higher FISM. In Table 2, compared one-stage and two-stage, two-stage achieves lower FID values and higher FSIM vales than one-stage.

Table 1. Quantitative results corresponding to different methods use CUHK dataset.

Criterion	dataset	MWF	MRF	cGAN	CA-GAN	SCA-GAN	Our model
FID	CUHK	87.0	68.2	43.2	39.7	36.2	34.4
FSIM	CUHK	71.4	70.4	71.1	71.2	71.4	72.7

Table 2. Quantitative results corresponding to our methods one-stage and two-stage use CUHK dataset. One-stage: Generate directly in one step, input the original image output sketch. In order to highlight the advantage of the two-stage generation model.

Criterion	dataset	one-stage	two-stage
FID	CUHK	44.3	34.4
FSIM	CUHK	69.8	72.7

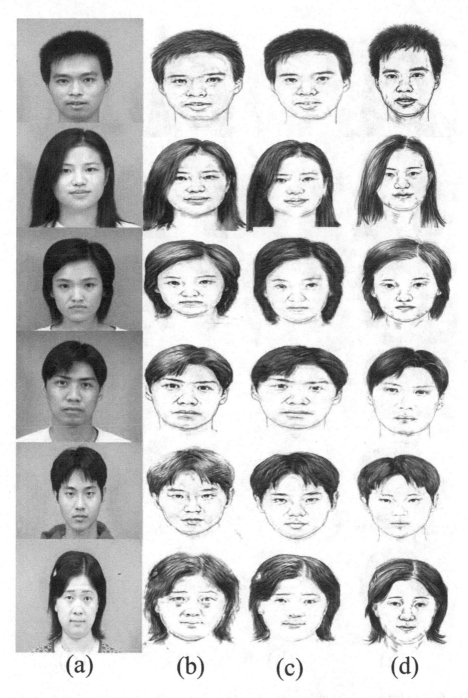

(a)   (b)   (c)   (d)

Fig. 4. (a) photos; (b) one-Stage results. Omitting a stage, directly using the results of one stage test; (c) two-Stage results(output from our method); (d) sketches by our artist

(a)   (b)   (c)   (d)

Fig. 5. (a) photos; (b) stage 1 results; (c) stage 2 results(our model); (d) sketches by our artist

5   Conclusion

In this paper, we proposed based on arts drawing steps constrain conditional generation adversarial networks (CGAN). In the first stage, we take the original face photos which concat noise z as input, generating stage 1 low resolution synthesize sketches. In the second stage take the stage 1 results and original face photo as inputs, yielding stage 2 high resolution synthesize sketches, which can express more natural textures and details. Proved by qualitative and quantitative results, our approach dramatically improves the realism of the synthesized face photos and sketches than most previous methods. Also show that two-stage generation is better than the one-stage generation. Future work includes extending this framework to three-stage, four-stage etc.

References

1. Chen, T., Cheng, M.M., Shamir, A., Shamir, A., Hu, S.M.: Sketch2photo: internet image montage. In: ACM SIGGRAPH Asia (2009)
2. Cho, K., et al.: Learning phrase representations using RNN encoder-decoder for statistical machine translation. Computer Science (2014)
3. Di, X., Patel, V.M.: Face synthesis from visual attributes via sketch using conditional VAEs and GANs. arXiv preprint arXiv:1801.00077 (2017)
4. Eitz, M., Richter, R., Hildebrand, K., Boubekeur, T., Alexa, M.: Photosketcher: interactive sketch-based image synthesis. IEEE Comput. Graph. Appl. **31**(6), 56 (2011)
5. Gao, X., Zhong, J., Jie, L., Tian, C.: Face sketch synthesis algorithm based on e-hmm and selective ensemble. IEEE Trans. Circuits Syst. Video Technol. **18**(4), 487–496 (2008)
6. Goodfellow, I.J., et al.: Generative adversarial nets. In: International Conference on Neural Information Processing Systems (2014)
7. Han, Z., Tao, X., Li, H.: Stackgan: text to photo-realistic image synthesis with stacked generative adversarial networks. In: Computer Vision and Pattern Recognition (2016)
8. Isola, P., Zhu, J.Y., Zhou, T., Efros, A.A.: Image-to-image translation with conditional adversarial networks. In: Computer Vision and Pattern Recognition (2016)
9. Iwashita, S., Takeda, Y., Onisawa, T.: Expressive facial caricature drawing. In: IEEE International Fuzzy Systems Conference (1999)
10. Johnson, J., Alahi, A., Li, F.F.: Perceptual losses for real-time style transfer and super-resolution. In: European Conference on Computer Vision (2016)
11. Uhl Jr., R.G., Lobo, N.D.V., Kwon, Y.H.: Recognizing a facial image from a police sketch. In: IEEE Computer Society Conference on Computer Vision and Pattern Recognition (1996)
12. Jun, Y., Shengjie, S., Fei, G., Dacheng, T., Qingming, H.: Towards realistic face photo-sketch synthesis via composition-aided GANs. In: Computer Vision and Pattern Recognition (2017)
13. Konen, W.: Comparing facial line drawings with gray-level images: a case study on PHANTOMAS. In: von der Malsburg, C., von Seelen, W., Vorbrüggen, J.C., Sendhoff, B. (eds.) ICANN 1996. LNCS, vol. 1112, pp. 727–734. Springer, Heidelberg (1996). https://doi.org/10.1007/3-540-61510-5_123

14. Koshimizu, H., Tominaga, M., Fujiwara, T., Murakami, K.: On kansei facial image processing for computerized facial caricaturing system PICASSO. In: IEEE International Conference on Systems (1999)
15. Krizhevsky, A., Sutskever, I., Hinton, G.E.: Imagenet classification with deep convolutional neural networks. In: Advances in Neural Information Processing Systems, pp. 1097–1105 (2012)
16. Li, C., Wand, M.: Precomputed real-time texture synthesis with markovian generative adversarial networks. In: Leibe, B., Matas, J., Sebe, N., Welling, M. (eds.) ECCV 2016. LNCS, vol. 9907, pp. 702–716. Springer, Cham (2016). https://doi.org/10.1007/978-3-319-46487-9_43
17. Lin, Z., Lei, Z., Xuanqin, M., David, Z.: FSIM: a feature similarity index for image quality assessment. IEEE Trans. Image Processing **20**(8), 2378 (2011). A Publication of the IEEE Signal Processing Society
18. Liu, Q., Tang, X., Jin, H., Lu, H., Ma, S.: A nonlinear approach for face sketch synthesis and recognition. In: IEEE Computer Society Conference on Computer Vision and Pattern Recognition (2005)
19. Liu, S., Yang, J., Huang, C., Yang, M.H.: Multi-objective convolutional learning for face labeling. In: Proceedings of the IEEE Conference on Computer Vision and Pattern Recognition, pp. 3451–3459 (2015)
20. Lucic, M., Kurach, K., Michalski, M., Gelly, S., Bousquet, O.: Are GANs created equal? A large-scale study. In: Advances in Neural Information Processing Systems, pp. 700–709 (2018)
21. Pathak, D., Krahenbuhl, P., Donahue, J., Darrell, T., Efros, A.A.: Context encoders: feature learning by inpainting. In: IEEE Conference on Computer Vision and Pattern Recognition (2016)
22. Phillips, P.J., Moon, H., Rauss, P., Rizvi, S.A.: The FERET evaluation methodology for face-recognition algorithms. In: Proceedings of IEEE Computer Society Conference on Computer Vision and Pattern Recognition, pp. 137–143. IEEE (1997)
23. Shrivastava, A., Pfister, T., Tuzel, O., Susskind, J., Wang, W., Webb, R.: Learning from simulated and unsupervised images through adversarial training. In: Computer Vision and Pattern Recognition (2016)
24. Tang, X., Wang, X.: Face photo recognition using sketch. In: International Conference on Image Processing (2002)
25. Tang, X., Wang, X.: Face sketch synthesis and recognition. In: Proceedings Ninth IEEE International Conference on Computer Vision, pp. 687–694. IEEE (2003)
26. Wang, L., Sindagi, V.A., Patel, V.M.: High-quality facial photo-sketch synthesis using multi-adversarial networks, pp. 83–90 (2018)
27. Wang, N., Zha, W., Jie, L., Gao, X.: Back projection: an effective postprocessing method for GAN-based face sketch synthesis. Pattern Recogn. Lett. **107**, S0167865517302180 (2017)
28. Wang, X., Tang, X.: Face photo-sketch synthesis and recognition. IEEE Trans. Pattern Anal. Mach. Intell. **31**(11), 1955–1967 (2009)
29. Wang, X., Gupta, A.: Generative image modeling using style and structure adversarial networks. In: Leibe, B., Matas, J., Sebe, N., Welling, M. (eds.) ECCV 2016. LNCS, vol. 9908, pp. 318–335. Springer, Cham (2016). https://doi.org/10.1007/978-3-319-46493-0_20
30. Yoo, D., Kim, N., Park, S., Paek, A.S., Kweon, I.S.: Pixel-level domain transfer. In: Leibe, B., Matas, J., Sebe, N., Welling, M. (eds.) ECCV 2016. LNCS, vol. 9912, pp. 517–532. Springer, Cham (2016). https://doi.org/10.1007/978-3-319-46484-8_31

31. Zhang, S., Ji, R., Hu, J., Gao, Y., Lin, C.W.: Robust face sketch synthesis via generative adversarial fusion of priors and parametric sigmoid. In: IJCAI, pp. 1163–1169 (2018)

32. Zhao, H., Xia, S., Zhao, J., Zhu, D., Yao, R., Niu, Q.: Pareto-based many-objective convolutional neural networks. In: Meng, X., Li, R., Wang, K., Niu, B., Wang, X., Zhao, G. (eds.) WISA 2018. LNCS, vol. 11242, pp. 3–14. Springer, Cham (2018). https://doi.org/10.1007/978-3-030-02934-0_1

33. Zhou, H., Kuang, Z., Wong, K.Y.K.: Markov weight fields for face sketch synthesis. In: IEEE Conference on Computer Vision and Pattern Recognition (2012)

34. Zhou, Y., Berg, T.L.: Learning temporal transformations from time-lapse videos. In: Leibe, B., Matas, J., Sebe, N., Welling, M. (eds.) ECCV 2016. LNCS, vol. 9912, pp. 262–277. Springer, Cham (2016). https://doi.org/10.1007/978-3-319-46484-8_16

35. Zhu, J.Y., Park, T., Isola, P., Efros, A.A.: Unpaired image-to-image translation using cycle-consistent adversarial networks. In: IEEE International Conference on Computer Vision (2017)

Research and Implementation
of Vehicle Tracking Algorithm
Based on Multi-Feature Fusion

Heng Guan[1]([⊠]), Huimin Liu[1], Minghe Yu[2], and Zhibin Zhao[1]

[1] School of Computer Science and Engineering, Northeastern University,
Shenyang 110819, China
guanheng@stumail.neu.edu.cn
[2] School of Software, Northeastern University, Shenyang 110819, China

Abstract. The main task of multi-object tracking is to associate targets in diverse images by detected information from each frame of a given image sequence. For the scenario of highway video surveillance, the equivalent research issue is vehicles tracking, which is necessary and fundamental for traffic statistics, abnormal events detection, traffic control et al. In this paper, a simplified and efficient multi-object tracking strategy is proposed. Based on the position and intersection-over-union (IOU) of the moving object, the color feature is derived, and unscented Kalman filter is involved to revise targets' positions. This innovative tracking method can effectively solve the problem of target occlusion and loss. The simplicity and efficiency make this algorithm applicable for the perspective of real-time system. In this paper, highway video recordings are explored as data repository for experiments. The results show that our method outperforms on the issue of vehicle tracking.

Keywords: Video surveillance · Multiple object tracking · Data association

1 Introduction

Multi-object tracking is an important research issue in computer vision. It has a wide range of applications, such as video surveillance, human-computer interaction, and driverless driving. The main task of multi-object tracking is to correlate the moving objects detected in the video sequence and plot the trajectory of each moving object, as shown in Fig. 1. These trajectories will contribute to abnormal event report, traffic control and other applications.

In the practical application of high-speed scenes, most of the surveillance cameras are installed on the roadside, so in the process of video surveillance, vehicles often occlude each other. When the target vehicle reappears in the video after occultation, it is likely recognized as new, which results in the loss of the original target. A simplified and efficient correlation algorithm is proposed in this paper. After calculating the correlation degree of each object in different images by its position, motion and color information, we match and merge those objects with a high degree as an identified target.

© Springer Nature Switzerland AG 2019
W. Ni et al. (Eds.): WISA 2019, LNCS 11817, pp. 72–78, 2019.
https://doi.org/10.1007/978-3-030-30952-7_8

Fig. 1. Vehicle detection and tracking on a highway

The main contributions of this paper are listed below:

- A vehicle location prediction model is proposed to accurately predict the location when occluded.
- A data association method based on multi-feature fusion is proposed for data association for simplification and efficiency.
- Tracking algorithm is improved on real data sets as experiments.

2 Related Work

From RCNN [1] series to the SSD [2] and YOLO [3] series, many achievements appear in object detection. As YOLOv3 is an end-to-end object detection method, it runs very fast [4, 5].

For video analysis, multi-object tracking has been a research focus which consists of Detection-Free Tracking (DFT) and Detection-Based Tracking (DBT) [6]. Because the vehicle in the video is tracked continuously, DBT is considered in this paper.

Based on DBT, many researchers propose different solutions [7–9]. However, these methods do not involve image features, the results are not acceptable for occlusion.

CNN is also introduced for multi-object tracking [10–13]. Based on SORT [8, 14, 15] uses CNN to extract the image features of the tracking target. Although CNN provides more precision, it shows powerlessness in real-time computation. Because the performances on single object tracking in complex scenarios have been improved [16–18], [19] transforms the multi-object tracking problem into multiple single object tracking problems, and also achieves good performance.

For location revision, Kalman filter [20] can filter the noise and interference and optimize the target state. However, Kalman filter is only applicable to Gauss function and deals with linear model.

Based on the above works, this paper proposes a method on multi-object tracking by correlating the color and the location features. It predicts the location when the target is lost in an image, and then applies the unscented Kalman Filter for calibration.

3 Data Association

Given an image sequence, we employ object detection module to get n detected objects in the t-th frame as $S^t = \{S_1^t, S_2^t, \ldots, S_n^t\}$. The set of all the detected objects from time t_s to t_e is $S^{t_s:t_e} = \{S_1^{t_s:t_e}, S_2^{t_s:t_e}, \ldots, S_n^{t_s:t_e}\}$, where $S_i^{t_s:t_e} = \langle S_i^{t_s}, S_i^{t_s+1}, \ldots, S_i^{t_e}\rangle (i \in n)$ is the trajectory of the object i-th from t_s to t_e. The purpose of multi-object tracking is to find the best sequence of states of all objects, which can be modeled by using MAP (maximum a posteriori) according to the conditional distribution of the states of all observed sequences, defined as:

$$\bar{S}^{t_s:t_e} = \arg\max_{S^{t_s:t_e}} P(S^{t_s:t_e}|O^{t_s:t_e}) \tag{1}$$

For more accuracy, prediction on the object position according to the historical trajectory is necessary. Equations (2) and (3) show the prediction method.

$$S_i^t = \left\langle Loc_{S_i^t}, MotionInfo_{S_i^t}, Features_{S_i^t} \right\rangle \tag{2}$$

$$\begin{cases} Loc_{S_i^t} = \left[\left(x_{S_i^t}, y_{S_i^t} \right), \left((x+w)_{S_i^t}, (y+h)_{S_i^t} \right) \right] \\ MotionInfo_{S_i^t} = \left[\bar{v}_{S_i^t}, \bar{a}_{S_i^t}, k \right] \\ Features_{S_i^t} = \left[color_{S_i^t}, type_{S_i^t} \right] \end{cases} \tag{3}$$

$Loc_{S_i^t}$ is the location of i-th object in t-th frame, including upper left coordinates $\left(x_{S_i^t}, y_{S_i^t} \right)$ and lower right coordinates $\left((x+w)_{S_i^t}, (y+h)_{S_i^t} \right)$. $MotionInfo_{S_i^t}$ is motion information of i-th object in t-th frame, including average velocity $\bar{v}_{S_i^t}$, average acceleration $\bar{a}_{S_i^t}$ and motion direction k; $Features_{S_i^t}$ is characteristic information of i-th object in t-th frame, including color features $color_{S_i^t}$ and object class $type_{S_i^t}$.

Kalman filter is used to calibrate the position. However, Kalman filter algorithm is applicable to linear model and does not support multi-object tracking. Therefore, UKF is used for statistical linearization called nondestructive transformation. UKF first collects n points in the prior distribution, and uses linear regression for non-linear function of random variables for higher accuracy.

In this paper, three data are used for data matching – location information, IOU (Intersection over Union) and color feature. When calculating the position information of each S_i^{t-1} and S_i^t, we use the adjusted cosine similarity to calculate the position similarity of S_i^{t-1} and S_i^t. Formula for calculating position similarity $L_{confidence}$ between S_i^{t-1} and S_i^t is defined as (4).

$$L_{confidence} = \frac{\sum_{k=1}^{n} \left(Loc_{S_i^t} - \overline{Loc}_{S_i^t} \right)_k \left(Loc_{S_i^{t-1}} - \overline{Loc}_{S_i^{t-1}} \right)_k}{\sqrt{\sum_{k=1}^{n} \left(Loc_{S_i^t} - \overline{Loc}_{S_i^t} \right)_k^2} \sqrt{\sum_{k=1}^{n} \left(Loc_{S_i^{t-1}} - \overline{Loc}_{S_i^{t-1}} \right)_k^2}} \tag{4}$$

The object association confidence between S_i^{t-1} and S_i^t is shown as:

$$\text{Confidence}\left(S_i^{t-1}, S^t\right) = \left\{c_1\left(S_i^{t-1}, S_1^t\right), c_2\left(S_i^{t-1}, S_2^t\right), \ldots, c_n\left(S_i^{t-1}, S_n^t\right)\right\} \tag{5}$$

Thereby $Loc_{S_i^t}$ is:

$$Loc_{S_i^t} = Loc_{S^t_{\text{maxIndex}\left(\text{Confidence}\left(S_i^{t-1}, S^t\right)\right)}} \tag{6}$$

4   Experiments

The real high surveillance videos are used as the experimental data set, and three different scenes are intercepted: ordinary road section, frequent occlusion and congestion. Each video lasts about 2 min with a total of 11338 pictures annotated according to the MOT Challenge standard data set format.

The performance indicators of multi-object tracking show the accuracy in predicting the location and the consistency of the tracking algorithm in time. The evaluation indicators include: MOTA, combines false negatives, false positives and mismatch rate; MOTP, overlap between the estimated positions and the ground-truth averaged over the matches; MT, percentage of ground-truth trajectories which are covered by the tracker output for more than 80% of their length; ML, percentage of ground-truth trajectories which are covered by the tracker output for less than 20% of their length; IDS, times that a tracked trajectory changes its matched ground-truth identity [6].

Firstly, we implement the basic IOU algorithm to calculate the association of targets. The comparison is between IOU, SORT and IOU17. In the experiment, our method sets $T_{minhits}$ (shortest life length of the generated object) as 8, T_{maxdp} as 30; the same to SORT. IOU17 sets $T_{minhits}$ as 8, σ_{iou} as 0.5; the objective detecting accuracy is set as 0.5 and the results are shown in Tables 1, 2 and 3.

Table 1. Accuracy for ordinary section

Method	MOTA($\uparrow$)	MOTP($\uparrow$)	MT($\uparrow$)	ML($\downarrow$)	IDS($\downarrow$)
IOU	43.1	76.1	22.8%	30.7%	238
IOU+ position prediction	52.5	77.2	24.6%	26.8%	178
IOU+ color feature	53.2	76.7	23.2%	27.8%	184
IOU+ position prediction + color feature	55.9	78.6	26.6%	24.5%	147
SORT [8]	48.6	75.5	21.2%	29.4%	266
IOU17 [9]	51.4	76.8	23.1%	27.6%	192

It is illustrated from Tables 1, 2 and 3 that the method with predicted position and color features we proposed in this paper outperforms other methods. On frequently occlusion scenario, predicted position method greatly improves the accuracy. In congestion sections, color features are more helpful in accuracy than location prediction.

Table 2. Accuracy for frequently occlusion section

Method	MOTA(↑)	MOTP(↑)	MT(↑)	ML(↓)	IDS(↓)
IOU	32.1	73.3	13.1%	48.3%	482
IOU+ position prediction	39.7	76.6	16.3%	40.8%	314
IOU+ color feature	37.6	75.4	15.2%	42.5%	357
IOU+ position prediction + color feature	42.8	78.6	18.6%	38.7%	286
SORT [8]	34.1	76.0	14.6%	46.7%	477
IOU17 [9]	36.9	78.6	16.2%	41.3%	376

Table 3. Accuracy for congestion sections

Method	MOTA(↑)	MOTP(↑)	MT(↑)	ML(↓)	IDS(↓)
IOU	38.2	75.3	18.6%	36.3%	369
IOU+ position prediction	43.3	76.5	19.8%	32.4%	287
IOU+ color feature	45.5	77.3	21.6%	30.2%	245
IOU+ position prediction + color feature	47.9	79.3	23.2%	29.5%	212
SORT [8]	39.2	76.8	17.3%	34.2%	385
IOU17 [9]	42.8	77.8	19.6%	33.6%	302

The method in this paper does not predict the position in the case of amble because of the large deviation. However, the color feature doesn't change in amble. For amble, the color feature will make the accuracy significantly increase.

SORT [8] and IOU17 [9] only use the position information of the target to correlate the data over image feature and position prediction. In the case of frequent congestion and occlusion, the vehicle can't be tracked effectively because of the decrease of the accuracy of the object detection and the occlusion of the tracking target.

5   Conclusions

Target loss happens when occlusion and other events occur in a video surveillance. In this paper, we use linear regression to analyze the vehicle's historical trajectory, and then predict the position of the vehicle when it disappears in a video frame, and recognize the target when the vehicle appears again by the predicted position and color feature. Experiments show that the algorithm also shows it sufficiency for occlusion and congestion.

This method only uses color feature as extracted image feature. Although the computation is light, the deviation for color similarity of targets exits. For the future work, we will research shallow CNN to extract image features to enhance the performances in terms of efficiency and differentiation.

Acknowledgement. This research is supported by the National Key R&D Program of China under Grant No. 2018YFB1003404.

References

1. Girshick, R., Donahue, J., Darrelland, T., et al.: Rich feature hierarchies for accurate object detection and semantic segmentation. In: IEEE Conference on Computer Vision and Pattern Recognition, CVPR 2014, Columbus, pp. 580–587. IEEE (2014)
2. Liu, W., et al.: SSD: Single Shot MultiBox Detector. In: Leibe, B., Matas, J., Sebe, N., Welling, M. (eds.) ECCV 2016, Part I. LNCS, vol. 9905, pp. 21–37. Springer, Cham (2016). https://doi.org/10.1007/978-3-319-46448-0_2
3. Redmon, J., Divvala, S., Girshick, R., et al.: You only look once: unified, real-time object detection. In: IEEE Conference on Computer Vision and Pattern Recognition, CVPR 2016, Las Vegas, pp. 779–788. IEEE (2016)
4. Redmon, J., Farhadi, A.: YOLO9000: better, faster, stronger. In: IEEE Conference on Computer Vision and Pattern Recognition, CVPR 2017, Hawaii, pp. 6517–6525. IEEE (2017)
5. Redmon, J., Farhadi, A.: YOLOv3: an incremental improvement. https://arxiv.org/abs/1804.02767. Accessed 15 Mar 2019
6. Luo, W., Xing, J., Milan, A., et al.: Multiple object tracking: a literature review. https://arxiv.org/abs/1409.7618. Accessed 15 Jan 2019
7. Jérôme, B., Fleuret, F., Engin, T., et al.: Multiple object tracking using K-shortest paths optimization. IEEE Trans. Pattern Anal. Mach. Intell. 33(9), 1806–1819 (2011)
8. Bewley, A., Ge, Z., Ott, L., et al.: Simple online and realtime tracking. https://arxiv.org/abs/1602.00763. Accessed 18 Feb 2019
9. Bochinski, E., Eiselein, V., Sikora, T.: High-speed tracking-by-detection without using image information. In: IEEE International Conference on Advanced Video and Signal Based Surveillance, AVSS 2017, Lecce, pp. 1–6. IEEE (2017)
10. Chu, Q., Ouyang, W., Li, H., et al.: Online multi-object tracking using CNN-based single object tracker with spatial-temporal attention mechanism. In: IEEE International Conference on Computer Vision, ICCV 2017, Venice, pp. 4846–4855. IEEE (2017)
11. Leal, T., Laura, F.C.C., Schindler, K.: Learning by tracking: Siamese CNN for robust target association. In: Computer Vision and Pattern Recognition Conference Workshops, pp. 33–40 (2016)
12. Son, J., Baek, M., Cho, M., et al.: Multi-object tracking with quadruplet convolutional neural networks. In: IEEE Conference on Computer Vision and Pattern Recognition, CVPR 2017, Hawaii, pp. 5620–5629. IEEE (2017)
13. Zhao, H., Xia, S., Zhao, J., Zhu, D., Yao, R., Niu, Q.: Pareto-based many-objective convolutional neural networks. In: Meng, X., Li, R., Wang, K., Niu, B., Wang, X., Zhao, G. (eds.) WISA 2018. LNCS, vol. 11242, pp. 3–14. Springer, Cham (2018). https://doi.org/10.1007/978-3-030-02934-0_1
14. Wojke, N., Bewley, A., Paulus, D.: Simple online and realtime tracking with a deep association metric. https://arxiv.org/abs/1703.07402. Accessed 22 Feb 2019
15. Chu, Q., Ouyang, W., Li, H., et al.: Online multi-object tracking using CNN-based single object tracker with spatial-temporal attention mechanism. In: IEEE International Conference on Computer Vision, ICCV 2017, Venice, pp. 4836–4845. IEEE (2017)
16. Bertinetto, L., Valmadre, J., Golodetz, S., et al.: Staple: complementary learners for real-time tracking. In: IEEE Conference on Computer Vision and Pattern Recognition, CVPR 2016, Las Vegas, pp. 1401–1409. IEEE (2016)

17. Fan, H., Ling, H.: Parallel tracking and verifying: a framework for real-time and high accuracy visual tracking. In: IEEE International Conference on Computer Vision, ICCV 2017, Venice, pp. 5487–5495. IEEE (2017)
18. Li, Y., Zhu, J.: A scale adaptive Kernel correlation filter tracker with feature integration. In: Agapito, L., Bronstein, M.M., Rother, C. (eds.) ECCV 2014, Part II. LNCS, vol. 8926, pp. 254–265. Springer, Cham (2015). https://doi.org/10.1007/978-3-319-16181-5_18
19. Chu, P., Fan, H., Tan, C.C., et al.: Online multi-object tracking with instance-aware tracker and dynamic model refreshment. In: IEEE Winter Conference on Applications of Computer Vision, WACV 2019, Hawaii, pp. 161–170. IEEE (2019)
20. Julier, S.J., Uhlmann, J.K.: New extension of the Kalman filter to nonlinear systems. In: Signal Processing, Sensor Fusion, and Target Recognition VI, vol. 3068, pp. 182–194. International Society for Optics and Photonics (1997)

An Enterprise Competitiveness Assessment Method Based on Ensemble Learning

Yaomin Chang[1,2], Yuzheng Li[1,2], Chuan Chen[1,2(✉)], Bin Cao[3], and Zhenxing Li[3]

[1] School of Data and Computer Science, Sun Yat-sen University, Guangzhou, China
{changym3,liyzh23}@mail2.sysu.edu.cn, chenchuan@mail.sysu.edu.cn
[2] Guangdong Key Laboratory for Big Data Analysis and Simulation of Public Opinion, Sun Yat-sen University, Guangzhou, China
[3] HuaXia iFinance (Beijing) Information Technology Co., Ltd., Beijing, China
bincao@hxifin.com

Abstract. It is of great significance to assess the competitiveness of enterprises based on big data. The current methods cannot help corporate strategists to judge the status quo and prospects of enterprises' development at a relatively low cost. In order to make full use of big data to evaluate enterprise competitiveness, this paper proposes an enterprise competitiveness assessment method based on ensemble learning. The experimental results show that our method has a significant improvement in the task of the enterprise competitiveness assessment.

Keywords: Enterprise competitiveness assessment · Data mining · Ensemble learning · Machine learning

1 Introduction

Effective and quick assessment of enterprise competitiveness can create huge economic value. It can help people identify more competitive enterprises in bank credit risk management and capital investment, thus to achieve a better allocation of funds. Traditional methods of enterprise assessment are mainly based on the tedious and lengthy investigation, analysis and report, which suffer from strong subjectivity, massive cost and poor generalization.

With the help of the rapid development of data mining techniques, artificial intelligence is widely used in lots of fields [3,9]. A large number of models armed with machine learning methodology have proven to be efficient, reliable

The work described in this paper was supported by the National Key Research and Development Program (2016YFB1000101), the National Natural Science Foundation of China (61722214,11801595), the Natural Science Foundation of Guangdong (2018A030310076), and CCF Opening Project of Information System.

W. Ni et al. (Eds.): WISA 2019, LNCS 11817, pp. 79–84, 2019.
https://doi.org/10.1007/978-3-030-30952-7_9

and powerful. However, there remain several major challenges in assessing enterprise competitiveness with artificial intelligence technology. Firstly, it is difficult to extract important features to evaluate the enterprise, while a huge amount of data is generated in business operations. In addition, it requires a good combination with the time series analysis techniques to consider the enterprises' data of the past years. Moreover, unlike traditional machine learning researches, which have relatively well-formulated problems and objective functions, there cannot be a standard to assess enterprise competitiveness comprehensively. Therefore, it's hard to select an appropriate target to train the model.

In this paper, we propose a method called Enterprise Assessment with Ensemble Learning (EAEL) to assess the enterprise competitiveness based on ensemble learning. Firstly, this method extracts enhanced features from the corporate annual financial data, basic registration information and the national macroeconomic data. Subsequently, we apply the idea of ensemble learning to train the model in four different dimensions, including profitability, operation competency, liquidity and growth opportunity, to achieve a more comprehensive enterprise competitiveness assessment. Finally, we evaluate and validate our model on a real-world dataset collected from Chinese A-share stock market. The experiments demonstrate that our method based on ensemble learning has a significant improvement compared with traditional methods.

2   Related Work

In the subject of management science, methods of assessing enterprise competitiveness are mainly listed into two categories: qualitative analysis and quantitative assessment. Qualitative methods such as Michael Porter's Five Forces Model [7] studied the factors that have a great influence on enterprise competitiveness, which is based on domain-relevant knowledge. However, these approaches rely heavily on the subjective judgment of experts and are so costly. Therefore, a larger number of researches turn to use financial data as indicators of enterprise competitiveness, which is more objective and quantifiable. Edward Altman [2] proposed the Z-Score model after a detailed investigation of the bankrupt and non-bankrupt enterprises. This work selected 5 indicators from 22 financial ratios by mathematical statistics methods and thus to make the model simpler, more effective and more intuitive. Although the Z-Score model has a long history, it still has a good performance on enterprise competitiveness assessment in the past 25 years, according to Vineet Agarwal's research [1].

3   Methodology

In this section, we introduce the feature engineering techniques used in our method, as well as the details of our ensemble learning sub-models.

3.1   Feature Engineering

Incremental Features. Since the enterprises' data from the past few years can reflect their development trend, we extract *Incremental Features* from original data under the management advice. *Incremental Features* are defined as the difference and the growth ratio between the current period and the previous period, both on a yearly and quarterly basis. *Quarterly Incremental Features* can reflect the effect of the enterprise's short-term business strategy and its development trend. On the other hand, *Yearly Incremental Features* can exclude the influence of seasonal factors, thus reflecting the long-term competitiveness of enterprises.

Unit Features. In addition to finding large companies with outstanding market performance, we also need to identify smaller companies that have high growth potential. Competitive enterprises of smaller size can perform well in the market with fewer employees. Less competitive enterprises of larger size can still have a large number of market share in the same industry, despite that their per capital benefit and asset utilization level have already located in a lagging position. To achieve a better evaluation on small and competitive enterprises, we construct *Unit Features* that are defined as the ratio of each feature to the number of employees, total assets, gross liability and book value of equity, which intuitively reflect the corresponding output on these certain features.

3.2   Ensemble Learning

Based on the previous researches [4–6], we propose Enterprise Assessment with Ensemble Learning (EAEL) to assess the competitiveness of enterprises. Our method contains two specific sub-models, i.e., the n-year prediction model and the annual series prediction model, as well as a collective model that combines the predictions of the former two sub-models.

XGBoost. XGBoost [4] is a scalable end-to-end tree boosting system that is used widely by data scientists to solve many machine learning problems in a variety of scenarios [10]. XGBoost is adopted in our EAEL method, where we use the data of the enterprise and the macro-economy x_i to predict a target assessment of this enterprise's competitiveness y_i. The original objective function is Eq. (1).

$$Obj^t(X,Y) = \Sigma_{i=1}^n L(y_i, \hat{y}_i^{t-1}) + \Omega(f_t) \tag{1}$$

where $\hat{y}_i^{t-1}$ represents the model's prediction of round $t-1$ in the training phase, and $L(y_i, \hat{y}_i^{t-1})$ is the self-defined cost function between y_i and $\hat{y}_i^{t-1}$. $\Omega(f_t)$ is the regularization term to control the complexity of the model, i.e., the L2 norm of the leaf scores.

To approximate the objective function, XGBoost performs the second-order Taylor expansion on Eq. (1), using both g_i (the first derivative) and f_i (the second derivative) in Eq. (2).

$$Obj^t(X,Y) \simeq \Sigma_{i=1}^n [L(y_i, \hat{y}_i^{t-1} + g_i f_t(x_i) + \frac{1}{2} h_i f_t^2(x_i)] + \Omega(f_t) \tag{2}$$

The n-year Prediction Model. As we have collected several years data of each enterprise, we use Ent_i^k and C_i^k to denote the ith enterprise's data and its corresponding competitiveness in year k, respectively. The n-year prediction model refers to the idea that we use the data Ent_i^k to predict the enterprise's market performance in the nth years, i.e., C_i^{k+n}. The n-year prediction model tries to dig out the implicit relationship between current year data and the enterprise's future competitiveness.

The Annual Series Prediction Model. In order to mine the annual sequence information of each company, we use time-based linear regression that performs on each feature to figure out the enterprise's development trend over the last few years. We take the results of linear regression as the features of the enterprises' general performance over the past years and train a new XGBoost model.

Collective Model. According to the research on model stacking [8], taking the output of multiple sub-models as the input features of the collective model can improve the generalization and fitting ability of the model. Therefore, we train T of the n-year prediction model and the annual series prediction model each and use their predictions as new features to train the collective XGBoost model.

4   Experiment

4.1   Experiment Setup

The following experiments use a real dataset crawled from different sources, including Chinese A-share stock markets, the Chinese National Bureau of Statistics and the Chinese National Bureau of Commerce and Industry. Totally the financial data and registration information of 3573 enterprises and the macroeconomic data are collected, covering from 2011 to 2018. Firstly we annotate each record in four dimensions (profitability, operation competency, liquidity and growth opportunity) according to experts' advice, turning this assessment problem into a classic binary classification problem. Then we split our dataset into training set, test set and validation set by 6:2:2. To evaluate the proposed method, we compare the predicted result with the traditional Z-Score model in all dimensions. Additionally, several verification experiments are presented to illustrate the effectiveness and necessity of feature engineering and different ensemble methods.

4.2   Experiment Results

Comparative Experiments with Traditional Methods. In this experiment, we compare our proposed method EAEL with the traditional method in different dimensions (profitability, operation competency, liquidity and growth opportunity). As the Z-Score model can only get numeric scores of enterprise competitiveness, we rank the scores predicted by the Z-Score model in descending order, to which we assign corresponding binary labels. The experimental results are listed in Table 1. We can see from Table 1 that the predictive ability of our EAEL method certainly outperforms the Z-Score model in all dimensions, especially on the comprehensive metrics F_1-score.

Table 1. Comparison results with traditional methods.

Dimension	Model	Accuracy	Precision	Recall	F_1-score
Profitability	Z-Score	0.5791	0.6421	0.6703	0.5765
	EAEL	**0.6939**	**0.7069**	**0.8569**	**0.6775**
Operation Competency	Z-Score	0.4654	0.4234	**0.6027**	0.4590
	EAEL	**0.6592**	**0.6350**	0.5117	**0.6528**
Liquidity	Z-Score	0.5219	0.5258	0.6436	0.5144
	EAEL	**0.6394**	**0.6338**	**0.6832**	**0.6386**
Growth Opportunity	Z-Score	0.4631	0.3772	**0.6145**	0.4620
	EAEL	**0.6684**	**0.6156**	0.3408	**0.6407**

Analysis of Feature Engineering. In this experiment, we try to verify how our feature engineering techniques affect the model's performance. In short, we only display the experimental results in the dimension of profitability. We use different kinds of features to train the model and the results are listed in Table 2. We can figure out that different techniques improve the performance of enterprise competitiveness assessment with a different degree. When we apply all feature engineering techniques to the training data, we can train the best-performing model.

Table 2. The effects of different feature engineering techniques

Data	Accuracy	Precision	Recall	F_1-score
Data (raw)	0.5641	0.5110	0.7914	0.4568
Data (with Unit Features)	0.6724	0.6854	**0.8631**	0.6485
Data (with Incremental Features)	0.6432	0.6584	0.8541	0.6141
Data (with Unit & Incremental Features)	**0.6939**	**0.7069**	0.8569	**0.6775**

Analysis of Ensemble Learning. In this experiment, we train the three sub-models mentioned in Sect. 3.2 and compare their predictions in the dimension of profitability. From Table 3 we can see that the performance of the n year prediction model decreases as the hyperparameter n increases. That suggests that the correlation between the current year's data and the enterprise's competitiveness in future years is declining. And the annual series prediction model is slightly worse than the 3-year prediction model. It also indicates that the competitiveness of enterprises is more related to the enterprises' data in recent years, while the long series of enterprises' data may disturb the model and reduce its evaluation performance.

Table 3. The performance of different sub-models

Model	Accuracy	Precision	Recall	F_1-score
The 1-year prediction model	0.6939	0.7069	0.8569	0.6775
The 2-year prediction model	0.6664	0.6726	0.8328	0.6503
The 3-year prediction model	0.6459	0.6604	0.8258	0.6203
The annual series prediction model	0.6456	0.6508	**0.9444**	0.5676
The collective model	**0.6964**	**0.7319**	0.8047	**0.6905**

5 Conclusion

In the era of big data, it is significantly important to fully explore the implicit information of the enterprises' data. In this paper, we propose an ensemble learning-based method EAEL to assess enterprise competitiveness. Firstly, this method performs feature engineering to get informative *Incremental Features* and *Unit Features*. Then, EAEL trains three sub-models to predict enterprise competitiveness. Finally, experimental results demonstrate that our method has significant improvement in enterprise competitiveness assessment.

References

1. Agarwal, V., Taffler, R.J.: Twenty-five years of the Taffler z-score model: does it really have predictive ability? Account. Bus. Res. **37**(4), 285–300 (2007)
2. Altman, E.I.: Financial ratios, discriminant analysis and the prediction of corporate bankruptcy. J. Finan. **23**(4), 589–609 (1968)
3. Carneiro, N., Figueira, G., Costa, M.: A data mining based system for credit-card fraud detection in e-tail. Decis. Support Syst. **95**, 91–101 (2017)
4. Chen, T., Guestrin, C.: XGBoost: a scalable tree boosting system. In: KDD-16, pp. 785–794 (2016)
5. Friedman, J.H.: Greedy function approximation: a gradient boosting machine. Ann. Stat. **29**, 1189–1232 (2001)
6. Opitz, D., Maclin, R.: Popular ensemble methods: an empirical study. J. Artif. Intell. Res. **11**, 169–198 (1999)
7. Porter, M.E.: The five competitive forces that shape strategy. Harv. Bus. Rev. **86**(1), 78–93 (2008)
8. Wolpert, D.H.: Stacked generalization. Neural Networks **5**(2), 241–259 (1992)
9. Wu, S., Ren, W., Yu, C., Chen, G., Zhang, D., Zhu, J.: Personal recommendation using deep recurrent neural networks in netease. In: ICDE-16, pp. 1218–1229 (2016)
10. Xi, Y., Zhuang, X., Wang, X., Nie, R., Zhao, G.: A research and application based on gradient boosting decision tree. In: Meng, X., Li, R., Wang, K., Niu, B., Wang, X., Zhao, G. (eds.) WISA 2018. LNCS, vol. 11242, pp. 15–26. Springer, Cham (2018). https://doi.org/10.1007/978-3-030-02934-0_2

A Method for Duplicate Record Detection Using Deep Learning

Qing Gu, Yongquan Dong[✉], Yang Hu, and Yali Liu

School of Wisdom Education, Jiangsu Normal University,
Xuzhou 221000, China
gdq@qq.com, tomdyq@163.com, 380703693@qq.com,
1204240062@qq.com

Abstract. As the scale of data in the database is growing rapidly, there exists the increasing number of duplicate information, which leads to data redundancy. Effective detection of duplicate records is of great significance for improving data quality and is an important issue in the field of data cleaning. At present, traditional duplicate record detection approaches use discrete features to build classification models which cannot express the semantic information well, and the classification accuracy needs to be improved. The deep learning model has good performance in feature extraction and classification. Therefore, this paper proposes a novel two-stage duplicate record detection model based on deep learning. In the first stage, the doc2vec model based on deep learning is applied to attribute vectorization construction, and the vector of record comparison pairs is constructed by concatenating attribute vectors. The second stage uses a Convolutional Neural Networks (CNN) to build a classification model for all record comparison vectors. The experimental results show that the proposed method is superior to the traditional duplicate record detection method.

Keywords: Duplicate record detection · Deep learning · Data cleaning

1 Introduction

As the scale of data stored in computers grows rapidly, there exists a large number of duplicate records. Duplicate record detection is a challenging problem because different data sources may represent the attribute values of an entity in different ways or even provide conflicting data [1]. It is also called entity recognition, which refers to detect duplicate records by using some scientific methods. And it is the research focus in the field of data cleaning. At the very beginning, it was proposed by Newcombe et al. [2] and is gradually used in official statistics, medicine and other fields. In big data environment, it is of great significance to detect duplicate records effectively for further data cleaning and data quality improvement.

The problem of duplicate record detection has been studied for many years. In order to improve the accuracy, researchers have proposed a series of duplicate record detection methods based on supervised learning. For example, the method of support vector machine (SVM) adopted in [3]. However, the above methods are not sufficiently expressive on semantic information.

© Springer Nature Switzerland AG 2019
W. Ni et al. (Eds.): WISA 2019, LNCS 11817, pp. 85–91, 2019.
https://doi.org/10.1007/978-3-030-30952-7_10

Deep learning has made great progress with the in-depth development of computer technology. Feature extraction and text classification models based on deep learning have also begun to appear in the field of Natural Language Processing (NLP). In text features extraction, Mikolov proposed the deep learning model word2vec [4] in 2013, a method of transforming each word in text into a vector representation, which effectively improves the traditional discrete text representation vector and makes up the insufficient for the unrelated between semantics. Later, the doc2vec model was extended to represent the distribution vector of text of arbitrary length. In classification, many scholars use CNN to get good results [5]. It does not require manual extraction of features, as it simplifies the complex text feature extraction steps, and can get better results when dealing with a lot of data. It can be seen that the deep learning model is superior to the traditional model in feature extraction and text classification.

According to the advantages of deep learning, this paper proposes a novel two-stage duplicate record detection model based on deep learning, In the first stage, the doc2vec model is applied to attribute the vectors of record comparison pairs. The second stage uses a CNN to construct a classification model. The experimental results show that the proposed method can significantly improve the performance of the duplicate record detection model compared with the traditional models.

2   Technical Principle

2.1   Doc2vec

Doc2vec also called Paragraph Vector. The model consists of two structures [6]. The PV-DM (Distributed Memory version of Paragraph Vector) model represents each paragraph as a vector and predict the next word through successive paragraphs between these paragraphs. The PV-DBOW (Distributed Bag of Words version of Paragraph Vector) model only uses paragraph vectors to predict words in a particular size window in the current paragraph.

The doc2vec model adds a new sentence vector in the input layer, which is different from the word2vec. When a new one needs to be given, the word vector in the prediction stage model and the softmax weights parameter projected to the output layer are fixed. Through the above steps, a vector of sentences can be obtained. Doc2vec is an unsupervised learning algorithm whose structure retains its semantic advantage over the word bag model. Through practice test, the doc2vec model has effects in calculating text similarity [7] and emotion classification [8].

2.2   Convolutional Neural Networks (CNN)

CNN can be regarded as a special feedforward neural network. In the process of text feature extraction and classification, high-recognition text features are extracted by the advantages of convolutional neural network in image preprocessing to reduce the workload of feature extraction. The CNN adds a convolutional layer and a pooling layer on the basis of the traditional neural network, which can effectively solve the problem of network layer limitation of the traditional neural network. At the same time, feature vectors of the same dimension can be generated through the pooling layer even if the input sentences are of different lengths. Finally, the pooled layer output is passed to the fully connected layer for classification.

3   Two-Stage Duplicate Record Detection Model Based on Deep Learning

The two-stage duplicate record detection model proposed in this paper uses doc2vec to complete the vectorization construction of the record comparison pair in the first stage, and uses CNN to classify the record comparison pairs in the second stage.

3.1   Constructing the Vector of Record Comparison Pair

In the first stage, we mainly construct a vector for each record comparison pair using the doc2vec model. First, we split the entire data set according to attributes. Second, we use the doc2vec model to get a vector for each attribute value in each record. Then, the vectors of all attributes of each record are concatenated together, and the vector of each record comparison pair is obtained, that is, each record comparison pair is constructed into a one-dimensional vector matrix.

3.2   Constructing a Classification Model

In the second stage, we use the CNN to build a classification model. Assuming that the dimension of each record comparison pair obtained by the first stage is k, the information of the input layer is a vector matrix of 1 * k. Assuming that the size of the convolution kernel is h, the vector matrix of the input layer passes through the convolution kernel of 1 * h, and a feature matrix with a column number of 1 is obtained. After the convolution, an activation function f is used to obtain the eigenvalue as the output of the convolutional layer. The next pooling layer uses the max pooling method, which is to extract the most representative value from the previously obtained feature matrix. Finally, the information extracted by the pooling layer is connected to the activation function sigmoid in a fully connected manner to obtain the classification result. The structure is shown in Fig. 1.

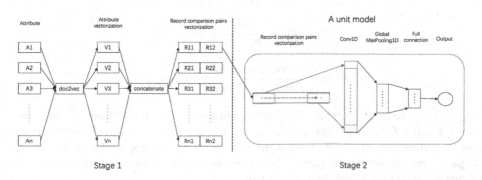

Fig. 1. The structure of the two-stage duplicate record detection model

4 Experimental Study

In order to prove the effectiveness of the proposed method, we did three experiments by using different text vector construction methods and different text classification methods.

4.1 The Data Set

The data set used Core data, which contains 3154 references and a total of 302 different references. The experiment selected four attributes of the data set, whose names and meanings are shown in Table 1. In order to get more accurate results, we did the work of balancing the data set in advance. A random function was used to obtain 22228 duplicate record comparison pairs and 22228 non-repeating record comparison pairs as data sets.

Table 1. Attribute name and its meaning

Name	Meaning
title	title name of the literature
book_title	source journal of the literature
authors	all author names of the literature
date	year of publication

4.2 Evaluation Standard

In the experiments, we use the classical evaluation standards proposed by Ananthakrishna et al. [9]: Accuracy, Precision, Recall and F1, whose corresponding calculation formulas are as

$$Accuracy = \frac{TP + TN}{TP + TN + FP + TN} \times 100\% \tag{1}$$

$$Precision = \frac{TP}{TP + FP} \times 100\% \tag{2}$$

$$Recall = \frac{TP}{TP + FN} \times 100\% \tag{3}$$

$$F1 = \frac{2 \times Precision \times Recall}{Precision + Recall} \times 100\% \tag{4}$$

where TP denotes the number of predict duplicate and actual duplicate record comparison pairs; FP denotes the number of predict duplicate and actual not duplicate record comparison pairs; FN denotes the number of predict not duplicate and actual duplicate record comparison pairs; TN denotes- the number of predict not duplicate and actual not duplicate record comparison pairs.

4.3 Experimental Results and Analysis

The first experiment uses TF-IDF to construct the vector of each attribute, and calculate the cosine similarity between the same attributes [10], then concatenate attribute for the cosine similarity between the record comparison pairs. Finally use the SVM to construct the classification model. The second experiment uses doc2vec to construct the vector of each attribute, and calculate the cosine similarity between the same attributes, then concatenate attribute for the cosine similarity between the record comparison pairs. Finally use the SVM to construct the classification model. The third experiment uses the vector representation of each attribute obtained in experiment 2, and the vector of record comparison pairs is constructed by concatenating attribute vectors, and finally the classification model is constructed by using CNN to compare all the records.

Experiment 1 and experiment 2 directly use the TF-IDF model and the doc2vec model in Gensim [11] to construct the attribute vector. We split the data set to the training set and test set randomly into 7:3, and extract 30% as a validation set in the training set. The SVM model is trained on training set. Finally, the test set is classified on the trained SVM model. In the third experiment, the classification model is constructed by the CNN. Though the experiment, we find that when the number of epochs is 180, the model has the highest accuracy and stability on the validation set. So we use 180 as the number of epochs of the model. The performance comparison of the three experiments on test set is shown in Fig. 2.

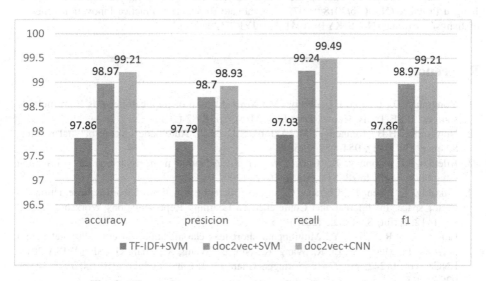

Fig. 2. The performance comparison of the three experiments

The experimental results show that Experiment 2 with doc2vec is much higher in four indexes than experiment 1. Meanwhile, experiment 3 with the two-stage repeated record detection method proposed in this paper is better in four indexes than the other two methods, which verifies the superiority of our method in the duplicate record detection.

5 Conclusion

Nowadays, there is a large amount of duplicate information in the database, which causes waste of resources. In order to improve data quality, it is necessary to effectively detect duplicate records. Based on the application of deep learning in the field of NLP, we propose a new model of duplicate record detection based on deep learning. The advantages of deep learning in feature extraction and classification make up for the deficiencies of the traditional discrete feature construction classification model, and can express semantic information. Experimental studies show that the proposed method can significantly improve the performance of the duplicate record detection model.

In the future we will deepen input information from the entire record comparison pairs to the same attribute of each record comparison pairs to achieve better results.

Acknowledgement. This work is supported by the National Natural Science Foundation of China (No. 61872168, 61702237), the 13th Five-year Plan Project for Education Science of Jiangsu Province (No. C-b/2018/01/07), Postgraduate Research & Practice Innovation Program of Jiangsu Province (No. SJKY19_1941, No. JYKTZ201712).

References

1. Yongquan, D., Dragut, E.C., Weiyi, M.: Normalization of duplicate records from multiple sources. IEEE Trans. Knowl. Data Eng. **31**(4), 769–782 (2019)
2. Newcombe, H.B., Kennedy, J.M., Axford, S.J., et al.: Automatic linkage of vital records. Science **130**(3381), 954–959 (1959)
3. Bilenko, M., Mooney, R., Cohen, W., et al.: Adaptive name matching in information integration. IEEE Intell. Syst. **18**(5), 16–23 (2003)
4. Mikolov, T., Chen, K., Corrado, G., et al.: Efficient estimation of word representations in vector space. In: International Conference on Learning Representations Workshop Track, pp. 1–12. eprint, Scottsdale (2013)
5. Liu, J., Cui, R., Zhao, Y.: Multilingual short text classification via convolutional neural network. In: Meng, X., Li, R., Wang, K., Niu, B., Wang, X., Zhao, G. (eds.) WISA 2018. LNCS, vol. 11242, pp. 27–38. Springer, Cham (2018). https://doi.org/10.1007/978-3-030-02934-0_3
6. Le, Q., Mikolov, T.: Distributed representations of sentences and documents. In: International Conference on Machine Learning, vol. 32(2), pp. 1188–1196 (2014)
7. Lee, S., Jin, X. Kim, W.: Sentiment classification for unlabeled dataset using Doc2vec with JST. In: 18th Annual International Conference on Electronic Commerce: e-Commerce in Smart connected World, pp. 1–5. Association for Computing Machinery, Suwon (2016)

8. Maslova, N., Potapov, V.: Neural network Doc2vec in automated sentiment analysis for short informal texts. In: Karpov, A., Potapova, R., Mporas, I. (eds.) SPECOM 2017. LNCS (LNAI), vol. 10458, pp. 546–554. Springer, Cham (2017). https://doi.org/10.1007/978-3-319-66429-3_54

9. Ananthakrishna, R., Chaudhuri, S., Ganti, V.: Eliminating fuzzy duplicates in data warehouses. In: 28th International Conference on Very Large Databases, pp. 586–597. Morgan Kaufmann, San Francisco (2002)

10. Elmagarmid, A.K., Ipeirotis, P.G., Verykios, V.S.: Duplicate record detection: a survey. IEEE Trans. Knowl. Data Eng. **19**(1), 1–16 (2007)

11. Rehurek, R., Sojka, P.: Gensim–statistical semantics in Python. In: EuroScipy, Pairs (2011)

Case Facts Analysis Method
Based on Deep Learning

Zihuan Xu, Tieke He(✉), Hao Lian, Jiabing Wan, and Hui Wang

State Key Laboratory for Novel Software Technology,
Nanjing University, Nanjing 210093, China
hetieke@gmail.com

Abstract. With the rapid development of artificial intelligence, it is a general trend to apply artificial intelligence to the judicial field. The use of deep learning to analyze judicial data can assist judicial decisions, and can also assist in legal counseling. Artificial intelligence can promote the intelligence and automation of the judiciary and improve the efficiency of the judicial process. The analysis of judicial texts has always been a research hotspot of judicial intelligence. Using deep learning to analyze judicial texts and screening similar cases according to the similarity of cases can provide objective and effective assistance for judicial decisions. This paper designs a method for analyzing case facts based on deep learning and recommending similar cases. The method implements word segmentation on the case facts of the training set cases and trains the word vector. Then, extracting k keywords of the 'fact' field for each training case. The representative central word vector of the case is constructed according to the keywords. The representative central word vector for each test case is constructed in the same way, and the training case to be recommended for the test case is the one which has the most similar vector to the vector of the test case. We used this method to implement the recommendation of judicial cases in experiment, and designed a verification experiment to evaluate the accuracy of the recommendation. The recommendation accuracy rate reached 91.3%.

Keywords: Judicial cases analysis · Case recommendation · Word segmentation · Word2vec · TextRank

1 Introduction

The judiciary has begun to invest in integrating artificial intelligence and big data into the judicial process. Using artificial intelligence to analyze big data to provide a reference for the judgment, which is an assurance that judgment does not include the subjective consciousness of the judge.

The research of judicial intelligence has two main directions. One is about legal reasoning [11] and legal ontologies [3], the other is about assisting judicial decisions, including legal information retrieval [10], penalty recommendation [4] and so on. The focus of this paper is on the second direction, which is about case recommendation and assisting judicial decisions. The judicial structure [5] has

© Springer Nature Switzerland AG 2019
W. Ni et al. (Eds.): WISA 2019, LNCS 11817, pp. 92–97, 2019.
https://doi.org/10.1007/978-3-030-30952-7_11

a complex structure including the facts of the case, the origin of the evidence and some other elements. The judicial record is a special instrument formed in the process of handling case and consist of case fact description, accusation, prison term, law article, and some other documents. In general, analyzing cases and implementing case recommendations need various kinds of information. This paper proposes a case recommendation method that only uses facts of cases. We use the deep learning method Word2vec [9] to learn facts of cases and train deep learning models. Constructing feature vector for each case to achieve case recommendation.

The contributions of this paper are as follows. This paper designs a method for analyzing case facts based on deep learning and recommending similar cases. The recommendation accuracy reached 91.3%. This paper combines some existed methods with our own method to build a case recommendation methodology.

The rest of this paper is organized as follows. Section 2 introduces our approach. Section 3 explains the dataset we use. Section 4 reports the experiment. The evaluation of the experiment is in Sect. 5. We discuss the related work in Sect. 6. Section 7 concludes this paper.

2 Methodology

In this paper, we propose a methodology to analyze case facts based on deep learning and recommend similar case. The method is divided into two parts: the training method and the recommendation method. The training method uses the Word2vec model to train the corpus. The corpus is the facts of the cases extracted from the training set, word segmentation is implemented on the facts, and then word2vec is used to train these words. Then word vectors are generated. Finally it constructs a representative central word vector for each case in the training set based on the word vectors generated. The recommendation method constructs a representative central word vector for the predicted case in a similar way. The case in training set that is recommended for the predicted case is the one which has the most similar representative central word vector to the vector of the predicted case. The framework of our methodology is shown in Fig. 1.

2.1 Training Method

a. Extract case facts from judicial cases in training set.
b. Implement word segmentation on case facts and generate the corpus.
c. Use Word2vec to train word vectors.
d. Use TextRank [8] to extract the first k keywords of case facts.
e. Construct a representative central word vector for each case.

$$\beta = \sum_{i=1}^{k} \frac{a_i}{k} \tag{1}$$

We use β to represent the representative central word vector of the case, and α_k represents the word vector corresponding to the kth keyword.

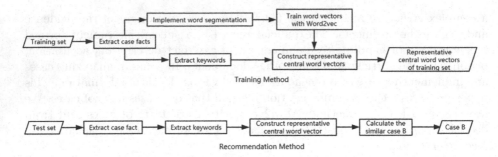

Fig. 1. Framework of the case analysis and recommendation methodology

2.2 Recommendation Method

a. Extract case fact from a single judicial case A in test set.
b. Extract keywords with TextRank algorithm.
c. Construct a representative central word vector for case A. The construction method is similar to step 5 of the training method. The calculation result is the representative central word vector of the case A.

$$\gamma = \sum_{i=1}^{k} \frac{a_i}{k} \tag{2}$$

We use γ to represent the representative central word vector of the case A, and α_k denotes the word vector corresponding to the kth keyword.
d. Calculate the similar case B. After the previous steps, each case in the training set has an n-dimensional vector corresponding to it. The method calculates the distance between the vector of each judicial case in the training set and the vector of the predicted case A. The formula is as follows.

$$distance = \sqrt{\sum_{i=1}^{n} (\gamma[i] - \beta[i])^2} \tag{3}$$

Where $\gamma[i]$ represents the ith element of the n-dimensional vector γ of case A, and $\beta[i]$ represents the ith element of the n-dimensional vector β of a case in training set. We can compare and then get the shortest distance. When the distance is the shortest, the case corresponding to the vector β is the similar case B.
e. Recommend the case B calculated in step 4.

3 Data Set

We use the CAIL2018 data set which contains 155,000 pieces of training data and 33,000 pieces of test data [13]. The data set is in the .json file format. Every piece of data is a criminal case. Each row of the data set corresponding to a case and all of them are stored in the type of dictionary. The information used in our paper contains the 'fact', the 'accusation' and the 'relevant articles' field. The 'fact' field is the case factual description, the 'accusation' field is the accusation of the suspect, and the 'relevant articles' field is the relevant legal articles about the case. In our experiment, we randomly selected 8000 cases in the training data, randomly selected 500 cases in the test data. These cases are distributed in 'theft', 'intentional injury', 'fraud' and several other accusations.

4 Experiment

4.1 Training Process

a. Extract case facts from judicial cases in training set. The CAIL2018 dataset is in json file format. Each row of the data set is corresponding to a case and all of them are stored in the type of dictionary. The training set has 8000 cases (i.e., 8000 lines). The program reads each line in order, decodes the JSON string into Python's dictionary, and extracts the 'fact' field in each dictionary.
b. Implement word segmentation on case facts and generate the corpus. In our experiment, we chose 'jieba' as the word segmentation tool.
c. Use Word2vec to train word vectors.
d. Use TextRank algorithm to extract the first k keywords of case facts.
e. Construct a representative central word vector for each case. Firstly, we search for the word vectors corresponding to the k keywords of a case in the word vectors library generated in step 3. Secondly, all the filtered vectors are added, and then divided by the number k (k is the number of keywords of the case). The calculation result is the representative central word vector of the case.

5 Evaluation

We designed an experiment to evaluate the recommendation accuracy of our recommendation method. From two perspectives, we classify the recommendation accuracy into the accusation accuracy and the articles accuracy. The cases in data set include not only the 'fact' field, but also the 'accusation' field and the 'relevant articles' field, which respectively represent the accusation of the suspect and the relevant legal articles about the case. We compare the 'accusation' field and the 'relevant articles' field of the recommended case A with the same fields of the recommendation case B. If there is any shared element, we think the case recommendation is successful. The specific steps to evaluate the accusation accuracy are as follows.

Step 1. There are 500 pairs of recommended and recommendation cases. For each pair of cases, we extract their 'accusation' field. The extracted content is a pair of lists of accusations, represented by list1 and list2. There may be only one accusation in a list, or there may be multiple accusations.

Step 2. Use the method of intersection to determine whether the two lists share the same element, and if so, we claim that the case recommendation is successful.

The experiment result shows that 458 of the 500 cases are successfully recommended. The accusation accuracy of case recommendation reaches 91.6%.

The method of evaluating the articles accuracy is the same as the method of evaluating the accusation accuracy. We only need to replace the 'accusation' field with the 'relevant articles' field. The experiment result shows that 455 of the 500 cases are successfully recommended. The articles accuracy of case recommendation reaches 91%.

The recommendation accuracy of our method is the average of the accusation accuracy and the articles accuracy. As shown in Table 1, the recommendation accuracy of our method reaches 91.3%.

Table 1. Case recommendation accuracy

Accusation accuracy	Articles accuracy	Case recommendation accuracy
91.6%	91.0%	91.3%

6   Related Work

Case recommendation is an important part of judicial intelligence. Providing judges with recommendations for similar cases can provide an objective reference for judges' judgments, and promote the fairness of judiciary. Thus, it is very important to recommend the correct case to the judge. Collaborative Filtering Recommendations [6] (CF) is the most popular recommendation method [2]. CF is consist of online collaborative and offline filtering. Online collaboration finds items of interest to users based on online resources, while offline filtering filters information that is worthless. Collaborative filtering recommendations can be divided into user-based collaborative filtering [12], item-based collaborative filtering [7], and model-based collaborative filtering [1]. User-based collaborative filtering is based on users. By analyzing user data, it finds similar user groups to achieve recommendations. Item-based collaborative filtering is based on items. It analyzes the similarity of items and recommends the items that meet the user preferences for the target users. Model-based collaborative filtering recommends items to target users based on existing user information and several items. It is commonly used in the field of natural language processing.

7   Conclusion

We design a case facts analysis method based on deep learning in this paper. The method uses the facts of judicial cases as training data, uses Word2vec to train the deep learning model, and recommends similar case for a given case based on the model. We implemented this idea through experiments. And we designed an experiment to evaluate the accuracy of our case recommendation method. The result shows that the recommendation accuracy rate reaches 91.3%.

There are 8,000 cases in our training set. These cases are distributed in 'theft', 'intentional injury', 'fraud' and several other accusations. And the cases are generated in different years. It should be pointed out that the accusations we selected did not cover all the criminal accusations. If the number and types of training set cases are more, the model trained will have a better performance.

Acknowledgment. The work is supported in part by the National Key Research and Development Program of China (2016YFC0800805) and the National Natural Science Foundation of China (61772014).

References

1. Breese, J.S., Heckerman, D., Kadie, C.: Empirical analysis of predictive algorithms for collaborative filtering. In: UAI, pp. 43–52 (1998)
2. Desrosiers, C., Karypis, G.: A Comprehensive survey of neighborhood-based recommendation methods. In: Ricci, F., Rokach, L., Shapira, B., Kantor, P.B. (eds.) Recommender Systems Handbook, pp. 107–144. Springer, Boston, MA (2011). https://doi.org/10.1007/978-0-387-85820-3_4
3. Gray, P.N.: The ontology of legal possibilities and legal potentialities. In: LOAIT, pp. 7–23 (2007)
4. He, T., Lian, H., Qin, Z., Zou, Z., Luo, B.: Word embedding based document similarity for the inferring of penalty. In: Meng, X., Li, R., Wang, K., Niu, B., Wang, X., Zhao, G. (eds.) WISA 2018. LNCS, vol. 11242, pp. 240–251. Springer, Cham (2018). https://doi.org/10.1007/978-3-030-02934-0_22
5. Helfer, L.R.: The politics of judicial structure: creating the united states court of veterans appeals. Conn. L. Rev. **25**, 155 (1992)
6. Herlocker, J.L., Konstan, J.A., Riedl, J.: Explaining collaborative filtering recommendations. In: CSCW, pp. 241–250 (2000)
7. Linden, G., Smith, B., York, J.: Amazon.com recommendations: item-to-item collaborative filtering. IEEE Internet Comput. **7**(1), 76–80 (2003)
8. Mihalcea, R., Tarau, P.: TextRank: bringing order into text. In: EMNLP (2004)
9. Mikolov, T., Chen, K., Corrado, G., Dean, J.: Efficient estimation of word representations in vector space. arXiv preprint arXiv:1301.3781 (2013)
10. Nanda, R., Adebayo, K.J., Di Caro, L., Boella, G., Robaldo, L.: Legal information retrieval using topic clustering and neural networks. In: COLIEE@ ICAIL, pp. 68–78 (2017)
11. Prakken, H.: Modelling reasoning about evidence in legal procedure. In: ICAIL, pp. 119–128 (2001)
12. Sarwar, B.M., Karypis, G., Konstan, J.A., Riedl, J., et al.: Item-based collaborative filtering recommendation algorithms. In: WWW, vol. 1, pp. 285–295 (2001)
13. Xiao, C., et al.: Cail 2018: a large-scale legal dataset for judgment prediction. arXiv preprint arXiv:1807.02478 (2018)

The Air Quality Prediction Based on a Convolutional LSTM Network

Canyang Guo[1], Wenzhong Guo[1], Chi-Hua Chen[1], Xin Wang[2],
and Genggeng Liu[1(✉)]

[1] College of Mathematics and Computer Science,
Fuzhou University, Fuzhou, China
liugenggeng@fzu.edu.cn
[2] College of Intelligence and Computing, Tianjin University, Tianjin, China

Abstract. In recent years, the rapid development of industrial technology has been accompanied by serious environmental pollution. In the face of numerous environmental pollution problems, particulate matter (PM2.5) which has received special attention is rich in a large amount of toxic and harmful substances. Furthermore, PM2.5 has a long residence time in the atmosphere and a long transport distance, so analyzing PM2.5 distributions is an important issue for air quality prediction. Therefore, this paper proposes a method based on convolutional neural networks (CNN) and long short-term memory (LSTM) networks to analyze the spatial-temporal characteristics of PM2.5 distributions for predicting air quality in multiple cities. In experiments, the records of environmental factors in China were collected and analyzed, and three accuracy metrics (i.e., mean absolute error (MAE), root mean square error (RMSE), and mean absolute percentage error (MAPE)) were used to evaluate the performance of the proposed method in this paper. For the evaluation of the proposed method, the performance of the proposed method was compared with other machine learning methods. The practical experimental results show that the MAE, RMSE, and MAPE of the proposed method are lower than other machine learning methods. The main contribution of this paper is to propose a deep multilayer neural network that combines the advantages of CNN and LSTM for accurately predicting air quality in multiple cities.

Keywords: Air quality prediction · Convolutional neural network ·
Long short term memory · PM2.5

1 Introduction

In recent years, with the development of industrial development, the problem of environmental pollution has become more and more serious, which has attracted significant attention. There are many factors that cause environmental pollution, such as sulfur dioxide, nitrogen oxides, fine particles (PM2.5) and so on. PM2.5 is the main cause of smog among them [1]. PM2.5 can stay in the atmosphere for a long time, and it also can enter the body by breathing, accumulating in the trachea or lungs and affecting the health of the body [2]. PM2.5 is larger than viruses and smaller than bacteria. It is easy to carry toxic substances into the human body [3]. For the

W. Ni et al. (Eds.): WISA 2019, LNCS 11817, pp. 98–109, 2019.
https://doi.org/10.1007/978-3-030-30952-7_12

environment and human health, the threat from PM2.5 is enormous. Therefore, the prediction and control of PM2.5 concentration are quite important issues.

This paper proposes a deep multilayer neural network model combining convolutional neural network (CNN) and long short-term memory (LSTM) to predict PM2.5 concentration. The model is able to predict the future PM2.5 concentration data based on the past PM2.5 concentration data. This study collects the environmental data from January 2015 to December 2017 in nine cities (i.e., Ningde City, Nanping City, Fuzhou City, Sanming City, Putian City, Quanzhou City, Longyan City, Zhangzhou City, and Xiamen City) in the Fujian Province of the People's Republic of China as a training set, and the environmental data from January to October 2018 as a testing set. For the evaluation of the proposed method, mean absolute error (MAE), root mean square error (RMSE) and mean absolute percentage error (MAPE) are used as accuracy metrics. Experimental results show that the proposed method is superior to other machine learning methods.

This paper is organized as follows. Section 2 presents a literature review on air quality prediction. Section 3 presents the prediction methods based on deep learning techniques. Section 4 describes the data processing and gives the practical experimental results and discussions. Section 5 summarizes the contributions of this study and discusses the future work.

2 Literature Reviews

In 2013, the World Health Organization's International Agency for Research on Cancer (IARC) published a report stating that PM2.5 is carcinogenic to humans and is considered a universal and a major environmental carcinogen [4]. Therefore, the prediction and control of PM2.5 are particularly important issues for air quality maintenance and urban development. At present, the air quality prediction methods are mainly classified into two categories: (1) the mechanism models based on the atmospheric chemical modes are called deterministic models [5, 6]; (2) the statistical models based on machine learning algorithms are called machine learning models [6, 7].

Gu et al. [8] designed a new picture-based predictor of PM2.5 concentration (PPPC) which employs the pictures acquired using mobile phones or cameras to make a real-time estimation of PM2.5 concentration. Although this method can estimate the PM2.5 concentration more accurately, it can just evaluate the current PM2.5 concentration and cannot predict the future PM2.5 concentration. Mahajan et al. designed a PM2.5 concentration prediction model that combined a neural network based hybrid model and clustering techniques like grid-based clustering and wavelet-based clustering [9]. The main focus is to achieve high accuracy of prediction with reduced computation time. A hybrid model was applied to do a grid-based prediction system for clustering the monitoring stations based on the geographical distance [9]. In 2014, Elangasinghedeng et al. [10] proposed the complex event-sequence analyses of PM10 and PM2.5 in coastal areas by using artificial neural network models and k-means clustering method. The study presented a new approach based on artificial neural network models, and the k-means clustering method was used to analyze the relationships between the bivariates of concentration–wind and speed–wind direction for extracting source performance signals from the time series of ambient PM2.5 and PM10 concentrations [10]. Li et al. [11] proposed a LSTM network for air pollutant concentration prediction, and Tsai et al. [12] showed the way of air pollution prediction based on RNN and LSTM. The method collected PM2.5 data from 66 stations in Taiwan from 2012 to 2016 and establishes LSTM and RNN models for precise prediction of air quality. Yu et al. [13] predicted the concentration of PM2.5 through the Eta-Community Multiscale Air Quality (Eta-CMAQ) model. The method is based on the chemical composition of PM2.5 and has certain references, but the cost of chemical analysis is too expensive. Verma et al. [14] proposed the use of a bi-directional LSTM model to predict air pollutant severity levels ahead of time. The models are robust and have shown superiority over an artificial neural network model in predicting PM2.5 severity levels for multiple stations in New Delhi City [14].

Because of the small particle size of PM2.5, it stays in the atmosphere for a long time and the transport distance is long. Therefore, PM2.5 concentrations have very close relationships with time and space. The proposed method based on a CNN and a LSTM network to perfectly extract the spatial-temporal characteristics of PM2.5 distributions for air quality prediction.

3   Prediction Methods Based on Deep Learning Techniques

The concepts and processes of CNNs, LSTM networks, and convolutional long short-term memory (ConvLSTM) networks are presented in the following subsections.

3.1   Convolutional Neural Networks

Deep neural networks have achieved remarkable performance at the cost of a large number of parameters and high computational complexity [15]. A convolutional neural network is a feedforward neural network that contains convolutional computation and has a deep structure. The difference between CNN and the fully connected neural network is the weight sharing. CNN [16, 17] has two advantages: (1) the number of weights is reduced, and the amount of training is greatly reduced; (2) spatio features can be effectively extracted. The network model can process multi-dimension data. In this paper, the input convolutional layer data is a 5×5 two-dimensional matrix. As shown in Fig. 1, the variables x_1 to x_{25} are inputs, and the variables w_1 to w_4 are convolution kernels which function to filter data and extract features. The variables h_1 to h_{16} are feature maps obtained after convolution.

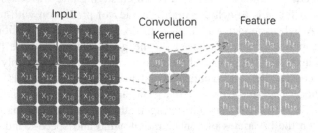

Fig. 1. A two-dimensional convolutional operation

3.2 Long Short-Term Memory Networks

A recurrent neural network (RNN) [18] is an artificial neural network that has a tree-like hierarchical structure, and the nodes of RNN recursively input information in the order in which they are connected. A LSTM [19, 20] network, a special RNN, differs from RNN in learning long-term dependencies. The repeating module in a conventional RNN contains only a single layer (shown in Fig. 2(a)), and the repeating module in a LSTM network contains four interacting neural network layers.

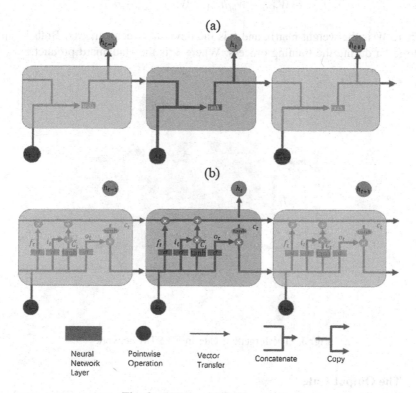

Fig. 2. RNN and LSTM networks

The LSTM network can remove or add information to the cell state and manage it by the gate structure. The LSTM network includes forgetting gates, input gates and output gates. The function σ in the module represents a sigmoid function, and the formula is as shown in Eq. (1). The sigmoid layer outputs a number between 0 and 1, which represents how much each component should pass the threshold. The value of "1" means that all ingredients pass, and the value of "0" means that no ingredients are allowed to pass.

$$S(t) = \frac{1}{1 + e^{-t}} \qquad (1)$$

3.2.1 The Forgetting Gate

In Fig. 3, the forgetting gate decides which information to discard. The formula is shown in Eq. (2).

$$f_t = \sigma(W_{xf}x_t + W_{hf}h_{t-1} + W_{cf} \circ c_{t-1} + b_f) \qquad (2)$$

Where W is the weight matrix and b is the deviation vector matrix. Both W and b need to learn during the training process. Where $\circ$ is the Hadamard product.

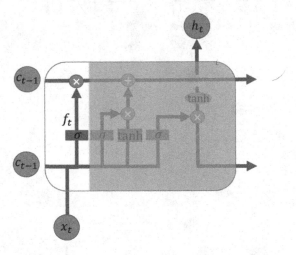

Fig. 3. The forgetting gate in a LSTM network

3.2.2 The Output Gate

In Fig. 4, the input gate determines which information to remember. The formulas are shown in Eqs. (3), (4) and (5).

$$i_t = \sigma(W_{xi}x_t + W_{hi}h_{t-1} + W_{ci} \circ c_{t-1} + b_i) \qquad (3)$$

$$\tilde{C}_t = \tanh(W_{xc}x_t + W_{hc}h_{t-1} + b_c) \qquad (4)$$

$$C_t = f_t \circ C_{t-1} + i_t \circ \tilde{C}_t \qquad (5)$$

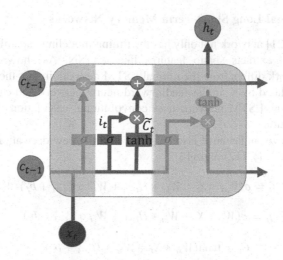

Fig. 4. The output gate in a LSTM network

3.2.3 The Input Gate
In Fig. 5, the input gate decides which information to update. The formulas are shown in Eqs. (6) and (7).

$$o_t = \sigma(W_{xo}x_t + W_{ho}h_{t-1} + W_{co} \circ c_t + b_o) \tag{6}$$

$$h_t = o_t \circ \tanh(c_t) \tag{7}$$

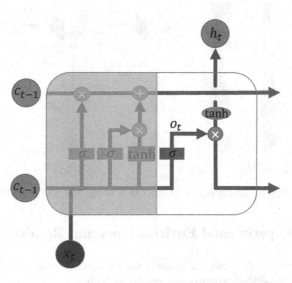

Fig. 5. The input gate in a LSTM network

3.3 Convolutional Long Short-Term Memory Networks

The ConvLSTM [21] network not only has the timing modeling capabilities of a LSTM network, but also extracts spatio features like a CNN. As shown in Fig. 6, the ConvLSTM network differs from the normal LSTM network in that the internal LSTM is internally calculated by a similar feedforward neural network and can be called FC-LSTM [21]. A ConvLSTM network uses convolutional calculations instead of fully connected calculations.

The derivation formulas have also changed, and the new derivations are shown in Eqs. (8), (9), (10), (11), (12) and (13).

$$f_t = \sigma(W_{xf} * X_t + W_{hf} * H_{t-1} + W_{cf} \circ C_{t-1} + b_f) \tag{8}$$

$$i_t = \sigma(W_{xi} * X_t + W_{hi} * H_{t-1} + W_{ci} \circ C_{t-1} + b_i) \tag{9}$$

$$\tilde{C}_t = \tanh(W_{xc} * X_t + W_{hc} * H_{t-1} + b_c) \tag{10}$$

$$C_t = f_t \circ C_{t-1} + i_t \circ \tilde{C}_t \tag{11}$$

$$o_t = \sigma(W_{xo} * X_t + W_{ho}H_{t-1} + W_{co} \circ C_t + b_o) \tag{12}$$

$$H_t = o_t \circ \tanh(c_t) \tag{13}$$

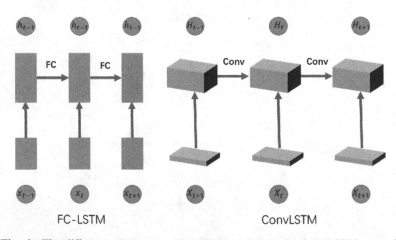

Fig. 6. The differences between a FC-LSTM network and a ConvLSTM network

4 Practical Experimental Environments and Results

This section illustrates the selected features for air quality prediction in Subsect. 4.1 and discusses the practical experimental results in Subsect. 4.2.

4.1 Practical Experimental Environments

For training and testing the air quality prediction methods, this study collected the environmental data in nine cities in the Fujian Province of the People's Republic of China from January 2015 to October 2018. The environmental factors include 7 dimensions which are air quality index (AQI), PM2.5, PM10, SO2, NO2, CO and O3; the AQI reflects the degree of air pollution. Seven environmental factors (i.e., AQI, PM2.5, PM10, SO2, NO2, CO and O3) at the t-th timestamp are elected as the inputs of neural networks, and the parameter of PM2.5 at the $(t + 1)$-th timestamp is elected as the output of neural networks. The mean squared error loss function is adopted for optimizing neural networks. In experiments, the environmental data from January 2015 to December 2017 is used as a training dataset, and the environmental data from January 2018 to October 2018 is used as a testing dataset.

For data pre-processing, if there are abnormal values or missing values in a record, the record will be deleted [22]. For data normalization, the data is processed by min-max normalization method [23] and represented by a number between 0 and 1. The number of records is N, and the value of the i-th record (x) can be normalized by Eq. (14).

$$z_i = \frac{x_i - \min_{1 \leq n \leq N}(x_n)}{\min_{1 \leq n \leq N}(x_n) - \min_{1 \leq n \leq N}(x_n)} \tag{14}$$

4.2 Practical Experimental Results and Discussions

For the evaluation of the proposed ConvLSTM method, multi-layer perception (MLP) neural networks [24–26], CNNs, LSTM networks are implemented and used to predict the air quality in the selected cities in Fujian Province. In order to compare the performance of each prediction method comprehensively and objectively, MAE, RMSE and MAPE were used as accuracy metrics. The value of the n-th actual data is defined as o_n, and the value of the n-th predicted data is defined as p_n. These three accuracy metrics can be estimated by Eqs. (15), (16) and (17), respectively. The practical experimental results based on these three accuracy metrics are shown in Tables 1, 2 and 3.

$$MAE = \frac{1}{N} \sum_{n=1}^{N} |o_n - p_n| \tag{15}$$

$$RMSE = \sqrt{\frac{\sum_{n=1}^{N} (o_n - p_n)^2}{N}} \tag{16}$$

$$MAPE = \frac{1}{N} \sum_{n=1}^{N} \frac{|o_n - p_n|}{o_n} \tag{17}$$

Table 1. The MAEs of each prediction method for each city

CITY	MLP	CNN	LSTM	CONVLSTM
Ningde	7.2652	7.2915	7.9911	**6.7214**
Nanping	6.0939	6.0738	6.4391	**5.3384**
Fuzhou	7.0248	7.1274	7.3394	**6.4475**
Sanming	6.2867	6.1960	6.3652	**5.4708**
Putian	8.4391	8.3612	8.4000	**8.1412**
Quanzhou	7.2165	7.4500	**6.8798**	6.9295
Longyan	7.0222	6.9505	6.7233	**6.3074**
Zhangzhou	7.4520	7.7403	7.6557	**6.7840**
Xiamen	6.3989	6.6250	6.2187	**5.9813**
Average	7.0221	7.0906	7.1125	**6.4579**

Table 2. The RMSEs of each prediction method for each city

City	MLP	CNN	LSTM	ConvLSTM
Ningde	9.7771	9.8717	10.7989	**9.5848**
Nanping	10.8600	10.9898	10.7856	**10.0890**
Fuzhou	9.3981	9.5205	9.6814	**8.8822**
Sanming	9.3716	9.2888	9.2222	**8.6397**
Putian	13.2202	12.9197	12.9752	**12.4293**
Quanzhou	10.7698	10.5632	10.3454	**10.0866**
Longyan	13.2247	12.9709	12.7514	**12.4693**
Zhangzhou	11.7122	11.5926	12.0828	**10.9533**
Xiamen	8.9357	8.9459	8.5964	**8.1712**
Average	10.8077	10.7404	10.8044	**10.1450**

Table 3. The MAPEs of each prediction method for each city

City	MLP	CNN	LSTM	ConvLSTM
Ningde	0.3952	0.3902	0.4271	**0.3322**
Nanping	0.3887	0.3906	0.4114	**0.3289**
Fuzhou	0.4118	0.4213	0.4238	**0.3442**
Sanming	0.3223	0.3186	0.3297	**0.2732**
Putian	0.4065	0.4268	0.4001	**0.3777**
Quanzhou	0.3455	0.3689	0.3313	**0.3272**
Longyan	0.3308	0.3305	0.3247	**0.2984**
Zhangzhou	0.2835	0.3047	0.2815	**0.2572**
Xiamen	0.3347	0.3614	0.3056	**0.2975**
Average	0.3577	0.3681	0.3595	**0.3152**

The MAEs from low to high are generated by ConvLSTM (6.4579), MLP (7.0221), CNN (7.0906) and LSTM (7.1125). Furthermore, the MAPEs from low to high are generated by ConvLSTM (0.3152), MLP (0.3577), LSTM (0.3595) and CNN (0.3681). Finally, the RMSEs from low to high are generated by ConvLSTM (10.1450), CNN (10.7404), LSTM (10.8044) and MLP (10.8077). From the comparison results, the performance of ConvLSTM is significantly better than the other methods, which proves that the superiority of the ConvLSTM network in predicting PM2.5 concentration.

A case study of air quality prediction by the proposed ConvLSTM network for each city is shown in Fig. 7. The actual records are illustrated as blue polylines, and the predicted records are expressed as orange polylines. In experiments, the predicted values of PM2.5 concentration by the proposed ConvLSTM network are roughly consistent with the actual values. Some large errors may be generally caused by human behaviors. For instance, a large number of fireworks and firecrackers are released during the Spring Festival and New Year's Eve, which causes the rising of PM2.5 concentration.

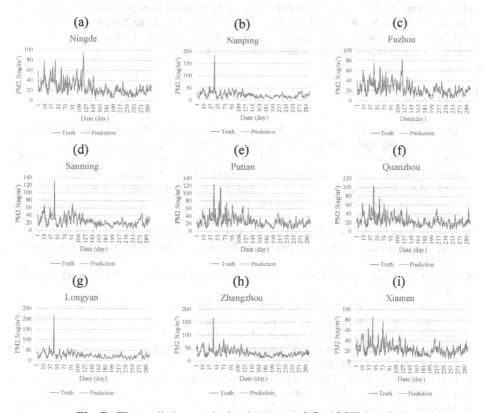

Fig. 7. The prediction results by the proposed ConvLSTM method

5 Conclusions and Future Work

A deep multi-layer neural network model based on CNN and LSTM (i.e., the ConvLSTM method) is proposed to analyze the spatio-temporal features for predicting air quality in multiple cities. A case study of the prediction of PM2.5 concentration in the Fujian Province of the People's Republic of China is given in this study, the proposed model estimate and predict the future concentration of PM2.5 in accordance with the past concentration of PM2.5. In experiments, the performances of each prediction method (e.g., MLP, CNN, LSTM, and ConvLSTM) were evaluated by MAE, MAPE and RMSE. The practical experimental results show that the proposed model combines the advantages of CNN and LSTM for analyzing the spatio-temporal features and improving the accuracy of PM2.5 concentration prediction.

In the future, this study can be applied to the prediction and control of air quality for other cities. Furthermore, the human behaviors can be detected and considered for the improvement of air quality prediction.

Acknowledgment. This work was supported in part the National Natural Science Foundation of China under Grants No. 61877010 and No. 11501114, and the Fujian Natural Science Funds under Grant No. 2019J01243. This research was partially supported by Fuzhou University, grant numbers 510730/XRC-18075 and 510809/GXRC-19037.

References

1. Querol, X., et al.: PM10 and PM2.5 source apportionment in the Barcelona metropolitan area, Catalonia, Spain. Atmos. Environ. **35**(36), 6407–6419 (2001)
2. Schwartz, J., Laden, F., Zanobetti, A.: The concentration-response relation between PM2.5 and daily deaths. Environ. Health Perspect. **110**(10), 1025–1029 (2002)
3. Bell, M.L., Francesca, D., Keita, E.: Spatial and temporal variation in PM2.5 chemical composition in the United States for health effects studies. Environ. Health Perspect. **115**(7), 989–995 (2007)
4. Badyda, A.J., Grellier, J., Dąbrowiecki, P.: Ambient PM2.5 exposure and mortality due to lung cancer and cardiopulmonary diseases in Polish cities. Adv. Exp. Med. Biol. **944**, 9–17 (2017)
5. Chan, C.K., Yao, X.: Air pollution in mega cities in China. Atmos. Environ. **42**(1), 1–42 (2008)
6. Kermanshahi, B.S., et al.: Artificial neural network for forecasting daily loads of a Canadian electric utility. In: Proceedings of the Second International Forum on Applications of Neural Networks to Power Systems, Yokohama, Japan (1993)
7. Fleming, S.W.: Artificial neural network forecasting of nonlinear Markov processes. Can. J. Phys. **85**(3), 279–294 (2007)
8. Gu, K., Qiao, J., Li, X.: Highly efficient picture-based prediction of PM2.5 concentration. IEEE Trans. Industr. Electron. **66**(4), 3176–3184 (2019)
9. Mahajan, S., Liu, H.M., Tsai, T.C., Chen, L.J.: Improving the accuracy and efficiency of PM2.5 forecast service using cluster-based hybrid neural network model. IEEE Access **6**, 19193–19204 (2018)

10. Elangasinghe, M.A., Singhal, N., Dirks, K.N., Salmond, J.A., Samarasinghe, S.: Complex time series analysis of PM10 and PM2.5 for a coastal site using artificial neural network modelling and k-means clustering. Atmos. Environ. **94**, 106–116 (2014)
11. Li, X., et al.: Long short-term memory neural network for air pollutant concentration predictions: method development and evaluation. Environ. Pollut. **231**, 997–1004 (2017)
12. Tsai, Y.T., Zeng, Y.R., Chang, Y.S.: Air pollution forecasting using RNN with LSTM. In: Proceedings of 2018 IEEE 16th International Conference on Dependable, Autonomous and Secure Computing, 16th International Conference on Pervasive Intelligence and Computing, 4th International Conference on Big Data Intelligence and Computing and Cyber Science and Technology Congress, Athens, Greece (2018)
13. Yu, S., et al.: Evaluation of real-time PM2.5 forecasts and process analysis for PM2.5 formation over the eastern United States using the Eta-CMAQ forecast model during the 2004 ICARTT study. J. Geophy. Res. Atmos. **113**, D06204 (2008)
14. Verma, I., Ahuja, R., Meisheri, H., Dey, L.: Air pollutant severity prediction using bi-directional LSTM network. In: Proceedings of 2018 IEEE/WIC/ACM International Conference on Web Intelligence, Santiago, Chile (2018)
15. Zhao, H., Xia, S., Zhao, J., Zhu, D., Yao, R., Niu, Q.: Pareto-based many-objective convolutional neural networks. In: Meng, X., Li, R., Wang, K., Niu, B., Wang, X., Zhao, G. (eds.) WISA 2018. LNCS, vol. 11242, pp. 3–14. Springer, Cham (2018). https://doi.org/10.1007/978-3-030-02934-0_1
16. Krizhevsky, A., Sutskever, I., Hinton, G.E.: ImageNet classification with deep convolutional neural networks. In: Proceedings of the 25th International Conference on Neural Information Processing Systems, Nevada, USA (2012)
17. Lawrence, S., Giles, C.L., Tsoi, A.C., Back, A.D.: Face recognition: a convolutional neural-network approach. IEEE Trans. Neural Netw. **8**(1), 98–113 (1997)
18. Mikolov, T., Karafiát, M., Khudanpur, S.: Recurrent neural network based language model. In: Proceedings of the 11th Annual Conference of the International Speech Communication Association, Chiba, Japan (2010)
19. Hochreiter, S., Schmidhuber, J.: Long short-term memory. Neural Comput. **9**(8), 1735–1780 (1997)
20. Baddeley, A.D., Warrington, E.K.: Amnesia and the distinction between long-and short-term memory. J. Verbal Learn. Verbal Behav. **9**(2), 176–189 (1970)
21. Shi, X., Chen, Z., Wang, H., Yeung, D.Y., Wong, W.K., Woo, W.C.: Convolutional LSTM network: a machine learning approach for precipitation nowcasting. In: Proceedings of the 28th International Conference on Neural Information Processing Systems, Montréal, Canada (2015)
22. Kotsiantis, S.B., Kanellopoulos, D., Pintelas, P.E.: Data preprocessing for supervised leaning. Int. J. Comput. Electr. Autom. Control Inf. Eng. **1**(12), 4091–4096 (2007)
23. Jain, Y.K., Bhandare, S.K.: Min max normalization based data perturbation method for privacy protection. Int. J. Comput. Commun. Technol. **3**(4), 45–50 (2014)
24. Haykin, S.: Neural Networks: A Comprehensive Foundation, 2nd edn. Prentice Hall, Upper Saddle River (1998)
25. Gardner, M.W., Dorling, S.R.: Artificial neural networks (the multilayer perceptron)—a review of applications in the atmospheric sciences. Atmos. Environ. **32**(14–15), 2627–2636 (1998)
26. Haykin, S.O.: Neural Networks and Learning Machines, 3rd edn. Prentice Hall, Upper Saddle River (2008)

Cloud Computing and Big Data

Core Solution Computing Algorithm of Web Data Exchange

Yuhang Ji[1,2(✉)], Gui Li[1,2], Zhengyu Li[1,2], Ziyang Han[1,2],
and Keyan Cao[1,2]

[1] Computer Engineering and Applications, Shenyang Jianzhu University,
Shenyang 110168, China
syjzjyh@163.com
[2] Faculty of Information and Control Engineering, Shenyang Jianzhu University,
Shenyang 110168, China

Abstract. Traditional Web data exchange research usually focuses on designing transformation rules but ignores the processing of the actual generated target data instances. Since the data instance is highly correlated with the schema and there are many duplicate elements in the source data instance, there is redundancy in the actual generated target data instance. In order to generate a target data instance solution that does not contain redundancy, under a given source-to-target exchange rule, a unified integration schema is designed firstly, and then, the instance block mechanism is introduced to analyze three mapping relationships of single homomorphism, full homomorphism and the isomorphism among the initial generated target data instances. According to the mapping relationship, three methods of instance selection, which are more compact, more informative and equivalence class processing, are proposed to remove redundant data instance in target data set and generate the core solution of target data instance. The experiment uses the data from the China Land Market Network to evaluate the performance of the data exchange core solution algorithm.

Keywords: Large web data · Data exchange · Redundant data processing · Core solution · Homomorphism relation · Schema integration

1 Introduction

Data exchange is very important in multi-source data integration. The original data exchange problem [1] was proposed by Fagin et al., the paper illustrates that data conversion typically takes source data as input and select the source data by a set of mapping rules (also known as tuple generation dependencies) to transform it into a target data set that satisfies a given schema mapping rule. On this basis, Web data exchange can be carried out at two levels, schema layer and instance layer [2]. The main work of schema layer is to design an accurate and complete set of mapping rules according to the attribute correspondence between source and target. The instance layer obtains and generates the target data set from multiple Web data sources according to the set of schema mapping rules between the given source and target database, and

© Springer Nature Switzerland AG 2019
W. Ni et al. (Eds.): WISA 2019, LNCS 11817, pp. 113–118, 2019.
https://doi.org/10.1007/978-3-030-30952-7_13

performs repeated data processing on the target data set to remove redundant data instances. However, traditional research on Web data exchange usually focuses on designing schema exchange rules, while ignoring the processing of the actual generated target data instance. In the data exchange scenario shown in Fig. 1, table A, B, C, and D represent the source database tables from different web sources. Table T1 and T2 represent the target database tables. The attribute correspondence of the source-to-target database tables and the given mapping rules are shown in Fig. 2.

Fig. 1. Source and target profile data

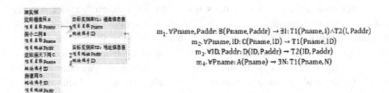

$m_1. \forall Pname, Paddr: B(Pname, Paddr) \rightarrow \exists I: T1(Pname, I) \wedge T2(I, Paddr)$

$m_2. \forall Pname, ID: C(Pname, ID) \rightarrow T1(Pname, ID)$

$m_3. \forall ID, Paddr: D(ID, Paddr) \rightarrow T2(ID, Paddr)$

$m_4. \forall Pname: A(Pname) \rightarrow \exists N: T1(Pname, N)$

Fig. 2. Attribute correspondence and mapping rules

The target data set of the target table T1, T2 in Fig. 1 can be obtained by given schema mapping rules, and these target data sets are called the canonical universal solution [2]. They basically satisfy the given mapping rules, but contain Multiple redundant tuples (gray background part). In this paper, T1 is obtained on the basis of the initial target data instance sets and T2 does not contain redundant target data instance sets, that is, the part with the white background in the target table T1 and T2. This part can be called the core universal solution. [2].

The contribution can be sum up as follows:

- We design an accurate integration schema and given the initial mapping rules.
- Three instance selection methods based on homomorphic relation are proposed to calculate data exchange core solution. Finally, a detailed experiment and theoretical analysis of the data exchange algorithm are carried out with the online data set, and the experimental results verify the performance of the algorithm in this paper.

2   Related Work

With the development of Web technology and the popularization of its application, Web data has become the main data source in various fields. Pichler et al. [3] started from the perspective of schema mapping, through the optimization of the mapping rules and converted them into executable scripts to calculate the kernel solution. However, when faced with large data exchange scenarios, this technique may result in a large amount of redundant data in the target database. In literature [6], it is considered that data exchange based on schema-level information only limits the ability to express semantics in data exchange, and cannot solve some fuzzy data exchange scenarios. In literature [9] and [10], data instances are used to reconstruct schema mapping, and constraints in the pattern are used to find natural associations and it hopes to find the kernel solution through better examples and mapping rules. The main difficulty with this method is the selection of data instances, which may be redundant due to the fact that different data instances may describe the same thing.

3   Methodology

3.1   The Achievement of the WEB Multisource Integration Mode

Because the relevance between data and mode is very high in the process of exchanging the web data, only when the mode integrates [2] and designs an initial exchange rule, the living example can be exchanged from source data into target data. The process of integration mode can be divided into two phases:

The Definition Phase of Integration Mode. Firstly, the data source to be exchanged and integrated should be determined. Be directed against different data sources, we should analyze their mode information. Analyze the universality and differences between the output modes of various part databases. Define the formal description of integration mode on the basis of meeting the needs of user data and obtain the all information which can be used in the follow-up procedures of integration mode.

The Definition Phase of Matching Relation Table. On the basis of the formal description of integration mode, we define the matching relation tables which include the matching relationship between output mode of source databases and integration mode of target databases in data table names, attribute names and operation names. Find certain mapping relationships which local in two different modes' elements. Input two modes as parameters and the output result is the mapping relationship between then which is the initial mapping rule set.

3.2   The Selection Method of Data Instance

Firstly, find all the instance block sets which content transfer rules of data. Obtain the irredundant instance block sets by using the homomorphism relationship between instance blocks to delete redundant instance blocks. Then, choose the instance blocks with higher accuracy to produce that by calculating the accuracy of data in instance blocks. According to the instance blocks, the basic features of core and popular solution can be defined. Choose and output the final target instance data.

Because there are different homomorphism relationships in instance blocks, this paper will use homomorphism to define what the "redundant" instance block is.

The Classification in Homomorphism. Given two instance set J and J', mapping $h : J \rightarrow J'$. When $J \rightarrow h(J), J' \rightarrow h(J')$, if $J * J' \rightarrow h(J) * h(J')$, we call that the mapping h is the homomorphism from J to J'.   Above all, there are two main reasons due to redundant instance blocks. The first one is that there is the epimorphism relationship between w and w' which makes w' more compact than w, and w is the redundant block which expressed by $w \prec w'$. The second one is that even we exclude the tuples included many uncertainties, the instance block w may still produce other instance block w' which exists with single homomorphism relationship by using other exchange rules and assignment, we call that w' has more information than w, w is the redundant block which expressed by $w \prec w'$.

Through the above steps, the most redundancies in initial solutions can be removed, but not enough to produce core solutions. When two biggest instance blocks are isomorphism to each other, the core solution only need to consider one of them.

In order to produce the instance block sets which can accurately calculate core solutions, the accuracy of each instance block w must be calculated first. Remove the instance blocks with lower accuracy until the change of accuracy $R(W) = \sum_{v \in v(w)} \frac{p(v)}{n}$ in each instance blocks are smaller than the given standard value $\left(P(v) = \frac{\sum_{i=1}^{v_j} Simv_i, v_i}{v_i} \right)$.

Then, choose the instance blocks with higher accuracy to make sure that there is no isomorphism relationship between instance blocks of the set. Finally, take advantage of these instance block sets to produce the core solutions. The accuracy of instance blocks can be calculated by the following formula.

4 Experiments

4.1 Experimental Setting and Dataset

The experimental data set used in this paper is from China Land and Market Network (www.landchina.com). To evaluate the performance of the algorithm, the real-world data is difficult to present all the problems, so the artificial data set is constructed by using the above data set. Three of these classified attribute data are chose to do the experiment. The information description of the data set includes 18052 estate records, 25308 item indicia records and 62371 address records.

In order to compare the superiority of this method and other similar works, we design the source databases which include 100k, 250k, 500k and 1M tuples. Compare our method with the computational algorithm of core solution in literature [3] and literature [4]. t1, t2 and t3 are used separately to represent our method and other two computational algorithms. Test the running time of our method and others aimed at core solution computing problems on the large instance tuples. We design five source databases with 10k data which are all from the real data set to prove through experiments that the target instances produced by core solution is less and better in quality than by standard solution. Every source database includes different degrees of "redundant" and the range is 0%–40%. We do the specific experiments with eight different mapping scenes. S1–S4 does not include self-join. SJ1–SJ4 includes self-join.

4.2 Experimental Results

As shown in Fig. 3, it is obvious that the time of the target core solutions in computing large data set is less and the efficiency of that is higher by using our method.

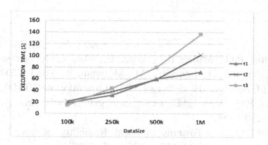

Fig. 3. Performance testing

By comparing the tuple numbers produced in the target database, the reduced generating tuple numbers percentage of the core solution compared with the universal solution is obtained. The Fig. 4 left shows the four project results of the target which do not include self-join and the core solution is more compact than the universal solution in all cases. As redundancy increases, this case becomes more apparent. Two hypothetical scenarios S1 and S2 follow the trend but not as significant as the two coverage scenarios S3 and S4, because the design of tgds in S1 and S2 often generates many duplicate tuples in the solutions and these tuples are deleted by core scripts and universal scripts. Figure 4 right shows the reduced percentage of four self-join solutions. Except SJ_1, the core solution is more compact than the universal solution in all cases. No *null* value is generated in the solution which the universal solution and the core solution coincide, because SJ_1 is the complete mapping.

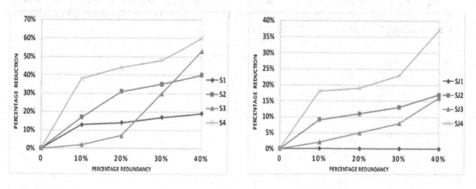

Fig. 4. Reduction of redundancy

5 Conclusion

In this paper, we proposed a data conversion kernel calculation method based on homomorphic relationship between target database instances. We used data from the China Land and Market Network (www.landchina.com) to build datasets and to evaluate the performance of the algorithm, three of these classified attribute data are chose to do the experiment. Experimental results prove the effectiveness of our method.

References

1. Fagin, R., Kolaitis, P.G., Miller, R.J., Popa, L.: Data exchange: semantics and query answering. In: Calvanese, D., Lenzerini, M., Motwani, R. (eds.) ICDT 2003. LNCS, vol. 2572, pp. 207–224. Springer, Heidelberg (2003). https://doi.org/10.1007/3-540-36285-1_14
2. Fagin, R., Kolaitis, P.G., Popa, L.: Data exchange: getting to the core. ACM Trans. Database Syst. **30**(1), 174–210 (2005)
3. Pichler, R., Savenkov, V.: Towards practical feasibility of core computation in data exchange. Theoret. Comput. Sci. **411**(7), 935–957 (2010)
4. Gottlob, G., Nash, A.: Data exchange: computing cores in polynomial time. In: ACM SIGMOD-SIGACT-SIGART Symposium on Principles of Database Systems. ACM (2006)
5. Cabibbo, L.: On keys, foreign keys and nullable attributes in relational mapping systems. In: International Conference on EDBT. DBLP (2009)
6. Sekhavat, Y.A., Parsons, J.: SEDEX: scalable entity preserving data exchange. IEEE Trans. Knowl. Data Eng. **28**(7), 1878–1890 (2016)
7. Mecca, G., Papotti, P., Raunich, S.: Core schema mappings: scalable core computations in data exchange. Inf. Syst. **37**(7), 677–711 (2012)
8. Cai, D., Hou, D., Qi, Y., Yan, J., Lu, Y.: A distributed rule engine for streaming big data. In: Meng, X., Li, R., Wang, K., Niu, B., Wang, X., Zhao, G. (eds.) WISA 2018. LNCS, vol. 11242, pp. 123–130. Springer, Cham (2018). https://doi.org/10.1007/978-3-030-02934-0_12
9. Arocena, P.C., Glavic, B., Ciucanu, R., Miller, R.J.: The iBench integration metadata generator. Proc. Very Large Data Bases **9**(3), 108–119 (2015)
10. Dong, X.L., Berti-Equille, L., Srivastava, D.: Integrating conflicting data: the role of source dependence. Proc. VLDB Endowment **2**(1), 550–561 (2018)
11. Han, Z., Jiang, X., Li, M., Zhang, M., Duan, D.: An integrated semantic-syntactic SBLSTM model for aspect specific opinion extraction. In: Meng, X., Li, R., Wang, K., Niu, B., Wang, X., Zhao, G. (eds.) WISA 2018. LNCS, vol. 11242, pp. 191–199. Springer, Cham (2018). https://doi.org/10.1007/978-3-030-02934-0_18

A Novel Adaptive Tuning Mechanism for Kafka-Based Ordering Service

Li Xu[✉], Xuan Ma, and Lizhen Xu

Department of Computer Science and Engineering,
Southeast University, Nanjing 211189, China
seu_lxu@163.com

Abstract. Thousands of resumes are sent to companies every year and it takes abundant time to authenticate the resumes. Blockchain (Hyperledger fabric) with its consensus algorithm, kafka-based ordering service, is a new solution to this issue, but it can not adapt to the dynamic workloads in real-time. This paper investigates an adaptive tuning mechanism based on feedback control theory to adjust the parameters connected to its consensus algorithm. In order to evaluate its efficiency, experiments have been done to compare the performance with the original kafka-based ordering service.

Keywords: Kafka-based ordering service · Hyperledger fabric · Adaptive mechanism

1 Introduction

Every year when it comes to graduate season, many well-known companies will receive thousands of resumes. It takes a lot of time and resources of the companies to identify the authenticity of the resumes. If the data of students' whole life cycle can be truly and untamable recorded such as awards record, certificates and grades, etc. a lot of time will be saved.

For that the blockchain [1, 2] is good at solving the problems of data security, sharing and reconciliation and so on, a lot of research has sprung up on the application industry of blockchain such as finance [3], medical care [4] and education [5, 6].

The sudden abundant data caused by students' collective activities requires fabric's consensus mechanism, kafka-based ordering service [7], should adapt to the current system workloads quickly. Therefore in this paper, a novel adaptive tuning mechanism (A-Kafka) contracting on adjustment of the parameters of the kafka-based ordering service based on feedback control theory [8] is proposed, with blockchain particularly Hyperledger fabric [9] utilized. In order to evaluate its efficiency, the proposed A-Kafka is studied on and analyzed in comparison with the original scheme of kafka (O-Kafka).

© Springer Nature Switzerland AG 2019
W. Ni et al. (Eds.): WISA 2019, LNCS 11817, pp. 119–125, 2019.
https://doi.org/10.1007/978-3-030-30952-7_14

2 Related Work

The consensus mechanism based on kafka is selected by Hyperledger fabric1.x, which has the characteristics of high throughput, low latency, extensibility, durability and reliability [10]. The kafka-based consensus mechanism contains kafka cluster and its associated zookeeper cluster, as well as ordering service nodes (OSN). Literature [11, 12] elaborated its consensus process:

Figure 1 shows the flow of kafka-based ordering service, when the transactions (e.g. tx1, tx2 ... txn) created by clients, they will be broadcasted toward OSNs into kafka cluster. When the configurations of batchSize which is the maximum message count of the block and batchTimeout which is the maximum delay time from the last block generated are met, OSNs will package previously received transactions and generate a new block and after that, the block will be presented to committer peers which validate transactions in every block.

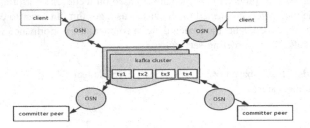

Fig. 1. The architecture of kafka-based ordering service

Currently, the configuration of O-Kafka about batchSize and batchTimeout could be set at startup [12], but the optimal policy configuration cannot be determined in real-time. Therefore, some researches related to the kafka-based consensus optimization have been carried out such as literature [13, 14].

3 Model

3.1 Model Representation

To describe the structure of A-Kafka, this paper models kafka-based ordering service into three stages: ordering, buffer and batching until configurations are met and committer validation.

A-Kafka continually evaluates the whole workload and the performance of consensus throughout a feedback control loop. It obtains information from kafka-based ordering service regarding the events such as pull transactions from the kafka cluster, buffer and cut block and process block (committer peers' validation) to adjust parameters of batchSize and batchTimeout.

3.2 Model Details

BatchTimeout

The batchTimeout should be large or small enough to meet the requirements of enough transactions batched or a busy state of the stage committer validation. MBV defined in this paper is the mean process time of buffer and batching and committer validation, and MOT defined is the mean time interval of OSNs pull transactions.

The Eq. (1) is the formula of MBV.

$$MBV_k = \alpha MBV_{k-1} + (1 - \alpha)BV_k \tag{1}$$

In the formula, α is the coefficient and $BV_k = T_{exe} - T_{start}$; T_{exe} is the instant time of complete validation of a block and T_{start} is the instant time when the OSN pulls a transaction from kafka cluster. MBV_{k-1} means the average value of the last T time intervals of MBV.

The Eq. (2) shows the formula of MOT.

$$MOT = \frac{t - t_0}{n} \tag{2}$$

In the formula, n represents the number of transactions that OSNs pulled in time interval $(t - t_0)$.

BatchSize

The adjustment method of batchSize is also should be large or small enough. The batchSize (BS) then could be calculated from the relationship of MBV and MOT and the Eq. (3) is presented below.

$$BS = \left\lceil \frac{MBV}{MOT} \right\rceil \tag{3}$$

3.3   Algorithms for A-Kafka

Algorithm 1 A-Kafka adjusting algorithm

Output: $BatchTimeout(BT)$, $BatchSize(BS)$
1: **on event:** OSN pulls a transaction **do**
2:    $T_{start} \leftarrow$ *the local clock timestamp*
3:    Number each transaction tx arrives OSN
4: **on event:**After committer peers validate transaction tx in block **do**
5:    **search** the transaction with the matched number
6:       $T_{exe} \leftarrow$ *the local clock timestamp*
7:       use T_{exe} and T_{start} compute MBV
8: **on event:** After OSNs pull tx from kafka-cluster **do**
9:    **call Algorithm 3:** BatchTimeout Adaptation
10: **on event:** After committer peers validate the last block **do**
11:    **call Algorithm 4:** BatchSize Adaptation

Algorithm 3 BatchTimeout Adaptaion

Input: MBV_k
Output: BT
1: compute MOT
2: **if** $MOT \geqslant MBV$ **then** $BS \leftarrow 0$
3: **else** $BT \leftarrow MBV_k$

Algorithm 4 BatchSize Adaptaion

Input: MBV, MOT
Output: BS
1: **if** fabric is startup **then** $BS \leftarrow default$
2: **else** $BS \leftarrow \lceil \frac{MBV}{MOT} \rceil$

---

4   Experiments

4.1   Objective

Latency (LT) and Throughput (TH) is widely used in performance evaluation. LT measures the mean time interval of a process from a transaction initially being created by clients to the transaction is confirmed by the blockchain. TH measures the numbers of transactions processed successfully number per second. What's more, $PER = \frac{TH}{LT*100}$ is utilized in this paper to evaluate the integral performance.

4.2 Experiments Design

The configuration of Hyperledger Fabric is fixed, the number of the nodes of zookeeper is fixed to 3, with that of the kafka nodes fixed to 4 and of the orderer nodes 3. And considering the influence factors of experimental performance are the size of transactions and the value of batchTimeout and batchSize. Therefore, the experiments are designed to those of fixed workloads and those of varying workloads.

4.3 Results and Analysis

Fixed-Workloads Experiments
The data collected from the fixed-workload experiments is shown from Tables 1 and 2. Obviously, the performance of A-Kafka is better than O-Kafka by observing variation of PER.

Table 1. Fixed-workloads results (5 clients, 8 KB)

Model	BS	BT(s)	LT(s)	TH(tx/s)	PER
O-Kafka	10	2	1.48	387.2	2.61
	10	30	1.53	395.0	2.58
	200	2	1.49	401.1	2.69
	200	30	0.73	536.9	7.39
A-Kafka			0.58	584.9	10.07

Table 2. Fixed-workloads results (50 clients, 8 KB)

Model	BS	BT(s)	LT(s)	TH(tx/s)	PER
O-Kafka	10	2	6.4	405.6	0.63
	10	30	5.5	413.0	0.75
	200	2	5.1	515.9	0.79
	200	30	3.7	838.6	1.01
A-Kafka			1.5	933.5	6.22

Varying-Workloads Experiments
Tables 3 and 4 present the results of the experiments of varying workloads with 5 and 50 clients engaged. We found that A-Kafka is more effective in experiments of varying workloads compared to those of fixed workloads by observing the change of PER.

To sum up, A-Kafka performs better than O-Kafka both in two types of experiments especially with varying workloads. To analyze, O-Kafka cannot cope with the changing workloads.

Table 3. Varying-workloads results (5 clients, 256B-8 KB)

Model	BS	BT(s)	LT(s)	TH(tx/s)	PER
O-Kafka	10	2	1.47	388.0	2.63
	10	30	1.39	389.9	2.80
	200	2	1.34	407.8	3.04
	200	30	0.54	542.6	10.01
A-Kafka			0.42	621.2	14.79

Table 4. Varying-workloads results (50 clients, 256B-8 KB)

Model	BS	BT(s)	LT(s)	TH(tx/s)	PER
O-Kafka	10	2	5.9	408.3	0.69
	10	30	5.0	416.6	0.83
	200	2	4.5	518.4	1.15
	200	30	3.2	841.9	2.63
A-Kafka			1.1	944.4	8.58

5 Conclusion

To address the difficulty in adapting to the dynamic workloads of the kafka-based ordering service in Hyperledger Fabric, this paper proposes an adaptive tuning mechanism A-Kafka. The results have shown that the effectiveness of this mechanism exceeds the O-Kafka, both in the experiments of fixed and varying workloads.

References

1. Nakamoto, S.: Bitcoin: A Peer-to-Peer Electronic Cash System (2006)
2. Garay, J., Kiayias, A., Leonardos, N.: The bitcoin backbone protocol: analysis and applications. In: Oswald, E., Fischlin, M. (eds.) EUROCRYPT 2015. LNCS, vol. 9057, pp. 281–310. Springer, Heidelberg (2015). https://doi.org/10.1007/978-3-662-46803-6_10
3. Treleaven, P.: Blockchain technology in finance. Computer **50**(9), 14–17 (2017)
4. Wang, X., Hu, Q., Zhang, Y., Zhang, G., Juan, W., Xing, C.: A kind of decision model research based on big data and blockchain in eHealth. In: Meng, X., Li, R., Wang, K., Niu, B., Wang, X., Zhao, G. (eds.) WISA 2018. LNCS, vol. 11242, pp. 300–306. Springer, Cham (2018). https://doi.org/10.1007/978-3-030-02934-0_28
5. Schmidt, P.: Blockcerts-an open infrastructure for academic credentials on the blockchain. In: MLLearning (2016)
6. The Melbourne Newsroom. https://about.unimelb.edu.au/newsroom/news/2017/october/university-of-melbourne-to-issue-recipient-owned-blockchain-records. Accessed 23 June 2019
7. Kafka, A.: Kafka 0.9. 0 Documentation. http://kafka.apache.org/documentation.html#-introduction. Accessed 13 Apr 2016

8. Lu, C., Stankovic, J.A., Sang, H.S., Gang, T.: Feedback control real-time scheduling: framework, modeling, and algorithms. Real-Time Syst. **23**(1/2), 85–126 (2002)
9. Cachin, C.: Architecture of the hyperledger blockchain fabric. In: Workshop on Distributed Cryptocurrencies and Consensus Ledgers, vol. 310 (2016)
10. Garg, N.: Apache Kafka. Packt Publishing Ltd, Birmingham (2013)
11. Christids, K.: A Kafka–based Ordering Service for Fabric. https://docs.google.com/documents/d/1vNMaM7XhOlu9tB_10dKnlrhy5d7b1u8lSY8akVjCO4. Accessed 2016
12. Hyperledger.org.: Hyperleger Fabric. https://hyperledger-fabric.readthedocs.io/en/latest/blockchain.html. Accessed 14 Apr 2019
13. Klaokliang, N., Teawtim, P., Aimtongkham, P., et al.: A novel IoT authorization architecture on hyperledger fabric with optimal consensus using genetic algorithm. In: 2018 Seventh ICT International Student Project Conference (ICT-ISPC), pp. 1–5. IEEE, Nakhonpathom, Thailand (2018)
14. Guo, Z., Ding, S.: Adaptive replica consistency policy for Kafka. In: MATEC Web of Conferences 2018, vol. 173, p. 01019. EDP Sciences (2018)

Author Name Disambiguation in Heterogeneous Academic Networks

Xiao Ma[1,2(✉)], Ranran Wang[1], and Yin Zhang[1]

[1] Zhongnan University of Economics and Law, Wuhan, China
cindyma@zuel.edu.cn, ran.ran.wang@foxmail.com, yin.zhang.cn@ieee.org
[2] State Key Laboratory of Digital Publishing Technology, Beijing, China

Abstract. In the real world, it is inevitable that some people share a name. However, the ambiguity of the author's name has brought many difficulties to the retrieval of academic works. Existing author name disambiguation works generally rely on the feature engineering or graph topology of the academic networks (e.g., the collaboration relationships). However, the features may be costly to obtain due to the availability or privacy of data. What's more, the simple relational data cannot capture the rich semantics underlying the heterogeneous academic graphs. Therefore, in this paper, we study the problem of author name disambiguation in the setting of heterogeneous information network, and a novel network representation learning based author name disambiguation method is proposed. Firstly, we extract the heterogeneous information networks and meta-path channels based on the selected meta-paths. Secondly, two meta-path based proximities are proposed to measure the neighboring and structural similarities between nodes. Thirdly, the embeddings of various types of nodes are sampled and jointly updated according to the extracted meta-path channels. Finally, the disambiguation task is completed by employing an effective clustering method on the generated paper related vector space. Experimental results based on well-known Aminer dataset show that the proposed method can obtain better results compared to state-of-the-art author name disambiguation methods.

Keywords: Name disambiguation ·
Heterogeneous information networks · Representation learning ·
Meta-paths

1 Introduction

In the physical world, the vocabulary is limited, but the population is constantly increasing. It is inevitable that some people share the same name, which poses a huge challenge for many applications, e.g., information retrieval and bibliographic data analysis [3,5,15].

The problem of name disambiguation is to identify *who is who*, which is a very important issue. For example, when we search a name like "Michael Jordan" in google scholar, many papers will be returned. However, the papers are not

© Springer Nature Switzerland AG 2019
W. Ni et al. (Eds.): WISA 2019, LNCS 11817, pp. 126–137, 2019.
https://doi.org/10.1007/978-3-030-30952-7_15

from the same "Micheal". With more and more scholars and papers included in databases like google scholar, DBLP, Aminer, etc., the name disambiguation problem becomes much more difficult and important.

Most of existing solutions for the name disambiguation task rely on the biographical features such as name, address, affiliation, email, homepage, and etc. [2,3,6]. However, the person-related information may not be easily obtained due to the availability or security of data. Therefore, some researchers have considered to employ the relational data, and solve the name disambiguation problem by using graph topological features [15,16]. Saha et al. [16] solve the name disambiguation task by extracting link information from a collaboration network. The only used information is the graph topology. By following their work, Zhang et al. [15] consider both the person-person collaboration graph and the person-document bipartite graph. As far as we know that, existing link-based method mainly rely on some simple relational data. However, the real bibliographic information networks are generally heterogeneous information networks [11], which contains multiple entities (i.e., author, paper, venue, topic) and multiple relationships (i.e., writing, publishing, collaborating). Simple link information are not sufficient to capture the rich semantics of the bibliographic graph.

To tackle this problem, in this paper, we consider to solve the author name disambiguation problem in the setting of heterogeneous bibliographic networks without considering the private information of researchers [8]. Inspired by the work of network representation learning [9], especially the heterogeneous information network representation learning [10], a **meta-path channel** based **heterogeneous network representation learning method (Mech-RL)** is proposed to solve the author name disambiguation problem[1]. The embeddings of each type of nodes are directly learned from the whole heterogeneous graph, which is different from the existing works which transformed the heterogeneous graph into some simple subgraphs [4] or a homogeneous graph [10]. The Mech-RL propose two meta-path based similarities to sampling and update the node embeddings, which makes it more flexible to learn the embeddings of different types of nodes in different embedding spaces than other meta-path based representation learning methods.

In the graph representation, the name disambiguation task becomes a clustering task of the low dimensional paper vectors, with the objective that each cluster contains papers pertained to a unique real-life person [15]. The contributions are: (1) The problem of author name disambiguation is solved in the setting of heterogeneous bibliographic graphs without considering the private information of researchers. By employing both the relational features and paper-related textual features, we propose a novel heterogeneous network representation learning method to obtain the embeddings of multiple types of nodes. (2) The textual

---

[1] The concept of meta-path channel is similar to the color channel in image processing. For example, an RGB picture can be viewed as a combination of three color channels i.e., red, green, and blue. Similarly, a heterogeneous information network can also be considered as a combination of meta-path channels which contains multiple meta-path instances with respect to different meta-paths.

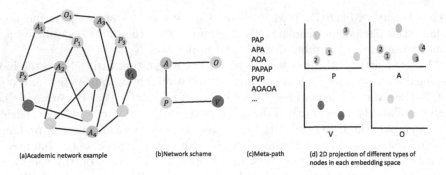

Fig. 1. A toy example: the procedure of the heterogeneous bibliographic networks representation learning.

features of papers are used to refine the initialization of the representation vectors, which increases the accuracy of the final node embedding. Two meta-path based proximity measures are proposed to evaluate the neighboring and structural similarities of node embedding in the heterogeneous graphs, respectively. (3) A real-world bibliographic dataset named Aminer based experiment results show the superior of our proposed method compared to the state-of-the-art name disambiguation approaches.

The rest of this paper is organized as follows. Section 1 gives the introduction of the related works. The background and preliminaries are introduced in Sect. 2. Our proposed Mech-RL method is detailed in Sect. 3. Section 4 includes our experimental results and analysis. Section 5 concludes this study with future work.

2 Background and Preliminaries

Given an author name a, $P^a = \{p_1{}^a, p_2{}^a, \ldots, p_N{}^a\}$ denotes the set of articles published by a. Each paper $p_i{}^a \in P^a$ is represented by a set of features, i.e., the title of the article, keywords, abstracts, venues (indicated by V), collaborators (represented by A), and affiliations (indicated by O). The features of $p_i{}^a$ are divided into two types: $X_{text}^{p_i{}^a} = \{title^{p_i{}^a}, keywords^{p_i{}^a}, abstract^{p_i{}^a}\}$ and $X_{entity}^{p_i{}^a} = \{A, O, V\}$ represent the text features and linkage features of paper, respectively.

Definition 1 (Author Name Disambiguation) [17]. *Given an author a, the task of author name disambiguation is to find a mapping ψ which divides the paper sets P^a into L disjoint subsets w.r.t. X.*

$$\psi(P^a) \xrightarrow{X} \{P_1{}^a, P_2{}^a, \ldots, P_L{}^a\} \tag{1}$$

where each paper subset $P_i{}^a(P_i{}^a \subseteq P^a)$ represents the paper collection of the i-th author whose name is a in the real world, and L represents the number of authors whose name is a.

Definition 2 (*Heterogeneous Network Representation Learning*). *For a given network $G(V, E)$, V and E represent the set of nodes and edges, respectively. $|V|(|V| > 1)$ and $|E|(|E| > 1)$ denote the node types and edge types. As we can see that, A Heterogeneous Information Network (HIN) is a directed graph, which contains multiple types of entities and links. The goal of heterogeneous information network representation learning is to find a mapping ψ_1 which will output a d_i dimensional vector λ_j^i to represent each node instance v_j^i of the i-type nodes in the heterogeneous networks.*

$$\psi_1(v_j^i) \to \lambda_j^i \tag{2}$$

Where v_j^i is the j-th node in the i-type node set. d_i represents the dimension of the projection space of the i-type node. The procedure of the heterogeneous bibliographic networks representation learning is shown as Fig. 1(a) to (d).

Definition 3 (*Meta-path*) [10]. *A meta-path Π is defined on an HIN schema $T_G = (M, R)$, and is denoted in the form of $M_1 \xrightarrow{R_1} M_2 \xrightarrow{R_2} \ldots \xrightarrow{R_L} M_{L+1}$, which defines a composite relation $R = R_1 \circ R_2 \circ \ldots \circ R_L$ between type M_1 and M_{L+1}, where $\circ$ denotes the composition operator on relations.*

For example, the network schema of the academic network (Fig. 1(a)) is shown as Fig. 1(b). The meta-paths (Fig. 1(c)) can be extracted as shown in Fig. 1(c) based on Definition 3.

Definition 4 (*Meta-path based First-order Proximity*). *Given a meta-path Π, if the node type of the starting point and end point is the same, we assume that the starting point and end point of any instances of the meta-path are similar.*

For example, in Fig. 1(a), $P_1 A_1 P_2$ is an instance of meta-path PAP. P_1 and P_2 represent two papers written by the same author in a heterogeneous information network, which indicates that papers P_1 and P_2 are similar. As we can find that nodes P_1 and P_2 are also closer than other nodes in Fig. 1(d).

Definition 5 (*Meta-path based Second-order Proximity*). *For any two instances of a given meta-path Π, if the starting point and end point of the instances are the same, we assume that the center nodes of the two instances are similar.*

For example, in Fig. 1(a), $P_1 A_1 P_2$ and $P_1 A_2 P_2$ are two instances of the meta-path PAP, which share the same starting point P_1 and end point P_2. It can be seen from this figure that nodes A_1 and A_2 are structurally similar. Besides, nodes A_1 and A_2 are also closer than other nodes in Fig. 1(d).

3 The Proposed Approach

3.1 Framework Description

The framework of the name disambiguation module is shown in Fig. 2. The input data is divided into two sections, one is the text information X_{text} and the other

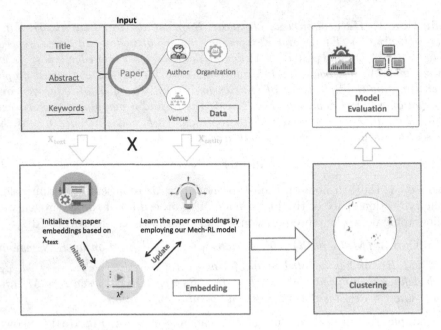

Fig. 2. The framework

is the entity relationship X_{entity}. Firstly, the paper embeddings are initialized based on X_{text}. Secondly, the paper embedding is updated and learned based on the X_{entity} by employing the proposed Mech-RL method. Thirdly, the generated paper embeddings are clustered into disjoint clusters, and each cluster contains documents pertained to a unique real-life person. Finally, the disambiguation results is evaluated.

3.2 Extracting Paper Embedding from Text Information

As shown in Fig. 2, the textual features we are interested in are the title, keywords, abstract of the papers. Instead of adopting a random initialization of the paper embeddings, we rely on the textual features to refine the initialization. The intuition is that the text features can be used to help distinguish different research topics, which will increase the accuracy of the final node embeddings.

Given a paper $p_i^a \in P^a$ which is published by the author named 'a', all its related text features like title, keywords and abstract are stored in the document doc_i^a. Then we employ Doc2vec [7] to initialize the paper embeddings based on the textual features extracted above.

3.3 The Mech-RL Algorithm

In order to learn the embedding of various types of nodes in heterogeneous networks, we propose a meta-path channel based heterogeneous network representation learning algorithm (Mech-RL). We use the meta-path based first-order

---

Algorithm 1. Mech-RL algorithm

Input:
    heterogeneous information network G(V,E);
    meta-path set S;
    embedding dimension of i-type node d_i;
    number of iterations $itera$
Output:
    latent node embedding λ $(\lambda^i \subseteq \lambda, \lambda^i \in \mathbb{R}^{|V| \times d_i})$
1: initialize λ;
2: **for** $it < itera$ **do**
3:     **for** each $s \in S$ **do**
4:         sampling(G(V,E)) $\xrightarrow{s}$ meta-path instances set M_s
5:         **if** $s_{endpoint} \in V_i$ **then**
6:             **for** each $m_s \in M_s$ **do**
7:                 update λ^i in the feature matrix corresponding to the endpoint of the
                path instance m_s
8:             **end for**
9:         **end if**
10:        **if** $s_{center} \in V_i$ **then**
11:            **for** each $m_s^j, m_s^k \in M_s$ **do**
12:                **if** the endpoints of m_s^j and m_s^k are same **then**
13:                    update λ^i in the feature matrix corresponding to the center of the
                   path instance m_s^j and m_s^k
14:                **end if**
15:            **end for**
16:        **end if**
17:     **end for**
18: **end for**
19: **return**

---

and second-order proximities to measure the similarities, and extensively incorporates the semantics underlying the heterogeneous bibliographic networks. The pseudo code of the Mech-RL algorithm is shown in Algorithm 1. The input of the algorithm is the heterogeneous network $G(V, E)$, the set of meta-paths S, the dimension d_i of i-type nodes, and the number of algorithm iterations $itera$. The output is the set of the latent embedding of various node types. According to the node type, λ can be divided into several sub-matrixes, i.e., λ^i, which represents the embedding of the i-type nodes. d_i is the dimensions of λ^i, where the row of λ^i represents a i-type node.

Specifically, traversing all the meta-path channels to learn all kinds of semantic of various type nodes is our core idea (line 3–17). For each meta-path channel, meta-path based first-order proximity was used to learn the representations of the end nodes (line 5–9), meta-path based second-order proximity was used to learn the representations of the center nodes (line 10–16).

3.4 The Objective Functions

In this section, we will introduce the objective functions of our Mech-RL method. Before going deep into the detail, let's first provide a brief background on the general definition of the objective function in network representation learning.

Given two nodes M_j and M_k whose embeddings are λ_j and λ_k. Their similarity in the embedding space can be calculated as the inner product of vectors, denoted as $S_{jk} = \lambda_j^T \lambda_k$ [15]. The larger S_{jk} is, the more similar they are. Suppose another node M_t is less similar to M_j than M_k, there will be $S_{jk} > S_{jt}$ in the latent embedding space. That is to say, the distance between nodes M_j and M_k should be smaller than the one between nodes M_j and M_t. Specifically, we model $S_{jk} > S_{jt}$ using the logistic function $\sigma(x) = \frac{1}{1+e^x}$. Therefore, the objective function which describes the relationships of nodes in the embedding space can be defined as follows:

$$\forall (M_j, M_k, M_t) \in G, maxpro(\sigma(S_{jk} - S_{jt})|\lambda_j, \lambda_k, \lambda_t) \tag{3}$$

Since we need to evaluate the similarity between heterogeneous nodes in our work, two meta-path based proximities are incorporated into the definition of the objective functions. Specifically, S'_{jk} represents the meta-path based first-order proximity, and S''_{jk} represents the meta-path based second-order proximity. Following the idea of Eq. 3, we minimize the sum of negative log-likelihood objective functions which describes the similarity of nodes can be defined as follows:

$$OBJ_M = min - \ln \sigma(S'_{jk} - S'_{jt}) + \gamma_1 Reg(M) \tag{4}$$

$$OBJ_N = min - \ln \sigma(S''_{jk} - S''_{jt}) + \gamma_2 Reg(N) \tag{5}$$

3.5 Embedding Update

We employ the stochastic gradient descent [1] to update the node embeddings. Given a meta-path MNM, the instance (M_j, N_j, M_k) will be used to update the embedding of the M_j. According to the meta-path based first-order similarity, M_k is the positive training sample and M_t is the negative one. We need to make sure that λ_k^M is closer to λ_j^M than λ_t^M in the embedding space.

Suppose there are two new instances (M_j, N_k, M_k) and (M_{j1}, N_t, M_{k1}). According to the meta-path based second-order similarity, N_k and N_t are the positive and negative training samples for N_j. We need to update λ_j^N, λ_k^N, λ_t^N correspondingly.

The meta-path based first-order proximity is updated as follows:

$$\lambda_j^M = \lambda_j^M - \alpha \frac{\partial OBJ_M}{\partial \lambda_j^M}$$

$$\lambda_k^M = \lambda_k^M - \alpha \frac{\partial OBJ_M}{\partial \lambda_k^M} \tag{6}$$

$$\lambda_t^M = \lambda_t^M - \alpha \frac{\partial OBJ_M}{\partial \lambda_t^M}$$

Table 1. Statistics of the dataset

Author name	#Real author	#Related papers	#Labeled papers
G Li	9	1440	362
Fei Liu	12	2827	467
Jian Huang	18	1810	237
Dan Wang	24	2678	281
Jing Zhou	34	1985	348
Bing Chen	50	1389	501

Where α is the learning rate. Similarly, the meta-path based second-order proximity is updated as follows:

$$\lambda_j^N = \lambda_j^N - \alpha \frac{\partial OBJ_N}{\partial \lambda_j^N}$$
$$\lambda_k^N = \lambda_k^N - \alpha \frac{\partial OBJ_N}{\partial \lambda_k^N} \quad (7)$$
$$\lambda_t^N = \lambda_t^N - \alpha \frac{\partial OBJ_N}{\partial \lambda_t^N}$$

4   Experiments

In this section, we conduct series of experiments to evaluate the performance of the proposed method on Aminer dataset [14].

4.1   Datasets

The dataset comes from the Aminer published by the Open Academic Data Challenge 2018[2], in which ambiguous authors come from various fields. We randomly select six ambiguous names. The statistics of the dataset is shown as Table 1. For example, there are 18 authors who share the same name "Jian Huang". The number of papers published by "Jian Huang" is 1810, of which 237 papers have been labeled that are published by which real "Jian Huang". Note that, only the labeled data can be used to evaluate the performance of comparative methods.

4.2   Training Examples Sampling

The meta-paths that we are interested in are listed in Table 2. All the samplings and updating are guided by the selected meta-paths. Each instance of the meta-path channel is used as the basis for directly updating the embeddings of neighbors or structural similar nodes in the embedding space.

---

[2] https://www.biendata.com/competition/scholar2018.

Table 2. Selected meta-paths and their explanations

Meta-path	Explanation
APA	Two authors collaborate on a paper
PAPAP	Two papers published by two authors who have collaborated before
OAPAO	Two cooperative units
PVPAP	One-hop neighbor of the paper
PAOAP	Two papers published by authors from the same organization
PAP	Two papers written by the same author
PVP	Papers published in the same venue

For example, given a meta-path APA, the meta-path instances (A_1, P_2, A_2), (A_3, P_3, A_4), (A_1, P_1, A_2) can be extracted from Fig. 1(a). As for (A_1, P_2, A_2), A_2 and A_4 are positive and negative training sample for A_1 based on the first-order proximity, while P_1 and P_3 are positive and negative training sample for P_2 based on the second-order proximity.

4.3   Comparative Methods

In the experiments, DeepWalk [9], LINE [13], NDAG [15] are selected as the comparative methods. Besides, two versions of meta-path channel based network representation learning methods are also compared. Mech-RL1 randomly initializes the embeddings for each type of nodes and the textual features are not considered during the embedding procedure. Mech-RL incorporates both the textual features and topological features to learning the representations of heterogeneous graph. Note that, after obtaining the paper embeddings for all the comparative methods, a hierarchical agglomerative clustering (HAC) method [12] is employed to generate the paper clusters for each unique person.

4.4   Methodology and Metrics

The cluster size L is determined by the number of ambiguous names in the real world. For example, there are 18 authors named "Jian Huang", therefore the number of clusters is set to 18. Each of the comparative method are run three times and we average the results. The number of iteration is set to 30. The dimension of the learned vectors is set to 20. Macro-F1 score is used to evaluate the effectiveness of the comparative methods. The Macro-F1 score is computed as follows:

$$Macro - F1 = \frac{1}{L} \sum_{i=0}^{L} \frac{2 \times Precison_i \times Recall_i}{Precison_i + Recall_i} \tag{8}$$

where $Precison_i$ and $Recall_i$ are the precision and recall metrics of the i-th cluster.

Table 3. Comparison results on Macro-F1 values (Dimension = 20)

Name reference	NDAG	DeepWalk	LINE	Mech-RL1	Mech-RL
G Li	0.364	0.022	0.452	0.560	**0.597**
Fei Liu	0.297	0.182	0.097	0.511	**0.532**
Jian Huang	0.509	0.024	0.290	0.627	**0.638**
Dan Wang	0.475	0.207	0.401	0.593	**0.610**
Jing Zhou	0.642	0.094	0.383	0.678	**0.693**
Bing Chen	0.370	0.165	0.399	0.514	**0.540**
Average	0.443	0.116	0.337	0.581	**0.602**

4.5 Experimental Results Analysis

Table 3 shows the performance comparison of name disambiguation between our proposed method and other competing methods for 6 different name references. As we can conclude that our proposed Mech-RL method achieves the best disambiguation results among all the comparisons. DeepWalk generates the worst performing results. In the sampling stage, DeepWalk adopts the random walk method to generate node sequences, which is not stable and affect the results of the representation learning. Besides, as DeepWalk can only be adopted to solve the homogeneous graphs embedding, which has destroyed the rich semantics underlying the heterogeneous bibliographic networks.

Compared to LINE, the average performance of Mech-RL has been improved by around 78.6%. LINE also introduces two proximity measures which evaluates the neighboring similarities and structural similarities. However, as has been demonstrated by the comparative experiments, our proposed meta-path based first-order and second-order proximities are more powerful in describing the relationships between entities in heterogeneous information networks.

Among the competing methods, NDAG is the most similar one to our proposed method. NDAG constructs three subnetworks, i.e., the A-A, A-P, and P-P networks. These subnetworks are processed separately, which ignored the heterogeneous features of the whole bibliographic information networks. Mech-RL learns the embeddings of different types of nodes by jointly learning and updating according to the meta-path channels and fully considers the heterogeneous topological features. That's the reason why Mech-RL shows superior performance over NDAG.

In order to verify the contributions of each component of the disambiguation framework, we also compare the results of Mech-RL and Mech-RL1. The difference between Mech-RL and Mech-RL1 lies in the initialization, where the former considers the textual features in initialization and the latter just randomly initializes the node embeddings. As we can see from the table, the disambiguation results of Mech-RL is better than Mech-RL1 for all the selected name references, which demonstrates that the textual features are meaningful in initialization.

5   Conclusion and Future Work

In this paper, we propose a novel heterogeneous information network representation learning based solution to address the author name disambiguation problem. We rely on the textual features of papers to refine the initialization of the paper embedding. The relational features are considered in updating the embeddings of various types of nodes. The author name disambiguation problem is competed by employing a hierarchical clustering algorithm on the generated paper embeddings. The experiment results demonstrate the effectiveness of our Mech-RL method. In the future, we will explore how to apply our method to other applications and validate our work on some other available datasets.

Acknowledgements. This research is funded by the National Natural Science Foundation of China under grant No. 61802440 and No. 61702553. We are also supported by the Fundamental Research Funds for the Central Universities, ZUEL: 2722019JCT037 and the Opening Project of State Key Laboratory of Digital Publishing Technology.

References

1. Bottou, L.: Large-scale machine learning with stochastic gradient descent. In: Lechevallier, Y., Saporta, G. (eds.) Proceedings of COMPSTAT'2010, pp. 177–186. Physica-Verlag HD, Heidelberg (2010)
2. Bunescu, R.C., Pasca, M.: Using encyclopedic knowledge for named entity disambiguation. In: Proceedings of the Conference of 11th Conference of the European Chapter of the Association for Computational Linguistics, Trento, Italy, EACL 2006, 3–7 April 2006
3. Cen, L., Dragut, E.C., Si, L., Ouzzani, M.: Author disambiguation by hierarchical agglomerative clustering with adaptive stopping criterion. In: Proceedings of the 36th International ACM SIGIR Conference on Research and Development in Information Retrieval, pp. 741–744. ACM (2013)
4. Chen, H., Perozzi, B., Hu, Y., Skiena, S.: HARP: hierarchical representation learning for networks. In: Thirty-Second AAAI Conference on Artificial Intelligence (2018)
5. Ferreira, A.A., Gonçalves, M.A., Laender, A.H.: A brief survey of automatic methods for author name disambiguation. ACM Sigmod Rec. **41**(2), 15–26 (2012)
6. Han, X., Sun, L., Zhao, J.: Collective entity linking in web text: a graph-based method. In: Proceedings of the 34th International ACM SIGIR Conference on Research and Development in Information Retrieval, SIGIR 2011, Beijing, China, 25–29 July 2011, pp. 765–774 (2011)
7. Le, Q., Mikolov, T.: Distributed representations of sentences and documents. In: International Conference on Machine Learning, pp. 1188–1196 (2014)
8. Li, Y., Li, C., Chen, W.: Research on influence ranking of Chinese movie heterogeneous network based on PageRank algorithm. In: Proceedings of the 15th International Conference on Web Information Systems and Applications, pp. 344–356 (2018)
9. Perozzi, B., Al-Rfou, R., Skiena, S.: DeepWalk: online learning of social representations. In: Proceedings of the 20th ACM SIGKDD International Conference on Knowledge Discovery and Data Mining, pp. 701–710. ACM (2014)

10. Shi, C., Hu, B., Zhao, X., Yu, P.: Heterogeneous information network embedding for recommendation. IEEE Trans. Knowl. Data Eng. **31**, 357–370 (2018)
11. Shi, C., Li, Y., Zhang, J., Sun, Y., Philip, S.Y.: A survey of heterogeneous information network analysis. IEEE Trans. Knowl. Data Eng. **29**(1), 17–37 (2017)
12. Steinbach, M., Karypis, G., Kumar, V., et al.: A comparison of document clustering techniques. In: KDD Workshop on Text Mining, Boston, vol. 400, pp. 525–526 (2000)
13. Tang, J., Qu, M., Wang, M., Zhang, M., Yan, J., Mei, Q.: Line: large-scale information network embedding. In: Proceedings of the 24th International Conference on World Wide Web, pp. 1067–1077. International World Wide Web Conferences Steering Committee (2015)
14. Tang, J., Zhang, J., Yao, L., Li, J., Zhang, L., Su, Z.: ArnetMiner: extraction and mining of academic social networks. In: Proceedings of the 14th ACM SIGKDD International Conference on Knowledge Discovery and Data Mining, pp. 990–998. ACM (2008)
15. Zhang, B., Al Hasan, M.: Name disambiguation in anonymized graphs using network embedding. In: Proceedings of the 2017 ACM on Conference on Information and Knowledge Management, pp. 1239–1248. ACM (2017)
16. Zhang, B., Saha, T.K., Al Hasan, M.: Name disambiguation from link data in a collaboration graph. In: IEEE/ACM International Conference on Advances in Social Networks Analysis and Mining (ASONAM), pp. 81–84. IEEE (2014)
17. Zhang, Y., Zhang, F., Yao, P., Tang, J.: Name disambiguation in AMiner: clustering, maintenance, and human in the loop. In: Proceedings of the 24th ACM SIGKDD International Conference on Knowledge Discovery and Data Mining, pp. 1002–1011. ACM (2018)

Is Bigger Data Better for Defect Prediction: Examining the Impact of Data Size on Supervised and Unsupervised Defect Prediction

Xinyue Liu[1,2] and Yanhui Li[1,2(✉)]

[1] State Key Laboratory for Novel Software Technology,
Nanjing University, Nanjing, China
[2] Department of Computer Science and Technology,
Nanjing University, Nanjing 210023, China
yanhuili@nju.edu.cn

Abstract. Defect prediction could help software practitioners to predict the future occurrence of bugs in the software code regions. In order to improve the accuracy of defect prediction, dozens of supervised and unsupervised methods have been put forward and achieved good results in this field. One limiting factor of defect prediction is that the data size of defect data is not big, which restricts the scope of application with defect prediction models. In this study, we try to construct *bigger* defect datasets by merging available datasets with the same measurement dimension and check whether bigger data will bring better defect prediction performance with supervised and unsupervised models or not. The results of our experiment reveal that larger-scale dataset doesn't bring improvements of both supervised and unsupervised classifiers.

Keywords: Defect prediction · Supervised · Classifier · Data size

1 Introduction

Fixing software defects is a very difficult and time-consuming job for software practitioners during the development of a large-scale software project [3], at the start of which we need to identify potential locations of these defects. In last decades, software defect prediction has been proposed to locate defect-prone code regions [7,9,11,18,19], most of which are constructed by supervised classifiers (e.g. trained with label information in the training set and tested on the disjoint test set). Generally, these supervised prediction models have helped a lot in software test and alleviate debug burden of software engineers. In most of previous studies, it has been reported that these prediction models have a promising performance in defect prediction.

Recently, unsupervised prediction defect has drawn more attention from academic fields. Zhou et al. [24] pointed out that a simple unsupervised prediction

© Springer Nature Switzerland AG 2019
W. Ni et al. (Eds.): WISA 2019, LNCS 11817, pp. 138–150, 2019.
https://doi.org/10.1007/978-3-030-30952-7_16

defect model (simple module size models) has a prediction performance comparable or even superior to most of the existing defect prediction models in cross-project scenario. Yang et al. [21] showed that in effort-aware just in time defect prediction settings, many simple unsupervised models have a better performance compared with the state-of-the-art supervised models.

One limiting factor of defect prediction is that the data size of defect data extracted from the real-world software projects is not big, which prevents defect prediction models from employing popular methods of analyzing and mining big data and restricts the scope of application with defect prediction models. In this study, we try to construct *bigger* defect datasets by merging available datasets with the same measurement dimension and check whether bigger data will bring better defect prediction performance or not.

In detail, we introduce traditional supervised classifiers and the state-of-the-art supervised classifier (simple module size model [24]) to test their performances on the bigger datasets we collected. Besides, due to the curiosity about if the promotion of the scale of dataset would boost the performances of supervised classifiers, we will organize a comparative experiment between raw smaller datasets and merged bigger datasets. The results of our experiment reveal that larger-scale dataset doesn't bring improvements of supervised classifiers, and simple module size model is also comparable to traditional supervised models on bigger datasets.

The rest of this paper is organized as follows. Section 2 introduces the background. Section 3 describes the preparations, including studied dataset, classifiers and evaluation measures. Section 4 presents the experimental results. Section 5 summarizes the threats to the validity. Section 6 concludes the paper.

2   Background and Related Work

There are two kinds of defect predict models: **supervised** and **unsupervised**. Supervised models are widely used and proved effective in prior work [22]. While unsupervised models could classify objects directly without training process, so it's time-saving and easy to implement. Clustering and simple module size model are two common unsupervised techniques used in software field.

According to the dataset based on, we also could divide defect prediction into two scenarios, one of them is within-project defect prediction (**WPDP**). In WPDP, the dataset used to train classifiers and the one used for prediction come from a same project. The other scenario is cross-project defect prediction (**CPDP**). In practice, some companies may find it's hard to collect enough data in their project to train the classifier [25], so it's essential for them to seek for CPDP solutions, which could utilize data from a different source project. In this study, we will leverage both supervised and unsupervised models, and organize experiments on both WPDP and CPDP.

3   Experimental Setup

In this section, we will introduce our preparations for the experiment. In the following statement, you will have a primary understanding about how we collect dataset, select classifier and other similar things.

3.1   Collected Datasets

We use data from 15 projects of three groups - AEEEM, Eclipse and JURECZKO. The data in one group have unified metrics, so they are easy to be merged up and conduct further testing. Each project has both code metrics and clear defect label. More specifically, each file's defect label is marked as 1 (buggy file) or 0 (clean file). Table 1 describes the 43 data sets used. The first to the third columns respectively list the group name, project name and version number. For each project, the fourth to the sixth columns respectively list the number of modules, the number of metrics, and the percent of defective modules.

Table 1. Information of collected data sets.

Group	Project	Version	#Modules	#Metrics	%Defective
AEEEM	JDT core	-	997	31	20.66%
	Equinox	-	324	31	39.81%
	Lucene	-	691	31	9.26%
	Mylyn	-	1862	31	13.16%
	PDE	-	1497	31	13.96%
Eclipse	eclipse	2.0,2.1,3.0	6729–10593	31	10.8%–14.8%
JURECZKO	ant	1.3,1.4,1.5,1.6,1.7	125–745	20	10.9%–26.2%
	camel	1.0,1.2,1.4,1.6	339–965	20	3.83%–35.5%
	jedit	3.2,4.0,4.1,4.2,4.3	272–492	20	2.2%–33.1%
	lucene	2.0,2.2,2.4	195–340	20	46.7%–59.7%
	poi	1.5,2.0,2.5,3.0	237–442	20	11.8%–64.4%
	synapse	1.0,1.1,1.2	157–256	20	10.2%–33.6%
	velocity	1.4,1.5,1.6	196–229	20	34.1%–75.0%
	xalan	2.4,2.5,2.6,2.7	723–909	20	15.2%–98.8%
	xerces	1.2,1.3,1.4,init	162–588	20	15.2%–74.2%

In this study, we try to figure out how these defect prediction models perform on datasets with larger scale, so all the versions in a project are merged into one file (they have the same measurement dimension). Group AEEEM is an exception, because the projects of it don't process series of versions for us to merge. Hence, we merge all the projects in AEEEM into one file, so this one could be seen as test on CPDP, while others are on WPDP. All these merged files are called **combination** in the following description. In contrast, original raw files are called **individual**. All the merged files are listed in **Table 2**.

3.2   Classifiers

In this study, we leverage three common supervised learning models and a simple module size model. They are DNN (Deep Neural Network), RF (Random Forest), LR (Logistic Regression) and ManualDown. Compared with other frequently-used supervised learning models such as Linear Regression or Decision Tree, the models we chose are more suitable and representative to be applied on our dataset. Notions and implement details are presented below in sequence.

Table 2. Information of merged datasets.

Merged Project	#Modules	#Metrics	%Defective
AEEEM	5371	31	15.88%
eclipse	17999	31	13.43%
ant	1567	20	21.06%
camel	2784	20	20.19%
jedit	1749	20	17.32%
lucene	782	20	56.01%
poi	1378	20	51.31%
synapse	635	20	25.51%
velocity	639	20	57.43%
xalan	3320	20	54.4%
xerces	1643	20	39.81%

Deep Neural Network. Deep Neural Network (DNN) is a classical machine learning technique and is widely used in a variety of learning scenarios. In our study, we build DNN model with the help of tensorflow. The DNN we use equips 2 hidden layers and 15 neural nodes in each layer. We select AdamOptimizer provided by tensorflow as training algorithm and Cross Entropy as loss function.

Random Forest. Random forest (RF) is an ensemble learning method usually used for classification. Nowadays it has been used more and more frequently in software engineering field study, and have been proved to be effective by former related work [5,15]. In our experience, we leverage the package in R named **randomForest** to implement the model. The number of trees grown (a parameter in RF function) is set 100, which is a apropos value after our testing.

Logistic Regression. Logistic regression (LR) is a technique borrowed by machine learning from the field of statistics. As a typical machine learning model, logistic regression is also chosen as one of traditional defect prediction model. we use function **glm** in R package to the model, which is used to fit generalized linear models.

ManualDown. ManualDown is a simple module size model coming from the idea that we could use module size in a target project to predict bugs' location. Project with larger module is thought to be more defect-prone. Since ManualDown do not need any data from the source projects to build the models, they are free of the challenges on different data metric sets of the source and target project data. In particular, they have small calculation demand and are easy to implement. Furthermore, previous studies show that module size has an intense confounding effect on the associations between code metrics and defect-proneness [2,23]. We hence include ManualDown as a baseline model for following comparison.

3.3   Prediction Setting

In order to decrease test error and estimate the accuracy of a predictive model in practice, we apply one round of 5-fold cross-validation for each test in our study. Firstly, we partition all the instances (or called module) in the data set into 5 complementary subsets, each pass one of them is chose as testing set, others as training set. After 5 passes, we could obtain 5 performance indexes, which reflect classifier's performance and constitute box plots exhibited in Sect. 4.

3.4   Evaluation Measures

There are numerous evaluation measures in defect prediction field. We mainly consider effort-aware scenarios and select three typical evaluation measures to evaluate our prediction models' performance.

AUC (The area under ROC curve). In Machine Learning, we usually count on AUC for evaluation when it comes to a classification problem. AUC is the area under ROC curve, which reflects classification model's performance variation at various thresholds settings. Hence AUC could fairly tell model's capability of distinguishing between classes. AUC is between 0 and 1 (inclusive). The closer AUC is to 1, the better the model performs to distinguish between instances with buggy and clean.

F1 (F1-Score). F1 (F1 score) is the harmonic mean of precision and recall:

$$F_1 = (\frac{recall^{-1} + precision^{-1}}{2})^{-1} = 2 \cdot \frac{precision \cdot recall}{precision + recall}$$

By the way of harmonic mean, F1 combines precision and recall to give a comprehensive description of how much model is able to distinguish two classes. If you get a good F1 score, it means that your false positives and low false negatives is low, so you correctly identify real buggy instances. A F1 score is also between 0 and 1 (inclusive), and is considered perfect when its value is close to 1.

CE (Cost-Effectiveness). During defect prediction process, software practitioners prefer to check those files with high defect-proneness and small module size due to limited resources. Therefore, we could rank files in descending order using their probability of being buggy. The effort-aware ranking effectiveness of classifiers is usually measured by cost-effectiveness (CE) curve, which is a common evaluation measure in the field of defect prediction. And it's widely used in prior work [12,16]. CE could be calculated by the following formula introduced by Arisholm et al. [1]:

$$CE_\pi = \frac{Area_\pi(M) - Area_\pi * (Random)}{Area_\pi(Optimal) - Area_\pi * (Random)}$$

Where $Area_\pi(M)$ is the area under the curve of model M (likewise to Optimal and Random) for a given π. π is a cut-off varying from 0 to 1, and it indicates the percentage of cumulative LOC we take into account. A larger CE_π represents a better ranking effectiveness. In this work, we discuss CE_π at $\pi = 0.2$.

3.5   Analysis Methods

In order to make the results in Sect. 4 more convincing, we introduce two practical analysis methods to help us analyze in detail.

Scott-Knott Test. Scott-Knott (SK) test [17] is an analysis method used for group classifiers into statistically distinct ranks. When a plot appears before our eyes, it's hard to identify if there is significant distinction between classifiers, and SK test could help on it. The SK test recursively ranks the given classifiers through hierarchical clustering analysis. It clusters the given classifiers into two group based on evaluation indicators and recursively executes until there is no significant distinct group created [4]. The SK test used in classifiers' comparation can be found in prior works [4,8,13]. In this study, we use SK test in 95% confidence level.

Win/Tie/Loss. Win/Tie/Loss result is another analysis method which is useful in performance comparison between different techniques and has been widely used in prior works [10,14]. Once we get classifiers' performance evaluation data, Wilcoxon signed-rank test [20] and Cliff's delta δ [6] could be conducted to compare their performance. If one outperforms another based on the Wilcoxon signed-rank test (p <0.05), and there is distinct difference between the two based on Cliff's delta δ ($\delta \geq 0.147$), we mark test as a 'Win'. In contrast, test is marked as a 'Loss' if p <0.05 and $\delta \leq$ -0.147. Otherwise, the case is marked as a 'Tie'. The Win/Tie/Loss result shows if one of them outperforms the other one actually in all conditions.

4   Experimental Results

This section will provide a detailed description of our experimental results. We focus on how different defect predict models perform on combination and individual datasets, and answer the following research question:

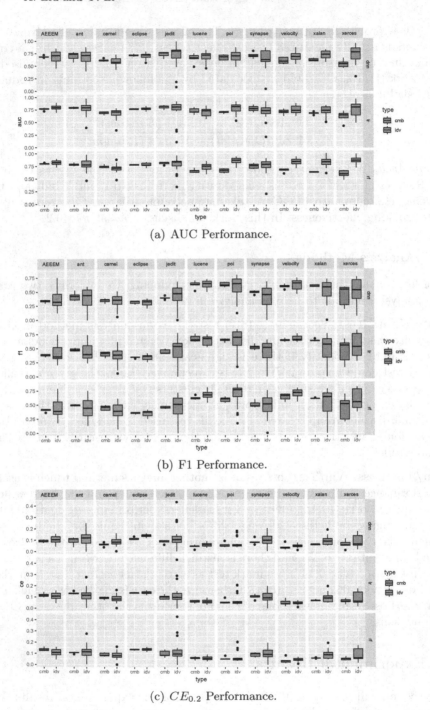

(a) AUC Performance.

(b) F1 Performance.

(c) $CE_{0.2}$ Performance.

Fig. 1. Comparision between combinations and individuals on AUC, F1 and $CE_{0.2}$ (Color figure online)

4.1   RQ1: Does Improvement of the Size of Dataset Promote the Effect of Defect Prediction?

The answer we give is NO. Generally speaking, training on the combination of a project doesn't outperform the one on the individual versions evaluated by AUC, F1 and $CE_{0.2}$. And the result is convincingly supported by Win/Tie/Loss.

Figure 1 shows an overview of the comparison between prediction on combinations and individuals. The boxplots show the distribution of evaluation measures (AUC, F1 and $CE_{0.2}$) of each classifier in the studied datasets. Green boxes represent the performances on merged files. Blue boxes represent the performances on individual files. Generally speaking, the results of the two don't exist significant discrepancy. But if you observe the plot more carefully, you could find that almost all the blue boxes are slightly higher than green boxes. Take the second plot in Fig. 1 as an example, except the performance on lucene with LR, all the blue boxes are higher than green boxes. This trend means that, surprisingly, the merge operation on the data set slightly reduces the performance of the classifier. This may be due to the hidden differences in different versions of a project.

Table 3. Information of Win/Tie/Loss indicator.

Index	Win	Loss	Tie	Relation	Result
AUC	11	1	21	win + tie >loss	Not significantly bad
F1	3	1	29	win + tie >loss	Not significantly bad
$CE_{0.2}$	1	1	31	win + tie >loss	Not significantly bad

In order to have a more detailed observation, we also apply the Win/Tie/Loss indicator to help analyzing. The Win/Tie/Loss result manifests whether combination is significantly better or not when compared with individual. Table 3 displays the details of Win/Tie/Loss result - it gives the number of Win/Tie/Loss in our test. Take the first row of table for illustration, it shows that performances of test on individuals wins 11 times, losses 1 time and ties 21 times against test on combinations. (11 merged files, 3 classifiers, so 33 competitions in total) This result manifests that our prior observation based on Fig. 1 is correct: individuals' performances slightly surpass the combination ones. In general, combination is not significantly better than individual on all three measure indexes. This result evinces the combination of software engineering dataset does not achieve distinct improvement.

We have considered many possibilities why increase in dataset scale didn't achieve distinct improvement, and the most possible reason comes to heterogeneity between different projects and different versions. Due to heterogeneous nature, increase in sample size doesn't provide supervised model with consistent and valid information, no matter on WPDP or on CPDP. So, it's hard to elevate classifiers' performances.

In summary, combination is not significantly better than individual on AUC, F1 and $CE_{0.2}$, which is supported by Win/Tie/Loss evaluation.

4.2 RQ2: Does ManualDown Outperforms Other Supervised Techniques?

In addition to differences between combinations and individuals, we also want to make a thorough inquiry on different performances of the four defect prediction model - DNN, RF, LR and ManualDown. Our question is if ManualDown could outperform other three typical supervised learning model, and it will be answered on both the merged and independent datasets.

First let's inspect the performance on independent datasets, and the three box plots in Fig. 2 show our effort. The models from left to right on the x-axis are in turn DNN, LR, ManualDown and RF. Three pictures represent the performance on AUC, F1 and $CE_{0.2}$. Our approach is to calculate the evaluation measure values of individual version and put them into box plot, so we can clearly discern the discrepancy between different model.

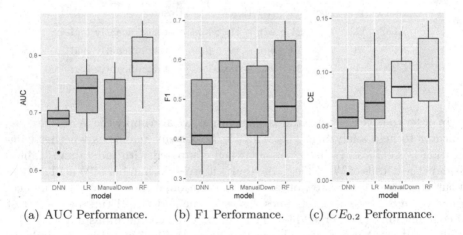

(a) AUC Performance.   (b) F1 Performance.   (c) $CE_{0.2}$ Performance.

Fig. 2. Comparision between classifiers on individuals

To explain with more detail, taking picture 1 for example. **ant** is one of the projects in the study, and firstly, we have a 5-fold DNN training and test on each version of **ant** (**ant1.3, ant1.4, ant1.5, ant1.6, ant1.7**) get 5 mean AUC values out of it. Then we calculate the mean of these values from different versions, so we get a synthesis AUC value of **ant**. In our experience we have prepared 11 projects, therefore we could procure 11 synthesis AUC value, which constitute the leftmost box in the plot. Similarly we could complement this box plot by using other models to have 5-fold training and test on the same dataset.

Besides, Scott-Knott (SK) test is also applied here to depict if distinct difference exists among models. In Fig. 2, clusters with high SK values have been tinted into carnation, while others are staying gray. From the three pictures we can tell that RF outperforms other models. And RF is significantly distinct from DNN, LR and ManualDown on AUC, and from DNN, LR on $CE_{0.2}$ through SK test. As a result, we could draw a conclusion that ManualDown doesn't outperform other supervised techniques on independent datasets.

Now let's pay our attention on the performs on merged datasets, and this is showed on Fig. 3. Unlike the experience on independent datasets, taking picture 1 for example. Firstly, we combine all the instances in different versions of a project. Hence, we get 11 combined datasets (because we have 11 projects). Then we have a 5-fold DNN training and test on each combined dataset and calculate AUC value of it, which make up the leftmost box in the picture 1 plot. In a similar way it's easy to draw up the performance box plot on $CE_{0.2}$ and F1 on merged datasets.

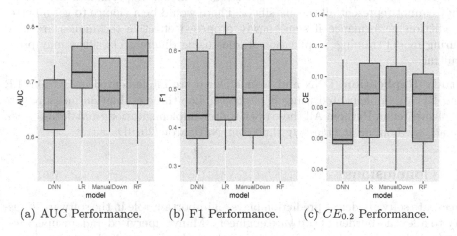

(a) AUC Performance.   (b) F1 Performance.   (c) $CE_{0.2}$ Performance.

Fig. 3. Comparision between classifiers on combinations

From Fig. 3, it's conspicuous that there doesn't exist obvious difference among the performance on these defect predict models. On the other hand, along with the increase of the instance's number in one dataset, the performance of model which has week performance on independent datasets makes a measly progress. And this lead to differences between different models is less obvious. In summary, ManualDown doesn't outperform other supervised techniques on merged datasets as well, and ManualDown is not significantly worse than traditional supervised models. Considering the complexity of traditional models, ManualDown is undoubtedly a predict method with more practical significance based on us experience.

In conclusion, ManualDown is also comparable to traditional supervised models on bigger combination datasets.

5 Threats to Validity

Project Selection. In this study, we select 15 open-source projects which have been used in prior works. These projects equip large enough data scale and regular metric information, which is conducive to train classifiers. Whereas limits still exist since these projects only come from 3 groups, thus there may be a different result if more diverse data is introduced in the experiment. So, it's necessary to replicate our study in the future with a wider variety of datasets.

Classifier Selection. The classifiers we select in this work are DNN, RF and LR. Although these are common investigated in defect prediction literatures, they cannot represent all the classifiers. There would be a chance to get a better prediction performance if some newly proposed effective techniques are used during model training. Replication studies using different classifiers may prove fruitful.

Study Replication. DNN and ManualDown are implemented in Python. LR and RF are implemented using R packages. The three evaluation measures are implemented in Python. All these open source implementations and datasets can be accessed online at https://github.com/NJUaaron/2019DataTest.

6 Conclusions

Accurate software defect prediction plays an important role in the software industry to alleviate burden of software engineers. Many supervised and unsupervised methods have been proposed in prior works and are proved effective in the literature. However, for supervised methods, there is a limiting factor of defect prediction - the size of data used for training is not big, which restricts the scope of application with defect prediction models in practice. In this study, we construct bigger defect datasets by merging available datasets with same measurement dimension and check whether the promotion in data size will lift defect prediction performance or not. Meanwhile, ManualDown, one of simple module size models, is introduced as a baseline model to measure supervised models' performances.

In the experience, we test DNN, RF, LR and ManualDown on individual files and merged files. Their prediction performances are evaluated by AUC, F1 and $CE_{0.2}$, and the experimental results are analyzed by SK test and Win/Tie/Loss technique. In summary, our conclusions are as follow:

- Performance on larger-scale dataset is not significantly better than performance on raw smaller dataset under AUC, F1 and $CE_{0.2}$. More precisely, the increase in the size of dataset even makes the classifier perform worse, although the degree of deterioration is not distinct.
- There is not significant difference between performances of ManualDown and other supervised techniques. In other words, classical supervised models cannot outperform simple module size model on our merged bigger data.

Acknowledgement. The work is supported by National Key R&D Program of China (2018YFB1003901) and the National Natural Science Foundation of China (Grant No. 61872177).

References

1. Arisholm, E., Briand, L.C., Johannessen, E.B.: A systematic and comprehensive investigation of methods to build and evaluate fault prediction models. J. Syst. Softw. **83**(1), 2–17 (2010)
2. Emam, K.E., Benlarbi, S., Goel, N., Rai, S.N.: The confounding effect of class size on the validity of object-oriented metrics. IEEE Trans. Softw. Eng. **27**(7), 630–650 (1999)
3. Erlikh, L.: Leveraging legacy system dollars for e-business (2000)
4. Ghotra, B., Mcintosh, S., Hassan, A.E.: Revisiting the impact of classification techniques on the performance of defect prediction models. In: International Conference on Software Engineering (2015)
5. Ibrahim, D.R., Ghnemat, R., Hudaib, A.: Software defect prediction using feature selection and random forest algorithm. In: International Conference on New Trends in Computing Sciences (2017)
6. Romano, J., Kromrey, J.D., Coraggio, J.: Exploring methods for evaluating group differences on the NSSE and other surveys: are the t-test and Cohen's d indices the most appropriate choices? (2006)
7. Jing, X.Y., Ying, S., Zhang, Z.W., Wu, S.S., Liu, J.: Dictionary learning based software defect prediction (2014)
8. Khalid, H., Nagappan, M., Shihab, E., Hassan, A.E.: Prioritizing the devices to test your app on: a case study of Android game apps. In: ACM SIGSOFT International Symposium on Foundations of Software Engineering (2014)
9. Kim, S., Zimmermann, T., Whitehead Jr., E.J., Zeller, A.: Predicting faults from cached history. In: International Conference on Software Engineering (2008)
10. Kocaguneli, E., Menzies, T., Keung, J., Cok, D., Madachy, R.: Active learning and effort estimation: finding the essential content of software effort estimation data. IEEE Trans. Softw. Eng. **39**(8), 1040–1053 (2013)
11. Lee, T., Nam, J., Han, D.G., Kim, S., In, H.P.: Micro interaction metrics for defect prediction (2011)
12. Ma, W., Lin, C., Yang, Y., Zhou, Y., Xu, B.: Empirical analysis of network measures for effort-aware fault-proneness prediction. Inf. Softw. Technol. **69**(C), 50–70 (2016)
13. Mittas, N., Angelis, L.: Ranking and clustering software cost estimation models through a multiple comparisons algorithm. IEEE Trans. Softw. Eng. **39**(4), 537–551 (2013)

14. Nam, J., Fu, W., Kim, S., Menzies, T., Tan, L.: Heterogeneous defect prediction. IEEE Trans. Softw. Eng. **PP**(99), 1 (2015)
15. Pushphavathi, T.P., Suma, V., Ramaswamy, V.: A novel method for software defect prediction: hybrid of FCM and random forest. In: International Conference on Electronics & Communication Systems (2014)
16. Rahman, F., Devanbu, P.: How, and why, process metrics are better. In: International Conference on Software Engineering (2013)
17. Scott, A.J., Knott, M.: A cluster analysis method for grouping means in the analysis of variance. Biometrics **30**(3), 507–512 (1974)
18. Wang, J., Shen, B., Chen, Y.: Compressed c4.5 models for software defect prediction. In: International Conference on Quality Software (2012)
19. Wang, S., Liu, T., Tan, L.: Automatically learning semantic features for defect prediction (2016)
20. Wilcoxon, F.: Individual comparisons of grouped data by ranking methods. J. Econ. Entomol. **39**(6), 269 (1946)
21. Yang, Y., et al.: Effort-aware just-in-time defect prediction: simple unsupervised models could be better than supervised models. In: Proceedings of the 2016 24th ACM SIGSOFT International Symposium on Foundations of Software Engineering, FSE 2016, pp. 157–168. ACM, New York (2016)
22. Zhang, J., Xu, L., Li, Y.: Classifying Python code comments based on supervised learning. In: Meng, X., Li, R., Wang, K., Niu, B., Wang, X., Zhao, G. (eds.) WISA 2018. LNCS, vol. 11242, pp. 39–47. Springer, Cham (2018). https://doi.org/10.1007/978-3-030-02934-0_4
23. Zhou, Y., Xu, B., Leung, H., Chen, L.: An in-depth study of the potentially confounding effect of class size in fault prediction. ACM Trans. Softw. Eng. Methodol. **23**(1), 1–51 (2014)
24. Zhou, Y., et al.: How far we have progressed in the journey? an examination of cross-project defect prediction. ACM Trans. Softw. Eng. Methodol. **27**(1), 1:1–1:51 (2018)
25. Zimmermann, T., Nagappan, N., Gall, H., Giger, E., Murphy, B.: Cross-project defect prediction a large scale experiment on data vs. domain vs. process. In: Proceedings of the Joint Meeting of the European Software Engineering Conference & the ACM SIGSOFT Symposium on the Foundations of Software Engineering (2009)

Using Behavior Data to Predict the Internet Addiction of College Students

Wei Peng, Xinlei Zhang$^{(\boxtimes)}$, and Xin Li

East China Normal University, Shanghai, China
wpeng@admin.ecnu.edu.cn, xinleizhang1997@gmail.com, xinli@stu.ecnu.edu.cn

Abstract. Internet addiction refers to excessive internet use that interferes with daily life. Due to its negative impact on college students' study and life, discovering students' internet addiction tendencies and making correct guidance for them timely are necessary. However, at present, the research methods used on analyzing students' internet addiction are mainly questionnaire and statistical analysis which relays on the domain experts heavily. Fortunately, with the development of the smart campus, students' behavior data such as consumption and trajectory information in the campus are stored. With this information, we can analyze students' internet addiction level quantitatively. In this paper, we provide an approach to estimate college students' internet addiction level using their behavior data in the campus. In detail, we consider students' addiction towards internet is a hidden variable which affects students' daily time online together with other behavior. By predicting students' daily time online, we will find students' internet addiction levels. Along this line, we develop a linear internet addiction (LIA) model and a neural network internet addiction (NIA) model to calculate students' internet addiction level respectively. And several experiments are conducted on a real-world dataset. The experimental results show the effectiveness of our method, and it's also consistent with some psychological findings.

Keywords: Student behavior data · Internet addiction ·
Neural network

1 Introduction

Internet addiction disorder refers to excessive internet use that interferes with daily life [1]. Some research shows that the addiction towards the internet has a negative impact on college students, such as the backwardness of study [8], health [1], social relationship [11] and so on. Therefore, it's necessary to discover students' addiction tendencies towards internet and make correct guidance for them.

At present, related works of internet addiction are concentrated on psychological fields. Such works focus on the causes, the influence of internet addiction

© Springer Nature Switzerland AG 2019
W. Ni et al. (Eds.): WISA 2019, LNCS 11817, pp. 151–162, 2019.
https://doi.org/10.1007/978-3-030-30952-7_17

and etc. There are few works on calculating internet addiction level quantitatively. Besides, the methods used for analyzing are mainly questionnaires and statistical analysis which is cumbersome and relays on the domain experts heavily. Therefore it's necessary to develop a method to explore students' internet addiction level quantitatively and automatically.

Fortunately, with the development of the smart campus, students' behavior data are collected, such as the access data, consuming data and so on. With this data, It's possible to analyze students' internet addiction level quantitatively.

In this paper, We propose an approach to estimate students' internet addiction level using their behavior data. Currently, there is not a method to evaluate students' addiction level precisely. Therefore, we can calculate students' internet addiction level through another task. In detail, we consider that the student' s internet addiction level is a hidden variable which will affect students' daily time online. Besides, student's behavior data such as consuming data, the internet access gap, etc. reflect student's daily activities which may also influence the time they spend online. By predicting students' daily time online with their behavior and internet addiction level, we will get students' internet addiction level. Along this line, we propose a linear internet addiction (LIA) model and neural network internet addiction (NIA) model to capture the relationship between students' behavior data, internet addiction and the time they spend online every data. Furthermore, both of the models take the regularity of students' behavior into consideration. Finally, we conduct extensive experiments on a real-world dataset from a Chinese college, and the experimental results demonstrate the correctness and effectiveness of the model we propose. And the results are also consistent with some psychological findings.

2   Related Works

The main related work of this paper can be divided into two parts: internet addiction analysis and campus data mining.

Internet Addiction Analysis. Internet addiction analysis is a research direction in the psychological field. Some works are focusing on the causes of internet addiction. Researchers found that interpersonal difficulties, psychological factor, social skills, etc. are all reasons of internet addiction [1,3,7]. Other works aim at finding the influence of internet addiction. Upadhayay et al. [8] claimed that excessive use of the internet would lead the drawback of study. He et al. [5] explored internet addiction's influence on the sensitivity towards punishment and award.There are also some works about the inner mechanism of forming internet addiction. Zhang et al. [12] focused on the inner reason of family function's negative influence on internet addiction. Zhao et al. [13] notice that stressful life event make users feel depressed, which causes the user addicted to the internet.

Campus Data Mining. Campus data mining refers to solving problems in campus with data mining method. Zhu et al. [14] propose an unsupervised method to calculate students' procrastination value with their borrow info in the library. Guan et al. [4] predict students' financial hardship through their smart card usage, internet usage and students' trajectories on campus(Dis-HARD model).There is also some work aiming at analyzing students' studying process and improving their performance in class. For example, Burlak et al. [2] identify if a student is cheating in an exam by analyzing their interactive data with online course systems such as start time, end time, IP address, access frequency, etc. Above all, to the best of our knowledge. There is no work on analyzing internet addiction using students' daily behavior. And we are the first to analyze internet addiction based on their behavior data with data mining method.

3   Preliminaries

Internet Addiction is an abstract concept in psychological field, so it's hard to give a measurable definition to internet addiction. To solve this problem, we first make a reasonable assumption about internet addiction. Then, based on this assumption, we calculate internet addiction value using students' behavior data.

3.1   Internet Addiction Assumption

Psychological research [6] shows that most college students are addicted to internet. And we mentioned that internet addiction refers to excessive use of internet interfering with daily life. Therefore students with different internet addiction level are very likely to spend different time online. Besides, different behaviors show the different activities in school, which in turn also lead to different online time. And students of different gender or department will also have some differences in internet use.

Based on such fact, we assume that internet addiction is a hidden factor which may influence students' daily time online together with their behavior and profile information. Therefore we will learn such factor by modeling how students' internet addiction and behaviors influence daily online time. To simplify the problem, we also assume students' internet addiction level will not change in a semester.

3.2   Problem Formulation

Since we don't have any label about internet addiction level, we can't use supervised method to study students' internet addiction value. Thus we need to estimate it through some known data. Based on our assumption that internet addiction value is a hidden variable which may affect the time students spend online, the value can be learned by predicting students' daily online time.

Formally, we define a_u as the internet addiction level of student u. Daily time online sequence of student u during a period T is represented as $\{T_u(t)\}$. And the daily behavior sequence of u during the same period is represented as

$\{B_u(t)\}$ We also define the personal profile information of student u as $\{p_u\}$. Our task is to model the relationship $\{a_u, p_u, B_u(t)\} - > \{T_u(t)\}$ which is how students' behaviors and internet addiction influence their daily time online. Then the internet addiction level a_u can be calculated from this model. Note that the t above is in the set T.

4   Linear Internet Addiction (LIA) Model

In this chapter, we first introduce how we use linear model to reveal the relationship of $\{a_u, p_u, B_u(t)\} - > \{T_u(t)\}$. Then to strengthen the model, we take the regularity of students' behaviors into consideration.

4.1   Naive LIA

Based on the internet addiction assumption, behavior is a factor which will influence students' online time. The impact of behavior on online time is not different in individuals, so every student shares this weight vector. We deal with the different kind of personal attributes the same way. And we suppose that different internet addiction level is the only reason which causes different time online with the same behavior and personal attributes. Here comes our naive linear internet addiction model.

$$y_u(t) = wx_u(t) + a_u$$

$y_u(t)$ represents the duration student u spend online at time t. $x_u(t)$ refers to the combination of behavior vector and personal attributes of student u at time t, and w is the weight vector of that combined vector. a_u here is the internet addiction level of student u. Our task is to find the value of a_u and w that minimize the loss function, that is:

$$argmin_{w,a_u} \sum_{u \in U} \sum_{t \in T} (y_u(t) - w^T x_u(t) - a_u)^2 + \lambda||w||^2 + \mu \sum_{u \in U} a_u^2$$

The item $\lambda||w||^2$ is used to prevent the model from overfitting. $\mu \sum_{u \in U} a_u^2$ can adjust the weight between behavior and internet addiction.

4.2   LIA with Regular Behavior

College student usually have a fixed curriculum, therefore, their behavior has some regularity every week, which will also lead to the regularity of the time they spend online. Take student u as an example, courses on Monday are kind of boring, so he spends a lot of time surfing the internet. However, courses on Tuesday are hard which means he must pay attention to the class, so he may not surf the internet in class. Based on such a fact, it's necessary to take the regular online time into consideration.

So we modify our linear internet addiction model by add an item $d_u(\pi(t))$ to represent the regular online time of student u at time t. Due to the character of college study, they perform similar online habit every week. So here $\pi(t)$ means which day of time t is of the week it belongs to, and $d_u(x)$ means the regular online time of the x-th day of the week. Here comes our new model :

$$y_u(t) = wx_u(t) + a_u + d_u(\pi(t))$$

For the convenience of calculation, we define $x_{2u}(t)$ as 8 dimension vector with the first item one standing for the internet addiction, others being a one-hot representation of the week. The formula above is equal to

$$y_u(t) = w^T x_u(t) + w_u^T x_{2u}(t)$$

with x_{2u} being equal to :

$$(1, \pi_1(t), \pi_2(t), \pi_3(t), \pi_4(t), \pi_5(t), \pi_6(t), \pi_7(t)) \tag{1}$$

$$\pi_i(t) = \begin{cases} 1, & \text{if } \pi(t) = i; \\ 0, & \text{otherwise} \end{cases}$$

Our task is to find a suitable w and w_u that will minimize the loss function, the first item of w_u is the internet addiction level of student u:

$$argmin_{w,w_u} \sum_{u \in U} \sum_{t \in T} (y_u(t) - w^T x_u(t) - w_u^T x_{2u}(t))^2 + \lambda ||w||^2 + \mu \sum_{u \in U} ||w_u||^2$$

Similarly, we add $\lambda ||w||^2$ to prevent the formula from overfitting, and the formula $\mu ||w_u||^2$ to adjust the weights between behavior, personal attributes and internet addiction level, regular habit.

5   Neural Network Internet Addiction (NIA) Model

In this section, we develop a neural network internet addiction (NIA) model to represent the non-linear influence of students' behaviors on their online time.

5.1   Network Structure

The neural network consists of two part: the public part and the private part. We use the public part to represent that the effect of behavior and personal attributes on daily online time is not different in individuals, and the weight matrix and threshold vector of this part will update every iteration. Because the internet addiction level and regular behavior are different in individuals, we use a private part to depict such characteristic. Every student has its own weight matrix and threshold vector, and the parameter will only be updated when the corresponding student's data is used as input. the private input of student u at time t is the same as vector (1). To ignore the influence of regular behavior, we can also keep only the first item of vector (1).

The structure of the network is shown as Fig. 1

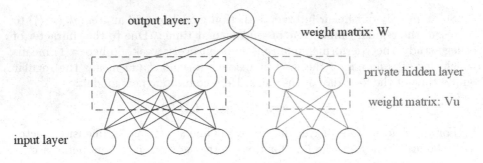

Fig. 1. Neural network internet addiction model.

5.2 Internet Addiction Calculation

After the neural network training is completed, the sum of the contribution that internet addiction gives to the private hidden units is the value of students' internet addiction level. We will calculate the internet addiction value as below:

$$a_u = \sum_{j=1}^{q_u} V_{uij}$$

q_u stands for the number of private hidden layer unit. i is the corresponding index of internet addiction in the private part input vector, and here the index is one. V_u is the matrix which connects the input layer and hidden layer of the private part. V_{uij} represents the i-th row and the j-th column value of matrix V_u.

6 Experiments

6.1 Data Description

Our data comes from a Chinese college, including students' consuming records in the school restaurant and internet access records. Besides it also includes the personal attributes information of students such as department, gender and age.

The consumption records consists of students' id, the time, place and amount of one consumption. Students have

Table 1. Features used in experiments

Type	Feature	Dimension	Representation
Profile	Gender	2	One-hot
	Department	61	
	Age	8	
Consumption	Dining amount	24	Statistical value
	Snack amount	24	
	Shower amount	24	
	Deposit amount	24	
	Total amount	24	
	Frequency	1	
Internet	Wifi access time	24	Statistical value
	Internet access gap	1	

various consumption behavior, therefore we first divide the places into different categories, and then extract the consuming amount on dining, snack, shower, deposit and total consuming amount per hour from the consuming records. We also count students' daily consumption frequency.

In addition, Because students can access the internet using campus wifi only when they get authenticated, based on the authentication record, we extract the time student accesses the campus wifi per hour. Similarly, Each time a student visits a website, a connection record is generated. When the visit is completed, there will be a disconnected record. Based on these records, we can extract the student's actual online time and the average gap between two internet access per day. We also represent every student with one-hot method using their profile information.

Due to some reason, we don't have students' internet access records in dormitory and library. Considering that students' activities are mainly centralized around classrooms and canteens as well as some college student activity centers. In class, students need to listen to the teachers at most time, and at the restaurant, they always play with a phone to kill time. Therefore, the actual online time we extract is mainly about entertainment. Intuitively, the entertainment time is suitable to be used to calculate the internet addiction level.

We choose the records of undergraduate students enrolled in 2016 from September 1st, 2018 to November 11th, 2018. After dropping students with records less than 35 days, there are 2341 students. The first 50 records of each student are used for training and the left records are used for testing. Students' profile representation and daily behavior vector is shown in Table 1.

6.2   Internet Addiction Calculation

LIA and NIA model can study the internet addiction level by predicting students' online time every day. To show the correctness of our model, we conduct several experiments.

For each model, we conduct three experiments. The first experiment removes the internet addiction and regular behavior part of LIA and NIA model, and predicts students' daily online time using students' behavior data and profile information, which is considered as a baseline. The second experiment only takes the internet addiction into consideration. The last experiment takes internet addiction and regular behavior into consideration.

For the linear model, the value of λ is set to 0.6, and μ is set to 0.4. For neural network model, the activation function of hidden layer is $f(x) = x$, and the activation function of output layer is $f(x) = tanh(x)$. In addition, the number of public hidden layer units is 10, and the number of private hidden layer units is 2. The learning rate is set to 0.01, and the number of the epoch is 40. The MSE performance of each method is shown as Table 2. Note that 'ia−' refers to the baseline experiments, 'ia' represents the second experiment, and 'ia+' stands for the third experiment.

From the results in Table 2, we know that no matter of linear model or neural network model, the prediction accuracy increases with our internet addiction assumption. Such results guarantee the correctness of our internet addiction assumption. However, adding the assumption of regular behavior, the accuracy

Table 2. Regression results

feature / model	ia-	ia	ia+
LIA	0.000052	**0.000048**(7.7%)	0.000049(5.8%)
NIA	0.000075	**0.000071**(5.3%)	**0.000071**(5.3%)

doesn't improve compared to the results without such an assumption. One possible reason is that there is some volatility in students' behavior, however. LIA and NIA are not able to model it. Generally, the results of the neural network model are worse than the linear model. Maybe it's because the linear model is strong enough to represent the relationship between students' behavior, internet addiction and online time.

6.3 Internet Addiction Verification

In this section, we devise regression and classification tasks to verify the correctness of the internet addiction value learned from the model we proposed.

Based on our assumption, internet addiction is a hidden variable which will influence students' daily time online. Therefore the learned internet addiction value should be a useful feature to predict students' online time. We devise two tasks to verify the truth of our learned internet addiction value.

The aim of the regression task is to predict students' daily online time. The baseline experiment takes the daily behavior vector and profile information as the input. The contrast experiment predicts the daily online time using students' internet addiction value, daily behavior vector and profile information. For the classification task, it is similar to the regression task. The aim of the classification task is to predict which online time interval it belongs to. The experiment settings are the same as the regression task. The method used in the regression task and classification task consists of Decision tree (DT), support vector machine (SVM), k-nearest neighbors (KNN), random forest (RF), Adaboost, gradient boosting decision tree (GBDT) [10], bagging and extremely randomized trees (ET). MSE is used as the evaluation method for the regression task, and f1-score for the classification task.

Table 3. Regression task

feature / model	ia-	ia(LIA)	ia(NIA)
DT	0.000069	**0.000057**(17.4%)	0.000066(4.3%)
SVM	0.003064	0.003360(-9.7%)	**0.002948**(3.8%)
KNN	0.000061	**0.000056**(8.2%)	0.000061(0%)
RF	0.000042	**0.000040**(4.8%)	0.000042(0%)
Adaboost	0.000078	**0.000069**(11.5%)	0.000084(-7.7%)
GBDT	0.000042	**0.000040**(4.8%)	0.000042(0%)
Bagging	0.000044	**0.000041**(6.8%)	0.000043(2.3%)
ET	0.000065	**0.000059**(9.2%)	0.000063(3.1%)

Table 4. Classification task

feature / model	ia-	ia(LIA)	ia(NIA)
DT	0.964367	**0.998905**(3.6%)	0.998016(3.5%)
SVM	0.968950	0.968950(0%)	0.968950(0%)
KNN	0.964855	0.975332(1.1%)	**0.982747**(1.9%)
RF	0.972281	**0.981374**(0.9%)	0.981024(0.9%)
Adaboost	0.968936	**0.968950**(0%)	0.98821(0%)
GBDT	0.970570	0.972571(0.2%)	**0.972747**(0.2%)
Bagging	0.971519	**0.999587**(2.9%)	0.998647(2.8%)
ET	0.962745	**0.967695**(0.5%)	0.967396(0.5%)

Note that 'ia—' refers to the baseline experiment, 'ia(LIA)' stands for the experiment with the internet addiction value learned by naive LIA, and 'ia(NIA)' represents the experiment with the best internet addiction value learned by NIA without regular behavior consideration.

From Table 3, we observe that for the regression task, the SVM model get a huge mean square error. One possible may be that it is not suitable for this task, so we will ignore the SVM results in the discussion below. After adding internet addiction value calculated by LIA, all of the prediction accuracy lifts. And after adding internet addiction value calculated by NIA, some methods still get promotion. For the classification task, no matter which internet addiction value is added to the behavior vector, except for the effect of the SVM and Adaboost methods has not changed, the effect of all other methods has been evidently improved (Table 4).

Generally speaking, after adding the internet addiction value calculated by LIA or NIA, both of regression and classification task get a remarkable promotion, which shows the effectiveness of the internet addiction value learned by the model we propose.

6.4   Internet Addiction Analysis

To show the internet addiction situation in college, we analyze the distribution of internet addiction. Because the naive LIA model has the best prediction accuracy when studying students' internet addiction value, the following analysis is based on the value calculated by naive LIA.

Internet Addiction Distribution. Figure 2(a) illustrates the number of students with respect to the calculated internet addiction value. The greater internet addiction value is, the more serious students' addiction towards the internet is. The internet addiction distribution is similar to a normal distribution. To show the distribution of internet addiction value clearly, we delete the value greater

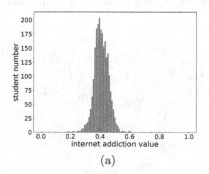
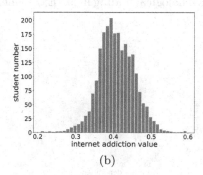

(a)   (b)

Fig. 2. Internet addiction distribution. (a): Number of students with different levels of dependency. (b): Some students with value greater than 0.6 or less than 0.2 are deleted.

than 0.6 or less than 0.2, which comes Fig. 2(b). If we define internet addiction less than 0.4 is normal, from Fig. 2(b), we observe that most of the students are addicted to the internet with different levels.

Internet Addiction Differences Among Groups. To reveal the differences of internet addiction between genders, we count the average internet addiction value of different genders. And we also count the average online time of different genders, Fig. 3 shows that girls spend more time on the internet than boys.

However, boys are more addicted to the internet than girls. Such a result is consistent with a finding in the psychological field. Wei et al. investigate the internet addiction situation of the college student in Hubei Polytechnic University using questionnaires. They point out that boys are usually not good at communication, and the way of communication with the network as the medium is easier to control, that is, they can improve the quality and

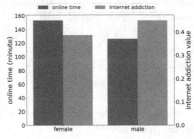

Fig. 3. Differences of online time and internet addiction between different genders.

quantity of communication in this way, which meets their needs of communication. Besides, Girls are better than boys in time management ability and deal with network use time more reasonably. So boys are more addicted to the internet than girls [9]. The consistency with the findings of psychology further proves the correctness of the internet addiction value we learned.

Figure 4(a) illustrate the average internet addiction level of different department. In general, except the internet addiction level of a few departments is extremely high, it fluctuates around 0.4. Further, we statistically analyze the differences in internet addiction level among students in different disciplines. In Fig. 4(b), we can observe that there is no significant difference in internet addiction level among students in different disciplines. The result is also consistent with the psychological finding in [9]. Experimental conducted by Wei et al. demonstrate

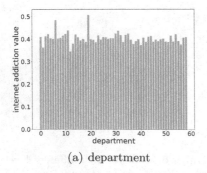

(a) department

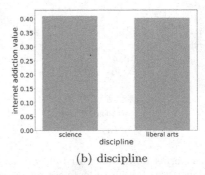

(b) discipline

Fig. 4. Differences of internet addiction among different departments and disciplines

the difference in internet addiction is not significant. The consistent result with psychological findings is also evidence of the effectiveness of the internet addiction value we learned.

Effect of Internet Addiction on Online Time. To show the role internet addiction plays when predicting students' online time, we extract students' daily behavior, and then we conduct two two-classification experiments using decision tree method: one predicts online time interval with daily behavior, and the other predicts online time interval with daily behavior and internet addiction value. Since the whole tree is too big to be put here, we select two representative branches. Note that all the values are normalized.

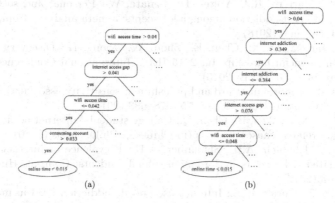

Fig. 5. Decision tree with behavior and internet addiction value. (a) predict with behavior data, (b) predict with behavior data and internet addiction value learned through naive LIA

From Fig. 5(a), we know that wifi access time and average internet access gap are important features when predicting online time. It is consistent with our intuitive thinking that less wifi access time and long internet access gap will cause less online time. Figure 5(b) illustrates that the internet addiction value is critical for predicting daily online time. Particularly, in this branch, the relatively low internet addiction value is a reason leading to short online time.

7   Conclusion

In this paper, we estimate college students' internet addiction level quantitatively using their behavior data in the campus. Specifically, we define the internet addiction value as a hidden variable which will affect students' online time, and formulate the problem as a regression problem. Along this line, we propose a linear internet addiction (LIA) model and a neural network internet addiction (NIA) model. Both of the models take students' regular behavior into consideration. Finally, we conduct excessive experiments on a real-world dataset from a Chinese college, the results demonstrate the effectiveness of our model, and the analysis results are consistent with some psychological findings, which also verify the correctness of the model we propose.

References

1. Internet addiction disorder. https://en.wikipedia.org/wiki/Internet_addiction_disorder (2019). Accessed 10 Apr 2019
2. Burlak, G.N., Hernandez, J.A., Ochoa, A., Munoz, J.: The use of data mining to determine cheating in online student assessment. In: Electronics, Robotics and Automotive Mechanics Conference (CERMA 2006), vol. 1, pp. 161–166. IEEE (2006)
3. Fumero, A., Marrero, R.J., Voltes, D., Peñate, W.: Personal and social factors involved in internet addiction among adolescents: a meta-analysis. Comput. Hum. Behav. **86**, 387–400 (2018)
4. Guan, C., Lu, X., Li, X., Chen, E., Zhou, W., Xiong, H.: Discovery of college students in financial hardship. In: 2015 IEEE International Conference on Data Mining, pp. 141–150. IEEE (2015)
5. He, W., et al.: Abnormal reward and punishment sensitivity associated with internet addicts. Comput. Hum. Behav. **75**, 678–683 (2017)
6. Liu, W., Bao, X.Y., Wen, B., Chen, Q.: College students' internet addiction investigation and related causes analysis (in Chinese). Ph.D. thesis (2010)
7. Malak, M.Z., Khalifeh, A.H., Shuhaiber, A.H.: Prevalence of internet addiction and associated risk factors in Jordanian school students. Comput. Hum. Behav. **70**, 556–563 (2017)
8. Upadhayay, N., Guragain, S.: Internet use and its addiction level in medical students. Adv. Med. Educ. Pract. **8**, 641 (2017)
9. Wei, Y.Y., Huang, G.S., Xie, Z.B., Wang, J.H., et al.: A study on the relationship between college students' network dependence and loneliness—taking Hubei institute of technology as an example (in Chinese). J. Liuzhou Vocat. Techn. Coll. **3**, 38–43 (2018)
10. Xi, Y., Zhuang, X., Wang, X., Nie, R., Zhao, G.: A research and application based on gradient boosting decision tree. In: Meng, X., Li, R., Wang, K., Niu, B., Wang, X., Zhao, G. (eds.) WISA 2018. LNCS, vol. 11242, pp. 15–26. Springer, Cham (2018). https://doi.org/10.1007/978-3-030-02934-0_2
11. Xue, Y., et al.: Investigating the impact of mobile SNS addiction on individual's self-rated health. Internet Res. **28**(2), 278–292 (2018)
12. Zhang, Y., Qin, X., Ren, P.: Adolescents' academic engagement mediates the association between internet addiction and academic achievement: the moderating effect of classroom achievement norm. Comput. Hum. Behav. **89**, 299–307 (2018)
13. Zhao, F., Zhang, Z.H., Bi, L., Wu, X.S., Wang, W.J., Li, Y.F., Sun, Y.H.: The association between life events and internet addiction among Chinese vocational school students: the mediating role of depression. Comput. Hum. Behav. **70**, 30–38 (2017)
14. Zhu, Y., Zhu, H., Liu, Q., Chen, E., Li, H., Zhao, H.: Exploring the procrastination of college students: a data-driven behavioral perspective. In: Navathe, S.B., Wu, W., Shekhar, S., Du, X., Wang, X.S., Xiong, H. (eds.) DASFAA 2016. LNCS, vol. 9642, pp. 258–273. Springer, Cham (2016). https://doi.org/10.1007/978-3-319-32025-0_17

Detection of Entity-Description Conflict on Duplicated Data Based on Merkle-Tree for IIoT

Yan Wang[1,2], Hui Zeng[1], Bingqing Yang[1], Zhu Zhu[1(✉)], and Baoyan Song[1]

[1] College of Information, Liaoning University, Shenyang 110036, China
zhuzhuzz@126.com
[2] State Key Laboratory of Robotics, Shenyang Institute of Automation, Chinese Academy of Sciences, Beijing 110016, China

Abstract. Inferior data processing in IoT has always been one of the research hotspots, especially entity-description conflict on duplicated data. Moreover, for the fluctuation of industrial production data, the existing entity resolution technology is not applicable. Therefore, this paper proposes an entity resolution (ER) based on Merkle-tree for massive industrial production data (IDEM). Experiments show that the IDEM can improve the recognition accuracy. Compared with other ER methods the commonly used, the overall accuracy of IDEM is about 14.85% higher.

Keywords: Entity-description conflict on duplicated data · IIoT · Merkle-tree

1 Introduction

The key to the detection of entity-description conflict on duplicated data is entity resolution [1, 2]. For the fluctuation of the industrial production data, traditional ER technology will greatly increase the false positive rate of recognition results of it. Therefore, this paper proposes IEDM algorithm, which is a progressive, lightweight detection algorithm of entity-description conflict on duplicated data.

- This paper analyzes the characteristics of massive industrial production data for the first time, and proposes a lightweight detection algorithm of entity-description conflict on duplicated data.
- This paper uses the idea of Merkle-tree to propose a Pro M-tree structure to ensures efficiency while improving accuracy.
- This paper proposes a chain structure called "St-Chain" for supporting the block operation in progressive hash coding.

W. Ni et al. (Eds.): WISA 2019, LNCS 11817, pp. 163–168, 2019.
https://doi.org/10.1007/978-3-030-30952-7_18

2 Related Work

There is common threshold-based ER algorithm in [3]. The threshold of the method is difficult to determine, and there is an error in the similarity. There are ER technologies using clustering on entity similarity graph in [4–6], which eliminate entity match conflict problem using clustering on graph. The ER technology using clustering on graph needs to select the appropriate clustering center, otherwise it will increase the false positive rate. There are ER algorithms based on machine learning in [7, 8]. Although the ER technology based on machine learning improves the accuracy of ER, it increases the computational complexity and is difficult to apply on edge nodes. In addition, existing ER technologies are mostly for text data, such as text entity resolution [9].

3 Entity Resolution Based on Merkle-Tree

3.1 Related Data Structure

This paper transforms the Merkle-tree and uses the modified Merkle-tree to ensure the recognition efficiency. If there is one different node in the Merkle-tree, the whole tree will be different. This paper transforms its structure. The modified structure uses a progressive method to calculate attribute hash values one by one, as shown in Fig. 1.

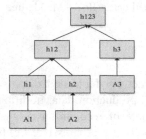

Fig. 1. Pro M-tree

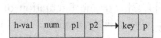

Fig. 2. St-Chain

The h1, h2, h3 represent the hash value of the attribute value, respectively. The intermediate nodes h12 represent the total hash value of its child nodes.

According to the hash values, all the tuples of the original data set are divided into some blocks. The tuples in different blocks must not be the same entity. Therefore, this paper introduces the following chain structure. In Fig. 2, the h-val represents the hash value stored by the node in the Pro M-tree. The num represents the number of tuples of the same hash value. The p1 is a pointer to the next block. The p2 is a pointer to the tuple in the block. The key holds the tuple ID. The p is a pointer to a subsequent tuple.

3.2   Standardization of Fluctuation Data

The traditional calculation methods cannot solve the fluctuation data in industrial production data. Therefore, this paper proposes a new data estimation method. The specific process is as follows. First, the cosine of each vector a(i) and b(j) is calculated from the cosine formula. A cosine vector v(key) is obtained. Second, sum the elements in each vector v(key) and calculate the average avg. Next, calculate the maximum of avg. Then calculate the average value of each attribute in the standard entity. Finally, take the corresponding average value of each attribute as its standard value. Where vector a(i) is a tuple in raw data set, $i = 1, 2, 3......n$. b(key) is a matrix composed of standard entities. $b(key) = \{b(j)|b(j) \in S, j = 1,2,3...k\}$, key: standard entity name. S is the sample set of standard entities, that is, there is no entity name conflict in S.

3.3   Determination of Attribute Sensitivity

Obviously, each attribute's ability to distinguish entities is different. In this paper, information entropy is used to determine the Attribute Sensitivity. After the above standardization, the larger the information entropy of the attribute, the more diverse the attribute value is, and the stronger the ability to distinguish the entity. Information entropy formula is as follows.

$$\sum_{i=1}^{n} p_i * \log_2(1/p_i)$$

3.4   Entity Resolution Algorithm Based on Pro M-Tree and Analysis

The specific algorithm is as shown in Algorithm 1. Where $Attr = \{attr(i)|i = 1,2,3...q\}$, attr(i) represents the i-th attribute. $Hash = \{hash(attr(i)(id))|i = 1,2,3...q$, id: the ID of the tuple in the data table$\}$. Algorithm 1 is mainly divided into two parts. Part 1: initialize St-Chain. First, hash encoding the first attribute, and then sort the hash value to simplify subsequent operations. Finally, the tuple is matched according to the hash value. Part 2: reconstructing St-Chain by iterating through the above operations.

Algorithm 1. ER Algorithm Based on Pro M-tree

Input： Processed data set D containing n industrial production data
Output： St-Chain with m entities

1.   for(each tuple t∈D)
2.   hash(attr(1)(id));
3.   end for
4.   Execute Shell Sort on hash(attr(1)(id));
5.   Allocate a memory block K(i), store (hj=0, num=0, p1=null, p2=null); // initialize St-Chain
6.   String s = hash(1)(id);   // Here hash(1)(id) refers to the first hash after sorting
7.   for (each hash (attr(i)(id)) ∈ Hash)
8.   if (hash (attr(i)(id)) == s)
9.   Link the memory block to the horizontal chain behind block K(i); Update K(i);
10.   else Link the memory block to the vertical chain below block K(i); s = hash(attr(i)(id));
11.   end if
12. end for
13. for(each attr(i)∈ Attr)
14.   for(each element∈ vertical chain)
15.   if(num != 1)
16.   for(each element∈ horizontal chain)
17.   hash(attr(i)(id)); hash(i)(id)= hash(attr(i-1)(id),attr(i)(id));
18.   end for
19.   Repeat steps 4-12;
20.   else Directly link num=1 to the new St-Chain;
21.   end if
22.   end for
23. end for

According to the description of Algorithm 1, The time complexity of the algorithm is in the interval $[n^{1.3}, n^{2.3})$.

4   Experiment and Analysis

The threshold-based ER algorithm and the ER algorithm based on entity similarity graph clustering are the two most classic ER algorithms. The paper selected the two typical algorithms of them: Part and ERC used to be compared with the IDEM.

4.1   Data Sets for Experiments

Two UCI datasets and one synthetic dataset are used for experiments. The two UCI datasets are Steel plate faults and Wine, respectively. They are real-world datasets. The synthetic dataset is generated by the generation tool according to the rules in [10]. It has 30000 pieces of data, each of which contains 6 attributes.

4.2 Analysis of Results

This section evaluates the performance of the IDEM mainly from the aspects of accuracy, F-measure and efficiency.

As can be seen from the Fig. 3, with the increasing of the size of the dataset, the accuracy of the IDEM algorithm is gradually increasing. The accuracy of the existing methods is basically unchanged.

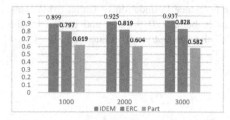
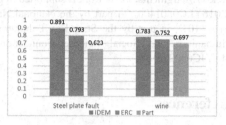

(a) Accuracy comparison on synthetic dataset (b) Accuracy comparison on real-world datasets

Fig. 3. Accuracy comparison chart

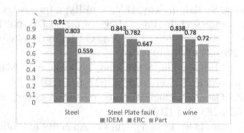

Fig. 4. F1-score comparison chart

As shown in Fig. 4, compared with the other two ER algorithms, the F1-score of the IDEM is about 7.53% and 22.17% higher, respectively. The overall average is 14.85% higher.

As shown in Table 1, the IDEM algorithm has no advantage in recognition efficiency, but the gap between the advantages and disadvantages is not large.

Table 1. Comparison of running time

Data sets	IDEM	ERC	Part
	Runtime(s)	Runtime(s)	Runtime(s)
Synthetic dataset	327	319	308
Steel plate faults	43	40	32
Wine	15	11	8

5   Conclusion

This paper analyzes the characteristics of massive industrial production data for the first time, and proposes a lightweight detection algorithm of entity-description conflict on duplicated data. The experimental results show IDEM can ensure the recognition efficiency while improving the accuracy.

Acknowledgments. This work is supported in part by the National Key R&D Program of China (2016YFC0801406), in part by the National Natural Science Foundation of China (61472169, 61871107), in part by the State Key Laboratory of Robotics (2019-O22), and in part by the Humanities and Social Science Foundation of the Ministry of Education of China (18YJC630276).

References

1. Li, J.Z., Liu, X.M.: An important aspect of big data: data usability. J. Comput. Res. Dev. **50** (6), 1147–1162 (2013)
2. Li, J.Z., Wang, H.Z., Gao, H.: State-of-the-art of research on big data usability. J. Softw. **27** (7), 1605–1625 (2016)
3. Altwaijry, H., Mehrotra, S., Kalashnikov, D.V.: QuERy: a framework for integrating entity resolution with query processing. In: Proceedings of the VLDB, pp. 120–131(2015)
4. Sun, C.C., Shen, D.R., Kou, Y., Nie, T.Z., Yu, G.: Entity resolution oriented clustering algorithm. J. Softw. **27**(9), 2303–2319 (2016)
5. Zhang, W.P., Li, Z.J., Li, R.H., Liu, Y.H., Mao, R., Qiao, S.J.: MapReduce-based graph structural clustering algorithm. J. Softw. **29**(3), 627–641 (2018)
6. Wang, H.Z., Li, J.Z., Gao, H.: Efficient entity resolution based on subgraph cohesion. Knowl. Inf. Syst. **46**(2), 285–314 (2016)
7. Yang, M., Nie, T.Z., Shen, D.R., Kou, Y., Yu, G.: An entity resolution approach based on random forest. J. Integr. Technol. **7**(02), 57–68 (2018)
8. Sun, C.C., Shen, D.R., Kou, Y., Nie, T.Z., Yu, G.: A genetic algorithm based entity resolution approach with active learning. Front. Comput. Sci. **11**(1), 147–159 (2017)
9. Liu, J., Cui, R., Zhao, Y.: Multilingual short text classification via convolutional neural network. In: Meng, X., Li, R., Wang, K., Niu, B., Wang, X., Zhao, G. (eds.) WISA 2018. LNCS, vol. 11242, pp. 27–38. Springer, Cham (2018). https://doi.org/10.1007/978-3-030-02934-0_3
10. Chen, J.X.: Commonly Used Chart Data Manual for Steelmaking, 2nd edn. Metallurgical Industry Press, Beijing (2010)

Anti-money Laundering (AML) Research: A System for Identification and Multi-classification

Yixuan Feng[1], Chao Li[3], Yun Wang[2], Jian Wang[2],
Guigang Zhang[2(✉)], Chunxiao Xing[3], Zhenxing Li[4],
and Zengshen Lian[4]

[1] Department of Engineering Physics, Tsinghua University,
Beijing 100084, China
fengyxl6@mails.tsinghua.edu.com
[2] Institute of Automation, Chinese Academy of Sciences, Beijing 100190, China
yuna9789@outlook.com,
{jian.wang, guigang.zhang}@ia.ac.cn
[3] Research Institute of Information Technology, Beijing National Research
Center for Information Science and Technology, Department of Computer
Science and Technology, Institute of Internet Industry, Tsinghua University,
Beijing 100084, China
{li-chao, xingcx}@tsinghua.edu.cn
[4] Beijing AgileCentury Information Technology Co., Ltd.,
Beijing 100085, China
{lizhenxing, lianzengshen}@agilecentury.com

Abstract. Anti-money laundering (AML) is of great significance for the integrity of financial system in modern society because of the huge amounts of money involved in and close relationships with other types of crimes. The typical AML system includes FinCEN for America which began in 1993 and AUSTRAC for Australia. This paper aims to develop a suspicious behavior detection and categorization system based on money transaction data collected from the simulator called Paysim. For preparation work, in order to facilitate crimes classification in the second step, the associated criminal activities were divided into five categories according to their different characteristics shown in the money transaction process. Then on the basis of the transaction data, a user profile was created and new features concerning both individual parties and network effect were extracted from the profile. With combined features, two models were developed for detection and classification using supervised learning methods separately. The results show good accuracy and recall rate which are most valued in reality. Meanwhile, the models display good robustness to further adjustment for practical use. Finally two models were connected in series, and the result shows a relatively good overall performance and verifies the feasibility of the system, as it provides more choices for users to decide which model (or both) to apply according to different situations in practice.

Keywords: AML · Features extraction · Classification

© Springer Nature Switzerland AG 2019
W. Ni et al. (Eds.): WISA 2019, LNCS 11817, pp. 169–175, 2019.
https://doi.org/10.1007/978-3-030-30952-7_19

1 Introduction

Money laundering refers to the action of legalizing illegal profits by disguising and cleaning the original source and nature of the black money through financial institutions like commercial banks, investment banks and insurance companies. ML (money laundering) has been a critical crime risk to almost all the countries for decades not only because that the significant amounts of money involved in can seriously undermine one country's financial system and encourage other kinds of crimes, but also owing to the complex processes and sophisticated and quickly evolving methodologies which make the detection work always hard to meet the need. Fortunately, the development of artificial intelligence technology has offered a chance to enhance the efficiency of AML (anti-money laundering) detection system and discover the newly emerging ML patterns and trading rules timely to combat these threats.

Generally AML cam be divided to three stages: prevention, detection & reporting and punishment. Preventative strategies include public education for AML, comprehensive information required while opening accounts and making transactions, promulgation of laws like <Anti-Money Laundering Regulations of Financial Institutions>. Detection & reporting indicates the detection of suspicious financial transactions with the help of human intelligence and data mining technology which is the point of this paper. Punishment refers to the financial and criminal penalties for ML criminals after detected.

For the detection & reporting part, generally the online monitoring system works first for suspicious detection and sends the report concerning target group to human for further investigation and judgement. To our best knowledge, most of the research on AML suspicious behavior detection mainly focuses on developing algorithms to distinguish potential illegal transactions from legal ones so far. Yet we think it should benefit further to develop a system that can identify illegal transactions and also make a reliable prediction on what kind of crimes the ML action is related with based on the characteristics of transaction pattern it displays, for it can offer a reliable reference for human analysts and reduce the labor cost to some extent. And it's the main purpose of our research.

2 Related Work

The application of information technology for AML work was first put forward in the 1990s. Senator et al. [1] made a detailed introduction on FAI (FinCEN Artificial Intelligence System) system which was designed to evaluate all kinds of financial businesses to identify ML and other criminal actions by using the rule-based method. And the rules were mainly settled by the knowledge and experience of experts which contributed to its high accuracy and disadvantage as well, as it was not capable enough of matching the quickly evolving ML approaches. Therefore after that further researches were done to propose improved detection systems for higher level of accuracy, automation, flexibility, etc. For example, Tang et al. [2] suggested a detection algorithm based on SVM instead of the preset rules and the result proved the improvement on the rate of misreport.

The improvement work includes two main branches. One considers to develop advanced algorithms for a better understanding of individual customers based on their own information. For example, Wang et al. [3] developed a decision tree method for ML identification based on four attributes extracted from companies' customer profile: industry, location, business size and products customer bought and the result proved the model's effectiveness. Kingdon et al. [4] proposed to develop a multidimensional adaptive probabilistic matrix for each bank account on the basis of its transaction behavior and make judgement according to ones' own behavior pattern. Yet the system is not always helpful for AML detection for unusualness is not equal to suspiciousness or illegalness. Zhang et al. [5] introduced wavelet analysis (Haar and bior3.7) to measure the suspicious level of individuals according to the time and amount series of transactions. Li et al. [6] proposes the credit score for analysis.

3 Algorithm of AML

The algorithm of AML system is shown in Table 1. One thing to note is that when we employed the model 2 separately to estimate its performance, the train set we used was 80% of A and test set the rest 20%.

Table 1. Algorithm of AML system

Algorithm Framework of our system
Input: The set of transaction information data A; **Output:** S', the identification of suspicious transactions and classification of crime categories on the test set A_2 of A; 1: Creating a user profile P based on A; 2: Extracting additional features from P concerning the initiator and recipient of every transaction and adding them to the original dataset A; 3: Splitting A into train set A_1 (80%) and test set A_2 (20%); 4: Training model 1 on A_1 using the methods of logistic regression, multi-layer perception and gradient boosting respectively, and testing it on A_2, gaining the suspicious transaction set S; 5: Collecting all the fraudulent transactions as set F; 6: Training the model 2 on $F\backslash S$ and testing it on S, gaining the suspicious transaction set S' labeled with the related crime category; 7: **return** S';

4 Experiments Preparations

4.1 The Sample Data

The lack of public available datasets is common for financial services and especially in the domain of money transactions due to the privacy protection. Fortunately there are several types of simulators generating synthetic datasets based on real data in order to resemble the normal operation of transactions. In the paper we adopted the money transaction data created by a simulator called Paysim. The sample it relied on was real transactions extracted from one-month financial logs in an African country provided by a multinational company. And a few modifications were done to make it more realistic and reliable.

4.2 The Classification of Crimes

Money laundering has a close relationship with other organized crimes such as drug trafficking, smuggling, terrorism, corruption, etc. According to their trading natures, the crimes were classified to five categories listed in Table 2. The result of classification would function as labels when the second model was trained.

Table 2. Five main categories of crimes related with money laundering

Category	Example	General characteristics
1	Illegal fund-raising, pyramid selling	A large number of accounts involved in, decentralized cash-in (each amount is small), centralized cash-out
2	Commercial bribery, power corruption	Huge amounts of money involved in, the number of accounts is limited, a long time period
3	Drug crimes, gambling crimes, mafia crimes	A relatively large number of accounts and huge amounts of money involved in, possible foreign factors
4	Illegal banks (underground banks)	Transactions with high frequency, possible foreign factors, relatively huge amounts of money involved in
5	Illegal activities with POS	Transactions with high frequency, each amount is relatively small, the type of transaction is debt in most cases

5 Multi-classification Model

5.1 Single Model Training

To make the system more useful and practical, further research has been done and the second model was developed to predict the most probable crime categories for the fraudulent transactions. Features used for model training are the same with suspicious detection model part, and the label was changed to *fraud_catg*.

For multi-classification, another three supervised learning techniques were used here: SVM (support vector machine), LR and MLP. 1047 pieces of fraudulent transactions were used for the training and testing processes. 80% of them were spilt to train set and 20% for test set. The numbers of the crimes with category 1, 2, 3, 4, 5 are 302, 214, 282, 203 and 46 respectively. Indicators included confusion matrix, precision rate, recall rate and F1.

The confusion matrices are shown as Eqs. (1), (2), (3). And the precision rate, recall rate, F1 are shown in Table 3.

Table 3. Performance indicators

Model	Precision rate /%	Recall rate /%	F1 /%
SVM	88.13	87.88	87.89
LR	87.53	87.68	87.59
MLP	90.42	90.59	90.48

$$M_{SVM} = \begin{pmatrix} 54 & 1 & 0 & 4 & 0 \\ 3 & 29 & 8 & 2 & 0 \\ 1 & 5 & 53 & 1 & 0 \\ 1 & 1 & 2 & 38 & 0 \\ 0 & 0 & 0 & 0 & 7 \end{pmatrix} \tag{1}$$

$$M_{LR} = \begin{pmatrix} 53 & 3 & 0 & 3 & 0 \\ 1 & 31 & 7 & 3 & 0 \\ 3 & 4 & 52 & 1 & 0 \\ 2 & 3 & 0 & 37 & 0 \\ 0 & 0 & 0 & 0 & 7 \end{pmatrix} \tag{2}$$

$$M_{MLP} = \begin{pmatrix} 54 & 2 & 0 & 3 & 0 \\ 1 & 35 & 4 & 2 & 0 \\ 1 & 5 & 54 & 0 & 0 \\ 2 & 2 & 0 & 37 & 0 \\ 0 & 0 & 0 & 0 & 7 \end{pmatrix} \tag{3}$$

The main parameters for three models are as follows.

SVM: *kernel* = 'rbf', *decision_function_shape* = 'ovo', C = 100.
LR: C = 100, *multi_class* = 'multinomial', *solver* = 'sag'.
MLP: *activation* = 'relu', *max_iter* = 500, *hidden_layer_size* = (100,).

The model performed well on both training and test dataset. The result shows that the classification for crimes done in preparation work is reasonable and practical for the model training. It's a good starting place for multi-classification problem for AML research. Currently one limitation is that with the development of economy and crime 'technology', the classification is supposed to change frequently to maintain good accuracy for the prediction of crime types.

5.2   Models in Series

After estimating the accuracy of two models separately, now we tested the performance of models in series. Figure 1 shows an overview of the system. After the extraction and combination of features for preparations, the model 1 is used to lock on the suspicious transaction set S from the test set. Then model 2 is trained on the fraudulent transaction set F (the fraudulent transactions which have been detected by model 1 are excluded here in order to make the prediction by model 2 more convincing). And then model 2 is employed to make classification of crime types on the set S. Finally the result is reported to intelligence analysts for further investigation and judgment.

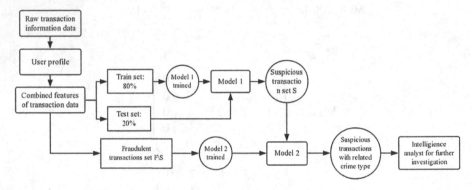

Fig. 1. System overview

6   Conclusion and Future Work

By using the supervised learning methods, this paper developed a system with two models for money laundering detection and crime category classification based on the transaction data. And the result indicates that each model is well-appreciated for the sample data and shows good robustness for different training methods and when models are combined together, it can give a valuable reference for manual inspection. In terms of the limitations of our research, further work is needed to improve the algorithms and robustness of the system.

Acknowledgement. Our work is supported by NSFC (61872443) and supported by CCF Opening Project of Information System.

References

1. Senator, T.E., Goldberg, H.G., Wooton, J., Cottini, M.A., Khan, A.F.U., Klinger, C.D., et al.: Financial crimes enforcement network AI system (FAIS) identifying potential money laundering from reports of large cash transactions. AI Mag. **16**(4), 21–39 (1995)
2. Tang, J., Yin, J.: Developing an intelligent data discriminating system of anti-money laundering based on SVM. In: International Conference on Machine Learning & Cybernetics. IEEE (2005)
3. Wang, S.N., Yang, J.G.: A money laundering risk evaluation method based on decision tree. In: IEEE 2007 International Conference on Machine Learning and Cybernetics. IEEE (2007)
4. Kingdon, J.: AI fights money laundering. IEEE Intell. Syst. **19**(3), 87–89 (2004)
5. Zhang, C.H., Zhao, X.H.: Research on time series of suspicious financial transactions based on wavelet analysis. Mod. Manage. Sci. **7**, 102–104 (2009). (in Chinese)
6. Li, R., Zhao, C., Li, X., Zhang, G., Zhang, Y., Xing, C.: Comparative analysis of medical P2P for credit scores. In: Meng, X., Li, R., Wang, K., Niu, B., Wang, X., Zhao, G. (eds.) WISA 2018. LNCS, vol. 11242, pp. 307–313. Springer, Cham (2018). https://doi.org/10.1007/978-3-030-02934-0_29

Research of Query Verification Algorithm on Body Sensing Data in Cloud Computing Environment

Yanfang Gao[1,2], Lan Yao[1(✉)], and Jinyue Yu[1]

[1] Northeastern University, Shenyang 110169, Liaoning, China
yaolan@mail.neu.edu.cn
[2] Neusoft Corporation, Shenyang 110179, Liaoning, China

Abstract. Cloud computing enables users with limited resources to accomplish complex computation. Meanwhile, the massive human physiological data generated by Wireless Body Area Network (WBAN) can be benefited by it through being stored in cloud servers. However, cloud storage of human physiological data is challenged by some issues, while one of them is soundness verification. Although local WBAN user is not powerful enough to compute the result on whole data repository, he may have concern on the correctness of the results from cloud which he pays for. To address this problem, this paper proposes a DMB+ tree data structure, and designs a query and verification algorithm for human physiological data in cloud computing environment based on DMB+ tree. By reducing the number of layers returned from the query tree, the algorithm reduces the height of the tree when verifying, thus improving the efficiency of query validation. The experimental results show that the performances in time and network traffic of DMB+ query verification are superior to Merkle tree and digital signature algorithm.

Keywords: Cloud computing · Verifiable computing · Query verification · WBAN · Human physiological data

1 Introduction

To help medical facilities collect bio-data from patients or prospective patients with long-term diseases, for example diabetic, to build their personal records and assist diagnosis, WBAN is applied and collected bio-data are stored in Cloud for the sake of the large size. However, cloud servers are not completely trustworthy. Once attackers compromise cloud servers, data and computation results may be falsified by them if users cannot prove the authenticity of the data stored on cloud servers, which often has a serious impact on medical diagnosis or message integrity. Therefore, it is necessary to provide users with an efficient approach to verify the accuracy and integrity of the results, so as to reduce the loss from malicious attacks.

At present, computational verification algorithms and soundness of human medical data have drawn a lot of researching attention regarding query verification. The research on verification algorithms of data query results are composed of three main categories. Merkle Hash Tree [1, 2] is proposed by Merkle. It guarantees the integrity

© Springer Nature Switzerland AG 2019
W. Ni et al. (Eds.): WISA 2019, LNCS 11817, pp. 176–188, 2019.
https://doi.org/10.1007/978-3-030-30952-7_20

of query results through Merkle Hash Tree. The data structure of Merkle Hash Tree has the advantage of one signature and multiple authentication. Data signature algorithm [3–5] is mainly used to prove the authenticity of data. There are DSA and RSA digital signature algorithms for verification. Some early query verification algorithms, such as those in reference [6], proposed a verification scheme using digital signatures, but this scheme cannot avoid large computational load, nor can it verify the integrity of the results. Pang et al. [7] proposed a new signature verification method. This method makes data records into a combination and signs the combination, which can solve the problem of correctness and integrity of query verification. Cheng [8] extends Pang's method to multi-dimension. By dividing and sorting multi-dimension data, a digital signature chain is constructed to complete data query and verification. Some other methods are based on probability by sampling and across-verification, mainly in reference to challenge-response method. Ateniese first proposes the data possession (PDP) [9] scheme, which verifies the data integrity by introducing the challenge-response protocol and homomorphic tags, while the computational overhead is high.

2 Verification Strategy Design

The main application scenarios of this paper are as follows: data owners, as some commercial application clients with massive human physiological data recorded by wearable health equipment, in order to enable system users to retrieve their own data, and to ensure the correctness and integrity of the data result set queried by users, and to provide users with the correct proof of query. In order to provide accurate information of human physiological data, cloud computing platform needs to support a verification object, so that users can verify the accuracy and integrity of query results even if they are weak in computation. Because these users come from different sources, there is no common trusted third party, so the query verification scheme designed in this paper does not rely on trusted third party. Therefore, the preliminary idea of the query and verification scheme for human physiological data which is listed below.

(1) Only a trusted key center is involved to issue public key certificates instead of trusted functions or data center.
(2) Data owners can process their own data before uploading data to the cloud, and once the data is stored, they cannot change it at will.
(3) Users are eligible to query data. When the query results are returned to users, they will receive them, as well as verification objects. Users apply public keys to decrypt and verify the verification objects to confirm the accuracy and integrity of the result set.

Based on the above strategy, this paper proposes a new query verification algorithm, which is a trade-off between flexibility and density of retrieval results. In the data processing stage, digital signature is introduced to reduce data returned by the verification object, so as to reduce the time complexity of the verification and ensure the accuracy and integrity.

2.1   A Query Index Structure – DMB+ Tree Design

Although Merkle hash tree solves the problem of large computation of digital signature algorithm, it still shows its incapability in time complexity and complex tree structure, and is not adaptive to the query verification of massive data. Therefore, DMB+ tree is proposed in this paper. DMB+ tree is a data structure that combines the hash concept of Merkle hash tree with B+ tree. The goal to design it is efficient query and verification on massive human physiological data.

Formal Definition of DMB+ Tree. Before defining the DMB+ tree proposed in this paper, we need to preprocess the data according to the characteristics of the data. Records are denoted in general form: $r\ (key,\ a_1,\ a_2\ldots\ldots a_n)$.

key is the key of the record r. For each data recorded, there is a unique key as the identification of the record. a_i is an attribute of record r, for example, blood pressure or respiratory data.

The precursor and after-cursor of a record r are defined here for subsequent data connection.

Definition 1: Precursor r_p: The precursor of record r is a record satisfying: $r_p.key < r.key$ and no $r_x \in r$ makes $r_p.key < r_x.key < r.key$.

Definition 2: After-cursor r_s: The aftercursor of record r is a record satisfying: $r. key < r_s.key$ and no $r_x \in r$ makes $r. key < r_x.key < r_s.key$.

First, records are accessed as data blocks, each of which stores the same number of records. Hash linked list can guarantee the integrity of the precursor of block b and the link of block b, and ensure that the data of the whole linked list will not be tampered with by malicious users at will. The preparation of defining a hash list (*HChain*) are the keywords of the data block and the representation of the Hash list. They are shown below.

Definition 3: $b.key$: $b.key$ is the key of data block b only if it satisfies: for any $r_i. \in b$, there is no $r_i.key < b.key$.

Definition 4: $HChain(b)$: $HChain(b) = H(H(b)\|HChain(b_P))$.

The above definitions are describing the relationship between records, or data blocks. The formal definition of n-order DMB+ tree is given below.

(1) If the DMB+ tree only has root nodes, the number of stored recorded data is less than n. Then the node has at least one keyword for retrieval, one hash value, at most N keywords and one hash value, and the root node has no pointer.

(2) If the node number of DMB+ tree is greater than 1, that is, the layer number of tree is greater than 1, and the stored records are greater than n, then the root node has at least two keywords for retrieval, one hash value and one pointer. It has at most n keywords, 1 hash value and n pointers.

(3) For all non-leaf nodes, each node has at least $\lceil n/2 \rceil$ keywords, $\lceil n/2 \rceil$ pointers, up to n keywords, 1 hash value and n pointers. Each non-leaf node also stores the signature of the node.

(4) All leaf nodes are located in the same layer of the DMB+ tree, where all data are stored, and each leaf node serves as a basic data block for data storage. A data

block is connected with its precursor and after-cursor through a hash chain, and each leaf node also stores the *HChain* value of that node. Users can retrieve data according to the head nodes of leaf nodes.

DMB+ Tree. The construction of DMB+ tree is divided into two steps. Firstly, using bottom-up method, the data is stored in the leaf nodes, and then the upper nodes are constructed layer by layer according to the leaf nodes through the root nodes. Suppose we need to construct a DMB+ tree with OD out-degrees and N records, and construct the algorithm of DMB+ tree as shown in Algorithms 1 and 2.

Algorithm 1 Pseudo code to build DMB+ tree leaf
```
Input: OD, n
Output: head
level=⌊log_OD n⌋+1;   // Tree layer number
m=⌈i/OD⌉;             // Number of leaf nodes
For (i=1 to n);
<ti.key,ti.record>←ri;// Generate a tuple ti based on ri
Ni=<<Ni.Ki>,<Ni.Fi>,Ni.hash,Ni.next>;//Set representation
                                      //of leaf nodes
F::<ti.key,ti.record>→<Nm.Ki,Nm.Fi>;//Store ri data in Nm
Nm.Next→Nm+1;        // The pointer points to the next node
Nm.Hash=HChain(Nm.hash ||H(ri));
End For;
Return head.
```

Algorithm 2 Pseudo code to build DMB+ tree
```
Input: OD, n, head;    // Input Out-degree, Number of
                    nodes, //Head Nodes
Output: root;
level=⌊log_OD n⌋+1; // Tree layer number
m=⌈i/OD⌉;              // Number of leaf nodes
For(j=level to 1):
Node N→Nj;     // N is the j-level header node
While (node ≠ null):
N=<<Ni.key>,<Ni.children>,Ni.hash,Ni.sign>;
            //Representation of non-//leaf nodes by sets
F'::MIN(<Ni.children>.Ki>) →Ni.Ki;//Generating nodes from
                            //child nodes and non-leaf nodes
N.Hash=HChain(N.hash||ri.key);
Nm.sign=SIGN(Nm.hash);
End while;
End For.
```

When constructing a DMB+ tree, the data are stored in the linked list in the form of data blocks according to algorithm 1. Then, the linked list is used as the leaf node of the tree, and the non-leaf nodes are constructed upward through the root. Then the signature of the root node and the root node is returned. When queried, relevant data will be searched from top to bottom to leaf nodes by binary search according to the keywords stored in each node. The instant of a DMB+ tree is shown in Fig. 1.

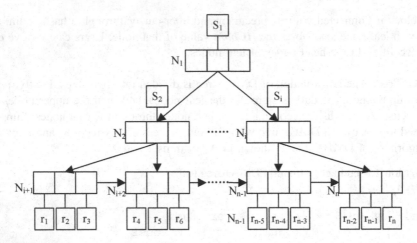

Fig. 1. DMB+ tree.

The data owner is the party who structures the data into a DMB+ tree when uploading the data to the cloud. In addition to the basic data, there are signatures for tree nodes and hash values for each node in the cloud server. These signatures and hash values have been stored in VO before the user queries and they are returned to the user with the query results for user's verification computation.

The following example is regarding medical data recorded by wearable body sensors. Each record contains treatment data and personal information of patient, as shown in Tables 1 and 2.

Table 1. Patient personal information.

Name	User ID	Sex	Age	Address
Zhang San	103142	Male	32	Hebei***
Li Si	103143	Male	44	Shanxi***

Table 2. Patient human physiological data.

User ID	Medical record	Blood pressure	Heartbeat	Time
103142	Heart disease......	**	**	2017-10-02 16:32:12
103143	Hypertension......	**	**	2015-10-29 09:42:58

Tables 1 and 2 show the medical information from a hospital. For the same user ID, the recording time of each data is unique. The data owner extracts the time field of each data as a key and stores other fields in the DMB+ tree.

The tree node out-degree has a direct impact on the tree depth. When the tree out-degree increases, the tree depth will decrease. Then, the query of DMB+ tree and the amount of data returned to VO will be reduced accordingly, thus reducing the time complexity of VO verification. According to reference [10–12], if the storage space is B bytes and the key word of tree is k bytes, the most effective way should be:

$$OD = (B/k) - 1 \qquad (1)$$

In the pre-processing stage of human physiological data, the following processing algorithm is used to generate a set ω, so that human physiological data can match the storage mode of DMB+ tree nodes in the form of a set. The preprocessing algorithm is as follows.

Algorithm 3: Medical data preprocessing

```
Input: r 1······ri;
Output: ω;
Init(ω);    // Initialize set ω
For (n= 1 to i):
    r ←r;
    w = <r.date,r.record>;  //Input data w, extract the
    ω ← w;                  // time field of w insert w intoω
End for;
Sort(ω);   // Sort the data in ω
Return ω;
```

By set ω, the keyword size k and the storage size B of data block can be computable. Formula 1 computes the output of the DMB+ tree used to store the data, and the data can be stored in the DMB+ tree nodes by mapping method as shown in Formula 2.

$$Fw \; :: \; <w.date, w.record> \rightarrow <ti.key, ti.record> \qquad (2)$$

After these procedures, the data owner can use the key to construct the DMB+ tree locally, and store each record in the leaf node in the form of formula 2. After building the DMB+ tree, the data owner needs to send the processed data to the server. The data is sent in blocks. The data and signature are sent separately so as to facilitate the later verification algorithm

2.2 Query Verification Algorithms Based on DMB+ Tree

The basic idea for verification is as follows:

(1) Cloud service providers query according to keywords and return user query results and minimum verification subtree as VO to verify.
(2) The DMB+ tree is reconstructed from bottom to top according to the VO returned during user verification.
(3) Verify the hash value of the root node according to the public key, and verify the correctness of the returned result through the signature of VO root node.

As shown in Fig. 2, the user queries the data from P0 to P1. When the cloud server returns VO, it only needs to return the minimum verification subtree containing P0 to

P1, that is, the shadow part of the graph, where the black node is the node independent of the query result. Because the cloud server only needs to return the sub-tree with the node N_2 as the root node as VO, but does not need to return the root node of the DMB+ tree. This reduces the data returned by the verification object, and reduces the unnecessary signature verification overhead in the process of user verification.

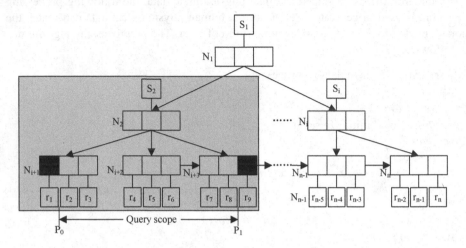

Fig. 2. VO construction of DMB+ tree query verification.

When the user receives the query result and verification object returned by the cloud server, the user needs to verify the query result according to the data returned. Because only the data on leaf nodes and the signatures and hash values of some non-leaf nodes are returned, users need to rebuild a new DMB+ tree from bottom to top to get the signature of the hash values of root nodes. Then, by verifying whether the signature is consistent with the signature returned by the verification object, we check whether the query results are correct. Specific verification algorithm is as follows.

Algorithm 4: User authentication algorithm
```
Input: VO(Signroot, N1......Nn)//data set R and VO
Output: True/False
For (i=1 to n)
Build _DMB+_leaf(Ni);   //Building leaf node based on
                        //algorithm 1
h=Ni.Hash;
If (HChain(Ni)≠Ni.hash)//udge if hash values are same
Return false;
Break;
End If;
End for;
Root=Build_DMB+ tree(head);  //Building tree based on
                             //algorithm 1, return root
If (Sign(Root.hash)≠VO.Signroot)
Return false;
Else
Return true;
End If.
```

2.3   Performance Analysis of Algorithm

Compared with Merkle tree verification, MB+ tree adopts an algorithm that generates hash values from multiple records as a data block. The MB+ tree reduces the number of hash computations compared with Merkle tree that each data generates a hash value. At the same time, the height of the returned tree is reduced in verification, which reduces the operation of the verifier to verify the signature.

If the out-degree of DMB+ tree is OD and the tree height is h, then the number of keywords n satisfies the inequality:

$$n \geqslant 1 + OD \cdot \sum\nolimits_{i=1}^{h} 2(OD+1)^{i-1} = 2(OD + h - 1) \tag{3}$$

From above, we can get the relationship between tree height h and out-degree OD:

$$h \leqslant log_{(OD+1)}(\frac{n+1}{2}) \tag{4}$$

If the query selection rate is $\in [0,1]$, and the height of the returned verification tree is h', then h' satisfies the following formula:

$$h' \leqslant log_{(OD+1)} \sigma + h \tag{5}$$

In this algorithm, the computational performance of human physiological data verification is improved by reducing the number of layers returned by the verification tree. Compared with Merkle tree which needs to return the verification information of the whole tree, DMB+ tree only needs to return the subtree of the whole tree structure to complete the verification of the query results. It turns out that the DMB+ tree algorithm improves the query verification efficiency when the query rate is lower.

3   Implementation and Evaluation of Algorithm

3.1   Experimental Environment

This experimental system is based on the open source experimental system provided by the research team of Texas University. In this paper, Pantry in the system is used as a prototype system to complete the algorithm experiment.

The experimental system is divided into 3 parties: cloud server, verification user and data owner. Tencent cloud server is selected as the experimental server; and the data set is the patient information, case information and human body data collected by a hospital. For privacy propose, we replace all patients' names with fake ones. Firstly, these data are filtered manually, and the data structure is shown in Tables 3 and 4.

Table 3. Patient information.

Name	Age	Sex	Address	Medical record
Zhang Xi	15	Male	Liaoning****	Heart disease, symptoms ***
Li Lin	37	Female	Liaoning****	Hypertension, symptoms ***
Wu Wen	43	Male	Shanxi****	Diabetes mellitus, symptoms ***

Table 4. Medical data.

Name	Heartbeat	Time
Li Lin	72	2017-06-01-22:45:49
Li Lin	77	2017-01-09-13:34:12
Zhang Xi	73	2018-07-12-07:12:45
Wu Wen	93	2018-05-14-00:01:23

Because these physiological data and patient information are processed separately in the database, it is necessary to connect these data first. Connecting the information of the same patient can reduce the dimension of the data while ensuring the uniqueness of each data. In the experimental environment, the data set is reduced to four dimensions, as shown in Table 5.

Table 5. Medical data.

Name	Medical record	Heartbeat	Time
Li Lin	Hypertension, symptoms***	72	2017-06-01-22:45:49
Li Lin	Hypertension, symptoms***	77	2017-01-09-13:34:12
Zhang Xi	Heart disease, symptoms***	73	2018-07-12-07:12:45
Wu Wen	Diabetes mellitus, symptoms***	93	2018-05-14-00:01:23

3.2 Verification Computing Scheme

The three parties, data owner, user and cloud server conduct the algorithm in four stages: data upload, authorization, request service and user verification. Figure 3 shows the flow of the scheme.

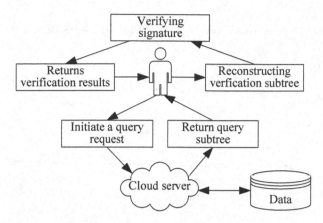

Fig. 3. Flow of request query service and user verification.

In Fig. 3, the data owner has the right to authorize the user. Data owners can encrypt their data with a unique key and then authorize it with the public key of the user (data visitor). At the same time, the scheme does not rely on trusted third party management, but allows users to authorize access to data visitors. Cloud servers can provide query verification services for users. They will return the response data according to the user request, and the user will verify the verification object locally.

3.3 Performance Analysis of Query Verification

In order to test the efficiency of DMB+ tree proposed in this paper on human physiological data sets, digital signature method and Merkle tree are applied to verify the data query with SHA-1 hash function. For query verification, the cost of client includes two parts: one is the communication cost of receiving the data sent by server for constructing Verification Object (VO), the other is the time cost for the object verification.

In the experiments, 999978 physiological data are involved. Data are stored in cloud servers in the form of digital signatures, DMB+ trees and Merkle trees. The single record size is 192 bytes and the keyword size is 19 bytes. The out-degree of Merkle tree are 2 and out-degree DMB+ tree are 10.

In order to verify the communication efficiency of human physiological data in the query verification algorithm based on DMB+ tree designed in this paper, we construct an experimental environment. By storing human physiological data in blocks, and building DMB+ tree on this basis, users can access data sets remotely and make query requests for data sets. The server searches the data through DMB+ tree and returns the corresponding result set. This experiment compares the cost of digital signature, DMB+ tree and Merkle tree in communication during verification. Figure 4 shows the comparison of the three algorithms in communication.

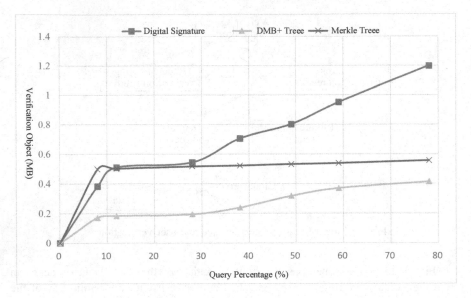

Fig. 4. Communication cost.

In Fig. 4, the communication cost of the three algorithms in different query selection rates is plotted. It illustrates that the communication cost of digital signature algorithm in verification is the greatest. Digital signature algorithm ensures the accuracy of data by adding signatures to data, but at the same time increases the amount of data. Therefore, in verification, a large number of signatures are returned, which adds a large amount of additional communication overhead. At the same time, when the query rate is between 5% and 40%, DMB+ tree can save 45% to 70% of the communication overhead of Merkle tree. This is mainly because the server can dynamically build VO, and it does not need to be built into the root node. When the query selection rate is low, it usually only needs to build to the level 2 and 3 nodes of the tree. However, due to the signature of the intermediate nodes in the DMB+ tree, the space increases. When the query selection rate reaches more than 60%, the DMB+ tree needs to be constructed to the level 4 and 5 nodes of the tree, and the communication cost of the DMB+ tree increases rapidly. But in the actual scenario, the query rate of users is less than 40%. Usually, the query rate of users is only 5% or even smaller. Therefore, DMB+ tree can greatly improve the communication cost of query verification in practical applications, and it can more effectively perform computational verification services with less bandwidth provided by cloud servers.

Figure 5 compares the time cost of the three algorithms to verify the query results by verifying the objects on the client side. Experiments show that after the cloud server returns the query results completely, the user verification and computation time of the query data results is eliminated, which eliminates the impact of network transmission delay. Then, in this period of time, the advantages and disadvantages of the three algorithms are compared through the overhead.

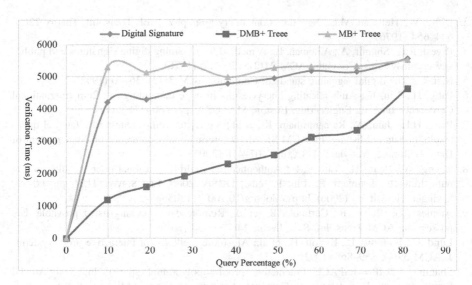

Fig. 5. Verification time cost in client end.

As shown in Fig. 5, the time cost of digital signature algorithm is relatively large, because a large number of signature operations are needed when querying more data. As a result, the growth of time is faster. Merkle tree needs to construct a complete Merkle+ tree every time, and it also needs a lot of signature and hash operations to verify. Because DMB+ tree can dynamically construct verification objects, it reduces the number of hash operations and verification time. When the query selection rate is between 5% and 40%, it can reduce 45% to 70%.

4   Conclusion

Although the research on verifiable computing has achieved more results, it is still in the theoretical stage, and there is no algorithm that can be fully used and applied in practical medical data query results. Based on this motivation, this paper designs a verification algorithm according to the characteristics of medical sensing data, and realizes the verification algorithm model in the experimental system with solid data from a hospital. As an application case of the proposed algorithm model, it has been proved with the feasibility of the computation verification of the model through the results of the actual computation task. Compared with existing algorithms, this method has advantages in medical data query verification.

References

1. Merkle, R.C.: Protocols for public key cryptosystems. In: IEEE Symposium on Security and Privacy, Oakland, CA, USA, pp. 122–134 (1980)
2. Merkle, R.C.: A certified digital signature. In: Brassard, G. (ed.) CRYPTO 1989. LNCS, vol. 435, pp. 218–238. Springer, New York (1990). https://doi.org/10.1007/0-387-34805-0_21

3. Diffie, W., Hellman, M.E.: New directions in cryptography. IEEE Trans. Inf. Theory **22**(6), 644–654 (1976)
4. Rivest, R.L., Shamir, A., Adleman, L.: A method for obtaining digital signatures and public-key cryptosystems. Commun. ACM **21**(2), 120–126 (1978)
5. Nist, C.: The digital signature standard. Commun. ACM **35**(7), 36–40 (1992)
6. Pang, H., Tan, K.: Authenticating query results in edge computing. In: 20th International Conference on Data Engineering, Boston, MA, USA, pp. 560–571 (2004)
7. Pang, H.H., Jain, A., Ramamritham, K., et al.: Verifying completeness of relational query results in data publishing. In: ACM SIGMOD International Conference on Management of Data, Baltimore, Maryland, USA, pp. 407–418 (2005)
8. Cheng, W., Pang, H., Tan, K.-L.: Authenticating multi-dimensional query results in data publishing. In: Damiani, E., Liu, P. (eds.) DBSec 2006. LNCS, vol. 4127, pp. 60–73. Springer, Heidelberg (2006). https://doi.org/10.1007/11805588_5
9. Ateniese, G., Burns, R., Curtmola, R., et al.: Remote data checking using provable data possession. ACM Trans. Inf. Syst. Secur. **14**(1), 1–34 (2011)
10. Lund, C., Fortnow, L., Karloff, H., et al.: Algebraic methods for interactive proof systems. J. ACM **39**(4), 859–868 (1992)
11. Shamir, A.: IP = PSPACE (interactive proof = polynomial space). In: 31st Annual Symposium on Foundations of Computer Science, St. Louis, Missouri, USA, vol. 1, pp. 11–15 (1990)
12. Wu, J., Ni, W., Zhang, S.: Generalization based privacy-preserving provenance publishing. In: Meng, X., Li, R., Wang, K., Niu, B., Wang, X., Zhao, G. (eds.) WISA 2018. LNCS, vol. 11242, pp. 287–299. Springer, Cham (2018). https://doi.org/10.1007/978-3-030-02934-0_27

Analysis of Macro Factors of Welfare Lottery Marketing Based on Big Data

Cheng Li[1,2(✉)], Hua Shao[1], Tiancheng Zhang[1], and Ge Yu[1]

[1] School of Computer Science and Engineering, Northeastern University,
Shenyang 110169, China
2103789@qq.com, 289973335@qq.com,
{tczhang, yuge}@mail.neu.edu.cn
[2] Liaoning Welfare Lottery Issuance Center, Shenyang 110015, China

Abstract. In order to analyze various factors of welfare lottery marketing in China objectively, we utilize several regression models to analyze the macro factors of the annual sales of welfare lottery. Afterwards, we adopt different CART decision tree models to analyze the micro factors of the daily sales of welfare lottery. The experimental results show that the learned rules provide an objective basis for scientific selection of marketing strategies.

Keywords: Welfare lottery · Marketing strategy · Macro-factor analysis · Micro-factor analysis

1 Introduction

China welfare lottery is an important means of reallocating of national resources and raising funds for the development of public welfare [1], which plays a very crucial role in maintaining social stability, promoting social harmony and promoting the development of social welfare service [2]. We should objectively analyze the current competitive situation of welfare lottery industry in China at various levels and adopt diversified measures to optimize the competitive strategy of welfare lottery issuance and management, to promote the comprehensive and coordinated development of welfare lottery and social welfare undertakings, and embark on long-term development road [3].

There are few scientific analyses on the relationship between welfare lottery marketing and macroeconomic and micro-behavior. Li et al. focus on guessing sports lottery and draw the following conclusions from the quantitative analysis of social factors related to it and other lotteries: guessing sports lottery marketing are more sensitive to the positive response of economic level; factors such as education level and male population proportion have a positive effect on its sales; its sales will increase significantly in the World Cup year, but there is no significant relationship between the sales volume and the football level of different provinces [4]. As a result, the decline in the rate of public welfare funds will reduce the overall sales of sports lottery and public welfare funds; online lottery sales will increase the sales of competitive guessing lottery, but the role is relatively small.

© Springer Nature Switzerland AG 2019
W. Ni et al. (Eds.): WISA 2019, LNCS 11817, pp. 189–198, 2019.
https://doi.org/10.1007/978-3-030-30952-7_21

In order to analyze the price factors of threaded steel, Chen et al. utilize scatter plots and trend lines to determine the influential factors and apply a multivariate linear regression model to make analysis [5]. It's worth mentioning that the multicollinearity between independent variables was eliminated by ridge regression analysis. The experimental results show that the model has high fitting degree and has certain practicability. In order to effectively improve the identification of HFMD, Yun et al. propose a realtime, automatic and efficient prediction system based on multi-source data (structured and unstructured data), and gradient boosting decision tree(GBDT) [6]. Besides, pruned CART tree was used to recognize the defects of manmade board [7]. Liu et al. construct CART tree based on Gini index, then the generated CART tree is pruned by cost complexity algorithm, which is proved to be effective in alleviating the overfitting problem.

In this paper, the macro and micro factors of welfare lottery marketing are modeled, and the historical data are used to verify them. Through the comparison of regression analysis methods, we can find that the accuracy of macro-factor model is higher, and the related factors reflect the changes of macro-sales situation more completely. Secondly, we conduct extensive experiments to evaluate the effectiveness of CART regression and the results demonstrate that it basically achieves the predictive accuracy that can be used for reference. Internal environmental factors do have an impact on micro-factors. Some analysis conclusions provide objective basis for scientific adjustment of marketing management strategies.

2 Macro-Factor Analysis

Macro factors are the factors that affect the annual sales of welfare lottery, such as population, GDP and PCDI. The reason of choosing regression analysis is that not only judge the influence of single attribute by unified regression and eliminate multi-collinearity problems by ridge regression, but also judge whether the factors affecting the change of lottery macro-sales are reasonable by the prediction accuracy of the model.

2.1 Linear Regression Model Design

Regression analysis can be divided into unified regression analysis and multivariate regression analysis. In practical applications, there are usually two or more independent variables that affect the dependent variables. If the relationship between multiple independent variables and dependent variables is linear, it is multivariate linear regression. The general model of multivariate linear regression analysis is as follows:

$$y = \beta_0 + \beta_1 x_1 + \beta_2 x_2 + \ldots + \beta_n x_n + \varepsilon \tag{1}$$

where $x_1, x_2, \ldots, x_n$ are n independent variables, y is dependent variable, $\beta_0, \beta_1, \beta_2, \ldots, \beta_n$ are $n + 1$ unknown parameters, β_0 is regression constant, $\beta_1, \beta_2, \ldots, \beta_n$ are regression coefficients, and ε is random error.

Sample data are selected from the annual reports of Liaoning Province from 2011 to 2018, including population data, GDP data, PCDI data, regional data and annual sales data of lottery. The selection of independent variables is realized by drawing scatter plots and fitting curves. In the model, $X1$ is population data, $X2$ is GDP data, $X3$ is PCDI data, $X1$-$X3$ is independent variable, Y is annual sales data of lottery, and is the target variable.

Table 1 lists the evaluation criteria for each fitting curve. The fitting degree R^2 of each fitting curve is greater than 0.45, which proves that the fitting effect of the model is good. Therefore, all three external factors are taken as explanatory variables.

Table 1. Fitting curve information

Column	Row	R^2	Root Mean Squared Error	P Value (Significance)
Y	X1	0.762853	18557.789	<0.0001
Y	X2	0.813275	1.647	<0.0001
Y	X3	0.451283	288228.719	<0.0001

The significance P value of each fitting curve in Table 1 is less than 0.05. The multivariate linear regression models of the above three variables are established by using the ordinary least squares (OLS) and the generalized least squares (GLS), which are model 1 and model 2, respectively. The goodness of fit is obtained as shown in Table 2. The table lists R value, R^2 value, adjusted R^2 value and standard error. The fitting degree of the two models is $R^2 = 91.2\%$. It shows that the fitting effect of the multivariate linear regression model obtained by the least square method is better.

Table 2. Goodness of fit test

Model	R	R^2	adjustedR^2	Standard Error
Model1	0.954	0.912	0.910	11284.494
Model2	0.954	0.912	0.909	11284.991
a. Predictors: (constants), X1, X2, X3				
b. Dependent variable: Y				

Table 3(a) and (b) are regression coefficients, standard deviations and T values obtained by using OLS and GLS models respectively. Combining the above information, the regression model of OLS is as follows:

$$Y = -41160 + 95.6518X1 + 0.0006X2 + 19280X3 \tag{2}$$

The regression model of GLS is as follows:

$$Y = -40650 + 95.0065X1 + 0.0006X2 + 19110X3 \tag{3}$$

Table 3. Least square estimation results

(a) OLS coefficient, standard deviation, t value and VIF

Coefficient[a]

Coefficient	B	Standard deviation	t value	VIF
Constant	−41160	5811.957	−7.081	
X1	95.6518	12.996	7.360	4.566
X2	0.0006	0.000	5.184	5.319
X3	19250	2058.506	9.351	1.416

(b) GLS coefficient, standard deviation, t value and VIF

Coefficient[a]

Coefficient	B	Standard deviation	T value	VIF
Constant	−40650	5889.699	−6.903	
X1	95.0665	13.072	7.272	4.587
X2	0.0006	0.000	5.198	5.348
X3	19110	2076.837	9.203	1.146

2.2   Ridge Regression Analysis Model Design

R. Frisch put forward the concept of multicollinearity in 1934, which pointed out that in the linear regression model variables if there is high precision correlation makes the model distortion or lower accuracy [8]. Table 3 shows that the variance expansion factor VIF of the independent variables $X1$ and $X2$ of OLS and GLS models are greater than 4, which indicates that there is a certain degree of multi-collinearity between $X1$ and $X2$.

Hoerl first proposed Ridge Regression Analysis in 1962 and developed it systematically in cooperation with Kennard after 1970. Ridge regression analysis improves the least square method to solve the collinearity problem between independent variables in linear regression analysis [9]. Multivariate linear regression analysis model can be written as $Y = \beta X + \varepsilon$, and parameter estimation $\widehat{\beta} = (X'X)^{-1}X'Y$ can be obtained by least square method. When $|X'X| \approx 0$, there are multiple collinearities between independent variables. Ridge regression analysis adds a normal number matrix αI to the function of least squares method, where $\alpha > 0$, I is a unit matrix, the estimator of ridge regression is obtained as follows:

$$\widehat{\beta}(\alpha) = (X'X + \alpha I)^{-1}X'Y \tag{4}$$

Formula 4 shows that the estimation result of ridge regression is that of least square method when α approaches zero; the effect of collinearity will gradually decrease with the increase of α, and the ridge regression estimation result is 0 at $\alpha \to \infty$, so α can not be too large.

The regression coefficients of different α in ridge regression analysis are shown in Table 4. Set the range of α to [0.1,1.9], and increase the step size by 0.2 each time.

From the data in the table, we can see that there is a certain collinearity between the dependent variables *X1* and *X2*, but it has little influence on the model.

Table 4. Choosing the variation of different α regression coefficients

α	X1	X2	X3	R^2
0.1	0.4861	1.1218e-03	6.1920e+03	0.87159954
0.3	0.4868	1.1216e-03	6.1841e+03	0.87159952
0.5	0.4876	1.1213e-03	6.1762e+03	0.87159948
0.7	0.4883	1.1211e-03	6.1684e+03	0.87159941
0.9	0.4890	1.1208e-03	6.1606e+03	0.87159931
1.1	0.4897	1.1205e-03	6.1528e+03	0.87159917
1.3	0.4906	1.1202e-03	6.1451e+03	0.87159901
1.5	0.4912	1.1120e-03	6.1373e+03	0.87159881
1.7	0.4919	1.1197e-03	6.1296e+03	0.87159859
1.9	0.4926	1.1195e-03	6.1219e+03	0.87159834

Fitting degree of ridge regression adjustment $R^2 = 0.872$, the ridge regression equation is as follows:

$$Y = -19645 + 0.4861X1 + 0.00012X2 + 6121.9X3 \tag{5}$$

2.3 Experimental Result

Table 5 is a comparison table of fitting degree between OLS model, GLS model and ridge regression model. The comparison shows that the fitting degree of OLS and GLS model is higher than that of ridge regression model. And the prediction accuracy is very high, and the prediction results are very ideal, which shows that these two models have high practical value in predicting the sales of lottery through macro factors.

Table 5. Comparison table of fitting degree of three model

Model	R2
OLS	91.2%
GLS	91.2%
ridge regression	81.8%

3 Micro-Factor Analysis

Micro-factor analysis refers to the analysis of the impact of weather, holidays and sports events on welfare lottery daily sales. The daily sales data of lottery tickets not only have a large amount of data, but also fluctuate greatly, so we cannot directly apply linear regression model in this case, which motivates us to select decision tree algorithm for analysis. Decision tree algorithm can not only construct classifier efficiently and solve the problem of large amount of data calculation, but also indirectly prove whether the factors affecting the daily sales of lottery tickets are considered comprehensively and reasonably. More importantly, the classification results of decision tree are relatively understandable, and the potential law represented in this model can provide some objective basis for optimizing the daily sales management of lottery tickets.

3.1 Cart

Decision tree is a kind of tree structure similar to flow chart. The top level of the tree is the root node. Each branch represents an output of the feature or attribute, and each leaf node stores a classification mark [10]. Decision tree learning algorithm usually chooses the best feature recursively, and divides the training data according to the feature, so that each sub-data set has the best classification process. Here, the optimal feature or attribute is calculated by the branch metric function. There are many kinds of branch metric functions, the most common Gini and Entropy.

CART decision tree is a binary decision tree, which can be used for classification or regression [11]. We use CART classification tree when the predicted result is discrete and use CART regression tree when the predicted result is continuous. The attribute with minimum Gini and Entropy is used as the optimal branch attribute in CART classification tree. The more the Gain and Gini values are, the better the effect of this attribute as split attribute is.

For sample S, the Gini calculation of CART classification tree is as follows:

$$Gini(S) = 1 - \sum P_i^2 \tag{6}$$

where p_i is the frequency of the first category appearing in the classification result of sample set S.

For sample D, which can be divided into m classes, the Entropy of CART classification tree is calculated as follows:

$$Info(D) = - \sum_{i=1}^{m} p_i \log_2(p_i) \tag{7}$$

where p_i is the probability of the first category appearing in the classification result of sample set D, and $\log_2(p_i)$ is the probability of choosing the classification. The information Gain can be calculated by formula 7. According to the value of an attribute, S is divided into n subsets, and the information gain is calculated as follows:

$$Gain(S) = Info(S) - \sum_{i=1}^{n} Info(S_i) \tag{8}$$

The mean square error (MSE) and mean absolute error (MAE) are used as attribute branch metric in CART regression tree. For data sets with n records, the MSE formula is expressed as follows:

$$MSE = \frac{1}{n} \sum_{i=1}^{n} (y_i - \hat{y}_i)^2 \tag{9}$$

where n is the total number of data, y_i is the actual value, and $\hat{y}$ is the predicted value. For data sets with n records, MAE calculations are as follows:

$$MAE = \frac{1}{n} \sum_{i=1}^{n} |y_i - \hat{y}_i| \tag{10}$$

where n is the total number of data, y_i is the actual value, and $\hat{y}$ is the predicted value.

3.2 Model Data Processing

Sample data were selected from the daily sales data of Heping District and Jianping County from 2017 to 2018. Heping District is the urban area of Shenyang City, as well as Jianping County is a county subordinate to Chaoyang City, where agriculture and animal husbandry account for nearly half of the total economic output. Chaoyang City and Shenyang City are two prefecture-level cities under Liaoning Province. The original data include date, regional grade, weather condition, holiday, date for lottery drawing, sports events, total sales of districts and counties. After data processing, the highest daily temperature, average temperature, maximum wind power, average wind power, holidays, date for lottery drawing and sports events are input vectors. The percentage value of daily sales of districts and counties to annual average value of districts and counties is the forecast target. We utilize CART regression tree and CART decision tree simultaneously and divide the percentage of daily sales of districts and counties to annual average of districts and counties into eight intervals: minimal, very small, small, slightly small, slightly large, large, very large and maximal, with a difference of 10% between adjacent intervals.

3.3 Experimental Design and Result

In experiment 1, We utilize CART regression tree analysis to predict the grade of percentage value of daily sales of districts and counties. The decision tree is pruned by controlling the depth of the decision tree, the number of nodes and the size of leaves. The tree depth is controlled by setting the maximum depth (max_depth) and the minimum sample number of nodes (min_samples_leaf). To evaluate the effectiveness, we adopt the evaluation methodology and measurement MSE and MAE. The prediction accuracy of attribute setting and different metrics is shown in Table 6.

Table 6. Prediction accuracy of CART regression tree

max_depth	min_samples_leaf	MSE score	MAE score
10	10	0.652648	0.660533
20	10	0.651752	0.651469
5	20	0.669257	0.669059
10	20	0.679681	0.665328
20	20	0.679681	0.665328
5	30	0.670415	0.670860
10	30	0.673488	0.671653
20	30	0.673488	0.671653

As can be seen from Table 6, the highest accuracy of MSE and MAE is 0.679681 and 0.671653 respectively when different attribute values are set. The corresponding prediction model can be used for reference.

In experiment 2, we utilize CART classification tree to predict the grade of percentage value of daily sales of districts and counties. Gini and Entropy were used as two attribute measurement methods. The tree depth is controlled by setting the maximum depth (max_depth) and the minimum sample number of nodes (min_samples_leaf). The experimental accuracy under different attribute values and different metrics is shown in Table 7.

Table 7. Prediction accuracy of CART classification tree

max_depth	min_samples_leaf	Gini score	Entropy score
10	10	0.477352	0.456446
20	20	0.477352	0.445993
5	5	0.452962	0.452962
10	10	0.487805	0.459930
20	20	0.487805	0.459930
5	5	0.477352	0.456446
10	10	0.477352	0.480836
20	20	0.477352	0.480836

As can be seen from Table 7, the maximum accuracy of Gini and Entropy is 0.487805 and 0.480836 respectively when max_depth and min_samples_leaf are set differently.

Different from other prediction models, Decision tree algorithm can generate understandable rules. To better understand and utilize the rules, the following 2 interesting rule explanations are given found by decision tree algorithm:

(a) Node #23 (gini = 0.333). The rule found by the algorithm is that when award_date is less than 1.5, region_grade > 2.0 and avg_temp > 15.75, the proportion of daily sales to annual sales is large. It means that Jianping County (on behalf of county districts) sells 110% to 120% of the annual average daily sales when the average temperature is over 15.75 C on the non-lottery-drawing date. Non-lottery-drawing days are also easier to sell in rural areas when temperatures are high.

(b) Node # 42 (gini = 0.533), the rules found by the algorithm are: when award_date > 1.5, region_grade < 2.0, Holiday > 1.0, high_wind < 4.5 and high_wind < 3.5, the proportion of daily sales to annual sales average is very small. It means that the daily sales of Heping District (representing the urban area) are less than 70% of the average annual sales when lottery is drawing, vacation, maximum wind power is greater than 3.5 grade. It can be understood that if the wind is too strong in the city, lottery tickets will sell very badly even at lottery-drawing day of the holidays.

4   Conclusion

In this paper, we utilize several different regression models to analyze the macro factors of welfare lottery annual sales. The experiment proves the effectiveness of proposed model, of which the prediction accuracy reaches 91%, and the results demonstrate that there is a positive linear relationship between population, GDP, per capita disposable income and annual sales of lottery. In addition, we utilize several different CART decision tree models to analyze the sales volume of welfare lottery daily sales. The accuracy of the regression tree algorithm reaches 67%. we can also observe that some cases, such as bad weather, have completely opposite effects on sales in different areas of cities and counties. These discovery rules are not only interesting, but also provide objective basis for the selection of welfare lottery sales management strategies.

Acknowledgement. This work was partially supported by the National Natural Science Foundation of China under Grant (Nos. U1811261, 61602103).

References

1. Dajuan, J.: Characteristic analysis of welfare lottery game archives. China Civ. Aff. **630**(09), 37–38 (2018)
2. Ning, G.: Computer network security management in welfare lottery management. Sci. Technol. Econ. Market **01**, 192–194 (2018)
3. Yang, L.: Optimized analysis of the competitive strategy of welfare lottery issuance management in China. Money China (Academic Edition) **05**, 41–42 (2018)
4. Gang, L., Yangzhi, L.: Research on developing strategies for the Toto in China. J. Sports Sci. **38**(9), 21–36 (2018)
5. Haipeng, C., Xuwang, L., Xuanjing, S., et al.: Analysis and prediction on rebar price based on multiple linear regression model. Comput. Sci. **44**(B11), 61–64 (2017)
6. Xi, Y., Zhuang, X., Wang, X., Nie, R., Zhao, G.: A research and application based on gradient boosting decision tree. In: Meng, X., Li, R., Wang, K., Niu, B., Wang, X., Zhao, G. (eds.) WISA 2018. LNCS, vol. 11242, pp. 15–26. Springer, Cham (2018). https://doi.org/10. 1007/978-3-030-02934-0_2
7. Chuanze, L., Longxian, C., Dawei, L., et al.: Defect recognition of wood-based panel surface using pruning decision tree. Comput. Syst. Appl. **27**(11), 168–173 (2018)

8. Hongliang, Y., Xiaoqin, M., Hao, W., et al.: Stock market trend prediction algorithm based on morphological characteristics and causal ridge regression. Comput. Eng. **42**(02), 175–183 (2016)

9. Xiangdong, Z., Kun, W., Hui, Q.: Fault tolerant regression model for sensing data. Comput. Sci. **43**(02), 140–143 (2016)

10. Gang, L., Wei, J.: Research on Parkinson UPDRS prediction model based on GBDT. Comput. Technol. Dev. **29**(01), 216–220 (2019)

11. Breimann, L.J., Friedmann, R.O., Sone, C.: Classification and Regression Tress. Wadsworth, Belmont (1984)

Information Retrieval

An Inflection Point Based Clustering Method for Sequence Data

Ying Fan[1,2(✉)], Yilin Shi[1,2], Kai Kang[1,2], and Qingbin Xing[1,2]

[1] Administration Residency Research Center of the Ministry of Public Security PRC, Beijing 100070, China
13910088751@139.com
[2] Nanjing Fiberhome StarrySky Co., Ltd., Nanjing 210019, Jiangsu, China

Abstract. With rapid development of information technology, large amount of sequential data has been accumulated. How to extract business-related knowledge from sequential data set has become an urgent problem. The data pattern of sequential data is similar to that of traditional relation data, but there is temporal association between different attributes. It makes the traditional distance measurement method fails, which leads to the difficulty of existing clustering methods to apply. Aiming at above problems, the concept of data inflection point is introduced, and a method of measuring the dissimilarity of inflection points is proposed to realize inflection points marking, leveraging density-based clustering. Further, data sequence clustering method, called DSCluster, is proposed by applying frequent item set mining algorithm Apriori to the marked inflection point sequence set. Theoretical analysis and experimental results show that DSCluster method can effectively solve the problem of sequence data clustering. It can fully take into account the sequence of data related features extracting, the algorithm is effective and feasible.

Keywords: Sequential data · Inflection point · Inflection point similarity · Sequence pattern

1 Introduction

With the construction and improvement of population management business systems at all levels, a large amount of business data is stored in the system. Sequence data is an important part of the system. Sequence data has attracted much attention for its fixed pattern structure, the data contains temporal associations and complex relationships. The data of birth, death, immigration, and migration in each province and city in the country is typical sequence data. How to mine and analyze the sequence data has become an important problem to be solved in the application of household registration management big data. In recent years, sequence data analysis has attracted the continuous attention of researchers [1–3].

As the basic function of data mining [4], clustering plays an important role in the research and application of data mining. In terms of clustering mining, researchers have proposed a series of clustering algorithms, however, existing clustering algorithms are mainly based on the traditional relational table data and use Euclidean distance to

© Springer Nature Switzerland AG 2019
W. Ni et al. (Eds.): WISA 2019, LNCS 11817, pp. 201–212, 2019.
https://doi.org/10.1007/978-3-030-30952-7_22

measure the differences between individual data. Although the sequential data is stored in a relational table structure, the sequential data is more concerned with the evolutionary characteristics of the attribute values at different times. Directly applying the traditional clustering algorithm based on Euclidean distance to the sequential data set will lead to the following deficiencies:

(1) The association among attribute dimensions in individual data sequence was separated.
(2) The distance measurement function lacks the ability to measure the similarity and heterogeneity of individual records from the perspective of individual record change trend.

According to the characteristics of sequence data structure, this paper introduces the definition of inflection point, and puts forward the calculation method of sequence data difference based on inflection point pattern. On this basis, the idea of density clustering is used to realize the classification and annotation of inflection points of sequence data, and a data sequence clustering method based on frequent item sets called DSCluster is further proposed to realize the mining and extraction of Sequence Data.

The main contributions of the paper include:

(1) The concept of slope of sequence data is introduced, and the method of extracting inflection point of data sequence based on slope is proposed to transform data sequence to inflection point sequence.
(2) A method for calculating the similarity of inflection points is proposed, and a density clustering method is used to classify and discretize the inflection points.
(3) Frequent item set mining method is used to realize frequent inflection point set mining of annotated inflection point sequence set and discovery of clustering pattern of sequence data.

The organizational structure of the paper is as follows: The Sect. 2 introduces the related work; The Sect. 3 describes the problem and introduces relevant concepts and definitions; In Sect. 4, the similarity measurement method of inflection points is proposed, and density clustering is used to realize the classification of inflection points; The Sect. 5 designs experiments to verify the effectiveness of the proposed method; Sect. 6 summarizes the full text and looks forward to the follow-up work.

2   Related Work

Clustering [5–12] is a basic function of data mining. Its purpose is to divide a large number of unknown annotated data sets into multiple different categories according to the data characteristics existing within the data, so that the data within the categories are relatively similar and the data similarity between the categories is relatively small. Clustering belongs to unsupervised learning. For more than ten years, researchers have proposed a series of clustering methods, including clustering algorithm based on partition, hierarchy, density and more. Density-based clustering is widely used for its ability to find clusters of arbitrary shapes and its insensitivity to noise data. DBSCAN algorithm is the representative algorithm [7]. The algorithm introduces the definition of core point from the perspective of data distribution in the domain of data objects, and extends the concept of density linkage, and describes data clustering as the maximum set of data objects with density linkage.

The focus of clustering algorithm is to calculate the similarity between sample data objects, also known as the distance between samples. Generally, appropriate similarity measurement methods are selected according to the pattern characteristics of data objects. The commonly used distance measurement methods include Hamming distance, Manhattan distance, Euclidean distance, Jacard coefficient cosine similarity distance, entropy-based KL distance [13] and so on.

Compared with Euclidean distance, Hamming distance and Manhattan distance are not robust enough in similarity measurement, which is likely to result in the amplification or erasing of individual attribute differences. Euclidean distance, as the most commonly used data object similarity measurement method, has been widely used, but it is mainly applicable to the relational data with numerical attribute type, fixed mode and moderate attribute number. Jacard coefficient is mainly used to measure the similarity of category attribute data objects. Cosine similarity distance, the vector is drawn into the vector space according to the coordinate value, and their similarity is evaluated by calculating the cosine value of the Angle between two vectors. It is mainly applicable to the calculation of text similarity. Two word vectors representing the text are established based on the word cutting results of two texts, and the cosine values of these two vectors are calculated to represent the similarity of the two texts.

The existing data distance measurement method does not meet the requirements of sequence data similarity measurement. Sequential data can be regarded as a special relational structure, and the attribute scale of sequential data is much larger than that of conventional relational data. Taking population data analysis as an example, if the monthly population birth data of a city are collected for two consecutive years, the attributes of generating sequence data reach 24, and the generation mechanism of each attribute is exactly the same. Common measurement methods such as Euclidean method are simply adopted to measure the similarity of different sequence data (corresponding to different enterprise energy use), which is difficult to characterize the internal evolution characteristics of the sequence. Due to the large attribute scale, there is a hidden danger that may cause dimensional disaster [14].

Therefore, it is necessary to establish a method for feature preprocessing and extraction of sequence data by combining the structural characteristics of sequence data pattern and the presentation and representation forms of internal features of sequence data. On this basis, a similarity measurement mechanism of key features of sequence data is constructed to support the clustering and mining analysis of sequence data.

Grouping approaches provide an abstract graph view that preserves the privacy of sensitive graph components, the major drawback is that the graphical structure of the workflow and the dependencies among the modules are destroyed, and it is even possible to add dummy dependencies.

3 Problem Statement and Definitions

Although sequence data can also be stored in the relational table mode, the main difference between them and conventional relational data is that each attribute field of sequence data has the same generation mechanism, and each field has continuity in time, and the characteristics shown by each field are similar. For example, the monthly product production of a typical manufacturer, the monthly birth data of each region, and the embedded/moved data are all typical sequence data. The sequence data set is defined as follows:

Definition 1. Sequence data. The form of sequence data is as follows: sd = {s_1, s_2, .., s_m}. Where m is a natural number, representing the number of sampling points of sequence data sd, s_i is real number ($1 \leq i \leq$ m). The value corresponding to the i-th sampling point.

Definition 2. Sequential data set. For a given number of sampling points m, the sequence data set S is defined as follows: S = {sd_1, .., sd_n}, where sd_i is the sequence data with the number of sampling points m.

Each sampling point in the sequence data describes the same type characteristic values of the objective object. The sequence data set S is regarded as a relational table containing n records, with m attributes for each record, which uses Euclidean distance to measure the distance between any two sequences as the analysis basis of sequence similarity or dissimilarity. Note that the characteristic values trend of different sampling points in the sequence, as well as the similarity and correlation of the trend characteristics of different sequence data, are more important for the analysis and application of sequence data. To characterize this trend, the sequence point slope and sequence inflection point are defined as follows:

Definition 3. Sequence point slope. A sequence sd with a given number of sampling points m, for the sampling point s_i ($1 \leq i \leq$ m − 1) in sd, the slope k(i) of the sequence sampling point si is defined as follows:

(1) If si <> 0, then k(i) = $(s_i + 1 - s_i)/s_i$.
(2) Else, k(i) = MAX, where MAX is the upper limit value set.

Definition 4. Sequence inflection point. Given the sequence sd with the number of sampling points m, if the slope of adjacent sampling points i and i + 1 in the sequence satisfies one of the following conditions:

(1) k(i) * k(i + 1) > 0, and $|k(i+1)/k(i)| > 1+\varepsilon$ or $|k(i)/k(i+1)| > 1+\varepsilon$, where $\varepsilon \in (0, 1)$.
(2) k(i) * k(i + 1) ≤ 0.

Then, the i + 1 sampling point of sequence sd is the inflection point, denoted by bp = {i + 1, k(i), k(i + 1)}.

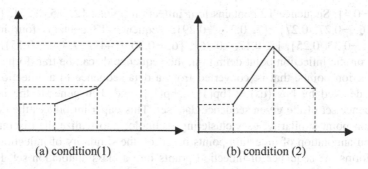

(a) condition(1)   (b) condition (2)

Fig. 1. Inflection point

The inflection point corresponds to the point in the sequence data where the value characteristics of sampling points change significantly. The inflection point satisfying the condition (1) corresponds to the situation shown in Fig. 1(a). At this point, both the slope $k(i)$ and $k(i + 1)$ are positive or negative, and the change rate satisfying the adjacent slope changes significantly. Parameter ε can be used to adjust the strength of the abrupt feature degree of the value of the sampling point within the sequence data. The larger the value of x is, the stronger the constraint requirement of inflection point is. The inflection point satisfying condition (2) corresponds to the case in Fig. 1(b). The value of slope $k(i)$ is opposite to that of $k(i + 1)$. Obviously, at the $i + 1$ sampling point, the sequence data changes significantly.

Table 1. Sequential data

	1	2	3	4	5	6	7	8	
T_1	10	12	8	11	14	12	15	9	...
T_2	6	9	11	13	15	11	14	10	...
T_3	12	8	10	12	14	9	13	15	...

According to the definition of the slope of the sequence point, it is easy to calculate the corresponding slope of the two sequences, as shown in Table 2.

Table 2. The slope of the sequence in Table 1

	1	2	3	4	5	6	7
T_1	0.2	−0.33	0.38	0.27	−0.14	0.25	−0.4
T_2	0.5	0.22	0.18	0.15	−0.27	0.27	−0.29
T_3	−0.33	0.25	0.2	0.17	−0.36	0.44	0.15

If the inflection point parameter is 0.3, it is easy to find that the sequence T1 contains 6 inflection points according to the definition of the inflection point: {2, 0.2, −0.33}, {3, −0.33, 0.38}, {4, 0.38, 0.27}, {5, 0.27, −0.14}, {6, −0.14, 0.25},

{7, 0.25, −0.4}; Sequence T2 contains four inflection points: {2, 0.5, 0.22}, {5, 0.15, −0.27}, {6, −0.27, 0.27}, {7, 0.27, −0.29}; Sequence T3 contains four inflection points: {2, −0.33, 0.25}, {5, 0.17, −0.36}, {6, −0.36, 0.44}, {7, 0.44, 0.15}.

Based on the inflection point definition, the sequence ds can be transformed into a set of inflection points, that is, converted from a data sequence to a inflection points sequence, denoted by bps (ds) = {bp(1), .., bp(l)}, and then generate the inflection point sequence set of the given sequence data set. This paper introduces the definition of inflection point similarity, uses clustering technology to realize the discrete stratification and annotation of inflection points based on the similarity of inflection points, and transforms the sequence of inflection points into a class annotation set. Frequent item set algorithm in data mining is used to mine patterns based on inflection points to realize pattern discovery of sequential data sets.

The general idea of the algorithm is shown in Fig. 2:

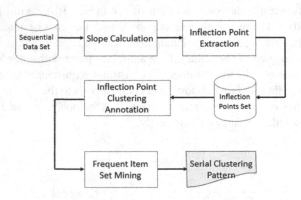

Fig. 2. Algorithm idea

The next section considers defining the measurement method of inflection point similarity from the perspective of inflection point pattern matching and proposes a method of mining sequence data pattern based on the characteristics of inflection point pattern.

4   Method

The difficulty in finding the clustering pattern of sequence data sets is that the internal timing of different individual sequences is dynamic, which is reflected in the internal inflection point characteristics of the sequence, and shows the difference of the rate of change of each inflection point and the duration.

DSCluster method mainly consists of extracting inflection points, discrete labeling inflection point sequence and frequent item set mining of inflection point sequence.

4.1   Inflection Point Classification Based on Inflection Point Similarity

Consider first measuring the similarity of different inflection points, so as to realize the conceptual layering and classification of inflection points from the inflection point variation feature angle.

The similarity for a given inflection point $bp_1 = \{i_x, k(i_x - 1), k(i_x)\}$ and $bp_2 = \left\{j_y, k\left(j_y - 1\right), k\left(j_y\right)\right\}$ is defined as follows:

Definition 5. Similarity of inflection point. The inflection point $bp_1 = \{i_x, k(i_x - 1), k(i_x)\}$, the inflection point $bp_2 = \left\{j_y, k\left(j_y - 1\right), k\left(j_y\right)\right\}$, the similarity is $Sbp(bp_1, bp_2)$, defines as follows: $Sbp(bp_1, bp_2) = a \times \left| \frac{k(i_x-1)-k(j_y-1)}{k(i_x-1)} \times \frac{k(i_x)-k(j_y)}{k(i_x)} \right| + b \times (1 - \frac{i_{x+1}-i_{x-1}}{l_{y+1}-l_{y-1}})$.

Where i_{x+1} is the sampling position of the nearest inflection point before bp_1 in the sequence, i_{x+1} is the sampling position of the nearest inflection point after bp_1 in the sequence. a, b $\in$ (0, 1), and a + b = 1, which is the weight.

The similarity between the two inflection points is mainly reflected in two aspects: the trend and magnitude of the slope before and after the inflection point are the same or different, and the duration of the corresponding inflection point. Therefore, the expression consists of two parts. The first item of the expression is the similarity of the slopes before and after the two inflection points, and the latter term corresponds to the similarity of the duration of the two inflection points. The smaller the overall value, the more similar the two inflection points are; otherwise, the more different the two inflection points.

Taking the sequence data described in Table 1 as an example, the inflection point parameters a = 0.5, b = 0.5, respectively calculate the similarity between the inflection point of T1{4, 0.38, 0.27} and the inflection point of T2{2, 0.5, 0.22}, {5, 0.15, −0.27}.

$$Sbp\left(\{4, 0.38, 0.27\}, \{2, 0.5, 0.22\}\right) = 0.5 \times \left| \frac{0.38 - 0.5}{0.38} \times \frac{0.27 - 0.22}{0.27} \right| + 0.5 \times (1 - \frac{5 - 3}{5 - 0}) = 0.329,$$

$$Sbp\left(\{4, 0.38, 0.27\}, \{5, 0.15, -0.27\}\right) = 0.5 \times \left| \frac{0.38 - 0.15}{0.38} \times \frac{0.27 - (-0.27)}{0.27} \right| + 0.5 \times (1 - \frac{5 - 3}{6 - 2}) = 0.855.$$

The similarity between the inflection point of the inflection point T1{4, 0.38, 0.27} and the inflection point of T2{2, 0.5, 0.22} is much greater than its similarity with the inflection point of T2{5, 0.15, −0.27}. This is consistent with the trend of the numerical distribution of the sequence data T1 and T2.

After generating the set of inflection points of S, extract the inflection points of the inflection point sequence to form an inflection point set:

$$IP(S) = \{p| \; p \text{ is the inflection point of the sequence ds, ds} \in S\}.$$

Density-based clustering of inflection points in IP(S) based on inflection point similarity (using DBSCAN algorithm [7], similarly defining core inflection point and density connection inflection point definition), the IP(S) is divided into a set of inflection point clusters, and the same inflection point features are marked on the inflection points belonging to the same cluster. Each sequence data in S is converted into an inflection point feature record in the order of the inflection points constituting the sequence. The process is as follows:

(1) Select the unlabeled inflection point $p \in$ IP(S). If p is the core inflection point, create a new cluster containing p and mark the cluster.
(2) Find all inflection points connected to p density and add them to the cluster to which p belongs.
(3) Repeat (1), (2) until all inflection points belong to a certain cluster.
(4) For ds $\in$ S, according to the order of the inflection points in ds, the inflection points are replaced by the clusters to which the inflection points belong, and the labeling sequence is formed.

Taking the data sequence of Table 1 as an example, assume that the 14 inflection points of T1, T2, and T3 are clustered into three clusters:

$C_1 = \{\{2, 0.2, -0.33\}, \{5, 0.27, -0.14\}, \{7, 0.25, -0.4\}, \{5, 0.15, -0.27\}, \{7, 0.27, -0.29\}, \{5, 0.17, -0.36\}\}$
$C_2 = \{\{2, 0.5, 0.22\}, \{4, 0.38, 0.27\}, \{7, 0.44, 0.15\}\}$
$C_3 = \{\{3, -0.33, 0.38\}, \{6, -0.14, 0.25\}, \{6, -0.27, 0.27\}, \{2, -0.33, 0.25\}, \{6, -0.36, 0.44\}\}$

Then T1, T2, T3 corresponding to the inflection point sequence can be expressed as $\{C_1, C_3, C_2, C_1, C_3, C_1\}$, $\{C_2, C_1, C_3, C_1\}$, $\{C_3, C_1, C_3, C_2\}$. The set of inflection points after forming the annotation is shown in Table 3:

Table 3. Inflection point sequence label set

T_1	C_1, C_3, C_2, C_1, C_3, C_1
T_2	C_2, C_1, C_3, C_1
T_3	C_3, C_1, C_3, C_2

4.2   Data Sequence Pattern Discovery Based on Frequent Patterns

Through the inflection point clustering based on the inflection point similarity, the classification of the inflection point is realized. The fine-grained difference between the data sequence and the change duration of the inflection point is described by a more abstract concept, which facilitates the extraction of subsequent sequence data clustering patterns.

Sequence data is different from common relational pattern data. Different individual records of relational pattern data have the same attribute pattern. The data pattern is mainly represented by the pattern of the overall statistical characteristics of the data set, and the pattern of the data sequence set is represented at two levels: (1) Within each individual data sequence, it exhibits temporal evolution characteristics; (2) Different individual data sequences may bring together some common feature patterns.

It is necessary to extract the pattern features of the annotation sequence data set from the above two levels. The annotation sequence data set shown in Table 3 has the characteristics of the transaction data set, and its pattern is expressed as the frequency and combination characteristics of different annotations in the individual sequence, which is consistent with the principle of frequent item set mining in data mining. Therefore, the classical algorithm Apriori method [4], which uses frequent item set mining, is applied to the cluster pattern discovery of the inflection point sequence set. The difference from the conventional Apriori method is that in the candidate frequent inflection point set generation process, the order relationship of each marked inflection point in the inflection point sequence needs to be maintained.

Data sequence pattern mining steps are as follows:

(1) Set the minimum support threshold for Apriori.
(2) Use the Apriori algorithm to obtain the frequent inflection point set C of the inflection point sequence set, and retain the longest frequent inflection point set in C (i.e., for x, y $\in$ C, if x is included in y, delete x).
(3) For each frequent inflection point set in C, the annotation is replaced by the average inflection point of the inflection point cluster corresponding to the annotation, and a cluster mode set of sequence data is formed.

Similarly, taking the sequence data shown in Table 1 as an example, the post-labeling inflection point sequence (sequence in Table 3) is used to perform frequent inflection point set mining using the Apriori algorithm. Assuming a minimum support degree of 0.6, a frequent inflection point set is generated as follows:

Frequent 2 inflection point set: C_1C_3, C_1C_2, C_1C_1, C_1C_3, C_3C_2, C_3C_1, C_3C_3, C_2C_1, C_2C_3, C_3C_1
Frequent 3 inflection point set: $C_1C_3C_1$, $C_1C_3C_2$, $C_3C_1C_3$, $C_2C_1C_3$, $C_2C_1C_1$, $C_2C_3C_1$
Frequent 4 inflection point set: $C_2C_1C_3C_1$

According to the frequent item set inclusion relationship, the final frequent inflection point set $\{C_2\,C_1\,C_3\,C_1, C_1\,C_3\,C_2, C_3\,C_1\,C_3\}$ can be obtained, corresponding to the pattern contained in the three sequence data.

5 Experimental Evaluation

In this part, the effectiveness of the proposed method is analyzed experimentally. Experimental data were obtained from UCI Knowledge Discovery Archive database (http://archive.ics.uci.edu/ml/datasets.html), combined with the application characteristics of the algorithm, the original data set is expanded to 10 and 20 dimensions respectively (use each record to randomly expand according to the trend of the original dimension value), and the basic information of the data is shown in Table 4. The experimental hardware platform is Intel(R) Core(TM) 2 CPU 1.80 GHz, 2 GB memory.

Table 4. Experimental data information

Data source	Name	Number of attributes	Number of records	Type
Blood Transfusion Data	DS1	10	748	Real
Gamma Telescope Data	DS2	20	1012	Real

Considering the essence of sequential pattern mining is to extract the typical group features of sequence data, each sequence pattern obtained by DSCluster mining corresponds to a sequence cluster. The density clustering algorithm DBSCAN is a clustering algorithm with good practicability. Therefore, the data sequence set is regarded as the regular relation pattern data, and the DBSCAN algorithm is used to cluster the data set. The clustering effect of DSCluster and DBSCAN is compared to verify the effectiveness of the proposed DSCluster method. The clustering results are compared with the F-measure index [15]. The larger the F-measure value, the better the clustering mode effect. In the experimental part, the parameters of the DBSCAN algorithm are set by the sampling data fitting method, and the minimum support threshold minsup of the Apriori algorithm is set to 0.3.

For the data sets DS1 and DS2, the DBSCAN clustering algorithm and the DSCluster algorithm are respectively run. As can be seen from Fig. 3, the F-measure value of the algorithm DSCluster is higher than the DBSCAN algorithm, especially for the data set DS2, the clustering effect of DSCluster is significantly better than the DBSCAN method. The reason is that the DSCluster algorithm extracts the inflection point features from a single data record (sequence) and measures the distance between records from the perspective of the feature similarity between the data records (sequences), which is insensitive to the data set attribute dimension, while the DBSCAN algorithm uses the Euclidean distance. As the dimension of the data set increases, the data distribution is sparse and the clustering effect is worse.

Figure 4 shows the clustering effect of the DSCluster method when ε is selected for different values. It is easy to see from Fig. 4. As the ε increases, the clustering effect of the algorithm increases gradually. When ε is 0.3, the F-measure reaches the maximum value. As the ε increases, the F-measure value decreases. The reason is that the ε adjustment controls the strength of the inflection point feature constraint. The larger the value, the more severe the inflection point constraint. The value is too small, the slope of the data change before and after the inflection point is not significant, resulting in a large number of non-inflection points being misjudged, and the value of ε is too large, the inflection point condition is too harsh, so that the true inflection point in the sequence is missed. It will also cause errors in the sequence data distance metric, resulting in a decrease in cluster quality.

Figure 5 analyzes the trend of the clustering effect of the proposed DSCluster method with the minimum support threshold minsup of the Apriori algorithm parameters. It is easy to verify that the larger or smaller minsup setting may cause the loss of cluster quality found in the sequence mode, because the minsup parameter is set too small, causing some non-representative features to be misjudged as frequent patterns; Too large, the frequent inflection point set condition is too harsh, which will cause some real sequence patterns to be misclassified as infrequent mode, resulting in the sequence pattern acquired by the mining does not conform to the true distribution of the data sequence. In practical applications, the sample inflection point sequence data can be used to analyze the inflection point statistical value method, and set the minimum support degree threshold parameter.

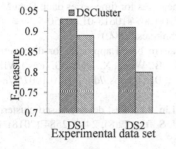

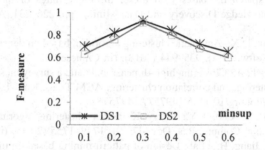

Fig. 3. Convergence quality comparison analysis (ε takes 0.3)

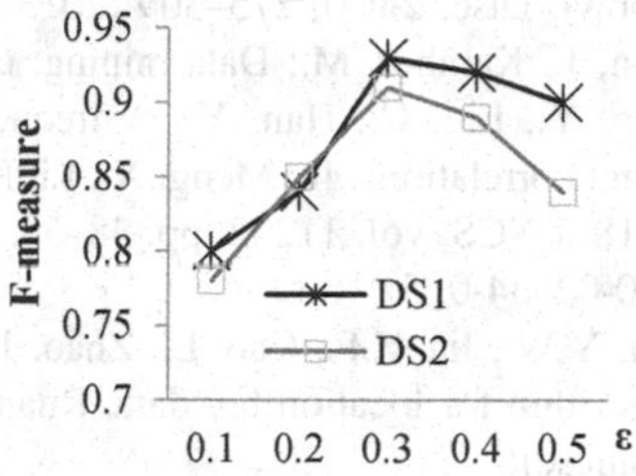

Fig. 4. The clustering effect of DSCluster algorithm changes with ε

Fig. 5. *DSCluster* clustering effect changes with *minsup*

6   Conclusion

The data object similarity measurement method based on Euclidean distance is difficult to adapt to the number of sequence data attributes and the time series correlation between attributes. This paper introduces the concept of data inflection point and designs the inflection point similarity measurement method. On this basis, the data sequence pattern mining method DSCluster is proposed, and the density clustering algorithm is used to realize the classification of the sequence data inflection point. For the category-type inflection point sequence, the frequent item set mining algorithm Apriori is used to implement data sequence pattern mining. The proposed method can effectively support the extraction of feature data of sequence data and realize the analysis of sequence data mining.

The next step will be to do some work: (1) Apply the proposed method to population big data applications to support business decision analysis (2) Combining the proposed method, study the pattern mining method of position trajectory sequence data.

References

1. Chen, Y.L., Wu, S.Y., Wang, Y.C.: Discovering multi-label temporal patterns in sequence databases. Inf. Sci. **181**(3), 398–418 (2011)
2. Wang, X., Mueen, A., Ding, H., Trajcevski, G., Scheuermann, P., Keogh, E.: Experimental comparison of representation methods and distance measures for time series data. Data Min. Knowl. Disc. **26**(2), 275–309 (2013). https://doi.org/10.1007/s10618-012-0250-5
3. Han, J., Kamber, M.: Data mining: concepts and techniques (2000)
4. Cao, Y., Liu, C., Han, Y.: A frequent sequential pattern based approach for discovering event correlations. In: Meng, X., Li, R., Wang, K., Niu, B., Wang, X., Zhao, G. (eds.) WISA 2018. LNCS, vol. 11242, pp. 48–59. Springer, Cham (2018). https://doi.org/10.1007/978-3-030-02934-0_5
5. Yu, Y.W., Jia, Z.F., Cao, L., Zhao, J.D., Liu, Z.W., Liu, J.L.: Fast density-based clustering algorithm for location big data. Ruan Jian Xue Bao/J. Softw. **29**(8), 2400–2584 (2018). (in Chinese)
6. Kaufman, L., Rousseeuw, P.J.: Finding groups in data: an introduction to cluster analysis. In: DBLP (1990). http://dblp.org/rec/books/wi/KaufmanR90.html. https://doi.org/10.1002/9780-470316801
7. Ester, M., Kriegel, H.P., Sander, J., Xu, X.: A density based algorithm of discovering clusters in large spatial databases with noise. In: Proceedings of the 2nd International Conference on Knowledge Discovery and Data Mining, pp. 226–231. AAAI Press, Menlo Park (1996)
8. Chen, Z.W., Chang, D.X.: Automatic clustering algorithm based on density difference. Ruan Jian Xue Bao/J. Softw. **29**(4), 935–944 (2018). (in Chinese)
9. Kriegel, H.P., Zimek, A.: Clustering high-dimensional data: a survey on subspace clustering, pattern-based clustering, and correlation clustering. ACM Trans. Knowl. Disc. Data **3**(1), 1–58 (2009). https://doi.org/10.1145/1497577.1497578
10. Liang, B.Y., Bai, L., Cao, F.Y.: Fuzzy K modes clustering algorithm based on new dissimilarity measure. Comput. Res. Dev. **47**(10), 1749–1755 (2010). (in Chinese)
11. Zhang, K., Li, C., Jiang, H., et al.: Design of pattern-mining based on time series. Comput. Eng. Appl. **51**(19), 146–151 (2015). (in Chinese)
12. Ni, W.W., Cheu, G., Wu, Y.J., Sun, Z.H.: Local density based distributed clustering algorithm. J. Softw. **19**(9), 2339–2348 (2008). (in Chinese)
13. Tan, P.-N., Steinbach, M., Kumar, V.: Introduction to Data Mining. Addison-Wesley, Boston (2005)
14. Beyer, K., Goldstein, J., Ramakrishnan, R., Shaft, U.: When is "nearest neighbor" meaningful? In: Beeri, C., Buneman, P. (eds.) ICDT 1999. LNCS, vol. 1540, pp. 217–235. Springer, Heidelberg (1999). https://doi.org/10.1007/3-540-49257-7_15
15. van Rijsbergen, C.J.: Information Retrieval, 2nd edn. Butterworths, London (1979)

Efficient Large-Scale Multi-graph Similarity Search Using MapReduce

Jun Pang[1,3](✉), Minghe Yu[2,5], and Yu Gu[4]

[1] School of Computer Science and Technology, Wuhan University of Science
and Technology, Wuhan 430065, Hubei, China
pangjun@wust.edu.cn
[2] Guangdong Province Key Laboratory of Popular High Performance Computers,
Shenzhen University, Shenzhen 518061, China
[3] Hubei Province Key Laboratory of Intelligent Information Processing
and Real-Time Industrial System, Wuhan 430065, Hubei, China
[4] School of Computer Science and Engineering, Northeastern University,
Shenyang 110819, Liaoning, China
guyu@mail.neu.edu.cn
[5] College of Software, Northeastern University, Shenyang 110819, Liaoning, China
yuminghe@mail.neu.edu.cn

Abstract. A multi-graph is a set consisting of multiple graphs. Multi-graph similarity search aims to find the multi-graphs similar to the query multi-graphs from the multi-graph datasets. It plays important role in a wide range of application fields, such as finding similar drugs, searching similar molecule groups and so on. However, existing algorithms of multi-graph similarity search are memory-based algorithms, which are not suitable for the large amount of multi-graph scenarios. In this paper, we propose a parallel algorithm based on the MapReduce programming model to solve the problem of the large-scale multi-graph similarity search. Our proposed algorithm consists of two MapReduce jobs, one for indexing and the other for filtering and validation. Specially, we adapt the localization strategy to further improve the performance of our algorithm, which not only reduces the communication cost, but also mitigates the load imbalance. Extensive experimental results show that our algorithm is effective and efficient.

Keywords: Multi-graph · Similarity · Large-scale · MapReduce

1 Introduction

A multi-graph is modeled as a set of several graphs [1–3]. It is a powerful model for describing the complicated structures of real world entities. For example, an image I_1 can be denoted as a multi-graph [1], as shown in Fig. 1. The target of multi-graph similarity search is to find out all multi-graphs similar to the query multi-graphs from multi-graph datasets [4]. The existing research works of the multi-graph mainly focus on the analysis rather than the similarity search, including the positive and unlabeled multi-graph classification [1], the supervised multi-graph classification [2,3,5,6] and the semi-supervised multi-graph classification [7]. However, the multi-graph similarity search has important practical

© Springer Nature Switzerland AG 2019
W. Ni et al. (Eds.): WISA 2019, LNCS 11817, pp. 213–225, 2019.
https://doi.org/10.1007/978-3-030-30952-7_23

value in finding drugs with similar properties, searching similar molecule groups, and so on [4]. It just has not attracted enough attentions in the field of graph data management, where the existing research works of graph similarity search mainly focus on two aspects: (1) finding all data graphs similar to the query graphs from a given collection of data graphs [8–13]; (2) exploring all substructures that satisfy the query conditions from a big graph [14–16].

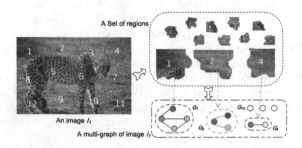

Fig. 1. Multi-graph representation of an image I_1. The image I_1 is regarded as a set of several regions. A region is represented as a graph [17], where a superpixel inside the region is denoted by a vertex, and the adjacency relationship between two superpixels is represented as an edge. A superpixel denotes a cluster of pixels.

The above two type of approaches cannot be directly used to solve the problem of multi-graph similarity search because of different structures. Specially, the research objects of above works are graphs, whose dissimilarities are measured by the graph edit distance [8,9] or distance metric based on max common subgraphs. Based on these dissimilarity metrics, efficient indices and filter conditions are proposed to reduce the search spaces. Because the multi-graphs have more complex structure than the graphs, the multi-graph similarity search needs different similarity measures, filters and indexes.

The works of the multi-graph similarity search are rarely reported. Pang et al. propose an efficient memory-based method, called MGSS, to solve the problem of multi-graph similarity search [4]. MGSS adopts the filtering-and-verify framework: it firstly calculates a small set of candidate results based on the proposed filters; then it verifies the candidate results to acquire the final query results based on the improved Kuhn-Munkras algorithm [4]. Furthermore, an incremental inverted index is proposed to speed up the query process. Extensive experimental results verify the effectiveness and efficiency of MGSS.

However, MGSS fails to process the large-scale multi-graph data because it is a memory-based algorithm. In real life, with the advent of the big data age, there are more and more modeled multi-graphs. As a mature parallel programming model, MapReduce is widely adopted to solve big data problems in academia and industry [18]. Therefore, we propose a parallel algorithm of MGSS based on MapReduce, called MR-MGSS, to solve the problem of the large-scale multi-graph similarity search. MR-MGSS consists of two MapReduce jobs, one for indexing and the other for filtering and validation. Experimental results show that MR-MGSS has two main shortcomings: (1) the communication costs of job1 are huge; (2) the load of the map tasks of job2 is very imbalanced. So we further adopt a localization strategy to improve MR-MGSS.

The main contributions of this paper are as follows:

1. Based on MGSS and the MapReduce programming model, a parallel algorithm MR-MGSS is proposed to solve the problem of large-scale multi-graph similarity search. In addition, the time complexity of MR-MGSS is analyzed.
2. A localization idea is utilized to enhance the MR-MGSS. Optimal MR-MGSS, i.e. MR-MGSS$^+$, not only reduces the communication cost, but also solves the load imbalance problem to some extent. Moreover, we compare the time complexities of MR-MGSS and MR-MGSS$^+$.
3. Extensive experimental results on large-scale synthetic data sets show that the MR-MGSS$^+$ algorithm is effective and efficient.

The rest organization structures of this paper are as follows. The related works are introduced in Sect. 2. The problem of the large-scale multi-graph similarity search is formalized in Sect. 3. The MR-MGSS algorithm based on MapReduce and MGSS is proposed in Sect. 4. Our MR-MGSS is improved in Sect. 5. In Sect. 6 we discuss the experimental results on multiple synthetic datasets, and verify the effectiveness and efficiency of our algorithm. At last, we conclude the whole paper.

2 Related Works

The related research works mainly include the multi-graph similarity search and analysis and distributed graph similarity query.

2.1 Multi-graph Similarity Search and Analysis

Many research works about the multi-graph analysis have been reported. Wu et al. [1] propose a learning framework, called puMGL, to solve positive and unlabeled multi-graph classification problem. Wu et al. [2,3] propose a approach gMGFL and a boosting approach to solve the problem of multi-graph classification. Pang et al. [5] propose a parallel approach based on MapReduce and the gMGFL to solve the problem of large-scale multi-graph classification. Pang et al. [7] propose a approach MGSSL to solve the problem of semi-supervised multi-graph classification. Pang et al. [6] expand the multi-graph model to the super-graph model, which contains more structure information.

There are few works about the problem of multi-graph similarity search. Pang et al. propose a distance of multi-graphs and the MGSS algorithm to solve the problem of multi-graph similarity search [4]. In order to reduce the searching space, many filters are conducted, including a number filter, a size filter, a complete edge filter and a lower bound filter. Moreover, the Kuhn-Munkras algorithm, i.e. KM, is improved to speed up the processing of verification. Also, a multi-layer inverted index is proposed to further improve the query efficiency.

This inverted index is incremental and can adapt to the dynamic changes of the data sets. However, the MGSS algorithm is not suitable for solving the problem of large-scale multi-graph similarity search because it needs to load all data into the memory.

2.2   Distributed Graph Similarity Query

Existing techniques of the distributed graph similarity query focus on a single big graph with a large scale of nodes and edges [19,21,22], e.g. the techniques of node similarity join [19], and the techniques of pattern matching [20]. As the best of our knowledge, techniques of the distributed query in a collection of very many graphs or multi-graphs hardly have been reported.

In literature [19], a new computing method of SimRank, called Delta-SimRank is proposed for the dynamic network. Delta-SimRank not only can satisfy these features of distributed computing, but also is suitable to solve the problem of node similarity join. Extensive experimental results show that Delta-SimRank is more efficient than SimRank. In literature [20], an algorithm based on Pregel is proposed to calculate the maximal matching of graph. Experimental results verify the scalability and efficiency of the proposed algorithm. Above techniques of similarity search for a big graph cannot be directly adopted to solve the problem of large-scale multi-graph similarity search because of the different structures of the multi-graphs.

3   Problem Statement

In this section, we formalize the problem of large-scale multi-graph similarity search.

Definition 1. *Multi-graph*
A multi-graph is a set of graphs represented as $MG = \{G_1, G_2, \cdots, G_{|MG|}\}$, where graph $G_i = (V_i, E_i, L_{G_i}, F_i)$, $1 \leq i \leq |MG|$. $|MG|$ denotes the number of the graphs contained by MG. MGS indicates a set of multi-graphs. For simplicity, MGS is called MS for short in the absence of ambiguity.

Definition 2. *Complete-bipartite-graph representation of a multi-graph pair*
A multi-graph pair (MG_1, MG_2) can be represented as a bipartite graph, called $CBG(MG_1, MG_2)$, where a vertex of X (or Y) denotes a graph of MG_1 (or MG_2) and an edge weight of $CBG(MG_1, MG_2)$ is the edit distance of two corresponding graphs.

Definition 3. *Distance between two multi-graphs*
The distance between MG_1 and MG_2 is the number of the minimum perfect match of CBG (MG_1, MG_2), called $GSD(MG_1, MG_2)$. If MG_1 is the same as MG_2, $GSD(MG_1, MG_2) = 0$. The larger are the differences between MG_1 and MG_2, the bigger is their $GSD(MG_1, MG_2)$.

Definition 4. *Large-scale multi-graph similarity search*
Given a large scale of multi-graph dataset MS, a multi-graph distance metric
$GSD()$, a query multi-graph MG_q, and a distance threshold τ, the target of large-
scale multi-graph similarity search is to find all multi-graphs whose distances with
MG_q are less than or equal to τ, i.e., $\{MG|GSD(MG, MG_q) \leq \tau, MG \in MS\}$.

4 Basic MR-MGSS Algorithm

In this section, we first briefly introduce the overview of MR-MGSS, and then
describe the main steps of MG-MGSS in detail.

4.1 Overview of MR-MGSS

The overview of MR-MGSS is shown as Fig. 2. At first, the first job *job1* reads
the multi-graph datasets from a hadoop distributed file system, i.e. HDFS, and
produce a multi-layered inverted index which is written into HDFS. Then the
second job *job2* reads the query graphs, multi-graph datasets and previous multi-
layered inverted index from HDFS, and responds to the query. In order to reduce
the search space, *job2* performs multiple filters proposed in MGSS, and verify
the candidate results to obtain the final results, which are written into HDFS.
The details of *job1* and *job2* are described in Sects. 4.2 and 4.3, respectively.

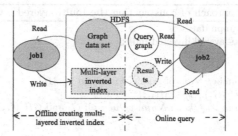

Fig. 2. Data flow illustration diagram of MR-MGSS.

4.2 Offline Creating Multi-layered Inverted Index

In MR-MGSS, a MapReduce job, i.e. *job1*, is utilized to construct a multi-layered
inverted index offline. The map and reduce functions are shown in Table 1. The
pseudocode of the algorithm of job1 is shown in Algorithm 1.

The map tasks of *job1* calculate the *number* of multi-graphs and the information which is needed to create the inverted index. The reduce tasks of *job1*
aggregate the outputs of the map tasks of job1 and produce the index. In detail,
a map task first catches a record of a multi-graph MG_i, calculates the Information Unit of MG_i, and then output a $< key, value >$ pair, where *key* is the

number of MG_i; value consists of the Information Unit and the content of MG_i. Reduce tasks divide the multi-graphs with the same *number* into groups which are written into HDFS. The Information Unit of MG is defined as the combination of the unique identifier of MG and the Information Units of all graphs contained by MG [4]. The Information Unit of a graph G is defined as a triple $GIU(G) =< graphId, size, branches >$, where $graphId$ denotes the identifier of G. Size indicates the sum of the numbers of the vertices and edges of G. A branch consists of a vertex of G and its adjacent edges.

Table 1. Map functions and reduce functions.

Job name	F	Input	Output
job1	M	$< id_{MG_i}, MG_i >$	$< Num_{MG_i},$ $MG_i.Info.Unit >$
	R	$< Num,$ $List(MG_i.Info.Unit) >$	$< Num,$ $List(MG_i.Info.Unit) >$
job2	M	$< Num,$ $List(MG_i.Info.Unit) >$	$< id_{MG_q}, (data_{MG_i},$ $GSD(MG_i, MG_q)) >$
	R	$< id_{MG_q}, List(data_{MG_i},$ $GSD(MG_i, MG_q)) >$	$< id_{MG_q}, List(data_{MG_i},$ $GSD(MG_i, MG_q)) >$

Algorithm 1: Approach of constructing the multi-layered inverted index

Input : Multi-graph datasets
Output: A multi-layered inverted index
1 //map phase:
2 Map(id, MG_i)
3   calculate the number Num of graphs contained by MG_i;
4   compute all $GIU(G_j)$, $G_j \in MG_i$;
5   emit($< key, value >$);//key : Num, value : all G_j and $GIU(G_j)$,
6 $G_j \in MG_i$.
7 //reduce phase:
8 Reduce($Num, List < MG_i.Info.Unit >$)
9   aggregate all graphs and Information Unit of multi-graphs with the same
   Num, which are written into HDFS.

Based on the cost model proposed in Literature [23], the total cost of a MapReduce job equals the sum of the cost of all mappers, the communication cost, and the calculating cost of all reducers, i.e., $Cos_t = Cost_m + Cost_c + Cost_r$, where $Cost_m$ denotes the cost of all mappers including the I/O and computing cost; $Cost_c$ indicates the total cost of communication, which is expressed in terms of the total amount of transmission; $Cost_r$ represents the cost of all reducers consisting of the calculating and I/O cost. Through analysis, for the *job1* of MR-MGSS, $Cost_m = O(|MS| + \sum_{j=1}^{MS} \sum_{i=1}^{MG_j}(|V_i|))$; $Cost_c = O(\sum_{j=1}^{MS} \sum_{i=1}^{MG_j}(|V_i|))$; $Cost_r = O(\sum_{j=1}^{MS} \sum_{i=1}^{MG_j}(|V_i|))$.

4.3   Online Query

MR-MGSS sets up another MapReduce job, i.e. $job2$, to answer the queries online. The map and reduce functions are also showed as Table 1. The pseudocode of the algorithm of $job2$ is described in Algorithm 2.

The map tasks of $job2$ read the inverted index as well as the query multi-graphs, and adopt the filter-and-verify framework to output query results, which are aggregated by the reduce tasks of $job2$ for returning the query results of each query multi-graph. Firstly, all query multi-graphs are read from HDFS and saved locally in each node in advance. Then, a map function is called to read a record, i.e. the corresponding index of MG_j, from HDFS every time and compare all the query multi-graphs to the index. If the comparison is successful, i.e. satisfying the number filter [4], other constraint conditions [4] are used to further filter. If all constraint conditions are met, the improved KM [4] is utilized to calculate the distance between the "query multi-graph" MG_{q_i} and the "data multi-graphs". If the calculated distance is not more than the threshold τ, a $< key, value >$ pair is exported, where key denotes the ID of MG_{q_i}; value represents ID of MG_j. Otherwise, noting is exported. For the reduce tasks, the query results for each query multi-graph are organized into a group, and written to HDFS.

Algorithm 2: Online query algorithm

   Input   : Multi-graph datasets and a multi-layered inverted index
   Output: Final query results
1 //map phase:
2 define $QMG = \emptyset$;
3 Setup( )
4   read the distance threshold ;
5   read and save the query multi-graphs to QMG from distributed caches ;
6 Map($Num, List < MG_i.Info.Unit >$)
7   **for** $\forall$ *a query multi-graph* $MG_j \in QMG$ **do**
8   **if** Num satisfies the number filter **then**
9   **for** $MG_i \in list < MGi.Info.Unit >$ **do**
10   **if** MG_i satisfies the size filter **then**
11   **if** MG_i satisfies the complete edge filter **then**
12   **if** MG_i satisfies the lower bound filter **then**
13   calculate $gsd_{ij} = GSD(MG_j, MG_i)$;
14   **if** $gsd_{ij} \leq \tau$ **then**
15   output pairs of $< key, value >$;//$key : id$ of MG_j,
                  $value : MG_i, gsd_{ij}$.

16 //reduce phase
17 Reduce($queryMG_j\_id, List < dataMG_i, gsd_{ij} >$)
18 aggregate and output the query results of the same query multi-graph to the HDFS.

Through analysis, for the $job2$ of MR-MGSS, $Cost_m = O(|MS| + |MAP| \times \sum_{j=1}^{|MS|} \sum_{i=1}^{|MG_j|}(|V_i|))$; $Cost_c = O(|Can|)$, where Can denotes the candidate result set; $Cost_r$ is the sum of the costs of calculating $|Can|$ multi-graph distances. Because the graph edit distance is a NP-Complete problem, the multi-graph edit distance is a NP-Complete problem, too.

5   Improved Algorithm MR-MGSS$^+$

Although MG-MGSS can answer the large-scale multi-graph similarity query, it has two main disadvantages: (1) The communication costs of $job1$ are huge; (2) the load of the map tasks of $job2$ is very imbalanced. The reasons are as follows. Firstly, because the multi-graphs are needed to conduct validation by $job2$, the map tasks of $job1$ output both the index and multi-graphs, whose number is huge. Therefore, the communications of $job1$ are very costly. Secondly, the number of times that each map task in job2 performs validation is inconsistent. The above disadvantages seriously affect the speed of MG-MGSS.

5.1   Basic Idea of MR-MGSS$^+$

To solve these problems, we propose an improved algorithm, called MR-MGSS$^+$, which adopts a localization strategy. The basic idea of MR-MGSS$^+$ is to locally read data before computing, and write the results into the locality; so that there are no or fewer long-distance communications among the nodes, which can greatly improve the efficiency of the algorithm. So the outputs of the map tasks of $job1$ are written into HDFS directly in MR-MGSS$^+$, which can not only save communication cost by avoiding a large amount of data communication among different nodes, but also solve the load imbalance problem of the map tasks of $job2$ to some extent.

5.2   MR-MGSS$^+$ Algorithm

MR-MGSS$^+$ mainly improves the $job1$ of MR-MGSS. The pseudocode of the improved algorithm of $job1$ is shown as Algorithm 3. Because the pseudocode of $job2$ of MR-MGSS$^+$ is the same as that of MR-MGSS, we do not list it.

Algorithm 3: Improved approach for constructing the multi-layered inverted index

Input   : Multi-graph datasets
Output: A multi-layered inverted index
1 //map phase:
2 Map(id, MG_i)
3   calculate the number Num of graphs contained by MG_i;
4   compute all $GIU(G_j)$, $G_j \in MG_i$;
5   write $< key, value >$ pairs into HDFS directly.
6   //key: Num, value: all G_j and $GIU(G_j), G_j \in MG_i$.

The map tasks read the multi-graphs, calculate the number of graphs contained by every multi-graph and the Graph Information Unit (line 3–4), which are written into HDFS with the form of $< key, value >$ pairs (line 5). Compare to MR-MGSS, the $job1$ of MR-MGSS$^+$ directly writes the outputs of the map tasks into HDFS instead of starting the reduce tasks. $Cost^+_{m\_job1} = O(|MS| + 2\sum_{j=1}^{|MS|} \sum_{i=1}^{|MG_j|}(|V_i|))$. $Cost^+_{c\_job1} = O(1)$. $Cost^+_{r\_job1} = O(1)$.

6   Experimental Evaluation

We compare the running time, online query time, scaleup and speedup of the MGSS, our MR-MGSS and improved MR-MGSS$^+$ over synthetic data sets because the real data sets are not publicly released.

NCI dataset: The NCI compound dataset is a benchmark dataset used for graph queries. Each molecule compound in the dataset can be modeled as a graph: atoms and bonds are represented as the vertices and the edges of the graph, respectively. In this paper, two small data sets ($NCI11$ and $NCI12$) and four large data sets ($NCI2$, $NCI3$, $NCI4$ and $NCI5$) are generated based on the NCI dataset with id 1 using the multi-graph construction method proposed in literature [2]. The statistical information is shown in Table 2.

Table 2. Statistics of data sets.

Dataset	Number of multi-graphs (thousand)	Size (G)
$NCI11$	20	0.06
$NCI12$	30	0.09
$NCI2$	1,000	3.01
$NCI3$	2,000	6.03
$NCI4$	3,000	9.04
$NCI5$	4,000	12.05

In this paper, MGSS algorithm is used as the baseline algorithm, and all experiments of serial algorithms are performed on a single computer. All experiments of parallel algorithms are based on a cluster of 33 nodes (1 master node and 32 slave nodes), and each computer is configured as follows: 2.93 GHz CPU, 4 GB RAM, 500 GB Hard disk, Redhat operating system and Hadoop 1.2.1. The average results of three experiments are reported in this paper.

6.1   Running Time

In this section, we compare the running time of our algorithms and the baseline algorithm on small and large-scale datasets, respectively. The experimental results are shown in Table 3, where the unit of measure is seconds.

Table 3 shows that both MR-MGSS and MR-MGSS$^+$ are faster than MGSS on datasets $NCI11$, $NCI12$, $NCI2$, and $NCI3$, where the experimental results of MGSS on the $NCI2$ and $NCI3$ datasets are not reported because a memory overflow occurs. However, our MR-MGSS and MR-MGSS$^+$ can run normally, and answer the query quickly because of adopting the MapReduce parallel programming framework. MR-MGSS$^+$ runs faster than MR-MGSS because of its localization strategy.

Table 3. Running time on small and large-scale datasets ($\tau = 5$).

Algorithm	$NCI11$	$NCI12$	$NCI2$	$NCI3$
MGSS	183.2	192.7	–	–
MR-MGSS	54.3	57.2	149.1	11803.5
MR-MGSS$^+$	50.0	50.2	51.6	266.5

6.2 Online Query Time

In this section, we examine the variations of the online query time of the optimized MR-MGSS algorithm to verify the effect of the locality strategy. The experimental results are shown in Table 4.

Table 4. Running time of MR-MGSS and MR-MGSS$^+$ over different datasets ($\tau = 5$).

Algorithm	$Time\ (S)$	$NCI2$	$NCI3$	$NCI4$	$NCI5$
MGSS	$job1$	69	106	174	207
	$job2$	80.1	11697.5	1427.4	46116.2
	Total	149.1	11803.5	1601.4	46323.2
MR-MGSS$^+$	$job1$	29.0	50.0	74.0	88.0
	$job2$	22.6	216.5	67.2	449.1
	Total	51.6	266.5	141.2	537.1

Table 4 shows that $job1$ and $job2$ of MR-MGSS$^+$ run faster than those of MR-MGSS on the same dataset. In the fastest case, MR-MGSS$^+$ is 86 times faster than MR-MGSS. Also, there are two main reasons: (1) $job1$ of MR-MGSS$^+$ saves much communication cost; (2) the loads of the map tasks of $job2$ of MR-MGSS$^+$ are more balanced than the ones of MR-MGSS.

6.3 Scale Up and Speed Up

The scaleup and speedup of MR-MGSS and MR-MGSS$^+$ are compared. Figure 3 reports the running time of MR-MGSS and MR-MGSS$^+$ with the synchronous changes of the dataset size and cluster size. Figure 4 demonstrates the running time of MR-MGSS and MR-MGSS$^+$ on $NCI3$ dataset with the increase of the cluster size. Figures 3 and 4 show that MR-MGSS$^+$ has good scaleup and speedup.

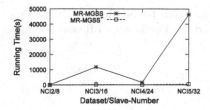

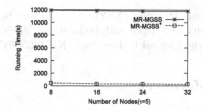

Fig. 3. Scaleup ($\tau = 5$).

Fig. 4. Speedup ($\tau = 5$, over $NCI3$ dataset).

6.4 Performance Evaluation with Different Thresholds

In this section, we test the running time of MR-MGSS and MR-MGSS$^+$ under different threshold conditions on $NCI3$ dataset, and the experimental results are shown in Fig. 5.

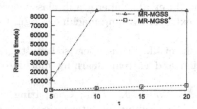

Fig. 5. Running time of MR-MGSS and MR-MGSS$^+$ with different distance thresholds.

Figure 5 shows under the same threshold condition, the MR-MGSS$^+$ algorithm runs faster than the MR-MGSS algorithm due to adopting the localization strategy. Moreover, the running time of MR-MGSS grows faster than MR-MGSS$^+$ because of the load imbalance problem of the map tasks of MR-MGSS.

7 Conclusion

In this paper, we propose a parallel algorithm MR-MGSS based on MGSS and the MapReduce programming framework to solve the problem of large-scale multi-graph similarity search query. Furthermore, the localization strategy is adopted to enhance our MR-MGSS algorithm, which not only reduces the communication cost, but also alleviates the imbalanced load of the map tasks. Extensive experimental results on large-scale synthetic datasets demonstrate that the optimized algorithm MR-MGSS$^+$ is effective and efficient.

Acknowledgment. The work is partially supported by the National Natural Science Foundation of China (No. 61702381, No. 61872070, No. 61772124), the Hubei Natural Science Foundation (No. 2017CFB196), Guangdong Province Key Laboratory of Popular High Performance Computers (No. 2017B030314073), Liao Ning Revitalization

Talents Program (XLYC1807158), the Scientific Research Foundation of Wuhan University of Science and Technology (2017xz015), and the Fundamental Research Funds for the Central Universities (N171605001).

References

1. Wu, J., Hong, Z., Pan, S., et al.: Multi-graph learning with positive and unlabeled bags. In: SDM, pp. 217–225 (2014)
2. Wu, J., Zhu, X., Zhang, C., et al.: Bag constrained structure pattern mining for multi-graph classification. IEEE Trans. Knowl. Data Eng. **26**(10), 2382–2396 (2014)
3. Wu, J., Pan, S., Zhu, X., et al.: Boosting for multi-graph classification. Trans. Cybern. **45**(3), 430–443 (2015)
4. Pang, J., Gu, Y., Yu, G.: A similarity search technique for graph set. J. Northeast. Univ. (Nat. Sci.) **38**(5), 625–629 (2017)
5. Pang, J., Gu, Y., Xu, J., et al.: Parallel multi-graph classification using extreme learning machine and MapReduce. Neurocomputing **261**, 171–183 (2017)
6. Pang, J., Zhao, Y., Xu, J., et al.: Super-graph classification based on composite subgraph features and extreme learning machine. Cogn. Comput. **10**(6), 922–936 (2018)
7. Pang, J., Gu, Y., Xu, J., et al.: Semi-supervised multi-graph classification using optimal feature selection and extreme learning machine. Neurocomputing **277**, 89–100 (2018)
8. Zheng, Z., Tung, A.K.H., Wang, J., et al.: Comparing stars: on approximating graph edit distance. In: Proceedings of International Conference on Very Large Databases (VLDB) Endowment, vol. 2, no. 1, pp. 25–36 (2009)
9. Wang, G., Wang, B., Yang, X., et al.: Efficiently indexing large sparse graphs for similarity search. IEEE Trans. Knowl. Data Eng. **24**(3), 440–451 (2012)
10. Zhao, X., Xiao, C., Lin, X., et al.: A partition-based approach to structure similarity search. In: Proceedings of International Conference on Very Large Databases (VLDB) Endowment, vol. 7, no. 3, pp. 169–180 (2013)
11. Zheng, W., Zou, L., Lian, X., et al.: Efficient graph similarity search over large graph databases. IEEE Trans. Knowl. Data Eng. **27**(4), 964–978 (2015)
12. Zhao, P.: Similarity search in large-scale graph databases. In: Zomaya, A.Y., Sakr, S. (eds.) Handbook of Big Data Technologies, pp. 507–529. Springer, Cham (2017). https://doi.org/10.1007/978-3-319-49340-4_15
13. Sun, Z., Huo, H., Chen, X.: Fast top-k graph similarity search via representative matrices. IEEE Access **6**, 21408–21417 (2018)
14. Roy, S.B., Eliassi-Rad, T., Papadimitriou, S.: Fast best-effort search on graphs with multiple attributes. In: Proceedings of International Conference on Data Engineering (ICDE), pp. 1574–1575 (2016)
15. Fang, Y., Cheng, R., Li, X., et al.: Effective community search over large spatial graphs. In: Proceedings of International Conference on Very Large Databases (VLDB) Endowment, vol. 10, no. 6, pp. 709–720 (2017)
16. Yu, W., Wang, F.: Fast exact CoSimRank search on evolving and static graphs. In: Proceedings of WWW, pp. 599–608 (2018)
17. Achanta, R., Shaji, A., Smith, K., et al.: SLIC superpixels compared to state-of-the-art superpixel methods. IEEE Trans. Pattern Anal. Mach. Intell. **34**(11), 2274–2282 (2012)

18. Viktor, M.S., Kenneth, C.: Big Data: A Revolution that Will Transform How We Live, Work and Think, pp. 9–10. Houghton Mifflin Harcourt, Boston (2013)

19. Cao, L., Cho, B., Kim, H., et al.: Delta-SimRank computing on MapReduce. In: Proceedings of International Workshop on Big Data, pp. 28–35 (2012)

20. Lim, B., Chung, Y.: A parallel maximal matching algorithm for large graphs using Pregel. IEICE Trans. Inf. Syst. **97–D**(7), 1910–1913 (2014)

21. Xiong, X., Zhang, M., Zheng, J., Liu, Y.: Social network user recommendation method based on dynamic influence. In: Meng, X., Li, R., Wang, K., Niu, B., Wang, X., Zhao, G. (eds.) WISA 2018. LNCS, vol. 11242, pp. 455–466. Springer, Cham (2018). https://doi.org/10.1007/978-3-030-02934-0_42

22. Jamour, F., Skiadopoulos, S., Kalnis, P.: Parallel algorithm for incremental betweenness centrality on large graphs. IEEE Trans. Parallel Distrib. Syst. **29**(3), 659–672 (2018)

23. Afrati, N.F., Sarma, D., et al.: Fuzzy joins using MapReduce. In: Proceedings of International Conference on Data Engineering (ICDE), pp. 498–509 (2012)

A Subgraph Query Method Based on Adjacent Node Features on Large-Scale Label Graphs

Xiaohuan Shan, Jingjiao Ma, Jianye Gao, Zixuan Xu,
and Baoyan Song[⊠]

School of Information, Liaoning University, Shenyang, China
bysong@lnu.edu.cn

Abstract. In the real world, various applications can be modeled as large-scale label graphs, which labels can be represented the feature of nodes. Subgraph query is a problem of finding isomorphism subgraphs that satisfy the query conditions, which has attracted much attention in recent years. Therefore, a subgraph query method based on adjacent node features on large-scale label graphs with fewer label categories was proposed in this paper. The method consists of two phases, which are offline index creation and online query. Firstly, we introduce a data structure to describe the information of nodes by scalar features. Secondly, we propose a three-level star structure features index include adjacent nodes' label category, the degree of star-centered and the number of each label, which is named adjacent node scalar features index (ANSF index). After that, according to the structure of query graph is star structure or not, we proposed two different processing strategies on the phase of online query. Experimental results show the efficiency and the effectiveness of the proposed method in index creation time, space usage and subgraph query.

Keywords: Adjacent node features · Feature index · Star structure · Subgraph query · Feature coding

1 Introduction

Large-scale label graphs may describe the information include the category attributes of entities, the association between entities and so on. In the real world, many networks can be abstracted into label graphs, such as protein interaction networks, social networks, etc. [1–5]. Meanwhile, the problem of efficient subgraph query on such graphs has become one of important research contents in the computer realm.

At present, most research of subgraph query on large-scale label graphs generally improves the query efficiency by creating indexes. Indexes which extract feature items according to structure features have a good query effect on the complex situation of feature items. On the contrary, when there are fewer label categories in the graph, the similarity among structures is high, and the feature items tend to assimilate, which leads to lower query efficiency.

However, large-scale graphs with fewer label categories exist widely in the real world. Taking Alipay social network as an example, the number of global users has researched one billion. Meanwhile, according to the integral the users are divided into

© Springer Nature Switzerland AG 2019
W. Ni et al. (Eds.): WISA 2019, LNCS 11817, pp. 226–238, 2019.
https://doi.org/10.1007/978-3-030-30952-7_24

three levels, which are ordinary user, gold user and platinum user. So the Alipay social network can be abstract into a graph with three label categories. In such a graph, nodes represent users, and there exists an edge between users who have a friend relationship. Meanwhile the users' levels are represented by label categories. Querying some structures in such graph, that is, querying some user groups with some kind of friend relationship, and then pushing specific financial information for a certain type of user group to obtain the maximum economic gain.

For the above problem of subgraph query, this paper makes an in-depth study of the subgraph query method on large-scale graph with fewer label categories. We propose a subgraph query method based on adjacent node features on large-scale label graphs utilizing the label feature information of each node and its adjacent nodes. The contributions of this paper are summarized as follows.

- We introduce a data structure to describe the information of nodes by scalar features. Meanwhile we introduce a structure that is star structure, and star-centered node as a special node, its feature information can describe the whole star structure.
- We propose a three-level star structure features index (ANSF index) include adjacent nodes' label categories, the degree of star center and the number of each label. And all star structures in the large-scale graph can be indexed.
- According to the structure of query graph is star structure or not, we propose two different processing strategies on the phase of online query, which are star subgraph query (SSQ) and non-star subgraph query (NSSQ).

The rest of this paper is organized as follows. In Sect. 2, we review related works. In Sect. 3, we explain the basic concepts while in Sects. 4 and 5 gives the details of the proposed novel index and subgraph query method. Experimental results and analysis are shown in Sect. 6. We finally conclude the work in Sect. 7.

2   Related Work

At present, subgraph query is the most common problem and has been studied extensively. Reference [6, 7] proposed the idea of accelerating the subgraph isomorphism on the data graph by using vertex relations. Ullmann [6] creates search tree according to the rule of node matching, which is used to prune and improve the efficiency of the determining isomorphism. But it needs recursive exhaustive, which leads to only applicable to small-scale graphs. VF2 [7] is optimized for Ullmann. It adopts the method of state space representation to determine the matching order, which prunes to narrow the scope of the state space quickly. However, the two algorithms ignore the information among nodes contained in the graph. BoostIso [8] uses the relationship among nodes to accelerate the isomorphism of subgraphs on the large-scale graph. It compresses the graph and makes the subgraph query on the hypergraph to narrow the query range. The inclusion relationship is used to set the matching order, and the pair of nodes to be matched is dynamically loaded according to the NLF filter [9]. However, for the case of fewer categories of labels, the compression of the graph is very expensive, and subgraph matching without index also takes a lot of time.

Treepi [10] and gIndex [11] propose algorithms based on feature path indexes. For building indexes, the two algorithms use different mining methods to organize the frequent items that meet the threshold. The time complexity of mining frequent subgraph algorithm is very high. Spath [12] creates reachability index, which extracts the shortest path in K neighborhood as the index items. In the phase of query, it decomposes the query graph and extracts the feature path. However, it is not suitable for processing the subgraph query of a single graph, and costly to calculate the shortest path. GraphQL [13] filters by neighbors of each node to minimize the number of candidate nodes. Because too much information is recorded in the filtering process, there is an increase in unnecessary memory overheads. DBrIndex [14] proposes an algorithm based on dual branches feature. The algorithm extracts the dual branches structure of each subgraph in the graph database to establish the inverted index. However, when the number of label categories is fewer, the dual branches structure will be too simple to result in a larger filter candidate set, and the query efficiency will reduce.

3   Preliminaries

In this section, we present the basic concepts used in the paper. Label graphs are usually used to represent graphs with attribute information. Subgraph matching of the label graph needs not only mapping of the structure, but also identical labels of corresponding nodes.

Definition 1 (Label Graph): Label graph is an undirected and unweight graph, which is represented as $G = (V, E, \Sigma, L)$ as shown in Fig. 1. Thereinto V is the set of nodes, and E is the set of undirected edges. Σ is the label set of G and L represents the node label, that $L(V) \in \Sigma$.

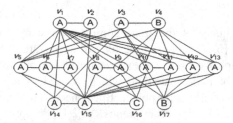

Fig. 1. Label graph G

As mentioned above, Alipay social network can be abstracted into a label graph G, as shown in Fig. 1, ordinary user, gold user and platinum user are represented as A, B and C respectively. The user v_3 with level A and user v_4 with level B are friends.

Definition 2 (Adjacent Node Scalar Feature): Given any node, the label category of this node and the information of its adjacent nodes are called adjacent node scalar feature. Thereinto the information of adjacent nodes includes the degree of this node, the label category and the number of its adjacent nodes.

Definition 3 (Star Graph): Star graph is a kind of label graph which has a depth of 1 and degree of n. And the node with a degree of n is called the star-centered node. Its formalization is represented as $sg_i = v|adj_A(v), adj_B(v), adj_C(v)...$, where v is the star-centered node, and $adj_l(v)$ represents the adjacent nodes of v.

The star graph is composed of star-centered node and its adjacent nodes, and the star-centered node can adequately describe the feature information of adjacent nodes. In the process of isomorphism detection, the scalar features of the star-centered nodes can be used as the basis for judging whether the matching of the node pairs is satisfied.

Definition 4 (Node Matching): Node $u \in G_q$, $v \in G$, if u and v are satisfied $L(v) = L_q(u)$ and $N_\Sigma(v) \geq N_\Sigma(v)$, $adj(v, \Sigma) \geq adj(u, \Sigma)$, then we call two nodes match. And $adj(v_i, \Sigma)$ represents the label set of v_i's adjacent nodes.

Definition 5 (Ordinary Decomposition): Given a query graph Gq, ordinary decomposition $\overline{D(Gq)} = \{\overline{sg_1}, \overline{sg_2}, ..., \overline{sg_t}\}$ is a subset of Gq decomposition, where $\overline{sg_i}(1 \leq i \leq t)$ represents a star graph composed of any node and all its adjacent nodes. For any $\overline{sg_i}$, there is $\bigcup \overline{sg_1}, ..., \overline{sg_{i-1}}, \overline{sg_{i+1}}, ..., \overline{sg_t} \neq Gq$. Ordinary decomposition results of query graph are not unique, and there are no duplicate subitems in the same decomposition set.

Definition 6 (Query Graph Normalization): The query graph G_q is decomposed from large to small. Add all the subgraphs to the query graph specification set and represent as $D(G_q) = \{sg_1, sg_2... sg_t\}$.

4 Adjacent Node Scalar Features Index

4.1 Number of Adjacent Node Label Category Encoding

The number of adjacent node label categories is one of the scalar features of star-centered node, and as the filtering feature directly, it will lead to the following problems: node categories are cluttered, which is not conducive to index construction, and the algorithm complexity is high. So we present a fixed length coding.

The commonly used hash coding will result in a large number of categories merging, and the cost of screening the number of adjacent nodes' label after merging is high. So according to the label category, we introduce coding bits d_i, the number of nodes with label 1 $N_l(v_i)$ is represented in decimal system. If the highest bit on the left is empty, the number is 0. The maximum number of bits *digit* (*l*) is the maximum coding bit $d_{max}(l)$. The calculation formulas are given as follow:

$$digit(l) = \lfloor lgN_l(v_i) \rfloor + 1 \qquad d_{max}(l) = \max\{digit(l_1), digit(l_2)...digit(l_t)\} \qquad (1)$$

4.2 Storage of Scalar Features

According to Definition 4, we classify star graphs by the label category of star-centered node and the feature information of adjacent nodes. The classification conditions are as follows: $L(v_i) = L(v_j)$; $adj(v_i, \Sigma) = adj(v_j, \Sigma)$; $N_l(v_i) = N_l(v_j)$, that $l \in \Sigma$.

To facilitate the construction of ANSF index, we introduce a data structure to describe the label categories and feature information of star-centered nodes. Firstly, we classify star-center nodes according to the star center label. Secondly, we distinguish the star graphs with different label categories of adjacent nodes under the current category. Then we classify the star graphs with current category according to the degree of the star-centered node, which is considered from constraining the number of adjacent nodes. Finally, we divide star graphs according to the number of adjacent nodes' label categories, and star graphs with the same label of star-centered node and adjacent node scalar features are classified into one group.

For example, in Fig. 1, the scalar features table of the label graph G is shown in Fig. 2. L and N_lable represent the label categories of star-centered nodes and theirs adjacent nodes. The degree is the degree of star-centered nodes. N_Σ is the coding of number of adjacent nodes' label categories. Star_id and Neighbor_id represent the ids of star-centered node and its adjacent nodes respectively.

L	N_lable	degree	N_Σ	Star_id	Neighbor_id
A	A	3	300	v_2	$v_1v_3v_{15},0,0$
				v_7	$v_1v_6v_{15},0,0$
		4	400	v_6	$v_1v_5v_{15}v_3,0,0$
				v_{14}	$v_1v_5v_{15}v_3,0,0$
A	AB	4	220	v_{13}	$v_1v_{15},v_4v_{17},0$
				v_{12}	$v_1v_{15},v_4v_{17},0$
		6	420	v_3	$v_5v_{14}v_{10}v_{11}, v_4v_{17},0$
				v_{11}	$v_1v_3v_{10}v_{15},v_4v_{17},0$
		7	610	v_5	$v_1v_2v_3v_{14}v_{15}v_6,v_4,0$
		12	110100	v_1	$v_2v_5v_6v_7v_8v_9v_{10}v_{11}v_{12}v_{13}v_{14},v_{17},0$
	AC	4	301	v_8	$v_1v_9v_{15},0,0,v_{16}$
		12	110001	v_{15}	$v_2v_5v_6v_7v_8v_9v_{10}v_{11}v_{12}v_{13}v_{14},0,0,v_{16}$
	ABC	5	311	v_9	$v_1v_3v_{15},v_4,v_{16}$
		6	411	v_{10}	$v_1v_3v_{11}v_{15},v_4,v_{16}$
B	A	5	500	v_{17}	$v_1v_3v_{11}v_{12}v_{13},0,0$
		7	700	v_4	$v_3v_5v_9v_{10}v_{11}v_{12}v_{13},0,0$
C	A	4	400	v_{16}	$v_3v_9v_{10}v_{15},0,0$

Fig. 2. Adjacent node scale feature

4.3 ANSF Index Construction

A third-level index is constructed based on the scalar features of star-centered nodes. We construct the first-level index by linked list, which realizes index from star-centered nodes' label categories to adjacent nodes' label categories. As the most widely used feature of nodes, we put degrees as the second-level index items, which is organized into B+ tree forest. It realizes the second-level index from adjacent nodes' label categories to the degrees of star-centered nodes. The third-level index is also constructed by B+ tree forest, which implements the index from the degrees of star-centered nodes to the number of adjacent nodes' labels, and the index value stores the star graphs under the corresponding category.

Star-Centered Label Category Index. The practical significance of the first-level index is to partition the star graphs according to label categories of star-centered and adjacent nodes.

Because we research the large-scale graph with fewer label categories, so the maximum possible number of combinations for each star-centered is few relatively. Such as Fig. 2, the combinations of adjacent label categories are "A", "AB", "AC", "ABC" when "L = A", as shown in Fig. 3.

Fig. 3. First-level index of label A

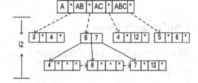

Fig. 4. Second-level index

Adjacent Node Degree Index. As one of the scalar features of adjacent nodes, the degree of star-centered node reflects the number of its adjacent nodes. We partition the star graphs based on the degree for the same adjacent nodes label. Because of the large number of star graphs and their star-centered degrees, so the adjacent node degree index is implemented by B+ tree. Each first-level index item points to the root node of the corresponding second-level index, so the second index contains many B+ trees. Each tree is indexed by the degrees of the star-centered nodes, and the order m is selected according to the amount of data. In our example, m is 3, and the number of keywords is 2. The leaf nodes' pointers point to the root nodes of the third-level index. Taking the label A as an example, the degrees containing in Table 1 with L = A are inserted into the B+ trees to form the second-level index. As shown in Fig. 4.

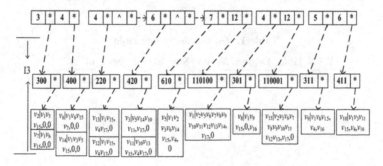

Fig. 5. Third-level index

Star-Centered Degree Index. When the number of adjacent nodes' label categories is unique, the three features of star-centered nodes' label, adjacent nodes' label and degrees of star-centered nodes can be used to represent this kind of star graph. On the contrary, the adjacent nodes' label categories are not unique, and the star-centered nodes are divided according to the number of adjacent nodes' label. Because each leaf node of the second-level index corresponds to the degree of a star-centered node and each degree points to the root of a tree, B+ trees of the number of adjacent nodes' label categories are formed, and the three-level index is finished. The index values store the star graphs satisfying the current partition, as shown in Fig. 5.

5 Subgraph Query

According to the structure of query graph is star structure or not, as shown in Figs. 6 and 7, we propose two different processing strategies on the phase of online query, which are star subgraph query (SSQ) and non-star subgraph query (NSSQ).

5.1 Star Subgraph Query

For the SSQ, we utilize the three-level star structure feature index to get the query results quickly.

Query Graph Features Extraction. We extract the same scalar features of the star-centered node from query graph. And we store the features in a sequential table, which contains the features of star-centered node and the ids of star-centered nodes and its adjacent nodes. We take Fig. 6 as an example. The label of star-centered node is A, the label categories of its adjacent nodes are A and B, and its degree is 6, the coding of the number of adjacent nodes' label categories is "510". The feature table of query graph Gq is shown in Table 1.

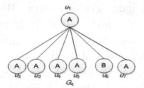
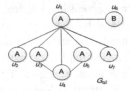

Fig. 6. Query graph Gq **Fig. 7. Non-star** query graph Gq_2

Table 1. Gq's scale feature table

L	N_lable	Degree	$N_\Sigma(v_i)$	Star_id	Neighbor_id
A	AB	6	510	u_1	$u_2u_3u_4u_5u_7, u_6, 0$

Query Based on ANSF Index. In the process of query, the feature index is traversed according to the features of adjacent nodes. According to star-centered node's label L and adjacent nodes' label N_lable, we traversal the first-level index to extract the items, which contains the labels of star-centered node and adjacent nodes. Then the items are found that are equal or greater than the degree of current star-centered from the second-level index. We find the items, which satisfy $N_\Sigma(v_i) \geq N_\Sigma(u_i)$ from the three-level index, and add the values into the intermediate results.

Theorem 1. If graph G contains graph Q, then Q contains any substructure of G.

According to Theorem 1, if query results contain query graph, then they contain any substructure, else no results are returned. Therefore, when the intermediate result of any index item is empty, it can be judged that there is no matching of the current query graph in the data graph, and no further query is required.

5.2   Non-star Subgraph Query

In addition, as for NSSQ, we decompose the query graph into some star graphs, and then get the intermediate results by the proposed index. Finally, we proposed a multi-join strategy with minimum cost to optimize the connection judgment and the minimum cost calculation. The join of intermediate results is realized effectively, and then the final results of subgraph query are obtained.

Query Graph Normalization. Because of the particularity of index structure, query graph can be directly filtered after it is split into star graphs by ordinary decomposition. E_{pi} denotes the unmatched edges contained in the query graph G_q of the decomposition set to be added, and E_r denotes the edges contained in the subitems of the decomposition set.

Query graph normalization decomposes query graph into minimum star subgraphs. It is proved that according to the definition of normalization, there is $E_r \cup E_r = E_{Gq}$. In the process of decomposition, if the current query graph has been decomposed into t star structures, then there is $E_{Gq} = \cup E_{r1}, \ldots, E_{rt} \cup E_p$. When the query graph G_q is decomposed completely, there is $E_p = 0$, and then $E_{Gq} = E_{r1} + \ldots + E_{rt}$. Because the edges of the query graph are fixed values, in the process of decomposition, the larger the item of E_{pi} is selected and added to the ordinary decomposition set, the smaller sum term is, that is, the smaller the total term t is.

Minimum Cost Connection. For non-star structure queries, in order to further reduce the operation cost, the two aspects of whether to execute the connection and calculating the minimum cost connection are optimized in this paper.

In order to reduce the connections of intermediate results, we try to filter the intermediate results, which do not have connection relations with star subitems in the query graph specification set, so the concept of graph model is introduced.

In practical applications, given a star subitem sequence, all possibility can be represented by graph model, one of which can be represented by a spanning tree of the graph. In order to realize the least cost connection, a pre-connected spanning tree is constructed based on the connectivity of graph model, and the intermediate results are filtered according to the order of spanning tree. It starts with the non-degree vertex of the spanning tree, scans whether the number of the common vertexes of the first group of intermediate results of the current vertex and the successor vertex are larger than the weights of the corresponding edges of the spanning tree, and if they are larger, continues to judge the number of the common vertexes of the first group of intermediate results of the successor vertex. Instead, it will prune and retrospect the query intermediate results of all successor vertexes. Then it will judge the second query intermediate result of the current vertex, and add each group of executable connections to the query intermediate result set until all the results are detected. Although connection preprocessing increases the scanning of relational tables, it omits most unnecessary operation time and effectively improves query efficiency for a large number of connections without common vertexes and satisfying conditions.

The MVP algorithm [15] based on graph model obtains the global minimum cost by calculating the local minimum cost of connection, and realizes the minimization of the size of connection results. The execution process is to select the least expensive connection from the graph model, and then merge the two relational vertexes into a new vertex until only one relational vertex is left in the graph, which represents the result of multi-join query. The formula for calculating the minimum cost of the algorithm is as follows:

$$\cos t(R_i \theta R_j) = 2(|R_1| + |R_j|) + (|R_1||R_j|JSF)[(||R_1|| + ||R_j||)JCF] \quad (2)$$

$$JSF = \frac{|R \theta S|}{|R| \times |S|} \quad JCF = \frac{||R \theta S||}{||R|| + ||S||} \quad (3)$$

When dealing with each set of query intermediate solutions set, the cost calculation is carried out, and two intermediate solutions with the minimum cos$t$ $(R_i \theta R_j)$ value in each set are selected to join, and the least cost two intermediate results are iteratively calculated and merged until only one result is left and the join result set is obtained.

6   Experiments

In this section, we present our experimental results by comparing our methods with some state-of-the-art algorithms. We evaluate our method in index construction time, storage overhead of the index and the efficiency of our query method in star query graph and non-star query graph.

6.1   Experimental Settings

All experiments are implemented in JAVA, and conducted on the computer of Intel Pentium (R) CPU (R) G3220 3.00 Hz processor, 16 GB memory and 500G hard disk.

Datasets. We use ego-Facebook and gemsec-Facebook real datasets and different scales of synthetic datasets. ego-Facebook dataset (node number is 4039, the number of edges is 882340) is the data of Facebook friends relationship. The gemsec-Facebook dataset (node number is 134833, the number of edges is 1380293) collected data on Facebook page (November 2017). Nodes represent pages, the edges represent the relationship of each page, and the dataset contains eight different types of pages, as the label categories in our experiments.

Synthetic datasets G_1, G_2, G_3, G_4 and G_5 are generated by R-MAT [16]. The number of nodes is 2k, 4k, 6k, 8k, 10k respectively, and the number of edges is 8 times nodes'. At the same time, each node is randomly assigned label "A", "B" or "C".

6.2   Experimental Analysis

Performance Analysis of ANSF Index. In this section, the experiments are compared with gIndex, DBrIndex and Spath algorithms from two aspects, which are index construction time and indexing storage overhead.

Figure 8(a) and (b) show the comparison of index construction time on synthetic and real datasets. As shown, each index construction time increase with the increasing scale of graphs. Among them, the gIndex index need mine frequent subgraph to construct index, which will cost a lot of time. DBrIndex and Spath algorithms need to extract feature structure of each subgraph, and there are containing relation among multiple extracting feature structures, which will directly lead to consume an amount of time to construct the inverted index. The proposed ANSF Index method in this paper extract the features of nodes to construct index, without the complex extracting process of DBrIndex, the shortest paths computation of Spath algorithm and mining frequent items of gIndex. Our method simplifies the process of feature extraction that reduces the index construction time effectively.

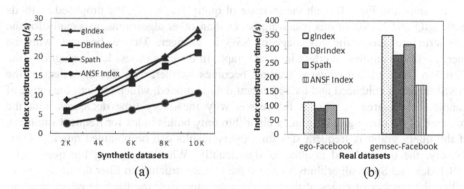

Fig. 8. Comparison of index construction time for different datasets

Figure 9(a) and (b) show the comparison of index storage overhead on different datasets. GIndex index need set threshold, regard frequent items bigger than the threshold as index items. Due to the non-frequent items are not stored, therefore the storage size of this index relatively low. DBrIndex utilizes the inclusion relation of the structure and regard dual branch structure as the index items. With the increasing of data, dual branch structure increases, so it needs more storage space. However, the dual branch structure exists repeatability and don't need to store many times, so the growth of index storage space is slowed. Spath index need set threshold D, according to different D, it need store multiple hops information of adjacent nodes for each node. With the increasing of D, the storage space is growing exponentially. The proposed ANSF Index in our paper, while also using inclusion relations as the index items, but because the label categories is less, and categories of index items are repetitive, which can control the size B + trees of our method effectively.

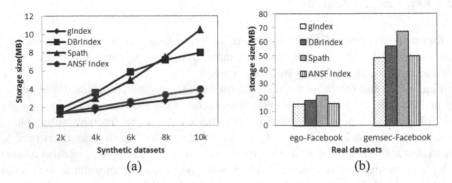

(a)   (b)

Fig. 9. Comparison of index storage overhead on different datasets

Performance Analysis of Query. In the experiments of evaluating query efficiency, we change the number of edges and structures of query graphs to analyze the query time.

As shown in Fig. 10, with the increase of query graph size, the proposed methods both SSQ and NSSQ in this paper are better than other algorithms, especially for the star structure query, the advantage of SSQ is stronger. That is because, with the increase of the number of edges in query graph, in the case of less label categories, the filtration range of star-centered features becomes smaller. The filtering effect of the ANSF index is enhanced and the redundant data is pruned, which reduces the range of candidate sets greatly. This is the reason why the query time decreases with the increase of the query graphs. gIndex algorithm only builds index for frequent subitems. If the query graph is frequent structure, query results can be obtained quickly. Conversely, the query speed is affected significantly. When processing the query with DBrIndex and Spath algorithms, they use the feature structure to filter the query graph. When the number of edges of the query graphs increases, the filtering effect weakens and the candidate set becomes larger. Therefore, the query time increases with the increasing number of edges of query graphs.

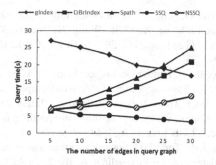

Fig. 10. Comparison of query time on ego-Facebook

7 Conclusion

In this paper, we proposed a subgraph query method based on adjacent node features on large-scale label graphs. Firstly, we propose a three-level star structure features index (ANSF Index) to retrieve all star structures in the large-scale graph. Secondly, according to the structure of query graph is star structure or not, we propose two different processing strategies on the phase of online query, which are SSQ and NSSQ. For the SSQ, we utilize the three-level star structure feature index to get the query results quickly. In addition, as for NSSQ, we decompose the query graph into some star graphs, and then get the intermediate results by the proposed index. Finally, we proposed a multi-join strategy with minimum cost to optimize the connection judgment and the minimum cost calculation. Experiments show that the proposed method can effectively support subgraph query on large-scale label graphs.

Acknowledgements. This work was supported by National Natural Science Foundation of China under Grant (Nos. 61472169, 61502215, 61802160, 51704138); The Key Research and Development Program of Liaoning Province (No. 2017231011); Shenyang City Young Science and Technology Innovation Talents Support Program (No. RC180244); Liaoning Public Opinion and Network Security Big Data System Engineering Laboratory (No. 04-2016-0089013); Scientific Research Fund of Liaoning Province Education Department (No. LYB201617).

References

1. Li, X.J., Yang, G.H.: Graph theory-based pinning synchronization of stochastic complex dynamical networks. IEEE Trans. Neural Netw. Learn. Syst. **28**(2), 427–437 (2017)
2. Calle, J., Rivero, J., Cuadra, D., et al.: Extending ACO for fast path search in huge graphs and social networks. Expert Syst. Appl. **86**, 292–306 (2017)
3. Sonmez, A.B., Can, T.: Comparison of tissue/disease specific integrated networks using directed graphlet signatures. BMC Bioinf. **18**, 135 (2017)
4. Ravneet, K., Sarbjeet, S.: A comparative analysis of structural graph metrics to identify anomalies in online social networks. Comput. Electr. Eng. **57**, 294–310 (2017)

5. Liu, W., Zhang, M., Niu, G., Liu, Y.: Weibo user influence evaluation method based on topic and node attributes. In: Meng, X., Li, R., Wang, K., Niu, B., Wang, X., Zhao, G. (eds.) WISA 2018. LNCS, vol. 11242, pp. 382–391. Springer, Cham (2018). https://doi.org/10.1007/978-3-030-02934-0_35
6. Ullmann, J.R.: An algorithm for subgraph isomorphism. J. ACM **23**(1), 31–42 (1976)
7. Cordella, L.P., Foggia, P., Sansone, C., et al.: A (sub) graph isomorphism algorithm for matching large graphs. IEEE Trans. Pattern Anal. Mach. Intell. **26**(10), 1367–1372 (2004)
8. Ren, X., Wang, J.: Exploiting vertex relationships in speeding up subgraph isomorphism over large graphs. Proc. VLDB Endow. **8**(5), 617–628 (2015)
9. Han, W.S., Lee, J., Lee, J.H.: TurboISO: towards ultrafast and robust subgraph isomorphism search in large graph databases. In: Proceedings of the 2013 ACM SIGMOD International Conference on Management of Data, pp. 337–348. ACM (2013)
10. Zhang, S., Hu, M., Yang, J.: TreePi: a novel graph indexing method. In: International Conference on Data Engineering, pp. 966–975. IEEE (2007)
11. Yan, X., Yu, P.S., Han, J.: Graph indexing: a frequent structure-based approach. In: SIGMOD 2004: Proceedings of the 2004 ACM SIGMOD International Conference on Management of Data, pp. 335–346 (2004)
12. Zhao, P., Han, J.: On graph query optimization in large networks. VLDB Endow. **3**(1–2), 340–351 (2010)
13. He, H., Singh, A.K.: Query language and access methods for graph databases. In: Proceedings of the ACM SIGMOD International Conference on Management of Data, pp. 125–160 (2010)
14. Zhang, Y.N., Gao, H., Zhang, W.: A dual branches feature based subgraph query algorithm. J. Comput. Res. Dev. **48**(z2), 114–123 (2011)
15. Meng, F.R., Zhang, Q., Yan, Q.Y.: Information entropy based algorithm for efficient subgraph matching. Appl. Res. Comput. **29**(11), 4035–4037 (2012)
16. Chakrabarti, D., Zhan, Y.P., Faloutsos, C.: R-MAT: a recursive model for graph mining. In: SIAM International Conference on Data Mining, Florida, pp. 442–446 (2004)

Semantic Web Service Discovery
Based on LDA Clustering

Heng Zhao[1], Jing Chen[1], and Lei Xu[2(✉)]

[1] Wuhan Digital Engineering, Wuhan 430205, China
[2] Department of Computer Science and Technology,
Nanjing University, Nanjing 210023, China
xlei@nju.edu.cn

Abstract. In recent years, with the exponential growth of Web services, how to find the best Web services quickly, accurately and efficiently from the large Web services becomes an urgent problem in Web service discovery. Based on the previous work, we propose a semantic Web service discovery method based on LDA clustering. Firstly, the OWL-S Web service documents are parsed to obtain the document word vectors. Then these vectors are extended to make the documents more abundant of semantic information. Moreover, these vectors are modeled, trained and inferred to get the Document-Topic distribution, and the Web service documents are clustered. Finally, we search the Web service request records or the Web services clusters to find Web services that meet the requirements. Based on the data sets of OWLS-TC4 and hRESTS-TC3_release2, the experimental results show that our method (LDA plus semantic) has higher accuracy (13.48% and 9.97%), recall (37.39% and 24.26%), F-value (30.46% and 23.58%) when compared with VSM method and LDA method.

Keywords: Web service discovery · Document parsing ·
Latent Dirichlet Allocation · Clustering · Semantic Web service

1 Introduction

Based on Service Oriented Architecture (SOA), Web service is a distributed calculating model, which realizes service invokes by the interactions among service providers, agencies and requesters. The number of Web services on the Internet increase rapidly. Different Web service providers have different descriptions of resources in Web services, resulting in a large number of Web services with similar functions and different expressions. So it becomes difficult for service requesters to find the required services correctly and efficiently from a large number of Web services.

At present, Web service discovery is the main way to solve the matching of user needs and service functions. Web service discovery first selects the corresponding matching method according to the user's requirements, matches the service stored in the service center, and then selects the target service set that meets the user's requirements. In the process of service discovery, the key step is to compare and analyze the service request of the user and the service description of the service agent one by one, and make a set of matching criteria according to the user demand.

W. Ni et al. (Eds.): WISA 2019, LNCS 11817, pp. 239–250, 2019.
https://doi.org/10.1007/978-3-030-30952-7_25

Web services mainly use WSDL (Web services Description Language) documents to describe services, while WSDL can only describe syntactic information, without functional semantic information, resulting in low precision and recall of service discovery.

Therefore, this paper puts forward a LDA (Latent Dirichlet Allocation) clustering method of semantic Web service discovery, which pretreatments OWL-S (OWL Web Ontology Language for Services) document firstly, then gets the initial term vectors of each document, and extracts and parses the OWL-S document corresponding ontology document. Next, through WordNet and Word2Vec, gets the words vector of the final document, and then uses the LDA model and Gibbs sampling algorithm to get the document-subject distribution, calculates the Jensen-Shannon Divergence (JS Divergence), sets these Web services into several clusters, and lookups query first memo databases have any records of the Web service request, document vocabulary of this query and result set of service discovery are also recorded.

The experiment shows that compared with the VSM (Vector Space Model) method based on TF-IDF, the method using LDA and semantic information has higher accuracy and F-value, especially better recall rates and acceptable efficiency.

2 Related Work

Current researches try to make the existing Web services with automaticity and intelligence, and put forward the concept of semantic Web services. There are mainly three kinds of semantic Web service matching, namely based on logic, based on non-logic, and based on mixed pattern.

Semantic Web service matching based on logic [1–3] can deduce and inference semantic information. The ontology corresponding to the service is obtained through semantic annotation of the Service Profile, and logical concepts and rules are extracted from the ontology. The relationship between concepts and rules are inferred by inference machine, and finally the matching set of service requests is obtained. Paolucci et al. [4] proposed the matching idea. It took the input and output parameters of the service description as the basis for matching, and divided the service matching degree into four categories: exact matching, plug-in matching, subsume inclusion and fail. It predefined formal expressions for each type of matching degree and calculated specific values by the shortest distance of concepts on the semantic classification tree. This matching method can guarantee the inevitability and precision of the results, but the implementation of the system largely depends on the integrity of the inference engine and inference rules adopted, and the flexibility and reliability are insufficient.

Non-logical semantic Web service matching [5, 6] is used to calculate the matching degree of service description between service requesters and service providers. The calculation criteria include syntax similarity, structural graph matching degree, and concept distance in ontology. Compared with the semantic Web service matching based on logic that explicitly utilizes the IOPE (Input, Output, precondition, Effect) semantics defined in logical rules, the semantic Web service matching based on non-logic implicitly utilizes information such as pattern matching, sub-graph matching and term frequency defined in Web service description. Wu et al. [7] proposed a semantic

Web service discovery method based on ontology and lexical semantic similarity, including the construction of Web service ontology corresponding to Web service, the similarity calculation of Web service. Although the method can quantify the service matching degree and improve the accuracy of service matching, it is difficult to propose a similarity calculation function that can adapt to all different application scenarios and requirements. At the same time, how to evaluate the advantages and disadvantages of a similarity calculation function is also a difficulty currently.

Semantic Web service matching based on mixed pattern is a combination of the above two, which can give full play to their respective advantages, reduce the influence brought by the defects, and improve the performance of Web service matching. Many researchers in industry have tried this method [8–11]. Klusch [12] presented a matching model based on logical reasoning and matching model based on similarity matching model OWLS-MX: first get five kinds of compatibility, and build the syntax of the similarity between the I/O, the experimental results show that the method improves the precision of service discovery and recall. Our method is also based on the mixed ways.

3   Parsing and Semantic Extension of Web Services

Web service parsing is an important prerequisite for Web service clustering and discovery, and the parsing results will directly affect the performance of service clustering and discovery. In order to make the document parsing more accurate and effective, combining the characteristics of web services and the semantic Web technology, we proposed a method to make web services parsing and semantic extension, known as PASEBLDA4WS (Parsing and Semantic extension Based on LDA for Web Services), as shown in Fig. 1.

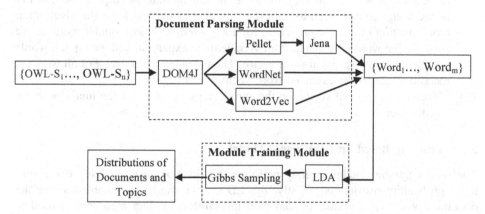

Fig. 1. Framework for PASEBLDA4WS

As can be seen from Fig. 1, the input is the collection of Web service documents $(OWL-S_1, OWL-S_2, ..., OWL-S_n)$, and the output is the document-topic distribution. The intermediate process includes two functional modules: document parsing module and model training module.

3.1 Parsing OWL-S Documents

OWL-S document is a kind of semantic annotation for Web services, including Service Profile, Service Model and Service Grounding. The document parsing module mainly completes the functions of OWL-S document parsing and semantic extension of service description. The specific process is as follows:

(1) Parsing the OWL-S file. Tool DOM4J is used to extract the service name, service description, input and output information of OWL-S document, and it conducts text preprocessing such as stop word and part of speech restoration for the service name and service description to obtain the initial vocabulary vector.

(2) Parsing OWL ontology files. By using Pellet inference machine and tool Jena to inference the OWL ontology file corresponding to OWL-S document, the concept vector related to input and output elements are obtained.

(3) Expanding WordNet semantics. Using Algorithm [13] to calculate the words in WordNet, which are more similar to the original word vectors than the threshold value and add them into the extended word vector.

(4) Expanding Word2Vec semantics. Word2Vec is an open-source, deep learning tool that Google launched in 2013 to convert a word into the vector form. After using Word2Vec, the processing of text content can be transformed into vector operation in vector space. In this paper, every article in the Wikipedia data set (version 2016.03.06, size 11.9 g) is converted into a line of text, and the punctuation and other contents are removed to obtain the training corpus. Then, neural network algorithm is used to train the language model and learn the distribution representation of each word. Then, for each word, the most similar word or the most similar word is searched from Wikipedia to expand it, and the top ten words are added to the expansion word vector. The extended word vector of all words is merged to obtain the extended word vector.

(5) Merging all the word vectors in the above steps, so as to get the final document word vector.

3.2 Training Based on LDA Topic Model

The model training module mainly performs the function of referencing document-topic probability distribution. Firstly, the LDA topic model[1] is established for the document vocabulary vector set, and then the Gibbs sampling algorithm is used to iteratively train the vocabulary vector set until the LDA topic model converges, and then the document-topic distribution in the document set is inferred.

---

[1] Latent Dirichlet allocation, https://en.wikipedia.org/wiki/Latent_Dirichlet_allocation.

The process of using LDA to generate documents in the corpus [14] is shown as follows:

(1) For each topic $k \in \{1, \ldots\ldots, K\}$, sampling generates word distribution for topic k: $\vec{\phi}_k \backsim Dirichlet\left(\vec{\beta}\right)$;
(2) For each document $m \in \{1, \ldots\ldots, M\}$, sampling generates topic distribution for document m: $\vec{\vartheta}_m \backsim Dirichlet(\vec{\alpha})$;
(3) For each word $n \in \{1, \ldots\ldots, N_m\}$ in document m, the generating process is shown as follows:

Choosing topic: in ϑ_m, sampling generates topic of word n in document m, $Z_{m,n}$ $\backsim Multinomial\left(\vec{\vartheta}_m\right)$;

Generating a word: sampling generates a word from the chosen topic, $W_{m,n} \backsim$ $Multinomial(\vec{\phi}_{Z_{m,n}})$.

LDA model [14] is shown in Fig. 2. K is the number of topics, M is the number of documents, N_m is the number of words in document m, $W_{m,n}$ is the nth word in document m, $\vec{\beta}$ is the *Dirichlet* priori argument of multinomial distribution of the word in each topic, $\vec{\alpha}$ is the *Dirichlet* priori argument of multinomial distribution of the topic in each document. $Z_{m,n}$ is the topic of the nth word in the mth document, $W_{m,n}$ is the nth word in the mth document. Two latent variables $\vec{\vartheta}_m$ and $\vec{\phi}_k$ are the distributions of topics under the mth document and words of the kth topic.

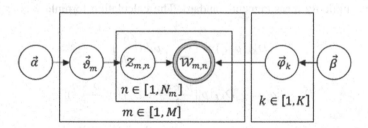

Fig. 2. Topic model of LDA

LDA topic model can extract the core semantics or features of objects from massive corpus. It is a probabilistic model for generating discrete data sets. It maps every document in the corpus to a probability distribution of a group of potential topics, and each potential topic corresponds to a probability distribution of words in the vocabulary set.

Gibbs sampling is a special case of MCMC (Markov Chain Monte Carlo), and the basic idea is as follows: select one dimension of the probability vector each time, sample the value of the current dimension with the variable values of other dimensions, and iterate continuously until the output parameter to be inferred converges.

We use LdaGibbsSampler[2] as the Gibbs sampling algorithm of LDA topic model for training and inference, until the model convergence, then evaluate the parameters of $Z_{m,n}$, $\vec{\vartheta}_m$ and $\vec{\varphi}_k$, finally get the document-topic distribution of the document vector collection. And the calculating formulas of $\vec{\vartheta}_m$ and $\vec{\varphi}_k$ are shown as follows.

$$\varphi_{k,t} = \frac{n_k^{(t)} + \beta_t}{\sum_{t=1}^{V} n_k^{(t)} + \beta_t} \tag{1}$$

$$\vartheta_{m,k} = \frac{n_m^{(k)} + \alpha_k}{\sum_{k=1}^{K} n_m^{(k)} + \alpha_k} \tag{2}$$

In formula (1) and (2), V is the number of words, K is the number of topics, $n_k^{(t)}$ is the number of word t assigned to topic k, $n_m^{(k)}$ is the number of words assigned to topic k in document m, β_t is the *Dirichlet* priori argument of multinomial distribution of the word t in topic k, α_k is the *Dirichlet* priori argument of multinomial distribution of the topic k in document m.

4   Clustering and Discovery of Semantic Web Services

We present an improved clustering algorithm, which combines LDA model and k-means algorithm. The Jensen-Shannon Divergence based on KL divergence [15] was used as the similarity measurement standard. The calculation formula was as follows:

$$D_{KL}(p||q) = \sum_{i=1}^{n} p_i \ln\left(\frac{p_i}{q_i}\right) \tag{3}$$

$$D_{JS}(p||q) = \frac{1}{2}[D_{KL}(p||\frac{p+q}{2}) + D_{KL}(q||\frac{p+q}{2})] \tag{4}$$

Where D(p||q) is the KL divergence, representing the information loss generated when the probability distribution p is used to fit the real distribution q, and p represents the real distribution and q represents the fitting distribution of p. The main reason for not using KL divergence as the similarity metric is that KL divergence is not symmetric. The specific clustering process is as follows: firstly, global variables are initialized, including randomly selecting the first initial clustering center; then calculate the distance between each document and the initial cluster center; then a vector with a large distance from all the existing initial clustering centers is calculated as the new initial clustering center. The JS divergence matrix is constructed, and the k-means clustering algorithm is called to obtain the clustering set and cluster center set of Web services.

---

[2] LdaGibbsSampler,   http://web.engr.illinois.edu/~mqian2/upload/projects/java/TextProcessor/ocessor/doc/textProcessor/LdaGibbsSampler.html.

We use a quad to represent the Web service request record, *WSRRecord* =<*RID*, *File-Name*, *ExtWordVec*, *WSResultSet*>, *RID* represents the ID number of the Web service record, *FileName* represents the name of the Web service request document, *ExtWordsVec* represents the document extension vocabulary vector of the Web service request document processed by the document parsing module, and *WSResultSet* represents the set of Web services that the Web service request has met the requirements. Memo database mainly includes query and storage function, which provides a fast and effective way for the query of Web service request.

The query function is used for a Web service request to find a request record for the Web service that already exists in the memo database, including a direct lookup for *FileName* and a matching lookup for *ExtWordVec*. Direct lookup of *FileName* is obtained by precise comparison of the *FileName* in all *WSRRecord* with the name of the document requested by the Web service; if the matching is successful, the set of Web services meeting the requirements can be extracted directly from the database. If it is not successful, then the matching lookup of *ExtWordVec* is required: first through document parsing module for the Web service request document parsing, expand vocabulary get the document vector, then with all the *ExtWordVec* and *WSRRecord* are matched. Since the results of the expansions of lexical semantic information vector after parsing are rich enough, we use the Jaccard's similarity coefficient [16] to measure the two lexical semantic matching degree of the vector, the specific formula is as follows:

$$Jaccard(p, q) = \frac{|p \cap q|}{|p| + |q| - |p \cap q|} \tag{5}$$

In formula (5), the numerator represents the intersection of all words of vector p and q, that is, the number of the same words, and the denominator represents the sum of their words minus the number of the same words, that is, the number of words in their union. And the result is a number inside $[0,1]$.

Web service matching module is a more accurate and complex search in the Web service clustering set, when the Web service set that meets the requirements is not found in the memo DB module. The search is mainly divided into the following two steps: matching the expanded vocabulary vector, and finding the Web service cluster with the largest similarity as the candidate set of Web services; in the Web service candidate set, the semantic distance between the extended vocabulary vector of the Web service request document and all the Web services is calculated, and the Web services whose semantic distances are less than the threshold are added to the Web service result set.

The search and matching process of Web service includes: firstly, initialize the distance between the Web service request and the nearest cluster center; then make the Web service request expanding the vocabulary vector and the cluster center; then calculate the JS divergence of these two vectors, update the minimum distance with the Web service request, and record the cluster center; so the candidate set of Web services is obtained through the cluster center with the smallest Web service request. Computing document-topic distribution of JS divergence with the request, and join these

Web services into the last result set, whose JS divergences are less than the threshold value.

When the corresponding records are not found in the memo DB module, by the Web service matching module, the Web service set that meets the requirements is obtained, and the result set is stored in the memo database in the format of *WSRRecord*.

5   Experiments and Evaluations

This paper conducts experimental evaluation on the proposed semantic Web service discovery method based on LDA clustering, with three research questions:

RQ1: How to determine the number of document topics when conducting Gibbs sampling training and inference on LDA topic model?

RQ2: Is it effective for the semantic Web service discovery method based on LDA clustering?

RQ3: Does the semantic Web service discovery method based on LDA clustering improve the performance of Web service discovery?

We use OWLS-TC4 as a test data set, comprising 1083 Web services in nine different domains. In addition, an associated set of each service query request, more commonly known as the factual basis for the query request, is also included in OWLS-TC4. More than 48 ontology files related to OWL-S files are included in OWLS-TC4, which plays an important role in semantic extension of OWL-S files.

In order to solve RQ1, we start from the fact basis of query request, and set the number of clustering as 9 according to the number of fields (9 fields are preset in the data set, namely $K2 = 9$), and considers the accuracy rate, recall rate and F-value of clustering results when $K1 = \{2,3, ...,50\}$, and $K1$ was determined by the experimental results. As shown in Fig. 3, when $K1 = 12$, the recall rate and F-value of clustering reached the maximum value of 0.547497869 and 0.619850749 respectively, and the accuracy reached its fourth value of 0.714238753. Since F-value is taken as the optimal standard, the value of $K1$ is finally chosen as 12.

To solve RQ2, we select the widely used VSM [17] methods, based on TF-IDF, as the contrast experiment. At the same time, in order to test the semantic expansion effect of LDA model and service description, we select the Web service discovery based on LDA clustering and the semantic Web service discovery based on LDA clustering, so as to compare these three methods in accuracies, recalls and F-values, the experimental results are shown in Figs. 4, 5 and 6.

In Figs. 4, 5 and 6, the horizontal axis shows the 42 service request, the longitudinal axis shows each service request under the three methods of accuracies, recalls and F-values, thus it can be seen that in most cases, the method based on the LDA+semantics in accuracy, recall and F-values is better than VSM method based on TF-IDF and the method based on LDA. And there is no value of 0, while the other two methods have the condition of 0value, which means that they cannot find the query results.

To solve RQ3, we compare the running time of the VSM method based on TF-IDF, the method based on LDA and our method, specifically including service parsing time, service clustering time and service matching time.

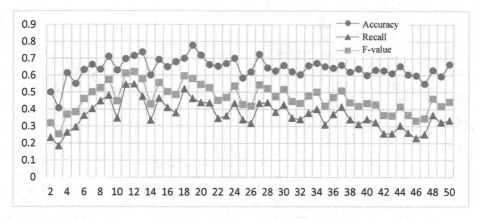

Fig. 3. Clustering effects of Web services with different K1 values (K2 = 9)

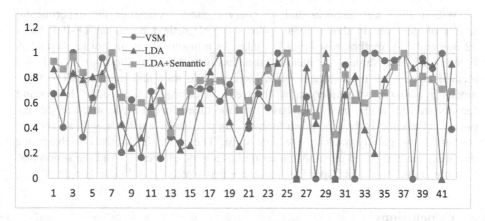

Fig. 4. Comparisons of Accuracies for experimental results

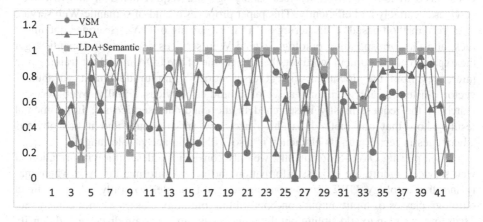

Fig. 5. Comparisons of Recalls for experimental results

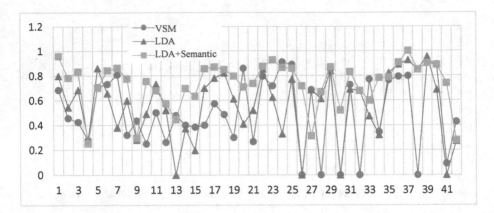

Fig. 6. Comparisons of F-values for experimental results

Within 1083 web services and 42 query request documents, the experimental results of the three methods are as follows: VSM method based on TF-IDF takes 22 min and 20 s; method based on LDA takes 28 min and 6 s; while our method takes 21 min and 13 s. Namely our method takes less time than the other two. LDA-based method takes more time since it needs more time to train and infer the LDA topic model. In addition, when querying all the web service request files, since all querying records are kept in the memo database, the time taken for semantic web service discovery is the same as the time taken for querying the database records, with an average cost of 57 ms, greatly reducing the time spent. Therefore, the semantic web service discovery method based on LDA clustering has a good efficiency. And the supervised learning is also useful in Paper [18] to handle text matching and classifying.

6 Conclusions

Users need to look for web services satisfying their requirements in the vast Web services accurately and efficiently. This paper proposes a kind of semantic Web service discovery method based on clustering of LDA, by parsing of OWL-S service document files, and the corresponding ontology extend Web services description semantics, get the expansion of document word set, then the theme of the LDA model for training and inferences from the document-subject distribution, again to document vocabulary vector clustering, and then find a set of candidate services. Finally, Web services matching the request mostly is searched in the candidate service set. Experimental results show that our method has high accuracy and F-value, and it has obvious advantages in recall rate, its efficiency is also within the acceptable range.

When parsing OWL-S documents, we only consider the description information of web services, ignore the process information and access information. And when analyzing the ontology OWL corresponding to OWL-S file, we only add the direct ontology classes of input-output concepts, sibling ontology classes are not considered. Therefore, we plan to add sibling ontology classes within a certain distance and density in the concept, and meanwhile reduce the direct ontology classes with distant relatives. And, using WordNet expand service description semantics, semantic similarity threshold values are determined by previous experimental experience value, is not necessarily applicable to the proposed approach, therefore, we plan for these two semantic similarity threshold experiment analysis, get the best results of semantic similarity threshold, improve the efficiency of web service discovery.

Acknowledgment. The work is supported by National Key R&D Program of China (2018YFB1003901), National Natural Science Foundation of China (61832009, 61728203).

References

1. Chakraborty, D., Perich, F., Avancha, S., et al.: DReggie: semantic service discovery for m-Commerce applications. In: 20th Symposium on Reliable Distributed Systems, pp. 28–31 (2001)
2. Zhang, C., Zhao, T., Li, W., et al.: Towards logic-based geospatial feature discovery and integration using web feature service and geospatial semantic web. Int. J. Geogr. Inf. Sci. **24** (6), 903–923 (2010)
3. García, J.M., Ruiz, D., Ruiz-Cortés, A.: Improving semantic Web services discovery using SPARQL-based repository filtering. Web Semant. Sci. Serv. Agents World Wide Web **17**, 12–24 (2012)
4. Paolucci, M., Sycara, K.: Autonomous semantic web services. IEEE Internet Comput. **7**(5), 34–41 (2003)
5. Deng, S.G., Yin, J.W., Li, Y., et al.: A method of semantic Web service discovery based on bipartite graph matching. Chin. J. Comput. **31**(8), 1364–1375 (2008)
6. Bernstein, A., Kiefer, C.: Imprecise RDQL: towards generic retrieval in ontologies using similarity joins. In: Proceedings of the 2006 ACM Symposium on Applied Computing, pp. 1684–1689 (2006)
7. Wu, J., Wu, Z.H., Li, Y., et al.: Web service discovery based on ontology and similarity of words. China J. Comput. **28**(4), 595–602 (2005)
8. Bianchini, D., De Antonellis, V., Melchiori, M., et al.: Semantic-enriched service discovery. In: Proceedings of 22nd International Conference on Data Engineering, p. 38 (2006)
9. Sangers, J., Frasincar, F., Hogenboom, F., et al.: Semantic web service discovery using natural language processing techniques. Expert Syst. Appl. **40**(11), 4660–4671 (2013)
10. Amorim, R., Claro, D.B., Lopes, D., et al.: Improving Web service discovery by a functional and structural approach. In: 20111 IEEE International Conference on Web Services, pp. 411–418 (2011)
11. Paliwal, A.V., Shafiq, B., Vaidya, J., et al.: Semantics-based automated service discovery. IEEE Trans. Serv. Comput. **5**(2), 260–275 (2012)
12. Klusch, M., Fries, B., Sycara, K.: OWLS-MX: a hybrid semantic web service matchmaker for OWL-S services. Web Semant. Sci. Serv. Agents World Wide Web **7**(2), 121–133 (2009)
13. Jiang, J.J., Conrath, D.W.: Semantic similarity based on corpus statistics and lexical taxonomy. In: Proceedings of International Conference Research on Computational Linguistics (ROCLING X), pp. 1–15 (1997)
14. Blei, D., Ng, A., Jordan, M.: Generative probabilistic model for collections of discrete data such as text corpora. J. Mach. Learn. Res. **3**, 993–1022 (2002)

15. Lamberti, P.W., Majtey, A.P., Borras, A., et al.: Metric character of the quantum Jensen-Shannon divergence. Phys. Rev. A **77**(5), 1–8 (2008)
16. Niwattanakul, S., Singthongchai, J., Naenudorn, E., et al.: Using of Jaccard coefficient for keywords similarity. In: Proceedings of the International MultiConference of Engineers and Computer Science, pp. 1–6 (2013)
17. Skoutas, D., Sacharidis, D., Simitsis, A., et al.: Ranking and clustering web services using multicriteria dominance relationships. IEEE Trans. Serv. Comput. **3**(3), 163–177 (2010)
18. Zhang, J., Xu, L., Li, Y.: Classifying Python code comments based on supervised learning. In: Meng, X., Li, R., Wang, K., Niu, B., Wang, X., Zhao, G. (eds.) WISA 2018. LNCS, vol. 11242, pp. 39–47. Springer, Cham (2018). https://doi.org/10.1007/978-3-030-02934-0_4

Research and Implementation of Anti-occlusion Algorithm for Vehicle Detection in Video Data

Yongqi Wu[1]([⊠]), Zhichao Zhou[1], Lan Yao[1], Minghe Yu[2], and Yongming Yan[3]

[1] College of Computer Science and Engineering, Northeastern University, Shenyang 110819, China
wuyongqi@stumail.neu.edu.cn
[2] College of Software, Northeastern University, Shenyang 110819, China
[3] Department of Video Big Data, DIXN Technology Co., Ltd., Shenyang 110004, China

Abstract. Object detection is an important branch of image processing and computer vision, which has become a hot research issue in recent years. Accurate target detection in a video is the foundation of intelligent surveillance system. Since the background scenario is dynamic and even especially complicated when vehicles occlude, the target detection accuracy is declined. Therefore, based on the bounding box regression algorithm, this paper constructs adjacent punishment mechanism to make the bounding box clear off other objects. The proposal weak confidence suppression is leveraged for the robustness of the detector when occlusion happens. Experiments show that the proposed method outperforms traditional methods on three different datasets.

Keywords: Object detection · Vehicle occlusion · Bounding box regression · Confidence suppression

1 Introduction

Vehicle detection, as an important part of intelligent transportation system, has always been a research hotspot to detect mobile vehicles in a video. It is the foundation of vehicle tracking, vehicle type identification, flow statistics, video velometer and other technologies. However, in practical applications, scenes often change dynamically. For example, if two vehicles are similar in exterior, it is likely to blur them. This detection result need to be further revised by the non-maximum suppression algorithm [1], otherwise the prediction box on the occluded vehicle is probably suppressed by the one on the adjacent vehicle causing a detection loss. Our work will improve the detection accuracy under occlusion circumstance.

The main contributions of the work are:

- Propose a bounding box regression algorithm that introduces the adjacent punishment mechanism, so that the bounding box frames its specified target and being clear off from adjacent targets.

© Springer Nature Switzerland AG 2019
W. Ni et al. (Eds.): WISA 2019, LNCS 11817, pp. 251–256, 2019.
https://doi.org/10.1007/978-3-030-30952-7_26

- A proposal weak confidence suppression algorithm is proposed to reduce the probability with which the proposal is suppressed by adjacent targets.
- Experiments are conducted on PASCAL VOC, UA-DETRAC and Highway-Vehicle datasets and show the efficiency of the proposed algorithm.

2 Related Work

Most of the early object detection algorithms are based on manual features. Malisiewicz [2] trains each sample through the SVM classifier based on the HOG feature. Felzenszwalb [3] proposes to use deformable part model to detect multi-class objects. In recent years, with the rise of deep learning, convolution neural network (CNN) has brought new ideas for object detection [4–10]. Networks such as SPP-net, R-FCN, and GoogleLeNet [11–13] can be used for object detection.

For vehicle detection, many researchers propose methods with efficiency and acceptable accuracy [14–16]. However, they have not considered occlusion. When the traffic flow on the road increases, the vehicles will occlude each other, which seriously affects the detection accuracy of vehicles if the above method are applied.

This paper improves the bounding box regression algorithm to make the proposal of the vehicle separated from adjacent targets, and proposes a proposal weak confidence suppression algorithm to avoid the proposal being mistakenly suppressed by adjacent targets, so as to improve the detection accuracy with occlusion.

3 Implementation Method

3.1 Bounding Box Regression and Its Improvement

Bounding box regression is used in object detection methods such as RCNN and Fast R-CNN to revise the position of the proposal and make it close to the designated target. Let P and G be the original proposal bounding box and true box represented by their center coordinates and width and height. Bounding box regression is to find a map from P to $\hat{G}$, while $\hat{G}$ is approximate to the ground-truth box and defined as:

$$f(P) = (\hat{G}) \quad \text{s.t.} \quad dist(G, \hat{G}) < dist(G, P) \tag{1}$$

However, this regression only makes the proposal as close as possible to its target, without considering the influence of adjacent objects. When occlusion occurs, as shown in Fig. 1(a), the small box is ground-truth of A and the big box is ground-truth of B. When A is partially occluded by B, A's proposal is likely to be misaligned due to the similarity of A and B as shown in Fig. 1(b). The dotted line box is A's proposal and it may drift as B's.

To solve this problem, the adjacent punishment mechanism is applied as the bounding box regression algorithm. In the detector training process, each proposal will not only approach its ground-truth and define it as positive term L_P, but also clear off from the ground-truth of adjacent objects and define it as negative term L_N. By

(a)The ground-truth boxes   (b) The proposal bounding box
of vehicle A and vehicle B   of Vehicle A drifts to that of vehicle B

Fig. 1. The proposal bounding boxes of the occluded vehicle

introducing the repulsive effect of adjacent objects on the proposal bounding box, the
detection accuracy can be improved by avoiding the proposal drifting to the near
similar objects when the target is occluded. The regression calculation is defined as:

$$L = L_P + L_N \tag{2}$$

The positive term L_P and negative term L_N are defined as:

$$L_p(p,g) = \sum_{i \in \{x,y,w,h\}} smooth_{L_1}(p_i - g_i) \tag{3}$$

$$smooth_{L_1}(x) = \begin{cases} 0.5x^2 & \text{if } |x| < 1 \\ |x| - 0.5 & \text{otherwise,} \end{cases} \tag{4}$$

Since $IoU(p, G_i) \in [0,1]$, the $smooth_{L_1}$ function is modified as follows:

$$L_N(p, G_n) = \sum_{i \in n} smooth_{L_1'}(IoU(p, G_i)) \tag{5}$$

$$smooth_{L_1'}(x) = \begin{cases} (0.5+x)^2 - 0.25 & x \leq 0.5 \\ \frac{|\ln(1-x)|}{2-x} - 0.5 & x > 0.5, \end{cases} \tag{6}$$

L_P is used to narrow the gap between proposals and ground-truth boxes, L_N is used for
repulsive effect on the proposal bounding box. G_n represents the ground-truth boxes set
of all objects except the target, and $IoU(p, G_i)$ represents the IoU between the proposal
and the ground-truth box.

3.2   Proposal Weak Confidence Suppression Algorithm

After the bounding box regression, a large number of proposals are generated near the target. In order to eliminate the false, the non-maximum suppression (NMS) algorithm is used to remove redundant proposals by the overlapping area (IoU), and reserve the proposal with the highest confidence for each target.

NMS algorithm selects the highest scored box b_m from the proposal set B, then removes the proposal whose IoU value with b_m is greater than the threshold N_t from B, and repeats until B is empty. However, there are some drawbacks in NMS. When dense occlusion occurs, the artificial N_t will matter the detection accuracy. If N_t is too large, it causes false detection. Contrarily, it loses detection.

Therefore, we improve NMS and propose a proposal weak confidence suppression algorithm. Instead of deleting proposals from the set B, the confidence S_i of the proposal b_i, which is determined by the polarity of IoU values of b_i and b_m is introduced to avoid false suppression and reduce the impact of N_t. Assuming that U_p represents the IoU between b_i and b_m, if U_p is greater than the threshold N_t, the confidence S_i of b_i is multiplied by the confidence attenuation coefficient α:

$$\alpha = -U_p * \ln(1 + U_p) + 1 \tag{7}$$

Its function is to reduce the confidence S_i. Conversely, if U_p is less than or equal to the N_t, the confidence S_i will not be changed. Repeat the process until set B is empty. Finally, the prediction box set D and the confidence set S_d is output.

4   Experiments and Results Analysis

We use two open source image datasets: the PASCAL VOC dataset and the UA-DETRAC dataset [17]. The PASCAL VOC dataset is used for PASCAL VOC Challenge Competition with a total of 1,659 vehicle images. The UA-DETRAC dataset is mainly taken in Beijing and Tianjin with 6,250 vehicle images. We select the peak time video of Shanghai-Hangzhou-Ningbo Expressway, frame it to obtain images, and employ 20 students to annotate images manually. Finally, the image and annotation information are constructed into Highway-Vehicle dataset, which contains 12,800 vehicle images.

In order to verify the performances, Faster R-CNN is applied as the detector and the VGG-16 network in Faster R-CNN is replaced by ResNet-101 network with stronger feature extraction. The related parameters are set as follows: the training learning rate is 0.001, the attenuation step size is 30000, the attenuation coefficient is 0.1, the training batch size is 128, and the detection confidence threshold is 0.5.

For the evaluation, if the IoU between the prediction box and its ground-truth box is greater than 0.5, it is a correct detection, otherwise it is a false detection. When the annotated vehicle is not detected, it is a detection loss. Using three image datasets, Faster R-CNN and its reformative method is verified by five-fold cross validation under the different IoU thresholds. The average precision (AP) of these two methods are shown in Table 1.

Table 1. Vehicle detection AP of two methods on three datasets

N_t	Method	PASCAL VOC	UA-DETRAC	Highway-Vehicle
0.2	Faster R-CNN	0.7353	0.6611	0.6134
	Proposed method	0.7587	0.6827	0.6447
0.3	Faster R-CNN	0.7521	0.6773	0.6353
	Proposed method	0.7763	0.7011	0.6724
0.4	Faster R-CNN	0.7603	0.6817	0.6407
	Proposed method	0.7833	0.7083	0.6752
0.5	Faster R-CNN	0.7581	0.6855	0.6378
	Proposed method	0.7805	0.6972	0.6718
0.6	Faster R-CNN	0.7413	0.6715	0.6232
	Proposed method	0.7705	0.6876	0.6633
0.7	Faster R-CNN	0.7154	0.6573	0.5946
	Proposed method	0.7428	0.6647	0.6285

The experimental results show that the detection performance on three image datasets is improved when the bounding box regression algorithm with adjacent punishment mechanism and the proposal weak confidence suppression algorithm is introduced. When the IoU threshold is 0.4, the proposed method yields the best performance. It is shown that the proposed method can effectively improve the detection accuracy and improve the accuracy and stability of the detector under the occlusion condition.

5 Conclusion

Vehicles have always been an important target for object detection. Due to the complexity of the actual road environment, such as vehicle occlusion, the accuracy and stability of object detection algorithms have been challenged. Compared with the traditional object detection algorithm, the proposed algorithm makes the proposal close to its designated target while clear from other nearby targets, and reduces the probability that the proposal is mistakenly suppressed by adjacent targets, so as to improve the detection performance of the detector. The experimental results show that the reformative Faster R-CNN is more efficient in vehicle detection.

Acknowledgements. This research is supported by the National Key R&D Program of China under Grant No. 2018YFB1003404.

References

1. Neubeck, A., Gool, L.: Efficient non-maximum suppression. In: International Conference on Pattern Recognition, vol. 3, pp. 850–855. IEEE, Hong Kong (2006)
2. Malisiewicz, T., Gupta, A., Efros, A.: Ensemble of exemplar-SVMs for object detection and beyond. In: IEEE International Conference on Computer Vision 2011, ICCV, vol. 1, no. 2. IEEE, Barcelona (2011)

3. Felzenszwalb, P., Girshick, R., Mcallester, D.A.: Visual object detection with deformable part models. Commun. ACM **56**(9), 97–105 (2010)
4. Zhao, H., Xia, S., Zhao, J., Zhu, D., Yao, R., Niu, Q.: Pareto-based many-objective convolutional neural networks. In: Meng, X., Li, R., Wang, K., Niu, B., Wang, X., Zhao, G. (eds.) WISA 2018. LNCS, vol. 11242, pp. 3–14. Springer, Cham (2018). https://doi.org/10.1007/978-3-030-02934-0_1
5. Krizhevsky, A., Sutskever, I., Hinton, G.: ImageNet classification with deep convolutional neural networks. In: Advances in Neural Information Processing Systems 2012, NIPS, pp. 1097–1105. IEEE, Lake Tahoe (2012)
6. Girshick, R., Donahue, J., Darrell, T., Malik, J.: Rich feature hierarchies for accurate object detection and semantic segmentation. In: Proceedings of the IEEE Conference on Computer Vision and Pattern Recognition 2014, CVPR, pp. 580–587. IEEE, Columbus (2014)
7. Girshick, R.: Fast R-CNN. In: Proceedings of the IEEE International Conference on Computer Vision 2015, ICCV, pp. 1440–1448. IEEE, Santiago (2015)
8. Ren, S., He, K., Girshick, R., Sun, J.: Faster R-CNN: towards real-time object detection with region proposal networks. In: Advances in Neural Information Processing Systems 2015, NIPS, pp. 91–99. IEEE, Montreal (2015)
9. Cai, Z., Fan, Q., Feris, Rogerio S., Vasconcelos, N.: A unified multi-scale deep convolutional neural network for fast object detection. In: Leibe, B., Matas, J., Sebe, N., Welling, M. (eds.) ECCV 2016. LNCS, vol. 9908, pp. 354–370. Springer, Cham (2016). https://doi.org/10.1007/978-3-319-46493-0_22
10. Lin, T., Girshick, R.: Feature pyramid networks for object detection. In: Proceedings of the IEEE Conference on Computer Vision and Pattern Recognition 2017, CVPR, pp. 2117–2125. IEEE, Hawaii (2017)
11. He, K., Zhang, X., Ren, S.: Spatial pyramid pooling in deep convolutional networks for visual recognition. IEEE Trans. Pattern Anal. Mach. Intell. **37**(9), 1904–1916 (2014)
12. Dai, J., Li, Y., He K.: R-FCN: object detection via region-based fully convolutional networks. In: Advances in Neural Information Processing Systems 2016, NIPS, pp. 379–387. IEEE, Barcelona (2016)
13. Szegedy, C., Liu, W., Jia, Y.: Going deeper with convolutions. In: Proceedings of the IEEE Conference on Computer Vision and Pattern Recognition 2015, CVPR, pp. 1–9. IEEE, Boston (2015)
14. Fan, Q., Brown, L., Smith, J.: A closer look at Faster R-CNN for vehicle detection. In: IEEE Intelligent Vehicles Symposium 2016, vol. IV, pp. 124–129. IEEE, Gothenburg (2016)
15. Song, H., Zhang, X., Zheng, B., Yan, T.: Vehicle detection based on deep learning in complex scene. Appl. Res. Comput. **35**(04), 1270–1273 (2018)
16. Lee, W., Pae, D., Kim, D.: A vehicle detection using selective multi-stage features in convolutional neural networks. In: International Conference on Control, Automation and Systems 2017, ICCAS, pp. 1–3. IEEE, Singapore (2017)
17. Wen, L., Du, D., Cai, Z.: UA-DETRAC: a new benchmark and protocol for multi-object detection and tracking. arXiv preprint arXiv:1511.04136 (2015)

Temporal Dependency Mining from Multi-sensor Event Sequences for Predictive Maintenance

Weiwei Cao[1,2(✉)], Chen Liu[1,2], and Yanbo Han[1,2]

[1] Beijing Key Laboratory on Integration and Analysis of Large-Scale Stream Data, North China University of Technology, Beijing, China
ml3611031336@163.com, {liuchen,hanyanbo}@ncut.edu.cn
[2] School of Computer Science,
North China University of Technology, Beijing, China

Abstract. Predictive maintenance aims at enabling proactive scheduling of maintenance, and thus prevents unexpected equipment failures. Most approaches focus on predicting failures occurring within individual sensors. However, a failure is not always isolated. The complex dependencies between different sensors result in complex temporal dependencies across multi anomaly events. Therefore, mining such temporal dependencies are valuable as it can help forecast future anomalies in advance and identifying the possible root causes for an observable anomaly. In this paper, we transform the temporal dependency mining problem into a frequent co-occurrence pattern mining problem and propose a temporal dependency mining algorithm to capture temporal dependency among multi anomaly events. Finally, we have made a lot of experiments to show the effectiveness of our approach based on a real dataset from a coal power plant.

Keywords: Predictive maintenance · Root causes · Temporal dependency · Frequent co-occurrence pattern

1 Introduction

Predictive maintenance aims to help anticipate equipment failures to allow for advance scheduling of corrective maintenance. It is usually performed based on an assessment of the health status of equipment [1, 2]. Thanks to the rapid development of IoT, massive sensors are deployed on industrial equipment to monitor health status. As a key phase for predictive maintenance, anomaly detection technologies have given us the ability to monitor anomalies from multivariate time-series sensor data or events [3].

In recent years, lot of researches have paid attention to the problem of anomaly detection/prediction from multivariate time-series sensor data. Anomaly detection within individual variables, referred to as "univariate anomaly", have already been extensively studied [4, 5]. However, it is much more challenging but common in the real applications to mine and analyze temporal dependencies among sensor data or "univariate anomaly" events. It means the possibility of finding new anomaly type or inferring the root cause of an anomaly [6–8].

© Springer Nature Switzerland AG 2019
W. Ni et al. (Eds.): WISA 2019, LNCS 11817, pp. 257–269, 2019.
https://doi.org/10.1007/978-3-030-30952-7_27

In Sect. 2, a simple example shows the temporal dependencies found among "univariate anomaly" events. Actually, in a complex industrial system, an anomaly/failure is not always isolated. Owing to the obscure physical interactions, trivial anomalies will propagate among different sensors and devices, and gradually deteriorate into a severe one in some device [9]. Mining such temporal dependencies are valuable as it can help forecast future anomalies/failures in advance and identifying the possible root causes for an observable device anomaly/failure [10].

In the paper, we try to propose an effective and explainable approach to predict the anomaly based on mining the temporal dependencies from multi-sensor event sequences. To reach this goal, we detect "univariate anomaly" events from sensor data and output multi-sensor event sequences. Then, we transform the temporal dependency mining problem into a frequent co-occurrence pattern mining problem. Next, a graph-based anomaly prediction model is built based on choosing and connecting the mined temporal dependencies for event predict in the experiment. Furthermore, a lot of experiments are done to show the effectiveness of our approach based on a real dataset from a coal power plant.

2   Motivation

An anomaly event carries much information about an anomaly like its occurrence time, sources and type. Here, we use a 4-tuple to depict an anomaly event: $e = $ *(timestamp, eventid, sourceid, type)*, where *timestamp* is the occurrence time of e; *eventid* is the unique identifier of e; *sourceid* is the unique identifier of the source sensor; and *type* is the type of anomaly event.

A time-ordered list of events from the same sensor construct an event sequence $E_i = \{e_1, e_2, \ldots, e_m\}$. All the event sequences construct the event space $\Theta = \{E_1, E_2, \ldots, E_n\}$. These events are not isolated with each other. They imply complex temporal dependencies.

Definition 1. Temporal Dependency: Let $A = \{E_1, E_2, \ldots, E_k\}, 1 \leq k < m$ be an event set contains k event sequences, $B = \{E_1', E_2', \ldots, E_h'\} 1 \leq h < m$ be another event set contains another h event sequences, and $A \cap B = \varnothing$. A temporal dependency [10, 11] is typically denoted as $A \xrightarrow{[t_1, t_2]} B$. It means that B will happen within time interval $[t_1, t_2]$ after A occurs.

Figure 1 shows a sample about temporal dependency among several anomaly events. There are several types of anomaly events have occurred on the different sensors. These events construct five event sequences which are shown in Fig. 1. In which, the event sequence $E^1 = \{e_i^1, i = 1, 2, 3, 4, 5, 6, 7\}$ is constructed by H-CF events, and the event sequences $E^3 = \{e_j^3, j = 1, 2, 3, 4, 5\}$ is constructed by the H-IAP event. According to the maintenance log, when every event e_j^3 in E^3 happen, the event e_i^1 in E^1 will happen within an average time lag $\Delta t = 18$ min. Thus, there is a temporal dependency between the H-CF event and H-IAP event, i.e. $H - CF \xrightarrow{18min} H - IAP$. Thus, if there is a H-CF event, we can predict that in the flowing 18 min, there will happen the H-IAP event.

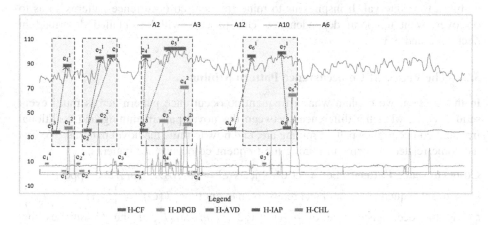

Fig. 1. A real case: Temporal dependencies among event sequences.

Besides the temporal dependency among two event sequences, there is temporal dependency among multi sequences. For example in Fig. 1, the part covered by a shadow contains three event sequences, E^1, $E^2 = \{e_k^3, k = 1, 2, 3, 4, 5, 6\}$ which is constructed by the H-DPGB events and E^3, when events in E^1 and E^3 occurs, the event in E^2 also occur within an average time lag $\Delta t = 61$ min. Thus, we can see that there is a temporal dependency between the H-CF event, H-IAP event and H-DPGB event, donated as $(H - CF, H - IAP) \xrightarrow{61min} H - DPGB$.

This case illustrates that if we can discovery such temporal dependency for multi events, we have chances of predicting the anomaly event or inferring the root cause of an observable anomaly. Thus, the goal of this paper is mining the temporal dependency from multi event sequences.

3 Temporal Dependency Mining

3.1 Overview

For now, there are lots of excellent techniques have been developed to detect the univariate events. The common ones include range-based approaches, outlier detection approaches [4, 5]. A range-based approach customizes value bounders for individual sensor based on inspectors' experiences, sensor/device instructions and so on. Outliers are widely known as the values which sufficiently deviate from most ones, the original outlier detection methods were arbitrary, but in recent years statistics techniques are used [12].

The main idea of mining the temporal dependency is to transform temporal dependency into a frequent co-occurrence pattern across multi event sequences. Essentially, a temporal dependency among the means that an event set B frequently occurs within time interval Δt after an event set A occurs. In other words, the temporal dependency is a relationship among the objects in a frequent co-occurrence pattern within a time interval. It inspires us to mine frequent co-occurrence patterns so as to discover event temporal dependencies. This process will be detailed described in Sects. 3.2 and 3.3.

3.2   The Frequent Co-occurrence Pattern Mining

In this section, we explain what a frequent co-occurrence pattern across multi event sequences is, what the differences between the novel pattern mining and traditional frequent co-occurrence pattern mining are, and how to mine the novel patterns. We first list some related concepts in mining the frequent co-occurrence patterns.

Co-occurrence Pattern: For a set of objects $\mathcal{O} = \{o_1, o_2, \ldots, o_k\}$ that appear in the same event sequence E_i, an object refers to an event type. $T(\mathcal{O})^{E_i} = \left\{t_{o_1^{E_i}}, t_{o_2^{E_i}}, \ldots, t_{o_k^{E_i}}\right\}$, $t_{o_j^{E_i}}$ is the occurrence time of o_j $(j = 1, 2, \ldots, m)$ in E_i, if the $\mathcal{O}$ satisfies that $max(T(\mathcal{O})^{s_i}) - min(T(\mathcal{O})^{s_i}) \leq \xi$, then we say that $\mathcal{O}$ is a co-occurrence pattern (CP), ξ is a user-specified threshold.

However, the challenge is how to identify the time lag between two event sets who has the temporal dependency. It actually reflects how long that a set of events will be affected by its related events. Unfortunately, most traditional frequent co-occurrence pattern mining algorithms cannot directly solve such problem. They only focused on the occurrence frequency of a group of unordered objects [13]. Hence, we try to design an algorithm to discover a constrained frequent co-occurrence pattern. Such pattern consists of two object groups, where intra-group objects are unordered and inter-group objects are time-ordered, and all objects span no more than Δt. We call such pattern as frequent co-occurrence pattern across multi event sequences.

Frequent Co-occurrence Pattern Across Multi Event Sequences: For a Co-occurrence pattern $\mathcal{O} = \{\mathcal{O}_{pre} \cup \mathcal{O}_{post}\}$ that occurs in a set of l event sequences $\{E_1, E_2, \ldots, E_l\}$. The $\mathcal{O}_{pre}$ and $\mathcal{O}_{post}$ will form the multi-dimensional co-occurrence pattern, donated as $MCP(O_{pre}, O_{post})$, if the $\mathcal{O}$ satisfies the following conditions: (1) every object $o_i \in O_{pre} \cup O_{post}$ comes from different event sequences; (2) the object in O_{post} always occurs after the occurrence of object in O_{pre}; (3) $max\{T(O_{post})\} - min\{T(O_{pre})\} \leq \Delta t$, in which Δt is the time lag, O_{pre} contains m events and O_{post} contains n events, i.e. $|O_{pre}| = m$ and $|O_{post}| = n$. Thus, the $MCP(O_{pre}, O_{post})$ also can be donated as $MCP_{m,n}(O_{pre}, O_{post})$. If the $MCP_{m,n}(O_{pre}, O_{post})$ have occurred more than k times in l event sequences $\{E_1, E_2, \ldots, E_l\}$, then the $MCP_{m,n}(O_{pre}, O_{post})$ will be regarded as the Multi-dimensional Frequent Co-occurrence Pattern, denoted as $FMCP(O_{pre}, O_{post})$ or $FMCP_{m,n}(O_{pre}, O_{post})$. In which, O_{pre} is the antecedent, O_{post} is the consequent, Δt is the time lag between them.

Based the above definition, it is obviously that, our FMCP mining task is significantly different from the traditional one. The difference is a MCP is supposed to be divided into two groups, where intra-group objects are unordered and inter-group objects are time-ordered. This time constraint raises the complexity of our task. Assume that the frequency of a FMCP $O = \{o_1, o_2, \ldots, o_m\}$ is l. To find out all valid divisions by traditional ideas, we have to count the frequency for any possible division of O. The number of possible divisions is $2 * (C_m^2 + \ldots + C_m^{\lceil m/2 \rceil})$, where $\lceil m/2 \rceil$ will return the closest integer greater than or equal to $m/2$, not to mention the number of object groups. Owing to this difference, our task is unable to be simply solved by the well-known generation and counting strategy.

3.3 FMCP Mining Algorithm

In this paper, we use $\gamma(A, B)$ to denoted the temporal dependency between event sequences, and we use the $\gamma(A, B).sup$ to donate the occurrence probability of B given the knowledge that A have occurred. It is used to filter the mined temporal dependency.

$$\gamma(A, B).sup = sup(O_{post}|O_{pre}) = \frac{freq(O_{post}|O_{pre})}{freq(O)}$$

In which, $freq(O_{post}|O_{pre})$ is the frequency of the occurrence of O_{post} after with O_{pre}, and $freq(O)$ is the frequency of O.

Assume that the occurrences threshold $freq_{min} = sup_{min}$, where sup_{min} is the threshold of support, and assume that FP is the set constructed by all the FMCP in Θ. All temporal dependency relationships whose satisfied that $sup > sup_{min}$ constitute the set R, then $\forall \gamma(E_i, E_j) \in R$ and there is and only one $TFCP(O_{pre}, O_{post}) \in FP$ satisfy that $\gamma(E_i, E_j)$ is the temporal dependency of O_{pre} and O_{post}, and vice versa.

Because of that $\gamma(E_i, E_j) \in R$, so $\gamma(E_i, E_j).sup \geq sup_{min}$, thus, the number of occurrences of E_j is over sup_{min} after the occurrence of E_i within a time range $\gamma(E_i, E_j).\Delta t$. In a conclusion, E_i and E_j be a FMCP, denoted as $TFCP(E_i, E_j)$, i.e. $TFCP(E_i, E_j) \in FP$. Every FMCP is constructed with antecedent and the consequent, and the item is the set is unique. Therefore, if there is $MCP(O'_{pre}, O'_{post}) \neq FMCP(O_{pre}, O_{post})$ and $FMCP(O'_{pre}, O'_{post}) \in P$, which satisfy that $\gamma(E_i, E_j)$ is the temporal dependency of O'_{pre} and O'_{post}, then $O'_{pre} = E_i, O'_{post} = E_j$, and the number of occurrences is $freq\left(FMCP\left(O'_{pre},\right.\right.$ $\left.\left.O'_{post}\right)\right) = \gamma(E_i, E_j).sup$, and time lag is $FMCP\left\langle O'_{pre}, O'_{post}\right\rangle.\Delta t = \gamma(E_i, E_j).\Delta t$. Then $TFCP\left(O'_{pre}, O'_{post}\right) = FMCP(O_{pre}, O_{post})$, but this is conflict. Thus, it is proofed that $\forall \gamma(E_i, E_j) \in R$, there is and only one $FMCP(O_{pre}, O_{post}) \in FP$ satisfy that $\gamma(E_i, E_j)$ is the temporal dependency of O_{pre} and O_{post}.

Therefore, if we can get all the FMCP in Θ, we can calculate support of temporal dependencies and filter the candidate temporal dependency sets that satisfy the conditions. Thus, based on the traditional method of frequent co-occurrence pattern mining, in this paper, we proposed an approach called as *ETD-mining* with a three-stage process, that is "Generation-Filter-Extension", to mine the FMCP.

Given a Frequent Co-occurrence Pattern $FMCP(O_{pre}, O_{post})$, the frequent number is donated as $freq(FMCP(O_{pre}, O_{post}))$, and the $freq(MCP(O_{pre}, O_{post})) = \gamma(O_{pre}, O_{post}).sup$. If the $freq_{min} = sup_{min}$, then for any temporal dependency $\gamma(E_i, E_j)$ that satisfied $\gamma(E_i, E_j).sup > sup_{min}$, there is $freq(MCP(E_i, E_j)) \geq freq_{min}$, it means that $MCP(E_i, E_j)$ is a FMCP. Thus, for any event $e \in E_i \cup E_j$, the occurrence number in some sequence satisfy $freq(e) \geq freq_{min}$. Therefore, the first step of mining FMCP is to find all the events whose occurrence number over $freq_{min}$, these events denoted as F^1. This step is consistent with the traditional method of frequent co-occurrence pattern mining.

According to the analysis of the above, any support more than the threshold value of event correlation, which incorporates the event set and the target event set can only consist of the event in F^1. Therefore, we can get the source through the events in combination F^1 events set and target set, and then to filter out support more than the threshold value of event temporal dependency.

The more the event sequences, the higher the cost of filter. If there is n_{seq} event sequences in Θ, and every event sequence contains m_i types of events, then in Θ there are $\dot{m} = \sum_{i=1}^{i=n_{seq}} m_i$ types of events. These events construct the temporal dependency sets can be denoted as $canP = \{\langle E_i, E_j \rangle\}$, then the number of $canP$ is $|canP| = C_m^2 * C_2^1 + \ldots + C_m^k * \left(C_k^1 + \ldots + C_k^{k-1}\right) + \ldots + C_m^m * \left(C_m^1 + \ldots + C_m^{m-1}\right)$. And for any $\langle E_i, E_j \rangle \in canP$, it is need to verify that if $\gamma(E_i, E_j).sup \geq sup_{min}$. It is obviously that the temporal dependency number need to be filter is very large.

Thus, we designed an extension strategy to avoid this problem based on the following Theorem.

Theorem 1: For any $FMCP(E_i, E_j)$ and the temporal dependency $\gamma(E_i, E_j)$ of event E_i, E_j. Assume that the frequency threshold is equal to the support threshold, i.e. $freq_{min} = sup_{min}$, then for any subset $E_i' \subseteq E_i$ and $E_j' \subseteq E_j$, they also be $FMCP\left(E_i', E_j'\right)$. Besides, when they satisfied that the time lag of temporal dependency $\gamma\left(E_i', E_j'\right).\Delta t \geq \gamma(E_i, E_j).\Delta t$, then there is $\gamma\left(E_i', E_j'\right).sup \geq \gamma(E_i, E_j).sup$.

Proof: It is obviously that the occurrence count number of E_j is $\gamma(E_i, E_j).sup$ within a time range $\gamma(E_i, E_j).\Delta t$ after the occurrence of E_i. For that $E_i' \subseteq E_i$ and $E_j' \subseteq E_j$, thus, the E_j' will occur at last $\gamma(E_i, E_j).sup$ times in a time range $\gamma(E_i, E_j).\Delta t$ after the occurrence of E_i'.

Based on the Theorem 1, we can inference that for any $\gamma(E_i, E_j)$ (the support has over the threshold), assume that $|E_i| > 1$, $|E_j| > 1$, the we can get the $FMCP(E_i, E_j)$ of $\gamma(E_i, E_j)$ by extending some $FMCP(\{\varepsilon_\alpha\}, \{\varepsilon_\beta\})$, in which $\varepsilon_\alpha \in E_i$ and $\varepsilon_\beta \in E_j$. We can compose the event in F^1 to get all the patterns such as $(\{\varepsilon_\alpha\}, \{\varepsilon_\beta\})$, then verify that if the pattern satisfied $\gamma(\{\varepsilon_\alpha\}, \{\varepsilon_\beta\}).sup \geq sup_{min}$, and filter the temporal dependencies whose support over the threshold. Thus, we can get $FMCP(\{\varepsilon_\alpha\}, \{\varepsilon_\beta\})$. Then, choose the remaining event in F^1 to extend the antecedent and consequent of $FMCP(\{\varepsilon_\alpha\}, \{\varepsilon_\beta\})$. At the same time, to verify that if the supports of temporal dependencies for antecedent and consequent are more than the threshold.

Algorithm 1 is the pseudocode of mining the temporal dependency. It first finds the events who had occurred more that sup_{min} times from the event space Θ, and put them into the F^1 (line 1–2). Then, modeling any two events into the pattern $(\varepsilon_\alpha, \varepsilon_\beta)$, and verifying that if the formed pattern is $TFCP_{1,1}$ (line 3–6). Next, it uses the *extend ()* function (line 18–26) to expand antecedents of $FMCP_{1,1}$ recursively (line 7–8). The extension process will break up until the extended pattern is not FMCP or there is no object in F^1 can be extended. Based on the result of extension for antecedents, we use *extend ()* function to expand consequents of $FMCP_{1,1}$ recursively (line 9–11). Finally, we put all the FMCP into the set P, the event relationship between antecedents and consequents components in P constitute the set R.

Algorithm1: *ETD-mining*

Input: Θ, event space which contains multi event sequences

$\delta_t \in [\Delta t_{min}, \Delta t_{max}]$, the time lag span

sup_{min} , *threshold of support*

Output: R, all significant temporal dependencies over Θ

1. *for each $seq_i \in \Theta$ // Generation stage*

2. *put the events into sup_{min}//whose occurrence number over sup_{min};*

3. *for each $\varepsilon_\alpha, \varepsilon_\beta \in F^1$*

4. *initialize the $E_i = \{\varepsilon_\alpha\}$, $E_j = \{\varepsilon_\beta\}$ and $P = \phi$;//P is FMCP*

5. *if(is FMCP(E_i, E_j, δ_t)) //Filter Stage*

6. *$P \leftarrow TFCP(E_i, E_j)$;*

7. *for each $\varepsilon \in F^1$ && $\varepsilon > E_i$ //ε is greater than E_i*

8. *extend $(E_i, \varepsilon, \delta_t)$; // Extension Stage, extend antecedents*

9. *for each $FMCP(E_p, E_q) \in P$*

10. *for each $\varepsilon'' \in F^1$ && $\varepsilon'' > E_j$*

11. *extend $(E_q, \varepsilon'', \delta_t)$; // extend consequents*

12. *For each $FMCP(E_u, E_v) \in P$*

13. *$R \leftarrow (E_u, E_v, FMCP(E_u, E_v).\Delta t, FMCP(E_u, E_v).freq)$; // The antecedents and consequents components of FMCP in P constitute the set R*

14. *return R;*

15. *isFMCP(E_i, E_j, δ_t):// judge that if E_i and E_j construct the $FMCP(E_i, E_j)$with the condition of δ_t;*

16. *if $(FMCP(E_i, E_j).\Delta t = max\{MCP(E_i, E_j).\Delta t\}$ && $FMCP(E_i, E_j).\Delta t \leq \delta_t)$*

17. *return true;*

18. *extend $\left((E_i, E_j), target, \varepsilon, \delta_t\right)$:*

19. *IF target == E_i*

20. $E_i \leftarrow \varepsilon$;

21. IF target $== E_j$

22. $E_j \leftarrow \varepsilon$;

23. IF (isTFCP (E_i, E_j, δ_t))

24. $P \leftarrow TFCP(E_i, E_j)$;

25. for each $\varepsilon' \in F^1 \&\& \varepsilon' > E_i$

26. extend $\left((E_i, E_j), target, \varepsilon', \delta_t \right)$;

---

4 Experiments

4.1 Experiment Dataset and Environment

Environments: The experiments are done on a PC with four Intel Core i5-6300HQ CPUs 2.30 GHz and 16.00 GB RAM. The operating system is Centos 6.4. All the algorithms are implemented in Java with JDK 1.8.5.

Dataset: There are totally 361 sensors deployed on 8 important devices. Each sensor generates one record per second, the dataset size is about 3.61 GB. And the dataset is divided into the training dataset and test dataset. The training set is from 2014-10-01 00:00:00 to 2015-03-31 23:59:59. The testing set is from 2015-04-01 00:00:00 to 2015-04-30 23:59:59.

We use real faults contained in the maintenance records of the plant power from 2014-07-01 00:00:00 to 2015-06-30 23:59:59 to verify our warnings.

4.2 Experiment Setup

Firstly, we conduct the temporal mining algorithm on the training dataset to find the temporal dependencies. We can get the temporal dependency quantity (TDQ) of the training data. We observe the variation trend of the TDQ under the different parameters of algorithm and different time of dataset.

Then, we built an anomaly prediction model with a directed graph over the anomaly events based on choosing and connecting their mined temporal dependencies. The graph is defined as $G = \langle V, E \rangle$, where V is the set of anomaly events, and $E \subseteq V \times V$ is a non-empty set of edges. Each direct edge $v_i \rightarrow v_j$ along with a weight, the weight is the time lag Δt, it means that if the anomaly event v_i occurs, we can predict that the anomaly event v_j will occur after a time lag Δt.

Based on the anomaly prediction graph, we conduct the experiment on test data set. We compare our anomaly prediction model with the other two typical approaches, they are the range-based approach [4], the outlier detection approach [5]. Once the associated anomalies are detected by them, they will make a warning of maintenance to the corresponding fault.

For evaluations, we consider the following performance metrics.

The Temporal Dependency Quantity (*TDQ*): The Number of temporal dependencies mined in the input data set;

Warning Time: Warning time is the difference between the timestamp an approach makes a warning of maintenance for a fault and the starting time of this fault;

Precision: Precision represents how many abnormities are accurate according to failure records;

Recall: It presents the probability of being able to classify positive cases, which is defined as following.

4.3   Experiment Result

Firstly, we conduct the experiment to verify how the value of TDQ changes under different data size with different time range when the parameter $sup_{min} = 0.8$ and $\Delta t = 10$. The data size increase from 1-month data to 6-months (the whole training set) dataset and each time for 1 month. This experiment mining the temporal dependences with no less than 0.8 probability (i.e., $sup_{min} = 0.8$). Table 1 shows the result of TDQ.

Table 1. Experiment results of TDQ under different data size ($sup_{min} = 0.8$, $\Delta t = 10$ min).

Time range	Data size	TDQ
2014.07 (1 month)	180 MB	4274
2014.07-2014.08 (2 months)	378 MB	4338
2014.07-2014.09 (3 months)	534 MB	4269
2014.07-2014.10 (4 months)	750 MB	4407
2014.07-2014.11 (5 months)	912 MB	4430
2014.07-2014.12 (6 months)	1128 MB	4954

The Table 1 shows that as the size of the data set increases, the total TDQ also shows an upward trend. However, it is clearly that there is no linear correlation between the TDQ and the size of the data set. Overall, the size of the data set has risen from one month to six months, and the TDQ generated is relatively close, always between 4,000 and 5,000. It is indicated that, as the time goes on increases, the TDQ does not rise rapidly, but increases relatively slowly and slowly, and may even be controlled within a certain range. The reason is that as the time goes on, the more temporal dependencies gradually enriched, but the growth rate trend is stable. This result indicated that our algorithm has a certain robustness.

Then, we compare the precision results and recall results of different methods. Notably, in this paper, we only consider the predict events with finally failures occurring both in training set and testing set. Figure 2 shows the final average results.

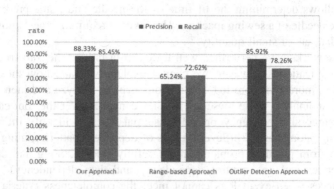

Fig. 2. Experiment Results of Precision and Recall under different approach

Figure 2 indicates that our methods performs well among the three methods, and achieves the highest accuracy and the second highest recall rate.

As shown in the Fig. 2, the precision of our approach is 88.33%, and the recall rate is 85.48%. The analysis of the intermediate results revealed that some of the repetitive anomaly propagation paths excavated in the training cannot be detected in the test set. For the same type of faults, the abnormal propagation path excavated in the test data set has changed compared to the training set. This phenomenon is essentially caused by caused by the character of uncertainty in the stream data.

The precision of the ranged-based approach and the outlier detection approach are 65.24% and 85.92%, and the recall of them are 72.62% and 78.26%. Analysis of the intermediate results revealed that the two approaches are based on the single sensor data. However, usually one fault may be caused by multiple anomalies, it can correspond to multiple isolated anomalous points. The signal sensor data-based detection method cannot detect such a fault. Our approach will find the correlation between multi sensors and then form the anomaly propagation paths, this helps us to find more hidden anomalies.

The above results show that the temporal dependency has an effective effect in constructing fault prediction logic and fault detection.

5   Related Works

Recently, researchers have designed several approaches dealing with the problem of anomaly prediction for the predictive maintenances. Several quantitative models ranging from simple linear discriminant analysis, more complex logistic regression analysis, and neural networks have been proposed for prediction [14]. Zhang et al. [1] presented a novel system that automatically parses streamed console logs and detects early warning signals for IT system failure prediction based on the LSTM neural network. Susto et al. [15] developed a multiple classifier machine learning methodology to deal with the unbalanced datasets that arise in maintenance classification problems. Baban et al. [16] used a fuzzy logic approach to develop a decision-making system that allows determining the lifetime of the needle and plane predictive maintenance of the needle of a sewing machine. However, it required expert knowledge and depends on datasets of small quantity.

The above works have performed well to detect or predict the univariate anomaly for the IoT applications. However, we cannot explain why and how these approach works. Besides, more and more IoT applications need to analyze and identify the root causes of the anomalies, while these approaches cannot answer the root causes.

Mining temporal dependency provide essential clues to identify the cause relationship among anomalies [17]. Several types of dependent pattern mining tasks have been induced from practical problems and carefully studied. Song and et al. mined activity dependencies (i.e., control dependency and data dependency) to discover process instances when event logs cannot meet the completeness criteria [7]. A dependency graph is utilized to mine process instances. However, the authors do not consider the dependency among events. Plantevit et al. presented a new approach to mine temporal dependencies between streams of interval-based events [8]. Friedberg et al. proposed a novel anomaly detection approach. It keeps track of system events, their dependencies and occurrences, and thus learns the normal system behavior over time and reports all actions that differ from the created system model [18].

In this paper, we introduce the temporal dependency into the predictive maintenance to improve the explanation of prediction approaches and discovery the root cause of anomaly.

6 Conclusion

In this paper, we try to propose an effective and explainable approach to predict the anomaly based on mining the temporal dependencies from multi-sensor event sequences. To reach this goal, we detect "univariate anomaly" events from sensor data and output multi-sensor event sequences. Then, we transform the temporal dependency mining problem into a frequent co-occurrence pattern mining problem. Furthermore, a lot of experiments have been done to show the effectiveness of our approach based on a real dataset from a coal power plant. But the speed of our method can also be improved. So, for future work, we are interested in parallel optimization algorithms to speed up our method.

Acknowledgement. This work is supported by "The National Key Research and Development Plan (No: 2017YFC0804406), Public Safety Risk Prevention and Control and Emergency Technical equipment", and the "Key Project of the National Natural Science Foundation of China No. 61832004 (Research on Big Service Theory and Methods in Big Data Environment)".

References

1. Zhang, K., Xu, J., Min, M.R.: Automated IT system failure prediction: a deep learning approach. In: International Conference on Big Data, pp. 1291–1300 (2016)
2. Sipos, R., Fradkin, D., Moerchen, F.: Log-based predictive maintenance. In: Proceedings of the 20th ACM SIGKDD International Conference on Knowledge Discovery and Data Mining, pp. 1867–1876. ACM, New York (2014)
3. Song, F., Zhou, B., Sun, Q.: Anomaly detection and explanation discovery on event streams. In: Business Intelligence for the Real-Time Enterprises (2018)
4. Ishimtsev, V., Bernstein, A., Burnaev, E.: Conformal k-NN anomaly detector for univariate data streams. arXiv: Machine Learning, pp. 213–227 (2017)
5. Siffer, A., Fouque, A., Termier, A.: Anomaly detection in streams with extreme value theory. In: Proceedings of the 23rd ACM SIGKDD International Conference on Knowledge Discovery and Data Mining, Canada, pp. 1067–1075. ACM (2017)
6. Qiu, H., Liu, Y., Subrahmanya, N.A., Li, W.: Granger causality for time-series anomaly detection. In: Proceedings of the IEEE International Conference on Data Mining, ICDM 2012, pp. 1074–1079 (2012)
7. Song, W., Jacobsen, H.A., Ye, C., Ma, X.: Process discovery from dependence-complete event logs. IEEE Trans. Serv. Comput. **9**(5), 714–727 (2016)
8. Plantevit, M., Robardet, C., Scuturici, V.M.: Graph dependency construction based on interval-event dependencies detection in data streams. Intell. Data Anal. **20**(2), 223–256 (2016)
9. Yan, Y., Luh, P.B., Pattipati, K.R.: Fault diagnosis of HVAC air-handling systems considering fault propagation impacts among components. IEEE Trans. Autom. Sci. Eng. **14**(2), 705–717 (2017)
10. Cao, Y., Liu, C., Han, Y.: A frequent sequential pattern based approach for discovering event correlations. In: Meng, X., Li, R., Wang, K., Niu, B., Wang, X., Zhao, G. (eds.) WISA 2018. LNCS, vol. 11242, pp. 48–59. Springer, Cham (2018). https://doi.org/10.1007/978-3-030-02934-0_5
11. Zeng, C., Tang, L., Zhou, W.: An integrated framework for mining temporal logs from fluctuating events. IEEE Trans. Serv. Comput. **12**(2), 199–233 (2019)
12. Domingues, R., Filippone, M., Michiardi, P., Zouaoui, J.: A comparative evaluation of outlier detection algorithms: experiments and analyses. Pattern Recogn. **74**, 406–421 (2018)
13. Yagci, A.M., Aytekin, T., Gurgen, F.S.: Scalable and adaptive collaborative filtering by mining frequent item co-occurrences in a user feedback stream. Eng. Appl. Artif. Intell. **58**, 171–184 (2017)
14. Warriach, E.U., Tei, K.: Fault detection in wireless sensor networks: a machine learning approach. In: 16th IEEE International Conference on Computational Science and Engineering, CSE 2013, Sydney, NSW, pp. 758–765 (2013)
15. Susto, G.A., Schirru, A., Pampuri, S.: Machine learning for predictive maintenance: a multiple classifier approach. IEEE Trans. Ind. Inform. **11**(3), 812–820 (2015)
16. Baban, C.F., Baban, M., Suteu, M.D.: Using a fuzzy logic approach for the predictive maintenance of textile machines. J. Intell. Fuzzy Syst. **30**(2), 999–1006 (2016)
17. Perng, C., Thoenen, D., Grabarnik, G., Ma, S., Hellerstein, J.: Data-driven validation, completion and construction of event relationship networks. In: Proceedings of the 9th ACM SIGKDD International Conference on Knowledge Discovery and Data Mining, pp. 729–734. ACM (2003)
18. Friedberg, I., Skopik, F., Settanni, G., Fiedler, R.: Combating advanced persistent threats: from network event correlation to incident detection. Comput. Secur. **48**, 35–57 (2015)

An Anomaly Pattern Detection Method for Sensor Data

Han Li[1,2], Bin Yu[1,2(✉)], and Ting Zhao[3]

[1] College of Computer Science, North China University of Technology, Beijing, China
lihan@ncut.edu.cn, yubin0574@qq.com
[2] Beijing Key Laboratory on Integration and Analysis of Large-Scale Stream Data, Beijing, China
[3] Advanced Computing and Big Data Technology Laboratory of SGCC, Global Energy Interconnection Research Institute, Beijing, China
zhaoting@geiri.sgcc.com.cn

Abstract. With the development of the Internet of Things (IOT) technology, a large number of sensor data have been produced. Due to the complex acquisition environment and transmission condition, anomalies are prevalent. Sensor data is a kind of typical time series data, its anomaly refers to not only outliers, but also the anomaly of continuous data fragments, namely anomaly patterns. To achieve anomaly pattern detection on sensor data, the characteristics of sensor data are analyzed including temporal correlation, spatial correlation and high dimension. Then based on these characteristics and the real-time processing requirements of sensor data, a sensor data oriented anomaly pattern detection approach is proposed in this paper. In the approach, the frequency domain features of sensor data are obtained by Fast Fourier Transform, the dimension of the feature space is reduced by describing frequency domain features with statistical values, and the high-dimensional sensor data is processed in time on the basis of Isolation Forest algorithm. In order to verify the feasibility and effectiveness of the proposed approach, experiments are carried out on the open dataset IBRL. The experimental results show that the approach can effectively identify the pattern anomalies of sensor data, and has low time cost while ensuring the high accuracy.

Keywords: Sensor data · Fourier Transform · Frequency domain feature · Isolation Forest · Anomaly pattern detection

1 Introduction

In recent years, as an information carrier based on Internet and traditional telecommunication network, the Internet of Things (IOT) enables all ordinary physical objects to interconnect and exchange information, and supports more intelligent physical object management. The rapid development of IoT technology can provide more abundant data. Based on these data, more abundant data analysis can be carried out to provide more accurate services. The environment of IoT is extremely complex, and there are many problems such as equipment failure, signal interference, abnormal transmission,

© Springer Nature Switzerland AG 2019
W. Ni et al. (Eds.): WISA 2019, LNCS 11817, pp. 270–281, 2019.
https://doi.org/10.1007/978-3-030-30952-7_28

etc. Therefore, the sensor data inevitably have anomalies. For abnormal sensor data, if it is not processed and are directly analyzed, there are two potential problems. Firstly, abnormal data will affect the accuracy of data analysis results, resulting in invalid decision-making. Secondly, if the anomalies hidden in the sensor data can't be identified as early as possible, it is not conducive to timely discovery of the physical world problems, and may cause unnecessary losses. Therefore, anomaly detection for sensor data is particularly important. Reliable anomaly detection can not only make data analysis decision more effective, but also detect anomaly sensors in time to reduce losses.

Hawkins [1] defines anomalies as distinct data in a data set, which makes one suspect that these data are not random deviations, but are generated by completely different mechanisms. Anomaly detection is a process of discovering abnormal data in data resources by using various data processing models and technologies. It is the premise and necessary link of discovering data anomalies and improving data quality. Traditional anomaly detection mostly aims at outliers, and is based on statistics, distance, density and clustering methods. But sensor data is mostly time series data, and there are not only outlier abnormalities, but also timing fragment abnormalities, that is, pattern abnormalities. Among them, anomaly pattern means that there is no anomaly at any data point of time series data (such as the data value is no more than the threshold), but the trend of this data fragment is obviously different from that of other similar data fragments. In addition, it is common to use multiple sensor devices to monitor the same physical entity. For example, various types of sensor data are usually combined to describe the operating conditions of industrial equipment or the environment. Thus, sensor data are always high dimensional. Similar to the single-dimensional time series, high-dimensional time series also has the problem of pattern anomalies, and efficiency is considered as one of the most important issues. If the traditional outlier detection methods are used to these high-dimensional time series, there will be a problem of high time cost, and it is difficult to judge the overall abnormal situation according to the abnormal situation of a single point.

Therefore, this paper considers the high-dimensional time series as the sensor data, and an anomaly pattern detection method is proposed for these sensor data to improve the efficiency under the premise of ensuring accuracy.

2 Related Work

In recent years, as a branch of data mining, anomaly detection is receiving more and more attention. Among them, according to the object of anomaly detection, it is mainly divided into anomaly detection of outliers and anomaly detection of time series data.

Outlier detection can be roughly divided into four categories: statistical-based [2], distance-based [3], density-based [4] and clustering-based [5] methods. The statistical-based approach is a model-based approach that firstly creates a model for the data and then evaluates it [6]. However, such methods need sufficient data and prior knowledge, and are more suitable for outlier detection of individual attributes. The distance-based method is similar to the density-based method. The distance-based method is based on the distance between the point and other points [7], and the density-based method is based on whether there are enough points in the neighborhood of the point [8]. These two methods are simple in thought, but generally require $O(n^2)$ time overhead. The cost is too high for large data sets, and the methods are also sensitive to parameters. Without proper parameters, the performance of the algorithm will be worse. Clustering-based method regards data that do not belong to any cluster as outliers [9]. Some clustering-based methods, such as K-means, have linear or near-linear time and space complexity, so this kind of algorithm may be highly effective for outlier detection. The difficulty lies in the selection of cluster number and the existence of outliers. The results or effects produced by different cluster numbers are completely different, so each clustering model is only suitable for a specific data type.

In the aspect of time series data anomaly detection, pattern anomaly detection algorithms for time series data have also made some achievements. Chen [10] et al. proposed the D-Stream clustering algorithm for clustering of time series data. Its main idea is to divide the data space into a series of grids in advance. By mapping the time series data to the corresponding grids, the results of grid processing can be obtained. However, this algorithm requires users to set more parameters in advance and the accuracy is low. Yan [11] et al. used the probability density function of data to re-express Euclidean distance, and obtained a probability measure to calculate the dissimilarity between two uncertain sequences. However, the detection effect of this algorithm depends on the size of the detection window. There is no suitable method to find the appropriate detection window size, and there are certain requirements for data. Cai [12] et al. proposed a new anomaly detection algorithm for time series data by constructing distributed recursive computing strategy and k-Nearest Neighbor fast selection strategy. However, the algorithm is effective for one-dimensional sensor data and does not consider multi-dimensional data.

Considering the problems of the above methods, this paper proposes an anomaly pattern detection method for high-dimensional sensor data. The method uses Fast Fourier Transform to transform time domain data into frequency domain data, and then realizes data dimensionality reduction through feature extraction. Finally, the anomaly patterns of sensor data are detected by using the spatial-temporal correlation characteristics of sensor data and the Isolation Forest algorithm.

3   Anomaly Pattern Detection Method for Sensor Data

3.1   Sensor Data Characteristics

Sensor data refers to the data collected by sensor devices, which is often used to continuously perceive the information of the physical world. Therefore, sensor data has many special characteristics. In order to identify the pattern anomalies of sensor data more pertinently, the main characteristics of sensor data are analyzed as follows:

(1) Time continuity: Sensor data is generated continuously by sensors. Generally, sensor data acquisition is carried out according to a certain frequency, so time continuity is one of the most basic characteristics of sensor data.
(2) Spatial-temporal correlation: Sensor data is usually used to perceive the information of the physical world, so the sensor data will be associated with the

physical world it perceives, and the association is specifically expressed as spatial and temporal correlation. That is to say, in different time or space environment, the sensor data collected by the same sensor may also be different.

(3) Data similarity: Based on the spatial-temporal correlation of sensor data, the sensor data collected by the same type of sensors in the similar time and space range have similarity. For example, different sensors which are set up in the same environment detect the similar environmental indicators at the same time. In addition, if the monitoring object of the sensor has similar behavior, the sensor data used to describe the behavior should also have similarity. For example, if users have similar electricity consumption patterns, their meter data fluctuations should be roughly similar. If there are abnormalities, there may be problems such as electricity theft. Therefore, it can be considered that similar sensor data have data similarity under similar time, space or behavior conditions.

(4) High-dimensionality: In the actual production environment, the single-dimensional sensor data has the problem of not being able to describe the complex physical world. Therefore, it is often necessary to combine various types of sensor data to describe the state of a physical entity, thus forming a high-dimensional sensor data. Taking the sensor data for monitoring the working conditions of thermal power generators as an example, the dimensions of the data are up to dozens.

3.2   The Proposed Approach

The goal of anomaly pattern detection method for sensor data is to quickly identify anomaly patterns in high-dimensional sensor data. The method is divided into two stages: data preprocessing and anomaly pattern detection.

The goal of data preprocessing stage is to reduce the data dimension for the accurate and fast detection of abnormal patterns on the premise of ensuring the characteristics of sensor data patterns. Briefly, the Fast Fourier Transform is used to transform the time series data into the frequency domain data, and then the characteristics of the frequency domain data are extracted for dimension reduction.

The goal of anomaly pattern detection stage is to improve the efficiency on the premise of ensuring accuracy. Briefly, based on the spatial-temporal correlation and data similarity of the sensor data, the sensor data with obvious difference from the pattern of the adjacent sensor data is found by comparing the time and space related sensor data. In order to solve the problem of fast processing of high-dimensional data, the method adopts the idea of ensemble learning and detects anomaly patterns based on Isolation Forest algorithm.

Data Preprocessing. Due to the time continuity and high-dimensionality of the sensor data, the sensor data segment consists of a large number of temporally consecutive high-dimensional data points. It is not only difficult to directly describe the pattern of the sensor data segments by these continuous high-dimensional data points, but also difficult to quickly identify the pattern abnormality. Therefore, the data preprocessing stage mainly focuses on the extraction of sensor data pattern features and the dimensionality reduction of data features. Figure 1 shows the workflow of the data preprocessing stage, which consists of feature extraction and feature reduction.

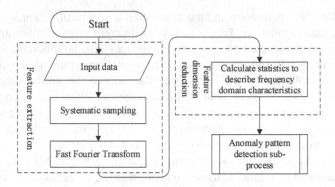

Fig. 1. Workflow of the data preprocessing stage

(1) Feature extraction

Time-frequency transform is used to discard the single difference of sensor data in time domain, and frequency domain features are used to describe the pattern characteristics of data fragments. In order to ensure the efficiency of anomaly pattern detection, Fast Fourier Transform (FFT) with low time complexity is used to transform time domain data to frequency domain data. Fast Fourier Transform makes full use of the symmetry and periodicity of exponential factors in Discrete Fourier Transform (DFT) formulas, and then obtains the results of Discrete Fourier Transform corresponding to these short sequences and combines them appropriately, so as to delete repetitive computation, reduce multiplication and simplify the structure. Compared with the time complexity of $O(n^2)$ in the Discrete Fourier Transform, the Fast Fourier Transform can reduce the time complexity to O (nlogn) level. In the case of a larger amount of data, the advantage of the Fast Fourier Transform in terms of time is more obvious. Assuming that the sensor data segment is $T = \{T_1, T_2, ..., T_i, ..., T_n\}$, and T_i is the data segment of the i-th dimension data. The extraction step of the sensor data feature is as follows: Firstly, the data fragments of all dimensions in T are sampled by equidistant sampling method, requiring that each dimension has 2^m sample points $\{t_1, t_2, ..., t_j, ..., t_2^m\}$, and t_j is the j-th sample point, and 2^m should be as close as possible to the number of original samples in this dimension. Secondly, performing Fast Fourier Transform on the sample data on the data segment of each dimension, and obtaining n frequency domain data sets $\{F_1, F_2, ..., F_i, ..., F_n\}$, and F_i is the frequency domain data of the i-th dimension data fragment in sensor data.

(2) Feature dimension reduction

In order to ensure that the frequency domain data can depict the time domain data as accurately as possible, the sample density of the time domain data is high. Therefore, the frequency domain data obtained by time-frequency transformation also has a high data density, that is, a large amount of data. For the purpose of ensuring the fast processing of high-dimensional data fragments, the method has to reduce the dimensionality of data. Because the amplitude of the sensor data at a certain frequency in the frequency domain space is directly related to the modulus of the result value of the fast

Fourier transform at that frequency. In order to reflect the concentration trend, the degree of dispersion and the maximum amplitude of the sensor data in the frequency domain space, the mean, variance and peak value of each dimension frequency domain data module of the sensor data fragment are selected as the frequency domain characteristics of the sensor data fragment.

Anomaly Detection. In this paper, the anomaly pattern of sensor data is defined as a pattern with "few but different" characteristics compared with other patterns. Based on the definition, the anomaly pattern does not only refer to the wrong pattern, but mainly emphasizes the specificity of the pattern. The pattern anomaly of high-dimensional sensor data can appear in any dimension, so it is necessary to detect the data of each dimension to identify the anomalies of the sensor data. Obviously, it would be more accurate to detect pattern anomalies in each data dimension, but the computational efficiency will be very low. Therefore, the goal of anomaly detection in this paper is to improve the efficiency of anomaly detection on the premise of ensuring accuracy.

Isolation tree is a random binary tree, which is similar to decision tree, can be used for data classification. When constructing binary tree, the isolation tree randomly selects attribute values, and identifies anomalies by the depth (also known as isolation depth) of the data in the binary tree. Therefore, the algorithm not only makes full use of the "few but different" characteristics of anomalies, but also has good processing performance. However, due to the randomness of the construction process of isolation tree, it is a weak classifier. Therefore, it is necessary to adopt ensemble learning method to improve the generalization performance of anomaly detection. In this paper, the anomaly detection is implemented by the Isolation Forest algorithm [13] based on bagging ensemble learning method and isolation tree algorithm. The workflow is shown in Fig. 2.

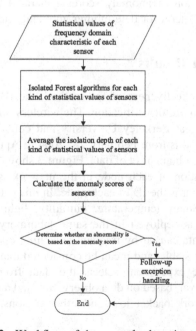

Fig. 2. Workflow of the anomaly detection stage

As shown in Fig. 2, anomaly detection consists of three steps: Firstly, the statistical values of each sensor are processed by the Isolation Forest algorithm to obtain the corresponding isolation depth $\{D_1, D_2, ..., D_i, ..., D_n\}$, where D_i is the isolation depth of the i-th sensor, including the average isolation depth $\{m_i, v_i, p_i\}$ of mean, variance and peak value. Secondly, calculating the total average isolation depth $\{d_1, d_2, ..., d_i, ... d_n\}$ according to the statistical values, where d_i is the isolation depth of the i-th sensor, and $d_i = (m_i + v_i + p_i)/3$, which is used to calculate anomaly score. Thirdly, the anomaly score is calculated for each sensor and the anomaly is judged according to the score. According to the definition of the anomaly score of the Isolation Forest algorithm, this paper defines the anomaly score $s(i, M)$ of the i-th sensor as 2^{-n}, where n is the ratio of the average isolation depth d_i of the i-th sensor to the average path length c (M) of the M isolation trees used to construct the isolation forest. The calculation method is as shown in Eq. (1), and the value of c(M) is calculated by Eq. (2).

$$s(i, M) = 2^{-\frac{d_i}{c(M)}} \tag{1}$$

$$c(M) = 2\ln(M - 1) + \xi - \frac{2(M - 1)}{M}, \text{where } \xi \text{ is Euler constant} \tag{2}$$

Based on the definition and calculation method of anomaly score, the anomaly score $s(i, M)$ has the following properties: Firstly, the range of anomaly score is [0,1], and the closer to 1, the higher the probability of anomaly. Secondly, if the anomaly scores of all samples are smaller than 0.5, it can be basically determined as normal data. Thirdly, if the anomaly scores of all samples are around 0.5, the data does not contain significant anomalous samples. According to the above properties of anomaly score, the anomaly of sensor data can be determined. It should be noted that anomalies are relative, and the distribution of anomaly scores generated by different data sets is different, so the specific criteria for determining anomalies are also different.

4 Experiments and Results

In order to evaluate the effectiveness of the method, the IBRL [14] (Intel Berkeley Research Lab) data set is used for verification. The wireless sensor network is deployed at the Intel Research Lab at Berkeley University and contains 54 Mica2Dot sensor nodes. The sampling period is from February 28, 2004 to April 5, 2004, and sampling is performed every 30 s to obtain a set of data. Figure 3 shows the deployment diagram of the network. The location of each node in the network is represented by a black hexagon. The white number is the ID of each node. In this network, each node collects four types of values, namely temperature, humidity, light and voltage. Since the wireless sensor network is deployed in the same laboratory, except for the sudden change of illumination data caused by frequent switching operations, other data sampling values are relatively stable, and it can be considered that the data obtained by the sensors are similar. The experiment selects the data from 08:00:00-24:00:00 on February 28, 2004. However, based on data observation and analysis, some observation data is lost due to network packet loss and other reasons. In order to ensure the reliability of the experiment, the missing values are interpolated based on the average or spatial correlation characteristics, and the values sampled in the time period of 15 s to 20 s per minute are used.

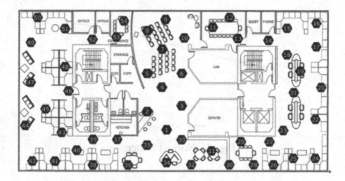

Fig. 3. Node distribution in the IBRL data set

To evaluate the performance of the abnormal pattern detection algorithm, some normal observation values were randomly modified to become abnormal data. To avoid the generality, the distribution of abnormal data should be different from the distribution of normal data, and the sample space should overlap as much as possible. In addition, the anomaly pattern should be a small probability event relative to the normal sample set collected by the non-faulty node. Therefore, the abnormal data generated by the simulation has a slight deviation from the normal sample data distribution. The details of the data set are shown in Tables 1 and 2. Table 1 shows the details of the 64-min data for a single experiment, and Table 2 shows the overall data details of this experiment.

Table 1. Data set per 64 min

Time	Number of normal sensors	Number of abnormal sensors	Number of normal samples	Number of abnormal samples
Every 64 min	52	2	3416	40

Table 2. Overall data set situation

Time	Total normal sensor cumulative value	Total anomaly sensor cumulative value	Total number of normal samples	Total number of anomaly samples
8:00-24:00	780	30	51240	600

In order to show the differences between the abnormal pattern and the normal pattern, the normal data of 8:00:17-8:30:17 is shown in Fig. 4.

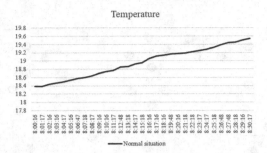

Fig. 4. Normal pattern

There are two kinds of abnormal patterns, which are mutation abnormalities and trend abnormalities, as shown in Fig. 5.

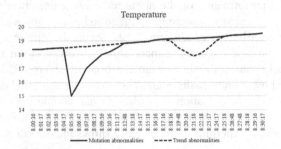

Fig. 5. Mutation abnormalities and Trend abnormalities

4.1 Evaluation Indicators

- Recall rate
  It is also known as True Positive Rate (TPR) or sensitivity (sensitivity). This indicator corresponds to the proportion of all abnormalities actually detected in the data set, that is, the ratio of correctly detected anomalies to all anomalies in the data set.
- False Positive Rate (FPR)
  It is also known as the false detection rate. The indicator corresponds to the ratio of the normal sample in the data set that is incorrectly judged to be abnormal, that is, the ratio of the sample misjudged as abnormal to all normal samples in the data set.
- Accuracy rate
  The indicator corresponds to the ratio of the correctly predicted samples in the data set, that is, the ratio of the correctly predicted samples to all samples in the data set.

4.2   Experimental Analysis

Since two sensors are selected to simulate 20 outliers every 64 min in the sensor data of 54 sensors, the experiment takes the two sensors with the largest abnormality score as sensors with abnormal pattern, and then observe whether the simulated anomaly sensor is correctly detected. Due to the randomness of the Isolation Forest, the experiment conducted 10 times on the data set. Table 3 shows the experimental results corresponding to each experiment.

Table 3. Experimental results

Data set time	Recall rate	False Positive Rate	Accuracy rate
8:00-24:00	0.8	0.0077	0.985
8:00-24:00	0.833	0.0064	0.987
8:00-24:00	0.8	0.0077	0.985
8:00-24:00	0.867	0.0051	0.99
8:00-24:00	0.833	0.0064	0.987
8:00-24:00	0.8	0.0077	0.985
8:00-24:00	0.8	0.0077	0.985
8:00-24:00	0.833	0.0064	0.987
8:00-24:00	0.8	0.0077	0.985
8:00-24:00	0.8	0.0077	0.985

The experimental results show that the recall rate of the results using the proposed method maintains above 80%, indicating that most sensors with abnormal pattern can be detected. At the same time, by observing the experimental data of the false-detection sensor, it is found that the misdetection of most sensors may be caused by the sudden change of illumination. Because the experimental data is collected under real conditions, there are cases of switching lights or curtains. For example, the No. 20 sensor by the window and the No. 11 sensor in the closed environment both have cliff-like rise or fall in the value of illumination attributes. As a result, the abnormal pattern is more obvious than the abnormal pattern simulated in this experiment, and the final calculated abnormality score is also higher, resulting in false detection.

4.3   Comparisons

In order to verify the efficiency of the proposed method, a comparative experiment is carried out between the commonly used anomaly detection algorithms DBSCAN, One Class SVM and the proposed method. In order to ensure that the anomaly can be detected normally, the statistical values after the fast Fourier transform are used as the data set of this comparative experiment. Table 4 shows the results.

Table 4. Efficiency of the three methods

Algorithm name	Run time/s	Accuracy rate
Isolation Forest	1.716 s	0.995
DBSCAN	4.400 s	0.935
One Class SVM	33.114 s	0.966

According to the above results, the advantage of the isolation forest algorithm in efficiency is relatively obvious on the basis of ensuring accuracy. Because DBSCAN and One Class SVM are density-based spatial clustering algorithms, which need to calculate the distance from the core point to the surrounding, and if the amount of data is larger, the computational complexity is higher.

5   Conclusions

Aiming at the pattern abnormality of sensor data, this paper analyses the characteristics of sensor data, including time continuity, spatial-temporal correlation, data similarity and high dimensionality. Considering these characteristics, an anomaly pattern detection method for sensor data is proposed under the consideration of efficiency and accuracy. In the approach, the frequency domain features of sensor data are obtained by Fast Fourier Transform, and then the dimension of feature space is reduced by describing frequency domain features with statistical values. Finally, the sensor with anomaly patterns is determined by anomaly detection using Isolation Forest algorithm. The experimental results show that the method not only has advanced accuracy, but also has obvious advantages in running time and can effectively identify the pattern anomalies of sensor data. In the future, the main work is to carry out more extensive experiments and to study anomaly pattern detection methods for more complex sensor data.

Acknowledgements. This paper is supported by the Scientific and Technological Research Program of Beijing Municipal Education Commission (KM201810009004) and the National Natural Science Foundation of China (61702014).

References

1. Hawkins, D.M.: Identification of Outliers. Springer, Netherlands (1980). https://doi.org/10.1007/978-94-015-3994-4
2. Xi, Y., Zhuang, X., Wang, X., et al.: A research and application based on gradient boosting decision tree. In: 15th International Conference on Web Information Systems and Applications, pp. 15–26 (2018)
3. Ramaswamy, S., Rastogi, R., Shim, K.: Efficient algorithms for mining outliers from large data sets. ACM SIGMOD Rec. **29**(2), 427–438 (2000)

4. Frank, R., Jin, W., Ester, M.: Efficiently mining regional outliers in spatial data. In: Papadias, D., Zhang, D., Kollios, G. (eds.) SSTD 2007. LNCS, vol. 4605, pp. 112–129. Springer, Heidelberg (2007). https://doi.org/10.1007/978-3-540-73540-3_7
5. Gaddam, S., Phoha, V., Balagani, K.: K-Means+ID3: a novel method for supervised anomaly detection by cascading K-Means clustering and ID3 decision tree learning methods. IEEE Trans. Knowl. Data Eng. **19**(3), 345–354 (2007)
6. Kasliwal, B., Bhatia, S., Saini, S., et al.: A hybrid anomaly detection model using G-LDA. In: 2014 IEEE International Advance Computing Conference, Gurgaon, pp. 288–293 (2014)
7. Zhang, Y., Du, B., Zhang, L., et al.: A low-rank and sparse matrix decomposition-based Mahalanobis distance method for hyperspectral anomaly detection. IEEE Trans. Geosci. Remote Sens. **53**(3), 1–14 (2015)
8. Huang, T., Zhu, Y., Zhang, Q., et al.: An LOF-based adaptive anomaly detection scheme for cloud computing. In: 37th Annual Computer Software and Applications Conference Workshops, Japan, pp. 206–211 (2013)
9. Münz, G., Li, S., Carle, G.: Traffic anomaly detection using K-Means clustering. In: 4th GI/ITG-Workshop MMBnet, Hamburg (2007)
10. Chen, Y.: Density-based clustering for real-time stream data. In: 13th ACM SIGKDD International Conference on Knowledge Discovery and Data Mining, California, pp. 133–142 (2007)
11. Yan, Q.Y., Xia, S.X., Feng, K.W.: Probabilistic distance based abnormal pattern detection in uncertain series data. Knowl. Based Syst. **36**(11), 182–190 (2012)
12. Cai, L., Thornhill, N., Kuenzel, S., et al.: Real-time detection of power system disturbances based on k-nearest neighbor analysis. IEEE Access **99**, 1–8 (2017)
13. Liu, F., Ting, K., Zhou, Z.H.: Isolation forest. In: 8th IEEE International Conference on Data Mining, Los Alamitos, pp. 413–422 (2008)
14. Intel Berkeley Research Lab dataset. http://db.csail.mit.edu/labdata/labdata.html. Accessed 18 Apr 2019

Natural Language Processing

A Sequence-to-Sequence Text Summarization Model with Topic Based Attention Mechanism

Heng-Xi Pan[1], Hai Liu[1,2(✉)], and Yong Tang[1]

[1] South China Normal University, Guangzhou 510631, China
liuhai@m.scnu.edu.cn
[2] Guangzhou Key Laboratory of Big Data Intelligent in Education,
Guangzhou 510631, China

Abstract. One of the limitation of automatic summarization is that how to take into account and reflect the implicit information conveyed between different text and the scene influence. In particularly, the generation of news headlines should under specific scene and topic in the field of journalism. Traditionally, Sequence-to-Sequence (Seq2Seq) with attention model has shown great success in summarization. However, Sequence-to-Sequence (Seq2Seq) with attention model is focusing on features of the text only, not on implicit information between different text and scene influence. In this work, we present a combination of techniques that harness scene information which reflects by word topic distribution to improve abstractive sentence summarization. This model combines word topic distribution of LDA topic model as an external attention mechanism to better text summarization result. This model contains an RNN network as an encoder and decoder part, encoder is used to embed original text in a low dimensional dense vector as previous works, and decoder uses attention mechanism to incorporate word-topic distribution and low dimensional dense vector of encoder. The proposed approach is evaluated by datasets of CNN/Daily Mail and citation. The result shows that it is better than the aforementioned methods.

Keywords: Text summarization · Attention mechanism · Topic model

1 Introduction

Summarization is the process for shortening a text document with software. Lots of technologies in machine learning and data mining can make a coherent summary with the major points of the original document to convey the most important information. It can efficiently solve the problem of finding main information in massive information. Currently, there are two general approaches to automatic text summarization: extraction and abstraction. Abstractive summarization attempts to use new sentences to describe the main idea of the original context, while extractive summarization extracts the important sentences of text to form the summarization [1]. Compared to extraction, abstraction is more close to the way how humans do summarization because of abstractive methods use new sentences to describe the main idea of the original context with the assumption that the new sentences are grammatically correct and human

© Springer Nature Switzerland AG 2019
W. Ni et al. (Eds.): WISA 2019, LNCS 11817, pp. 285–297, 2019.
https://doi.org/10.1007/978-3-030-30952-7_29

readable. Automatic text summarization has played an important role in a variety of Internet applications, such as the news headlines generation of news website, the product highlights of e-commerce website, and the content targeting of social media.

Automatic text summarization can be based on machine learning. For example, Kupiec [2] used naive Bayesian classification model to determine whether every sentence of the article belong to summary, Lin [3] considered the correlative of the summary and introduced a method of decision tree, Osborne [4] observed the association of various features and constructed a method of Long-Linear Models, Conroy and O'leary [5] used hidden Markov model to generate summary. Recently, with the rapid development of deep learning techniques, deep neural network models have been widely used for NLP tasks such as automatic text summarization, especially the attention based sequence-to-sequence (seq2seq) framework [6] prevails in automatic text summarization. On this foundation, Nallapati et al. [7] introduced Generator-Pointer mechanism to solve the problem of OOV and low frequency words; Kikuchi et al. [8] constructed a LenEmb and Lenlnit model to control the length of summarization. All these neural network models only focus on features of self-text to promote the quality of summary. However, these works neglect the problem of the implicit information conveyed between different text and the influence of scene. In fact, when people write an article, they will base on given background or scene. Therefore, it is important and useful to do summarization based on certain scene and thus is of interest to us in this paper.

As the idea mentioned above, particular scene may play an important role in summarization. As to a scene, we can use the word distribution of text topics to reflect it. Word distribution is the latent topic information of the document sets or corpus and thus these information reflect the background or scene related to the topic. Jiang et al. [9] attempted to incorporate the topic model and Machine Translation method to generate Chinese metrical poetry. These techniques which use the word distribution of topic model as the scene to assist Chinese metrical poetry generation has achieved remarkable results. Thus, it is feasible to develop a learning model that incorporates the word distribution of text topics with original content to generate summarization.

In this paper, we propose a learning model (topics based attention model) that incorporating the word distribution of text topics with original content to generate summary. Our proposed model not only involves the sequence-to-sequence (seq2seq) framework and the attention mechanism but also involves the topic model to generate word distribution as certain scene to assist the summarization. On the one hand, word distribution affects the attention of the original text. On the other hand, the attention mechanism can incorporate the word distribution of text topics with original text in semantic layer. Our model suits to apply to news headlines generation of the journalism with the reason that news headlines generation should under certain scene and remain models pay less attention on this problem. Our contributions are as follows:

1. We explore the implicit information conveyed between different text and the scene influence.
2. We present a novel method, which uses word topics distribution to affect the attention mechanism and assists decoder to produce next word in automatic text summarization.

3. In automatic text summarization, we illustrate that machine learning and deep learning can work together.

The proposed model remains competitive to existing model on the datasets of CNN/Daily Mail and citation. In experiments, we show that our proposed model outperforms the summarization of short text with the same quantity of word distribution. The rest of the paper is organized as follows. Section 2 contextualizes our model with background of abstractive text summarization. In Sect. 3, we describe topics based attention model in detail. We present the results of our experiments on two different datasets in Sect. 4. In Sect. 5, we conclude the paper with remarks on our future direction.

2 Background

Automatic text summarization has been deeply studied. Extractive or abstractive approaches were proposed to address this challenging task. In extractive summarization (Neto et al. [10]; Zajic et al. [11]; Wang et al. [12]; Filippova et al. [13]; Durrettet al. [14]; Nallapati et al. [15]) focused on the summarization that extracts the important sentences from the source input and combined them to produce a summary. However, extractive models can produce coherent summaries, but they can't use new words to create concise summaries. In recent years, neural networks have been exploited to summarization. Prior to deep learning, abstractive summarization has been investigated. (Chen et al. [16]; Miao and Blunsom [17]; Zhou et al. [18]; Paulus et al. [19]; See et al. [20]; Gehrmann et al. [21]) focused on the encoder-decoder architecture, where the encoder and decoder are constructed using either Long Short Term Memory (LSTM) (Hochreiter and Schmidhuber [22]) or Gated Recurrent Unit (GRU) (Cho et al. [23]), to promise results of the abstractive summarization. On this foundation, the state-of-the-art abstractive summarization models are based on sequence-to-sequence models with attention (Bahdanau et al. [24]). Kikuchi et al. [8] constructed a LenEmb and Lenlnit model to control the length of summarization. The pointing mechanism (Gulcehre et al. [25]; Gu et al. [26]) allowed a summarization system to both copy words from the source text and generate new words from the vocabulary. Reinforcement learning is exploited to directly optimize evaluation metrics (Paulus et al. [27]; Kryściński et al. [28]; Chen and Bansal [29]). However, very few methods have explored the performance of word distribution as the scene in summarization tasks. Topic model, such as Latent Dirichlet Allocation (LDA), is popular in machine learning. It can extract a set of latent topics from a text corpus. Each topic is a unigram distribution over words, and the high-probability words often present strong semantic correlation. Therefore, the word distribution can reflect the scene. In this paper, we explored the word distribution as scene to assist summarization tasks.

The definition of sentence summarization is to generate corresponding summarization for input sentences. Generally speaking, the input contains multiple sentences and consists of a sequence of m words which is represented as $X = (x_1, x_2, \ldots, x_m)$. The word is coming from a fixed vocabulary v of size $|v| = v$ and we represent each word as an indicator vector. A summarizer takes X as input and outputs a sequence

$Y = (y_1, y_2, \ldots, y_n)$. Words of summary come from the same vocabulary v. Here, we will assume the summarizer knows the fixed output length of the summary before generation. Summary is the shorter sentence and its length N < M. As the problem of the generation of summary, an extractive system tries to transfer words from the input.

$$\underset{m \in \{1, \ldots, M\}^N}{\arg \max} \; g(X, X_{[m_1, \ldots, m_N]}) \tag{1}$$

Here, g is a scoring function, $X_{[m_1, \ldots, m_N]}$ represents the summary. While a system is abstractive if it tries to find the optimal sequence from the set $\gamma \subset (\{0, 1\}^V, \ldots, \{0, 1\}^V)$.

$$\underset{Y \in y}{\arg \max} \; g(X, Y) \tag{2}$$

Abstractive summarization poses a more difficult generation challenge, the lack of hard constraints gives the system more freedom in generation and allows it to fit with a wider range of training data. [6] Traditionally, abstractive summarization regards the whole vocabulary size set y as the summary choice space. Such a large choice space will cause efficiency problem and it seems to look for a needle in a haystack without any purpose. We try to limit the whole vocabulary size set y under the certain scene which can be reflected through the word distribution of topic. Topic model can be used to mine the latent topic information in the document sets or corpus. These information reflect the scene related to the topic. In general, abstractive summary is generated based on set γ. But with the assistant of scene, the set γ will be restricted and more precise. Besides, the input, together with the word distribution can more fully describe the information of the text. Therefore, our model tries to find the optimal sequence as follow:

$$\underset{Y \in \gamma}{\arg \max} \; g([X, W_L], Y) \tag{3}$$

Here, W_L is the word distribution of original text. Our work focuses on the scoring function g. According to the conditional probability of a summary given the input and the word distribution, we can get $g([X, W_L], Y) \sim P(Y|X, W_L)$. So we can write this as:

$$p(Y|X, W)_L = \prod_{t=1}^{|y|} p(y_t | y_{<t}, Z_t) \tag{4}$$

Here, $y_{<t} = y_1, y_2, \ldots, y_{t-1}$ and Z_t is the context environment variable. Section 3 we will introduce the details of this equation.

3 Our Model

Suppose that the text content is represented as $X = (x_1, x_2, \ldots, x_m)$ and corresponding summarization as $Y = (y_1, y_2, \ldots, y_n)$. Typically, this conditional probability is factorized as the product of conditional probabilities of the next word in the sequence:

$$p(Y|X) = \prod_{t=1}^{|y|} p(y_t|y_{<t}, c_t)$$ (5)

Here, $y_{<t} = y_1, y_2, \ldots, y_{t-1}$ and c_t is the context vector.

In this work, we propose to incorporate the word distribution of text topics W_L with original content to generate summarization. The word distribution is obtained by LDA model. Thus, the product of conditional probabilities of the next word in the sequence:

$$p(Y|X, W_L) = \prod_{t=1}^{|y|} p(y_t|y_{<t}, Z_t)$$ (6)

Here, $y_{<t} = y_1, y_2, \ldots, y_{t-1}$ and Z_t is the context environment variable.

The subsequent sections focus on the context environment variable Z_t which is the vector that fuses the word distribution of the text with source sentences. As Fig. 1 shown, we propose the Topic based attention model to produce the context environment variable to assist the generation of summarization. It contains Encoder, acquisition of word distribution, Decoder with attention and word distribution. Compared with the Seq2Seq model based on attention mechanism, our model which adds the word distribution as the certain scene can more fully describe the information of the text and thus the generation of summarization is better. In the subsequent sections, we will introduce Encoder, acquisition of word distribution, Decoder with attention and word distribution.

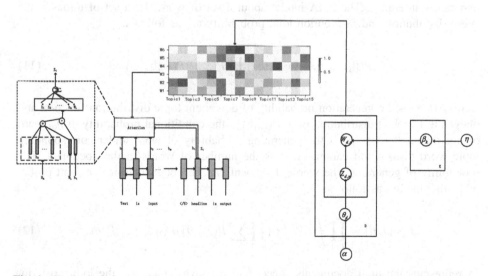

Fig. 1. Topic based attention Model. When generating each word of summarization, the attention mechanism fuses the content information of the whole text (H_m) which is obtained by bidirectional RNN with the word distribution (W_L) to produce the context environment variable Z_t. Then, we use the Z_t as feature vector to generate summarization. (See Sects. 3.1, 3.2 and 3.3 for more details).

3.1 Encoder

We use the bi-directional RNN as encoder. A bi-directional RNN processes the original text for both forward and backward directions with two RNNs. During the encoding process, the forward hidden states and backward hidden states can obtain as follows:

$$\overrightarrow{k_t} = \overrightarrow{LSTM}(x_t, k_{t-1}) \tag{7}$$

$$\overleftarrow{k_t} = \overleftarrow{LSTM}(x_t, k_{t+1}) \tag{8}$$

We use the long short-term memory (LSTM) networks as the recurrent unit. The hidden states finally can obtain as follows:

$$h_t = [\overrightarrow{k_t}, \overleftarrow{k_t}] \tag{9}$$

Each time of the hidden state vectors can be preserved and formed H_m, and H_m represents the content information of the whole text.

$$H_m = (h_1, h_2, \ldots, h_m) \tag{10}$$

3.2 The Acquisition of Word Distribution

Generally speaking, according the mind of Maximum Likelihood Estimate (MLE), we must get joint posterior distribution of all parameters which will be estimated before parametric inference. The LDA model about a set of words w, a set of topics z, the latent distribution and distribution joint probability are as follows:

$$P(\theta, z, w|\alpha, \beta) = P(\theta|\alpha) \prod_{n=1}^{N} P(z_n|\theta)P(w_n|z_n, \beta) \tag{11}$$

Here, $P(\theta|\alpha)$ is the generation probability of document topic distribution function and obeys the Dirichlet distribution of α, $P(z_n|\theta)$ is the conditional probability of n_{th} word of document, $P(w_n|z_n, \beta)$ is the generating probability of word under corresponding topic word distribution parameter. N is the number of words in the document. The probability of generating the whole document can be obtained from the joint probability distribution as follows:

$$P(w|\alpha, \beta) = \prod_{d=1}^{M} \int P(\theta_d|\alpha)(\prod_{n=1}^{N} \sum_{z_{dn}} P(z_{dn}|\theta_d)P(w_{dn}|z_{dn}, \beta))d\theta_d \tag{12}$$

M represents the total documents. According to Bayes' theorem, the joint posterior probability of (11) is

$$P(\theta, z | w, \alpha, \beta) = \frac{P(\theta, z | \alpha, \beta)}{P(w | \alpha, \beta)} \tag{13}$$

The denominator of (13) contains two parameters α and β. Therefore, it is complicated to estimate the joint posterior probability about the parameters and we can't obtain the analytical solution directly. In generally, there are two ways to solve the parameter inference problem of this formula. One is the Variational inference which used by the original author of LDA. The main idea of Variational inference is that using the Variational inference to calculate the lower bound of maximum likelihood function, then uses the intermediate variables to simplify the model, finally, get the estimated parameters α, β by EM algorithm. However, the disadvantage of this method is that the amount of computation is relatively large. Here, we will use Gibbs Sampling which is a concise and fast method to get the estimation of parameters. Gibbs Sampling is an algorithm of Markov chain Monte Carlo (MCMC). The main idea of the algorithm is that using a sample sequence $\{Y_1, Y_2, \ldots, Y_n\}$ which is generated by joint probability distribution $P(X_1, X_2, \ldots, X_n)$ with two or more variables to approach joint distribution. Here, we use Gibbs Sampling to sample the hide variable Z. Let w be the word of document set and Z be the distribution vectors of the topic. For a particular word, which topic does it belong to depends on the subordinate situation of other words to the current topics. This subordination process can be obtained as follows:

$$P(z_i = k | \overrightarrow{z_{\neg i}}, \overrightarrow{w}) \propto \frac{n_{m,\neg i}^{(k)} + \alpha_k}{\sum\limits_{j=1}^{k} (n_{m,\neg i}^{(j)} + \alpha_k)} \cdot \frac{n_{k,\neg i}^{(v)} + \beta_t}{\sum\limits_{v=1}^{V} (n_{k,\neg i}^{(v)} + \beta_v)} \tag{14}$$

Here, i is the word of the document, $n_{m,\neg i}^{(k)}$ is amount of k_{th} topic of other words except i_{th} word of document, $\sum\limits_{j=1}^{k} (n_{m,\neg i}^{(j)})$ is the total amount of other words, $n_{k,\neg i}^{(v)}$ is amount of other words of k_{th} topic except i_{th} word of document, $\sum\limits_{v=1}^{V} (n_{k,\neg i}^{(v)})$ is the total amount of words of k_{th} topic. After the ends of all iterations, the formula of θ and ϑ can be obtained as follows:

$$\theta_{k,j} = \frac{n_m^{(k)} + \alpha_k}{\sum\limits_{z=1}^{T} n_m^z + \alpha_z} \tag{15}$$

$$\vartheta_{k,t} = \frac{n_k^{(t)} + \beta_t}{\sum\limits_{v=1}^{V} n_k^{(v)} + \beta_t} \tag{16}$$

We choose maximum top V possibility words in a LDA model with k topics. Therefore, the topic distribution of a document is $\theta = (\theta_1, \theta_2, \ldots, \theta_k)$ and word distribution of all topics of a document is $\vartheta = (\vartheta_{1v}, \vartheta_{2v}, \ldots, \vartheta_{kv})$. We can obtain the word distribution of a document as follows:

$$W_L = f_\vartheta(\theta) \tag{17}$$

Here, $\theta = (\theta_1, \theta_2, \ldots, \theta_k)$ is the topic distribution of a document, f_ϑ is the function to get the most likely word distribution to reflect a certain scene.

3.3 Decoder

Our decoder is based on an RNN with LSTM:

$$h_t = LSTM(x_t, h_{t-1}) \tag{18}$$

We fuse the content information of the whole text (H_m) with the word distribution (W_L) to obtain the fusion vector (F), and then use the CNN [30] convolution kernel to extract the feature (S) of the fusion vector. Equations are shown as follows:

$$F = g(H_m, W_L) \tag{19}$$

$$S = f_{conv}(F) \tag{20}$$

Here, g is a connection function and f_{conv} is a convolutional function. The activation function can be sigmoid function, tanh function, ReLU function and so on. Here, the ReLU function is selected. We use the attention mechanism (Bahdanau et al. [24]) for reference. We use decoder hidden state h_t, feature S and fusion vector F to compute context environment variable Z_t of time step t. We calculate the context environment variable Z_t as the weighted sum of these summarized vectors:

$$Z_t = \sum_{j=1}^{T_{m,L}} \alpha_{tj} F_j \tag{21}$$

Here, α_t is the weight at the $t - th$ step for F_j computed by a softmax operation:

$$\alpha_{tj} = \frac{\exp(e_{tj})}{\sum_{k=1}^{T_{m,L}} \exp(e_{tk})} \tag{22}$$

Here, $e_{tj} = b(h_t, S_j)$, b is the alignment function. After context environment variable Z_t is calculated, the model updates the distribution of next word as follows:

$$h_t = \tanh(W_{ht}[h_t; Z_t] + b_{ht}) \tag{23}$$

$$p(Y|X, W_L) = soft \max(W_{ho}h_t + b_{ho}) \tag{24}$$

Here, W_{ht}, b_{ht}, W_{ho}, b_{ho} are learnable parameters.

4 Experiments

We evaluate the proposed approach on datasets of CNN/Daily Mail and citation. As a comparison, we use the same training procedures and the same datasets for both models.

4.1 Datasets

We train our model on a part of the CNN/Daily Mail dataset (Hermann et al. [31]; Nallapati et al. [7]), which contains articles (766 tokens on average) and summaries (53 tokens on average) and open citation dataset of AMiner [32] platform, which contains articles(152 tokens on average) and summaries (15 tokens on average). The citation dataset is a citation dataset published on the AMiner platform. The data is extracted from DBLP, ACM, MAG and other academic databases. We used scripts to obtain the data. CNN/Daily Mail dataset contains 267,113 training pairs, 10,883 validation pairs and 9,117 test pairs. Citation dataset contains 41,098 training pairs, 4,229 validation pairs and 1,902 test pairs.

4.2 Evaluating Indicator

For the evaluation of the text summaries, this paper uses the ROUGE (Recall-Oriented Understudy for Gisting Evaluation) automatic testing tool, which is proposed by Lin, and it is widely used in the field of automatic text summarization. The experiment uses the ROUGE-N index to measure the quality of the algorithm generated by the algorithm. The formula is defined as follows:

$$ROUGE - N = \frac{\sum_{S \in Ret} \sum_{gram_n \in S} Count_{match}(gram_n)}{\sum_{S \in Ref} \sum_{gram_n \in S} Count(gram_n)} \tag{25}$$

Here, the n represents the length of the n-gram setting. ROUGE-N is an indicator to evaluate the quality of the generated digest by the ratio of the number of n-gram. $\sum_{S \in Ref} \sum_{gram_n \in S} Count(gram_n)$ and the total n-gram number in the generated summary are calculating by the generated summary and the reference summary Ref. The larger the value, the more similar the generating summary and the reference abstract on the n-gram unit.

4.3 The Acquisition of Word Distribution

Latent Dirichlet Allocation (LDA [33]) is a generative probabilistic model of a corpus. It can be used to mine the latent topic information in the document set or corpus. These information reflects the background content related to the topic. We use LDA model to train our data sets. As to CNN/Daily Mail dataset, we set $\alpha = 0.15$, $\beta = 0.15$, citation dataset we set $\alpha = 0.10$, $\beta = 0.10$. For each topic's word distribution, it can consider that these words are related to the same theme and reflect scene. After training with LDA, word distribution can be obtained.

4.4   Results and Analysis

We trained our model on a Quadro M400 GPU with Tensorflow1.3. We set the dimension of word embedding to 80, the dimension of LSTM hidden state to 160. Each dataset train 35 epoch and the learning rate is 0.15. The performance of the model on the test set is compared with Seq-to-Seq+attn baseline model on CNN/Daily Mail dataset is shown on Table 1. The results of ROUGE-1, ROUGE-2, ROUGE-L of CNN/Daily Mail dataset quotes from See A et al. [20]. Table 2 shows the ROUGE scores of each model on citation. On CNN/Daily Mail dataset, in order to compared with the Seq-to-Seq+attn baseline model, we limit the vocabulary size to 150k.

Table 1. Experimental results on the summary tasks of CNN/Daily Mail dataset on various ROUGE metrics.

Model	CNN/Daily Mail		
	ROUGE-1	ROUGE-2	ROUGE-3
Seq-to-Seq+attn baseline (150k vocab)	30.49	11.17	28.08
Topic based Attention Seq2Seq (150k vocab)	30.74	12.08	28.76

Table 2. Experimental results on the summary tasks of citation dataset on various ROUGE metrics.

Model	Citation		
	ROUGE-1	ROUGE-2	ROUGE-3
Seq-to-Seq+attn baseline	32.26	14.16	28.76
Topic based Attention Seq2Seq	35.57	17.17	32.18

It is clear from Tables 1 and 2 that our model can improve the ROUGE-N values. With the same quantity of word distribution, on CNN/Daily Mail dataset, comparing to seq-to-seq+attn baseline model, ROUGE-1 increased by 0.25%, ROUGE-2 increased by 0.91% and ROUGE-L increased by 0.68%; On citation datasets, ROUGE-1 increased by 3.31%, ROUGE-2 increased by 3.01% and ROUGE-L increased by 3.42%.

Therefore, the results show that the word distribution as background content assists summarization plays an active role in summarization. Combining the original text with the word distribution, the information of the text can be described more completely and the summarization is better. At the same time, the same quantity of word distribution as background content assists summarization prefers the short text summarization and the effect of short text summarization is more special. The main reason is that compare to long text, the same quantity of word distribution can describe the information of the short text more completely.

5   Conclusion

In this work, we try to explore the impact of word distribution in summarization and the combination of the method of deep leaning with machine learning to promote the summary. We propose a deep learning model Topic based attention model. We apply the Topic based attention model for the task of abstractive summarization with promising results on two different datasets. Our model uses the word distribution as background content to assist summarization. As a part of our future work, we plan to focus on Dynamic topic models.

Acknowledgments. This work was partially supported by the following projects: Natural Science Foundation of Guangdong Province, China (Nos. 2016A030313441), research and reform project of higher education of Guangdong province (outcome-based education on data science talent cultivation model construction and innovation practice).

References

1. Hu, X., Lin, Y., Wang, C.: Summary of automatic text summarization techniques. J. Intell. **29**, 144–147 (2010)
2. Kupiec, J., Pedersen, J., Chen, F.: A trainable document summarizer. In: Advances in Automatic Summarization, pp. 55–60 (1999)
3. Lin, C.Y.: Training a selection function for extraction. In: Proceedings of the Eighth International Conference on Information and Knowledge Management, pp. 55–62. ACM (1999)
4. Osborne, M.: Using maximum entropy for sentence extraction. In: Proceedings of the ACL 2002 Workshop on Automatic Summarization, vol. 4, pp. 1–8. Association for Computational Linguistics (2002)
5. Conroy, J.M., O'leary, D.P.: Text summarization via Hidden Markov Models. In: Proceedings of the 24th Annual International ACM SIGIR Conference on Research and Development in Information Retrieval, pp. 406–407. ACM (2001)
6. Rush, A.M., Chopra, S., Weston, J.: A neural attention model for abstractive sentence summarization. In: Proceedings of the 2015 Conference on Empirical Methods in Natural Language Processing, pp. 379–389 (2015)
7. Nallapati, R., Zhou, B., dos Santos, C., et al.: Abstractive text summarization using sequence-to-sequence RNNs and beyond. In: Proceedings of the 20th SIGNLL Conference on Computational Natural Language Learning, pp. 280–290 (2016)
8. Kikuchi, Y., Neubig, G., Sasano, R., et al.: Controlling output length in neural encoder-decoders. In: Proceedings of the 2016 Conference on Empirical Methods in Natural Language Processing, pp. 1328–1338 (2016)
9. Jiang, R.Y., Cui, L., He, J., et al.: Topic model and statistical machine translation based computer assisted poetry generation. Chin. J. Comput. **38**(12), 2426–2436 (2015)
10. Neto, J.L., Freitas, A.A., Kaestner, C.A.A.: Automatic text summarization using a machine learning approach. In: Bittencourt, G., Ramalho, G.L. (eds.) SBIA 2002. LNCS (LNAI), vol. 2507, pp. 205–215. Springer, Heidelberg (2002). https://doi.org/10.1007/3-540-36127-8_20
11. Zajic, D., Dorr, B.J., Lin, J., et al.: Multi-candidate reduction: Sentence compression as a tool for document summarization tasks. Inf. Process. Manag. **43**(6), 1549–1570 (2007)

12. Wang, L., Raghavan, H., Castelli, V., et al.: A sentence compression based framework to query-focused multi-document summarization. In: Proceedings of the 51st Annual Meeting of the Association for Computational Linguistics. Long Papers, vol. 1, pp. 1384–1394 (2013)
13. Filippova, K., Alfonseca, E., Colmenares, C.A., et al.: Sentence compression by deletion with LSTMs. In: Proceedings of the 2015 Conference on Empirical Methods in Natural Language Processing, pp. 360–368 (2015)
14. Durrett, G., Berg-Kirkpatrick, T., Klein, D.: Learning-based single-document summarization with compression and anaphoricity constraints. In: Proceedings of the 54th Annual Meeting of the Association for Computational Linguistics. Long Papers, vol. 1, pp. 1998–2008 (2016)
15. Nallapati, R., Zhai, F., Zhou, B.: SummaRuNNer: a recurrent neural network based sequence model for extractive summarization of documents, pp. 3075–3081. AAAI (2017)
16. Chen, Q., Zhu, X., Ling, Z., et al.: Distraction-based neural networks for modeling documents. In: Proceedings of the Twenty-Fifth International Joint Conference on Artificial Intelligence, pp. 2754–2760. AAAI Press (2016)
17. Miao, Y., Blunsom, P.: Language as a latent variable: discrete generative models for sentence compression. In: Proceedings of the 2016 Conference on Empirical Methods in Natural Language Processing, pp. 319–328 (2016)
18. Zhou, Q., Yang, N., Wei, F., et al.: Selective encoding for abstractive sentence summarization. In: Proceedings of the 55th Annual Meeting of the Association for Computational Linguistics. Long Papers, vol. 1, pp. 1095–1104 (2017)
19. Paulus, R., Xiong, C., Socher, R.: A deep reinforced model for abstractive summarization. arXiv preprint: arXiv:1705.04304 (2017)
20. See, A., Liu, P.J., Manning, C.D.: Get to the point: summarization with pointer-generator networks. In: Proceedings of the 55th Annual Meeting of the Association for Computational Linguistics. Long Papers, vol. 1, pp. 1073–1083 (2017)
21. Gehrmann, S., Deng, Y., Rush, A.: Bottom-up abstractive summarization. In: Proceedings of the 2018 Conference on Empirical Methods in Natural Language Processing, pp. 4098–4109 (2018)
22. Hochreiter, S., Schmidhuber, J.: Long short-term memory. Neural Comput. 9(8), 1735–1780 (1997)
23. Cho, K., van Merrienboer, B., Gulcehre, C., et al.: Learning phrase representations using RNN encoder–decoder for statistical machine translation. In: Proceedings of the 2014 Conference on Empirical Methods in Natural Language Processing (EMNLP), pp. 1724–1734 (2014)
24. Bahdanau, D., Cho, K., Bengio, Y.: Neural machine translation by jointly learning to align and translate. In: ICLR (2015)
25. Gulcehre, C., Ahn, S., Nallapati, R., et al.: Pointing the unknown words. In: Proceedings of the 54th Annual Meeting of the Association for Computational Linguistics. Long Papers, vol. 1, pp. 140–149 (2016)
26. Gu, J., Lu, Z., Li, H., et al.: Incorporating copying mechanism in sequence-to-sequence learning. In: Proceedings of the 54th Annual Meeting of the Association for Computational Linguistics. Long Papers, vol. 1, pp. 1631–1640 (2016)
27. Paulus, R., Xiong, C., Socher, R.: A deep reinforced model for abstractive summarization. In: International Conference on Learning Representations (ICLR) (2018)
28. Kryściński, W., Paulus, R., Xiong, C., et al.: Improving abstraction in text summarization. In: Proceedings of the 2018 Conference on Empirical Methods in Natural Language Processing, pp. 1808–1817 (2018)

29. Chen, Y.C., Bansal, M.: Fast abstractive summarization with reinforce-selected sentence rewriting. arXiv preprint: arXiv:1805.11080 (2018)
30. Zhao, X., Zhang, Y., Guo, W., Yuan, X.: Jointly trained convolutional neural networks for online news emotion analysis. In: Meng, X., Li, R., Wang, K., Niu, B., Wang, X., Zhao, G. (eds.) WISA 2018. LNCS, vol. 11242, pp. 170–181. Springer, Cham (2018). https://doi.org/10.1007/978-3-030-02934-0_16
31. Hermann, K.M., Kočiský, T., Grefenstette, E., et al.: Teaching machines to read and comprehend, pp. 1693–1701 (2015)
32. Tang, J., Zhang, J., Yao, L., et al.: ArnetMiner: extraction and mining of academic social networks. In: ACM SIGKDD International Conference on Knowledge Discovery and Data Mining, pp. 990–998. DBLP (2008)
33. Blei, D.M., Ng, A.Y., Jordan, M.I.: Latent Dirichlet Allocation. J. Mach. Learn. Res. **3**, 993–1022 (2003)

Sentiment Tendency Analysis of NPC&CPPCC in German News

Ye Liang[✉], Lili Xu, and Tianhao Huang

Beijing Foreign Studies University, Beijing, China
liangye@bfsu.edu.cn

Abstract. The sentiment tendency analysis on news reports serves as a tool to study the main stream attitude towards a hot event. With the China's going-out strategy processing, we can effectively avoid the potential risks in the help of the in-depth study and interpretation of China's relevant policy in a certain country and region and the understanding of the local public opinion and the conditions of the people. We can not use the existing mature tools which use Chinese and English as the research object, so the sentiment tendency analysis to German of which relevant work is relatively absent need to be solved. On the basis of completing the basic work of German sentiment dictionary, degree adverb dictionary, negative word dictionary and stop word dictionary, this paper puts forward a set of calculating methods aiming at the sentiment tendencies in German, which is applied to the calculation of the sentiment tendencies related to NPC_CPPCC event in one of the most mainstream media in Germany, and the results of the calculation will be interpreted and analyzed.

Keywords: German news · Sentiment analysis · NPC&CPPCC

1 Introduction

With the China's national strength increasing, its influence on other countries is also increasing, meanwhile China's words and deeds are increasingly being paid attention, especially those countries, which have frequent communication with China, are always paying attention to China's words and deeds in the international community. They will do an in-depth interpretation of the subtle changes of China's attitude in an event in its domestic media in order to measure the possible impact to their politics, economy, culture, diplomacy, military and other aspects. For China, it is possible to test the gains and losses of China's international policy through the reports relating to China in other countries' media, which will help us to know others and ourselves when we formulate the international policies.

We selected Germany as our target country of sentiment analysis. As the biggest economy in Europe, Germany is not only an valuable companion in the strategy of "the Belt and Road", but also a pivotal supporting point in the context of China's going-out strategy. On account of China's increasing influence in international politics and economy, issues relevant to China become more and more frequent over time in German network media. The perspective and attitude of news editors have direct or

© Springer Nature Switzerland AG 2019
W. Ni et al. (Eds.): WISA 2019, LNCS 11817, pp. 298–308, 2019.
https://doi.org/10.1007/978-3-030-30952-7_30

indirect orienting influence on their audiences' emotion, which further impact the factual effect of China's two great strategy. Therefore, sentiment analysis on news relevant to China in German network is valuable for China both practically and theoretically.

2 Related Work

With the rapid development of the information technology, the Internet has become the main carrier of social public opinion. The major mainstream media also actively expand its news and reports, which are published on the network in order to enhance its audience coverage. After gaining the ability to acquire the real-time dynamic information in the network, the public opinion researchers will move on to the various forms of new media. Professor Hsinchun Chen of Arizona University has made ground-breaking work in the use of the network to obtain and process information for visualization of site content, network relationships and activity levels, and built the Dark Web platform based on the public information. This platform can be used in the real-time online monitoring of the terrorist activities, which made outstanding contributions in the field of national security [1]. Professor Xing Chunxiao from the Institute of Information Technology of Tsinghua University accumulates the data of interpersonal relationships and interactive exchange in the social platform, opening new development ideas to promote the research work like wisdom finance, wisdom medical care and so on [2].

The difficulty of public opinion research can be divided into two parts. First, it is necessary to identify the subject exactly from the massive network information data. Secondly, analyze the emotional tendencies of the identified content [3, 4]. The current research to the first part is more mature. So this paper will focus on the method of analyzing the emotional tendencies in German language.

2.1 Sentiment Analysis Based on Sentiment Dictionary

The sentiment tendency analysis, which is based on the sentiment dictionaries, will use a well-written sentiment dictionary for data analysis. The classic research results: the online comment tracking system built by Tong RM, which manually extracted the sentiment vocabulary in the film and television commentary and marked to build the sentiment dictionary. Scholar Kim S M divides the opinion into four parts: the subject, the holder, the statement and the emotion, using the seed word and the WordNet extension dictionary to calculate the sentiment tendencies. Some scholars expand the sentiment dictionary based on the general sentiment dictionary, adding a specific field of sentiment dictionary. For example, the sentiment analysis in the field of social networking sites, Ravi Parikh analyzes the information as a corpus that the users have published on Twitter. In China, the researchers carry out experiments in the mass of information according to "Sina microblog". Sentiment dictionary-based sentiment analysis method is more accurate. This method can make better use of human knowledge that has been mastered, but the construction of sentiment dictionary work is more time-consuming.

2.2   Sentiment Analysis Based on Machine Learning

Common used models which take used of the machine learning technology to analyze the text of sentiment tendencies are: central vector classification algorithm, Naive Bayesian classification algorithm, maximum entropy classification algorithm, K nearest neighbor classification algorithm and support vector machine classification algorithm.

In the study results of the English text, Bo Pang, Lillian Lee and others used the machine learning method to classify the text according to sentiment tendencies. They used three different classification methods to experiment, and the results show that the method of the support vector machine classification shows the best classification effect among these methods.

As the Chinese sentences need additional operation of word segmentation on the text before sentiment analysis, so the difficulty is greater than the English text, and the process is more complex. Tang Huifeng, Tan Songbo and others used different types of adjectives, adverbs, nouns and verbs as different textual features, and experimented with different machine learning methods.

In the study of sentiment analysis of uncommon talk language texts, Boiy and Moens used the machine learning method to do the sentiment classification with the texts of English, Dutch and French. The experimental results show that the accuracy rate of the classification of the English text can reach 83%, but the Dutch and French text only reach to 70%. It can be seen that the machine learning method in the sentiment analysis of the uncommon talk languages also shows a certain advantage, but because of the lack of the resources, accuracy is still to be further improved.

Based on the machine learning method, we can obtain the knowledge objectively and accurately, but it has a high dependence on training corpus and needs a long training period.

2.3   The Problem to Be Solved

In order to study the sentiment tendencies of the news related to China's event expressed in German network news and to improve the analysis accuracy, it is necessary to construct a sufficiently large and widely covered German sentiment dictionary. And then according to the sentiment dictionary, we extract the sentiment information and the evaluation of the object in the news, in order to achieve a sentiment tendency analysis system for the German language news. And then we can study and analyze the sentiment tendencies to NPC&CPPCC in the German network news media.

3   Acquisition and Preprocessing of News Data

3.1   Source of News Data

We selected one of the most popular media news website (Spiegel.de), which is among the best in the ranking, which have a large user base in Germany and have certain credibility in this country, meanwhile the news data obtained from these sites is more accurate and persuasive.

3.2   The Realization of Web Crawler

Using Java to compile the crawler program, we crawled more than 25 thousand news reports in the previous website, using those in a time span from January 1st, 2015 to April 15th, 2019. Through the data cleaning, topic recognition, topic classification, news anti-repetition, event detection, reporting relationship identification, topic tracking, theme extraction and other links, we extract 54 articles in German news reports related to China's NPC&CPPCC event from the above news.

3.3   News Data Preprocessing

German is a single-byte text that uses the Latin alphabet to express different grammatical semantics by attaching various components to a word that can no longer be divided into smaller morphemes. German uses the word as the basic unit and the words separated by space. A single word can express an independent meaning, without additional word segmentation operation.

4   Sentiment Analysis of News Tests Based on Sentiment Dictionary

In order to analyze the sentiment tendency of German news texts, a sentiment dictionary based sentiment analysis technique for German news texts is proposed. The analysis process is shown in Fig. 1.

4.1   Sentiment Analysis Method

The analysis of sentiment tendency includes two aspects: sentiment tendency and degree of sentiment tendency. In the field of journalism, sentiment tendency refers to the news editors' attitudes towards the news event, such as whether to support the news, expressing positive emotion or negative emotion. For example, when the news text contains commendatory terms such as "ankurbeln" and "appreciate", the news editors expresses positive attitudes. When the news text contains derogatory terms such as "unkoordiniert" and "drastisch", it expresses news editor is opposed to news events, the emotion is negative. The degree of sentiment tendency refers to the intensity of the direction in which sentiment tendency is expressed. In the field of journalism, news editors use different degrees of adverbs, modal words and punctuation to express different intensity of feelings. For instance, both "förderlich" and "zuversichtlich" are commendatory phrases, expressing positive emotions, "echt" is an adverb of degree, having "echt" or not expresses different degrees of the sentiment intensity, and the former is much higher than that of the latter. In order to distinguish between different degrees adverbs in emotional tendency, in addition to the sentiment dictionary, a degree adverb dictionary was established.

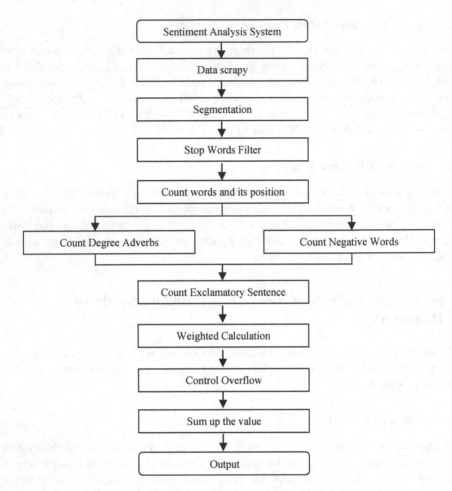

Fig. 1. Flow chart of sentiment analysis.

4.2 The Establish of the Sentiment Lexicon

Sentiment words refer to a collection of words, phrases, or sentences that contain emotion. It plays a very important role in the process of sentiment identification and analysis. The coverage of the sentiment lexicon greatly influences the judgment accuracy of sentiment orientation. In order to improve the coverage of the emotional lexicon, the lexicon of sentiment words and adverbs of degree are built. On the basis of commendatory and derogatory dictionary, the new dictionary adds six degrees of adverbs. The program uses the pre-written lexicon to calculate the sentiment orientation of the news text.

Sentiment Lexicon. With the help of German language students, the lexicon of sentiment word issued by CNKI was translated into German. Through artificial selection, culling and supplement and revision, the German sentiment lexicon was constructed, including 10 thousand sentiment words, 5000 commendatory terms and 5000 derogatory terms. The lexicon is mainly composed of adjectives and adverbs. As the language is constantly updated and enriched, the lexicon will continue to add new login words.

Degree Adverbs Lexicon. Adverbs of degree usually appear before emotional words. They are mainly used to modify the range and extent of emotional words. The news text often contains a large number of adverbs of degree, such as "extremely", "very", and "totally", etc. The degree adverbs modify the emotional intensity of the writers' attitudes and views. For example, "the weather is very good today". This statement uses degree adverb "very" to modify commendatory terms "good", expressing a strongly positive sentiment. It can be seen that the addition of adverbs of degree can make the text change greatly in emotional orientation, therefore it needs to be dealt with accordingly.

On the basis of "HowNet" sentiment dictionary, after translating, adding and modifying, the lexicon of adverb of degree was constructed. There are 47 German adverbs of degree and they are divided into five grades, extremely high, high, medium, low and extremely low. Each adverbs of degree has been given a corresponding weight according to the degree of intensity. (Shown in Table 1).

Table 1. Tht degree adverbs weight assignment table.

Level	Example	Translation	Number	Weight
High	wirklich, echt	Very, highly	19	2
Medium	viel, mehr	Enough	15	1
Low	leicht, wenig	A bit of	13	0.5

Negative Lexicon. The negative word is a kind of degree adverb, and it is the word that expresses negative meaning. When a negative word modifies an emotional word, the original sentimental tendency is transformed into an opposite emotion. Therefore, the negative words have the unique grammar meaning and influence in the text. For example in English, "this dress is not beautiful", in this statement, "beautiful" is a positive emotional word, and the negative word "not" modifies the phrase into a negative tendency, that's also true of German. There will be multiple negatives in the text, and multiple negatives may cause multiple changes in the sentimental orientation. When there is more than one negative word before sentimental words, if the number of negative words is odd, is still negative; if the number is even, it expresses positive meaning. In this paper, a negative vocabulary of 15 German negative words is constructed by referring to the negative word list proposed by Zhang Jinming, and the weights are set from −0.5 to −1 according to the degree of intensity. (Shown in Table 2).

Table 2. The negative words weight assignment table.

Level	Example	Translation	Number	Weight
Low	Bagaimana	Little	12	−0.5
High	Perlu, Tidak	Not, no	11	−1

4.3   The Calculation of the Sentiment Tendency

In order to calculate the sentiment tendency and the degree of the emotion of the news text, after the preprocessing of the news text, the program using the lexicon of sentimental words, degree adverbs and negative words dictionary to calculate and classify the sentiment tendency and emotional intensity.

The program automatically gets the news documents from database, after calculation, the score will be written to the property "score" of the according item.

In the process of preprocessing, the original text is divided into n sentences. The program scans these sentences one by one under the guidance of the pre-created sentiment lexicon, extracting the positive words and negative words in each sentence. If it is not a sentiment word, then go to the next candidate character word until the whole news text is processed. If it is a sentiment word, it will scan the words before and after the sentiment word, finding if there is an adverb of degree or negative words to emotional words, if any, read the corresponding weight of it and multiply it by the procedure as follow. The sentiment tendency and emotional intensity calculation formula of the phrase are shown in formulas (1) and (2).

$$O_{pi} = M_{wa} \times S_{pi} \tag{1}$$

$$O_{Ni} = M_{wa} \times S_{Ni} \tag{2}$$

In formula 1, S_{Pi} is the weight of the degree adverb, and M_{Wa} is the weight of the sentiment word W_a in the sentence. S_{Ni} is the weight of the negative sentiment word N_i.

Sentence may contain m positive sentiment words and n negative sentiment words. Therefore, the formula to calculate the sentiment tendency of the whole sentence is shown in formula (3).

$$O_{Si} = \sum_{i=1}^{m} O_{pi} - \sum_{i=1}^{n} O_{wi} \tag{3}$$

In order to calculate the sentiment tendency and the degree of the emotion of the news text, after the preprocessing of the news text, the program using the lexicon of sentimental words, degree adverbs and negative words dictionary to calculate and classify the sentiment tendency and emotional intensity.

$$O_{di} = \sum_{i=1}^{q} O_{Si} \tag{4}$$

Finally, the degree of sentiment orientation and the degree of the sentiment orientation of the news text Ni will be classified by the value of the final sentiment orientation value P. According to different values, the intensity of emotion is divided into 5 levels: extremely negative, negative, neutral, positive, and extremely positive. The specific method of classification is shown in the formula (5).

$$P_{di} = \begin{cases} > 0 & positive \\ = 0 & neutral \\ < 0 & negative \end{cases} \tag{5}$$

At last, some statistical scores will be calculated, including the number of positive, neutral and negative news, total average sentiment value and total sentiment variance, which can be used to help the data analysis.

4.4 The Choose of the Weight

Sentiment adverbs are divided into 6 different degrees, ranging between (0, 2). The weight of the negative adverbs is between (−1, 0). The closer to −1, the higher the negative degree is; the closer to 0, the lower degree of negation is. The weight of other adverbs of degree is between (0, 2), the closer to 2, the stronger the intensity is, the closer to 0, the weaker the intensity is. There are some artificial assumptions to select the weights shown in Tables 1 and 2 to design algorithms. First of all, most of the news scores shown in the result are negative, if the weights of the negative word are chosen too large, and it will lead to lower scores and bigger fluctuation, which is not conducive to the analysis of data. In addition, due to the differences between Chinese and German language, in German, the low and extremely low degree adverbs are very similar in the German language, so the distinction of the weights is changed from 0.5 to 0.25.

4.5 The Output of the Result

The result is stored in the database by updating the column "score".

5 Analysis of the Sentiment Tendency of the News on "NPC&CPPCC" in German

5.1 General Sentiment Distribution of News

Using the sentiment algorithm above, 69 German news texts have been calculated and analyzed. The results are divided into 3 levels according to the formula (5): the extremely positive, positive, neutral, negative and extremely negative. In the 69 news texts, 39 news show negative emotion tendency, accounting for 56%; 14 of the news show neutral sentiment tendency, accounting for 20%; 16 news texts show extremely negative sentiment tendency, accounting for 23%. The calculations show that the German government's overall mood towards the NPC&CPPCC of China tends to be negative (Fig. 2).

According to the calculation, it was found that the average emotion value was −1.7, and the emotion fluctuation was great. The extremes were 6 and −11, the maximum appeared in the news on March 4, 2019, and the minimum appeared on March 6, 2015.

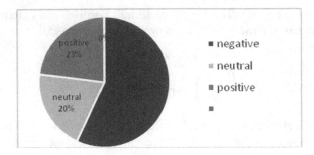

Fig. 2. Sentiment tendency distribution.

5.2   Time Series Analysis

In order to analyze the emotional trend of the news more comprehensively and accurately, through the statistical analysis of the emotional value of each news, the trend chart of sentiment value in time series is created. (Shown in Fig. 3).

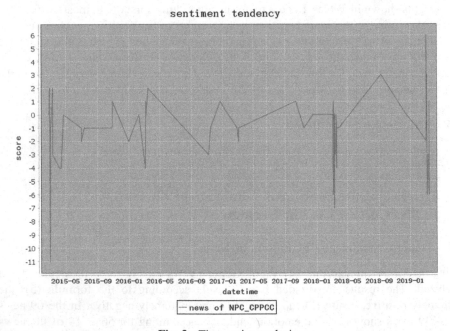

Fig. 3. Time series analysis.

As it shown in the Fig. 3, the sentiment tendency that the German news media expressed towards NPC&CPPCC shown a downward trend as time goes on. Statistics show that the emotional value of the news released in the early 2015 is mostly negative.

The analysis shows that, since 2015, the general sentiment tendency in news has been fluctuated below zero, but the intensity is relatively low, and is relatively stable.

After entering 2018, with the continuous development of events, the impact scope expanded gradually, the emotional fluctuation of the news event began to increase, showing an overall more negative or extremely negative direction. In general, with the continuous development of events, "NPC&CPPCC" constantly caused controversy and stronger dissatisfaction in German.

From the above analysis shows that, the German official media's attitudes toward "NPC&CPPCC" change as time goes by. In the early stage, the emotional tendency is neutral, in the process of the event, and the negative emotional tendency becomes more obvious.

5.3   Analysis of the Positive and Negative Content in the News

By analyzing the trend of sentiment tendency in chronological order, according to the emotional trend shown in Fig. 3, the original news corresponding to the position of the wave crest and trough was analyzed. Studying the sentences and factors that cause the change of emotional value, a summary of the news analysis of the position of peaks and troughs in the emotional trend in Fig. 3 is shown in Tables 3 and 4.

Table 3. Positive news content.

Date	Wave vrest	Translation
2019.3.4	Hoffnung auf Ende des Handelsstreits: Gipfel mit Trump und Xi könnte Ende März stattfinden	The Department of Defense said on Wednesday that it would complete the deployment of NPC&CPPCC in a few weeks

The above table shows that the "NPC&CPPCC" causes a positive emotion because of progress of the negotiation between two national leaders of China and USA. They are looking forward to the end of this dispute.

Table 4. Negative news content.

Date	Wave trough	Translation
2019.3.4	Kamis lalu saat kementerian pariwisata China menginstruksikan operator tur di Beijing untuk menghentikan penjualan perjalanan ke Korea Selatan dari 15 Maret	Hoping to end the trade dispute: Summit with Trump and Xi could take place in late March
2018.3.11	Verfassungsänderung in China: Xi für immer	Constitutional change in China: Xi forever

The above table shows that the "NPC&CPPCC" has aroused resentment among German media. The diplomatic relations between countries and South Korea have become extremely tense. As an onlooker, the official news media in German also expressed its attitude through the news. It is clearly that German's attitude towards the event is also negative.

Trough position has repeatedly appeared with the news of the Lotte Group in Korea, which shows that the deployment of "NPC&CPPCC" has a very negative impact on Lotte group. Lotte is the fifth largest enterprise in Korea, and China is its largest overseas market. Lotte has entered the Chinese market since 1994 and has invested in 24 provinces, including food, retail, tourism, petrochemical, construction and finance. Such as supermarkets, coffee shops, cinemas and so on. According to the news description, "By the Korean local time at 4 pm on March 8th, a total of 55 Lotte Mart supermarkets in China has closed". This also caused the South Korean Lotte Company's stock fell greatly, which has the adverse effect to the South Korean economy.

6 Summary and Prospect

On the basis of existing Chinese and English sentiment dictionaries, a German sentiment lexicon is constructed, and a sentiment analysis method based on German news is proposed. The validity of the proposed sentiment orientation analysis method is verified by the analysis of the news in the background of "NPC&CPPCC".

The German sentiment dictionary still has a large space to expand and improve, especially in the language contains some specific local idioms and slang. These words are not included in the existing sentiment lexicon. At the same time, with the continuous emergence of new words in the Internet, it brings new challenges to the maintenance of sentiment lexicon. Only keep up with the times and improve the system constantly, can be more conducive to the judgment of the sentiment tendency.

Acknowledgment. This work was supported in part by the Social Science Foundation of Beijing (No. 15SHA002), and the First-class Disciplines Construction Foundation of Beijing Foreign Studies University (No. YY19SSK02).

References

1. Chen, H.: Security informatics: the dark web experience. American Association for the Advancement of Science annual meeting, Vancouver, 18 February 2012
2. Zhang, J., Zhang, Y., Hu, Q., Tian, H., Xing, C.: A big data analysis platform for healthcare on apache spark. In: Xing, C., Zhang, Y., Liang, Y. (eds.) ICSH 2016. LNCS, vol. 10219, pp. 32–43. Springer, Cham (2017). https://doi.org/10.1007/978-3-319-59858-1_4
3. Yu, J., An, Y., Xu, T., Gao, J., Zhao, M., Yu, M.: product recommendation method based on sentiment analysis. In: Meng, X., Li, R., Wang, K., Niu, B., Wang, X., Zhao, G. (eds.) WISA 2018. LNCS, vol. 11242, pp. 488–495. Springer, Cham (2018). https://doi.org/10.1007/978-3-030-02934-0_45
4. Liang, Y.: Multilingual financial news retrieval and smart recommendation based on big data. J. Shanghai Jiaotong Univ. (Sci.) **21**(1), 18–24 (2016)

Ensemble Methods for Word Embedding Model Based on Judicial Text

Chunyu Xia, Tieke He[✉], Jiabing Wan, and Hui Wang

State Key Laboratory for Novel Software Technology,
Nanjing University, Nanjing 210093, China
hetieke@gmail.com

Abstract. With the continuous expansion of computer applications, scenarios such as machine translation, speech recognition, and message retrieval depend on the techniques of the natural language processing. As a technique for training word vectors, Word2vec is widely used because it can train word embedding model based on corpus and represent the sentences as vectors according to the training model. However, as an unsupervised learning model, word embedding can only characterize the internal relevance of natural language in non-specific scenarios. For a specific field like judicial, the method of expanding the vector space by creating a professional judicial corpus to enhance the accuracy of similarity calculation is not obvious, and this method is unable to provide further analysis for similarity in cases belonging to the same type. Therefore, based on the original word embedding model, we extract factors such as fines and prison term to help identify the differences, and attach the label of the case to complete supervised ensemble learning. The result of the ensemble model is better than any result of single model in terms of distinguishing whether they are the same type. The experimental result also reveal that the ensemble method can effectively tell the difference between similar cases, and is less sensitive to the details of the training data, the choice of training plan and the contingency of a single inaccurate training run.

Keywords: Natural language · Word embedding · Judicial text · Ensemble learning · Supervised learning

1 Introduction

In the past decade, artificial intelligence (AI) has become a popular term that covers all advanced smart technologies such as adversarial competition, computer vision and natural language processing, bringing amazing advances to computer-aided intelligence. As an extension of AI, Natural Language Processing (NLP) has achieved unprecedented success in dealing with machine translation, speech recognition, information retrieval and text similarity. Although NLP has evolved into different aspects of a smart society, intelligent justice lacks specific and accurate applications.

© Springer Nature Switzerland AG 2019
W. Ni et al. (Eds.): WISA 2019, LNCS 11817, pp. 309–318, 2019.
https://doi.org/10.1007/978-3-030-30952-7_31

In terms of intelligent justice, NLP can help the court save time by identifying similar cases and recommending relevant text [1]. The word embedding method overcomes the basic problem in the judicial text, namely natural semantics [13]. In addition, Word2vec-based technology can help with word embedding in a more efficient manner. Unfortunately, word embedding lacks the ability to overcome common pitfalls in natural language, such as polyphonic words. These common deficiencies can lead to accidental deviations, and accidental deviations should be eliminated in the field of judicial texts.

In contrast, ensemble methods for embedding words can help reduce accidental mistakes. Through experiments, we found that the set model performed well in terms of abnormal similarity due to inaccurate expression. More importantly, this model can be induced into a dimensionality reduction method. Projecting a high-dimensional vector into a three-dimensional vector simplifies calculations and shows excellent data fit.

2 Related Work

Looking back at the history of text similarity architectures, the classical text similarity architecture is based on specific differences between similar sentences. The definition of classical difference is the difference in the number and length of different words represented by two sentences, or in short, the ratio of intersection to union. Due to the rigid model and limited vocabulary, this classic building was replaced by a high-dimensional model based on a large corpus. In view of this model, the term frequency-inverse document frequency (TF-IDF) and bag of words (BOW) appeared. Their main idea is that if you create a corpus full of words, you can project each sentence into a high-dimensional vector. However, creating a corpus can only convert sentences into vectors without considering word order or synonyms. This problem actually has a fundamental impact on the emergence of word embedding.

Word embedding is a concept that describes the relationship between adjacent words and tends to predict the meaning of word. For example, a regular sentence can be converted into an inverted sentence, or even another sentence with completely different words can be reconstructed, albeit with the same meaning. Based on classical models or corpus models, it is hard to distinguish similarities and even categorize them as opposite vectors by mistake. Nevertheless, the word embedding can simulate real natural language scenes and produce mutually replaceable words.

In addition, in 2013, Mikolov proposed a new technique to train word embedding, which is called Word2vec [12]. Word2vec is a method of constructing a word embedding model after text training, with the options for continuous bag-of-words (CBOW) and skip-gram (SG) [10]. With the help of Word2vec, the concept of word embedding can be extended into other applications besides the NLP domain. Recent applications have trained word embedding of user actions like clicks, requests and searches to provide personal recommendations. Domains like E-commerce, E-business, and Market have utilized this approach to handle search rankings [4].

In terms of judicial text [5], in order to enhance the scalability of word embedding, relative research has added a professional corpus to trained word embedding. This method is to refine the word embedding model and improve the accuracy of comparing similar sentences. Although the vector space seems to work well in experiments due to the expansion of professional corpus, this solution is very poor when dealing with totally different categories of cases. The similarities in different cases may be high, not down to zero. More importantly, if the two cases are of the same type, the exact difference cannot be accurately stated.

3   Model Design

In order to avoid the shortcomings of word embedding, we need to convert unsupervised learning into supervised learning [2]. Word2vec is an unsupervised learning without tags [9]. An obvious problem with Word2vec is that the training model is only trying to cluster data features in the subconscious. Sometimes, when coincident feature learning happens to fit the essential differences in the category, the model can produce better results. More often than not, training models simply over-fitting features and creating some unreasonable boundaries. Therefore, we use supervised learning to address these shortcomings.

It is undeniable that we cannot supervise the word embedded model during Word2vec training. Instead, we can consider the original model as part of the new supervisory training model [14]. As a principle feature, fines and imprisonment are also important factors in dealing with similarities.

In this paper, we present an ensemble method for offsetting defects displayed in word embedding. The general idea is that we can use Word2vec to train word embedding in judicial texts and then combine the similarities of Word2vec calculations with other features [15]. By adjusting these features, we can adapt to the characteristics of different types of cases.

As described in the image, the entire process is derived from the cail data set. The first operation is to extract a single type of useful data from a composite data set, such as text words from facts, penalties from money, and prison terms from imprisonment. Once the data is ready, we can define different operations for them. In particular, the operation of text words is exactly the same as in the normal Word2vec model, and it is optional to train word embedding based on professional corpora. However, the most significant difference is the vector formed by the regular Word2vec model and other gaps, which are calculated by comparing the corresponding penalty and prison terminology [3]. The accusations in the figure reveal the basic idea that we can transform an unsupervised model into a supervised model. Ultimately, the entire data and tags can be injected into the deep neural network (DNN) model (Figs. 1 and 2).

DNN is an effective machine learning method that facilitates data fitting [6]. The ensemble DNN model we propose here is a DNN model that absorbs the advantages of unsupervised learning and supervised learning. By modifying the parameters that represent weights in specific situations, we can modify the

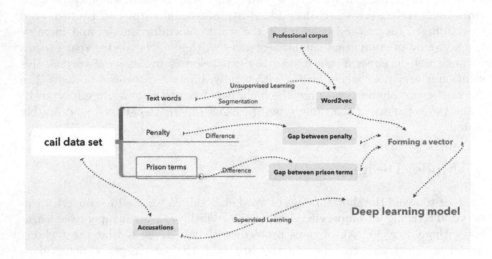

Fig. 1. Model architecture

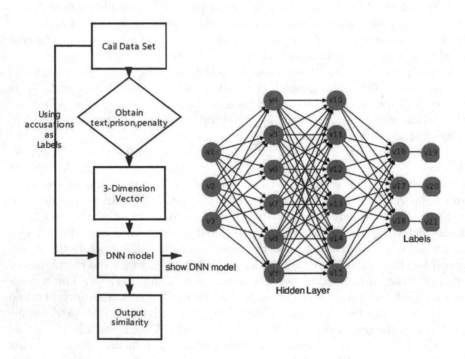

Fig. 2. Ensemble DNN model

model to accommodate minor changes. On the one hand, unsupervised learning can derive some features from a general point of view, which may be somewhat inaccurate but correspond to natural language. On the other hand, supervised learning can compensate for shortcomings by precise gradient descent and data fitting. In short, the model designed here takes into account the particularity of the similarity of judicial texts.

4   Evaluation and Results

When evaluating this model, we applied this model to two extreme cases and general cases. By experimenting with these situations, we can make some conclusions based on the results.

4.1   Case 1: Texts of Different Types with High Similarity

For instance, theft cases are very similar to burglary cases except one is on the premise of breaking into the room. However, we observed that most burglary cases have a prison term of more than 10 years, so the gap between the same burglary cases is relatively small. In this case, we can increase the weight of imprisonment and the results are obvious. Even if the text is linguistically similar, the gap between prison terms can seriously affect similarity, and the end result will decline.

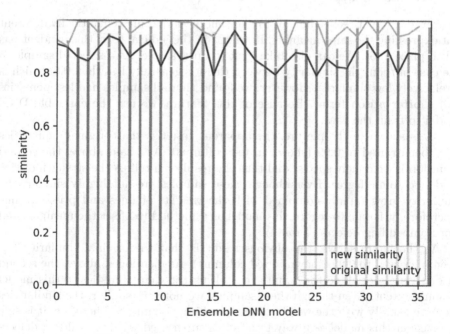

Fig. 3. Case 1 experiment (Color figure online)

We use a simple chart to illustrate this situation. The blue curve has the same trend as the orange curve. The consistency of the two curves means that the ensemble DNN model can be adapted to the original word embedding model and improved by means of other factors. More importantly, the green column is generally lower than the red column, which represents the effectiveness of the new model when dealing with confusing text (Table 1).

Table 1. Reduce the similarity of different types of text

	Low	High	Average
origin	0.91	0.99	0.96
ensemble	0.79	0.94	0.85

For cases with very similar categories, the initial similarity is very high, from the lowest value, the highest value and the average value. This is a highly error-prone situation. Based on this problem, our ensemble model can reduce the similarity in this case, and thus ensure the accuracy of the application in the actual scene (Fig. 3).

4.2   Case 2: Texts with Relatively Low Similarity but Belonging to the Same Type

In everyday life, it seems difficult to find sentences that are semantically different but expressing the same meaning. However, in the judicial text, this incident has taken place a lot. We can also take theft as an example. As a selected sample in the case of theft, the similarity between them is generally less than 0.8, which is a relatively low similarity. However, we found that all samples had less penalties and shorter prison terms. Because of this feature, we use the ensemble DNN model to train the data.

As is shown in the picture, the apparent result is that original similarities have been raised to 0.99 (shown in red columns). As I said above, the second situation is extremely special and thus the result can tell us the strength of this model to some degree. Even though these samples are not highly similar, the similarity grow rapidly combined with the weights of fines and prison terms. Therefore, it is safe to regard this model as a method to offset the unsupervised word embedding defects (Table 2).

As the figure shows, the obvious result is that the original similarity has been increased to 0.99 (shown in red columns). As mentioned above, the second case is very special, so the results can tell us the characteristics of this model to some extent. Even though these samples are not very similar, the similarities increase rapidly with the weight of fines and imprisonment. Therefore, it is safe to consider this model as a way to offset unsupervised word embedding defects (Fig. 4).

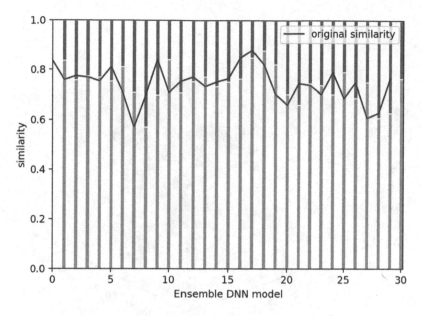

Fig. 4. Case 2 experiment (Color figure online)

Even in the same type of cases, occasionally the similarity is too low. The average similarity is only 0.78. In this case, we increased the similarity according to the prison terms and the penalty, and the final similarity can be increased to 0.97, which is regarded as a significant increase of 0.19.

Table 2. Raise the similarity of the same type of text

	Low	High	Average
origin	0.58	0.86	0.78
ensemble	0.94	0.99	0.97

4.3 General Case: High-Dimensional Space Mapping for Three-Dimensional Space

In addition to two extremely special cases, the normal situation of the model is the clustering problem in three-dimensional space. The general advantage of this model is dimensionality reduction. Word2vec represents a word embedding model in high-dimensional space, and the difference in high-dimensional space has been converted to a number between 0 and 1. This process is the first dimension reduction we know in the word embedding model (Fig. 5).

Besides, the overall DNN model combines this number with the gaps in other factors. This process also reduces complexity. The three-dimensional vector constitutes the input to the DNN model. By training this model, we can get the corresponding output to cluster relative types [7].

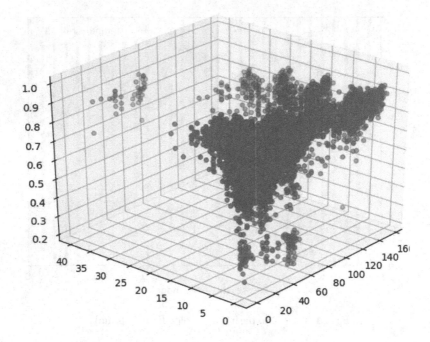

Fig. 5. General case experiment

Overall, for Case 1, we improved the accuracy of 0.29 because we increased the weight of the non-text indicators. For case 2, we only improved the accuracy of 0.12 due to its relatively high accuracy (Table 3).

Table 3. General increased accuracy

	Case1	Case2
origin-acc	0.65	0.86
ensemble-acc	0.94	0.98
variation	0.29	0.12

5   Discussion and Future Work

The ensemble method we propose here for word embedding is actually a way to reduce complexity and increase data fit. This model may be helpful in reducing accidental natural language similarities and enhancing the determinacy of reliability factors. The results also showed better applicability when dealing with sudden similar inaccuracies.

In terms of shortcomings, this is merely an attempt based on an ensemble approach that lacks sufficient reliability. The reconstructed vector is in three dimensions, and this reduced complexity may result in over-fitting of the data.

In addition, DNN models are sometimes less efficient, and several changes in parameters can have unpredictable effects.

For future work, when deciding the similarity of judicial texts, more attention should be paid to collecting useful factors [8]. We should consider choosing a more compatible model [11]. The new model should measure the importance of the different factors themselves during the training process. In addition, the stability of the new model is included when dealing with subtle changed parameters.

Acknowledgment. The work is supported in part by the National Key Research and Development Program of China (2016YFC0800805) and the National Natural Science Foundation of China (61772014).

References

1. Al-Kofahi, K., Jackson, P., Travers, T.E., Tyrell, A.: Systems, methods, and software for classifying text from judicial opinions and other documents, uS Patent 7,062,498, June 2006
2. Barlow, H.B.: Unsupervised learning. Neural Comput. **1**(3), 295–311 (1989)
3. Deng, L., Platt, J.C.: Ensemble deep learning for speech recognition. In: Fifteenth Annual Conference of the International Speech Communication Association (2014)
4. Grbovic, M., Cheng, H.: Real-time personalization using embeddings for search ranking at Airbnb. In: Proceedings of the 24th ACM SIGKDD International Conference on Knowledge Discovery & Data Mining, pp. 311–320. ACM (2018)
5. He, T., Lian, H., Qin, Z., Zou, Z., Luo, B.: Word embedding based document similarity for the inferring of penalty. In: Meng, X., Li, R., Wang, K., Niu, B., Wang, X., Zhao, G. (eds.) WISA 2018. LNCS, vol. 11242, pp. 240–251. Springer, Cham (2018). https://doi.org/10.1007/978-3-030-02934-0_22
6. Hinton, G., et al.: Deep neural networks for acoustic modeling in speech recognition. IEEE Signal Process. Magaz. **29**, 82 (2012)
7. Iwayama, M., Tokunaga, T.: Cluster-based text categorization: a comparison of category search strategies. In: Proceedings of the 18th Annual International ACM SIGIR Conference on Research and Development in Information Retrieval, pp. 273–280. Citeseer (1995)
8. Jaderberg, M., Vedaldi, A., Zisserman, A.: Deep features for text spotting. In: Fleet, D., Pajdla, T., Schiele, B., Tuytelaars, T. (eds.) ECCV 2014. LNCS, vol. 8692, pp. 512–528. Springer, Cham (2014). https://doi.org/10.1007/978-3-319-10593-2_34
9. Jain, A.K., Farrokhnia, F.: Unsupervised texture segmentation using gabor filters. Pattern Recognit. **24**(12), 1167–1186 (1991)
10. Levy, O., Goldberg, Y.: Neural word embedding as implicit matrix factorization. In: Advances in Neural Information Processing Systems, pp. 2177–2185 (2014)
11. Marti, U.V., Bunke, H.: Using a statistical language model to improve the performance of an hmm-based cursive handwriting recognition system. In: Hidden Markov Models: Applications in Computer Vision, pp. 65–90. World Scientific (2001)
12. Mikolov, T., Chen, K., Corrado, G., Dean, J.: Efficient estimation of word representations in vector space. arXiv preprint arXiv:1301.3781 (2013)

13. Tang, D., Wei, F., Yang, N., Zhou, M., Liu, T., Qin, B.: Learning sentiment-specific word embedding for twitter sentiment classification. In: Proceedings of the 52nd Annual Meeting of the Association for Computational Linguistics (Volume 1: Long Papers), vol. 1, pp. 1555–1565 (2014)
14. Turian, J., Ratinov, L., Bengio, Y.: Word representations: a simple and general method for semi-supervised learning. In: Proceedings of the 48th Annual Meeting of the Association for Computational Linguistics, pp. 384–394. Association for Computational Linguistics (2010)
15. Zhang, C., Ma, Y.: Ensemble Machine Learning: Methods and Applications, 1st edn. Springer, New York (2012). https://doi.org/10.1007/978-1-4419-9326-7

Multilingual Short Text Classification Based on LDA and BiLSTM-CNN Neural Network

Meng Xian-yan, Cui Rong-yi, Zhao Ya-hui[(⊠)], and Zhang Zhenguo

Yanbian University, 977 Gongyuan Road, Yanji, China
{cuirongyi, yhzhao}@ybu.edu.cn

Abstract. The rapid development of the Internet makes information resources have the characteristics of language diversity, and the differences between different languages can cause difficulties in information exchange. Therefore, the integration of multilingual data resources has important application value and practical significance. In this paper, we propose a novel model based on LDA and BiLSTM-CNN to solve the multilingual short text classification problem. In our method, we make full use of topic vectors and word vectors to extract text information from two aspects, in each language, to solve the sparse problem of short text features. Then, we use Long Short Term Memory (LSTM) neural network to capture the semantic features of documents and use Convolutional Neural Network (CNN) to extract local features. Finally, we cascade the features of each language and use SoftMax function to predict document categories. Experiments on parallel datasets of Chinese, English and Korean scientific and technological literature show that our proposed the short text classification model based on LDA and BiLSTM-CNN can effectively extract text information and improve the accuracy of multilingual text classification.

Keywords: Multilingual text classification · Linear discriminant analysis · Recurrent neural network · Convolutional Neural Network

1 Introduction

Multilingual text classification is a relatively new field compared to monolingual text classification, and research started late. Cross-language text classification, proposed by *Bel et al.* [1] refers to extending the existing text classification system from single language to two or more languages without manual intervention. At present, multilingual text categorization is mainly divided into dictionary-based, corpus-based, machine translation and other methods. *Amine et al.* [2] proposed a multilingual text classification method based on WordNet, which uses WordNet to represent training text as synset, thus obtaining the relationship between synset and category. *Ni et al.* [3] proposed a method of constructing multilingual domain topic model based on Wikipedia multilingual information. *Hanneman et al.* [4] proposed a full-text translation algorithm based on syntax to improve the accuracy of classification. *Liu et al.* [5] proposed a multilingual text categorization model based on autoassociative memory.

In this study, we propose a LDA-based BiLSTM-CNN network for multilingual text categorization to solve the barriers between different languages. The algorithm

© Springer Nature Switzerland AG 2019
W. Ni et al. (Eds.): WISA 2019, LNCS 11817, pp. 319–323, 2019.
https://doi.org/10.1007/978-3-030-30952-7_32

works as follows: Combining word vectors and topic vectors, we construct multilingual text representation from word meaning and semantics. We use BiLSTM network to capture the semantic features of multilingual text, and use convolution layer to further optimize features, finally SoftMax function is used to predict classification.

2 Extracting Text Vector

For texts in different languages, each language trains the word vector and the subject vector independently, constructing a text matrix. Taking Chinese text as an example, the word vector trained by Word2Vec is $\mathbf{w}_m$:

$$\mathbf{w}_m = [\mathbf{t}_{m,1}, \mathbf{t}_{m,2}, \cdots, \mathbf{t}_{m,N}] \in R^{D \times N} \tag{1}$$

$\mathbf{W}_m$ represents the word embedding vector of the m-th document, $\mathbf{t}_{m,i}$ represents the word vector of the i-th (i = 1, 2, …, N) word in the text m, N is the number of words in the sentence, and D is the word vector Dimensions. The topic vector trained with the LDA theme model is $\boldsymbol{\theta}_m$:

$$\boldsymbol{\theta}_m = (\theta_{m,1}, \theta_{m,2}, \ldots, \theta_{m,k}) \in R^K \tag{2}$$

$\boldsymbol{\theta}_m$ represents the topic vector of the m-th document, K is the number of topics, and is the same as the word vector dimension D, K = D, splicing the word vector and the topic vector to form a new input $\mathbf{W}$, containing the meaning and semantic features of the word, such as (3), where $\oplus$ is the stitching symbol.

$$\mathbf{W} = [\mathbf{w}_m \oplus \boldsymbol{\theta}_m]^{\mathbf{T}} \in R^{(N+1) \times D} \tag{3}$$

3 Extracting Text Vector

The proposed neural network model is shown in Fig. 1. The Chinese sub-network, the English sub-network and the Korean sub-network have the same network structure and different model parameters. The following mainly introduces the Chinese model.

For texts in different languages, the text matrix is constructed according to the way described in Sect. 2 and input into the neural network. Then two LSTM neural networks are used to extract text features from forward and backward directions. Convolution operation is performed on multiple convolution layers of different convolution kernels, wherein the convolution kernel window width k is the same as the output width of the BiLSTM layer, and the ith value of the convolution result vector s is calculated as a formula (4):

$$s_i = w * a_{i:i+h-1} + b, \, w \in R^{h \times k} \tag{4}$$

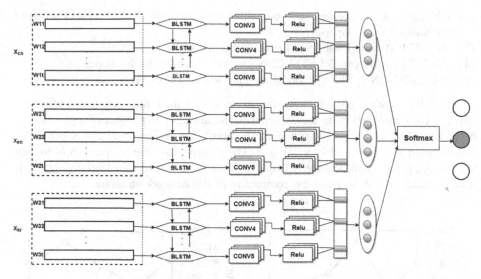

Fig. 1. BiLSTM-CNN Neural Network

Where a represents the output data of the BiLSTM layer, W represents the convolution kernel weight parameter for calculating the inner product with a, and b is the offset term. The activation layer selects f_{relu} as the activation function.

The maximum pooling strategy is adopted at the pooling level, and the maximum eigenvalue is taken, which can significantly reduce the parameters. Finally, the text representations of r_{cn}, r_{en}, and r_{kr} are cascaded into the SoftMax function to predict categories. The formula of SoftMax function is as follows:

$$y = soft\max(r_{cn}\|r_{en}\|r_{kr}) \tag{5}$$

4 Experimental Results and Analysis

4.1 Data Set

The corpus contains more than 90,000 abstracts of scientific and technical literature in Chinese, English and Korean, which are divided into 13 categories including biology, ocean and aerospace etc. The training and test sets are selected according to the ratio of 9:1 for each category.

4.2 Experiments and Analysis of Results for Multilingual Text Classification

This experiment mainly compares the following benchmark methods:

(1) Machine Translate (MT): Chinese and Korean abstracts are translated into English by machine translation tool Google, and classified with English text.

(2) CNN [6]: Neural network structure with multiple convolution kernels of different sizes and pooling layers.
(3) BiLSTM: Bidirectional LSTM neural network structure.
(4) BiLSTM-CNN: The BiLSTM-CNN method presented in this paper.
(5) T-BLSTM-CNN: The method of merging topic vectors proposed in this paper.

From Fig. 2, we can see that the dimension size of the topic vectors will affect the effect of the classification. In the experiment, the appropriate number of topics should be set according to the specific situation. The number of hidden layer nodes affects the classification accuracy as shown in Fig. 3. The results show that the number of hidden layer nodes is too small, and the network cannot have the necessary learning ability. If it is too much, it will increase the complexity of the network structure.

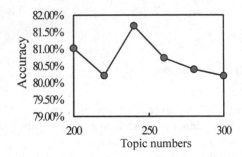

Fig. 2. Performance with topic numbers.

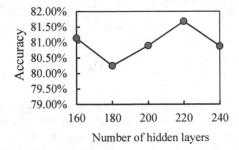

Fig. 3. Performance with number of hidden layers.

The cross-entropy and accuracy are used as performance indicators. The experimental results of each algorithm are shown in Table 1.

Table 1. Classification Accuracy rate for different methods

Algorithm	Cross entropy	Accuracy(%)
MT	1.12	78.15
TextCNN	0.74	78.50
BiLSTM	0.72	78.14
BiLSTM-CNN	0.68	80.29
T-BiLSTM-CNN	**0.57**	**81.68**
cn-T-BiLSTM-CNN	–	81.05
en-T-BiLSTM-CNN	–	83.85
kr-T-BiLSTM-CNN	–	80.14

From the experimental results, it can be seen that the method proposed in this paper is better than machine translation method, and this method has good expansibility. The classification accuracy of BiLSTM-CNN method proposed in this paper is about 2 higher than that of CNN and BiLSTM models. It shows that the optimization of sequence information by CNN can highlight the significant features of text. The classification accuracy of fusion topic vector method is significantly higher than other method, which shows that the combination of semantic information and word meaning information can fully express the characteristics. At the same time, it can be seen that the multilingual text categorization model based on T-BiLSTM-CNN can not only classify three parallel datasets correctly, but also achieve better classification results when only Chinese, English and Korean monolingual texts are input.

5   Conclusion

This paper presents a multilingual text categorization model, which can extract multilingual information more deeply and improve the accuracy of multilingual text categorization. At the same time, by incorporating the topic information into the neural network model, we can capture the feature information of the whole semantic level of the text, complement the word vector information, and extract more abundant text representation.

In the next step, different neural networks can be designed according to the characteristics of different languages to better extract text features of different languages and further enhance the effect of multilingual text categorization.

Acknowledgements. This research was financially supported by State Language Commission of China under Grant No. YB135-76 and Scientific Research Project for Building World Top Discipline of Foreign Languages and Literatures of Yanbian University under Grant No. 18YLPY13.

References

1. Bel, N., Koster, C.H.A., Villegas, M.: Cross-lingual text categorization. In: Koch, T., Sølvberg, I.T. (eds.) ECDL 2003. LNCS, vol. 2769, pp. 126–139. Springer, Heidelberg (2003). https://doi.org/10.1007/978-3-540-45175-4_13
2. Amine, B.M., Mimoun, M.: Wordnet based cross-language text categorization. In: IEEE/ACS International Conference on Computer Systems & Applications, pp. 845–855. Infocomp Journal of Computer Science, Amman, Jordan (2007)
3. Ni, X., Sun, J.T., Hu, J., et al.: Mining multilingual topics from Wikipedia. In: 18th International Conference on World Wide Web, pp. 1155–1156. ACM Press, Madrid (2009)
4. Hanneman, G., Lavie, A.: Automatic category label coarsening for syntax-based machine translation. In: Fifth Workshop on Syntax, Semantics and Structure in Statistical Translation, pp. 98–106. ACM Press, USA (2011)
5. Liu, J., Cui, R., Zhao, Y.: Multilingual short text classification via convolutional neural network. In: Meng, X., Li, R., Wang, K., Niu, B., Wang, X., Zhao, G. (eds.) WISA 2018. LNCS, vol. 11242, pp. 27–38. Springer, Cham (2018). https://doi.org/10.1007/978-3-030-02934-0_3
6. Kim, Y.: Convolutional Neural Networks for Sentence Classification. Eprint Arxiv (2014)

Data Privacy and Security

Random Sequence Coding Based Privacy Preserving Nearest Neighbor Query Method

Yunfeng Zou[1,2,3](✉), Shiyuan Song[1,2,3], Chao Xu[1,2,3],
and Haiqi Luo[1,2,3]

[1] State Grid Jiangsu Electronic Power Company Research Institute,
Nanjing 210036, China
yunfeng.zou@163.com
[2] Department of Computer Science and Engineering, Southeast University,
Nanjing 211189, China
[3] Nanjing Suyuan High Technology Company, Nanjing 210008, China

Abstract. Location based service has been widely used in people' life. It brings convenience to the people, in parallel with the risk of query user's location privacy disclosure. As a result, privacy preserving location based nearest neighbor queries witness its thriving in recent years. Private information retrieval (PIR) based solutions receives continuous attention for in privacy preserving for its merits in high level privacy protection strength and independence of the trusted third parties. However, existing PIR based methods fall short in high time consuming of encoding, querying efficiency and poor security to mode attacks. To address above issues, random sequence is introduced to encode POI data, which can resist mode attacks and reduce the time of data encoding. As a consequence, location privacy protection effectiveness is improved. Meanwhile, to accelerate query efficiency, a hash table structure is built at the server-side to store rules of POI distribution in the manner of space bitmap, which can position nearing POI quickly and reduce the I/O cost of database visiting efficiently. Theoretical analysis and experimental results demonstrates our solution's efficiency and effectiveness.

Keywords: Location based service · Location privacy ·
Private information retrieval · Random sequence

1 Introduction

The rapid development of mobile communication and spatial location technology promotes the rise of location based services (LBS) [1]. k-Nearest Neighbor (kNN) query is a query mode in location services, which refers to finding k POIs (points of interest, POI) closest to the current location of the inquirer. For example, inquirer wants to find k nearest restaurants or gas stations to his current position. The basic mode is that the inquirer submits its own location and query request to LBS server, and the server responses request and returns query result to the inquirer. During this process, the inquirer s will inevitably share their locations to an untrustable third

© Springer Nature Switzerland AG 2019
W. Ni et al. (Eds.): WISA 2019, LNCS 11817, pp. 327–339, 2019.
https://doi.org/10.1007/978-3-030-30952-7_33

party including the LBS service provider. There exists the risk of location privacy leakage.

Currently, existing solutions in privacy preserving location based nearest neighbor queries mainly rely on spatial confusion [2–5], data transformation [6–8], false location perturbation [9–11] and PIR (private information retrieval) technology [12–16]. Compared with the location privacy preserving technology mentioned above, PIR-based location privacy preserving querying technology can provide higher protecting strength. Based on PIR technology, inquirer can retrieve any data on an untrusted server without exposing the data information. PIR technology can compensate for the security at server side that cannot be guaranteed by traditional solutions, and provide a stronger location privacy protection. As a result, PIR technology has received continuous attention. A series of PIR-based location privacy preserving query algorithms have been proposed in recent years. Reference [14] constructs the index structure of POI dataset, and proposes a PIR based privacy retrieval algorithm. However, it spends a long time in regional division and preprocessing, and there is some difficulties to defend against pattern attacks. Reference [15] uses kd-tree, R-tree structure and Voronoi polygon to complete nearest neighbor query. However, there exists defects such as insufficient accuracy and expensive preprocessing cost. Besides, it is also not applicable to k-nearest neighbor query sceneory. Pattern attack is an important threat to location privacy protection. It means that the attacker uses the frequencies of different inquirers' database access to infer the inquirer's next target. In order to cope with pattern attacks, Reference [16] proposes an AHG (Aggregate Hilbert Grid) algorithm based on aggregated Hilbert grid, which introduces a query plan to improve query accuracy and protecting effectiveness against specific pattern attacks. However, it still has the following disadvantages:

(1) Hilbert curve-based POI encoding costs quite a lot of offline processing time at LBS server side.
(2) To cope with pattern attack, POI dataset is divided into multiple tables. In each round of the query, LBS server needs to scan the database multiple times, which leads to large processing overhead.
(3) Large number of pattern attacks facilitate attackers ability to achieve visiting frequency of different POI, which is the key of POIs' coding characteristics. Query contents can be deduced easily via leakage of coding characteristics.

To address these issues of AHG algorithm, a privacy-preserving kNN query method RSC_kNN is proposed based on random sequence and PIR technology. A random sequence is introduced to finish POI encoding, instead of the Hilbert curve. Auxiliary storage structure is built at LBS server side to reduce the cost of scanning the POI database. What's more, privacy protection effect to large amount of pattern attack is improved, as well as the query efficiency.

The main contributions of the paper are as follows:

(1) The random sequence is used instead of Hilbert curve to reduces the encoding time overhead to POI data set.
(2) A hash table structure is devised to cache some visited POIs in memory, which can reduce cost of POI database access and improve query performance.
(3) Random sequence has higher randomicity and non-repeatability, which provides higher security. Furthermore, it can be periodically updated to improve security of the system.

The paper is organized as follows: Sect. 2 summarizes related work. Section 3 describes the problem and related concepts. Section 4 introduces the algorithm of RSC_kNN. Section 5 analyzes the proposed algorithm and verifies its effectiveness. Finally, summarize the full text and look forward to the future work.

2 Related Work

2.1 PIR Technology

PIR is an important protocol proposed to protect server-side user data security. PIR protocol has the characteristics of private retrieval in the server-side database. As a result, it has become an important method to protect the location and privacy of neighbors. In order to facilitate understanding of the principle of PIR technology, the model of PIR technology is simplified as follows: suppose the database has n data blocks: $d_1, d_2, \ldots, d_n$. When the inquirer initiates a PIR protocol based data block request q(i) to the database in LBS server side, q(i) represents the i-th data block d_i in the database, the process of q(i) can be completed without any data block information leakage to potential adversary, including the LBS server. The so-called "private" means that the database server or other illegal attacker does not know the content of the user's interest during the server query phase, thereby ensuring the security of the query initiator's location data.

The PIR protocol can be divided into three categories: information-based security PIR, computation-based security PIR, and security-based PIR. Their security levels and application scenarios are different. Most of existing PIR based location privacy preserving methods belongs to computation-based security PIR of quadratic congruence. The quadratic congruence problem is the theoretical basis of computation-based security PIR technology. The quadratic congruence problem is a typical NP-hard problem, so it theoretically determines the security of quadratic congruence based PIR technology. Reference [8] demonstrates the security of PIR technology theoretically and proves that PIR technology can protect the location privacy of query initiators in the process of server-side data retrieval and improve system security effectively. Reference [14] builds an index structure to accelerate PIR based target POIs retrieval.

Reference [15] uses indexing structure of kd-tree and R-tree, in parallel with Voronoi polygon structure to realize POI data set division to improve query efficiency. However, voronoi polygon schema requires a large time overhead to finish POIs segmentation, and is only applicable to nearest neighbor query rather than k-nearest neighbor query mode. Reference [16] proposes the concept of "query plan" to deal with pattern attacks. However, the following problems exist: (1) The security of the algorithm depends on the encryption characteristics of the Hilbert curve and the security features of PIR protocol. In the face of large number of pattern attacks, the attacker can still calculate the encoding characteristics of POI data set, and then obtain the query content of the inquirer. (2) This algorithm encodes POI data and cooperates with a specific query plan to cope with pattern attack. However, the high complexity of the Hilbert curve in POI data encoding stage deteriorates the coding efficiency and increases the system time overhead.

2.2 AHG Algorithm

PIR based location privacy preserving technology has attracted continuous attention because of its high security and independence of trusted third parties. AHG algorithm is the most representative solution. The kNN query is implemented based on the query plan [17] by dividing POI data into multiple data tables to avoid pattern attacks and achieve strong location privacy protection.

AHG algorithm divides server-side POI data into three parts and stores them in three tables, which are DB_1, DB_2 and DB_3. DB_1 stores id of the POI point encoded by the Hilbert curve, DB_2 stores coordinates of each POI point, and DB_3 stores additional information of each POI point. The inquier initiates a kNN location based nearest neighbor query in the manner of submitting to LBS server an encrypted query request, which includes his current location Q, query content and the privacy request. After receiving the query request, LBS server searches for DB_1, DB_2 and DB_3 sequentially according to the inquirer's query request and return query results to the client in encrypted form.

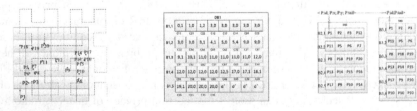

Fig. 1. POI space G encoded by Hilbert curve   **Fig. 2.** Illustration of DB1   **Fig. 3.** Illustration of DB2, DB3

Position P is in the form of <P: id; P: x; P: y; P: tail>. P: id denotes the identifier of the position P. P: x and P: y are the coordinates of the point P. P: tail stores the additional information of P. The coordinates of point P <P: id; P: x; P: x; P: y; P: tr> and additional information of P are stored in DB_2 and DB_3, respectively, where DB_2 is associated with DB_3 via the P: tr. Figure 1 illustrates a distribution map of POI data set. In order to facilitate storage and query, the entire space is divided into a grid with scale $g \times g$ (6×6 in Fig. 1). Hilbert curve code is applied to encode the figure G. DB_1 stores the POIs' id encoded by the Hilbert curve (as shown in Fig. 2). The grid C_{14} does not contain any POI points. The Hilbert curve passes through 3 POI points, namely P_1, P_2, P_3, so $C_{14} = (3, 0)$. If the last data block does not have corresponding grid, it will be filled with a fake point. As shown in Fig. 3, DB_2 stores the coordinates of the POI point encoded by the Hilbert curve, and the name and attribute of each POI is stored in DB_3.

When inquirer initiates a k-nearest neighbor query at location Q, LBS server first finds the grid in DB_1 which is nearest to Q and obtains corresponding k ids. Subsequently, the server searches the coordinates of these ids in DB_2. Secondly, the table DB_3 is scanned to get additional information of these k POIs. Finally, k POIs' coordinates and additional information is returned to the inquirer. Due to the encryption characteristics of the Hilbert curve, AHG algorithm can improve location privacy protection strength to some extent. However it still has the following drawbacks:

(1) In the POI coding phase on the LBS server, The high complexity of Hilbert curve leads to a large time overhead in DB1 construction.
(2) POI data is divided into three tables DB1, DB2 and DB3. For each query request initiated by the client, the server needs to retrieve three tables separately, which increases the query response time.
(3) The security of the algorithm depends on the encryption characteristics of Hilbert curve and PIR. However, in the face of a large number of pattern attacks, attackers may still deduce the coding characteristics of POI datasets on the LBS server side, and then obtain specific query content, which has security risks.

3   Problem Statement and Definitions

3.1   Problem Statement

PIR based privacy preserving location-based queries needs to solve the following problems:

(1) Server-side POI data set coding can achieve good efficiency under the premise of both data availability and privacy security.
(2) Accommodating accuracy of the query, the number of client queries to search database can be reduced to improve the query efficiency.
(3) Enhance the system's ability to cope with a large number of pattern attacks.

The PIR-based AHG algorithm has many drawbacks such as long encoding time, low query efficiency and weakness in coping with pattern attack. Aiming at these problems, our solution introduces random sequence instead of Hilbert curve to encode POI dataset. A new Hash table structure is designed to store encoded POI data rather than the table DB_1 in AHG. Correspondingly, a tailed querying algorithm is proposed to complete the server-side querying.

We still follow the POI spatial graph G structure adopted in AHG algorithm. The whole data space is divided into $g \times g$ grids. As shown in Fig. 4, in the coding phase in LBS server-side, information (including the number and id of POI points) of each grid is stored in the Hash table. Meanwhile, the POI data is divided into two tables, called DB_1 and DB_2. In this stage, an efficient random sequence generation algorithm is used to generate the storage order of POI data points. POI data is stored in DB_1 and DB_2 according the storage order. As shown in Fig. 5, DB_1 stores the coordinates of POI points and DB_2 stores their additional information. When the inquirer initiates the query plan QP at location Q, LBS server responses the request and return the result to the inquirer in an encrypted form.

Random number is a sequence of randomly generated permutations. Random sequence can play the role of blur and encrypted queries, and can also rearrange and encrypt databases. A random sequence is introduced to encrypt the database and corresponding queries to make query contents indistinguishable.

Even if the number of queries is different, it can resist the pattern attack because the querier and the query content can not be accurately inferred. In addition, the generation of random sequence has the characteristics of single parameter and relatively simple process.

3.2 Definitions

Concerning these problems of Γ - privacy model, a novel privacy-preserving provenance model is devised to balance the tradeoff between privacy-preserving and utility of data provenance. The devised model applies the generalization and introduces the generalized level. Furthermore, an effective privacy-preserving provenance publishing method based on generalization is proposed to achieve the privacy security in the data provenance publishing. Relevant definitions are as follows:

Definition 1. Query Plan QP [16]: QP specifies the number of times that each data table (DB1, DB2, DB3...) needs to be retrieved for each kNN query. It depends on the kNN algorithm, size of the POI data set and the distribution of data points in the data set.

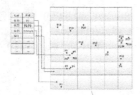

Fig. 4. Count of Hash table POI point

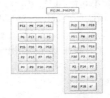

Fig. 5. POI data storage

Definition 2. POI Data Set Space Bitmap G [16]: G = <n, d, id>. N denotes the number of grids in graph G. D is the length of edges of each grid in the graph G. Symbol id is the identifier of the grid that can be uniquely distinguished from other grids after the POI data space bitmap is divided into g × g.

For example, the id of the grid in the lower left corner of Fig. 1 is G11.

Definition 3. POI Date Set S (size, array) [16]: Size denotes the number of POI points in S, and the specific information of all the points is stored in the array.

Hash table can realize fast positioning of POI data in the data query stage, reduce the number of database I/O needed in the query stage, and improve the query efficiency. Therefore, Hash table structure is designed to store POI data.

Definition 4. Hash table storage for POI data sets: The Hash table storage structure of POI dataset is as follows:

m = new Hash <G:id, <num, array>>. The key of the Hash table is used to store G: id. The value of the Hash table is used to store num and array where num and array represent number and id of the POI point, respectively.

As shown in Fig. 1, num and array of the grid G23 are <4, <5, 6, 7, 8>>, and it means that grid G23 contains four POI points P5, P6, P7, and P8.

4 Method

4.1 Algorithm RSC_KNN

In order to reduce the coding time of POI data set, random sequences is applied to encode the POI data, which can provide higher level security against large number of pattern attacks. Further, Hash table structure, in parallel with improved query strategy are leveraged to enhance query efficiency.

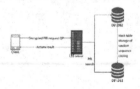

Fig. 6. Illustration of system flow

The system flow is shown in Fig. 6. The client sends the query request QP to the LBS server. At the server side, POI data is encoded by random sequence, and the encoding results are stored in the Hash table. When the server receives the client's kNN query request QP (QP contains client location information, query content and privacy requirements), it sequentially searches the hash table, DB1, DB2 to obtains k POI points including the id and the additional information, and returns the result to the client in an encrypted form.

4.2 Server-Side POI Coding

To cope with large number of pattern attacks. random sequences is introduced to encode the POI data in LBS server side. Compared with Hilbert curve, random sequence is more random and unpredictable which increases the difficulty for the attacker to obtain the client's retrieval target. At the same time, random sequence has a lower time complexity. Existing random sequence's time complexity is O(n), which is much lower than that of Hilbert curve. Data encoding work at server-side consists of two parts, to store corresponding POI information in Hash table storage and store them in the random storage.

Hash Table Constructing. The POI data set is traversed first, and the grid of the POI point is judged according to its coordinates <P: x, P: y>. The id of the Grid is used as Hash value. Hash table m is stored in system memory. It stores the number of POI points and POI point's id of each grid.

Algorithm 1 POI data set hash table storage
Input: Candidate POI point set S
Output: Hash table with POI point distribution information

---

1. New HashMap m();
2. For each POI in S /* Count one POI point at a time */
3.   G:id=($\lceil P{:}x/G{:}d \rceil$,$\lceil P{:}y/G{:}d \rceil$); /* Calculate the grid according to the coordinates of the POI point */
4.   m[G:id].num++;
5.   m[G:id].array.add(P:id);
6. Return m;

Although the encryption feature of the Hilbert curve adopted in AHG algorithm ensures that the encoding of POI data is more concealed, Hilbert curve has high complexity when POI dataset space map G is too large. It brings large time overhead. In addition, the attacker can still find some clues related to the POI point storage order in the database by analyzing a large number of query processes and then obtain the query content of some inquirer.

In order to reduce the complexity of POI data storage and processing overhead, and improve the effect of location privacy protection, random sequence is considered to encode POI data sets. Firstly, according to the size of POI data set S, the random sequence table is generated, which consists of no duplicate values with size S: size. POI data information is stored in table DB1 and DB2 in turn. Due to the low time complexity of random sequence, the overhead of POI data coding is reduced accordingly. The security of location privacy protection can be improved further by changing the encoding mode of POI data in the way of updating random sequence and data tables DB1 and DB2, periodically.

Algorithm 2 POI data set random storage

Input: Candidate POI point set S

Output: POI points stored in database with corresponding order

---

1. set up system timer/*Update a, DB_1 and DB_2 regularly*/
2. While（timer）
3.   New array() a/* Array a is used to store the order of POI points stored in the database. */
4.   a= Random number generation algorithm（S:size）/* Use a specific random number generation algorithm to create a random number sequence*/
5.   For each POI in S
6.   Store the coordinates of all POI points in the database DB_1 in the order of the random sequence table a;
7.   Store the additional information of all POI points in the database DB_2 in the order of the random sequence table a;

4.3 Server-Side Query Processing

Processing POI data at the server side can improve the security of location protection and query efficiency. When the inquirer initiates a kNN query request, the server-side processing steps list as follows:

(1) The nearest grid to Q is obtained according to the location of the query point Q, and the number of POI data stored in this grid is determined according to the Hash table. If not, the nearest neighbor grid of Q is extended until the nearest neighbor grid of Q is satisfied with kNN requirement.
(2) According to the nearest neighbor grid obtained in step (1), k POI points (P: id) are retrieved from the Hash table.
(3) According to the P: id obtained in step (2), database queries satisfying the query plan are completed sequentially, and the coordinates of POI points are retrieved by DB1. According to the P: tr pointer of POI points in DB1, DB2 is retrieved sequentially to obtain additional information of POI points.
(4) returns the query result including coordinates and additional information to the inquirer in an encrypted form.

Algorithm 3 Server-side query algorithm
Input: Corrdinates of location Q, query content and query value k
Output: POI points including the coordinates and additional information

---

1.   Obtain corrdinates of location Q:Q:x,Q:y;
2.   New array() Garray;/* Garrayis used to store Q's neighboring point*/
3.   findNearestNeighborG(){/*Obtain Q's neighboring grid*/
4.   G:id=($\lceil Q:x/G:d \rceil$,$\lceil Q:y/G:d \rceil$);//Determine grid which Q belongs to according Q:x,Q:y;
5.   If(Q is in a single grid){
6.   Garray.add(grid);/*Add this grid to the array of Q's neighboring gird*/}
7.   If(Q is at two or more grid junctions){
8.   Garray.add(grids bordering each other);/* Add these grids to Q's neighbor grid array*/}
9.   Search the hash table m to obtain the total number of POI points in the neighboring grid of Q;
10. While(count<k){
11. Extend Q's neighbor grid based on the Garray;
12.   Retrieve the hash table m and update count;}
13. New Vector() v1=Garray.P:id;/* Search the hash table m for the P:id of all POI points in Garray and save them in vector v1*/
14. New Vector() ret/*Save results*/
15. For each POI in v{
16.   Retrieving the coordinate information in DB$_1$ according to P: id;
17.   Ret[Pid].add(P:x,P:y); //Save the coordinate information of the point P in ret;
18.   Search the additional information of the P in DB$_2$ according to P:tr;
19.   Ret[Pid].add(P:tail);// Save the additional information of the point P in ret;}
20. Return ret;

4.4   Performance Analysis

This section mainly analyzes privacy protection strength and time complexity of our improved solution. In terms of location privacy preserving, our solution provides guarantee from two aspects, namely PIR and random sequence. PIR mechanism provides privacy retrieval database internally and request interface externally, which has high security strength. Random sequence reorders POI data sets to achieve double privacy protection for query process. In terms of time complexity, random sequence coding has lower time complexity than Hilbert coding, and the coding phase of POI data sets takes less time.

Compared with Hilbert curve, random sequence has the characteristics of randomness, unpredictability and non-reproducibility. At the same time, the process of retrieving database by PIR meets the basic privacy information retrieval constraints; POI data sets are also encrypted. Therefore, even through a large number of pattern attacks, attackers can not deduce the encryption rules of POI datasets, nor can they further crack PIR requests.

The time complexity of reordering POI data sets using random sequences in the server-side data coding stage is O (n), which is much lower than that of Hilbert curve coding O (n^2). In addition, RSC_kNN algorithm introduces the Hash table structure. The POI data information stored in DB_1 is stored in the memory through the Hash table. Each query uses only two tables, DB_1 and DB_2, which reduces the number of database I/O in the query stage, and further reduces the time cost of query processing.

In addition, the low complexity of random sequence makes the coding of server POI data set more efficient. The system can complete the re-coding of POI data set (update Hash table, DB1 table, DB2 table) by updating random sequence regularly or irregularly, thus further improving privacy protection strength.

5 Experimental Analysis

The experimental algorithm is implemented in Java and runs on Windows 10 platform with i7-3307 3.4 GHz processor and 8 GB memory. California's real POI data points were used in the experiment, including 105 K POI data points. The experiment compares our solution RSC_kNN with the AHG algorithm. The performance is mainly compared from the following aspects: G granularity selection of POI data bitmap, query time, coding time overhead of POI data set and system security in the face of pattern attack.

The grid size of POI data sets determined by the size of granularity (when POI data sets are divided into g * g grids), directly determines the number of grids, and affects the organizational structure of the database, thereby affecting the number of I/O queries to the database. Figures 7(a) and (b) show the change of query time under different granularity sizes when k = 1 and K = 5, respectively. Figure 8 describes the coding time overhead of POI data sets at system initialization.

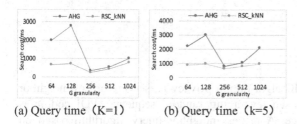

(a) Query time（K=1）   (b) Query time（k=5）

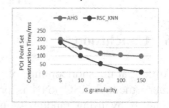

Fig. 7. Effect of granularity on query time overhead

Fig. 8. Coding time overhead for POI datasets

Assuming that the number of queries Q1 and Q2 accessing the database is C1 and C2 respectively, when C1 is not equal to c2, the attacker can determine the POI of the next query or even locate the location of the query according to the frequency of query requests sent to the server by the interceptor. This attack is called pattern attack. The experiment simplifies the pattern attack, and simulates the pattern attack by predicting the POI of the next query by analyzing the frequency characteristics of query requests for DB1 and DB2 in a certain number of different queries. Figures 9(a) and (b) show the security of the query system in the face of pattern attack without updating the random sequence and updating the random sequence in time, respectively.

As can be seen from Fig. 7, the response time of RSC_kNN algorithm system based on random sequence is shorter. At the same time, the AHG algorithm is sensitive to the choice of granularity size. From Fig. 8, we can see that the RSC_kNN method based on random sequence has less time overhead and faster speed in the coding phase of POI data sets. As shown in Fig. 9(a), in the face of a large number of pattern attacks, due to the randomness and unpredictability of random sequences, the attacker infers from the existing POI data access characteristics that the possibility of querying the next POI data to be accessed by the user is lower, and the RSC_kNN algorithm based on random sequences is more secure than the AHG algorithm. At the same time, because of the short offline time of POI data encoding of RSC_kNN algorithm based on random sequence, the POI data storage of LBS server can be updated by periodically updating random sequence. From Fig. 9(b), it can be seen that under the condition of periodically updating random sequence, the system can better cope with a large number of pattern attacks, and the degree of system privacy leakage tends to be stable with the increase of attacks. Safety is higher.

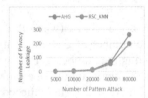   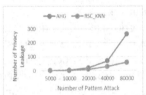

(a) System security (not update random sequence)   (b) System security (update ran dom sequence periodically)

Fig. 9. System security analysis

6 Conclusion

Aiming at the shortcomings of AHG algorithm, a privacy-preserving nearest neighbor query method RSC_kNN is proposed leveraging random sequence. POI coding is realized by using random sequence. A nearest neighbor query algorithm based on auxiliary storage structure is designed on LBS server side, which can improve the intensity of location privacy protection and query processing efficiency. Theoretical analysis and experiments verify the efficiency and privacy preservation of the proposed algorithm. Effectiveness of nursing.

Whether the granularity selection of bitmap G in POI dataset is reasonable or not has a certain impact on query performance. Large or small granularity may reduce query efficiency. The optimization of storage overhead on LBS server side and the granularity setting method of POI bitmap set G will be further studied in the future.

References

1. Mokbel, M.F.: Privacy in location-based services: state-of-the-art and research directions. In: International Conference on Mobile Data Management, p. 228. IEEE Computer Society (2007)

2. Gruteser, M., Grunwald, D.: Anonymous usage of location-based services through spatial and temporal cloaking. In: International Conference on Mobile Systems, Applications, and Services, DBLP, pp. 31–42 (2003)

3. Bu, G.G., Liu, L.: A customizable k-Anonymity model for protecting location privacy. In: ICDCS, pp. 620–629 (2004)

4. Xiao, Z., Meng, X., Xu, J.: Quality aware privacy protection for location-based services. In: Kotagiri, R., Krishna, P.R., Mohania, M., Nantajeewarawat, E. (eds.) DASFAA 2007. LNCS, vol. 4443, pp. 434–446. Springer, Heidelberg (2007). https://doi.org/10.1007/978-3-540-71703-4_38

5. Gedik, B., Liu, L.: protecting location privacy with personalized k-Anonymity: architecture and algorithms. IEEE Trans. Mob. Comput. 7(1), 1–18 (2007)

6. Wu, J., Ni, W., Zhang, S.: Generalization based privacy-preserving provenance publishing. In: Meng, X., Li, R., Wang, K., Niu, B., Wang, X., Zhao, G. (eds.) WISA 2018. LNCS, vol. 11242, pp. 287–299. Springer, Cham (2018). https://doi.org/10.1007/978-3-030-02934-0_27

7. Indyk, P., Woodruff, D.: Polylogarithmic private approximations and efficient matching. In: Halevi, S., Rabin, T. (eds.) TCC 2006. LNCS, vol. 3876, pp. 245–264. Springer, Heidelberg (2006). https://doi.org/10.1007/11681878_13

8. Khoshgozaran, A., Shahabi, C.: Blind evaluation of nearest neighbor queries using space transformation to preserve location privacy. In: Papadias, D., Zhang, D., Kollios, G. (eds.) SSTD 2007. LNCS, vol. 4605, pp. 239–257. Springer, Heidelberg (2007). https://doi.org/10.1007/978-3-540-73540-3_14

9. Man, L.Y., Jensen, C.S., Huang, X., et al.: SpaceTwist: managing the trade-offs among location privacy, query performance, and query accuracy in mobile services. In: IEEE, International Conference on Data Engineering, pp. 366–375. IEEE (2008)

10. Ni, W., Zhen, J., Chong, Z.: HilAnchor: location privacy protection in the presence of users' preferences. J. Comput. Sci. Technol. 27(2), 413–427 (2012)

11. Gong, Z., Sun, G.Z., Xie, X.: Protecting privacy in location-based services using K-anonymity without cloaked region. In: Eleventh International Conference on Mobile Data Management, pp. 366–371. IEEE (2010)

12. Williams, P., Sion, R.: Usable PIR. In: Network and Distributed System Security Symposium, NDSS 2008, San Diego, California, USA. DBLP (2008)

13. Yi, X., Paulet, R., Bertino, E.: Private information retrieval. J. ACM 45(6), 965–981 (1998)

14. Khoshgozaran, A., Shahabi, C., Shirani-Mehr, H.: Enabling location privacy; moving beyond k-anonymity, cloaking and anonymizers. Knowl. Inf. Syst. 26(3), 435–465 (2011)

15. Ghinita, G., Kalnis, P., Khoshgozaran, A., et al.: Private queries in location based services: anonymizers are not necessary. In: SIGMOD 2008, pp. 121–132 (2008)

16. Papadopoulos, S., Bakiras, S., Papadias, D.: Nearest neighbor search with strong location privacy. PVLDB 3(1), 619–629 (2010)

17. Gedik, B., Liu, L.: Location privacy in mobile systems: a personalized anonymization model. In: Proceedings of 2005 IEEE International Conference on Distributed Computing Systems, ICDCS 2005, pp. 620–629. IEEE (2005)

Multi-keyword Search
Based on Attribute Encryption

Xueyan Liu[(⊠)], Tingting Lu[(⊠)], Xiaomei He, and Xiaotao Yang

Northwest Normal University, Lanzhou 730070, China
liuxy@nwnu.edu.cn, 15343612736@163.com

Abstract. To address the problems of key leakage and large computation amount in current attribute-based searchable encryption (SE) schemes, this study proposes a multi-keywords ranked search scheme based on attribute encryption in cloud environment. Firstly, the scheme not only supports fine-grained search authorisation, but also has constant ciphertext length. Secondly, users' private keys are blinded before being submitted to the cloud server, This process ensures the security and confidentiality of users' keys. Thirdly, to reduce search complexity, this study uses $TF \times IDF$ rules to calculate the relevance scores of documents and given keywords, and then sorts the documents. Lastly, formal evidence shows that our scheme exhibits selective plaintext security and is indistinguishably secure against chosen keyword attacks. Experimental results demonstrate that our scheme is efficient.

Keywords: Attribute-based searchable encryption · $TF \times IDF$ rule ·
Fast searchable · Rank

1 Introduction

An SE scheme allows cloud users to perform a search over encrypted data. Song et al. [1] introduced the concept of SE. Farras et al. [2] proposed a public key scheme for searching keywords in ciphertext. Thereafter, other studies [3–7] have focus on enriching search functionality including multi-keyword search, dynamic search and result ranking. In the traditional SE schemes, however, the relationship between data owner and user is one-to-one encryption [8, 9]. In many cases, the data owner is expected to share his/her data in an expressive manner.

Attribute-based encryption (ABE) was introduced to solve this problem [10], ABE schemes are classified into two types of ABE scheme: key-policy ABE (KP-ABE) [11] and ciphertext-policy ABE (CP-ABE) [12]. Sun et al. [13] presented the first attribute-based keyword search (ABKS) scheme. Miao et al. [14] proposed a verifiable keyword search over encrypted data that uses third party audit to verify the search results by appending a signature to each file. However, inherent to ABE technology, ciphertext size in the aforementioned ABKS scheme is proportional to the number of attributes. Accordingly, search time is also proportional to the number of attributes. Emura et al. [15] presented a CP-ABE scheme with constant ciphertext length.

© Springer Nature Switzerland AG 2019
W. Ni et al. (Eds.): WISA 2019, LNCS 11817, pp. 340–346, 2019.
https://doi.org/10.1007/978-3-030-30952-7_34

The major contributions of this study are summarised as follows: (1) To reduce search complexity, this study uses *TF* × *IDF* rules [8] to calculate the relevance scores of documents and given keywords, and then ranks the documents. (2) The scheme introduces ABE technology to realise fine-grained access authority. (3) To ensure the security and confidentiality of users' keys, users' private keys are blinded before being submitted to the cloud server. (4) The proposed scheme is multi-value independent and has constant ciphertext length.

2   Construction of Proposed Scheme

2.1   System Model

Briefly speaking, the interaction between the four entities as described in Fig. 1. Data owner builds the index and encrypts files, and uploads them to cloud. TA is in charge of issuing the system public key, master secret key, attribute keys for users. CSP stores data uploaded by data owner and retrieve the stored files requested by user. Data user verifies the validity of search results and decrypts it.

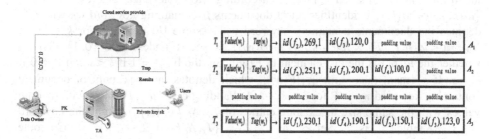

Fig. 1.   System model   **Fig. 2.**   Example model

Supposing that keyword set is $W = \{w_1, w_2, w_3\}$ and $F = \{f_1, f_2, f_3, f_4\}$ is file set in Fig. 2. If the authorized user generates correct trapdoor about interested keywords $W = \{w_2\}$, CSP seeks for the location of w_2 and returns the most top-3 files $C_{top-3} = \{\{f_2, f_1, f_4\}, Tag(w_2)\}$ to user. If $W = \{w_1, w_2, w_3\}$, CSP will seek for the location of w_2 with the lowest frequency and filter out the documents f_1, f_4. Only f_2 containing W. The correlation score of f_2 containing W is $369 + 251 + 150 = 670$. the documents $C_{top-1} = \{f_2\}$ that contain W and have the highest relevance are sent to users.

2.2   The Description of Scheme

Our scheme consists of seven algorithms.

Setup(1^k). TA picks two bilinear groups (G_1, G_2) of prime-order p, $g, h \in G_1$, $y \in {}_R Z_p^*$, and $t_{ij} \in {}_R Z_p^*$ ($i \in [1, n], j \in [1, n_i]$), $MAC : \{0,1\}^\lambda \times \{0,1\}^* \to \{0,1\}^l$, $\varphi : \{0,1\}^\lambda \times \{0,1\}^* \to \{0,1\}^\lambda$, and $H : \{0,1\}^* \to Z_p^*$ are hash functions. TA computes

$T_{ij} = g^{t_{ij}} (i \in [1,n] j \in [1,n_i])$, $Y = e(g,h)^y$, $\hat{h} = h^y$, and generates $PK = \{e, g, \hat{h}, Y, \varphi,$
$H, MAC, \{T_{ij}\} (i \in [1,n] j \in [1,n_i])\}$ as public parameter, $MSK = \{y, t_{ij}(i \in [1,n],$
$j \in [1,n_i])\}$ as master secret key. Note that $\forall S, S'(S \neq S')$, $\sum_{v_{ij} \in S} t_{ij} \neq \sum_{v_{ij} \in S'} t_{ij}$ is
assumed.

$KeyGen(PK, MSK, S)$. TA chooses $r_1 \in_R Z_p^*$, computes $K_1 = \hat{h} \left(g^{\sum_{v_{i,j} \in S} t_{ij}} \right)^{r_1}$,
$K_2 = g^{r_1}$, and sends $sk_u = \{K_1, K_2\}$ to a user linked to attribute set S.

$Encrypt(PK, W, F, \varepsilon k, \mathbb{A})$. Given the symmetric key εk, an AND-gate access policy
$\mathbb{A} = \{\mathbb{A}_1, \cdots, \mathbb{A}_n\}$, keywords $W = \{w_1, \cdots, w_L\}$ and files $F = \{f_1, \cdots, f_K\}$. Data
owner selects $a \in_R Z_p^*$, computes each weighted score $TF \times IDF_{lk}$ between w_l and f_k.
The T_l and A_l $(l \in [1, \mu], \mu \geq L)$ are built for each keyword w_l. T_l $(l \in [1, \mu])$ includes
two parts: $Value(w_l)$ and $Tag(w_l)$, where $Value(w_l) = g^{aH(w_l)}$ is to retrieve the keyword
location. $Tag(w_l) = \langle E_{\varepsilon k}(\rho_l) \| \delta_l \rangle$ is proof for each keyword, $\rho_l(l \in [1,L])$ is K bit
vectors in which the k-th dimension is 1 if document f_k contains the keyword w_l,
otherwise, set 0. Data owner computes $\delta_l = MAC(w_l, \rho_l)$, encrypts ρ_l using εk, i.e.
$E_{\varepsilon k}(\rho_l)$. A_l $(l \in [1, \mu], \mu \geq L)$ is built based on an ordered sequence of document
identifiers $id(f_k)$ that is ranked according to relevance scores. $F(w_l) =$
$\{id(f_{l1}), \cdots, id(f_{ls})\}$ is identifier set of documents f_k containing keyword w_l, where f_{lk}
denotes the documents containing keyword w_l, Assume $|F(w_l)| = s$. For $k \in [1,s]$,
$N_{lk} = (<id(f_{lk}) \| TF \times IDF_{lk} \| \beta > \oplus \varphi(w_l))$ is secret nodes, $\beta \in \{0,1\}$ denotes
whether the $id(f_{lk})$ is the last element in $F(w_l)$. Set the bit $\beta = 1$ if $k < s$ and $\beta = 0$ if
$k = s$. If $s < max$, $max = max_{l=1}^L |F(w_l)|$, which denotes the maximum of document
containing keyword $w_l \in W$ for $l = 1, \cdots, L$. Padding other $\mu - L$ arrays with random
values with the same size as existing entries of $T_l, A_l (l \in [1, \mu])$. Thus, the keyword
index table is set as $I_w = \{T_l, A_l | l \in 1, 2, \cdots, \mu\}$. Compute
$C_k \leftarrow E_{\varepsilon k}(f_k)(k = 1, \cdots, K)$, $C = \{C_1, C_2, \cdots, C_K\}$ is encrypted file set. Set
$\hat{C}_1 = \left(\prod_{v_{ij} \in A} T_{ij} \right)^a$, $\hat{C}_2 = g^a$, $\hat{C}_3 = \varepsilon k \cdot e(g, \hat{h})^a$, $\hat{C}_4 = g^{aH(\varepsilon k)}$, the ciphertext is
$CT = \{\hat{C}_1, \hat{C}_2, \hat{C}_3, \hat{C}_4\}$. Finally $\langle I_w, CT, C \rangle$ is uploaded to CSP.

$Trapdoor(sk, W)$. Data user selects $r_2 \in_R Z_p^*$ and generates trapdoor of his interested
keywords $W' = \{w_1, \cdots, w_t\}(t \leq L)$ as: $Trap = \left\langle \{\tau_l = \varphi(w_l)\}_{l \in [1,t]}, tk_1 = K_1^{r_2} tk_3 = \right.$
$\left. (\hat{h})^{r_2 \sum_{l=1}^{t} H(w_l)}, tk_2 = K_2^{r_2 H(w_l)} \right\rangle$. Finally, user sends $Trap$ to CSP.

$Test(I, Trap)$. The CSP calculates $e(\hat{C}_1, tk_2) \cdot e(\hat{C}_2, tk_3) = e\left(\prod_{l=1}^{t} Value(w_l), tk_1 \right)$ as
follows:

$$e(\hat{C}_1, tk_2) \cdot e(\hat{C}_2, tk_3) = e\left(\left(\prod_{v_{ij} \in \mathbb{A}} T_{ij}\right)^a, (g^{r_1})^{r_2 \sum_{l=1}^{t} H(w_l)}\right) \cdot e\left(g^a, (\hat{h})^{r_2 \sum_{l=1}^{t} H(w_l)}\right)$$

$$= e(g,g)^{a\, r_1\, r_2 \sum_{l=1}^{t} H(w_l) \sum_{v_{ij} \in \mathbb{A}} t_{ij}} \cdot e(g,h)^{ayr_2 \sum_{l=1}^{t} H(w_l)}$$

$$e\left(\prod_{l=1}^{t} Value(w_l), tk_1\right) = e\left(\prod_{l}^{t} g^{aH(w_l)}, \left[\hat{h}\left(g^{\sum_{v_{ij} \in S} t_{ij}}\right)^{r_1}\right]^{r_2}\right)$$

$$= e(g,h)^{ar_2y \sum_{l}^{t} H(w_l)} \cdot e(g,g)^{ar_1r_2 \sum_{l}^{t} H(w_l) \sum_{v_{ij} \in S'} t_{ij}}.$$

Initialize $F(w_l)$ as an empty posting list, decrypt: $A_l[k] \oplus \varphi(w) = (id(f_k), TF \times IDF_{lk}, \beta)$ $l = \{1, 2, \cdots, t\}$, if $\beta = 1$, add $id(f_{lk})$ to $F(w_l)$; Otherwise, stop searching the array. Supposing that the least frequency keyword is w_1, $F(w_1) = \{f_{11}, \cdots, f_{1c}\}$ and $|F(w_1)| = c$, so that the document that does not contain keyword w_1 can be filtered, CSP computes the relevant scores between $\{w_l\}(l \in [2, t])$ and $\{f_{1k}\}_{k \in [1,c]}$ though the equation, ranking the retrieval relevant score. The relevant order file can be ranked in to $C_{top-s}(W')$. Assume $C_{top-t}(W') = \{C_{u1}, \cdots, C_{ut}\}$, the search result Results $= \left\{\{Tag(w_l)\}_{l \in [1,t]}, CT, C_{top-t}(W')\right\}$ returned by CSP is sent to user.

Decrypt(CT, sk). A authorized data user can compute $\dfrac{e(\hat{C}_1, K_2) \cdot \hat{C}_3}{e(\hat{C}_2, K_1)} =$

$$\frac{\varepsilon k \cdot e(g,g)^{a\, r_1 \sum_{v_{ij} \in S} t_{ij}} \cdot e(g, h^y)^a}{e(g,g)^{a\, r_1 \sum_{v_{ij} \in S} t_{ij}} \cdot e(g, h^y)^a} = \varepsilon k$$ and $\hat{C}_2^{H(\varepsilon k)} = \hat{C}_4$; otherwise, rejects.

Verify. The data user calculates as follows: decrypts $Tag(w_l) l \in [1, t]$ using the symmetric εk to get the string ρ_l, computes $MAC(w_l, \rho_l) \overset{?}{=} \delta_l$, where $\rho = \rho_1 \wedge \rho_2 \wedge \cdots \wedge \rho_t$ and u_j-th dimension of ρ is 1. If not, reject.

3 Security Analysis

Theorem 1. Assume that the scheme in [15] is selectively CPA-secure, then the basic CP-ABE scheme is also selectively CPA-secure [16].

Theorem 2. The proposed scheme is computationally infeasible to selective keywords attract under the DBDH assumption.

Theorem 3. TA chooses a $r_1 \in {}_R Z_p^*$, and adds it to user secret key $K_1 = \hat{h}\, g^{r_1 \sum_{v_{ij} \in S'} t_{ij}}$, which cannot be separated by collusion users. Thus, the scheme can resist collusion attract.

4 Performance Comparison

In this section, we compare the performance of our scheme with the techniques in literature [12, 14]. N denotes the number of attributes in system, n denotes the number of attributes in access policy, m denotes the number of user's attributes, l_1 denotes the number of keywords extracted from files, l_2 denotes the number of query keywords. E_1, E_2 denotes exponential operation in G_1, G_2, respectively. E_P denotes pairing operation. Tables 1 and 2 show that our scheme is more efficient.

Table 1. Computation overhead comparison

	Index	Trapdoor	Search	Decryption
[12]	$(2n+6)E_1 + E_2 + E_P$	$(2m+2)E_P + 3E_1 + 2mE_2$	$4E_P + E_1 + H$	–
[14]	$(n+l_1)E_1 + E_2$	$(2N+1)E_1 + 2E_2$	$(N+1)E_P + E_2$	–
Ours	$4E_1 + E_2$	$3E_1 + l_2H$	$3E_2$	$E_1 + 2E_2$

We perform an experimental evaluation by using a large real-world dataset. We implement the experiment using C on lenovo AMD A8-6410 APU with AMD Randeon R5 Graphics 2.00 GHz and 8 GB RAM, Windows10, based on Paring-Based Cryptography PBC library [17]. The comparison results as illustrated in Figs. 3, 4, 5 and 6.

Table 2. Performance comparison

	Traditional ABKS scheme	Our scheme
Encryption	$\{C_1, \cdots, C_n\}$	$\{C_1, C_2, C_3, C_4, C_5,\}$
Secret key	$\{K_1, \cdots, K_n\}$	$\{K_1, K_2\}$
Trapdoor	$\{tk_1, \cdots, tk_n\}$	$\{tk_1, tk_2, tk_3, tk_4\}$
Search	$\{e(C_i, tk_i) \vert i = (1, \cdots, n)\}$	$e(\hat{C}_1, tk_2) \cdot e(\hat{C}_2, tk_3) = e(Value(w_l), tk_1)$
Decryption	$\{e(C_i, K_i) \vert i = (1, \cdots, n)\}$	$e(\hat{C}_1, K_2) \cdot \hat{C}_3 / e(\hat{C}_2, K_1)$

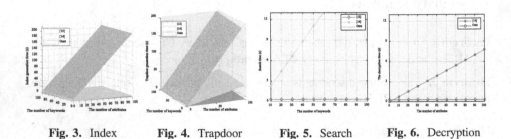

Fig. 3. Index   **Fig. 4.** Trapdoor   **Fig. 5.** Search   **Fig. 6.** Decryption

5   Conclusion

The proposed ABKS scheme with ranking results can perform single and multiple keyword search by introducing ABE technology to realise fine-grained access authority. In the trapdoor generation process, trapdoors are blinded by random values to ensure the security and confidentiality of users' key. In addition, the access policy in this scheme is AND gate and multi-value independent. This feature enables the scheme to have constant ciphertext size, and achieve efficient keyword search. A message authentication code can be used to verify the results returned by a credential service provider. In particular, our security analysis illustrates that the proposed scheme can realise selective plaintext security and is indistinguishably secure against chosen keyword attacks. Moreover, an experiment demonstrates that our scheme is more efficient than others schemes whereas experience demonstrates that our scheme is more efficient than others schemes. In our future work, we intend to explore dynamic file updating.

References

1. Song, X.D., Wagner, D., Perrig, A.: Practical techniques for search on encrypted data. In: Proceedings of the 2000 IEEE Symposium on Security and Privacy, pp. 44–55. IEEE Press (2000)
2. Farras, O., Ribes-Gonzalez, J.: Provably secure public-key encryption with conjunctive and subset keyword search. Int. J. Inf. Secur., 1–16 (2019)
3. Li, H., Liu, D., Jia, K., et al.: Achieving authorized and ranked multi-keyword search over encrypted cloud data. In: 2015 IEEE International Conference on Communication (ICC), pp. 7450–7455. IEEE (2015)
4. Wang, C., Cao, N., Ren, K., Lou, W.: Enabling secure and efficient ranked keyword search over outsourced cloud data. IEEE Trans. Parallel Distrib. Syst. 23, 1467–1479 (2012)
5. Chen, Z., Wu, C., Wang, D., Li, S., Ostrovsky, R.: Searchable symmetric encryption with efficient pairing, constant ciphertext and short trapdoor. In: PAISI 2012, pp. 176–189. Springer (2012)
6. Xia, Z., Wang, X., Sun, X., Wang, Q.: A secure and dynamic multi-keyword search over encrypted cloud data. IEEE Trans. Parallel Distrib. Syst. 27(2), 340–352 (2016)
7. Li, H., Liu, D., Jia, K., Lin, X.: Achieving authorized and ranked multi-keyword search over encrypted cloud data. In: 2015 IEEE International Conference on Communications (ICC), pp. 7450–7455. IEEE (2015)
8. Jiang, X.X., Yu, J., Yan, J.B., Hao, R.: Enabling efficient and verifiable multi-keyword ranked search over encrypted cloud data. Inf. Sci. 403–404, 22–41 (2017)
9. Wang, X., Hu, Q., Zhang, Y., Zhang, G., Juan, W., Xing, C.: A kind of decision model research based on big data and blockchain in eHealth. In: Meng, X., Li, R., Wang, K., Niu, B., Wang, X., Zhao, G. (eds.) WISA 2018. LNCS, vol. 11242, pp. 300–306. Springer, Cham (2018). https://doi.org/10.1007/978-3-030-02934-0_28
10. Sahai, A., Waters, B.: Fuzzy identity-based encryption. In: Cramer, R. (ed.) EUROCRYPT 2005. LNCS, vol. 3494, pp. 457–473. Springer, Heidelberg (2005). https://doi.org/10.1007/11426639_27
11. Li, S., Xu, M.: Attribute-based searchable encryption. Chin. J. Comput., 05–26 (2016)

12. Cui, J., Zhou, H., Zhong, H., Xu, Y.: AKSER: attribute based keyword search with efficient revocation in cloud computing. Inf. Sci. **423**, 343–352 (2018)
13. Sun, W., Yu, S., Lou, W., Hou, Y.T., Li, H.: Protecting your right: attribute-based keyword search with fine-grained owner-enforced search authorization in the cloud. In: 2014 Proceedings IEEE INFOCOM, pp. 226–234 (2014)
14. Miao, Y.B., Ma, J.F., Jiang, Q., Li, X., Sangaiah, A.K.: Verifiable keyword search over encrypted cloud data in smart city. Comput. Electr. Eng. **65**, 90–101 (2017)
15. Emura, K., Miyaji, A., Nomura, A., Omote, K., Soshi, M.: A ciphertext-policy attribute-based encryption scheme with constant ciphertext length. In: Bao, F., Li, H., Wang, G. (eds.) ISPEC 2009. LNCS, vol. 5451, pp. 13–23. Springer, Heidelberg (2009). https://doi.org/10.1007/978-3-642-00843-6_2
16. Li, J., Sha, F., Zhang, Y., Huang, X., Shen, J.: Verifiable outsourced decryption of attribute-based encryption with constant ciphertext length. Secur. Commun. Netw. **2017**, 1–11 (2017)
17. Lynn, B.: Pbc library (2006). http://cryptostanford.edu/pbc

Private Trajectory Data Publication for Trajectory Classification

Huaijie Zhu[1,2]([✉]), Xiaochun Yang[3], Bin Wang[3], Leixia Wang[3],
and Wang-Chien Lee[4]

[1] School of Data and Computer Science, Sun Yat-Sen University, Guangzhou, China
zhuhuaijie@mail.sysu.edu.cn
[2] Guangdong Key Laboratory of Big Data Analysis and Processing,
Guangzhou, China
[3] School of Computer Science and Engineering, Northeastern University,
Shenyang, China
{yangxc,binwang,wleixia}@mail.neu.edu.cn
[4] The Pennsylvania State University, State College, USA
wlee@cse.psu.edu

Abstract. Trajectory classification (TC), i.e., predicting the class labels of moving objects based on their trajectories and other features, has many important real-world applications. Private trajectory data publication is to anonymize trajectory data, which can be released to the public or third parties. In this paper, we study *private trajectory publication for trajectory classification (PTPTC)*, which not only preserves the trajectory privacy, but also guarantees high TC accuracy. We propose a private trajectory data publishing framework for TC, which constructs an anonymous trajectory set for publication and use in data services to classify the anonymous trajectories. In order to build a "good" anonymous trajectory set (i.e., to guarantee a high TC accuracy), we propose two algorithms for constructing anonymous trajectory set, namely Anonymize-POI and Anonymize-FSP. Next, we employ *Support Vector Machine* (SVM) classifier to classify the anonymous trajectories. Finally, the experimental results show that our proposed algorithms not only preserve the trajectory privacy, but also guarantee a high TC accuracy.

Keywords: Trajectory classification · Private trajectory · Classification accuracy

1 Introduction

With the wide deployment of GPS and RFID technologies, a tremendous amount of trajectory data have been generated and used for real-world applications. *Trajectory classification (TC)*, a fundamental mining task, is essential for trajectory data analytic. TC, aiming to predict the class labels of moving objects based on their trajectories and derived features, has many important applications, such as

© Springer Nature Switzerland AG 2019
W. Ni et al. (Eds.): WISA 2019, LNCS 11817, pp. 347–360, 2019.
https://doi.org/10.1007/978-3-030-30952-7_35

city and transportation planning; road construction, design, and maintenance; and traffic congestion detection.

Although a lot of research works [8,9,20–22,24] on TC have been reported in the literature, to the best knowledge of the authors, there is no existing work on privacy preserving trajectory data for TC. To classify trajectory data, the privacy of users whose trajectories being collected may be jeopardized through the necessary analysis of trajectory data. If the trajectory data owner outsources the original trajectory data (without being preprocessed using any privacy protecting schemes) to some data mining vendors (or researchers) to develop predictive models for TC, the users privacy is exposed. For example, trajectory data may be explored to predict some symptoms of illness. Table 1 shows some sample trajectories collected from patients with different symptoms of illness. As shown, those records only contains record id RID, trajectory, and the illness symptoms of patients, where RID is a unique random ID, corresponding to an individual. The first record in the table indicates that patient with $RID = 1$ has visited location p_1, p_4, p_3 and p_6 at time-stamps 2, 4, 5 and 8, respectively. Assume that an adversary has the knowledge that the patient Bob has been to location p_1 and p_6 at time-stamps 2 and 8, respectively. By matching the records, the adversary may find that Bob has an HIV. This example shows that publishing user trajectory data (without any anonymization) may cause serious privacy threats. While many research studies [1,5,10,13,15] have worked on the issues of private trajectory data publication, to the best knowledge of the authors, these works have not taken the TC accuracy into consideration.

Table 1. Patients' trajectory data

RID	Trajectory	Symptoms of illness
1	$(p_1, 2) \to (p_4, 4) \to (p_3, 5) \to (p_6, 8)$	HIV
2	$(p_4, 1) \to (p_6, 2) \to (p_7, 6) \to (p_2, 7) \to (p_5, 8) \to (p_3, 10)$	Flu
3	$(p_2, 2) \to (p_7, 3) \to (p_3, 6) \to (p_6, 8) \to (p_5, 9)$	Flu
4	$(p_6, 5) \to (p_7, 7) \to (p_3, 10)$	HIV
5	$(p_6, 3) \to (p_1, 4) \to (p_7, 6) \to (p_8, 9)$	Fever

In this paper, we study the problem of *private trajectory publication for trajectory classification* (PTPTC), which not only protects the trajectory privacy, but also guarantees a high TC accuracy. For protecting the trajectory privacy, we consider the classical Trusted Third Party framework. The data holder sends the original trajectory data to the trusted third party, and the trusted third party produces anonymous trajectories to a classifier server to train a classifier for TC. As a trade-off between user privacy and TC accuracy is expected, a major challenge faced in addressing PTPTC lies in a potential conflict of interests, i.e., the trajectory classifier requires as precise data as possible, while the users do not want to disclose their exact movements.

In order to well address the issues of private trajectory data publication for TC problem, we first analyze "what affects the accuracy of TC the most for a classifier". As we known, feature selection is crucial for improving the TC accuracy. Effective TC depends on the discriminative features. Firstly, based on the similarity of Points of Interest (POIs) (i.e., as one of features) of trajectories, we propose a trajectory similarity, called trajectory POI similarity. Based on this similarity, we propose a (k, γ)-anonymized trajectory set. Accordingly, we design an algorithm, called Anonymize-POI, to construct a (k, γ)-anonymized trajectory set. Notice that *frequent sequential patterns* are considered as the best features in most of the-state-of-the-art methods [2,8,9,17,20], as it preserves the order of visiting sequence of trajectories. Similarly, the frequent sequential pattern as the feature is also an alternative way in our classifier model, which tips us regarding how to maintain frequent sequential patterns while protecting the trajectory privacy. Thus, we propose a novel *trajectory similarity* function based on frequent sequential patterns. Accordingly, we define (k, γ)-anonymized trajectory set for trajectory privacy preservation. Meanwhile, we propose an effective *FSP-based (k, γ)-anonymized trajectory set constructing* algorithm, called Anonymize-FSP. After anomynizing all the trajectories, i.e., satisfying the k-anonymity, we employ *Support Vector Machine* (SVM) classifier to classify the anonymous trajectories. Finally, experimental results show that the proposed algorithms not only preserve trajectory privacy, but also guarantee a high TC accuracy.

In summary, the primary contributions summarized as follows made in this paper are four-fold:

- We formalize the private trajectory data publication for TC problem. To the best of our knowledge, this work is the first attempt to tackle the PTPTC problem.
- According to the *trajectory similarity* based on POIs, we propose (k, γ)-*anonymized trajectory set* for trajectory privacy preservation. For privacy preserving in TC, we design an algorithm called *Anonymize-POI*, for constructing a POI-based (k, γ)-anonymized trajectory set.
- Alternatively, we propose a novel *trajectory similarity* function based on *frequent sequential patterns* and we propose an algorithm, called Anonymize-FSP, for constructing FSP-based (k, γ)-anonymized trajectory set.
- At last, experimental result shows that our algorithms are efficient to protect trajectory privacy while guaranteeing high TC accuracy. Meanwhile, our study shows that Anonymize-FSP achieves a higher TC accuracy than Anonymize-POI.

2   Related Work

In recent years, in order to preserve user privacy in trajectory data, a lot of research studies have been reported in the literature. Generally speaking, research on trajectory privacy can be classified into trajectory privacy in location-based Service and trajectory publication.

Trajectory Privacy in Location-Based Service. In LBS, a lot of research works [11,12] study the spatial cloaking [5,25] techniques to protect trajectory privacy. Then, a distortion-based approach [12] has been proposed to aim at overcoming the drawbacks of the group-based approach. It not only requires querying users to report their locations to the location Another way to ensure k-anonymity is to use individuals' historical footprints instead of their real-time locations [15]. A footprint is defined as a user's location collected at some point of time. Similar to the previous two approaches, a fully-trusted location anonymizer is placed between users and LBS service providers to collect users' footprints. The concept mix-zones [3] have been extended to LBS for preserving trajectory privacy. Without relying on a trusted third party to perform anonymization, a mobile user can generate fake spatial trajectories, called dummies, to protect its privacy [19].

Trajectory Privacy in Trajectory Publication. In this section, we introduce privacy-preserving techniques for trajectory data publication. The anonymized trajectory data can be released to the public or third parties for answering spatio-temporal range queries and data mining. The clustering-based approach [16] utilizes the uncertainty of trajectory data to group k co-localized trajectories within the same time period to form a k-anonymized aggregate trajectory. Also, the generalization-based algorithm [10] first generalizes a trajectory data set into a set of k-anonymized trajectories. After that, generalization-based algorithm [14] with Enhanced l-Diversity [18] is proposed. Another direction is to use differential privacy techniques [7].

In summary, existing privacy-preserving techniques for spatial trajectory publication only support simple aggregate analysis, such as range queries. None of these private trajectory data publication works take TC into consideration.

3 Preliminaries

In this section, we introduce the background needed for our work, and then formulate our PTPTC problem.

3.1 Background and Definitions

In this work, we focus on TC. Give a set of trajectories $T = \{tj_1, tj_2, \ldots, tj_{num\_tra}\}$, with each trajectory associated with a class label $c_i \in C = \{c_1, \ldots, c_{num\_cat}\}$, where the trajectory, *trajectory sequences* and some other features are defined as follows:

Definition 1 *(Trajectory). A trajectory tj is a sequence of spatio-temporal points. Formally, $tj = \{ID, (x_1, y_1, t_1), \ldots, (x_n, y_n, t_n)\}$ ($t_1 < \ldots < t_n$), where ID represents a trajectory, and time t_i ($1 \leq i \leq n$) denotes that user passes by location (x_i, y_i) at time t_i.*

Definition 2 *(Point of Interests, POIs). POI is a 2-tuple (ID, loc), where ID represents the ID of trajectory which this POI belongs to, and loc = (x, y) represents the geography coordinates of this POI.*

Since the points denoting the same POI may have different position values, all the position values in a certain range are normalized to denote the same POI.

Definition 3 *(Trajectory Sequence, TS). TS means the sequence of POIs in accordance with the order of time-stamp t_i.*

Table 2. Trajectory data

RID	POI sequence	Sequential pattern	Maximum sequential subPattern
tj_1	$\langle p_1, p_2, p_5 \rangle$	$\langle p_1, p_2, p_5 \rangle$	$\langle p_1, p_2, p_5 \rangle$
tj_2	$\langle p_1, p_2, p_3, p_5 \rangle$	$\langle p_1, p_2, p_5 \rangle, \langle p_3, p_5 \rangle$	$\langle p_1, p_2, p_5 \rangle, \langle p_3, p_5 \rangle$
tj_3	$\langle p_4, p_2, p_6, p_5 \rangle$	$\langle p_4, p_5 \rangle, \langle p_6, p_5 \rangle$	$\langle p_4, p_5 \rangle, \langle p_6, p_5 \rangle, \langle p_2, p_5 \rangle$
tj_4	$\langle p_1, p_7, p_8, p_5 \rangle$	*null*	$\langle p_1, p_5 \rangle$
tj_5	$\langle p_6, p_4, p_3, p_5 \rangle$	$\langle p_4, p_5 \rangle, \langle p_6, p_5 \rangle, \langle p_3, p_5 \rangle$	$\langle p_4, p_5 \rangle, \langle p_6, p_5 \rangle, \langle p_3, p_5 \rangle$

In order to define the frequency of TS, we give some necessary notations. Assume there are two TSs $\alpha = \langle a_1, a_2, \ldots; a_m \rangle$ and $\beta = \langle b_1; b_2; \ldots; b_n \rangle$. β contains α, denoted as $\alpha \sqsubseteq \beta$, iff $\exists k_1; k_2; \ldots; k_m, 1 \le k_1 < k_2 < \ldots < k_m \le n$ and $a_1 = b_{k1} \wedge a_2 = b_{k2} \wedge \ldots \wedge a_m = b_{km}$. T_α denotes the set of trajectories T which contains α, i.e., $T_\alpha = \{tj | tj \in T \wedge \alpha \sqsubseteq tj\}$. In a set of TSs, a TS α is *maximal* if α is not contained in any other sequence. We now define *frequent sequential pattern* in Definition 4.

Definition 4 *(Frequent sequential pattern, FSP). Trajectory sequence ξ is a frequent sequential pattern if $\theta_\xi = |T_\alpha|/|T| \ge \theta_0$ and ξ is maximal, where θ_ξ is the frequency of ξ, and θ_0 is the minimum support threshold $(0 < \theta_0 < 1)$. The set of frequent sequential patterns is denoted as $\Re$.*

FSPs are extracted form the set of trajectories T. Each trajectory may share different numbers of FSPs.

Example 1. Table 2 illustrates the FSPs. As shown, with minimum support set to 25%, i.e., a minimum support must contain at least 2 trajectories. Notice that the length of one FSP is not less than 2. Sequence $\langle p_1, p_2, p_5 \rangle$, $\langle p_3, p_5 \rangle$, $\langle p_6, p_5 \rangle$ and $\langle p_4, p_5 \rangle$ are the FSPs which satisfy the minimum support. In detail, $\langle p_1, p_2, p_5 \rangle$ is supported (contained) by trajectories 1 and 2. The other FSPs contained in the other four trajectories are shown in the third column of Table 2.

To propose the private trajectory data publication methods, we define *(k, γ-anonymized trajectory set* as follow.

Definition 5 *((k, γ)-anonymized Trajectory Set). A trajectory set F is (k, γ)-anonymized trajectory set, if F consists at least k trajectories and the similarity of any two trajectories is not less than γ.*

3.2 Problem Definition

(*User Identification Problem* (Attack Model)). Given a trajectory database, trajectory tj_i exists *user identification problem* iff attacker utilizes the background knowledge to find who has visited some locations at certain time to identify the user of tj_i.

In this paper, we study the PTPTC problem. Based on this attack model, our goal is to preserve the trajectory privacy to prevent the attackers identifying the user of any trajectory in published trajectory set T. For the sake of preventing the attackers identifying the user of any trajectory, we focus on how to construct the good (k, γ)-anonymized trajectory set.

4 Private Trajectory Data Publication for Trajectory Classification Approach

In this section, we propose private trajectory data publication for TC method, based on (k, γ)-anonymized trajectory set. The proposed method consists of two main phases: (1) constructing the anonymous trajectory set; and (2) classifying the anonymous trajectories.

4.1 Constructing the Anonymous Trajectory Set

In this section, we propose the algorithms for constructing an anonymous trajectory set. In order to build a good anonymous trajectory set for guaranteeing high TC accuracy, we first analyze the essence of TC. One of the most important requirements for effective TC is to identify discriminative features. As the extensively used features, i.e., POIs and FSPs, we propose two algorithms for constructing anonymous trajectory set, namely *Anonymize-POI* and *Anonymize-FSP*.

Anonymize-POI Algorithm. According to POIs contained in a trajectory, we define the *trajectory POI similarity*.

Definition 6 *(Trajectory POI similarity). For two trajectories, the number of POIs shared in them is M, whereas the total number of POIs contained in them is N. Therefore, the trajectory POI similarity is M/N, denoted by Sim_POI.*

Based on trajectory POI similarity, we construct the (k, γ)-anonymized trajectory set, as defined in Definition 5. Thus, such a (k, γ)-anonymized trajectory set preserves the trajectory privacy according to the k-anonymity property. Accordingly, we develop an algorithm for constructing the POI-based (k, γ)-anonymized trajectory set, called Anonymize-POI. Anonymize-POI performs the following three steps: (i) obtain all the POIs; (ii) according to the similarity function, group all the trajectories; and (iii) anonymize the trajectories in each group.

In order to measure a trajectory, we first extract all the POIs, via the function getPOI. We consider the POIs in this paper, which consists of starting point,

---

Algorithm 1. Anonymize algorithm

Input: A group of trajectories g, and anonymous value k;
Output: A group of anonymous trajectory set g';
1 if $g.size >= k$ then
2   for Each trajectory tj in g do
3   Randomly select one trajectory to replace it and attach original ID;
4   Put the generated trajectory into g';

5 else
6   for $|g|$ to k do
7   Randomly choose one trajectory from g;
8   Construct a new trajectory via copying the chosen trajectory and attach a new id;
9   Put the generated trajectory into g';

10 Return g';

---

ending point, staying point and turning point. Our privacy protection of entire trajectory goal is by protecting these sensitive points. After processing all the POIs in each trajectory, we utilize the trajectory POI similarity to measure and cluster the trajectories. Since how many groups to be finally obtained and clustered is unknown, we adopt a hierarchical clustering method to cluster the trajectories (using a bottom-up strategy). Finally, we anonymize the trajectories in each group. Algorithm 1 presents the pseudo-code of Anonymize. The main idea of Anonymize algorithm is as follows. If the size of group g is not less than the anonymous value k, we randomly use one of the trajectories to replace each trajectory and keep the original trajectory id ID; Otherwise, we construct $k - |g|$ fake trajectories.

Let's show an running example for illustration of the Anonymize-POI algorithm. Recall the trajectories in Table 2, assuming that the similarity threshold γ is 0.5, and the anonymous value $k = 2$. According to the POI similarity function, the groups obtained by the clustering method are $\{1, 2\}$, $\{3, 5\}$, $\{4\}$. For group $\{1, 2\}$, $\{3, 5\}$, final anonymous set $\{2, 2\}$, $\{5, 5\}$ are obtained using function Anonymize. While for $\{4\}$, we construct a fake trajectory with id 6 and form the group $\{6, 4\}$. Therefore, the final 2-anonymized trajectory set is $\{2, 2\}$, $\{5, 5\}$ and $\{4, 4\}$.

Anonymize-FSP Algorithm. Since POI similarity does not capture well the features of trajectory, using POI similarity does not guarantee high precision of TC. Notice that, good features play a key role in TC. As many state-of-the-art TC algorithms explore *frequent sequential pattern*, which preserves the order of visiting sequence of trajectories, we explore the frequent sequential patterns in designing a good similarity to serve the group function. Thus, we propose a novel trajectory similarity function, which takes the frequent sequential patterns into consideration, called Sim_FSP, as shown in Definition 7.

Definition 7 *(Trajectory FSP similarity). For two trajectories tj_i, $tj_k \in T$ and a minimum support threshold θ, the number of FSPs shared in R_{tj_i} and R_{tj_k} is M, whereas the total number of FSPs contained in tj_i and tj_k is N. Then, the trajectory FSP similarity is M/N, denoted by Sim_FSP.*

Recall the trajectories data in Table 2. $Sim\_FSP(tj_1, tj_2)$ is 0.5 according to Sim_FSP equation, $Sim\_FSP(tj_3, tj_5)$ is 0.5, and the similarity between trajectory 4 and other trajectories is 0. From this example, the FSP similarity function is not a good choice for grouping the trajectories, which leads the special case that the similarity score between two trajectories is 0. There exists two situations: (a) some trajectories only contain a subset of FSP (not the whole FSP), but the subset is not the maximal; (b) the sequential patterns shared in two trajectories are not supported by enough number of trajectories (i.e., the support degree is not high). For example, trajectory 4 contains the subset $\langle p_1, p_5 \rangle$ of FSP $\langle p_1, p_2, p_5 \rangle$. For the first situation, we give a new concept, *maximal FSP subset (MFS)*, which is the maximum subset of the FSP with respect to a trajectory. With regard to trajectory 4, $\langle p_1, p_5 \rangle$ is the maximal FSP subset for itself. The maximal FSP subsets for other trajectories are shown in the rightmost part of Table 2. For two subsets of the same FSP, we think they are similar in some const. Thus, we utilize a function to measure the similarity between two maximal FSP subsets, r_i and r_k, which satisfy $r_i \subseteq r_k$ or $r_k \subseteq r_i$. The similarity function between r_i and r_k is as follow:

$$Sim(r_i, r_k) = \begin{cases} \frac{|r_i \bigcap r_k|}{|r_i \bigcup r_k|}, & r_i \subseteq r_k || r_k \subseteq r_i \\ 0, & else \end{cases} \tag{1}$$

According to Eq. 1, the similarity score between $\langle p_1, p_5 \rangle$ and $\langle p_1, p_2, p_5 \rangle$ is 2/3. Thanks to maximal FSP subset similarity function, it is not necessary for each trajectory to have FSPs and the subsets of FSP can also contribute to the similarity computation. Based on the similarity between two maximal FSP subsets, we propose another trajectory similarity function, as defined below.

Definition 8 *(Trajectory MFS similarity). For two trajectories tj_i, $tj_k \in T$ and a minimum support threshold θ_0, maximal FSP subset obtained in tj_i is denoted by RS_{tj_i}, and the maximal FSP subset obtained in tj_k is denoted by RS_{tj_k}. Then, the trajectory MFS similarity is defined in Eq. 2.*

$$Sim\_MFS(t_i, t_j) = \frac{\sum_{r_i \in RS_{tj_i}, r_k \in RS_{tj_k}} Sim(r_i, r_k)}{|R_{tj_i} \bigcup R_{tj_k}|} \tag{2}$$

Example 2. As shown in Table 2, according to trajectory MFS similarity, the similarity score between trajectory 1 and 2 is 1/2, the similarity score between trajectory 1 and 4 is 1/3 and the similarity score between trajectory 3 and 5 is $\frac{1+1}{1+1+2} = 1/2$.

Accordingly, based on trajectory MFS similarity, we design another (k, γ)-anonymized trajectory set. Then we construct (k, γ)-anonymized trajectory sets from original trajectories. Accordingly, we develop an effective algorithm for constructing *FSP-based (k, γ)-anonymized trajectory set*, called Anonymize-FSP. The Anonymize-FSP can be divided into four steps: (1) obtaining the POIs for each trajectory; (2) mining all the FSPs and find the maximal FSP subsets for each trajectory using VMSP algorithm [6]; (3) grouping all the trajectories according to the similarity function; and (4) anonymizing the trajectories in each group.

4.2 Classifying the Anonymous Trajectories

After the third trust party anonymizes the original trajectories, classifier server utilize the SVM model [4] to classify the anonymous trajectories.

5 Experiments

5.1 Experimental Setup

In this section, we evaluate the proposed algorithms, which are implemented in Java on an Intel Core 8 Duo CPU E7500 2.93 GHz PC with 6 GB RAM. The experiments are conducted using both synthetic and real datasets. Synthetic dataset (including 2012 trajectories) is generated using the software Network-based Generator of Moving Objects on website "http://iapg.jade-hs.de/personen/brinkhoff/generator/" based on Oldenburg dataset. The real dataset [23] contains 17621 GPS trajectories. Each trajectory is labelled as *driving*, *by bus*, *riding* and *walking*. All datasets are normalized in order to design the experiments.

In order to investigate the efficiency and accuracy of the proposed algorithms, the experiments are carried out by varying various parameters, which are summarized in Table 3. In each experiment, we test one parameter at a time (by fixing the other parameters at their default values (i.e., in bold)). The metrics in our experimental study include *classification accuracy* on anonymous trajectories and *constructing time* of anonymous trajectories. For TC accuracy, we compared three algorithms, namely withoutAnonymize, Anonymize-POI and Anonymize-FSP.

Table 3. Parameter ranges and defaults values

Parameter	Range
k (anonymous value)	5, **7**, 9, 11, 13
Similarity threshold δ	0.4, 0.5, **0.6**, 0.7, 0.8
A minimum support threshold θ_0	0.005, **0.006**, 0.007, 0.008, 0.01

5.2 Classification Accuracy of Anonymous Trajectories

In this section, we evaluate the accuracy of the proposed anonymity algorithms, including withoutAnonymize, Anonymize-POI and Anonymize-FSP. We use SVM to classify the trajectories. We measure the TC accuracy corresponding to three different parameters: (a) anonymous value k; (b) similarity threshold γ and (c) minimum support threshold θ.

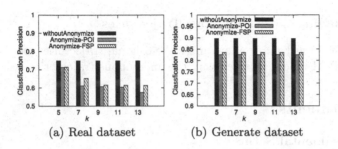

(a) Real dataset   (b) Generate dataset

Fig. 1. Classification accuracy vs. k value

Effect of Anonymous Value k. In this experiment, we compare the accuracy of classifying the anonymous trajectories constructed by three algorithms under different anonymous k values. The results for classifying different anonymous trajectories are depicted in Fig. 1. Figure 1(a) shows the result on Real dataset. As shown, the accuracy of classifying the trajectories without anonymizing (i.e., withoutAnonymize) is more than 0.75. Obviously, it is not affected by the k values. The accuracy of classifying the anonymous trajectories generated by Anonymize-FSP algorithm is nearly 0.65. Therefore, the accuracy results of trajectories using withoutAnonymize and Anonymize-FSP are very close. This is because the Anonymize-FSP algorithm not only protects the privacy of the trajectories, but also maintains some good TC features for TC. As expected, the TC accuracy become worse when the anonymous value k increases. This shows that if we want to preserve more privacy about the anonymous data (i.e., k increases), the TC accuracy becomes lower. The results obtained on Generate dataset are shown in Fig. 1(b). Compared to the accuracy results on Real dataset, the TC accuracy is higher using different anonymization algorithms, as the generated trajectories are distributed more uniformly than Real data. The observed trend and conclusion on Generate dataset are consistent with the results on Real dataset.

Effect of Similarity Threshold γ. Figure 2 compares the TC accuracy using different anonymization algorithms by varying the similarity threshold γ. For Real dataset, the results in Fig. 2(a) show the superiority of Anonymize-FSP algorithm over Anonymize-POI. It can be also seen that the TC accuracy using

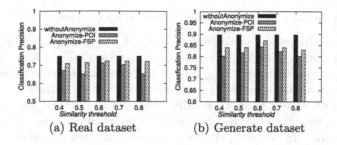

Fig. 2. Classification accuracy vs. similarity threshold

without Anonymize is not affected by increasing the similarity threshold. By increasing the similarity threshold, the accuracy of classifying the anonymous trajectories generated by the proposed two algorithms first becomes higher, and then becomes lower.

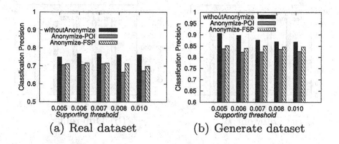

Fig. 3. Classification accuracy vs. minimum support threshold

Effect of Minimum Support Threshold θ. We then compare the accuracy of classifying different anonymous trajectories by varying minimum support threshold and show the results in Fig. 3. The accuracy results on Real dataset are shown in Fig. 3(a), where the TC accuracy of the trajectories using withoutAnonymize is more than 0.75, and it is affected when changing different minimum support thresholds. The TC accuracy using Anonymize-FSP algorithm is 0.7 under different minimum support thresholds. Moreover, it becomes higher first and then become lower when the minimum support threshold is increasing. Moreover, the Anonymize-FSP shows the superiority to Anonymize-POI for constructing the anonymous trajectories. While the TC accuracy results on Generate dataset in Fig. 3(b) is higher than on Real dataset under each minimum support threshold. This is because the FSPs can be a good feature when classifying the trajectories which are uniformly generated.

5.3 Constructing Time of Anonymous Trajectories

Finally, we evaluate the constructing time of the proposed anonymization algorithms, including withoutAnonymize, Anonymize-POI and Anonymize-FSP. Figure 4(a) shows the constructing time of the anonymous trajectories by varying different anonymous values. It can be seen that the constructing time using Anonymize-POI is less than that using Anonymize-FSP, as Anonymize-FSP needs a lot of time to mine the FSPs and MFSs when constructing the anonymous trajectories. In addition, the time of these two anonymization algorithms changes relatively stable when increasing the k values. The constructing time with respect to different similarity thresholds is depicted in Fig. 4(b). It shows similar results on these two different parameters.

In summary, by comparing the TC accuracy and constructing time, the Anonymize-FSP shows the superiority of Anonymize-POI on constructing the anonymous trajectories for TC, but it takes more time to construct anonymous trajectories.

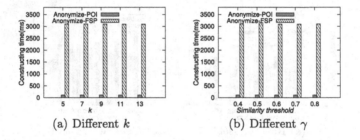

(a) Different k (b) Different γ

Fig. 4. Constructing time vs. k value

6 Conclusion

In this paper, we study the problem of *private trajectory data publication for trajectory classification*, which not only protects the trajectory privacy, but also guarantees high classification accuracy. We propose a private trajectory data publishing framework for trajectory classification, which first constructs the anonymous trajectory set, and then classifies the anonymous trajectory. In order to build a good anonymous trajectory set, we propose two algorithms for constructing anonymous trajectory set, namely *Anonymize-POI* and *Anonymize-FSP*. At last, comprehensive performance evaluation is conducted to validate the proposed ideas and demonstrate the accuracy and effectiveness of the proposed algorithms. The experimental results show that the proposed algorithms not only preserve the trajectory privacy, but also guarantee a high classification accuracy.

This work may lead towards several new directions for future work, e.g., using encryption techniques and differential privacy.

Acknowledgment. This work was supported by the Postdoctoral fund (2018M-643307), Key R&D Program of Guangdong Province (2018B010107005, 2019B010-120001), and the National Natural Science Foundation of China (Nos. 61532021, 61572122, U1736104).

References

1. Abul, O., Bonchi, F., Nanni, M.: Never walk alone: uncertainty for anonymity in moving objects databases. In: ICDE, pp. 376–385 (2008)
2. Agrawal, R., Srikant, R.: Mining sequential patterns. In: ICDE, pp. 1–12 (1995)
3. Beresford, A.R., Stajano, F.: Location privacy in pervasive computing. IEEE Pervasive Comput. **2**(1), 46–55 (2004)
4. Chang, C.C., Lin, C.J.: LIBSVM: a library for support vector machines. ACM Trans. Intell. Syst. Technol. **2**(3), 1–27 (2011)
5. Chow, C.Y., Mokbel, M.F.: Privacy of spatial trajectories. In: Zheng, Y., Zhou, X. (eds.) Computing with Spatial Trajectories. Springer, New York (2011). https://doi.org/10.1007/978-1-4614-1629-6_4
6. Fournier-Viger, P., Wu, C.-W., Gomariz, A., Tseng, V.S.: VMSP: efficient vertical mining of maximal sequential patterns. In: Sokolova, M., van Beek, P. (eds.) AI 2014. LNCS (LNAI), vol. 8436, pp. 83–94. Springer, Cham (2014). https://doi.org/10.1007/978-3-319-06483-3_8
7. He, X., Cormode, G., Machanavajjhala, A., Procopiuc, C.M., Srivastava, D.: DPT: differentially private trajectory synthesis using hierarchical reference systems. In: VLDB, pp. 1154–1165 (2015)
8. Lee, J.G., Han, J., Li, X.: TraClass: trajectory classification using hierarchical region-based and trajectory-based clustering. VLDB **1**, 1081–1094 (2008)
9. Lee, J.G., Han, J., Li, X., Cheng, H.: Mining discriminative patterns for classifying trajectories on road networks. TKDE **23**(5), 713–726 (2011)
10. Nergiz, M.E., Atzori, M., Saygin, Y.: Towards trajectory anonymization: a generalization-based approach. Trans. Data Privacy **2**(1), 52–61 (2009)
11. Palanisamy, B., Liu, L.: MobiMix: protecting location privacy with mix-zones over road networks. In: ICDE, pp. 494–505 (2011)
12. Pan, X., Meng, X., Xu, J.: Distortion-based anonymity for continuous queries in location-based mobile services. In: SIGSPTAIL, pp. 256–265 (2009)
13. Terrovitis, M., Mamoulis, N.: Privacy preservation in the publication of trajectories. In: MDM, pp. 65–72 (2008)
14. Wu, J., Ni, W., Zhang, S.: Generalization based privacy-preserving provenance publishing. In: Meng, X., Li, R., Wang, K., Niu, B., Wang, X., Zhao, G. (eds.) WISA 2018. LNCS, vol. 11242, pp. 287–299. Springer, Cham (2018). https://doi.org/10.1007/978-3-030-02934-0_27
15. Xu, T., Cai, Y.: Exploring historical location data for anonymity preservation in location-based services. In: INFOCOM, pp. 547–555 (2008)
16. Xin, Y., Xie, Z.Q., Yang, J.: The privacy preserving method for dynamic trajectory releasing based on adaptive clustering. Inf. Sci. **378**, 131–143 (2017)
17. Yan, X.: CloSpan: mining closed sequential patterns in large datasets. In: SAIDM, pp. 166–177 (2003)
18. Yao, L., Wang, X., Wang, X., Hu, H.: Publishing sensitive trajectory data under enhanced l-diversity model. In: MDM (2019, to appear)
19. You, T.H., Peng, W.C., Lee, W.C.: Protecting moving trajectories with dummies. In: MDM, pp. 278–282 (2008)

20. He, Z., Gu, F., Zhao, C., Liu, X., Wu, J., Wang, J.: Conditional discriminative pattern mining. Inf. Sci. **375**, 1–15 (2017)
21. Zheng, Y.: Computing with Spatial Trajectories. Springer, New York (2011). https://doi.org/10.1007/978-1-4614-1629-6
22. Zheng, Y.: Trajectory data mining: an overview. ACM Trans. Intell. Syst. Technol. **6**, 29 (2015)
23. Zheng, Y., Li, Q., Chen, Y., Xie, X., Ma, W.Y.: Understanding mobility based on GPS data. In: MOBIQUITOUS, pp. 312–321 (2008)
24. Zheng, Y., Liu, L., Wang, L., Xie, X.: Learning transportation mode from raw GPS data for geographic applications on the web. In: WWW (2008)
25. Zhu, H., Yang, X., Wang, B., Lee, W.C.: Range-based obstructed nearest neighbor queries. In: SIGMOD, pp. 2053–2068 (2016)

Adaptive Authorization Access Method for Medical Cloud Data Based on Attribute Encryption

Yu Wu, Nanzhou Lin, Wei Song$^{(\boxtimes)}$, Yuan Shen, Xiandi Yang,
Juntao Zhang, and Yan Sun

School of Computer Science, Wuhan University, Wuhan 430072, China
songwei@whu.edu.cn

Abstract. The outsourcing storage of medical big data encrypted in the cloud can effectively alleviate the problem of privacy disclosure, but cipher text storage will lead to the inconvenience of data access, which brings new challenges to medical big data shared access. The existing flexible authorization solutions for encrypted data are mainly based on methods such as CP-ABE, which requires the data owner to define the data access strategy, while in reality, the patient is the data owner of the medical data. At the same time, the existing scheme does not support access control authorization in emergency scenarios, and in medical big data applications, when patients can not authorize data users to access cipher text medical data, it will lead to unpredictable consequences. According to the application requirements of encrypted medical big data shared service in cloud environment, an adaptive authorization access method based on attribute encryption is proposed to realize flexible and secure medical data access authorization in normal and emergency situations. Experimental results demonstrate that our scheme is efficient.

Keywords: Attribute encryption · Adaptive access control · Privacy protection · Medical big data

1 Introduction

The development of cloud data enables medical institutions to provide high-quality, convenient and universal medical services. After collecting physiological data from the medical Internet of things, data is transmitted to the Medical big data Center for storage and disease diagnosis. In order to protect the privacy of patients, medical documents need to be encrypted before transmission to prevent eavesdropping on the public domain [1]. In order to realize the authorized sharing service of encrypted medical big data, patients, as the owners of medical data, formulate access policies to protected encrypted medical data and define authorization attributes and relationships. Only users with appropriate attribute keys (health care workers) have the right to decrypt the ciphertext. This encryption method is called attribute-based encryption [2]. There may be an emergency in the medical system, such as a car accident or a sudden collapse of the patient, and the first aid personnel on the scene need to access the electronic medical records of the patient in the process of emergency treatment. However, first aid

© Springer Nature Switzerland AG 2019
W. Ni et al. (Eds.): WISA 2019, LNCS 11817, pp. 361–367, 2019.
https://doi.org/10.1007/978-3-030-30952-7_36

personnel often do not have access to encrypted medical files, hampering emergency care for patients' lives. Therefore, a flexible access authorization method is required in practical medical applications to solve the problem of flexible access authorization for medical data in this emergency.

Aiming at the flexible access authorization problem of encrypted medical data in the cloud environment above, this paper proposes an adaptive access control method based on KP-ABE [3] technology. Under normal circumstances, patients have absolute control over medical data and authorize data users to access personal encrypted medical data in the cloud. In case of emergency, considering that patients are unable to perform data authorization operations, data users contact emergency contacts to negotiate access rights to encrypted medical data, and abnormal authorized access services of medical data in emergency are recorded by the cloud for audit use.

2 Related Work

Aiming at the data access requirements in emergency situations in medical big data applications, Brucker et al. [4] proposed a break-glass access model, which can still access when the system crashes. However, these studies [4, 5] only proposed a framework, but did not implement specific security data access authorization scheme.

Lattice-based cryptography system has the characteristics of simple calculation and high security. Ajtai [6] proposed a scheme to construct ciphers based on lattice problems, which has the advantages of high execution efficiency and strong security.

The research group has carried out research work on the privacy protection of medical big data. Wang et al. [7] proposed a method of Medical data encryption in cloud computing.

3 The Proposed Scheme

Cloud platforms provide an important support for the storage of medical data. But there are security problems, so an adaptive access control method is needed. Under normal circumstances, the data user is authorized to access the data by the patient, but in case of emergency, the patient cannot authorize the data in time, endangering the patient's life safety. For this special situation, this system model allows medical staff to negotiate access to encrypted medical data with emergency contact person after authentication in case of emergency (patients are unable to conduct data authorization due to illness). In order to ensure the accessibility of ciphertext data in emergency scenario, patients also need to set up emergency contact person and share the encryption key to them. In advance, and patients and emergency contact person can negotiate secret parameters together, so that in case of emergency, emergency contact person can reconstruct the key to decrypt patients' medical files through encryption key.

3.1   Scheme Model

The system model includes six entities: Key generation center, Medical institutions, Data owner, Cloud service provider, Data user and Emergency contact. The characteristics and functions of each entity are described as follows.

(1) Key generation center: responsible for generating system public parameters and creating master system key MSK. Meanwhile, the key generation center generates key pairs for patients and medical institutions, and generates file encryption parameter matrix A for patients, which is sent to patients and medical institutions through secure channels.

(2) Medical institutions: they are composed of various hospitals with medical capacity. A medical institution manages its staff and provides medical services to patients. After the registration of a medical institution, KGC generates public-private key pairs for the medical institution, and securely transmits the private key to the medical structure. A medical institution generates a set of attributes for its medical staff to describe their data access characteristics and generate an attribute key for them.

(3) Data owner: in order to protect the security and privacy of medical data, medical data is considered as a resource that is completely managed by patients (data owners). In the process of providing medical services to patients, medical institutions will send corresponding personal electronic medical documents to patients, who will encrypt the medical documents and store them in the cloud. Patients assign access attribute set to encrypted personal medical data, and only authorized visitors, whose pre-allocated access attributes satisfy the access policy of the corresponding encrypted file, can successfully access the encrypted data.

(4) Cloud service provider: cloud service provider is responsible for storing the ciphertext of medical documents and the set of attributes formulated by patients, and responding to queries according to the access policies of medical institutions.

(5) Data user: data user (such as medical staff of a hospital) registers with medical institutions to obtain attribute keys. Data consumers send data access requests to cloud service providers to obtain encrypted medical files and decrypt them using attribute keys.

(6) Emergency contact: the patient and the emergency contact negotiate the secret parameter y in advance. When patients are in a state of normal authorization, emergency contact use secret parameters restore the encryption key Ψ together with the users of the data.

3.2   Description of Proposed Scheme

Key generation center create system public parameters PP and master key MSK according to the security parameter 1^k. The public parameter PP is public in the whole system, and MSK is stored secretly by the key generation center.

GlobalSetup$(1^k) \rightarrow (PP, \text{MSK})$. The key generation center operates the GlobalSetup algorithm. The key generation center randomly sets the hash function $H_1: \{0, 1\}^* \rightarrow Z_p^*$, $H_2: \{0, 1\}^* \rightarrow \mathcal{K}$ and Generate symmetric encryption pair

SEnc/SDec in security key space K. Then the key generation center sets the random number $\eta \in Z_p^*$, g, g_1, g_2, $g_3 \in \mathbb{G}$ and compute bilinear pairs $Y = e(g_1, g_2)^\eta$. Finally, the key generation center sets common parameters $PP = (g,\ g_1,\ g_2,\ g_3,\ Y,\ H_1,\ H_2,$ SEnc/SDec) and master key $MSK = \eta$.

When a medical organization is registered as the ith medical institution. After the key generation center verifies the identity, it distributes the identity identification MI_i to each medical institution and generates the corresponding PK_i and SK_i.

MiKeyGen $(MI_i, MSK) \to (PK_i, SK_i)$. The algorithm is executed in the key generation center, then it randomly sets α_i, β_i, $\gamma_i \in Z_p^*$ and generates public key constituent element $pk_{i,1} = g^{\alpha_i}$, $pk_{i,2} = g^{\beta_i}$. Private key constituent element $sk_{i,1} = g_1^{\alpha_i}$, $sk_{i,2} = \beta_i$, $sk_{i,3} = g_2^\eta g_3^{\gamma_i}$, $sk_{i,4} = g_1^{\gamma_i}$, $sk_{i,5} = g_1^{\alpha_i \cdot \gamma_i}$.

When a medical user registers as the jth user in a medical institution. After verifying the user's identity, medical institutions generate an identity identifier $PID_{i,j} \in \mathbb{G}$ for patients, $HID_{i,j} \in \mathbb{G}$ for medical person. According to the role of the medical staff, assign attribute sets $\{attr_k\}_{k \in [\varphi]}$. The key generation center generates attribute keys $SK_{i,j}$ and $PK_{i,j}$ for each user.

UserKeyGen$(MI_i, SK_i, HID_{i,j}, \{attr_k\}_{k \in [\varphi]}) \to (PK_{i,j}, SK_{i,j})$. Medical institutions randomly sets $\gamma'_{i,j}$, $t \in Z_p^*$ and $\gamma_{i,j} = \gamma_i + \gamma'_{i,j}$. When a data owner registers as a patient P_i. Key generation center will generate public-private key pairs for patients according to file encryption parameter matrix and X solution set. The patient will embed the set of attributes in the key.

OwnKeyGen$(PID_i, A, X, \{attr_p\}_{p \in [\varphi]}) \to (PPK_i, PSK_i)$. File encryption parameter matrix A and A set of solutions to $AX = 0$, $X = \{\vec{x} | A\ \vec{x} = 0\}$. $PPK_i = g^{H_2(A)}$, $psk_{i,1} = g_1^{H_1(\vec{x_i})}$, $psk_{i,2} = g_2^{H_1(PID_i) \cdot \beta_i}$, $psk_{i,3} = g_2^{\alpha_i} g_3^{H_1(attr_p)}$, then $PSK_i = (psk_{i,1}, psk_{i,2}, psk_{i,3})$.

DepKeyGen$(PID_{i,j}, SK_{i,j}) \to DK_{i,j}$. When the patient P_i sets up an emergency contact, the corresponding key DK_i is generated for the emergency contact. The patient randomly sets parameter $\lambda \in Z_p^*$, then calculate $DK_{i,1} = (psk_{i,1})^\lambda = (g_1^{x_i})^\lambda$, $DK_{i,2} = (psk_{i,2})^\lambda = \left(g_2^{H_1(PID_i) \cdot \beta_i}\right)^\lambda$, $DK_{i,3} = (psk_{i,3})^\lambda = \left(g_2^{\alpha_i} g_3^{H_1(attr_p)}\right)^\lambda$.

PatientGen$(PID_i, PPK_i, PSK_i, DK_i, file) \to (Kf, y)$. Patient calculate bilinear pairs $E = e(PPK_i, PSK_i)$, then $Kf = H_2(E, PID_i, H_1(file))$, and use diffie-hellman key exchange protocol to negotiate the secret parameter $y = H_1 g^{PSK_i \cdot DK_i}$ with the emergency contact for emergency.

Enc$(Kf, file, A, \vec{x_i}) \to (CT, \Psi)$. When the patient completes the encryption key, symmetric encryption method is adopted to encrypt the file. Encryption key $\Psi = (Kf + \vec{x_i})A$, ciphertext $CT = SEnc(\Psi, file \| 0^\varpi)$.

PropertyMap$(ap, AP_H, \rho) \to 1/0$. Normally ap maps to AP based on implicit rules. Under normal circumstances, if the attribute judgment result returns 1, then the normal decryption algorithm is carried out.

NorDec$(1/0, HID_{i,j}, CT, \Psi) \to file/\bot$. If the attribute determination result is 1, the patient decrypts the medical file with the encryption key and sends it to the corresponding medical personnel through the secure channel; Output $\perp$ if the attribute determines that the result is 0.

EcpDec$(Kf,\ HID_{i,j},\ PSK_i,\ SK_{i,j},\ DK_i) \to$ file. The patient pre-negotiates the encryption key pair $((Kf + \vec{x} + y + r)A,\ rA)$ in the event of an emergency with the emergency contact. In case of emergency, the key generation center generates a group password $\vec{x'}(\vec{x'} \in X)$ for medical personnel. Send $Kf + \vec{x'}$ to Emergency contact. Emergency contact generation $(Kf + \vec{x'} + y)A$ and returned to the medical staff. The medical staff obtains rA from the key generation center to generate the key $(Kf + y + r)A$, and the medical staff decrypts the medical file.

4 Experiments and Verification

In order to verify the efficiency of the adaptive authorized access method of medical cloud data based on attribute encryption, the A^{2}MAE scheme proposed in this paper is implemented by JAVA parsed method cipher library (jPBC). Four groups of comparative experiments were conducted to compare the efficiency of this scheme with the existing IOT scheme [8], ABEC scheme [9] and IPSD scheme [10] in key generation algorithm, encryption algorithm, normal and emergency decryption algorithm (Fig. 1).

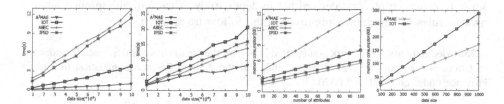

Fig. 1. Key generation   **Fig. 2.** Encryption   **Fig. 3.** Nor-decryption   **Fig. 4.** Em-decryption

The time consumption of key generation and encryption is tested in different data scales. After implementing key generation and encryption phases of IOT scheme, ABEC scheme and IPSD scheme respectively, the experiment conducted a comparative analysis on the time consumption of each phase of the four schemes. In the key generation stage, as shown in Fig. 2, with the continuous increase of data size, IOT scheme uses a large number of bilinear pairwise operations, leading to a greatly increased key generation time. ABEC scheme and IPSD scheme increase the number of keys to enhance security, but at the same time the key generation time significantly increases. When the data volume is large enough, the performance of our scheme will be better.

In the encryption phase, it can be seen from Fig. 3 that our scheme is more stable in terms of time consumption than IOT scheme. ABEC scheme and IPSD scheme show a significant increase in time consumption with the increase of data size. In general, our scheme is more effective under the condition of ensuring certain safety. For the decryption phase, normal decryption is compared with ABEC scheme, IOT scheme and IPSD scheme for communication consumption. The communication consumption in emergency declassification is compared with IOT scheme.

In the case of normal decryption, as shown in Fig. 3, memory consumption increases linearly as the number of attributes increases. When the number of attributes was 10, our scheme memory consumption was only 1.41 KB. ABEC scheme uses a lot of factorial memory consumption, and our scheme is slightly better than IOT scheme and IPSD scheme. It can be seen that our scheme consumes less decryption memory under normal circumstances. In the case of emergency decryption, due to non-attribute decryption, we only compare the emergency decryption of IOT scheme as shown in Fig. 4. With the increase of data volume, our scheme performs better than the IOT scheme in the case of emergency.

5 Conclusion

For satisfying complex requirements of cloud medical data access control, this paper proposes a medical cloud data based on attribute encryption adaptive grant access method, this method achieved under normal conditions and data access in an emergency. Experimental results show that this method has shorter time and higher performance than other methods on the premise of ensuring safety, but in the scheme attribute matching phase is derived not in-depth study, Therefore, the implicit authorization rules will be further improved in the follow-up research.

Acknowledgements. This work is supported in part by the National Natural Science Foundation of China under grants 61572378, U1811263, and the Natural Science Foundation of Hubei Province under grant 2017CFB420.

References

1. Song, W., Wang, B., Wang, Q., Peng, Z., Lou, W.: Tell me the truth: practically public authentication for outsourced databases with multi-user modification. Inf. Sci. **387**, 221–237 (2017)
2. Sahai, A., Waters, B.: Fuzzy identity-based encryption. In: Cramer, R. (ed.) EUROCRYPT 2005. LNCS, vol. 3494, pp. 457–473. Springer, Heidelberg (2005). https://doi.org/10.1007/11426639_27
3. Goyal, V., Pandey, O., Sahai, A., Waters, B.: Attribute-based encryption for fine-grained access control of encrypted data. In: Proceedings of CCS, pp. 89–98 (2006)
4. Brucker, A.D., Petritsch, H.: Extending access control models with break-glass. In: Proceedings of SACMAT, pp. 197–206 (2009)
5. Wang, X., Hu, Q., Zhang, Y., Zhang, G., Juan, W., Xing, C.: A kind of decision model research based on big data and blockchain in eHealth. In: Meng, X., Li, R., Wang, K., Niu, B., Wang, X., Zhao, G. (eds.) WISA 2018. LNCS, vol. 11242, pp. 300–306. Springer, Cham (2018). https://doi.org/10.1007/978-3-030-02934-0_28
6. Ajtai, M.: Generating hard instances of lattice problems. In: Proceedings of ACM Symposium on Theory of Computing, pp. 99–108 (1996)

7. Wang, B., Song, W., Lou, W., Thomas Hou, Y.: Privacy-preserving pattern matching over encrypted genetic data in cloud computing. In: Proceedings of INFOCOM, pp. 1–9 (2017)

8. Yang, Y., Zheng, X., Guo, W., Liu, X., Chang, V.: Privacy-preserving smart IoT-based healthcare big data storage and self-adaptive access control system. Inf. Sci. **479**, 567–592 (2019)

9. Hui, C., Deng, R.H., Li, Y., Guowei, W.: Attribute-based storage supporting secure deduplication of encrypted data in cloud. IEEE Trans. Big Data **99**, 1 (2017) ·

10. Han, J., Susilo, W., Yi, M., Zhou, J., Man Ho Allen, A.: Improving privacy and security in decentralized ciphertext-policy attribute-based encryption. IEEE Trans. Inf. Forensics Secur. **10**(3), 665–678 (2015)

Dummy-Based Trajectory Privacy Protection Against Exposure Location Attacks

Xiangyu Liu[1,2](✉), Jinmei Chen[1], Xiufeng Xia[1], Chuanyu Zong[1],
Rui Zhu[1], and Jiajia Li[1]

[1] School of Computer Science, Shenyang Aerospace University,
Shenyang 110136, Liaoning, China
liuxy@sau.edu.cn
[2] School of IT and Business, Wellington Institute of Technology,
Lower Hutt 5010, New Zealand
Xiangyu.Liu@weltec.ac.nz

Abstract. With the development of positioning technology and location-aware devices, moving objects' location and trajectory information have been collected and published, resulting in serious personal privacy leakage. Existing dummy trajectory privacy preserving method does not consider user's exposure locations, which causes the adversary can easily exclude the dummy trajectories, resulting in a significant reduction in privacy protection. To solve this problem, we propose a dummy-based trajectory privacy protection scheme, which hides the real trajectory by constructing dummy trajectories, considering the spatio-temporal constraints of geographical environment of the user, the exposure locations in trajectory and the distance between dummy trajectories and real trajectory. We design a number of techniques to improve the performance of the scheme. We have conducted an empirical study to evaluate our algorithms and the results show that our method can effectively protect the user's trajectory privacy with high data utility.

Keywords: Trajectory · Data publishing · Dummy · Privacy protection

1 Introduction

In recent years, with an increasing popularity of positioning technology and location-aware devices, the location and trajectory information of moving objects have been collected and published. Many new applications have emerged because of the mining and analysis of trajectory information. For example, Investors can make business decisions by analyzing trajectory information of users in a specific area, such as where to build a mall. At the same time, government agencies can optimize the design of traffic management systems and traffic routes by analyzing vehicle trajectories in cities. Although publishing of trajectory information plays a significant role in its mobility-related decisions, it also causes serious threats to personal privacy. If a malicious

The work is partially supported by Key Projects of Natural Science Foundation of Liaoning Province (No. 20170520321) and the National Natural Science Foundation of China (Nos. 61502316, 61702344).

W. Ni et al. (Eds.): WISA 2019, LNCS 11817, pp. 368–381, 2019.
https://doi.org/10.1007/978-3-030-30952-7_37

adversary obtains trajectory information, he can get privacy information of user through data mining technology [1–3], such as: home address, hobbies, living habits and health conditions, sensitive relationship etc. Therefore, the privacy protection of trajectory information has been widely concerned by scholars at home and abroad.

The dummy based trajectory privacy protection method has been widely used in practical research due to its simplicity, no need for trusted third-party entities, and the ability to retain complete trajectory information. However, existing dummy based trajectory protection method does not consider user's exposure location when generating the dummy trajectory, which may reduce the effect of anonymous protection or even directly reveal user's true trajectory. Figure 1(a) shows an example of user's real trajectory $tr = \{(l_1, t_1), (l_2, t_2), (l_3, t_3), (l_4, t_4), (l_5, t_5)\}$, he posts a dynamic through Weibo in location l_2 at time t_2, showing that he is now in location l_2, this information can be obtained by the adversary. When the user exploits existing dummy based trajectory privacy protection scheme to protect his real trajectory, adversary can use this exposure location l_2 to identify some false trajectories. As shown in Fig. 1(b), there are two dummy trajectories generated by algorithm in [14] and a real trajectory. Since the trajectory d_2 does not pass the location l_2 at t_2, adversary can identify it as a false trajectory, so the probability of identifying real trajectory becomes $\frac{1}{2}$, which is greater than anonymous requirement of $\frac{1}{3}$, resulting in user trajectory privacy leakage. We define this attack model as exposure location attack (the specific definition is given later).

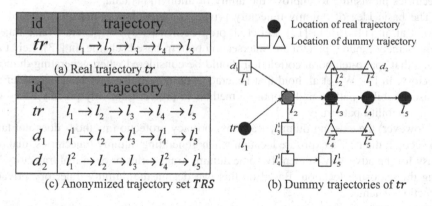

id	trajectory
tr	$l_1 \rightarrow l_2 \rightarrow l_3 \rightarrow l_4 \rightarrow l_5$

(a) Real trajectory **tr**

id	trajectory
tr	$l_1 \rightarrow l_2 \rightarrow l_3 \rightarrow l_4 \rightarrow l_5$
d_1	$l_1^1 \rightarrow l_2 \rightarrow l_3^1 \rightarrow l_4^1 \rightarrow l_5^1$
d_2	$l_1^2 \rightarrow l_2 \rightarrow l_3 \rightarrow l_4^2 \rightarrow l_5^1$

(c) Anonymized trajectory set *TRS*

● Location of real trajectory
□△ Location of dummy trajectory

(b) Dummy trajectories of **tr**

Fig. 1. A real trajectory *tr*, dummy trajectories of *tr* generated by random algorithm and the anonymized trajectory set *TRS* of *tr*

Aiming at this problem, we propose a dummy based trajectory privacy protection scheme. The basic idea is to construct $k - 1$ dummy trajectories that are similar to real trajectory and contain exposure locations to hide the real trajectory. It is worth mentioning that the trajectory privacy protection means to protect both whole real trajectory not to be re-identified and sensitive locations (locations of real trajectory except exposure locations) not to be exposed. Figure 1(c) shows the trajectory set after adding dummy trajectories, each dummy trajectory contains the exposure location l_2. The experimental results show that our algorithms can effectively protect the user's real trajectory.

The rest of this paper is organized as follows. Section 2 reviews some related works and Sect. 3 provides preliminaries. Section 4 presents the main steps of DTPP, including generate dummy locations and construct dummy trajectories. Section 5 is devoted to the experimental results. Finally, Sect. 6 concludes this paper.

2   Related Work

Trajectory privacy protection is mainly classified into following categories: dummy trajectories, trajectory k-anonymity and trajectory suppression. In [6] Yarovoy et al. proposed two algorithms to generate anonymity groups that satisfies the novel k-anonymity. In [14], Moreale et al. proposed a method based on spatial generalization and k-anonymity to transform the original GPS trajectory to achieve anonymity in trajectory dataset. In [7], Abul et al. presented a method named Never walk Alone (*NWA*) to achieve (k, δ)-anonymity through trajectory clustering and space translation. In [8], Huo et al. proposed an approach called You Can Walk Alone (*YCWA*) to protect trajectory privacy through generalization of stay points.

The trajectory suppression method is to selectively publish trajectory data by removing sensitive or frequently accessed locations. In [9], Terrvitis et al. deviced a data suppression technique, which protect privacy while keeping the posted data as accurate as possible. In [10], Zhao et al. proposed two methods based on frequency in trajectories publishing to improve the utility of anonymous data.

The basic idea of dummy trajectory privacy protection was first put forward by Kido et al. in [11, 15]. In [12], Lei et al. proposed two ways to generate false trajectories, namely random pattern and intersection pattern-based scheme. In [13], Lei et al. argued that spatio-temporal correlation should be considered when generating dummy trajectory. In [4], Wu et al. hold that generating dummy trajectory should considering the user mobility pattern and propose a method to protect trajectory privacy based on gravity mobility pattern.

However, the existing dummy trajectory privacy protection methods does not take into account the user's exposure location when generating dummy trajectories, making it easy for the adversary to identify some dummy trajectories or even real trajectory by using the exposure location. Based on this, we design the dummy trajectory privacy protection scheme.

3   Preliminaries

In this section, we define the concepts and notations used throughout the paper. In this paper, a location is a point of interest on the map (e.g. hospital, restaurant, store, bank, etc.), it can be expressed as $l = (x, y)$, where x is the longitude and y is the latitude of location l, we directly use l to represent the location. A trajectory tr is a sequence of n locations, it can be expressed as $tr = \{(l_1, t_1), (l_2, t_2), \ldots (l_n, t_n)\}$, where (l_i, t_i) represents the user checked in the location l_i at t_i, the dummy trajectory is described as $d = \{(l'_1, t_1), (l'_1, t_2), \ldots (l'_n, t_n)\}$. The length of trajectory tr is denoted by $|tr|$. The trajectory set formed by tr and $k - 1$ dummy trajectories is defined as $TRS = \{d_1, d_2, \ldots d_{k-1}, tr\}$.

Exposure Location: Given real trajectory $tr = \{(l_1, t_1), \ldots, (\lambda, t_\lambda), \ldots (l_n, t_n)\}$, if the user posts λ to the social network at time t_λ which all people can know it, we say λ is a exposure location of tr.

It is worth mentioning that there may be multiple exposure locations defined in the trajectory, we use $EL = \{(\lambda_1, t_{\lambda_1}), \ldots (\lambda_m, t_{\lambda_m})\}$ ($m < n$) to represent the set of exposure locations of tr, other locations in trajectory except exposure locations are sensitive locations. As shown in Fig. 1(a), l_2 is the exposure location, and other locations $\{l_1, l_3, l_4, l_5\}$ in the trajectory are sensitive locations.

Exposure Location Attack: Assume TRS be trajectory set with respect to tr, and the exposure location set of tr is EL. The adversary utilizes the exposure location $(\lambda_i, t_{\lambda_i}) \in EL$ as priori knowledge to attack the trajectory set TRS, when a trajectory in TRS does not pass the exposure location λ_i at time t_{λ_i}, it can be identified by adversary as a false trajectory, resulting in user trajectory privacy leakage. We define this type of attack as exposure location attack.

For example, as shown in Fig. 1(b), the trajectory d_2 is identified as a false trajectory because it does not pass the location l_2 at time t_2.

Trajectory Leakage Rate (TE): Given the trajectory set TRS with respect to tr, the adversary uses his background knowledge to identify the false trajectories of TRS and the probability of predicting user's real trajectory is defined as follows:

$$TE = \frac{1}{|TRS| - |TRS'|}$$

where $|TRS'|$ indicates the number of false trajectories identified by the adversary.

Average Location Leakage Rate (LE): Given the trajectory set TRS with respect to tr, the length of tr is n, and the location set containing the real and dummy locations of TRS at t_i is L_i, location leakage rate of anyone location in trajectory tr at t_i is defined as $\frac{1}{|L_i|}$, so average location leakage rate is defined as follows:

$$LE = \frac{1}{n} \sum_{i=1}^{n} \frac{1}{|L_i|}$$

(p, k)-anonymity: Given the trajectory set TRS with respect to tr, and anonymity threshold p, k, if the location leakage rate of anyone location in trajectory tr is not greater than $\frac{1}{p}$, the trajectory leakage rate is not more than $\frac{1}{k}$, we say that the trajectory set TRS satisfies (p, k)-anonymity.

For example, as shown in Fig. 1(c), the trajectory set satisfies $(2, 3)$-anonymity.

Location Distance: Given two locations l_i and l_j, the location distance is defined as Euclidean distance between them.

$$dist(l_i, l_j) = \sqrt{(l_i \cdot x - l_j \cdot x)^2 + (l_i \cdot y - l_j \cdot y)^2}$$

Trajectory Distance: Given two trajectories $tr_1 = \{(l_1, t_1), (l_2, t_1), \ldots (l_n, t_n)\}$, $tr_2 = \{(l'_1, t_1), (l'_2, t_2), \ldots (l'_n, t_n)\}(1 \le i \le n)$, the trajectory distance is defined as follows:

$$TDist(tr_1, tr_2) = \sum_{i=1}^{n} dist\left(l_i, l'_i\right)(1 \le i \le n)$$

4   Dummy-Based Trajectory Privacy Protection Scheme

In this section, we present dummy-based trajectory privacy protection algorithm (denoted as DTPP). The main idea of algorithm 1 is to select $k - 1$ dummy trajectories to form anonymous trajectory set with real trajectory. In this paper, we hold the view to protect both trajectory and location privacy. Therefore, each dummy trajectory needs to be similar to the real trajectory in shape as much as possible, and can increase the number of dummy locations for sensitive locations in the trajectory set. Algorithm DTPP first obtains a list $Cand_{tr}$ of candidate dummy trajectories sorted by $Score$ in descending order (line 2), where $Score$ is a heuristic function that measures the impact on both trajectory similarity and location diversity. A dummy trajectory with higher $Score$ indicates that more trajectory similarity and location diversity would be achieved by its generation. Then DTPP runs a loop (lines 3–5) while $k - 1$ dummy trajectories have been generated, and it attempts to select a dummy trajectory with highest $Score$, which is selected from the top one of $Cand_{tr}$. After getting $k - 1$ dummy trajectories, DTPP examines whether the $|L_i|$ of sensitive location satisfies the location anonymity threshold p, if not, it indicates that the dummy locations of this location cannot make the trajectory set satisfy (p, k)-anonymity (lines 7–9). So we suppress the location to ensure the user's location privacy. Finally DTPP returns the anonymous trajectory set (line 10). The details will be introduced in the followings.

Algorithm1: Dummy-based trajectory privacy protection algorithm (DTPP)

Input: Real trajectory tr, exposure location set EL, anonymity threshold (p, k) trajectory distance threshold (α, β),

Output: Trajectory set TRS

---

1. $TRS \leftarrow \emptyset$;
2. $Cand_{tr} = DTC(tr, \alpha, \beta, EL)$;//a list of candidate dummy trajectories sorted by $Score$ in descending order;
3.   **while** $|TRS| \ne k - 1$ **do**
4.   Select the top one from $Cand_{tr}$ to TRS;
5.   Update $Cand_{tr}$;
6. $TRS = TRS \cup tr$;
7.   **for** each sensitive location set L_i in TRS **do**
8.   **if** $|L_i| < p$ **then**
9.   Delete L_i;
10. **return** TRS;

Metric for a Dummy Trajectory. In this section, we consider a goodness metric for a dummy trajectory. We use $d = \{l'_1, l'_2, \ldots l'_n\}$ to represent a dummy trajectory, the effect of a dummy trajectory is summarized by "trajectory similarity", denoted by $Sim(d, tr)$, and the "location diversity", denoted by $Div(d, TRS)$. To maximize the effect of dummy trajectory, we designed a heuristic function as shown in Eq. (1) to select the dummy trajectory.

$$Score = Sim(d, tr) \times Div(d, TRS) \tag{1}$$

Trajectory similarity $Sim(d, tr)$ is defined as Eq. (2), it shows how similar the dummy trajectory is to the real trajectory, and we measure it by the standard deviation of the location distance between the dummy trajectory and real trajectory. In order to match the semantics, we take it countdown. The larger the $Sim(d, tr)$, the more similar dummy trajectory is to the real trajectory.

$$Sim(d, tr) = \frac{1}{\sqrt{\frac{1}{n}\sum_{i=1}^{n}\left(dist(l_i, l'_i) - \frac{1}{n}\sum_{i=1}^{n} dist(l_i, l'_i)\right)^2}} \tag{2}$$

The larger the $|L_i|$ of sensitive location in the TRS, the lower the location leak rate. We use location diversity $Div(d, TRS)$ to measure the effect of dummy trajectory to LE. It is defined as Eq. (3), it indicates the proportion of location sets with increasing amounts after adding the dummy trajectory. Where $div(l'_i, TRS_i)$ is the change of location set at time t_i after adding the dummy location l'_i, there is a change of 1, no change is 0.

$$Div(d, TRS) = \frac{\sum_{i=1}^{n} div(l'_i, TRS_i)}{|tr|} \tag{3}$$

4.1 Generating Dummy Trajectory Candidate Set

In this section, we present Algorithm 2 to generate dummy trajectory candidate set. We propose to construct a dummy trajectory through connecting dummy locations. Suppose a trajectory with s sensitive locations, which generates m dummy locations at each sensitive location, if the enumeration method is used to generate dummy trajectories, there exists m^s trajectories, the number of dummy trajectories increases exponentially with the number of sensitive locations. Considering that the real trajectory has spatio-temporal characteristics, spatio-temporal reachability should be satisfied between adjacent locations of the dummy trajectory. Therefore, we present to model the dummy trajectory candidate set as a directed graph according to whether the adjacent locations of the dummy trajectories is reachable, and formalize it as $G = \{V, E\}$, which will greatly reduce the number of dummy trajectory candidates. V is a set of locations, E is a set of edges. Spatio-temporal reachability is judged by formula (4), where t_i and t_{i+1} represent the timestamps of accessing locations l_i and l_{i+1} respectively, v_{max} is the user's maximum speed. Obviously, if the formula (4) does not hold, it means that the user could not attend location $v_{i+1,j}$ before t_{i+1} when he starts moving from t_i.

$$\frac{dist\left(v_{ij}, v_{i+1,j}\right)}{v_{max}} \leq (t_{i+1} - t_i)(1 \leq i \leq n, 0 \leq j \leq |Cand_{l_i}|) \tag{4}$$

As shown in Fig. 2, there is a trajectory directed graph of dummy trajectory candidate set for real trajectory in Fig. 1(a), wherein the location v_{20} is an exposure location. If the adjacent locations are reachable, there is an edge between them. From the directed graph, we can get all possible trajectories.

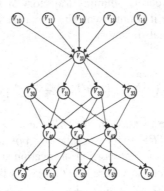

Fig. 2. Trajectory directed graph

Algorithm 2: Generate dummy trajectory candidate set(DTC)

Input: Real trajectory tr, trajectory distance threshold (α,β), exposure location set EL, location anonymity threshold p

Output: Dummy trajectory candidate set $Cand_{tr}$

1. $Cand_{tr} \leftarrow \emptyset$;
2. $Cand_l = DLC(tr, \beta, EL, p)$;
3. $L_0 = l_0 \cup Cand_{l_0}$;
4. $G=TDG(tr, Cand_l)$; // directed graph with spatiotemporal characteristics
5. **for** each location l_i in L_0 **do**
6.   Performs a Depth-First Search(DFS) taking l_i as a starting point on the G, and insert trajectories into TR;
7.   **for** each trajectory tr_i in TR **do**
8.   **if** $\alpha \leq TDist(tr, tr_i) \leq \beta$ **then**
9.   Calculate Score=Score(tr_i);
10.   Insert <tr_i,Score(tr_i)> into $Cand_{tr}$;
11. **return** $Cand_{tr}$;

In Algorithm 2, each dummy trajectory uses the location l_0 or a dummy location of $Cand_{l_0}$ as starting point and terminates with l_n or a location in $Cand_{l_n}$, DTC performs a Depth-First Search (DFS) taking $l_i \in \{l_0 \cup Cand_{l_0}\}$ as a starting point on the G, then can get all dummy trajectory candidates (lines 5–6). If the distance between dummy trajectory and real trajectory conforms to trajectory distance threshold (α, β), calculates *Score* of the dummy trajectory, and inserts $<tr_i$, Score $(tr_i)>$ to the dummy trajectory candidate set $Cand_{tr}$ (lines 7–10). Finally, returns $Cand_{tr}$ (line 11).

Algorithm 3: Trajectory directed graph(TDG)

Input: Real trajectory tr, dummy location candidate set $Cand_l$
Output: Trajectory directed graph G

1. $G \leftarrow \emptyset$;
2. v_{max} is the max speed of tr;
3. **for** (i=0; i<$|tr|$; i++) **do**
4. $\quad V_i = l_i \cup Cand_{l_i}$;
5. **for** (i=0; i<$|tr|$-1; i++) **do**
6. $\quad$ **for each** location v_{ij} in V_i **do**
7. $\quad\quad$ **for each** location $v_{i+1,j}$ in V_{i+1} **do**
8. $\quad\quad\quad$ Calculate $dist(v_{ij}, v_{i+1,j})$;
9. $\quad\quad\quad$ **if** $\frac{dist(v_{ij}, v_{i+1,j})}{v_{max}} \leq (t_{i+1} - t_i)$ **then**
10. $\quad\quad\quad\quad$ Insert $E<v_{ij}, v_{i+1,j}>$;
11. $\quad\quad\quad\quad$ $G_{ij} \leftarrow < v_{ij}, E < v_{ij}, v_{i+1,j} >>$;
12. $\quad\quad$ Insert G_{ij} into G_i;
13. $\quad$ Insert G_i into G;
14. **return** G;

The algorithm TDG first merges l_i and $Cand_{l_i}$ into V_i to obtain all points in the directed graph (lines 3–4). Then it iterates all points and judges the spatiotemporal reachability between adjacent locations, if it is satisfied, there exists an edge $E\langle v_{ij}, v_{i+1,j}\rangle$, and finally returns the directed graph G (lines 5–13).

4.2 Generating Dummy Location Candidate Set

In this section, we propose algorithm DLC to generate dummy location candidate set. In the real world, the adversary can obtain the map information from Internet, thus, he can easily exclude the dummy locations according to the geographic feature of the area the dummy locations belong to. For example, if the adversary have captured a location of user is a lake, he can derive that it is a dummy location. So we advocate using real and meaningful locations on the map as dummy locations to protect privacy.

In order to improve the operational efficiency, we propose to divide the map based on grid. The grid increment is set to 2β according to the trajectory distance threshold (α, β). As shown in Fig. 3(a), if the sensitive location is just at the center of the grid, only need to inquire a grid; as shown in Fig. 3(b), if the sensitive location is not in the center of the grid, need to demand four grids, the time is greatly shortened.

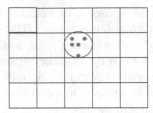 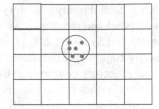

(a) sensitive location at the center of grid (b) sensitive location not at the center of grid

Fig. 3. Query grids according to sensitive location

Algorithm 4: Generate dummy location candidate set(DLC)

Input: Real trajectory tr, trajectory distance threshold β, exposure location set EL, location anonymity threshold p

Output: Dummy location candidate set $Cand_l$

1. $Cand_l \leftarrow \emptyset$;
2.   Divide the map into grids with a size of 2β;
3.   **for** each location l_i in tr **do**
4.   **if** l_i contains in EL **then**
5. $Cand_{l_i} = \emptyset$;
6.   Insert $Cand_{l_i}$ to $Cand_l$;
7.   **else**
8.   Query grids according to coordinate;
9.   **for** each location l'_i in girds **do**
10.   **if** $dist(l_i, l'_i) \leq \beta$ **then**
11.   Insert l'_i to $Cand_{l_i}$;
12.   **if** $|Cand_{l_i}| < p$ **then**
13.   delete l_i in tr;
14.   **else**
15.   Insert $Cand_{l_i}$ to $Cand_l$;
16. **return** $Cand_l$;

In algorithm DLC, the whole map of California is uniformly divided into grids with size of 2β (line 2). For each location in the trajectory, if it is an exposure location, its dummy location candidate set is $\emptyset$; if not, query grids according to location coordinate and if the location l'_i of grids satisfies $dist(l_i, l'_i) \leq \beta$, add this location to the candidate set $Cand_{l_i}$ (lines 3–11). If $|Cand_{l_i}| < p$, it indicates that the region where l_i belongs to is sparse, and cannot generate enough dummy locations to anonymize it. So we suppress it to ensure the user's location privacy (lines 12–13).

5   Experiments

In this section, we provide extensive experiments to evaluate our methods. The user's trajectory data comes from two real datasets:*Gowalla* and *Brightkite*, we also obtain the map data of California, which contains 21,047 nodes and 21,692 edges. In this paper, we select 5,000 trajectories of 5,000 users from two datasets respectively for experiment. Table 1 shows the statistics of the experimental data.

We first conduct experiments to obtain the optimal trajectory distance threshold of the algorithm (DTPP), then evaluate the performance of DTPP by comparing with the algorithm in [4], denoted by GM, and the algorithm in [14], denoted by SM. Finally, we evaluate the influence of the parameters only involved in our algorithm. All programs were implemented in Java and performed on a 2.33 GHz Intel Core 2 Duo CPU with 4 GB DRAM running the Windows 7 operating system. The location anonymity threshold p is set $[6, 27]$ (default value 15), the trajectory anonymity threshold k is set $[3, 9]$(default value 3), and the number of exposure location set EL is set $[1, 4]$ (default value 1). We obtain the optimal trajectory distance threshold by testing the trajectory similarity, average location leakage rate (LE) and running time of DTPP at different trajectory distances.

Table 1. Statistics of datasets

	Gowalla	Birghtkite
Number of users	5000	5000
Number of trajectories	5000	5000
Number of locations	42683	38916
Average length of trajectory	8.536	7.7832
Total time of trajectories (h)	14734	12176
Total distance of trajectories (km)	40560	37916
Maximum speed of users (km/min)	1.13	1.22

Figure 4 shows that: (1) with the trajectory distance increases, the trajectory similarity and the average location leakage rate gradually decreases, the running time gradually increases. This is because DLC generate more dummy locations when trajectory distance is larger. (2) When the trajectory distance is 2 km, the average location leakage rate is very large although the trajectory similarity is high, the *Gowalla* even reaches 60%; when the trajectory distance is greater than 6 km, the trajectory similarity is reduced, but the average location leakage rate is basically stable at 20%. (3) When the trajectory distance is less than 6 km, the running time of the algorithm is within 4 s. So we set the trajectory distance threshold (α, β) to (3, 6). (4) From the comparison of the two datasets, the *Gowalla* dataset is better than *Birghtkite* in terms of trajectory similarity and average positional leakage rate, but the running time is longer than *Birghtkite*, that is due to the location distribution density in *Gowalla* dataset is slightly higher than *Birghtkite*.

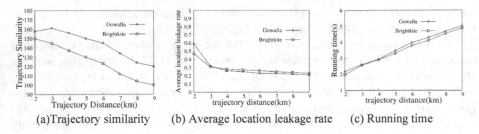

(a)Trajectory similarity   (b) Average location leakage rate   (c) Running time

Fig. 4. Trajectory similarity, average location leakage rate and running time with varying trajectory distance

We compare the performance of the algorithms DTPP, GM and SM by three metrics: trajectory leakage rate, trajectory similarity and running time.

It can be seen from Figs. 5, 6 and 7 that: (1) The trajectory similarity of DTPP is about 5% lower than GM, this is because DTPP considers the exposure locations, resulting in a difference in the location distance between the dummy trajectory and the real trajectory larger than GM, but it is about 4 times higher than SM. (2) When there is an exposure location in real trajectory, the adversary cannot identify any false trajectory because DTPP considers this exposure location; while the adversary can uniquely identify the real trajectory in GM due to it does not intersect with the real trajectory when generating dummy trajectories; SM randomly generates a dummy trajectory, so it may contain the exposure location, but the average trajectory leakage rate is as high as about 50%. (3) The running time of DTPP is almost the same as RM, which is lower than 5 s, but much lower than GM. This is because the algorithm GM generates dummy trajectories by using the enumeration method, so it's running time increases exponentially with the number of locations. DTPP greatly reduces the dummy trajectory candidate set by constructing a reasonable data structure, which saves a lot of time. Combining the above points, it can be concluded that the algorithm DTPP can maintain high trajectory similarity and consume little running time while considering the exposure locations, so the algorithm DTPP can effectively protect the user's real trajectory.

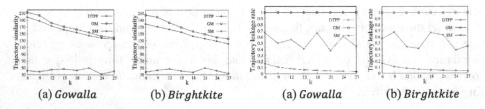

(a) Gowalla   (b) Birghtkite   (a) Gowalla   (b) Birghtkite

Fig. 5. Trajectory similarity with varying k **Fig. 6.** Trajectory leakage rate with varying k

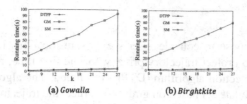

(a) *Gowalla*   (b) *Birghtkite*

Fig. 7. Running time of different protection models

Next, we evaluate the influence of the parameters only involved in our algorithm. First, evaluate the impact of the number of exposure locations on trajectory similarity and running time. Then use the location suppression ratio to measure the impact of the location anonymity threshold p on the algorithm. The location suppression ratio is defined as the percentage of locations that are suppressed.

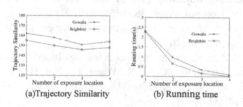

(a)Trajectory Similarity   (b) Running time

Fig. 8. The effect of $|EL|$ on DTPP

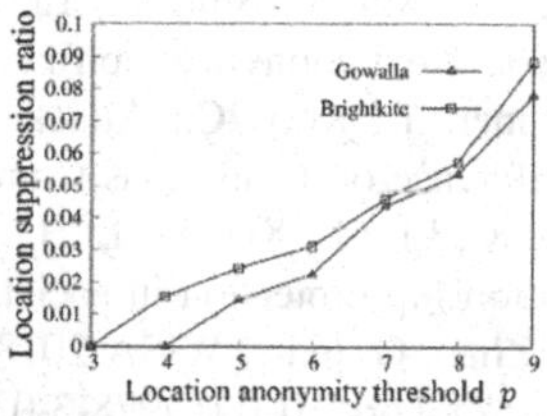

Fig. 9. The effect of p on DTPP

It can be seen from Fig. 8 that: (1) when the number of exposure location $|EL|$ is between 1 and 3, trajectory similarity decreases with increasing the number of exposure location, while $|EL| > 3$, the trajectory similarity increases. This is because when the number of exposure location is increased to a certain extent, the overlapping portions between the dummy and real trajectory become more, so the trajectory similarity is correspondingly improved. (2) The running time of the algorithm DTPP decreases greatly with the increase of the number of exposure location. When $|EL| > 2$, the running time is lower than 2 s, this is because the more the number of exposure location, the fewer dummy locations need to be generated, so the overall time consumption is very small.

As shown in Fig. 9, the location suppression ratio increases with the increase of the location anonymity threshold p, but the location suppression ratio does not exceed 1%, indicating that the algorithm does not need to suppress too many locations to satisfy the location anonymity threshold.

6 Conclusions

In this paper, we propose to consider user's exposure locations in the case of using dummy trajectories to protect trajectory privacy for the first time, and based on this, a trajectory privacy protection algorithm (DTPP) is studied. The algorithm generates the dummy location candidate set based on girds, constructs the trajectory directed graph to store the dummy trajectory candidate set, and establishes a heuristic rule to select the dummy trajectory, which making dummy trajectories have better trajectory similarity while protecting real trajectory. Experiments based on real trajectory datasets show that the algorithm DTPP can effectively protect the trajectory privacy against exposure location attacks.

References

1. Gao, S., Ma, J., Sun, C., Li, X.: Balancing trajectory privacy and data utility using a personalized anonymization model. Netw. Comput. Appl. **38**, 125–134 (2013)
2. Fechner, T., Kray, C.: Attacking location privacy: exploring human strategies. In: ACM Conference on Ubiquitous Computing, pp. 95–98 (2012)
3. Liu, X., Li, M., Xia, X., Li, J., Zong, C., Zhu, R.: Spatio-temporal features based sensitive relationship protection in social networks. In: Meng, X., Li, R., Wang, K., Niu, B., Wang, X., Zhao, G. (eds.) WISA 2018. LNCS, vol. 11242, pp. 330–343. Springer, Cham (2018). https://doi.org/10.1007/978-3-030-02934-0_31
4. Wu, Q., Liu, H.X., Zhang, C., Fan, Q., Li, Z.Q., Wang, K.: Trajectory protection schemes based on a gravity mobility model in IoT. Electronics **8**(2), 148–166 (2019)
5. Li, F.H., Zhang, C., Niu, B., Li, H., Hua, J.F., Shi, G.Z.: Efficient scheme for user's trajectory privacy. J. Commun. **36**(12), 114–123 (2015)
6. Yarovoy, R., Bonchi, F., Lakshmanan, S., Wang, W.H.: Anonymizing moving objects: how to hide a MOB in a crowd? In: Kersten, M.L., Novikov, J. (eds.) EDBT 2009, pp. 72–83. ACM Press, New York (2009)
7. Abul, O., Bonchi, F., Nanni, M.: Never walk alone: uncertainty for anonymity in moving objects databases. In: 24th IEEE International Conference on Data Engineering, pp. 215–226. IEEE Press, Washington (2008)
8. Huo, Z., Meng, X., Hu, H., Huang, Y.: You Can Walk Alone: trajectory privacy-preserving through significant stays protection. In: Lee, S.-g., Peng, Z., Zhou, X., Moon, Y.-S., Unland, R., Yoo, J. (eds.) DASFAA 2012. LNCS, vol. 7238, pp. 351–366. Springer, Heidelberg (2012). https://doi.org/10.1007/978-3-642-29038-1_26
9. Terrovitis, M., Mamoulis, N.: Privacy preservation in the publication of trajectories. In: 9th International Conference on Mobile Data Management (MDM), pp. 65–72. IEEE (2008)
10. Zhao, J., Zhang, Y., Li, X.H., Ma, J.F.: A trajectory privacy protection approach via trajectory frequency suppression. J. Comput. **37**(10), 2096–2106 (2014)
11. Kido, H., Yanagisawa, Y., Satoh, T.: An anonymous communication technique using dummies for location-based services. In: International Conference on Pervasive Services (ICPS), pp. 88–97. IEEE (2005)
12. Lei, P.R., Peng, W.C., Su, I.J., et al.: Dummy-based schemes for protecting movement trajectories. J. Inf. Sci. Eng. **28**(2), 335–350 (2012)

13. Lei, K.Y., Li, X.H., Liu, H., Pei, Z.X., Ma, J.F., Li, H.: Dummy trajectory privacy protection scheme for trajectory publishing based on the spatiotemporal correlation. J. Commun. **37** (12), 156–164 (2016)
14. Moreale, A., et al.: Movement data anonymity through generalization. IEEE Trans. Data Privacy **3**, 91–121 (2010)
15. Kido, H., Yanagisawa, Y., Satoh, T.: Protection of location privacy using dummies for location-based services. In: 21st International Conference on Data Engineering Workshops, p. 1248. IEEE (2005)

Privacy Protection Workflow Publishing Under Differential Privacy

Ning Wu[1,2], Jiaqiang Liu[1,2], Yunfeng Zou[1,2(✉)], Chao Xu[1,2], and Weiwei Ni[1,2]

[1] State Grid Jiangsu Electronic Power Company Research Institute,
Nanjing 210036, China
yunfeng.zou@163.com
[2] Department of Computer Science and Engineering, Southeast University,
Nanjing 211189, China

Abstract. The workflow has been widely used in data quality assessment, error data location and other fields. As data sharing deepens, so does the need to share data lineages. The topology of the lineage workflow contains private information that includes the data generation process, that is, the privacy of the lineage workflow structure, which directly exposes the structure privacy leakage of the lineage workflow. There are the following deficiencies in the privacy protection methods of the lineage workflow structure: (1) The privacy protection method based on the restricted release has a weak theoretical foundation and can only qualitatively measure the privacy protection effect of the lineage workflow structure; (2) Focusing on the maintenance of the local mapping relationship of modules, the maintenance of the key path of the lineage workflow is weak. Aiming at the above problems, this paper proposes a privacy protection method PPWP-DP for the lineage workflow structure, which satisfies the differential privacy. Key path and key path priority concepts are introduced. On this basis, the θ-project projection algorithm is proposed to reduce the degree of the lineage workflow. At the same time, according to the user's preference for key path priority, the maintenance of high priority key path reachability is achieved. The concept of oi-sequence is introduced to extract the structure characteristics of the lineage workflow, and add Laplacian noise to the oi-sequence to satisfy the differential privacy constraint. Adjust the global sensitivity of the oi-sequence after noise addition by the θ-project algorithm to reduce the Laplacian noise scale. Finally, the perturbed oi-sequence is used to reconstruct the lineage workflow for publication, which realizes the workflow privacy security and the maintenance of key path accessibility. Theoretical analysis and experiments verify the effectiveness of the proposed algorithm.

Keywords: Lineage workflow · Privacy security · Structural privacy · Differential privacy

W. Ni et al. (Eds.): WISA 2019, LNCS 11817, pp. 382–394, 2019.
https://doi.org/10.1007/978-3-030-30952-7_38

1 Introduction

Data lineage describes the generation principle and evolution process of data [1]. The data lineage plays an important role in data quality assessment, scientific experimental data reproduction, error data location, and data recovery [2, 3]. With the increasing widespread distribution of lineage workflows, the issue of privacy protection for lineage workflows is increasingly attracting researchers' attention, and the owners of lineage workflows want to protect certain sensitive information from potential attackers. Sensitive information in the lineage workflow can be divided into three categories [3–5]: (1) Intermediate data and parameter information during the execution of the lineage workflow; (2) Input-output mapping relationship of modules in the lineage workflow; (3) Topological relationship among modules in the lineage workflow. The above three types of sensitive information correspond to data privacy, module privacy and structural privacy of the lineage workflow [6].

At present, the privacy of the lineage workflow structure mainly adopts the protection method based on the deletion edge and aggregation module [7]. The main idea is to delete sensitive dependencies or aggregation modules in the lineage workflow according to the background knowledge of the workflow owner to achieve the purpose of protecting the privacy of the structure, and in the process of deleting sensitive dependencies, focusing on maintaining the local mapping relationship of the module input and output; The following defects existing:

(1) Based on the restricted release method of hiding sensitive dependencies, the theoretical basis of privacy processing technology is weak, and there is no strict privacy model and mathematical theory support. Causing the inability to quantitatively measure the privacy and security of the lineage workflow structure;
(2) The existing methods mainly maintain the local mapping relationship between the input and output of the module, and lack the maintenance of the key path describing the evolution process of the data item, which result in the poor availability of workflow.

To solve above problems, a lineage workflow publishing method PPWP-DP based on differential privacy is proposed. Based on the concepts of key path and key path priority, the θ-project algorithm is designed to reduce the degree of the lineage workflow, and the high-priority key path reachability is maintained according to the user's preference for key path priorities. The concept of oi-sequence is proposed and extract the structure characteristics of the lineage workflow. The global sensitivity of the oi-sequence after adding noise is adjusted by θ-project, so that the oi-sequence after the disturbance satisfies the ε-differential privacy model constraint while reducing the noise scale. Finally, the perturbed oi-sequence is used to reconstruct the lineage workflow for publication, which realizes the protection of workflow structure privacy and the maintenance of key path accessibility.

The main contributions of the paper are as follows:

(1) Introduce the concept of key path and key path priority. On this basis, the θ-project projection algorithm is proposed to reduce the degree of the lineage workflow. At the same time, according to the user's preference for the key path priority, the high

priority key path reachability is maintained. The concept of oi-sequence is devised to realize structure feature extraction from lineage workflow.

(2) The θ-project algorithm makes the degree of the module in the lineage workflow not greater than θ, reduces the global sensitivity of the oi-sequence, as well as scale of Laplacian noise added to the oi-sequence while satisfying the ε-differential privacy constraint.

(3) This paper proposes a privacy protection method PPWP-DP for the lineage workflow structure that satisfies the differential privacy model constraint, and realizes the purpose of quantitatively measuring the privacy security degree of the lineage workflow through ε value; And realizing the release of the oi-sequence reconstructed lineage workflow after the disturbance, while maximizing the accessibility of the key path; verifying the effectiveness of the proposed method through experiments on the public dataset.

The paper is organized as following: The second chapter introduces the research status of the privacy protection of the lineage workflow in recent years. The third chapter describes the problem and proposes the solution. The fourth chapter introduces the concept of the key path and oi-sequence of the lineage workflow, and proposes the privacy release method of the lineage workflow structure based on the ε-differential privacy. The fifth chapter analyzes and verifies the effectiveness of the proposed method. Finally, summarizing the full text and looking forward to the follow-up work.

2 Related Work

In recent years, the protection of lineage workflow privacy has received continuous attention, and researchers have done a lot of work in the field of lineage workflow privacy protection. Literature [3] proposes a module privacy protection model based on L-Diversity, which generates a visible view by hiding part of the attribute set, so that for any set of inputs, the attacker can guess that the probability of correct output of the module is not more than $1/\Gamma$. Literature [7] uses the technique of deleting sensitive edges and aggregating adjacent modules to protect the sensitive edges of the lineage workflow. Literature [8] uses differential privacy protection technology to implement module privacy protection by adding noise attributes. Literature [9] proposes a method of privacy protection for the lineage workflow based on view and access control. First, dividing the roles of workflow users and generating different security views for each role to achieve privacy protection. [10] proposed a semantic-based workflow protection mechanism for the lineage, which realizes the representation and execution of the privacy mechanism by defining privacy protection and related analysis terms as ontology. [11] proposed a hierarchy protection workflow protection framework based on semantic representation and distributed execution. [12] introduced the concept of closure in the lineage workflow, and solved the problem of private information dissemination in the lineage workflow containing both public and private modules by hiding the closure between private groups.

The differential privacy model has been widely used in the field of graph privacy protection with good privacy security. The application of the differential privacy model in the graph is mainly divided into two aspects [19]: edge differential privacy and node differential privacy. Since the global sensitivity of the node difference is too large, the edge difference is wider than the node difference in the field of graph privacy protection. The dk-graph model [17] can effectively extract the structural features of the graph, and has a central role in the edge differential privacy model protection; but when d is greater than 3, the global sensitivity is large, resulting in poor usability of the graph. [20] proposed a differential privacy model protection technique based on sampling and smoothing sensitivity, which reduces the global sensitivity of the differential privacy model by sampling and reduces the noise scale.

3 Problem Description and Related Concepts

3.1 Problem Description

The existing privacy protection methods for the lineage workflow structure are mainly based on the privacy protection technology of the deletion edge and aggregation module. The privacy protection technology based on the deletion edge and aggregation module is simple, lacks strict privacy model and mathematical principle support, and the theoretical foundation is weak. It is difficult to quantitatively measure the security level of the lineage workflow after privacy protection. The evolution of certain data items in the lineage workflow is critical to the user and is referred to as the key path. Maintaining the accessibility of key paths in the privacy protection of the lineage workflow structure is important for improving the availability of the lineage workflow. In summary, the privacy protection of the lineage workflow structure needs to meet the following requirements: (1) The privacy protection method is based on a strict privacy protection model and can measure the degree of privacy security quantitatively; (2) Maintain the availability of key paths that describe the evolution of data items in the lineage workflow.

3.2 Related Definition

Definition 1. ε-differential privacy [14]: Data set D_1 and D_2 have at most one record different, range(F) is the range of the random function F. If F satisfies the following conditions, then F is said to satisfy the ε-differential privacy model:

$$\Pr[F(D_1) \in S] \leq \exp(\varepsilon) \times \Pr[F(D_2) \in S], \, S \subseteq \text{Range}(F)$$

Among them, ε is called privacy protection budget, and D_1 and D_2 are called adjacent data sets.

Definition 2. Global Sensitivity [17]: For the function f: $D \rightarrow Rd$, the input is the data set D and the output is the d-dimensional real vector. For any adjacent data set D_1 and D_2, the global sensitivity Δf is defined as:

$$\Delta f = \max_{D_1,D_2} \|f(D_1) - f(D_2)\|$$

Theorem 1 [16]. Let f: D → Rd be a query function. If method A satisfies the following conditions, then A is said to satisfy differential privacy:

$$A(D) = f(D) + [Lap(\frac{\Delta f}{\varepsilon})]^n$$

Among them, Lap($\Delta f/\varepsilon$) is a Laplace noise variable independent of each other, and the noise magnitude is proportional to Δf and inversely proportional to ε. Therefore, the greater the global sensitivity, to achieve the same level of privacy protection, the greater the amount of noise required.

Definition 3. Module and module degree d/out degree dout/in degree din: For module m, its input I = {a1, ..., at}, output O = {b1, ..., bn}; then module m can be abstracted as a mapping R: I → O. The degree of the module m is d = t+n, the degree of exit is dout = n, and the degree of entry is din = t.

Definition 4. adjacent modules If there is a dependency r: mi → mj between the modules mi and mj, then mj is called the adjacency module of mi, and <mi, mj> is called an adjacency pair.

From another point of view, if the input attribute of m_j is the output attribute of m_i, then m_j is the adjacency module of m_i. As shown in Fig. 1, m_2 and m_3 are adjacent modules of m_1.

Fig. 1. Schematic diagram of the adjacent module

4 PPWP-DP

The main idea of PPWP-DP (Privacy-Preserving Workflow Publishing Method Based on Differential Privacy) is: first introduce the concept of key path and key path priority, and further propose the projection algorithm θ-Project, so that the maximum degree of the lineage workflow module is not greater than θ. Meanwhile, the priority path with higher priority is maintained more likely than the priority path with lower priority. The concept of oi-sequence is introduced to realize the feature extraction of the lineage workflow. The global sensitivity of the oi-sequence is reduced by θ-project processing, so that the oi-sequence after the disturbance satisfies the differential privacy constraint while reducing the noise scale. Finally, the perturbed oi-sequence is used to reconstruct the lineage workflow for publication.

4.1 Lineage Workflow Projection Algorithm

The dependencies between modules in the lineage workflow characterize the flow of data in the lineage workflow. If the special dependencies between the modules cannot be maintained after the privacy protection process, the workflow may be meaningless. To describe the special dependencies among modules in the lineage workflow, the concept of the key path of the lineage workflow is introduced as follows.

Definition 5. path P: For the lineage workflow $W = \{M, I\}$, the modules $ms \in M$, $md \in M$, the module md depends on the module ms through the path $P = \{ms \rightarrow m1$ $m2 \rightarrow \ldots \rightarrow mk \rightarrow md\}$, and the modules in the path are not duplicated. Then P is called a path in W.

Some paths in the lineage workflow are important to the user, called the key path KP (KeyPath), and there can be multiple key paths in the lineage workflow. As shown in Fig. 2, the path $P = \{m_1, m_3, m_5, m_6\}$ is the key path KP of the lineage workflow W. In this paper, unless otherwise stated, the meaning of the key path is defined by 5.

Definition 6. key path priority: KP1 and KP2 are the two key paths in the lineage workflow. If the importance of KP_1 is greater than KP_2 according to user preference, the priority of the key path KP_1 is higher than the priority of KP_2, which is recorded as $KP_2 \propto KP_1$.

Reducing the global sensitivity of the algorithm is important work. Considering the method of graph projection to reduce the global sensitivity of edge differential privacy and to ensure the accessibility of key paths with higher priority, based on this, the θ-Project algorithm is proposed. See Algorithm 1 for specific algorithm steps.

Algorithm 1 θ-Project

Input: Lineage workflow W=<M,I>, degree threshold θ, stable dependency sequence Λ_{kp} and Λ_{rest}

Output: Lineage workflow Wθ with a maximum degree less than θ

---
1. $I^\theta = \varnothing$
2. $d(m)=0$ for each $m \in M$/* Set the degree of all modules in M to 0.*/
3. **for** $i=(m_j, \ m_{j+1}) \in \Lambda_{kp}$ **do**
4.   **if** $d(m_j)<\theta$ and $d(m_{j+1})< \theta$ **then**
5. $I^\theta=\{i\} \cup I^\theta$; $d(m_j)=d(m_j)+1$; $d(m_{j+1})=d(m_{j+1})+1$ /* Add dependencies on key paths */
6. **end for**
7. **for** $i=(m_t, \ m_{t+1}) \in \Lambda_{rest}$ **do**
8.   **if** $d(m_t)<\theta$ and $d(m_{t+1})< \theta$ **then**
9. $I^\theta=\{i\} \cup I^\theta$; $d(m_t)=d(m_t)+1$; $d(m_{t+1})=d(m_{t+1})+1$
10.   **end if**
11. **end for**
12. return $W^\theta= (M, \ I^\theta)$

The θ-Project algorithm requires that the ordering of the DAG graph edges be stable; the so-called edge sorting is stable: the DAG graph $G = (V, E)$ and $G' = (V', E')$ differ only by one node, but $\Lambda(G)$ and $\Lambda(G')$ are consistent, that is, the relative order of the two side-by-side sorts is consistent.

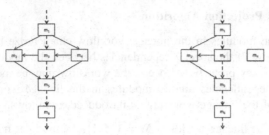

(a) Lineage workflow W   (b) W is processed by θ-Project to generate W'

Fig. 2. The lineage workflow W is processed before and after the θ-Project algorithm, where θ = 3, and the dotted line is the key path of the lineage workflow W.

4.2   oi-Sequence

The dk-sequence [17] is a powerful tool for describing the structural features of graphs, which describes the distribution of nodal degrees in the graph. To better describe the characteristics of the lineage workflow structure, the concept of oi-sequence is introduced as follows.

Definition 7. oi-sequence: For workflow W, its oi-sequence is represented by the set OIS: OIS = {ois_1, ois_2, …, ois_n}, where ois_i is a triple: ois_i = <d_{out}, d_{in}, count>, d_{out} is the degree of m_i, and d_{in} is the degree of m_j, where m_j is the adjacency module of m_i, and count is the number of occurrences of the adjoining pair whose degree is d_{out} and the degree of ingress is d_{in}.

Since the lineage workflow has initial input and final output, it is convenient to extract the oi-sequence, adding "source module" and "end module" to the lineage workflow. The lineage workflow in Fig. 2 is shown in Fig. 3 after adding "source module" and "end module".

Figure 3 shows the lineage workflow after the "source module" and "end module" are added to the lineage workflow $W^θ$. The right side of the figure is its corresponding oi-sequence. $W^θ$ has four sets of oi-sequences, which are denoted as set OIS($Wθ$) = {<1, 1, 3>, <2, 2, 3>, <2, 1, 1>, <1, 2, 1>}. Wherein <2, 2, 3> means that there are three adjacent pairs with an outdegree of 2 and an entry degree of 2, respectively <m_1, m_2> and <m_3, m_5>, <m_3, m_2>.

The oi-sequence describes the characteristics of the outflow and ingress distribution of the lineage workflow module, which can better realize the extraction of the characteristics of the lineage workflow.

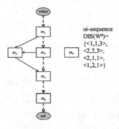

Fig. 3. The lineage workflow $W^θ$ adds "source module" and "end module" and its oi-sequence.

4.3 Privacy-Preserving Workflow Publishing Method Based on Differential Privacy

After acquiring the structural feature $OIS(W^\theta)$ of the lineage workflow W, the PPWP-DP considers using the Laplace noise perturbation $OIS(W^\theta)$ to generate an oi-sequence that satisfies the ε-differential privacy constraint. And reconstruct the lineage workflow for release. The pseudo code of the PPWP-DP complete algorithm is given.

The complete PPWP-DP algorithm steps are given below in Sects. 4.1 and 4.2.

Algorithm 2 PPWP-DP

Input: lineage workflow W, parameter θ, privacy budget ε, time-series oi-sequence set KPS

Output: lineage workflow Wr that satisfies the ε-differential privacy model constraint

---

1. $W^\theta = \theta\text{-}project(W, \theta)$ /* Reduce the maximum degree of W */
2. Compute $OIS(W^\theta)$ according to W^θ /* Generate oi-sequence according to W^θ*/

3. Using ε to perturb $OIS(W^\theta)$ and acquire ε-differentially private $\widetilde{OIS}(W^\theta)$ /* Theorem 1*/
4. DDS= Ø
5. compute KPOIS for keypaths
6. Compute DDS of W^θ /* Calculate the DDS of W^θ*/

7. rebuild keypaths according to DDS , KPOIS and $\widetilde{OIS}(W^\theta)$

8. update DDS and $\widetilde{OIS}(W^\theta)$
9. Create W_r with modules but no edges , stubs using DDS

10. For each $<d_{out}, d_{in}, count>$ in $\widetilde{OIS}(W^\theta)$
11.   For i=1 to count
12.   Choose any module m which has output degree d_{out} and n which has input degree d_{in}
13.   If m does not have free output stubs
14.   Choose u' whice has free output stubs
15.   m=m';
16.   End if
17.   If n does not have free input stubs
18.   Choose n' whice has free input stubs;
19.   n=n';
20.   End if
21.   Add edge from m to n;
22.   End for
23. End for
24. return W_r

---

5 Effectiveness Analysis

The lineage workflow W is processed by the θ-Project algorithm to generate W^θ, and the oi-sequence of W^θ is calculated to extract the feature of W^θ structure, and Laplace noise is added to the oi-sequence to satisfy the difference privacy model constraint. The global sensitivity of the oi-sequence and its proof are given below.

Theorem 2. The lineage workflow W is processed by the algorithm θ-Project to generate the lineage workflow as W^θ, the global sensitivity upper bound of the oi-sequence of the lineage workflow W^θ is $4\theta-3$.

Proof: e is a dependency between modules m_1 and m_2 added to the lineage workflow W^θ. Once e is added between m_1 and m_2, the degree of m_1 changes from d to d + 1, and the degree of m_2 changes from d' to $d' + 1$. Assume that the ingress of module m_1 is i and the outdegree of module m_2 is j. Then, after adding e, the count of $d - i + - j$ oi-sequences is increased by 1, and the count of $d - i + d' - j$ oi-sequences is decremented by one. Together with the newly added dependency, the oi-sequence change before and after adding e is $2(d - i + d' - j) + 1$. In the worst case, both d and d' take the maximum degree θ, then the global sensitivity of the oi-sequence is $4\theta - 2(i + j) + 1$. Since the minimum value of i and j can take 1, the upper bound of $4\theta - 2(i + j) + 1$ is $4\theta - 3$. The certificate is completed.

It can be known from Theorem 2 that the global sensitivity of the oi-sequence of the lineage workflow W is proportional to the maximum degree of the module in W. The θ-project algorithm reduces the maximum degree of the lineage workflow module, thus reducing the global sensitivity of the oi-sequence and reducing the Laplacian noise scale.

6 Experimental Analysis

The experimental data used two social network datasets on the Stanford Network Analysis Platform: wiki-vote (http://snap.stanford.edu/data/wiki-Vote.html), soc-sign-bitcoin-alpha (http://snap.stanford.edu/data/soc-sign-bitcoin-alpha.html). See Table 1 for details of the data set. The experimental environment is: Windows7 Ultimate, 64-bit operating system, Intel(R) Core(TM) i5-6500 3.2 GHz processor, 8 GB of installed memory.

Table 1. Experimental data set information

Data name	Nodes	Edges	Max degree	Average degree
wiki-vote	7115	103689	1167	29.15
soc-sign-bitcoinalpha	3783	24186	888	12.78

This section analyzes data availability from two perspectives: data distortion and path query reachability. The DKDP process was designed to compare with PPWP-DP. The natural logarithm LNMSE (Natural Logarithm of Mean Square Error) comparing the mean square error of the sequence after PPWP-DP and DKWP treatment was used as an experimental evaluation standard. The performance of the PPWP-DP method in critical path maintenance is verified by comparing the reachability of key path queries with the reachability of common path queries.

6.1   LNMSE Comparison Experiment

LNMSE is a measure used to evaluate the degree of difference between the estimator and the estimated amount. In this experiment, the difference between the true value of the oi-sequence and the post-disturbance value is evaluated using LNMSE. For DKWP, the dk- is evaluated using LNMSE. The difference between the true value of the sequence and the value after the disturbance. The calculation formula of the original sequence oi-s and the LNMSE of the perturbed sequence $oi \overset{\sim}{-} s$ is as follows:

$$LNMSE\left(oi - s, oi \overset{\sim}{-} s\right) = \ln\left(\frac{\sum\limits_{i=1}^{n}\left(\left(oi - s.count\right) - \left(oi \overset{\sim}{-} s.count\right)\right)^2}{n}\right)$$

In order to ensure the reliability of the results, multiple experiments are used to average the method to avoid errors. Specifically, for the same θ value, 100 times of LNMSE values are averaged for each method as the value of the final LNMSE.

Figure 4(a) shows the experimental results on the dataset wiki-vote. Since the θ value is independent of the DKWP method, the LNMSE value of the DKWP remains around 0.72. As the θ value increases, the value of the LNMSE of the PPWP-DP method gradually increases, indicating that the data distortion is intensified. Because the global sensitivity of the oi-sequence increases as the value of θ increases, the noise scale also increases, resulting in increased data distortion. For the same θ value, the LNMSE of the PPWP-DP method is smaller than the LNMSE of the DKWP method, indicating that the DPWP method reduces the data distortion and improves the data

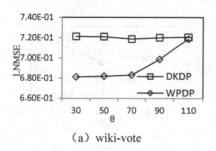

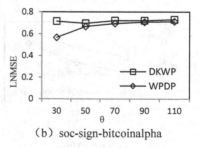

(a) wiki-vote   (b) soc-sign-bitcoinalpha

Fig. 4. LNMSE changes with θ size when privacy budget ε = 1

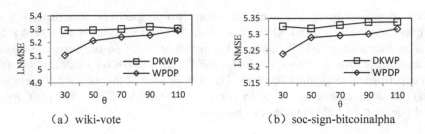

Fig. 5. LNMSE changes with θ size when privacy budget ε = 0.1

availability under the same ε value. Figure 4(b) shows the experimental results on the data set soc-sign-bitcoinalpha. Figure 5 are experimental results under the privacy budgets ε = 0.1. It can be seen that as the θ value increases, under the same conditions, the larger the value of LNMSE, the more severe the data distortion. At the same time, under the condition that the parameter θ and the privacy budget ε are the same, the LNMSE of the PPWP-DP is smaller than the LNMSE of the DKWP, indicating that the availability of the method PPWP-DP is higher than that of the DKWP.

6.2　Key Path Query

This section experimentally verifies the maintenance of key paths by the PPWP-DP. Two groups of queries were designed, 50 in each group. The first group of queries is the key path query, and the second group is the other path queries. The experiments is performed on the noise-disturbed wiki-vote dataset and the soc-sign-bitcoinalpha dataset. The success of the query means that module A and module B are connected by path p in the original workflow. If in the disturbed lineage workflow, module A and module B can still be connected by path p′, where p′ and p are the same. The definition of the query success rate SQR is given below:

$$SQR = \frac{success\_count}{sum} * 100$$

Among them, success_count is number of successful queries, and sum is e total number of queries (Fig. 6).

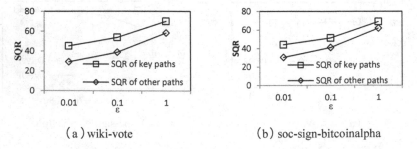

Fig. 6. SQR changes under different privacy budgets ε

7   Summary

Aiming at the problem that the existing lineage workflow structure privacy protection method can't quantitatively measure privacy security and ignore the maintenance of key path, this paper proposes a privacy protection method PPWP-DP that satisfies the ε-differential privacy constraint. Introducing the concept of key path and key path priority, the θ-project projection algorithm is designed to reduce the processing of the lineage workflow. The concept of oi-sequence is proposed to extract the features of the lineage workflow structure. The global sensitivity of the oi-sequence after adding noise is adjusted by θ-project, so that the oi-sequence after the disturbance satisfies the ε-differential while reducing the noise scale. Finally, the perturbed oi-sequence is used to reconstruct the lineage workflow for publication, which realizes the protection of workflow structure privacy and the maintenance of key path accessibility. After theoretical verification and experimental analysis, the proposed method can quantify the privacy security intensity and improve the privacy security while maximizing the accessibility of key paths.

Acknowledgment. Our work is supported by the National Natural Science Foundation of China (No. 61772131).

References

1. Simmhan, Y.L., Plale, B., et al.: A survey of data provenance in e-science. ACM SIGMOD Rec. **34**(3), 31–36 (2005)
2. Braun, U., Shinnar, A., et al.: Securing provenance. In: International Provenance & Annotation Workshop, vol. 4403, p. 752 (2008)
3. Davidson, S.B., Khanna, S., Roy, S., et al.: On provenance and privacy. In: International Conference on Database Theory, ICDT 2011, Uppsala, Sweden, 21–24 March 2011, Proceedings, DBLP, pp. 3–10 (2011)
4. Davidson, S.B., Khanna, S., Milo, T., et al.: Provenance views for module privacy (2011)
5. Wu, J., Ni, W., Zhang, S.: Generalization based privacy-preserving provenance publishing. In: Meng, X., Li, R., Wang, K., Niu, B., Wang, X., Zhao, G. (eds.) WISA 2018. LNCS, vol. 11242, pp. 287–299. Springer, Cham (2018). https://doi.org/10.1007/978-3-030-02934-0_27
6. Davidson, S.B., Roy, S.: Provenance: privacy and security (2017)
7. Davidson, S.B., Khanna, S., Milo, T.: To show or not to show in workflow provenance. In: Tannen, V., Wong, L., Libkin, L., Fan, W., Tan, W.-C., Fourman, M. (eds.) In Search of Elegance in the Theory and Practice of Computation. LNCS, vol. 8000, pp. 217–226. Springer, Heidelberg (2013). https://doi.org/10.1007/978-3-642-41660-6_10
8. Suriarachchi, I.: Addressing the limitations of Γ-privacy. School of Informatics and Computing, Indiana University (2014)
9. Chebotko, A., Chang, S., Lu, S., et al.: Scientific workflow provenance querying with security views 2008
10. Gil, Y., Cheung, W.K., Ratnakar, V., et al.: Privacy enforcement in data analysis workflows. Analysis (2008)
11. Gil, Y., Fritz, C.: Reasoning about the appropriate use of private data through computational workflows. In: AAAI Spring Symposium (2010)

12. Davidson, S.B., Milo, T., Roy, S.: A propagation model for provenance views of public/private workflows, pp. 165–176. Computer Science (2012)
13. Machanavajjhala, A., Gehrke, J., Kifer, D., et al.: L-diversity: privacy beyond k-anonymity. In: 22nd International Conference on Data Engineering. IEEE Computer Society (2006)
14. Dwork, C.: A firm foundation for private data analysis. Commun. ACM **54**(1), 86 (2011)
15. Dwork, C.: Differential privacy. In: Bugliesi, M., Preneel, B., Sassone, V., Wegener, I. (eds.) ICALP 2006. LNCS, vol. 4052, pp. 1–12. Springer, Heidelberg (2006). https://doi.org/10.1007/11787006_1
16. Dwork, C., McSherry, F., Nissim, K., Smith, A.: Calibrating noise to sensitivity in private data analysis. In: Proceedings of the 3rd Conference on Theory of Cryptography, New York, USA, pp. 265–284 (2006)
17. Mahadevan, P., Kroiukov, D., Fall, K., Vahdat, A.: Systematic topology analysis and generation using degree correlations. ACM SIGCOMM Comput. Commun. Rev. **36**(4), 135–146 (2006)
18. Tillman, B., Markopoulou, A., Butts, C.T., et al.: Construction of Directed 2K Graphs (2017)
19. Hay, M., Li, C., Miklau, G., et al.: Accurate estimation of the degree distribution of private networks. In: 2009 Ninth IEEE International Conference on Data Mining. IEEE Computer Society (2009)
20. Nissim, K., Raskhodnikova, S., Smith, A.: Smooth sensitivity and sampling in private data analysis. In: Proceedings of the Thirty-Ninth Annual ACM Symposium on Theory of Computing, STOC 2007, San Diego, California, USA, 11 June–13 June 2007, p. 75 (2007)
21. Zhang, X., Wang, M., Meng, X.: An accurate method for mining top-k frequent pattern under differential privacy. J. Comput. Res. Dev. **51**(1), 104–114 (2014)

Knowledge Graphs and Social Networks

A Cross-Network User Identification Model Based on Two-Phase Expansion

Yue Kou, Xiang Li, Shuo Feng, Derong Shen[(✉)], and Tiezheng Nie

Northeastern University, Shenyang 110004, China
{kouyue,shenderong,nietiezheng}@cse.neu.edu.cn,
2645544252@qq.com, fengshuo1989818@hotmail.com

Abstract. Cross-network user identification is a technique to infer the potential links among the shared user entities across multiple networks. However, existing methods mainly rely on a small set of seed users which might not get enough evidence for identification. In this paper, we propose a cross-network user identification model based on two-phase expansion. On one hand, in order to effectively solve the cold start problem, we propose a global seed expansion method to expand the seed set. On the other hand, we propose a local search range expansion method with the aim to ensure higher accuracy at a lower time cost. Experiments demonstrate the effectiveness of our proposed model.

Keywords: User identification · Cross-network · Global seed expansion · Local search range expansion

1 Introduction

With the increasing popularity of social networks, more and more people participate on multiple online social networks to enjoy their services. Therefore, the same user entities are often included by multiple social networks simultaneously. Cross-network user identification is a technique to infer the potential links among the shared user entities across multiple networks. As the foundation of many cross-network applications, such as recommendation [1] and link prediction, cross-network user identification has attracted increasing attention in recent years.

The goal of cross-network user identification is to find all the matched users among different networks according to the given set of seed users. Most state-of-the-art methods of cross-network user identification are iterative algorithms, also called propagation methods. Traditionally, they require a small set of identified users, also called seed users or seeds, as inputs. Then the other unknown users will be iteratively identified whether they refer to the same person. In each iteration, only the users with higher similarities are regarded as the new matched users. And they will be used as the new seeds for the next iteration.

However, existing methods for cross-network user identification mainly rely on the set of inputted seed users. When the set of seed users is large enough, cross-network user identification can be carried out efficiently. But in fact, the number of seed users is often very limited, resulting in the absence of evidence for identification. In addition,

W. Ni et al. (Eds.): WISA 2019, LNCS 11817, pp. 397–403, 2019.
https://doi.org/10.1007/978-3-030-30952-7_39

current methods usually regard all the unidentified users adjacent to the identified ones as candidates to be matched, which easily leads to higher computation cost.

In order to solve the issue, in this paper, we propose a cross-network user identification model based on two-phase expansion. Different from traditional models, by combining global seed expansion and local search range expansion, our model can be still effective to identify users in the absence of many seed users. More specifically, we make the following contributions:

(1) In order to effectively solve the cold start problem, we propose a global seed expansion method to expand the set of seed users.
(2) We propose a local search range expansion method with the aim to ensure higher accuracy at a lower time cost.
(3) Experiments demonstrate the effectiveness of our proposed model.

2 Related Work

The approaches for cross-network user identification mainly include: enumeration-based identification method and local expansion-based identification method.

As for enumeration-based identification method, it is to enumerate any pairs of users from different networks and identify whether they refer to the same entity. For example, in [2] a degree-based matching method is proposed, which firstly makes the users with higher degrees in networks become the candidates to match, and then considers the users with lower degrees as candidates. In [3], a semi-supervised multi-objective framework is proposed to build a multi-objective user matching model based on user attributes and user topics.

As for local expansion-based identification method, its basic idea is to use seed users as starting nodes, from which to expand the search range gradually. For example, in [4] a matching model based on stable marriage matching algorithm is proposed. In [5], a rule-based propagation algorithm is proposed, which firstly takes the neighbors of the matched users as candidates, and then matches them according to their attribute information. In [6], an energy model is proposed, which takes local consistency and global consistency into consideration for user identification. In [7], a user identification framework based on friend relationship and a semi-supervised propagation algorithm based on neighbors are proposed respectively. In [8], the users from neighbors within two hops of matching users are selected as candidates to improve the matching accuracy.

The differences between our work and existing work are as follows: (1) Most methods for cross-network user identification tend to rely on the small set of inputted seed users. We propose a global seed expansion method to expand the set of seed users, which can generate more evidence for identification. (2) The worst time cost of enumeration-based identification methods is $N * N$ (N is the number of users in the network). Most local expansion-based identification methods do not consider how to give a suitable expansion range. We propose a local search range expansion method with the aim to ensure higher accuracy at a lower time cost.

3   Model Overview

Definition 1 (Social Network). A social network is a graph $G(V, E)$, where V represents the set of users and E represents the set of relationships (edges) between users. We use G_s and G_t to represent the source network and the targeted network respectively.

Definition 2 (Seed Set). Given G_s and G_t, the seed set is denoted as A, where for each user pair $<v_i^s, v_j^t>$ in A (v_i^s and v_j^t are from V^s and V^t respectively), v_i^s and v_j^t refer to the same user.

Definition 3 (Cross-network User Identification). Given G_s, G_t and the seed set, the task of cross-network user identification is to predict whether a pair of users, v_i^s and v_j^t, refer to the same one.

We propose a cross-network user identification model based on two-phase expansion. As shown in Fig. 1, it includes two phases.

The first phase (see Sect. 4.1): Global seed expansion is used for solving the cold start problem. We first represent the nodes as vectors. Then, we select the nodes with more rich information as candidates. Finally, we regard the candidates with the similarities exceeding the threshold as the new seeds.

The second phase (see Sect. 4.2): In the phase of local search range expansion, the unidentified users will be matched iteratively. In each iteration, we choose a set of unidentified users in the source network as the ones to be identified. For each unidentified user v_i^s, we determine the search range in G_t based on the global optimal node and the local optimal node. Then the user in the target network having the highest similarity with v_i^s will be v_i^s's matching user.

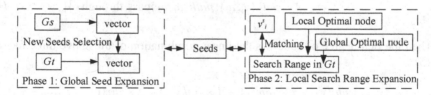

Fig. 1. Overview of cross-network user identification model based on two-phase expansion.

4   Proposed Solution

4.1   Global Seed Expansion

We propose a global seed expansion method to expand the seed set, which can effectively make up for the absence of many seed users. It includes the following steps.

Step 1. Node vectorization. In G_s and G_t, from each known seed node, its adjacent nodes are obtained layer by layer (we set the maximum number of layers to 6). Then each adjacent node v will be represented as a vector (Eq. 1). Here $d(v, a_i)$ means the distance between v and the known seed node a_i ($a_i \in A$, $i = 1 \sim |A|$). If v is not an adjacent node of a_i, then $d(v, a_i)$ is null.

$$vec(v) = \left(d(v, a_1), \ldots, d\left(v, a_{|A|}\right)\right) \tag{1}$$

Step 2. Calculate the rich degree of each node vectorized in Step 1. We quantify the rich degree of nodes from the following three aspects.

(1) The number of effective dimensions in a vector (denoted as dim). For a node v, the number of effective dimensions means the number of such $d(v, a_i)$ in its vector with a non-null value (Eq. 2).

$$dim(v) = |\{d(v, a_i) | d(v, a_i) \in vec(v) \wedge d(v, a_i) \text{ is not null}\}| \tag{2}$$

(2) The number of paths connecting two seeds (denoted as $links$): For a node v, sometimes it is between several pairs of seed nodes. The number of paths connecting two seeds means the number of paths which take the seeds as the endpoints (i.e. $path(a_i, a_j)$) and pass through v (Eq. 3).

$$links(v) = |\{path\left(a_i, a_j\right) | a_i, a_j \in A \wedge path\left(a_i, a_j\right) \text{passes through } v\}| \tag{3}$$

(3) The overlap of paths (denoted as dir): The value of dir is calculated based on the value of $links$. When a path considered in (2) contains another path which also takes the seeds as the endpoints and passes through v, dir will subtracts 1 from $links(v)$. Similarly, all the paths passing through v are considered. So as long as the above situation is true, dir subtracts 1 from $links(v)$ (Eq. 4).

$$\begin{aligned} dir(v) = links(v) - |\{path\left(a_i, a_j\right) | a_i, a_j, a_k \\ \in A \wedge path\left(a_i, a_j\right), path(a_i, a_k) \text{pass through } v\}| \end{aligned} \tag{4}$$

Then the rich degree of node v is the linear combination of the above three factors (Eq. 5).

$$info(v) = \alpha \times dim(v) + \beta \times links(v) + \gamma \times dir(v) \tag{5}$$

Step 3. Expand the set of seeds. We select the nodes with higher rich degree in G_s and G_t respectively. And then their similarities are calculated according to Eq. 6. The node pairs with higher similarities will be expanded to the current seed set.

$$sim\left(v_i^s, v_j^t\right) = vec(v_i^s) \times vec(v_j^t) / |vec(v_i^s)||vec(v_j^t)| \tag{6}$$

4.2   Local Search Range Expansion

As for the second phase, we propose a local search range expansion method with the aim to ensure higher accuracy at a lower time cost. Given G_s, G_t, a seed $<v_i^s, v_j^t>$ in A and an unidentified node v_z^s which is adjacent to v_i^s, our goal is to find the matching node of v_z^s in G_t. The pseudocode is shown in Algorithm 1.

Step 1. Push v_j^t into the queue, which is the initial global optimal node (Line 1–2).

Step 2. If the global optimal node remains unchanged, the algorithm converges and the process of matching ends (Line 3). Otherwise, pop a node from the queue (Line 4). If it is a global optimal node, its neighbors will be expanded and pushed into the queue. The global optimal node is re-selected (Line 5–8). Otherwise, it must be a local optimal node, then only the branch it is on will be expanded. Also the local optimal node on the branch is re-selected and pushed into the queue (Line 9–11). Finally, according to the matching scores (Eq. 7), the new global optimal node is re-selected (Line 12). The global optimal node is the node with the highest score in each expansion. Here $N(v_i^s)$ and $N\left(v_j^t\right)$ are the neighbor set of v_i^s and v_j^t respectively. The matching score of the node pair $<v_i^s, v_j^t>$ is relevant to the number of common neighbors of $N(v_i^s)$ and $N\left(v_j^t\right)$, and the maximum value of $\left|N(v_i^s)\right|$ and $\left|N\left(v_j^t\right)\right|$.

$$score\left(v_i^s, v_j^t\right) = \left|CN\left(N(v_i^s), N(v_j^t)\right)\right| / max\left(|N(v_i^s)|, |N(v_j^t)|\right) \tag{7}$$

Algorithm 1 Local search range expansion algorithm

Input: A pair of seed nodes $<v_i^s, v_j^t>$, an unidentified node v_z^s, G_s, G_t

Output: The matching node of v_z^s in G_t

Process:

```
1  queue_q.push(vᵗⱼ);
2  global_optimal_node← vᵗⱼ;
3  while global_optimal_node is changed
4    n←queue_q.pop();
5    if n is global_optimal_node
6      for each node m in N(n)
7        queue_q.push(m);
8        global_optimal_node←update(vˢz, global_optimal_node, Gs, Gt);
9    else
10     local_optimal_node←select(N(n), global_optimal_node, Gs, Gt);
11     queue_q.push(local_optimal_node);
12   global_optimal_node←update(vˢz, global_optimal_node, Gs, Gt);
   return global_optimal_node.
```

---

5 Experiments

Our experiments are conducted on real datasets: facebook_links (63731 nodes, 1634180 edges) and facebook_wall (63891 nodes, 366824 edges). To demonstrate the effectiveness of our model, we compare to the following methods. (1) MNA: Users are identified based on stable marriage matching algorithm. (2) TPA: Users are identified based on the local expansion from the second-layer neighbors of seeds. (3) OEES: Users are identified by combining local search range expansion and global seed expansion (our model).

The performance of different methods is illustrated in Figs. 2 and 3. It shows that OEES is significantly better than MNA and TPA when the seed nodes are less. With the increase of seed nodes, OEES is also better than MNA and TPA. Also the running time of OEES is lower than that of MNA and TPA.

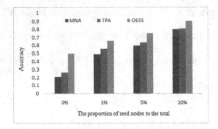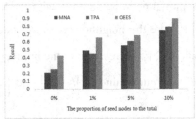

Fig. 2. Accuracy and recall of different methods.

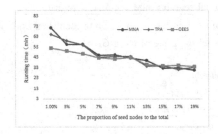

Fig. 3. The running time of different methods.

6 Conclusion

We propose a cross-network user identification model based on two-phase expansion. First, based on global seed expansion, the set of seed users is expanded. Second, by local search range expansion, nodes can be matched with higher accuracy at a lower time cost. Experiments demonstrate the effectiveness of our proposed model.

Acknowledgment. This work is supported by the National Key R&D Program of China (2018YFB1003404), the National Natural Science Foundation of China (61672142, U1435216) and the Program of China Scholarships Council (201806085016).

References

1. Yang, R., Han, X., Zhang, X.: A multi-factor recommendation algorithm for POI recommendation. In: Meng, X., Li, R., Wang, K., Niu, B., Wang, X., Zhao, G. (eds.) WISA 2018. LNCS, vol. 11242, pp. 445–454. Springer, Cham (2018). https://doi.org/10.1007/978-3-030-02934-0_41
2. Korula, N., Lattanzi, S.: An efficient reconciliation algorithm for social networks. PVLDB **7**(5), 377–388 (2014)
3. Liu, S., Wang, S., Zhu, F., Zhang, J., Krishnan, R.: HYDRA: large-scale social identity linkage via heterogeneous behavior modeling. In: SIGMOD 2014, pp. 51–62 (2014)
4. Kong, X., Zhang, J., Yu, P.: Inferring anchor links across multiple heterogeneous social networks. In: CIKM 2013, pp. 179–188 (2013)
5. Raad, E., Chbeir, R., Dipanda, A.: User profile matching in social networks. In: NBiS 2010, pp. 297–304 (2010)
6. Zhang, Y., Tang, J., Yang, Z., Pei, J., Yu, P.: COSNET: connecting heterogeneous social networks with local and global consistency. In: KDD 2015, pp. 1485–1494 (2015)
7. Zhou, X., Liang, X., Zhang, H., Ma, Y.: Cross-platform identification of anonymous identical users in multiple social media networks. IEEE Trans. Knowl. Data Eng. **28**(2), 411–424 (2016)
8. Zhang, Y., Fu, J., Xiao, C.: A local expansion propagation algorithm for social link identification. Knowl. Inf. Syst. (2018). https://doi.org/10.1007/s10115-018-1221-y

Link Prediction Based on Node Embedding and Personalized Time Interval in Temporal Multi-relational Network

Yuxin Liu, Derong Shen[✉], Yue Kou, and Tiezheng Nie

College of Information Science and Engineering,
Northeastern University, Shenyang, China
grimmjoyLiu@163.com,
{shenderong,kouyue,nietiezheng}@cse.neu.edu.cn

Abstract. Link prediction on temporal networks has a wide range of applications, such as facilitating individual relationship mining, user recommendation, and user behavior analysis. The traditional link prediction methods on temporal network only considered the structure of single-relational networks, which ignored the diversity of network link types and the influence between different link types. This paper proposes a temporal multi-relational network link prediction method combining personalized time interval and node embedding. Firstly, the node embedding is generated according to the structure of target network and auxiliary network which overcomes the defect of single network information sparseness; then, considering the diversity of link types, we construct the relationship formation sequence based on personalized time interval and the influence between different relationships for each link; next, the relationship formation sequence is modeled based on the Hawkes process, which takes product of the node embedding as the initial value to calculate the possibility of link formation. The method captures the dynamic characteristics and multi-relational property of the network, which is helpful to improve the accuracy of link prediction. Experimental results find that the proposed method has better performance and can be applied to large-scale networks.

Keywords: Link prediction · Multi-relational network ·
Personalized time interval · Node embedding · Relation influence

1 Introduction

In recent years, dynamic networks or named as temporal networks have achieved special attention, such as on social platforms, with the change of time, the interaction between users will form temporal social networks. Temporal networks can help people analyze the change of user behaviors or preferences, and temporal link prediction is a hot topic, which is mainly based on snapshots of networks formed at different times, and predicts the probability of link formation between node-pairs using temporal dynamics of graph topology [9].

Temporal link prediction has been widely used in many fields [1], including finding the association of documents in information retrieval, predicting the friend relationship

© Springer Nature Switzerland AG 2019
W. Ni et al. (Eds.): WISA 2019, LNCS 11817, pp. 404–417, 2019.
https://doi.org/10.1007/978-3-030-30952-7_40

between users in recommendation system [8], capturing the interaction between proteins in biological networks and so on. Existing link prediction methods of temporal networks focus on homogeneous networks, in fact, the relationships between node-pairs tend to be diversified. For example, in academic networks, authors form cooperative relationships with each other, also they form conference relationships because they participate in the same conference. So, the concept of temporal multi-relational social network is proposed by considering this situation. Given the target relationship type, the link prediction of temporal multi-relational networks is to predict the probability of link formation with the link type is the target relationship type.

At present, the existing work on temporal networks has the following limitations: (1) Only homogenous networks was constructed without considering the diversity of link types and the influence between different types of links. (2) They fixed the size of time interval, but in fact, the link formation rules among different node-pairs are different, and the uniform time interval will lead to the uneven distribution of information. (3) Most node embedding methods only considered the structure information of the target network. However, by incorporating the auxiliary network into embedding methods, the node embeddings can capture the structure information of multiple networks and improve the prediction precision.

The main contributions of this paper are: (1) We propose a node embedding method based on shared matrix factorization, which combines the structure information of target network and auxiliary network. (2) We propose the relationship formation sequence based on personalized time interval to combine dynamic and multi-relational property of networks. (3) We propose a method called personalized time interval link prediction (PTI-LP) which using Hawkes process to predict links in temporal multi-relational network. (4) Experiment is performed to demonstrate the effectiveness of our method.

The remainder of this paper is organized as follows. In Sect. 2 we mention some previous works related to ours. In Sect. 3 we introduce definitions and background. In Sect. 4 we describe our approach. In Sect. 5 we show its experimental evaluation. Finally, we summarize our findings in Sect. 6.

2   Related Work

Many methods of link prediction have been proposed in recent years. Newnam et al. [2] first proposed using common neighbors to solve the link prediction problem in static network in 2001. Liben-Nowell et al. [3] proposed the concept of snapshot graphs which dividing a temporal network into subgraphs according to time windows. Initial studies on link prediction focus on static and homogeneous networks, however many networks are heterogeneous and dynamic network, where nodes are connected by various types of links. Some of the studies [4, 5] consider temporal dynamics of the network in order to predict future link. Yu et al. [4] proposed a link prediction model LIST which characterizes the network dynamics as a function of time and integrates the spatial topology of network at each timestamp. Li et al. [5] proposed a novel link prediction framework SLIDE on dynamic attributed networks. SLIDE maintains and updates a low-rank sketching matrix to summarize all observed data, and leverage the sketching matrix to infer missing links. Some of the studies [6, 9–11] consider the multi-relational property of networks. Ozcan et al. [6] combined the correlations between different link types and the effects of different topological local and global similarity measures in different time periods to predict future links. Sett et al. [9] proposed a feature set called TMLP which combines dynamic of graph topology and history of interactions at dyadic level for link prediction in dynamic heterogeneous networks. Li et al. [10] used the probabilistic model to preserve the triangular and quadrilateral structures in directed multi-relational networks. Recent years, affected by language model Skip-Gram, many researchers used embedding method to predict future links [7, 12, 14]. Huang et al. [7] focused on heterogeneous network and proposed an embedding method based on meta-path similarity. Chao et al. [14] constructed subgraph correlation matrix and use subgraph embedding to generate node embeddings for link prediction.

Compared with the existing work, we use the auxiliary network and target network in node embedding method to improve the prediction accuracy. In order to solve the problem of uneven distribution of information, we create the personalized time interval instead of fixed time interval. In addition, we use word embedding and short text topic model GPU-DMM to construct topicSim relationship between nodes, which is more accurate and explainable than use coincidence of keywords.

3   Problem Formulation

A multi-relational network is a network that allows multiple relationships between node-pairs. In academic network, there are three type of links between two authors, i.e., co-author, conference and topicSim. Each link is assigned a label to denoting the type of relationship. We define a temporal multi-relational network on the basis of multi-relational network by considering the timestamps of link formation. The definitions used in this paper is as follows:

Temporal Multi-relational Network. Let $G = (V, L, E, T)$ be a temporal multi-relational network, V is the set of vertices, L is the set of labels denoting the relationship types. $E = \{E_1 \cup E_2 \ldots \cup E_L\}$ is the set of links, E_l is the link set of type l, we denote the element in E_l is a tiple (x, y, t) which x, y denote the vertices, t denotes the set of link formation time-stamps between x and y.

Relationship Formation Sequence. Give a link (x, y, t), the link type is l, R(x, y, l) returns a set of tuples $\{(1, I_1(x, y)), \ldots, (i, I_i(x, y))\}$, i is the event of type l, $I_i(x, y)$ is the set of timestamps which x and y participated in event i. For example, authors A and B participated in KDD conference in 2008 and 2013, also participated in ICDE conference in 2012, so R(A, B, conference) = $\{(\text{KDD}, \{2008, 2013\}), (\text{ICDE}, \{2012\})\}$.

Relationship Influence. For node x and y, the relationship types between them are l_i and l_j, the influence on relationship l_i from relationship l_j is denote as $r_{l_i,l_j}^{x,y}$.

Personalized Time Interval. We define a personalized time interval for each link (x, y, t) as follows:

$$\Delta t^{(x,y,l)} = \frac{t_{max}^{(x,y,l)} - t_{min}^{(x,y,l)}}{M} \qquad (1)$$

Where M is a predefined maximum number of time intervals, l is the link type, $t_{min}^{(x,y,l)}$ and $t_{max}^{(x,y,l)}$ denote the first and the last link formation time-stamps between x, y. For each link (x, y, t), based on personalized time interval $\Delta t^{(x,y,l)}$, we define the time-window sequence $W_l^{(x,y,l)} = \{w_1^{(x,y,l)}, w_2^{(x,y,l)} \ldots\}$, $w_i^{(x,y,l)}$ is a time window start at $t_{min}^{(x,y,l)} + \Delta t^{(x,y,l)} \times (i - 1)$, end at $t_{min}^{(x,y,l)} + \Delta t^{(x,y,l)} \times i$.

Link Prediction in Temporal Multi-relational Network. Given two nodes $x, y \in V$ and a target relationship type $l \in L$, the link prediction is to predict whether x, y will be connected by a link of type l or not.

4 Link Prediction Model in Temporal Multi-relational Network

In this paper, we proposed a link prediction model named PTI-LP (Personalized Time Interval Link Prediction) in temporal multi-relational network based on node embedding and personalized time interval. It can be applied to any multi-relational network, such as YOUTOBE network, academic network and so on. For the convenience of understanding, we take the academic multi-relational network as an example. The model is composed of three parts: multi-relational network preparation, node embedding generation and link prediction. The model framework is described in Fig. 1.

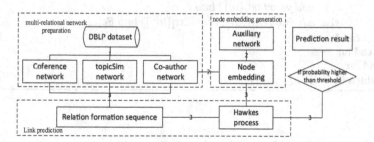

Fig. 1. PTI-LP model framework

4.1 Multi-relational Network Preparation

When generating the multi-relational network from academic dataset, we take the authors as nodes, if two authors collaboratively write a paper, they will connected by a co-author link, if two authors write paper(s) in the same conference, they will connected by conference link. So we denote the subgraph of G consisting of the co-author type of links as $G_1 = \{V, E_{author}\}$, the subgraph of G consisting of the conference type of links as

$G_2 = \{V, E_{conference}\}$. In addition, if two authors have similar research interests, they will connected by topicSim link. Existing works use the coincidence degree of paper keywords to determine whether two authors will form a topicSim link or not, however this method does not consider semantic information. As the title of paper can be regard as short text and each title corresponds to a topic, we use the topic model GPU-DMM combined with word embeddings to construct the topicSim network $G_3. = \{V, E_{topicSim}\}$. Algorithm 1 shows the main steps of topicSim network generation.

Algorithm 1: TopicSim network generation

Input:
- *nodelist*: a set of nodes, $nodelist = \{node_1, ... node_i, ..\}$, for example in academic dataset, $node_i$ contains $titlelist_i$ and $yearlist_i$, $titlelist_i$ is an array of paper titles, $yearlist_i$ is an array of public year of papers.
- Word embedding matrix W, topic number T, threshold ε

Output:
- $G_3 = \{V, E_{topicSim}\}$: topicSim network, each element in G_3 is a triple $(node_i, node_j, t)$
1:   **for** $node_i \in NodeList$ **do**
2:      **for** $title \in titlelist_i$ **do**
3:         $title \leftarrow id, doclist[id] = title$
4:      **end for**
5:   **end for**
6:   $topiclist = GPU\text{-}DMM(doclist, W, T)$
7:   initialize $G_3 = (V, E_{keyword})$
8:   **for** $node_i \in NodeList$ **do**
9:      **for** $node_j \in NodeList$ **do**
10:        **for** $id_x^i \in titlelist_i, id_y^j \in titlelist_j$ **do**
11:           **if** $topicList[id_x^i] == topicList[id_y^j] \&\& yearList_i[id_x^i] == yearlist_j[id_y^j]$。
12:              **if** $(node_i, node_j, t) \in E_{keyword}$
13:                 add $yearList_i[id_x^i]$ **into** t
14:              **else** add $(node_i, node_j, yearList_i[id_x^i])$ **into** $E_{keyword}$,
15:           **end for**
16:        **end for**
17:     **end for**
18:     **return** $G_3 = (V, E_{keyword})$

4.2   Node Embedding Method Based on Network Structure

In academic multi-relational network, if the target type is co-author, $G_1 = (V, E_{co\text{-}author})$ is the target network, considering nodes in different domains of networks have different activities, for example, in Fig. 2, author Bob cooperates with Tom in data mining domain, but in database domain, Bob cooperates with Anna, so it is possible that Bob will cooperates with Anna in future. If we only consider the information in data mining co-author network, the cooperation between Bob and Anna will be ignored. So in this paper, we obtain another co-author network from different domain, i.e. auxiliary network $G' = (V', E')$, from different datasets which is intersecting but not identical.

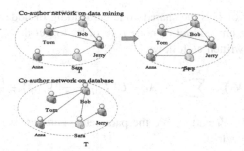

Fig. 2. The influence of database co-author network to data mining co-author network

The node embedding is generated by combining the structural information of target network and auxiliary network. Firstly, in Sect. 4.2.1, we use the SVD-based matrix completion to enrich the adjacency matrix of target network and auxiliary network, then we decompose matrices to obtain the node embedding in Sect. 4.2.2.

4.2.1   Adjacency Matrix Completion

Most real networks are sparse which only a small fraction of the link are presented in adjacency matrix. In order to enrich the structure information, we decompose matrices G and G' by singular value decomposition to get the left singular vectors L_1 and L', the singular matrices Σ_1 and Σ', the right singular vectors R_1 and R'. Then let $M_1 = L_1\Sigma_1R_1$ and $M' = L'\Sigma'R'$, M_1 and M' are the completed adjacency matrices. The process is shown in the Fig. 3.

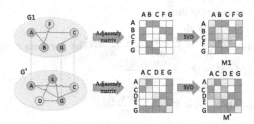

Fig. 3. Adjacency matrix completion

After adjacency matrix completion, we integrate all nodes of target network and auxiliary network, i.e. $V_{all} = V_1 \cup V'$, so that both networks have the same set of nodes V_{all} and different sets of edges, i.e. $G_1 = (V_{all}, E_{co-author})$, $G' = (V_{all}, E')$. Now, the adjacency matrices of network after node integration are $A_1 \in R^{|V_{all}| \times |V_{all}|}$ and $A_2 \in R^{|V_{all}| \times |V_{all}|}$, which have the same dimenson.

4.2.2   Node Embedding Based on Shared Matrix Factorization

Using the shared matrix factorization [13], the adjacency matrices are decomposed into a product of constant matrix U and $\{V_r\}_{r=1}^2$, where U is a constant matrix shared by all networks. The ith row of U is the shared node embedding of node i, and the ith row of V_r is the private node embedding of node i. in order to get the node embedding, the objective function is written as follows:

$$L(U, V_1, V_2) = \sum_{k=1}^{2} \left(\|A_k - UV_k^T\|_F^2 + \frac{\alpha}{2} \|V_k\|_F^2 \right) + \frac{\beta}{2} \|U\|_F^2 \tag{2}$$

where $U \geq 0$. Since U, V_1 and V_2 are the parameter matrices, we need to compute a partial derivative for each.

$$\frac{\partial L(U, V_1, V_2)}{\partial U} = \sum_{k=1}^{2} \left(UV_k^T - A_k \right) V_k + \beta U \tag{3}$$

$$\frac{\partial L(U, V_1, V_2)}{\partial V_1} = \left(UV_1^T - A_1 \right) U + \alpha V_1 \tag{4}$$

$$\frac{\partial L(U, V_1, V_2)}{\partial V_2} = \left(UV_2^T - A_2 \right) U + \alpha V_2 \tag{5}$$

The node embedding method is generated by Algorithm 2:

Algorithm2: Generate Node Embeddings

Input:
- Target network A_1, auxiliary network A_2, dimenson K
Output:
- Result of matrices U and V_i, i=1,2.
1:    **for** *time*=0 ; *time* < *step* **do**
2:        **for** j=1; $j \leq m$; j++ **do**
3:            **compute** $\frac{\partial L(U,V_1,V_2)}{\partial U}$, $\frac{\partial L(U,V_1,V_2)}{\partial V_1}$, $\frac{\partial L(U,V_1,V_2)}{\partial V_2}$ **by formula** (3) (4) (5)
4:            **update** U, $U = U - \lambda \frac{\partial L(U,V_1,V_2)}{\partial U}$
5:            **for** i=1; i <=2 **do**
6:                **update** V_i, $V_i = V_i - \lambda \frac{\partial L(U,V_1,V_2)}{\partial V_i}$
7:            **end for**
8:            **for** $m \in (0, |V|)$ **do**
9:                **for** $n \in (0, K)$ **do**
10:                    $U_{mn} = max(0, U_{mn})$
11:                **end for**
12:            **end for**
13:        **end for**
14:    **end for**
15:    **return** U, V_1, V_2

4.3 Link Prediction Based on Node Embedding and Personalized Time Interval

4.3.1 Relationship Formation Sequence Based on Personalized Time Interval

In this section, we construct the personalized time interval and the relationship formation sequence to capture the change of network topology. Given a link (x, y, t), the link type is l, it's personalized time interval is $\Delta t^{(x,y,l)}$. based on personalized time intervals, time windows are divided differently for each link. We assume the current moment is t_{now} and the total number of time windows is M, so the m-th time window of link (x, y, t) of type l is defined as

$$w_m^{(x,y,l)} = \left(t_{now} - \Delta t^{(x,y,l)} \times (M - m + 1), t_{now} - \Delta t^{(x,y,l)} \times (M - m)\right) \quad (6)$$

And, the relationship formation sequence on m-th time window is populated by

$$S_m(x,y,l) = \sum_{(i,I_i(x,y)) \in R(x,y,l)} \left|I_i(x,y) \cap w_m^{(x,y,l)}\right| \quad (7)$$

where $(i, I_i(x, y))$ is i-th event of type l, for example, if l is conference type, i is a conference name, such as AAAI, so $I_i(x, y)$ is the set of timestamps which x and y participation in AAAI at the same time. Then for a link (x, y, t), it's relationship formation sequence of link type l is $S(x,y,l) = \{S_1(x,y,l), \ldots S_m(x,y,l), \ldots S_M(x,y,l)\}$.

An example of relationship formation sequence is shown in Fig. 4, in academic multi-relational network, the relationship formation sequence of co-author link is generated as follows: author A and B connected by link (A, B, {2001, 2005, 2006, 2007}), the link type is co-author, let $M = 3$, so $\Delta t^{(x,y,l)} = 2$, and the time window sequence is {(2001, 2003), (2003, 2005), (2005, 2007)}, on 1-th time window, $S_1(A, B, \text{co-author}) = 1$, on 2-th and 3-th time window, $S_2(A, B, \text{co-author}) = 1$, $S_3(A, B, \text{co-author}) = 2$, so finally the relationship formation sequence of link (A, B, t) is $S(A, B, \text{co-author}) = (1, 1, 2)$.

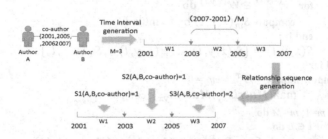

Fig. 4. The generation process of co-author relationship formation sequence

4.3.2 Relationship Influence Generation

In multi-relational network, different type of relationships might influence each other, and the influences on target relationship from other relationships are different. For example, for two authors, it is more likely to have the same research interests and form a co-author relationship than to meet at a meeting and form a co-author relationship. So we defined the relationship influence. Given two nodes x, y, the timestamps set of they connected by l_1 type of link is n_{l1}, the timestamps set of they connected by l_2 type of link is n_{l2}, so the influence of l_1 relationship on l_2 relationship is $r_{l1,l2}^{x,y} = \frac{|n_{l1} \cap n_{l2}|}{n_{l1}}$.

4.3.3 Link Prediction Based on Personalized Time Interval and Node Embedding

The Hawkes process [15] is a time-point process that models a discrete sequence of events, assuming that past event excites future event. In this paper, we consider the relationship formation sequences as a point process. Algorithm 3 shows the generation process of $\lambda_{x,y}^t$ which is the probability of nodes x, y connected by a link of type l. The basic probability $\mu_{(x,y)}$ is obtained by the product of node embeddings generated in Sect. 4.2.1, $\lambda_{x,y}^t$ is defined as follows:

$$\lambda_{x,y}^t = \mu_{(x,y)} + \sum_{m=1}^{M} \sum_{l \in L} S_m(x,y,l) r_{l,l_{target}}^{x,y} \kappa\left(\Delta t^{(x,y,l)} \times (M - m + 1)\right) \qquad (8)$$

where $\kappa(t) = exp(-\delta t)$.

Algorithm3: Link Prediction Based on Personalized Time Interval

---

Input:
- The multi-relational network $G=(V,L,E,T)$, node embedding matrix U,V_1, the target relationship l_{target}, threshold α
Output:
- $\tilde{E}$: The set of links which will form in future
1:    **for** (x,y,t) $\in$ E **do**;
2:        **for** $l \in L$ **do**
3:            **if** (x,,y,t)$\in E_l$**then**
4:                compute $\Delta t^{(x,y,l)}$, $W^{(x,y,l)} = \{w_1^{(x,y,l)}, w_2^{(x,y,l)} ...\}$
5:                **for** $w_1^{(x,y,l)} \in W^{(x,y,l)}$ **do**
6:                    compute $S_m(x,y,l)$ by formula (7)
7:                **end for**
8:                $S(x,y,l) = \{S_1(x,y,l), ... S_m(x,y,l), ... S_M(x,y,l)\}$
9:        **end for**
10:        compute $Matrix = UV_1^T$, $temp = 0$
11:        $\mu_{(x,y)} = max(Matrix_{xy}, Matrix_{yx})$, compute $r_{l,l_{target}}^{x,y}$
12:        **for** $m=1$; $m<M$ **do**
13:            **for** $l \in L$ **do**
14:                $temp += S_m(x,y,l) r_{l,l_{target}}^{x,y} \kappa(\Delta t^{(x,y,l)} \times (M - m + 1))$
15:            **end for**
16:        **end for**
17:        $\lambda_{x,y}^t \leftarrow \mu_{(x,y)} + temp$
18:        **if** $\lambda_{x,y}^t > \alpha$ **then**
19:            add (x,y,t) into $\tilde{E}$
20:    **end for**
21:    **return** $\tilde{E}$

---

5   Experiments

We experimentally evaluated the performance of our model PTI-LP on DBLP dataset, in which co-authorship information, conference information and paper information of six conferences (KDD, ICDE, SDM, PKDD, WSDM, PAKDD) in the field of data mining are selected to construct the temporal multi-relational network. We also construct the auxiliary network based on co-authorship information in the field of database in DBLP dataset. All tests were conducted on a computer server with a 64-bit, 8.0G of RAM Intel Core (3.30 GHz) CPU.

5.1   Evaluation Measures

We use the AUC score, precision and runtime to evaluate the performance of link prediction. At each time we pick an existing link and an non-existing link, if among n comparisons there are n_1 times the existing link have higher scores and n_2 times they have same score, the AUC value is $(n_1 + 0.5n_2)/n$. Given a set of node-pairs V_L, if m node-pairs will connected in the future, the precision is $m/|V_L|$. And the runtime reflects the complexity of the method.

5.2   Performance Evaluation

For comparison, we consider the following link prediction methods: TMLP [10], LPMR [1], PTI-LP-SN, and PTI-LP-SR. The PTI-LP-SN is the PTI-LP method which only considers the target network without auxiliary network. The PTI-LP-SN is the PTI-LP method which only considers single relationship, i.e., co-author relationship. In our experiment, we set the time interval number $M = 3$, the topic number is 25 and the node embedding dimension $k = 200$.

Comparison of AUC Score. Figure 5 shows that the proposed method PTI-LP has highest AUC score and proves that the method which combines multiple types of relationships between nodes and time information has better performance on link prediction.

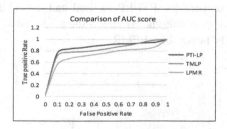

Fig. 5. The ROC curve

Impact of Topic Number. The number of topics plays a key role in the formation of the TopicSim network. As shown in Fig. 6, too many topics may introduce excessive noise into network, too few topics may lead to inadequate information extraction. Our method PTI-LP has the highest AUC score when the topic number is 25.

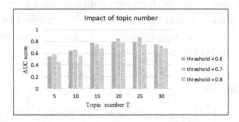

Fig. 6. Impact of the topic number

Impact of Auxiliary Network. As the TMLP cannot combine the auxiliary network for link prediction, this paper compares the LPMR, PTI-LP, and PTI-LP-SN. As shown in Fig. 7, method PTI-LP that combines both target network structure and auxiliary network structure outperforms than only consider one.

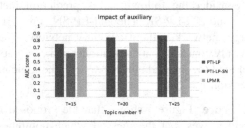

Fig. 7. Impact of auxiliary network structure

Impact of Time Interval Number. Since the LPMR method does not consider time information, we only compare the PTI-LP method and the TMLP method. Figure 8 shows that the number of time intervals M affects the predicted accuracy. When $M = 3$, PTI-LP performs slightly better than LPMR.

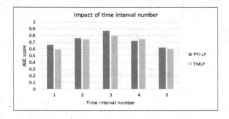

Fig. 8. Impact of time interval number

Runtime Comparison. In general, the other methods need to train a large number of parameters before the prediction, while the PTI-LP method uses the Hawkes process to predict without training. Figure 9 shows that PTI-LP is better than the comparison methods, so it is more suitable for the data sets which have more relationships.

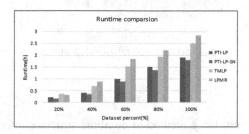

Fig. 9. Runtime comparison

Dimension Selection. We observe the impact of the dimension k node embeddings on the prediction results. Only PTI-LP, PTI-LP-SN and PTI-LP-SN use the shared matrix factorization, so we compare these three methods. As shown in Fig. 10, an appropriate dimension can reduce the running time and improve prediction accuracy, when k = 200, our method achieves the highest AUC score 0.836.

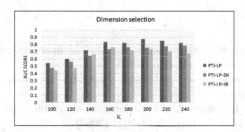

Fig. 10. Impact of dimension K

Impact of Relationship Number. We construct four networks based on same dataset, the AUTH network only contains co-author links, the AUTH-CONF contains co-author links and conference links, the AUTH-TSim network contains co-author links and TopicSim links, and the AUTH-CONF-TSim network contains all type of links. Figure 11 shows that PTI-LP has the highest AUC score on AUTH-CONF-TSIM, it proves the necessity to combing the multi-relational information with temporal dynamics of networks.

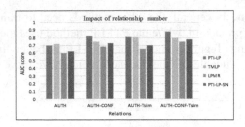

Fig. 11. Impact of relationship number

6 Conclusion

We propose a node embedding method to combine structure information of multiple networks, and propose a link prediction model based on personalized time interval in multi-relational network which can smooth the uneven distribution of information. The experiments show that our method has the best link prediction accuracy. In future, we may model the influence between nodes in link prediction.

Acknowledgment. This work is supported by the National Key R&D Program of China (2018YFB1003404), the National Natural Science Foundation of China (61472070, 61672142, 61602103) and the Program of China Scholarships Council (201806085016).

References

1. Dai, C., Chen, L., Li, B.: Link prediction in multi-relational networks based on relational similarity. Inf. Sci. **394–395**, 198–216 (2017)
2. Newman, M.E.J.: Clustering and preferential attachment in growing networks. Phys. Rev. E **64**(2), 025102 (2001)
3. Liben Nowell, D., Kleinberg, J.: The link-prediction problem for social networks. In: ACM CIKM, Louisiana, USA, pp. 1019–1031 (2003)
4. Yu, W., Cheng, W., Aggarwal, C.C., et al.: Link prediction with spatial and temporal consistency in dynamic networks. In: Twenty-Sixth International Joint Conference on Artificial Intelligence, Melbourne, Australia, pp. 3343–3349 (2017)
5. Li, J., Cheng, K., Wu, L., et al.: Streaming link prediction on dynamic attributed networks. In: ACM WSDM, Marina Del Rey, USA, pp. 369–377 (2018)
6. Ozcan, A., Oguducu, S.G.: Link prediction in evolving heterogeneous networks using the NARX neural networks. Knowl. Inf. Syst. **55**(2), 333–360 (2018)
7. Huang, Z., Mamoulis, N.: Heterogeneous information network embedding for meta path based proximity. CoRR abs/1701.05291 (2017)
8. Li, Y., Li, C., Chen, W.: Research on influence ranking of chinese movie heterogeneous network based on PageRank algorithm. In: Meng, X., Li, R., Wang, K., Niu, B., Wang, X., Zhao, G. (eds.) WISA 2018. LNCS, vol. 11242, pp. 344–356. Springer, Cham (2018). https://doi.org/10.1007/978-3-030-02934-0_32
9. Sett, N., Basu, S., Nandi, S., et al.: Temporal link prediction in multi-relational network. World Wide Web Web Inf. Syst. **21**(2), 395–419 (2018)

10. Liu, L., Li, X., Cheung, W.K., et al.: A structural representation learning for multi-relational networks. In: Proceedings of the Twenty-Sixth International Joint Conference on Artificial Intelligence, Melbourne, Australia, pp. 4047–4053 (2017)
11. Wang, Y., Gemulla, R., Li, H.: On multi-relational link prediction with bilinear models. In: Proceedings of the 32th AAAI Conference on Artificial Intelligence, New Orleans, USA, pp. 4227–4234 (2018)
12. Tang, J., Qu, M., Wang, M., et al.: LINE: large-scale information network embedding. In: International Conference on World Wide Web, Florence, Italy, pp. 1067–1077 (2015)
13. Yu, W., Aggarwal, C.C., Wang, W.: Modeling co-evolution across multiple networks. In: ACM SDM, San Diego, USA, pp. 675–683 (2018)
14. Cao, Z., Wang, L.: Link prediction via subgraph embedding-based convex matrix completion. In: Proceedings of the 32th AAAI Conference on Artificial Intelligence, New Orleans, USA, pp. 2803–2810 (2018)
15. Lemonnier, R., Scaman, K., Kalogeratos, A.: Multivariate hawkes processes for large-scale inference. In: Proceedings of the 31th AAAI Conference on Artificial Intelligence, California, USA, pp. 2168–2174 (2017)

A Unified Relational Storage Scheme for RDF and Property Graphs

Ran Zhang[1,2], Pengkai Liu[1,2], Xiefan Guo[1,2], Sizhuo Li[1,2], and Xin Wang[1,2(✉)]

[1] College of Intelligence and Computing, Tianjin University, Tianjin, China
{gxf_1998,wangx}@tju.edu.cn
[2] Tianjin Key Laboratory of Cognitive Computing and Application,
Tianjin University, Tianjin, China

Abstract. With the advance of Semantic Web and development of Linked Data, the scale of knowledge graphs has surged dramatically. On the one hand, RDF graph is a mainstream data model of the knowledge graph. On the other hand, property graphs are widely accepted in graph databases. How to manage large-scale RDF and property graphs in an interchangeable way has become popular in both academic and industrial communities. Thus we present an effective unified relational storage scheme, that can seamlessly accommodate both RDF and property graphs. Furthermore, we have implemented the storage schema on an open-source graph database to verify its effectiveness. Ultimately, our experimental results show that the proposed unified storage schema for both RDF and property graphs can effectively manage large-scale knowledge graphs, efficiently avoid data redundancy, and achieve high-performance queries.

Keywords: Knowledge graph · RDF · Property graph · Efficient storage

1 Introduction

Knowledge graphs have become the cornerstone of artificial intelligence. The construction and publishing of large-scale knowledge graphs in various domains have posed new challenges on the management of those graphs.

Currently, there are two mainstream data models of knowledge graphs, namely the RDF [8] (Resource Description Framework) model and the property graph model. The former has been standardized by the W3C (World Wide Web Consortium), and the latter has been widely accepted in industrial communities of graph databases. Unlike the relational database communities, however, the two models of knowledge graphs and their query languages have not yet been unified. For RDF graphs, their model has a profound mathematical foundation and relatively complete model characteristics, and with the Linked Data [11] initiative, an increasingly large number of RDF data have been published on

© Springer Nature Switzerland AG 2019
W. Ni et al. (Eds.): WISA 2019, LNCS 11817, pp. 418–429, 2019.
https://doi.org/10.1007/978-3-030-30952-7_41

the Semantic Web; whereas for property graphs, their model has built-in support for properties and several query languages, including Cypher [9], Gremlin [10], and PGQL [12]. Property graphs, which have not been standardized yet, have been widely recognized in industrial communities with the application of graph databases. Due to the hypergraph structure of RDF graphs, it has been demonstrated that RDF graphs are more expressive than property graphs. How to effectively manage both RDF and property graphs in a unified storage schema has become an urgent problem. In this paper, we firstly focus on the integration of RDF and property graphs at the storage layer.

The relational model has increasingly turned mature over several decades. It has concise and universal relational structures, and expresses the operations and constraints of relationships using relational algebra with strict mathematical definitions. Therefore, it can provide a solid theoretical foundation to store RDF and property graphs.

Our contributions can be summarized as follows:

(1) We propose a unified relational storage schema, that can seamlessly accommodate both RDF and property graphs.
(2) We then implement the storage schema on an open-source database, Agens-Graph, to verify its effectiveness and efficiency.
(3) To some extent, we manage RDF and property graphs in an interchangeable way and realize the interoperability between the two models.

The remainder of this paper is organized as follows: Sect. 2 introduces the related work and the formal definitions of RDF and property graphs are given in Sect. 3. The unified storage schema we proposed is illustrated in Sect. 4, the subsequent Sect. 5 describes the implementation on an open-source graph database with the experimental results. Finally, we conclude the paper by discussing future research directions in Sect. 6.

2 Related Work

The knowledge graph data model is based on the graph structure, with vertices representing entities and edges representing the relationships between those entities. This kind of general data representation can naturally depict the extensive connections between things in the real world.

2.1 The RDF Storage Schema

There are two typical approaches to designing RDF data management systems: relational approaches and graph-based approaches [18]. The relational approaches map RDF data to a tabular representation and then execute SPARQL queries on it while the latter approach is graph-based, which model both RDF and the SPARQL query as graphs and execute the query by subgraph matching using homomorphism [16].

Relationship-Based Knowledge Graphs Storage Management. Relational databases are still the most widely used database management system at present, and the storage scheme based on relational database is a main storage method of knowledge graphs data currently [1]. The triple table storage scheme directly stores RDF data; the horizontal table storage scheme [3,7,15] records all predicates and objects of a subject in each row; the property table storage scheme is a subdivision of the horizontal table, and the same subject will be stored in a table, which solves the problem of too many columns in the table; the vertical partitioning storage scheme creates a two-column table for each predicate [2]; the sextuple indexing storage scheme divides all six permutations of a triple into six tables [14]. Last but not least, DB2RDF [6] has been used to improve query performance recent years by creating entity-oriented storage structures that reduce the Cartesian product operations in queries.

Graph-Based Knowledge Graphs Storage Management. The advantage of graph-based approach is that it maintains the original representation of the RDF data as well as it enforces the intended semantics of SPARQL. The disadvantage, however, is that the cost of subgraph matching by graph homomorphism is NP-complete [18]. Systems such as that proposed by Bönström et al. [5], gStore [15,17], and chameleon-db [4] follow this approach.

2.2   The Property Graph Storage Scheme

A property graph is a directed, labeled, and attributed multi-graph. It means that the edges of a property graphs are directed, and both vertices and edges can be labeled and can have any number of properties, and there can be multiple edges between any two nodes [13]. Neo4j[1] is a native graph database that supports transactional applications and graph analytics, and it is currently the most popular property graphs database. Neo4j is also based on a network-oriented model where relations are first-class objects.

At present, the knowledge graph data model and the query language are not unified. The main reason for the surge of relational databases is that it has a precisely defined relational data model and a unified query language SQL. The unified data model and query language not only reduce the development and maintenance costs of the database management system, but also reduce the learning difficulty of users. Therefore, based on the existing work, we propose a unified relational storage scheme for RDF and property graph model.

3   Preliminaries

In this section, we provide the formal definitions of RDF triple, RDF graph, and property graph, which can be the basis for the transformations to relational tables in the document.

---

[1] https://neo4j.com/.

Definition 1 (RDF triple). *Let U, B and L be disjoint sets of URIs, blank nodes and literals, respectively. An RDF triple $(s, p, o) \in (U \cup B) \times U \times (U \cup B \cup L)$ states the fact that the resource s has the relationship p to the resource $o \in U$, or the resource s has the property p with the value $o \in L$, where s is called the subject, p the predicate (or property), and o the object.*

Definition 2 (RDF graph). *A finite set of RDF triples is called an RDF graph. Given an RDF graph T, we use $S(T)$, $P(T)$, and $O(T)$ to denote the set of subjects, predicates, and objects in T, respectively. For a certain subject $s_i \in S(T)$, we refer to the triples with the same subject s_i collectively as the entity s_i, denoted by $\mathsf{Ent}(s_i) = \{t \in T \mid \exists p, o \text{ s.t. } t = (s_i, p, o)\}$.*

We can use RDF Schema (RDFS) to define classes of entities and the relationships between these classes. For example, $(s, \mathsf{rdf:type}, C)$ declares that the entity s is an instance of the class C. Given an RDF graph T, we assume that for each subject $s \in S(T)$ there exists at least a triple $(s, \mathsf{rdf:type}, C) \in \mathsf{Ent}(s)$, denoted by $s \in C$. We believe that this assumption is reasonable since every entity should belong to at least one type in the real world.

Definition 3 (Property graph). *Let L and T be countable sets of node labels and relationship types, respectively [16]. A property graph is a tuple $G = (N, R, src, tgt, l, \lambda, \tau)$ where:*

- *N is a finite subset of N, whose elements are referred to as the nodes of G.*
- *R is a finite subset of R, whose elements are referred to as the relationships of G.*
- *$src: R \to N$ is a function that maps each relationship to its source node.*
- *$tgt: R \to NN$ is a function that maps each relationship to its target node.*
- *$l: (N \cup R) \times K \to V$ is a finite partial function that maps a (node or relationship) identifier and a property key to a value.*
- *$\lambda: N \to 2L$ is a function that maps each node id to a finite (possibly empty) set of labels.*
- *$\tau: R \to T$ is a function that maps each relationship identifier to a relationship type.*

4   The Unified Relational Storage Schema

Originally, we propose a unified relational storage schema for both RDF and property graphs. Then we elaborate on the specific rules for transforming RDF and property graphs into relational tables to effectively realize the storage integration.

4.1   Integration of RDF and Property Graphs in Relational Tables

As the representations of knowledge graph models, RDF and property graphs are relatively independent with expressivity difference, increasing the difficulty of the direct mapping. As shown in Fig. 1, we select the mature relational model as the physical storage model to realize the integration of RDF and property graphs.

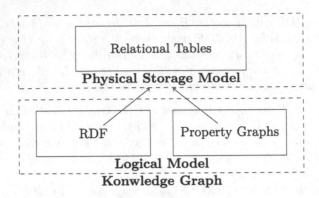

Fig. 1. The unified relational storage schema

4.2   Transforming RDF Graphs into Relational Tables

Since an RDF graph is defined as a finite set of triples, an RDF graph can be mapped into multiple relational tables. Mapping rules for an RDF graph to relational tables will be defined as follows.

RDF triples, by definition, will be formalized as $(s, p, o) \in (U \cup B) \times U \times (U \cup B \cup L)$. For simplicity, the namespace prefix of the resource and predicate URI names will be omitted in this paper (RDF built-in names is not omitted, such as rdf:type). Since the introduction of blank nodes will not make a fundamental change to the RDF data management method, the blank node in the RDF graph will be equated to the URI in this paper.

For three different forms of RDF triples, we define the basic mapping rules for RDF to relational tables as follows:

Rule 1. An RDF triple in the form of $\langle U_1 \rangle \langle$ rdf:type $\rangle \langle U_2 \rangle$, that the predicate of the RDF triple is $\langle$ rdf:type $\rangle$, then it can be expressed as a row with *id* (primary key) and *properties* in relational table U_2.

Rule 2. An RDF triple in the form of $\langle U_1 \rangle \langle U_2 \rangle \langle L \rangle$, that the object of the RDF triple is *literal*, then it can be expressed as a property $\{U_2 : L\}$ in properties of U_1.

Rule 3. An RDF triple in the form of $\langle U_1 \rangle \langle U_2 \rangle \langle U_3 \rangle$, that the subject, the predicate, and the object of the RDF triple are all URI, then it can be expressed as a row with *id* (primary key), *start* that is the foreign key referencing the *id* of U_1, *end* that is the foreign key referencing the *id* of U_3, and *properties* in relational table $U2$.

As shown in Fig. 2, most RDF graphs can be mapped to relational schemata according to the above basic rules.

In particular, the intersection of vertices and edges is not empty in RDF graphs. Specifically, the predicate can also act as the subject or the object of another RDF triple. We then propose a solution to implementation of RDF reification. In the relational schema, we artificially create a relational table

RDF Graph   **Relational Tables**

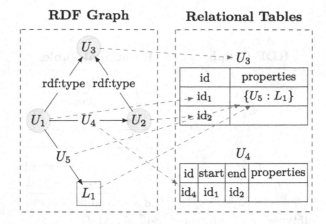

Fig. 2. The basic mapping from RDF graphs to relational tables

called "Edge_Vertex" with column *Vertexid* (primary key), column *Edgeid* that is the foreign key referencing the id of the edge, and column *properties*. The Edge_Vertex table stores edges as vertices to realize following relationships between edges and vertices or between edges and edges. Namely, as presented in Fig. 3, we use the dual storage to reserve the complete information of RDF in the relational model.

4.3 Transforming Property Graphs into Relational Tables

Property graphs also play a considerable role in knowledge graphs. In property graphs, an entity is represented as a vertex. Vertices and edges can have an arbitrary number of properties and can be categorized with labels. Labels are used to gather vertices and edges that have the same category. Furthermore, edges are directionally connected between two vertices, a start vertex and an end vertex.

We explore the transformation from property graphs to relational tables. For vertices and edges, we define the mapping rules for property graphs to relational tables as follows:

Rule 1. Labels can be represented as relational tables within vertices and edges of the same category.

Rule 2. Vertex tables have two columns, namely *id* (primary key) and *properties*.

Rule 3. Edge tables have four columns, namely *id* (primary key), *start* and *end* that are both the foreign keys referencing the *id* of vertex tables, and *properties*.

Rule 4. A vertex or an edge can be expressed as a row of the relational table.

According to the above rules, Fig. 4 visually shows the mapping from property graphs to relational tables.

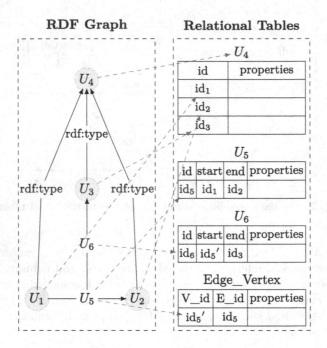

Fig. 3. The complete mapping from RDF graphs to relational tables

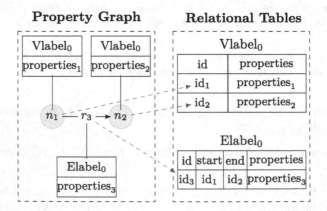

Fig. 4. The mapping from property graphs to relational tables

5   Experiments

We have conducted experiments on synthetic RDF datasets to verify the effectiveness and efficiency of our method. The database is deployed on a desktop computer that has an Intel i54520 CPU with 2 cores of 2.31 GHz, 8 GB memory, 512 GB disk, and 64-bit Centos7.0 as the OS.

We implemented the storage schema on AgensGraph v2.1.1[2]. AgensGraph is a new generation multi-model graph database for the modern complex data environment, that is very robust, fully-featured and ready for enterprise use. AgensGraph both supports relational tables and property graphs, and it has already realized the mapping from property graphs to relational tables. Consequently, RDF graphs are required to be imported into AgensGraph as relational tables.

As shown in Fig. 5, based on the existing storage mechanism, we extended the storage schema to accommodate RDF storage for AgensGraph with no effect to the original storage of relational tables and property graphs. According to the extension, RDF and property graphs can be stored and managed independently and compatibly in AgensGraph.

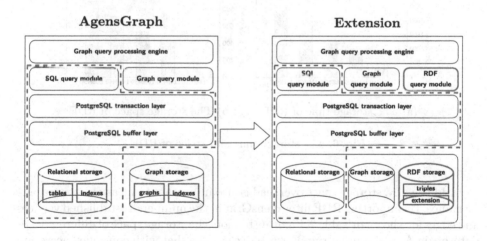

Fig. 5. The complete mapping from RDF graphs to relational tables

We generated five synthetic datasets using the LUBM (Lehigh University Benchmark), which is developed to facilitate the evaluation of Semantic Web repositories in a standard and systematic way, as a test sample imported into AgensGraph. LUBM consists of a university domain ontology, customizable and repeatable synthetic data, a set of test queries, and several performance metrics. The characteristics of each dataset are shown in Table 1.

---

[2] https://bitnine.net/.

Table 1. Characteristics of experimental datasets

Datasets	Sizes	Triple #
LUBM10	168.4M	1,316,700
LUBM20	358.0M	2,781,322
LUBM30	529.6M	4,107,812
LUBM40	709.0M	5,494,144
LUBM50	889.5M	6,888,642

Experiment 1: Storage Performance Analysis. We considered two indicators to evaluate the storage performance, namely storage time and storage space.

Storage time overhead is a significant indicator to evaluate the performance of the storage schema for importing RDF triples. Figure 6(a) shows the storage time to store RDF datasets of different sizes.

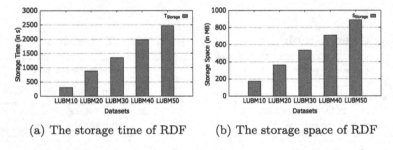

(a) The storage time of RDF   (b) The storage space of RDF

Fig. 6. The storage time and space of RDF

Additionally, storage space overhead is also important to measure storage performance. By importing RDF into AgensGraph, the number of established vertices and edges are shown in Table 2. Figure 6(b) plots the storage space of different RDF datasets in AgensGraph. From Fig. 6(b) we can see that with continued accretion

Table 2. Number of vertex and edge

Datasets	Vertex#	Edge#
LUBM10	219,680	673,602
LUBM20	463,296	1,422,567
LUBM30	684,222	2,101,605
LUBM40	915,519	2,810,798
LUBM50	1,147,136	3,524,142

of the size of RDF data, the storage scheme can significantly reduce the spatial storage of knowledge graphs and the redundancy of data storage.

Experiment 2: Interoperability of RDF and Property Graphs. LUBM provides 14 SPARQL query statements to measure the performance. Therefore, we tested them on AgensGraph to realize the interoperability of RDF and property graphs. For instance, Query number, answer, and query time of LUBM50 are shown in Table 3. From Table 3, we found the storage schema can effectively achieve the interoperability.

Table 3. Query Results of LUBM50

QueryNo.	Answer	Time/ms
Q1	4	8.438
Q2	130	9,099.834
Q3	6	32.527
Q4	34	29.932
Q5	678	105.774
Q6	519,842	17,578.083
Q7	67	56.935
Q8	7,790	5,515.660
Q9	13,639	10,193.622
Q10	4	43.472
Q11	0	25.351
Q12	0	0.561
Q13	0	10.739
Q14	393,730	546.531

Experiment 3: Comparison Between Import Methods. To verify the effectiveness of the storage schema, we compared the unified relational storage schema (Our-Method) with importing RDF graphs as property graphs (Agens-Grpah) on storage time. From the experimental results, as shown in Fig. 7, the storage time and storage space are positively correlated with the size of the datasets. The efficiency of the proposed relational storage schema has increased hundreds of times with the roughly equivalent storage space, which is valid for large-scale RDF storage.

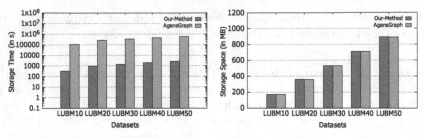

(a) The comparision of storage time   (b) The comparison of storage space

Fig. 7. The comparison between our method and AgensGraph

6   Conclusion and Outlook

In this paper, we have developed a unified relational storage schema of RDF and property graphs. On the one hand, we have solved the large-scale knowledge graph storage problem to some extent. On the other, the proposal of the unified storage schema promotes the integration of two mainstream data models of knowledge graph, playing an important role in the establishment of dominant knowledge graph databases.

The Unified data model not only lowers the development and maintenance cost of database management system, but also reduces the learning difficulty of users. Based on the unified storage schema, a unified query schema of Cypher and SPARQL needs to be proposed to realize a real sense of RDF to property graph interoperability. Therefore, it is an important research direction in the future to develop a unified knowledge graph query language with precise grammar and semantics. Furthermore, the research and development of distributed storage of large-scale knowledge graph data is still in its infancy, and the efficient algorithm of distributed queries is to be improved.

Acknowledgments. This work is supported by the National Natural Science Foundation of China (61572353), the Natural Science Foundation of Tianjin (17JCY-BJC15400), and the National Training Programs of Innovation and Entrepreneurship for Undergraduates (201810056090).

References

1. Wang, X., Zou, L., Wang, C.K., Peng, P.: Research on knowledge graph data management: a survey. Ruan Jian Xue Bao/J. Softw. (2018). (in Chinese)
2. Abadi, D.J., Marcus, A., Madden, S.R., Hollenbach, K.: SW-store: a vertically partitioned dbms for semantic web data management. VLDB J. **18**(2), 385–406 (2009)
3. Alexaki, S., Christophides, V., Karvounarakis, G., Plexousakis, D., Tolle, K.: The ICS-forth RDFSuite: managing voluminous RDF description bases. In: SemWeb (2001)

4. Aluc, G.: Workload matters: a robust approach to physical RDF database design (2015)
5. Bonstrom, V., Hinze, A., Schweppe, H.: Storing RDF as a graph. In: Proceedings of the IEEE/LEOS 3rd International Conference on Numerical Simulation of Semiconductor Optoelectronic Devices (IEEE Cat. No. 03EX726), pp. 27–36. IEEE (2003)
6. Bornea, M.A., Broekstra, J., Kampman, A., Van Harmelen, F.: Building an efficient RDF store over a relational database. In: Proceedings of the 2013 ACM SIGMOD International Conference on Management of Data, pp. 121–132. ACM (2013)
7. Broekstra, J., Kampman, A., van Harmelen, F.: Sesame: a generic architecture for storing and querying RDF and RDF schema. In: Horrocks, I., Hendler, J. (eds.) ISWC 2002. LNCS, vol. 2342, pp. 54–68. Springer, Heidelberg (2002). https://doi.org/10.1007/3-540-48005-6_7
8. Cyganiak, R., Wood, D., Lanthaler, M., Klyne, G., Carroll, J.J., McBride, B.: RDF 1.1 concepts and abstract syntax. W3C recommendation, 25 February 2014
9. Francis, N., et al.: Cypher: an evolving query language for property graphs. In: Proceedings of the 2018 International Conference on Management of Data, pp. 1433–1445. ACM (2018)
10. Khokha, M.K., Hsu, D., Brunet, L.J., Dionne, M.S., Harland, R.M.: Gremlin is the BMP antagonist required for maintenance of Shh and Fgf signals during limb patterning. Nat. Genet. 34(3), 303 (2003)
11. Li, W., Chai, L., Yang, C., Wang, X.: An evolutionary analysis of DBpedia datasets. In: Meng, X., Li, R., Wang, K., Niu, B., Wang, X., Zhao, G. (eds.) WISA 2018. LNCS, vol. 11242, pp. 317–329. Springer, Cham (2018). https://doi.org/10.1007/978-3-030-02934-0_30
12. van Rest, O., Hong, S., Kim, J., Meng, X., Chafi, H.: PGQL: a property graph query language. In: Proceedings of the Fourth International Workshop on Graph Data Management Experiences and Systems, p. 7. ACM (2016)
13. Rodriguez, M.A., Neubauer, P.: Constructions from dots and lines. Bull. Am. Soc. Inf. Sci. Technol. 36(6), 35–41 (2010)
14. Weiss, C., Karras, P., Bernstein, A.: Hexastore: sextuple indexing for semantic web data management. Proc. VLDB Endow. 1(1), 1008–1019 (2008)
15. Zou, L., Mo, J., Chen, L., Özsu, M.T., Zhao, D.: gStore: answering SPARQL queries via subgraph matching. Proc. VLDB Endow. 4(8), 482–493 (2011)
16. Zou, L., Özsu, M.T.: Graph-based RDF data management. Data Sci. Eng. 2(1), 56–70 (2017)
17. Zou, L., Özsu, M.T., Chen, L., Shen, X., Huang, R., Zhao, D.: GStore: a graph-based SPARQL query engine. VLDB J. Int. J. Very Large Data Bases 23(4), 565–590 (2014)
18. Özsu, M.T.: A survey of RDF data management systems. Front. Comput. Sci. 10(3), 1–15 (2016)

Construction Research and Application of Poverty Alleviation Knowledge Graph

Hongyan Yun[1], Ying He[2(✉)], Li Lin[1], Zhenkuan Pan[1],
and Xiuhua Zhang[1]

[1] College of Computer Science & Technology, Qingdao University,
Qingdao 266071, Shandong, China
[2] School of Electronic Information, Qingdao University, Qingdao 266071,
Shandong, China
Yunhy2001@163.com

Abstract. Based on the integration of multi-source data, an approach of domain-specific knowledge graph construction is proposed to guide the construction of a "people-centered" poverty alleviation knowledge graph, and to achieve cross-functional and cross-regional sharing and integration of national basic data resources and public services. Focusing on "precise governance and benefit people service", poverty alleviation ontology is constructed to solve semantic heterogeneity in multiple data sources integration, and provide an upper data schema for poverty alleviation knowledge graph construction. Karma modeling is used to implement semantic mapping between ontology concepts and data, and integrate multi-source heterogeneous data into RDF data. The RDF2Neo4j interpreter is developed to parse RDF data and store RDF data schema based on the graph database Neo4j. Based on visualization technology and natural language processing technology, Poverty Alleviation Knowledge Graph Application System is designed to achieve knowledge graph query and knowledge question answering function, which improved the application value of government data.

Keywords: An approach of knowledge graph construction ·
Poverty alleviation knowledge graph · Neo4j graph storage ·
Bayesian classification · Knowledge question answering

1 Introduction

There is a huge amount data spread across the Web and stored in databases that we can use to build knowledge graph. However exploiting this data to build knowledge graph is difficult due to the heterogeneity of the sources. Knowledge graph is a semantic network and a data structure that can store knowledge, and infer new knowledge to users. Knowledge graph organizes massive Internet data into knowledge network through deep semantic analysis and data mining, searches and displays knowledge intuitively. Knowledge graph has strong knowledge organization ability and semantic processing ability, which provides important guarantee for Internet intelligent search, big data analysis, intelligent question and answer, personalized recommendation and so on [1].

© Springer Nature Switzerland AG 2019
W. Ni et al. (Eds.): WISA 2019, LNCS 11817, pp. 430–442, 2019.
https://doi.org/10.1007/978-3-030-30952-7_42

The data source of general knowledge graph is mainly Internet Web pages, which identify entity data from Web pages and then summarize data patterns. For example, DBpedia uses structure data in Wikipedia entries as data sources, supports 125 languages and covers more than 10 million entities [2]; Zhishi.me uses structured data in Baidu Encyclopedia, Interactive Encyclopedia and Wikipedia as data sources, extracts entities and integrates entity information by structured information extraction method to build a Chinese general knowledge graph [3]. The Knowledge Workshop Laboratory of Fudan University developed and maintained CN-DBpedia, which is a large-scale open Encyclopedia Chinese knowledge graph covering tens of millions of entities and hundreds of millions of relationships. Domain knowledge graph focuses on entities and concepts of knowledge in specific fields or industries, such as agricultural knowledge graph, traditional Chinese medicine knowledge graph, financial knowledge graph, legal knowledge graph, etc. Domain knowledge graph can obtain complete knowledge for analysis and solution of domain problems.

At present, there is a lack of knowledge graph in the field of government big data application and the basic data of many government departments only exists in isolation, which fails to realize the interconnection and interoperability of data. Aiming at the lack of government poverty-alleviation knowledge graph, meeting the needs of precise governance and benefiting the people, this paper builds a "people-centered" poverty-alleviation knowledge graph by integrating national basic information data such as population, corporate enterprises and data resources in the fields of credit and social security. It will make efficient use of open government data on poverty alleviation, integrate isolated data nodes into knowledge base, and provide users with a friendly information service platform for poverty alleviation.

From the perspective of rapid integration of multi-source heterogeneous data, this paper proposes an approach to construct knowledge graph. A crawler system is constructed to obtain data and information related to poverty alleviation from local government poverty-alleviation websites. Fuse data from the poverty alleviation database published by the government to construct the poverty alleviation ontology that provides the upper data model for the construction of poverty-alleviation knowledge graph. Neo4j storage scheme for triple data is developed for structured and semi-structured. Graph storage method effectively improves the efficiency and scalability of data retrieval, and provides an important storage guarantee for further expansion and maintenance of knowledge graph. Based on entity recognition and Bayesian classification, an application system is designed to implement knowledge graph query and intelligent question-answer functions. This application system improves the application value of open government data.

2 An Approach of Building Knowledge Graph

The general knowledge graph covers the knowledge field and the data, and the construction technology covers many fields such as semantic technology, natural semantic processing, machine learning, deep learning, etc. The construction process is complex and the construction cycle is long [4]. Domain knowledge graph covers relatively less domain-specific knowledge and data, which shortens the construction cycle.

We propose an approach to build knowledge graph. We present the techniques using poverty alleviation domain as an example, though the general approach can be applied to other domains. Figure 1 shows the architecture of overall approach to construct knowledge graph for poverty alleviation, which can meet the needs of effective utilization and accurate poverty alleviation of government data in the field of poverty alleviation. The construction process of knowledge graph includes data acquisition, knowledge fusion and knowledge processing.

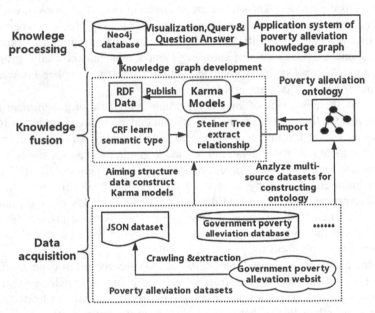

Fig. 1. An approach of Poverty-alleviation knowledge graph construction.

2.1 Data Acquisition

The data in the field of poverty alleviation is mainly composed of two parts: one is the structured data provided by government departments stored in MySQL database, the other requires finding relevant municipal governmental poverty alleviation websites and extracting the required information from those pages.

Based on Java crawler framework named WebMagic (https://gitee.com/flashsword20/webmagic), a crawler system is designed to support crawling at scale. The crawler system mainly includes downloader, page parser, scheduler and pipe component. The four elements are organized in the Spider container, and regular expressions are written to crawl the required website data through interactive and process execution. The workflow of this crawler system includes the following five steps. ① The initial URL of the reptile is the Shandong Provincial Poverty Alleviation Supply and Demand Platform (http://ax.sdfp.gov.cn/) and Chongqing Poverty Alleviation Open Website (http://www.cqfp.gov.cn/channels). ② The downloader uses Apache HttpClient as the download component to initiate a request for the initial URL to obtain the Web page object Page. ③ Page parser uses process() method in Processor to parse web pages, uses Jsoup to parse HTML pages into DOM trees, extracts useful information resources and discovers new seed URLs through CSS Selector. Aiming at government's poverty alleviation website, take the poor householders as the center to crawl the structured information in the website. ④ The scheduler manages the URLs to be fetched and delete-duplicate operations. ⑤ Pipeline is responsible for the processing of extraction results, mainly including saving data to files or databases.

After data extraction, we convert pages related to poor households into structured knowledge for storage. The extracted relational instances are defined in the following forms: <conceptual instance1, relational concepts, conceptual instance2>. The basic information of the poor households can be expressed in the form of multiple triples: <31185489099, reason of poverty, "illness">, <31185489099, detailed address, "Wenshangji Town, Chengwu County, Heze City>, <31185489099, helper, "Yuan Cunkui"> … etc.

2.2 Knowledge Fusion

Knowledge fusion solves the problem of data normalization and multi-source heterogeneous data integration. Construct domain ontology and Karma models (http://isi.edu/integration/karma/) to quickly integrate multi-source heterogeneous data. The integration scheme of multi-source heterogeneous data based on ontology and Karma modeling has been validated effectively in the fields of food security [5] and regional armed conflict event data (ACLED) integration [6]. Karma is an ontology-based data integration tool, which supports to import multiple data sources, including spreadsheets, relational databases, XML, CSV, JSON, etc. Through data cleaning and standardization, Karma builds a model or semantic description for each data source to integrate data across multiple data sources. Karma uses conditional Random Field (CRF) model to learn and recognize semantic types to complete mapping between data and ontology concepts. Under the constraint of ontology, Karma uses Steiner Tree algorithm to extract the relationship between data in data sources. Users can interact with the system to adjust the automatically generated model. In this process, users can convert data format according to their needs, standardize or reconstruct the data expressed in different formats. In order to support rapid modeling of big data, small batch data can be imported to build Karma model. Karma supports to export R2RML (RDB to RDF Mapping Language) model and then directly apply the model to the complete big data set [7], which can be published into a unified RDF data, so as to achieve the purpose of rapid modeling. Poverty alleviation ontology constructing will be described in Sect. 3. Karma modeling will not be described in this paper.

2.3 Knowledge Processing

This paper builds an interpreter based on the RDF data in poverty alleviation field, and stores RDF data into Neo4j graph database (introduced in Sect. 4 of this paper), so as to improve the expansibility, maintainability and the ability of visual analysis and processing of the knowledge graph. This paper will describe design and implementation of application system of poverty-alleviation knowledge graph in Sect. 5.

3 Poverty Alleviation Ontology

Ontology formally describes the semantic relationship between concepts in the domain, which provides a unified perspective for treating domain knowledge and solving domain problems [8]. Domain ontology can provide the upper data model for knowledge graph, and it is an important part of knowledge graph. By analyzing poverty alleviation data sets of different municipal government departments, main class, data property and object property are extracted to construct poverty alleviation ontology. Some extracted triples are shown as follows. By analyzing extracted triples, define main classes and properties of poverty alleviation ontology are shown in Table 1.

Table 1. Defined main classes and properties of Poverty-alleviation ontology.

Class/Attribute name	Class/Property	Meaning
Helper	Class	Helper people
Poor households	Class	Poor households
Family	Class	Poor family
GovHelp	Class	Support project
has_helper	ObjectProperty	
has_govhelp	ObjectProperty	
has_family	ObjectProperty	
has_name	DataProperty	Name of poor household
has_id	DataProperty	ID of poor household
has_address	DataProperty	Address of poor household
has_phone	DataProperty	Phone of poor household
has_starttime	DataProperty	Start time of support
has_endtime	DataProperty	End time of support
has_proname	DataProperty	Name of support project
has_protype	DataProperty	Type of support project

{<poor household, has a family, family information>; <poor household, has a householder, name>; <poor household, has family address, address information>; <poor household, has phone, phone number>; <poor household, has a helper, helper name>; <poor household, has a support project, support project name>}

{<helper, start time to support, start time> ;<helper, end time to support, end time>}

{<GovHelp, has a name of support project, name of support project>; <GovHelp, has a type of support project, type of support project>; <GovHelp, has information of support project, information of support project>}

{<family, has family introduce, introduce information>; <family, has a family ID, family ID>; <family, has number of students, number>; <family, has information of annual income, annual income>; <family, has house information, house information>}

Gradually refine concepts and its relationship, construct poverty alleviation ontology is shown in Fig. 2. Data property in the graph is identified by dotted lines with arrows, and object property is identified by solid lines with arrows.

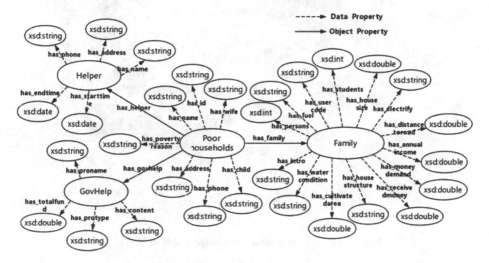

Fig. 2. Poverty alleviation ontology.

4 Knowledge Graph Building and Storage

Poverty alleviation ontology defines the terminology for representing nodes and edges of the knowledge graph. The key decision in the design of knowledge graph is the desire to record provenance data for each node and edge of the graph [9]. Neo4j is one of the current mainstream graph databases and has the characteristics of high performance, high reliability, scalability and transaction support. Neo4j follows the data model of attribute graph, which solves the problem of performance degradation when a large number of connection queries are made in relational databases. The graph-based traversal algorithm enables it to traverse nodes and edges at the same speed. Its traversal speed has nothing to do with the magnitude of graphics, so it can also perform well in big data.

In the view of the complexity and dynamics of data relations and the extension and maintenance of knowledge graph, this paper uses Neo4j graph database to store knowledge graph. The representation of knowledge graph is a unified set of triples. Poverty-alleviation knowledge graph is imported into Neo4j as shown in Fig. 3.

The RDF2Neo4j interpreter is designed to import structured data into Neo4j graph database. The specific steps are as follows: ① Export multiple tables from MySQL database into CSV format. ② Using Cypher statement to load several database tables, each table is regarded as a kind of data type, each record as different instance data of this type, encapsulating corresponding attributes and attribute values according to requirements, creating nodes in the knowledge graph. ③ According to the foreign key

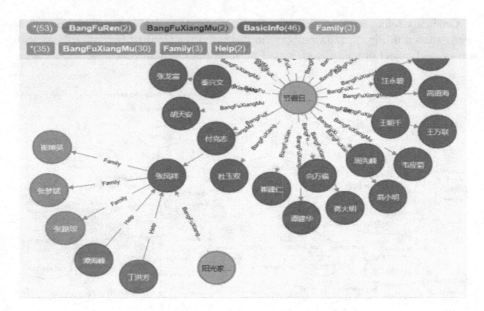

Fig. 3. Poverty-alleviation knowledge graph in Neo4j.

association between different tables, it maps to the relation type of different types of nodes, and creates relations in the knowledge graph.

5   Poverty Alleviation Knowledge Graph Application System

Data visualization technology presents data intuitively and interactively. Combine knowledge graph with visualization can accurately analyze and display data [10]. The poverty-alleviation knowledge graph application system has four-layer architecture, including data layer, data query layer, business logic layer and presentation layer, as shown in Fig. 4. The data layer stores poverty alleviation data in Neo4j graph database. Data query layer mainly uses Spring Boot framework to build micro services, provides RESTful interface, uses Spring Data module to operate Neo4j graph database, and writes Cypher statements to interact with Neo4j graph database. Business logic layer mainly deals with data, and further encapsulates returned data by calling data query layer to complete data statistical analysis and format specification. The business logic is compiled according to functions that need to be completed, and the encapsulated data is transmitted to the presentation layer. The data exchange format is JSON. Presentation layer passes received data to the front-end for rendering, and uses Echarts components and HTML5 to visualize front-end pages of the system. The application system of poverty alleviation knowledge graph has designed and implemented functions including display, query (inquiry based on the names of poor households, projects and relationships among poor households) and intelligent knowledge question and answer for users. This section elaborates on development of knowledge graph query and question-and-answer function.

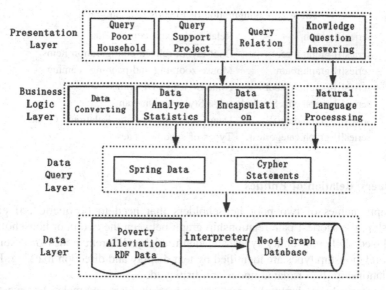

Fig. 4. P architecture of Poverty-alleviation knowledge graph application system.

5.1 Query Names of Poor Households and Support Projects

Users input the name of the poor household head and click on Query button to display a one-dimensional relationship. Different entity types are identified by different colors nodes, and different relation types are identified by text display and direction identification. The main types of relationships are support projects, help people, family information and so on. By using support project as the keyword to query knowledge graph, the multi-dimensional relationship of support project entity is returned, which can clearly show the relationship between different poor households. The steps to implement the query function are as follows.

Use ECharts to Build the Oriented Layout. Echarts is a powerful, interactive charting and visualization library offering an easy way to add intuitive, interactive and highly customizable charts to your need (http://echarts.apache.org/). ECharts defines a variety of icon styles and functions. According to users' actual needs, developers use corresponding JS code to send requests to the background and render returned data. In HTML, the ECharts container is created first. The option object is set in the container and the related code of force-oriented graph is introduced. The title, type, icon, node style and callback function of the graph are specified. The main configurations of ECharts used in knowledge graph visualization design are shown in Table 2.

Send Requests. The front-end uses AJAX in JQuery library to send asynchronous requests. It can refresh local data without refreshing web pages, parse and transmit returned data. Obtain query conditions entered by users and send POST requests to the background, call different methods to query, render returned JSON data in a specific format. For example, user query one-dimensional relationship with "a poor household as the center" and multi-dimensional relationship with "a poverty alleviation project as the center".

Table 2. Configuration of Echarts relation graph.

series[i]-graph	Relational graph components
series[i]-graph.force	Boot layout-related configuration items
series[i]-graph.roam	Mouse zooming and panning roaming
series[i]-graph.symbol	Node Styles in Diagrams
series[i]-graph.edgeSymbol	Edge Styles in Diagrams
series[i]-graph.data[i]	Data Format of Specific Relational Graph
series[i]-graph.categories[i]	Types of specific nodes

5.2 Query Relation of Entities

Users input names of two poor householders that need to be queried. If exists a relationship, the shortest path relationship graph between the two poor households will be displayed. Different entity types are identified by different nodes of colors, and different relationship types are identified by text display and direction marking. If there is no relationship, two isolated nodes are displayed.

The system obtains inputted names of two householders and makes a request to the server. The business logic layer of the server receives name for verification. After verification, call the method of data query layer to transfer into parameters. The shortest path function shortestPath() of Neo4j graph database is used to query the path between two poor householders and return result set. The system parses and processes the result set, transfers JSON data encapsulated in a specific format to ECharts component rendering, and finally displays it in graph.

5.3 Knowledge Question Answering Design

Intelligent Knowledge Question Answering (KQA) refers to semantic understanding and parsing of a given natural language question, and then querying the knowledge base to get the answer. For a natural language problem, the challenge is how to map question to knowledge base (corresponding data query statement representation) and convert natural language questions into understandable expressions in databases.

The main processes of question processing include question segmentation, information extraction and text classification. We design a knowledge question answering system as shown in Fig. 5, which consists of constructing question training set, pre-processing natural language questions input by users, including word segmentation, named entity recognition, using Naive Bayesian classification method to classifiers question, putting the results of question text preprocessing into the classifier model, and calculating the short text classification of question sentences. Execute queries corresponding to categories to get answers from knowledge graph queries and implementation process includes the following four steps.

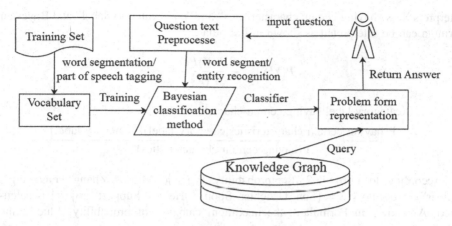

Fig. 5. Architecture of question answering system.

Focusing on the "people-centered" knowledge graph of poverty alleviation, the training set of poverty alleviation knowledge related issues is constructed. Using Java open source Chinese word segment tool named Ansj (https://github.com/NLPChina/ansj_seg), question training set is segmented and part-of-speech tagged to generate a vocabulary set. User-defined dictionary can be loaded at the same time to make word segmentation more accurate.

Question Text Preprocessing. Ansj is used in word segment. Questions are divided into fine-grained word forms to express, identify names, determine central word of the questions, and ensure that the answers to the questions are obtained for central word.

Abstract Classification of Problems. Naive Bayesian method is simple to implement and has high prediction efficiency. It is widely used in the field of text classification. In order to classify question sentences, we use Bayesian classifier to extract feature vectors of text sentences through training set and One-hot to calculate sentence similarity, so that classifier can effectively deal with natural language question sentences and accurately identify the intention of user questions.

According to Bayesian Classification Calculation to Get Question Categories. According to setting query rules, query sentences corresponding to the categories is executed to match knowledge points from knowledge graph to answer user's questions. Take question "张凤祥的帮扶人是谁 (Who is Zhang Fengxiang's helper?)" as an example, the QA process includes the following four main steps. ① Construct training set of question in poverty alleviation field, and use Ansi toolkit to participle and part-of-speech annotate training set. A vocabulary set of {之子 (son), 之女 (daughter), 配偶 (spouse), 帮扶人 (helper), 帮扶项目 (GovHelp)} is constructed. ② Use Ansj toolkit to participle question text, we can get {"张凤祥","的","帮扶人","是","谁"} ({"Zhang Fengxiang's", "Helper", "is", "who"})and confirm that the central word of the question is "张凤祥 (Zhang Fengxiang)" through person name entity recognition. ③ According to Bayesian classification, the most probabilistic category is the category of the current problem. From the perspective of machine learning, Bayesian formula (Formula 1) interprets X as "having certain characteristics" and Y as "category label" and Bayesian formula can be interpreted as shown in Formula 2.

$$P(Y|X) = \frac{P(X|Y)P(Y)}{P(X)} \tag{1}$$

P(category label|having certain characteristics)

$$= \frac{P(\text{having certain characteristics}|\text{category label})P(\text{category label})}{P(\text{having certain characteristics})} \tag{2}$$

According to Formula 2, the probability P (S |("Who is Zhang Fengxiang's Helper") of category S = {Son, Daughter, Spouse, Helper, Support Project} is calculated. As shown in Formula 3, the maximum category of probability value is the category of question.

$$P\big(S\big|(\text{"Who is Zhang Fengxiang's Helper"})\big)$$
$$= \frac{P(\text{"Who"}|S) * P(\text{"is"}|S) * P(\text{"Zhang Fengxiang's"}|S) * P(\text{"Helper"}|S) * P(S)}{P(\text{"Who is Zhang Fengxiang's Helper"})} \tag{3}$$

According to the calculation, the problem category S = {帮扶人 (Helper)} is obtained, and the knowledge storage ("张凤祥 (Zhang Fengxiang)", Helper, "谭海峰 (Tan Haifeng))" and ("张凤祥 (Zhang Fengxiang)", Helper, "丁洪芳 (Ding Hongfang))" in the knowledge map of poverty alleviation is matched and queried. The exact

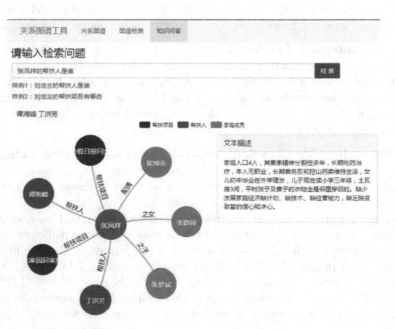

Fig. 6. Knowledge question answering query.

answer to the question is {谭海峰 (Tan Haifeng), 丁洪芳 (Ding Hongfang)} through the query sentence. This question is searched in the application system of poverty alleviation knowledge graph as shown in Fig. 6.

6 Conclusion

In this paper, a knowledge graph construction approach is proposed. According to this approach, poverty-alleviation knowledge graph is constructed. This paper analyzes poverty alleviation data to construct poverty alleviation ontology to solve the problem of semantic heterogeneity and provide the upper data model for the construction of poverty-alleviation knowledge graph. Construct Karma models to integrate multi-source data and publish unified RDF data. The RDF2Neo4j interpreter is designed to parse and store RDF data into Neo4j. It provides important data support for big data analysis and government precise poverty alleviation decision-making and provides storage guarantee for the expansion and maintenance of knowledge graph.

Based on Spring Boot framework, visualization technology, natural language processing technology and Bayesian classification algorithm, poverty alleviation knowledge graph application system is designed, which provides users with knowledge graph query and intelligent knowledge question-and-answer functions.

Next, we need to collect poverty alleviation data of different regions, so that the poverty-alleviation knowledge graph application system can provide more accurate reference value in the analysis of poverty alleviation data and the government's precise poverty alleviation decision-making.

Acknowledgment. This work was supported by National Key Research &Development Program of China (No. 2016YFB1001103).

References

1. Juanzi, L.I., Lei, H.O.U.: Reviews on knowledge graph research. J. Shanxi Univ. (Nat. Sci. Ed.) **40**(3), 454–459 (2017)
2. Li, W., Chai, L., Yang, C., Wang, X.: An evolutionary analysis of DBpedia datasets. In: Meng, X., Li, R., Wang, K., Niu, B., Wang, X., Zhao, G. (eds.) WISA 2018. LNCS, vol. 11242, pp. 317–329. Springer, Cham (2018). https://doi.org/10.1007/978-3-030-02934-0_30
3. Ruan, T., Sun, C.-l., Wang, H-f., Fang, Z.-j.: Construction of traditional Chinese medicine knowledge graph and its application. J. Med. Informatics **37**(4), 8–13 (2016)
4. Shijia, E., Lin, P., Xiang, Y.: Automatical construction of Chinese knowledge graph system. J. Comput. Appl. **36**(4), 992–996, 1001 (2016)
5. Yun, H., Xu, J., Knoblock, C.A., Xu, R.: Research and application of multi-source data integration based on ontology. Int. J. u- e- Serv. Sci. Technol. **9**(9), 75–88 (2016)
6. Yun, H.-y., Huang, C., Yu, X.-y., Sui, Y., Hu, G.: Exploiting semantics for conflict event data integration. J. Qingdao Univ. (Nat. Sci. Ed.) **29**(3), 47–52 (2017)
7. Knoblock, C.A., Szekely, P.: Exploiting semantics for big data integration. AI Mag. (S0738-4602) **36**(1), 25–38 (2015)

8. Yun, H., Xu, J., Guo, Z., Wei, X.: Modeling of marine ecology ontology. J. Comput. Appl. **34**(4), 1105–1108 (2015)
9. Szekely, P., et al.: Building and using a knowledge graph to combat human trafficking. In: Arenas, M., et al. (eds.) ISWC 2015. LNCS, vol. 9367, pp. 205–221. Springer, Cham (2015). https://doi.org/10.1007/978-3-319-25010-6_12
10. Zhou, Y., Zhou, M., Wang, X., Huang, Y.: Design and implementation of historical fig.s knowledge graph visualization system. J. Syst. Simul. **28**(10), 2560–2566 (2016)

Graph Data Retrieval Algorithm
for Knowledge Fragmentation

Wang Jingbin and Lin Jing[✉]

College of Mathematics and Computer Science, Fuzhou University,
Fuzhou 350116, China
527564460@qq.com

Abstract. In this era of big data, data are diversified, strongly connected, fragmented, dynamic, and combined with dynamic knowledge fragments to optimize the distributed storage of graphs and enable fast and efficient knowledge graph query problems. Presently, the distributed storage scheme of graph data has a large number of hop accesses between partitions, which leads to a long retrieval response time and is not conducive to fragment knowledge expansion. According to the characteristics of real-time inflow knowledge fragments and the storage structure and principles of graph databases, the Metis+ algorithm is proposed. The label graph is used as the initial initialization segmentation graph, and it is roughened to reduce the cutting of the large-weight edge. The weighted LND algorithm is proposed to run the balancing strategy for storage and assign the similar nodes and closely related nodes to the same partition to the greatest extent, which minimizes jump accesses between the partitions during retrieval.

Keywords: Knowledge graph · Graph database · Label graph ·
Knowledge fragmentation

1 Introduction

Presently, the relationship between the graph structure of strong connections and the dynamics of data puts an increasingly high load on storage technology [1]. The succession of problems accompanying the processing of massive amounts of data has prompted the development of large-scale knowledge graphs [2], and with their powerful semantic processing capabilities, new solutions to the problems have been made possible [3]. Combined with the structural characteristics of dynamic fragmentation knowledge, this study focuses on the advantages of a graph database in dealing with strong relational data and fuses loose data fragments to establish a knowledge graph network with strong links.

The core technology of distributed data storage is graph partitioning, which divides the graph into several subgraphs and then assigns these subgraphs to different computing nodes [4]. A large number of domestic and foreign experts have researched and improved on the partitioning methods. Combining the characteristics of the resource description framework (RDF) [5] graph data structure in the Semantic Web and parallel computing in a distributed environment, EAGRE [6] compresses the RDF data graph

W. Ni et al. (Eds.): WISA 2019, LNCS 11817, pp. 443–448, 2019.
https://doi.org/10.1007/978-3-030-30952-7_43

into an entity graph and uses the METIS algorithm to divide it. However, because the current distributed storage method of graph data is mostly based on static horizontal segmentation of the file, a query on it may require a large number of jump accesses between the cluster partitions when the graph is traversed, affecting the query performance.

This study aims to minimize the jump access between partitions, achieve an optimal storage to enable fast queries, improve the overall retrieval efficiency, and complete the real-time distributed storage of the dynamic graph database. For the demonstration, we constructed an urban safety knowledge graph in this study, taking into consideration the characteristics of the graph data and the load balancing requirements in distributed clusters. We applied the Metis+ algorithm to optimize the distributed storage process and the segmentation and storage of the original graph data. We then performed a balance strategy to store the real-time inflow of knowledge fragments. Finally, the effectiveness of the storage and retrieval algorithm was verified by experiments.

2 Proposed Method

This paper describes the application of the Metis+ algorithm on the constructed urban safety knowledge graph, which is mainly divided into three phases: (1) RDF data graph to graph database mapping; (2) graph data distributed partitioning; and (3) distributed dynamic knowledge fragment storage. The distributed partitioning of the graph data is further divided into two parts: the roughening combined with heavy edge matching (HEM) [7], an edge fusion algorithm, and the segmentation algorithm combined with the weighted leveled nested dissection (LND) algorithm.

2.1 Graph Roughening Combined with HEM Edge Fusion Algorithm

Suppose there are k partitions in the Neo4j [8] distributed cluster. The storage capacity of each partition is M, and thus, the total cluster capacity is kM. $P = |P(1), P(2), ..., P(k)|$ is the sum of all current partition storage states, and $|P(i)|$ indicates the total number of nodes in the partition with subscript i ($1 \leq i \leq k$).

When the graph $G_i = (V_i, E_i)$ is roughened to the next level graph $G_{i+1} = (V_{i+1}, E_{i+1})$, the matching is performed by selecting a larger weight, which can reduce more weights in the roughening graph. The method proceeds to find the maximum matching of edge weights, that is, to find the vertex V in all adjacent unmatched vertices of the vertex U to maximize the weight of the edge e_{uv}, and the algorithm complexity of the method is $O(|E|)$. Figure 1 shows an example where the label graph GL is initialized to the weighted undirected graph GL_0.

In Fig. 1, the left part shows a partial label graph GL, and the right side shows a weighted undirected graph GL_0. The total number of instance nodes is the weight $W(V_i)$ of the node V_i in the weighted undirected graph, and the total number of out-degrees and in-degrees of instances between the tags V_i and V_j is the weight $W(e_{ij})$ of the edge e_{ij} in the weighted undirected graph. The steps of the HEM edge fusion algorithm for the graph $GL_0 = (V_0, E_0)$ are as follows: (1) Fusion operation is performed on the vertices: found $max(W(e_{ij}))$ in all vertices of GL_0. At this time, the vertex

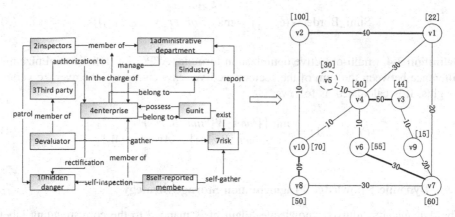

Fig. 1. GL conversion to GL_0

V_i, V_j are merged into a new vertex V_{ij}. (2) When the vertices are merged, the vertex weight is transformed as: $W(V_{ij}) = W(V_i) + W(V_j)$. (3) The edge e_{ij} connecting the vertices V_i, V_j is deleted. (4) Given the threshold θ, the vertex fusion operation is iteratively performed until $|V_m| < \theta$.

2.2 Multi-objective Weighted LND Segmentation Algorithm

After the vertexs roughening process, the original graph $GL_0 = (V_0, E_0)$ is coarsened to $GL_m = (V_m, E_m)$ by k steps. Because the hierarchical nesting partitioning LND algorithm does not consider the weight of the vertices, this study hopes that the total weight of the vertices of the graph can be divided into k partitions as evenly as possible when k-way partitioning is performed on the graph GL_m and the sum of the boundary weights in the sub-area is as large as possible.

Definition 3-1. (average vertex weight, Average_W) represents the vertex weight ideally assigned to each partition. It is calculated as follows:

$$Average\_W = [\sum_{i=0}^{n} W(V_{m\_i})]/k \tag{1}$$

Definition 3-2. (sum of vertex weights, Sum_W(V_i)) represents the sum of all vertex weights from 0 to label i (i $\geq$ 0). It is calculated as follows:

$$Sum\_W(V_i) = \sum_{k=0}^{i} W(V_k) \tag{2}$$

Definition 3-3 (the sum of the maximum edge weights, Sum_BorderW($e_{i,i+1}$)): To obtain the minimum cut edge, when the next hop label from the label i is selected as the vertex of i + 1, the accumulation with the largest edge weight should be selected first. It is calculated as follows:

$$\text{Sum\_BorderW}(e_{i,i+1}) = \max(Border\_W(e_{i,i+1\_j})) \tag{3}$$

Definition 3-4 (multi-objective optimization formula, APP(i, i + 1)): To minimize the difference between the sum of the accumulated vertex weights and the average vertex weights. It is calculated as follows:

$$APP(i, i+1) = \frac{\min\{\{Sum\_W[Sum\_BorderW(e_{i,i+1})]}{+ Sum\_W(V_i)\} - Average\_W\}} \tag{4}$$

2.3 Dynamic Knowledge Fragmentation Storage Strategy

The dynamically influxed knowledge fragment is mapped to the corresponding label set to determine if there is a corresponding label in the k partitions of the distributed cluster. If so, the knowledge fragments are stored in the partition corresponding to the label. Otherwise, the balancing policy for storage is implemented.

Definition 3-5 (Balance strategy): min(|P(i)|) is chosen. If there are multiple partitions that meet the requirements, one of them is randomly selected. The partition number index is returned according to the following formula.

$$Index = random(\{i|\min(|P(i)|), i \in |k|\}) \tag{5}$$

3 Experimental Settings and Result

The experiment used the open domain dataset provided by OpenKG.CN. This study used an urban area for the relevant resource sets. The basic parameters of the dataset are described in Table 1 and the search examples used in the experiment are shown in Table 2. The dataset covers urban air quality, meteorology, related diseases, traffic safety, risk hazards, and many other aspects. The experiment proves that $|V_m| < 100$ is a standard value suitable for ending the roughening process, so the threshold value $\theta = 100$ is selected for the edge fusion algorithm. After the dataset training, the key relationship contribution coefficients are selected as $\alpha = 0.7$, $\beta = 0.3$.

Table 1. Basic parameters of urban datasets

Dataset name	Document size (GB)	Pattern triples	Entity triples (10,000)	Classes	Attributes
Geography	6.20	186	4845	43	51
Meteorological	10.6	186	7596	43	51
Traffic	1.04	487	1253	18	43
Company	3.02	7768	2194	573	2355

Table 2. Search examples

Search	Keyword collection
Q1	世和, 重工业, 注塑机械, 起重机械, 环境污染
Q2	触电, 坍塌, 化学品泄露, 高风险, 中等风险
Q3	化工, 易燃性, 泄露, fire, 爆炸, 高风险, 中等风险
Q4	Aristotle, petrol station, wharf, thermal injury, 稳定性差, 中等风险
Q5	写字楼, 配电箱, 燃气具, 市安全监管局, 市消防监管局, 无防护, 飞溅物, 室内给排水不良, 低风险

This paper presents a comparative experiment, where the Metis+ algorithm was compared with file horizontal segmentation and the Metis algorithm on the keyword set scale Q1–Q5 with selected data sizes of 3 GB and 14 GB. To eliminate the influence of accidental factors, 10 results were retrieved for each example, and the highest and lowest values were removed to obtain the search average.

It can be seen from Figs. 2 and 3 that when the dataset size and the keyword set size are the same, the average search response time of the Metis+ algorithm is the least. When the data size is 3 GB, the average retrieval response time of the Metis algorithm is 1.12 times that of Metis+, and the file level division is 1.47 times that of Metis+. For a data size of 14 GB, the Metis algorithm is 1.7 times slower than Metis+ and file level division is nearly 2.5 times that of Metis+.

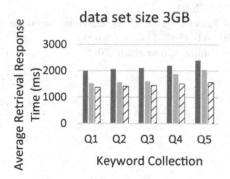

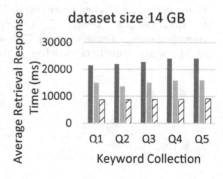

Fig. 2. Comparison of average retrieval response time for 3 GB data scale

Fig. 3. Comparison of average retrieval response time for 14 GB data scale

When the dataset sizes are the same, the average search response time of the Metis algorithm for Q5 is nearly 1.3 times that of Q1 and the horizontal file division is nearly triple. The Metis+ algorithm used in this study is only doubled. It can be seen from the experiment that Metis+ shows a slower rise in the average retrieval response time as the keyword set size is gradually increased.

4 Conclusion

The algorithm not only realizes the maximum allocation of the same type of nodes and closely related nodes to the same partition, but also minimizes the jump access between partitions during retrieval and improves the retrieval efficiency. In future research, we propose to improve the algorithm by considering the correlation between the semantics in the distributed storage stage of the graph data.

References

1. Pujara, J., Miao, H., Getoor, L., Cohen, W.: Knowledge graph identification. In: Alani, H., et al. (eds.) ISWC 2013. LNCS, vol. 8218, pp. 542–557. Springer, Heidelberg (2013). https://doi.org/10.1007/978-3-642-41335-3_34
2. Li, J.Z., Hou, L.: Reviews on knowledge graph research. J. Shanxi Univ. (Nat. Sci. Ed.) **40**(3), 454–459 (2017). (in Chinese)
3. Cai, D., Hou, D., Qi, Y., Yan, J., Lu, Y.: A distributed rule engine for streaming big data. In: Meng, X., Li, R., Wang, K., Niu, B., Wang, X., Zhao, G. (eds.) WISA 2018. LNCS, vol. 11242, pp. 123–130. Springer, Cham (2018). https://doi.org/10.1007/978-3-030-02934-0_12
4. Lasalle, D., Karypis, G.: A parallel hill-climbing refinement algorithm for graph partitioning. In: 45th International Conference on Parallel Processing (ICPP), pp. 236–241 (2016)
5. Leng, Y., Chen, Z., Zhong, F.: BRDPHHC: a balance RDF data partitioning algorithm based on hybrid hierarchical clustering. In: IEEE 12th International Conference on Embedded Software and Systems (ICESS), pp. 1755–1760 (2015)
6. Inokuchi, A., Washio, T., Motoda, H.: Complete mining of frequent patterns from graphs: mining graph data. Machin. Learn. **50**(3), 321–354 (2003)
7. Sun, L.Y., Leng, M., Deng, X.C.: Core-sorted heavy-edge matching algorithm based on compressed storage format of graph. CEA **47**(10), 41–45 (2011). (in Chinese)
8. Lal, M.: Neo4j Graph Data Modeling. Packt Publishing, Birmingham (2015)

A Method of Link Prediction Using Meta Path and Attribute Information

Zhang Yu$^{(\boxtimes)}$, Li Feng, Gao Kening, and Yu Ge

School of Computer Science and Engineering, Northeastern University,
Shenyang 110189, China
{zhangyu,lifeng,gkn,yuge}@mail.neu.edu.cn

Abstract. Heterogeneous information networks (HIN) contain different types of nodes and edges. Predicting the connection between nodes in HIN is a non-trivial problem. Meta paths connect multiple types of nodes through a set of relationships, and are used to describe the different semantics of connections between different types of nodes in HIN. Although several similarity measures based on meta paths have been proposed, the challenge of how to use the measures under different paths to predict links remains open. Besides, the attribute information of nodes and edges in HIN can also be used for link prediction. In this paper, we propose a framework that combines similarity measures of meta path with other attribute information, and formulate a supervised learning task to find the optimal parameters. Experiments on a real data set show that the method has good performance in the problem of link prediction.

Keywords: Link prediction · Meta path ·
Heterogeneous information network

1 Introduction

Link prediction is an important task in network analysis. It has mainly focused on predicting missing links. Most conventional methods of link prediction are designed for homogeneous networks [1,2]. In recent years, a considerable amount of attention has been devoted to the heterogeneous information networks [3,4] (HIN). The types of nodes and links of HIN are different. Link prediction on HIN is more complex, for the methods for homogeneous networks cannot be directly applied to HIN. The nodes and links in HIN have different semantics, which can be explored to reveal subtle relations among nodes. The meta path [3], connecting two objects through a sequence of relations between object types, is widely used to exploit rich semantic information. Although many meta path based methods have been proposed in the past couple of years, how to combine the similarity measured by different meta path is still an open issue. Moreover, many studies have shown that the attribute information, like node attributes or edge attributes, can improve the accuracy of link prediction effectively. We also focus on how to use the rich attribute information for link prediction in the paper.

© Springer Nature Switzerland AG 2019
W. Ni et al. (Eds.): WISA 2019, LNCS 11817, pp. 449–454, 2019.
https://doi.org/10.1007/978-3-030-30952-7_44

2   Problem Definition and Method

First we present some preliminary knowledge.

Definition 1. *Network Schema*. *The network schema is a meta template for a HIN, denoted as $T_G = (\mathcal{A}, \mathcal{R})$.*

Network schema can show the types of nodes and relations of a HIN clearly. Figure 1 shows the network schema of douban, which is a dataset we used in this paper. There are 6 different types of nodes in this network, including User (U), Group (G), Movie (M), Actor (A), Director (D) and Movie Type (T).

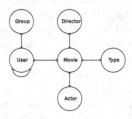

Fig. 1. The network schema of Douban network.

Definition 2. *meta path*. *A meta path P is a path defined on the network schema, denoted in the form $A_1 \xrightarrow{R_1} A_2 \xrightarrow{R_2} \dots \xrightarrow{R_l} A_{l+1}$, which defines a path template from A_1 to A_{l+1}.*

Given a meta path, several similarity measures can be defined for a pair of nodes, such as Path Count, which is the number of path instances between two nodes following the meta path. Random Walk, another straightforward measure, is the probability of the random walk that starts from one node and ends with the other one following the meta path. Besides, some researchers proposed several meta path based similarity measures on HIN, such as PathSim [3], PCRW [5] and HeteSim [6]. However, they do not consider how to combine the measures by different meta paths effectively. Furthermore, they only use the network structure to measure the similarity of two nodes and ignore the rich attribute information.

Definition 3. *link prediction in HIN*. *In a HIN, for a pair of nodes x and y, they have n similarity measures based on n meta paths, and they have m other attribute features. To predict whether x and y should be connected based on all the information above is called link prediction in HIN.*

According to the definition, we need to make the prediction by the node similarity based on different meta paths and other attribute information. First, we define a unified framework for this problem. Assume that node x and y have different similarity scores by different meta paths, and they can form a vector

$v(x, y) = [S_{P_1}, S_{P_2}, \ldots, S_{P_n}]$, where $S_{P_i}(i \in [1, n])$ is the similarity score defined on meta path P_i. We take other attributes of node x and y to form a attribute vector $\psi(x, y) = [a_1, a_2, \ldots, a_m]$. Then the unified measure of similarity can be defined as follows.

$$I(x, y) = f_\omega(v(x, y)) + f_\theta(\psi(x, y)) \tag{1}$$

where f_ω and f_θ is the combination function of $v(x, y)$ and $\psi(x, y)$ parameterized by ω and θ respectively. Although we use the simple linear combination function in the experiment, the two functions are still discussed separately.

For learning the weight ω and θ, we formulate a supervised task inspired by [7]. For a specific node x, we have a positive node set $D = \{d_1, d_2, \ldots, d_k\}$ and a negative node set $L = \{l_1, l_2, \ldots, l_k\}$, where each node in D will be connected to x and each node in L will be not. The method of selecting nodes for D and L will be discussed in the next section. Considering that for node x, the proximity from x to any positive node in D should be greater than to any negative node in L, we propose the following formulation.

$$\max_{\omega, \theta} F(\omega, \theta) = ||\omega||_2^2 + ||\theta||_2^2 + \lambda \sum_{d \in D, l \in L} h(I(x, l) - I(x, d)) \tag{2}$$

where h is the loss function, it returns a non-negative value. If $I(x, l) - I(x, d) < 0$, then $h(\cdot) = 0$, which means the constraint is not violated, while for $I(x, l) - I(x, d) > 0$, also $h(\cdot) > 0$. For convenience we use a differentiable function for h.

$$h(x) = \frac{1}{1 + exp(-x/b)} \tag{3}$$

To obtain the solution for Eq. 2, we find partial derivatives for ω and θ respectively. As for ω, the partial derivative of F can be written as follows:

$$
\begin{aligned}
\frac{\partial F(\omega, \theta)}{\partial \omega} &= 2\omega + \sum_{d,l} \frac{\partial h(I(x, l) - I(x, d))}{\partial \omega} \\
&= 2\omega + \sum_{d,l} \frac{\partial h(d_I)}{\partial d_I}(\frac{\partial I(x, l)}{\partial \omega} - \frac{\partial I(x, d)}{\partial \omega})
\end{aligned}
\tag{4}
$$

In Eq. 4, $d_I = I(x, l) - I(x, d)$, and we have the derivative of loss function h:

$$\frac{\partial h(d_I)}{\partial d_I} = \frac{1}{b}h(d_I)(1 - h(d_I)) \tag{5}$$

for the other part of Eq. 4, we have that:

$$\frac{\partial I(x, l) - \partial I(x, d)}{\partial \omega} = \frac{\partial f_\omega(v(x, l))}{\partial \omega} - \frac{\partial f_\omega(v(x, d))}{\partial \omega} \tag{6}$$

Similarly we can have the partial derivative of F for θ, then we can get the solution by using the general gradient descent method. We have several issues to explain. First, the optimization function in Eq. 2 is not convex, thus gradient descent methods will not always find the global minimum. We resolve this by using different starting points to find a good result. Second, we only consider one specific node x for the optimization. In practice, we can create a training set X, and Eq. 2 can be modified slightly as follows:

$$\max_{\omega,\theta} F(\omega,\theta) = ||\omega||_2^2 + ||\theta||_2^2 + \lambda \sum_{x \in X} \sum_{d \in D_x, l \in L_x} h(I(x,l) - I(x,d)) \tag{7}$$

3   Experiments

We used a dataset from Douban [8], a well known social media network in China. The network schema is shown in Fig. 1 and the detailed description of the dataset can be seen in Table 1. The dataset also includes 1068278 movie ratings, which can be seen as the attribute information.

Table 1. Statistics of Douban dataset.

Relations (A-B)	Number of A	Number of B	Number of (A-B)
user-movie	13367	12677	1068278
user-user	2440	2294	4085
user-group	13337	2753	570047
movie-actor	11718	6311	33587
movie-director	10179	2449	11276
movie-movie type	12676	38	27668

First we create a node set for training by choosing the nodes with more than 5 neighbors for each type. Then we create a positive and negative set for every node in the training set. We achieve this by selecting neighbors of the node randomly for positive set and selecting the nodes with no connection with the original node randomly for negative set. Second, we choose 6 different meta paths, including (1) UUU; (2) UMU; (3) UMAMU; (4) UGU; (5) UMDMU; and (6) UMTMU. Then we use 3 method of measuring the similarity of nodes by meta path for the experiment, including path count, random walk and PathSim. Moreover, we use the rating information to create 6 features, including (1) the rating numbers of two users; (2) the average rating scores of two users; (3) the number of movies that two users both rated; (4) the number of movies that two users rated same score.

In the experiment we found that the final ω and θ is relative small, so the overfitting problem can be ignored. We simply set $\lambda = 1$. Then we found that the choose of b in loss function h have little impact on the result, we set $b = 1$ for convenience.

We use 4 methods for comparison, including Common Neighbors (CN), Adamic/Adar (AA), Logistic Regression (LR) and Support Vector Machine

(SVM). And we use AUC for the comparison of all the methods. We first make a comparison for 3 methods of similarity measures of nodes by meta paths. We found that in LR, SVM and our algorithm, the random walk method achieve a better performance. Figure 2 show the ROC of the 3 methods by SVM. So we use the random walk method for the final report.

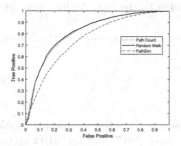

Fig. 2. The ROC of different measures by SVM.

In order to verify the effectiveness of our method (MA), five test environments with different scales of test set were selected randomly from 10% to 50% (step size is 10%). The experimental results are shown in Fig. 3. It can be seen that for HIN, it is difficult to get good prediction results by using the method for homogeneous network. Among all the methods, our approach achieves relatively better prediction results. When the network structure information is abundant (test set 10%), the SVM method will get the best results, but when the network structure information is not enough, the prediction effect of our method is better. This further shows that the generalization ability of our method is stronger.

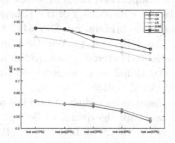

Fig. 3. The AUC value of 5 methods in 5 test environments.

4   Conclusion

Aiming at the link prediction problem in HIN, this paper first proposes a framework that combines the similarity under meta paths with the similarity of other attributes, then establishes an optimization function to obtain the optimal parameters, and deduces the optimization algorithm for the function. Finally, the experimental results show the effectiveness of the proposed method.

The combination function and the similarity measure of meta-path adopted in this paper are relatively simple. Next we will continue to try to use other combination functions and similarity measures, and refer to the weighted heterogeneous information network method [9] for more in-depth analysis.

Acknowledgments. This work is supported by the National Natural Science Foundation of China (U1811261).

References

1. Liben-Nowell, D., Kleinberg, J.: The link-prediction problem for social networks. J. Am. Soc. Inform. Sci. Technol. **58**(7), 1019–1031 (2007)
2. Lü, L., Zhou, T.: Link prediction in complex networks: a survey. Physica A **390**(6), 1150–1170 (2011)
3. Sun, Y., Han, J., Yan, X., Yu, P.S., Wu, T.: PathSim: meta path-based top-k similarity search in heterogeneous information networks. Proc. VLDB Endowment **4**(11), 992–1003 (2011)
4. Shi, C., Hu, B., Zhao, W.X., Philip, S.Y.: Heterogeneous information network embedding for recommendation. IEEE Trans. Knowl. Data Eng. **31**(2), 357–370 (2018)
5. Lao, N., Cohen, W.W.: Fast query execution for retrieval models based on path-constrained random walks. In: Proceedings of the 16th ACM SIGKDD International Conference on Knowledge Discovery and Data Mining, pp. 881–888. ACM (2010)
6. Shi, C., Kong, X., Huang, Y., Philip, S.Y., Wu, B.: HeteSim: a general framework for relevance measure in heterogeneous networks. IEEE Trans. Knowl. Data Eng. **26**(10), 2479–2492 (2014)
7. Backstrom, L., Leskovec, J.: Supervised random walks: predicting and recommending links in social networks. In: Proceedings of the Fourth ACM International Conference on Web Search and Data Mining, pp. 635–644. ACM (2011)
8. Shi, C., Zhang, Z., Luo, P., Yu, P.S., Yue, Y., Wu, B.: Semantic path based personalized recommendation on weighted heterogeneous information networks. In: Proceedings of the 24th ACM International on Conference on Information and Knowledge Management, pp. 453–462. ACM (2015)
9. Li, Y., Li, C., Chen, W.: Research on influence ranking of Chinese movie heterogeneous network based on PageRank algorithm. In: Meng, X., Li, R., Wang, K., Niu, B., Wang, X., Zhao, G. (eds.) WISA 2018. LNCS, vol. 11242, pp. 344–356. Springer, Cham (2018). https://doi.org/10.1007/978-3-030-02934-0_32

TransFG: A Fine-Grained Model for Knowledge Graph Embedding

Yaowei Yu, Zhuoming Xu[✉], Yan Lv, and Jian Li

College of Computer and Information, Hohai University, Nanjing 210098, China
{ywyu, zmxu, ylv, jli}@hhu.edu.cn

Abstract. Although concepts and instances in a knowledge graph (KG) are distinguished, TransC embeds concepts, instances, and various relations into the same vector space, which leads to the following problems: (1) The same instance in different triples that model different relations between instances is represented as the same vector, resulting in improper representation of different properties possessed by this instance; (2) Multiple instances not belonging to one concept may be located in the sphere representing this concept, resulting in an inaccurate modeling of the instanceOf relations between these instances and the concept. Based on TransC, this paper proposes a fine-grained KG embedding model called TransFG. TransFG embeds concepts, instances, and relations into different vector spaces and projects the instance vectors from the instance space to the concept space and the relation spaces through dynamic mapping matrices. This causes the projected vectors of the same instance in different triples to have different representations and the projected vectors of multiple instances belonging to the same concept to be spatially close to each other; otherwise they are far away. Experiments on the YAGO39K and M-YAGO39K datasets show that on the triple classification task, TransFG outperforms TransC and other typical KG embedding models in terms of accuracy, precision, recall and F1-score in most cases, and on the link prediction task, TransFG outperforms these compared models in terms of MRR and Hits@N in most cases.

Keywords: Knowledge graph embedding · Vector space · Mapping matrix · Triple classification · Link prediction

1 Introduction

A knowledge graph (KG) [1, 2] is a multi-relational graph consisting of entities (represented as nodes) and relations (represented as edges) between entities. The facts in KG are normally expressed as RDF triples. KGs like YAGO [3] and DBpedia [4, 5] have been widely used in knowledge-based applications such as question answering.

KG embedding [1, 2] is aimed at embedding entities and relations into continuous, low-dimensional vector space for efficiently performing downstream tasks such as link prediction and triple classification. According to the different types of scoring functions, there are two categories of KG embedding models: translational distance models and semantic matching models [1, 2]. TransE [6] is a representative of translation distance models. TransE has flaws in dealing with multi-mapping relations (one-to-many,

© Springer Nature Switzerland AG 2019
W. Ni et al. (Eds.): WISA 2019, LNCS 11817, pp. 455–466, 2019.
https://doi.org/10.1007/978-3-030-30952-7_45

many-to-one, and many-to-many). Based on TransE, many improved models have been proposed, for example, TransH [7], TransR [8], and TransD [9]. HolE [10], DistMult [11], and ComplEx [12] are typical semantic matching models.

Lv et al. [13] pointed out that existing KG embedding models fail to distinguish between concepts and instances, leading to some problems. Hence, they proposed a new model called TransC, which distinguishes between concepts and instances, and divides the triples in a KG into three disjoint subsets: the instanceOf triple set, the subClassOf triple set, and inter-instance relation triple set. However, TransC embeds concepts, instances, and various relations into the same vector space, which leads to the following problems: (1) The same instance in different triples that model different inter-instance relations is represented as the same vector, resulting in improper representation of different properties possessed by this instance; (2) Multiple instances not belonging to one concept may be located in the sphere representing this concept, resulting in an inaccurate modeling of the instanceOf relations between these instances and the concept.

Based on TransC, this paper proposes a fine-grained KG embedding model called TransFG. TransFG embeds concepts, instances, and relations into different vector spaces and projects the instance vectors from the instance space to the concept space and the relation spaces through dynamic mapping matrices. This causes the projected vectors of the same instance in different triples to have different representations and the projected vectors of multiple instances belonging to the same concept to be spatially close to each other; otherwise they are far away. We used two typical KG downstream tasks, triple classification and link prediction, to compare and evaluate TransFG, TransC, and several KG embedding models including TransE, TransH, TransR, TransD, HolE, DistMult, and ComplEx on the YAGO39K and M-YAGO39K datasets [13]. The experimental results show that on the triple classification task, TransFG outperforms TransC and other typical embedding models in terms of accuracy, precision, recall and F1-score in most cases, and on the link prediction task, TransFG outperforms these compared models in terms of MRR (the mean reciprocal rank of all correct instances) and Hits@N (the proportion of correct instances in the top-N ranked instances) in most cases.

2   TransFG: A Fine-Grained Model

In this section, we expatiate on our proposed TransFG model. We first briefly explain the basic idea of the model, then describe the model in detail, and finally explain the method of model training.

2.1   Basic Idea of TransFG

Like TransC, TransFG divides KG triples into three types: the instanceOf triples, the subClassOf triples, and inter-instance relation triples, and defines different loss functions for each type of triples. The main difference between TransFG and TransC is the improvement of the representations of the instanceOf triples and inter-instance relation triples. This is mainly achieved by embedding concepts, instances, and various relations between them into different spaces and applying corresponding mapping matrices.

For the representation of the `instanceOf` triples, TransFG projects instance vectors from the instance space to the concept space through dynamic mapping matrices. Let us use Fig. 1 to explain the representation of this type of triples. The meanings of the mathematical symbols in the figure are listed in Table 1. As shown in the figure, triangles, such as **e** and **f**, and pentagrams, such as **b**, denote different instance vectors belonging to two different concepts c_i and c_j, and $s_i(\mathbf{p}_i, m_i)$ and $s_j(\mathbf{p}_i, m_j)$ are two different concept spheres, respectively. For three `instanceOf` triples (e, r_e, c_i), (f, r_e, c_i), and (b, r_e, c_j), TransFG projects **e**, **f**, and **b** from the instance space to the concept space through the mapping matrices $\mathbf{M}_{ei}$, $\mathbf{M}_{fi}$, and $\mathbf{M}_{bj}$ and obtains the projected vectors $\mathbf{e}_\perp$, $\mathbf{f}_\perp$, and $\mathbf{b}_\perp$. If the three triples are positive triples (i.e., they exist in the KG), $\mathbf{e}_\perp$ and $\mathbf{f}_\perp$ are located in the sphere $s_i(\mathbf{p}_i, m_i)$, and $\mathbf{b}_\perp$ is located in sphere $s_j(\mathbf{p}_j, m_j)$. The loss functions for the `instanceOf` triples are then defined using the relative positions between the projected vectors and the concept spheres.

For the representation of the `subClassOf` triples, TransFG directly uses the corresponding method in TransC [13], that is, the loss functions for the `subClassOf` triples are defined using the relative positions between the two concept spheres.

For the representation of inter-instance relation triples, just like TransD [9], TransFG projects instance vectors from the instance space to the relation spaces through the corresponding mapping matrices. The loss functions for inter-instance relation triples are then defined using the projected vectors.

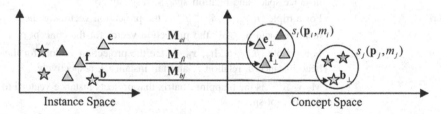

Fig. 1. Representation of the `instanceOf` triples in TransFG.

2.2 The TransFG Model

In this subsection, we describe the TransFG model in detail. We first list the mathematical symbols used to describe the model, and then define the loss functions for the three different types of triples: the `instanceOf` triples, the `subClassOf` triples, and inter-instance relation triples.

The mathematical symbols introduced in TransC's paper [13] and in our paper are listed in Table 1, where the symbols in the first eleven rows are introduced in [13].

Table 1. Mathematical symbols introduced in [13] and in our paper.

Symbols	Meanings
$\mathcal{KG} = \{\mathcal{C}, \mathcal{I}, \mathcal{R}, \mathcal{S}\}$	$\mathcal{KG}$ is a knowledge graph, where $\mathcal{C}$ denotes the concept set, $\mathcal{I}$ the instance set, $\mathcal{R}$ the relation set, and $\mathcal{S} = \mathcal{S}_e \cup \mathcal{S}_c \cup \mathcal{S}_l$ the set of triples existing in $\mathcal{KG}$ (i.e., the set of *positive triples*)
$\mathcal{R} = \{r_e, r_c\} \cup \mathcal{R}_l$	$\mathcal{R}$ is a relation set, where r_e is the `instanceOf` relation, r_c the `subClassOf` relation, and $\mathcal{R}_l$ the set of other relations
$\mathcal{S}_e = \{(i, r_e, c)_k\}_{k=1}^{n_e}$	$\mathcal{S}_e$ is the set of the positive `instanceOf` triples, where $i \in \mathcal{I}$ is an instance and $c \in \mathcal{C}$ a concept
$\mathcal{S}_c = \{(c_i, r_c, c_j)_k\}_{k=1}^{n_c}$	$\mathcal{S}_c$ is the set of the positive `subClassOf` triples, where both c_i and c_j are concepts
$\mathcal{S}_l = \{(h, r, t)_k\}_{k=1}^{n_l}$	$\mathcal{S}_l$ is the set of the positive inter-instance relation triples, where $h, t \in \mathcal{I}$ are the head instance and tail instance of a triple, respectively, and $r \in \mathcal{R}_l$ is an inter-instance relation
$\mathcal{S}' = \mathcal{S}'_e \cup \mathcal{S}'_c \cup \mathcal{S}'_l$	$\mathcal{S}'$ is the set of all negative triples, where $\mathcal{S}'_e, \mathcal{S}'_c, \mathcal{S}'_l$ are the sets of negative triples corresponding to $\mathcal{S}_e, \mathcal{S}_c, \mathcal{S}_l$, respectively
ξ, ξ'	$\xi \in \mathcal{S}$ is a positive triple, and $\xi' \in \mathcal{S}'$ is a negative triple
$\mathbf{i}$	$\mathbf{i} \in \mathbb{R}^n$ is the vector of the instance i in the triple (i, r_e, c)
$\mathbf{h}, \mathbf{r}, \mathbf{t}$	$\mathbf{h}, \mathbf{r}, \mathbf{t}$ are the vectors of the head instance h, relation r, and tail instance t in the inter-instance relation triple (h, r, t)
$s_i(\mathbf{p}_i, m_i)$	Sphere $s_i(\mathbf{p}_i, m_i)$ denote concept $c_i \in \mathcal{C}$, where $\mathbf{p}_i \in \mathbb{R}^k$ and m_i are the center and radius of the concept sphere, respectively
$\gamma_e, \gamma_c, \gamma_l$	$\gamma_e, \gamma_c, \gamma_l$ are the margins separating the positive triples and the negative triples in the three types of triples
k, n, z	k, n, z denote the dimensions of the vectors in the concept space, instance space and relation spaces, respectively
$\mathbf{i}_p, \mathbf{p}_p$	For a triple (i, r_e, c), $\mathbf{i}_p \in \mathbb{R}^n$ is the projection vector for the instance i and $\mathbf{p}_p \in \mathbb{R}^k$ the projection vector for the concept c
$\mathbf{h}_p, \mathbf{r}_p, \mathbf{t}_p$	For a triple (h, r, t), $\mathbf{h}_p, \mathbf{r}_p, \mathbf{t}_p$ are the projection vectors for the head instance h, relation r, and tail instance t, respectively
$\mathbf{M}_{ic}$	$\mathbf{M}_{ic} \in \mathbb{R}^{k \times n}$ is the mapping matrix that projects instance vector $\mathbf{i}$ to the concept space
$\mathbf{M}_{rh}, \mathbf{M}_{rt}$	$\mathbf{M}_{rh}, \mathbf{M}_{rt} \in \mathbb{R}^{z \times n}$ are the mapping matrices that projects the head instance and tail instance vectors $\mathbf{h}$ and $\mathbf{t}$ to the relation spaces
$\mathbf{i}_\perp$	$\mathbf{i}_\perp \in \mathbb{R}^k$ is the projected vector obtained by projecting the instance vector $\mathbf{i}$ to the concept space
$\mathbf{h}_\perp, \mathbf{t}_\perp$	$\mathbf{h}_\perp \in \mathbb{R}^z, \mathbf{t}_\perp \in \mathbb{R}^z$ are the projected vectors obtained by projecting the head instance vector $\mathbf{h}$ and tail instance vector $\mathbf{t}$ to the relation spaces, respectively

InstanceOf Triple Representation. For an `instanceOf` triple (i, r_e, c), TransFG learns the instance vector $\mathbf{i}$ of instance i, the projection vector $\mathbf{i}_p$ for i, the center vector $\mathbf{p}$ of the sphere representing concept c, and the projection vector $\mathbf{p}_p$ for c. The vectors $\mathbf{i}_p$ and $\mathbf{p}_p$ are used to construct the mapping matrix $\mathbf{M}_{ic} \in \mathbb{R}^{k \times n}$. TransFG projects $\mathbf{i}$ from the instance space to the concept space through the mapping matrix, which is defined as Eq. (1).

$$\mathbf{M}_{ic} = \mathbf{p}_p \mathbf{i}_p^\top + \mathbf{E}^{k \times n} \tag{1}$$

where $\mathbf{E}^{k \times n}$ is an identity matrix used to initialize the mapping matrix. As can be seen from Eq. (1), each mapping matrix is determined by an instance and a concept. Hence, TransFG uses different matrices to project the same instance in different `instanceOf` triples, and the projected vectors are also different. The projected vector $\mathbf{i}_\perp$ of $\mathbf{i}$ is defined as Eq. (2).

$$\mathbf{i}_\perp = \mathbf{M}_{ic}\mathbf{i} \tag{2}$$

In TransFG, the `instanceOf` relations are represented using the relative positions between the projected vectors and the concept spheres. For an `instanceOf` triple (i, r_e, c), if it is a positive triple, then $\mathbf{i}_\perp$ should be inside the concept sphere of c to represent the `instanceOf` relation between i and c. If $\mathbf{i}_\perp$ is outside the concept sphere, the instance embedding and concept embedding need to be optimized. The loss function is defined as Eq. (3).

$$f_e(i, c) = ||\mathbf{i}_\perp - \mathbf{p}||_2 - m \tag{3}$$

SubClassOf Triple Representation. For a `subClassOf` triple (c_i, r_c, c_j), the concepts c_i and c_j are represented by the spheres $s_i(\mathbf{p}_i, m_i)$ and $s_j(\mathbf{p}_j, m_j)$. There are four relative positions between the two spheres, as illustrated in Fig. 2 (this figure is taken from [13]). As shown in Fig. 2(a) and described in [13], if (c_i, r_c, c_j) is a positive triple, the sphere s_i should be inside the sphere s_j to represent the inclusion relation between the two concepts, which is the optimization goal.

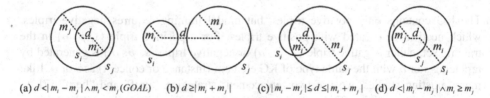

(a) $d < |m_i - m_j| \wedge m_i < m_j$ (GOAL)   (b) $d \geq |m_i + m_j|$ (c) $|m_i - m_j| \leq d \leq |m_i + m_j|$ (d) $d < |m_i - m_j| \wedge m_i \geq m_j$

Fig. 2. Four relative positions between concept spheres s_i and s_j. (Source: Figure 2 in [13])

If the two spheres are separate from each other (as shown in Fig. 2(b)) or intersect (as shown in Fig. 2(c)), the two spheres need to get closer via optimization. The loss function is thus defined as Eq. (4) [13].

$$f_c(c_i, c_j) = ||\mathbf{p}_i - \mathbf{p}_j||_2 + m_i - m_j \tag{4}$$

where $||\mathbf{p}_i - \mathbf{p}_j||_2$ denotes the distance d between $\mathbf{p}_i$ and $\mathbf{p}_j$ of the two spheres.

If s_j is inside s_i as shown in Fig. 2(d), we need to reduce m_i and increase m_j. The loss function is therefore defined as Eq. (5) [13].

$$f_c(c_i,\ c_j) = m_i - m_j \tag{5}$$

Inter-instance Relation Triple Representation. Just like TransD [9], for an inter-instance relation triple $(h,\ r,\ t)$, TransFG learns six vectors: the head instance vector $\mathbf{h}$, relation vector $\mathbf{r}$, tail instance vector $\mathbf{t}$, projection vector $\mathbf{h}_p$ for h, projection vector $\mathbf{r}_p$ for r, and projection vector $\mathbf{t}_p$ for t. The projection vectors $\mathbf{h}_p$, $\mathbf{r}_p$, and $\mathbf{t}_p$ are used to construct the mapping matrices $\mathbf{M}_{rh}$ and $\mathbf{M}_{rt}$, which are defined as Eqs. (6) and (7) [9].

$$\mathbf{M}_{rh} = \mathbf{r}_p \mathbf{h}_p^\top + \mathbf{E}^{z \times n} \tag{6}$$

$$\mathbf{M}_{rt} = \mathbf{r}_p \mathbf{t}_p^\top + \mathbf{E}^{z \times n} \tag{7}$$

where $\mathbf{E}^{z \times n}$ is an identity matrix. TransFG projects $\mathbf{h}$ and $\mathbf{t}$ from the instance space to the corresponding relation space through the mapping matrices, obtaining the projected vectors $\mathbf{h}_\perp$ and $\mathbf{t}_\perp$. The vectors $\mathbf{h}_\perp$ and $\mathbf{t}_\perp$ are defined as Eq. (8) [9]:

$$\mathbf{h}_\perp = \mathbf{M}_{rh}\mathbf{h}, \quad \mathbf{t}_\perp = \mathbf{M}_{rt}\mathbf{t} \tag{8}$$

The loss function is then defined as Eq. (9) [9].

$$f_r(h,\ t) = \ ||\mathbf{h}_\perp + \mathbf{r} - \mathbf{t}_\perp||_2^2 \tag{9}$$

Finally, similar to other embedding models [6–9, 13], we enforce constraints as $||\mathbf{h}||_2 \le 1$, $||\mathbf{t}||_2 \le 1$, $||\mathbf{r}||_2 \le 1$, $||\mathbf{p}||_2 \le 1$, $||\mathbf{h}_\perp||_2 \le 1$, $||\mathbf{t}_\perp||_2 \le 1$ in our experiments.

2.3   Model Training

The KG contains only positive triples, but model training requires negative triples, which need to be created with positive triples. For a positive triple $(s,\ p,\ o)$ in the training set, either negative triple $(s',\ p,\ o)$ or negative triple $(s,\ p,\ o')$ is generated by replacing s or o with the same type of KG element (instance or concept) as s or o. Like many existing studies, we use two replacement strategies, "unif" and "bern" [7], to generate negative triples. The replacement strategy "unif" means replacing the subjects or the objects in positive triples with the same probability, while "bern" means replacing the subjects or the objects with the different probabilities for reducing false negative labels. Each positive or negative triple is indicated by a label.

Just like TransC [13], we define the margin-based ranking loss $\mathcal{L}_e$ for the instanceOf triples as Eq. (10) [13].

$$\mathcal{L}_e = \sum_{\xi \in \mathcal{S}_e} \sum_{\xi' \in \mathcal{S}_e'} \max(0,\ \gamma_e + f_e(\xi) - f_e(\xi')) \tag{10}$$

Similarly, the margin-based ranking loss $\mathcal{L}_c$ for the subClassOf triples and the margin-based ranking loss $\mathcal{L}_l$ for inter-instance relation triples are defined as Eqs. (11) and (12) [13].

$$\mathcal{L}_c = \sum_{\xi \in \mathcal{S}_c} \sum_{\xi' \in \mathcal{S}_c'} \max(0, \ \gamma_c + f_c(\xi) - f(\xi')) \tag{11}$$

$$\mathcal{L}_l = \sum_{\xi \in \mathcal{S}_l} \sum_{\xi' \in \mathcal{S}_l'} \max(0, \ \gamma_l + f_r(\xi) - f_r(\xi')) \tag{12}$$

The overall ranking loss $\mathcal{L}$ is therefore defined as Eq. (13) [13].

$$\mathcal{L} = \mathcal{L}_e + \mathcal{L}_c + \mathcal{L}_l \tag{13}$$

The goal of model training is to minimize the overall ranking loss using stochastic gradient descent (SGD).

3 Experimental Evaluation

3.1 Experimental Design

Evaluation Tasks. We used two typical KG downstream tasks, triple classification and link prediction, to compare and evaluate our TransFG and several KG embedding models including TransC [13], TransE [6], TransH [7], TransR [8], TransD [9], HolE [10], DistMult [11], and ComplEx [12]. We used Accuracy, Precision, Recall and F1-score as the evaluation metrics for the triple classification task, while we used MRR and Hits@N as the evaluation metrics for the link prediction task.

Implementation. The program code of TransC directly uses its C++ code published in [13] (cf. https://github.com/davidlvxin/TransC). The program code of TransFG was generated by modifying the loss function calculation module and model training module of TransC's code. These codes were used to generate the corresponding experimental results. We copied the results of the other models from [13], as both experiments used the same experimental datasets and parameter settings.

Datasets. We used the same experimental datasets YAGO39K and M-YAGO39K as in [13]. YAGO39K was built in [13] by randomly extracting triples from YAGO, consisting of 39 types of relations including `instanceOf`, `subClassOf`, and inter-instance relations. As stated in [13], M-YAGO39K was formed based on YAGO39K by generating some new triples using the transitivity of the IS-A relations. We trained the TransC and TransFG models using the training set of YAGO39K, and obtained the best parameter configurations on YAGO39K and on M-YAGO39K through the validation sets of the two datasets, respectively. The triple classification task was evaluated on the test sets of the two datasets, respectively, while the link prediction task was evaluated on the test set of YAGO39K.

3.2   Experimental Results of Triple Classification

Triple classification is a binary classification task that determines whether a given triple is a positive triple or not. When the triple classification task is performed by TransFG, we set a threshold δ_r for each relation r. The threshold is achieved by maximizing the classification accuracy on the validation set. For each triple in the test set, TransFG uses the loss function defined for the triple to calculate the score. If the score is less than the threshold, the triple is classified as positive; otherwise it is classified as negative.

In our experiments, we set the parameters of TransC according to the best configurations given in [13]. For TransFG, we select the learning rate for SGD among $\{0.1, 0.01, 0.001\}$, the dimension of concept, instance and relation vectors k, n, z among $\{20, 50, 100\}$, and the margins γ_e, γ_c, γ_l among $\{0.1, 0.3, 0.5, 1\}$. The best configurations of TransFG are then determined by the classification accuracy on the validation set. The best configurations on both YAGO39K and M-YAGO39K datasets are: $\lambda = 0.001$, $k = 100$, $n = 100$, $z = 100$, $\gamma_e = 0.1$, $\gamma_c = 0.1$, $\gamma_l = 1$, and taking L_1 as dissimilarity. We train the TransC and TransFG models for 1,000 rounds.

The experimental results of performing the `instanceOf`, `subClassOf`, and inter-instance relation triple classification tasks on the experimental datasets are shown in Tables 2, 3, and 4, respectively, where "P" stands for Precision, "R" Recall, and "F1" F1-Score, and "unif" and "bern" are the two replacement strategies described earlier. Observing these results, we have the following findings:

1. From Table 2, we can find that TransC and TransFG perform slightly worse than other models on the `instanceOf` triple classification task on YAGO39K. We agree with the viewpoint in [13]: Since the `instanceOf` triples account for the majority (53.5%) in YAGO39K, the `instanceOf` relation is trained too many times in the other models, resulting in almost the best performance on this task, but the performance on other triple classification tasks is not good. It is worth noting that TransC and TransFG achieve relatively balanced performance on all three triple classification tasks. TransFG achieves slightly worse performance than TransC under the "unif" strategy, but it achieves better performance than TransC under the "bern" strategy.

Table 2. Results (%) of the `instanceOf` triple classification on the two datasets.

Model	YAGO39K				M-YAGO39K			
	Accuracy	P	R	F1	Accuracy	P	R	F1
TransE	82.6	83.6	81.0	82.3	71.0↓	81.4↓	54.4↓	65.2↓
TransH	82.9	83.7	81.7	82.7	70.1↓	80.4↓	53.2↓	64.0↓
TransR	80.6	79.4	**82.5**	80.9	70.9↓	73.0↓	66.3↓	69.5↓
TransD	83.2	84.4	81.5	82.9	72.5↓	73.1↓	71.4↓	72.2↓
HolE	82.3	86.3	76.7	81.2	74.2↓	81.4↓	62.7↓	70.9↓
DistMult	**83.9**	**86.8**	80.1	**83.3**	70.5↓	86.1↓	49.0↓	62.4↓
ComplEx	83.3	84.8	81.1	82.9	70.2↓	84.4↓	49.5↓	62.4↓
TransC (unif)	80.3	81.6	80.0	80.8	85.5↑	88.4↑	81.9↑	85.0↑
TransC (bern)	79.8	83.3	74.5	78.6	85.4↑	86.2↑	**84.3**↑	85.2↑
TransFG (unif)	80.2	82.4	78.6	80.4	85.5↑	88.3↑	82.0↑	85.0↑
TransFG (bern)	81.7	83.8	75.5	79.4	**85.9**↑	**88.4**↑	82.2↑	**85.2**↑

Table 3. Results (%) of the `subClassOf` triple classification on the two datasets.

Model	YAGO39K				M-YAGO39K			
	Accuracy	P	R	F1	Accuracy	P	R	F1
TransE	77.6	72.2	89.8	80.0	76.9↓	72.3↑	87.2↓	79.0↓
TransH	80.2	76.4	87.5	81.5	79.1↓	72.8↓	92.9↑	81.6↑
TransR	80.4	74.7	91.9	82.4	80.0↓	73.9↓	92.9↑	82.3↓
TransD	75.9	70.6	88.8	78.7	76.1↑	70.7↑	89.0↑	78.8↑
HolE	70.5	73.9	63.3	68.2	66.6↓	72.3↓	53.7↓	61.7↓
DistMult	61.9	68.7	43.7	53.4	60.7↓	71.7↑	35.5↓	47.7↓
ComplEx	61.6	71.5	38.6	50.1	59.8↓	65.6↓	41.4↑	50.7↑
TransC (unif)	83.0	77.2	93.7	84.6	83.1↑	77.6↑	93.2↓	84.7↑
TransC (bern)	83.8	78.1	93.9	85.3	84.5↑	**80.8**↑	90.4↓	85.3↑
TransFG (unif)	82.8	75.7	**96.5**	84.8	83.1↑	76.5↑	**95.5**↓	84.9↑
TransFG (bern)	**84.5**	**78.6**	95.2	**86.1**	**84.7**↑	78.7↑	94.1↓	**85.7**↓

Table 4. Results (%) of inter-instance relation triple classification on YAGO39K.

Model	YAGO39K			
	Accuracy	P	R	F1
TransE	92.1	92.8	91.2	92.0
TransH	90.8	91.2	90.3	90.8
TransR	91.7	91.6	91.9	91.7
TransD	89.3	88.1	91.0	89.5
HolE	92.3	92.6	91.9	92.3
DistMult	93.5	93.9	93.0	93.5
ComplEx	92.8	92.6	93.1	92.9
TransC (unif)	93.5	94.4	92.7	93.6
TransC (bern)	93.8	**94.9**	92.7	93.8
TransFG (unif)	93.1	94.0	92.7	93.4
TransFG (bern)	**94.4**	94.7	**93.3**	**94.0**

2. From Tables 2, 3, and 4, we can find that TransFG performs better than TransC on all three triple classification tasks in terms of almost all evaluation metrics, except for the Recall value of the `instanceOf` triple classification task on M-YAGO39K, the Precision value of the `subclassOf` triple classification task on M-YAGO39K, and the Precision value of the inter-instance triple classification task on YAGO39K. This suggests that TransFG does better than TransC in terms of the representation of multiple different properties possessed by an instance and the modeling of the `instanceOf` relations. TransFG also performs better than all other models on all three triple classification tasks in terms of all evaluation metrics, except for the `instanceOf` triple classification task on YAGO39K.

3. From Tables 2 and 3, we can find that in most cases of performing the `instanceOf` and `subClassOf` triple classification tasks, TransFG performs better

on the M-YAGO39K dataset than on the YAGO39K dataset, which indicates that TransFG can handle the transitivity of the IS-A relations very well.

4. From Tables 2, 3, and 4, we can find that in most cases of performing all three triple classification tasks on the both datasets, TransFG performs better under the "bern" strategy than under the "unif" strategy, which indicates that the "bern" strategy can reduce false negative labels more effectively than the "unif" strategy.

Based on the above findings, we can conclude that the performance of TransFG performing the triple classification task is generally better than that of TransC and other compared models.

3.3   Experimental Results of Link Prediction

Link prediction is aimed at predicting the missing head or tail for an inter-instance triple. When the link prediction task is performed by TransFG, for each triple in the test set we first replace the head instance and the tail instance with all instance in $\mathcal{I}$ one by one, thereby obtaining so-called *corrupted triples* (two corrupted triples per replacement). Then we obtain the scores by calculating the loss function defined for the corrupted triples, and finally rank the instances in $\mathcal{I}$ in ascending order of the scores. Note that a corrupted triple may also exist in the KG, so such a triple that exists in the KG should be regarded as a correct prediction (a positive triple). Like existing works [6–10, 12, 13], our experiments also use two common evaluation settings "Raw" and "Filter". The "Raw" setting means that the corrupted but positive triples are not filtered out, while the "Filter" setting means that these triples are filtered out.

In our experiments, we set the parameters of TransC according to the best configurations given in [13]. For TransFG, the parameters are selected in the same way as on the triple classification task as described in Sect. 3.2. The best configurations of TransFG are then determined according to the Hits@10 on the verification set. The best configurations on the YAGO39K dataset are: $\lambda = 0.001$, $k = 100$, $n = 100$, $z = 100$, $\gamma_e = 0.1$, $\gamma_c = 1$, $\gamma_l = 1$, and taking L_1 as dissimilarity. We train the TransC and TransFG models for 1,000 rounds.

The experimental results of performing the link prediction task on YAGO39K are shown in Table 5. Observing these results, we have the following findings:

1. TransFG performs slightly worse than DisMult, but is the same as TransE and better than all other models, in terms of MRR (in the "Raw" setting).
2. TransFG outperforms TransC and all other models in terms of MRR (in the "Filter" settings) and Hits@N, which indicates that TransFG can represent multiple different properties possessed by an instance very well.
3. TransFG performs better under the "bern" strategy than under the "unif" strategy, which indicates that the "bern" strategy can reduce false negative labels more effectively than the "unif" strategy.

Based on the above findings, we can conclude that the performance of TransFG performing the link prediction task is generally better than that of TransC and other compared models.

Table 5. Results of link prediction for inter-instance relation triples on YAGO39K. Hist@N uses the results in the "Filter" evaluation setting.

Model	YAGO39K				
	MRR		Hits@N (%)		
	Raw	Filter	1	3	10
TransE	0.114	0.248	12.3	28.7	51.1
TransH	0.102	0.215	10.4	24.0	45.1
TransR	0.112	0.289	15.8	33.8	56.7
TransD	0.113	0.176	8.9	19.0	35.4
HolE	0.063	0.198	11.0	23.0	38.4
DisMult	**0.156**	0.362	22.1	43.6	66.0
ComplEx	0.058	0.362	29.2	40.7	48.1
TransC (unif)	0.087	0.421	28.3	50.0	69.2
TransC (bern)	0.112	0.420	29.8	50.2	69.8
TransFG (unif)	0.105	0.404	28.3	50.2	69.4
TransFG (bern)	0.114	**0.475**	**32.5**	**52.1**	**70.1**

4 Conclusions

Based on TransC, this paper proposes a fine-grained KG embedding model TransFG that embeds concepts, instances, and relations into different vector spaces and projects instance vectors from the instance space to the concept space and the relation spaces through dynamic mapping matrices. We conducted experimental evaluation through link prediction and triple classification on datasets YAGO39K and M-YAGO39K. The results show that TransFG outperforms TransC and other typical KG embedding models in most cases, especially in terms of the representation of multiple different properties possessed by an instance and the modeling of the `instanceOf` relations.

References

1. Wang, Q., Mao, Z., Wang, B., Guo, L.: Knowledge graph embedding: a survey of approaches and applications. IEEE Trans. Knowl. Data Eng. **29**(12), 2724–2743 (2017)
2. Rosso, P., Yang, D., Cudré-Mauroux, P.: Knowledge graph embeddings. In: Sakr, S., Zomaya, A.Y. (eds.) Encyclopedia of Big Data Technologies. Springer, Cham (2019). https://doi.org/10.1007/978-3-319-63962-8_284-1
3. Suchanek, F.M., Kasneci, G., Weikum, G.: Yago: a core of semantic knowledge. In: Proceedings of the 16th International Conference on World Wide Web, WWW 2007, pp. 697–706. ACM (2007). https://doi.org/10.1145/1242572.1242667
4. Lehmann, J., Isele, R., Jakob, M., et al.: DBpedia - a large-scale, multilingual knowledge base extracted from Wikipedia. Semant. Web **6**(2), 167–195 (2015)
5. Li, W., Chai, L., Yang, C., Wang, X.: An evolutionary analysis of DBpedia datasets. In: Meng, X., Li, R., Wang, K., Niu, B., Wang, X., Zhao, G. (eds.) WISA 2018. LNCS, vol. 11242, pp. 317–329. Springer, Cham (2018). https://doi.org/10.1007/978-3-030-02934-0_30

6. Bordes, A., Usunier, N., García-Durán, A., Weston, J., Yakhnenko, O.: Translating embeddings for modeling multi-relational data. In: Advances in Neural Information Processing Systems 26: 27th Annual Conference on Neural Information Processing Systems 2013, NIPS 2013, pp. 2787–2795 (2013). http://papers.nips.cc/paper/5071-translating-embeddings-for-modeling-multi-relational-data

7. Wang, Z., Zhang, J., Feng, J., Chen, Z.: Knowledge graph embedding by translating on hyperplanes. In: Proceedings of the Twenty-Eighth AAAI Conference on Artificial Intelligence, AAAI 2014, pp. 1112–1119. AAAI Press (2014). https://www.aaai.org/ocs/index.php/AAAI/AAAI14/paper/view/8531

8. Lin, Y., Liu, Z., Sun, M., Liu, Y., Zhu, X.: Learning entity and relation embeddings for knowledge graph completion. In: Proceedings of the Twenty-Ninth AAAI Conference on Artificial Intelligence, AAAI 2015, pp. 2181–2187. AAAI Press (2015). https://www.aaai.org/ocs/index.php/AAAI/AAAI15/paper/view/9571

9. Ji, G., He, S., Xu, L., Liu, K., Zhao, J.: Knowledge graph embedding via dynamic mapping matrix. In: Proceedings of the 53rd Annual Meeting of the Association for Computational Linguistics and the 7th International Joint Conference on Natural Language Processing of the Asian Federation of Natural Language Processing, vol. 1, pp. 687–696. The Association for Computer Linguistics (2015). https://www.aclweb.org/anthology/P15-1067

10. Nickel, M., Rosasco, L., Poggio, T.: Holographic embeddings of knowledge graphs. In: Proceedings of the Thirtieth AAAI Conference on Artificial Intelligence, AAAI 2016, pp. 1955–1961. AAAI Press (2016). https://www.aaai.org/ocs/index.php/AAAI/AAAI16/paper/view/12484

11. Yang, B., Yih, W.-T., He, X., Gao, J., Deng, L.: Embedding entities and relations for learning and inference in knowledge bases. In: 3rd International Conference on Learning Representations, ICLR 2015. Conference Track Proceedings (2015). https://arxiv.org/pdf/1412.6575

12. Trouillon, T., Welbl, J., Riedel, S., Gaussier, É., Bouchard, G.: Complex embeddings for simple link prediction. In: Proceedings of the 33rd International Conference on Machine Learning, ICML 2016, JMLR Workshop and Conference Proceedings, vol. 48, pp. 2071–2080. JMLR.org (2016). http://proceedings.mlr.press/v48/trouillon16.html

13. Lv, X., Hou, L., Li, J., Liu, Z.: Differentiating concepts and instances for knowledge graph embedding. In: Proceedings of the 2018 Conference on Empirical Methods in Natural Language Processing, EMNLP 2018, pp. 1971–1979. Association for Computational Linguistics (2018). https://aclweb.org/anthology/papers/D/D18/D18-1222/

Adjustable Location Privacy-Preserving Nearest Neighbor Query Method

Linfeng Xie[1,2(✉)], Zhigang Feng[1,2], Cong Ji[1,2], and Yongjin Zhu[1,2]

[1] Jiangsu Frontier Electric Technology Co., LTD,
Nanjing 211189, Jiangsu, China
15905166617@139.com
[2] School of Computer Science and Engineering,
Southeast University, Nanjing 211189, China

Abstract. Location-based services facilitate the daily life of the people, nevertheless, they also bring about the problem of privacy preserving. Privacy preserving methods without anonymity server, for example, Coprivacy, attract increasing concerning from researchers for their simple and reliable structure and the avoidance of high cost of communication and computing resulting from the using of cloaking area. The drawbacks of Coprivacy are the high cost of communication and computing and the uncontrollability during query period. A feedback based incremental nearest neighbor query method (FINN) is propose to solve the problem. The user sends feedback to the server according to the query, and the server chooses POIs to send to the user according to the feedback. Theoretical analysis and experimental results show that FINN can improve the performance of the system significantly while ensures user's anonymous requirements.

Keywords: Location-based services · Location privacy preserving ·
Feedback information · Incremental nearest neighbor query

1 Introduction

The rapid development of mobile communication and spatial positioning technology has promoted the rise of location-based services (LBS). K nearest neighbor query is an important query service of location services. It refers to finding K target objects (POI, point of interest) nearest to a query's current location, such as finding K restaurants or gas stations nearest to the query. This service requires the querier to provide the service provider with its exact location to obtain the query results. Real-time location information contains user behavior patterns. With the increasing attention to individual information security, the security of user location has been paid more and more attention. Sending location to service providers may lead to the leakage of privacy information such as identity and behavior patterns of individual users. How to realize k-nearest neighbor query without revealing the location privacy of individual users has become a hot topic in the field of privacy-sensitive location services.

At present, the main idea of location privacy preserving query is to hide the location of the inquirer and submit the hidden location and query request to the LBS server.

© Springer Nature Switzerland AG 2019
W. Ni et al. (Eds.): WISA 2019, LNCS 11817, pp. 467–479, 2019.
https://doi.org/10.1007/978-3-030-30952-7_46

The server completes the query processing of the hidden location information, and feeds the query results back to the initiator for filtering the target results. The main hiding technologies include spatial obfuscation [10, 11], data transformation [12–14], location perturbation and Private Information Retrieval [5, 13, 14]. Spatial obfuscation enlarges the location of the query to a generalized region containing the location and submits it to the LBS server. The query results are screened out by the query or trusted third party from the returned candidate solutions. Data transformation achieves the privacy protection of the query location by transforming the location of the query and the target object into another data space for query processing. In location disturbance, the query is directed to L. BS servers submit query requests for specific false locations until they return the results satisfying their query accuracy and privacy security requirements. Most PIR technologies are based on the quadratic congruence problem, and use location keys to provide strong privacy protection intensity, but there are problems such as large computational load and high communication cost. These methods have different emphasis on privacy protection intensity, query accuracy and query processing performance. The specific comparison is shown in Table 1.

Table 1. Comparison of location privacy protection query technologies

	Spatial cloaking	Data transformation	Location perturbation	PIR
Protecting strength	Controllable	Controllable	Uncontrollable	Controllable
Query cost	Higher	Higher	High	High
Query accuracy	Exact	Not exact	Exact	Exact
Dependence on the trusted third party	Yes	Yes	No	Yes

From the perspective of query mode architecture, spatial cloaking and data transformation technology mostly rely on trusted third party (acting as anonymizing servers) to participate in query processing in online or offline mode [1–4]. There are difficulties in implementation and the trusted third party is inclined to be the bottlenecks of the query system. Location perturbation technology uses the mode of direct interaction between query client and LBS server to realize privacy protection query, which has the advantage of not depending on trusted third party. The false point method is proposed in ref [6] and ref [7], which uses the false position point instead of the user's real location to initiate the query. SpaceTwist method is proposed in ref [8], in which the client specifies a location as the anchor node, and the server performs incremental nearest neighbor query on the anchor node until the critical condition is satisfied. Privacy protection query method Coprivacy [9] is proposed based on SpaceTwist. It implements K anonymity by querying the user to form an anonymous group, and can support privacy preserving query satisfying K anonymity.

Existing location perturbation based privacy-preserving query methods, just as the representative method SpaceTwist of them, have the following shortcomings.

(1) The number of iteration queries between the query client and the LBS server is uncontrollable. It results in large computation and communication overhead.

(2) The query client lacks the regulation mechanism for query efficiency and privacy security, which makes it difficult to meet the personalized regulation requirements of query efficiency and privacy security.

To solve the above problems, a feedback-based incremental nearest neighbor query method (FINN) is proposed. Based on location perturbation, the query client provides feedback information to the LBS server in each iteration of query between the query client and the LBS server to realize the existence of the LBS server. Guided query processing reduces the communication and computing overhead of location perturbation method, and realizes the adjustable query efficiency and location privacy protection effect.

The main contributions of this paper list as follows:

(1) A feedback angle-based control mechanism for query efficiency and privacy protection intensity is proposed to realize the dynamic and controllable adjustment of the query processing process of the server.

(2) By using feedback angle control mechanism, a nearest neighbor query method based on location perturbation to protect location privacy, FINN, is proposed to dynamically adjust query processing efficiency and privacy protection intensity.

(3) FINN method is realized, and experiments are devised to verify the effectiveness of the solution.

The organizational structure of the paper is as follows: Sect. 2 summarizes the related work and expounds the existing problems of SpaceTwist method. Section 3 gives the basic idea of FINN method and the calculation method of key parameter feedback angle; Sect. 4 elaborates the execution process of FINN method on client and LBS server respectively. Sections 5 and 6 carries out theoretical analysis and experiment on FINN method, respectively. Finally, it summarizes the full text and looks forward to the next step.

2 Related Work and Problem Description

Trusted third party is easy to become the bottleneck of system performance and the target of attacker's concentrated attack, so the location perturbation method which does not rely on trusted third party has its unique advantages.

2.1 SpaceTwist Method

SpaceTwist algorithm uses the query client to iteratively initiate queries about false locations (anchors). By analyzing the geometric relationship between the POI of the false locations feedback by the LBS server and the real locations of the queriers, it decides whether to continue the query process until the target query results are obtained. The algorithm introduces the concepts of query demand space γ and supply space τ. The demand space is a circular region with the center of the query's real location and the radius of its distance from current k nearest neighbor POI, and the supply space is a circular region with the center of anchor and the radius of its distance from the nearest return POI. In order to reduce communication overhead, several POIs are transmitted in one message. The querier maintains a heap W_k record of the k-nearest neighbor currently known by the querier. As shown in Fig. 1, q represents the querier and q' represents the anchor. The specific process is: (1) the demand space and the supply space are initialized to ∞ and 0, respectively; (2) the server continuously makes incremental nearest neighbor queries on the anchor and sends the query results to the querier; (3) when the query keeps judging whether the supply space fully contains the demand space, if execution (2) is not included. The query ends.

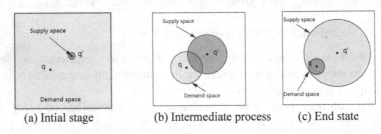

(a) Intial stage   (b) Intermediate process   (c) End state

Fig. 1. Demand space and supply space in SpaceTwist

2.2   Problem Description

Location perturbation-based method SpaceTwist has the disadvantage of high communication and computational overhead, and the intensity of location privacy protection and query efficiency can not be adjusted. How to dynamically adjust the intensity of privacy protection and query efficiency, and reduce the computational and communication overhead caused by unpredictable pseudo-location iteration queries are the main problems to be solved.

The high cost of query processing and the uncontrollable intensity of privacy protection in location perturbation-based query methods originate from the uncontrollable number of iteration query rounds and process between the query client and the LBS server. Considering that the query client can provide the LBS server with effective auxiliary information about iteration rounds, the unsupervised iteration can be changed into the guided iteration, and the dynamic strength of privacy protection and query efficiency can be realized.

3   Query Control Mechanism Based on the Feedback Angle

3.1   Basic Idea

In the SpaceTwist method, the LBS server iteratively queries the anchor until the critical condition is satisfied. Multiple rounds of communication are needed between the query and the LBS server, and the query needs multiple rounds of computation for the demand space and the supply space.

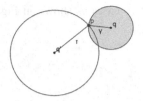
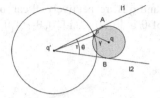

Fig. 2. Intermediate stage of SpaceTwist   **Fig. 3.** Illustration of feedback angel

As shown in Fig. 2, the gray area represents the current requirement space. Assuming that the W_k is full at this time, that is, the current k-nearest neighbor has been found, then the real k-nearest neighbor must be in this grey area. Because the server does not know this information, it will continue to iterate incremental queries on the anchor. If the distribution of POI adjacent to the anchor is uneven and the POI distributed outside the grey area is more intensive, more communication and computing overhead will be generated. In addition, in this process, queriers can not dynamically adjust their privacy protection intensity and query efficiency to meet their personalized needs. The analysis shows that the reason for the above situation is that the server only knows the end condition of the query and has no additional information to guide the query processing, which makes the query process uncontrollable.

Definition 1 Feedback angle θ. Feedback angle is the angle between the two tangent lines tangent to the outermost circle of the demand space through anchor node.

As shown in Fig. 3, tangent lines l_1 and l_2 are made to the outermost circle of the demand space through anchor point q', intersecting with the outermost circle of the demand space at points A and B, A q'B is the feedback angle.

The real k nearest neighbor of the query is in the grey area, and in the sector area where the anchor node is the vertex and the feedback angle is the angle. If the query node feeds back the information about this angle to the server, the server filters the query results that will be sent to the user according to the feedback angle, and only sends the query results that satisfy the specific conditions, it will reduce the communication overhead between the LBS server and the query, and thus reduce the computational overhead of the query. At the same time, the larger the feedback angle, the less likely the attacker will infer the real location of the query. Therefore, the query efficiency and privacy protection intensity can also be adjusted by adjusting the feedback angle.

3.2 Feedback Angle θ Computing

Without losing generality, the whole region space is assumed to be a rectangular region, with the point at the lower left corner of the region as the origin, the lower x-axis, and the left Y-axis as the coordinate system. For convenience, coordinates represent the longitude and latitude of nodes.

As shown in Fig. 4, there are two tangent points A and B. The anchor node q' is taken as the end point and qM is made along the positive direction of the x-axis. The qM is parallel to the x-axis. θ_1 and θ_2 are the angles of q'M rotating counterclockwise to q'A and q'B, respectively. θ can be divided into two cases as shown in Fig. 4. The range of θ is $[\theta_1, \theta_2]$ and $[0, \theta_1] \cup [\theta_2, 2\pi]$.

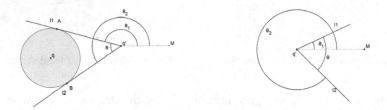

Fig. 4. The range of feedback angel θ

Assume that the attacker grasps the following information: (1) all POI locations; (2) anchor locations; (3) processing algorithm on client side; (4) query processing algorithm on server-side. The attacker can get information about the vertex and angle of the sector area, as well as the fact that the inquirer located on the angle bisector of the vertex. In order to ensure the location privacy of the query, the range of feedback angle θ is extended to both sides with scale α and β, that is, the range is changed to $[(1-\alpha)\theta_1, (1+\beta)\theta_2]$ or $[0, \theta_1(1+\alpha)] \cup [\theta_2(1-\beta), 2\pi]$.

The calculating process of θ is as follows: (1) make tangent line along the anchor nodes to the outermost circle of the demand space, and get tangent points A and B. (2) calculate the angles θ_1 and θ_2 between q'A, q'B and qM, and get range of θ (3) The specified range of θ is extended to both sides.

4   Feedback Angel Based Location Privacy Preserving Nearest Neighbor Query Method FINN

4.1   Client-Side Processing

The client is mainly responsible for judging the end of the query and calculating the feedback angle.

Definition 2 Query request Q. Queries initiated by queriers can be expressed as Q = {l, t, con}. Among them, l = {x, y} represents the location of the anchor point, x denotes the longitude of the anchor, y is the latitude of the anchor; t is the query time; con is query content.

The main process of FINN algorithm in client side include four steps: (1) sending query request Q to LBS server; (2) calculating the distance dist (p, q) between querier q and p for each POI point p sent by server, and putting <dist (p, q), p> into w_k until the cumulative number reaches k. At this time, it corresponds that the current known k nearest neighbor has been found, and the real k nearest neighbor of the querier just locates in the current demand space; (3) judging whether the anchor is outside the demand space, if dist (p, q) < γ, then calculating θ and send it to the server; (4) when the critical condition is satisfied, that is, when the supply space contains the demand space, end the query process. During the whole process, the client continues updating the demand space and supply space and updates w_k. The specific implementation process is shown in Algorithm 1.

Algorithm1. FINN_Client

Input: q, k, α, β

Output: k nearest neighbors of the querier

---

1. $\tau \leftarrow 0$;
2. $\gamma \leftarrow \infty$;
3. $W_k \leftarrow$ k pairs of <NULL, ∞>;
4. querier qsubmit the query request to LBS server;
5. while $\gamma + dist(p,q) > \tau$;
6.   receive message S from the server;
7.   for each p in S
8.   while $|w_k| < k$
9. $w_k \leftarrow < dist(p,q),\ p>$;
10.   end while;
11.   while $dist(q',q) < \gamma$
12.   if $dist(p,q) < \gamma$
13. $\tau \leftarrow dist(p_i,q)$;
14.   heap tuple $\leftarrow <p, dist(p,q)>$;
15.   end if;
16.   end while;
17.   Generate_θ;
18.   send θ to the server;
19.   if $dist(p,q) < \gamma$
20. $\tau \leftarrow dist(p_i,q)$;
21.   heap tuple $\leftarrow <p, dist(p,q)>$;
22.   end if;
23.   end for;
24. end while
25. Send end query request to server
26. return W_k;

4.2 Server-Side Processing

The server-side processing consists of four steps: (1) receiving anchor node q'; (2) sending back the query results before receiving θ; (2) after receiving θ, the server calculates the $\angle P\ q'M$ shown in Fig. 4 and sends p to the querier if $\angle P\ q'M \in$ range of θ; (4) receiving the request of the querier to ending the query. The specific implementation process is shown in Algorithm 2.

Algorithm2. FINN_Server

Input: the anchor q' , feedback angel θ

Output: query result p

---

1. receive query request;
2. while Not receiving request of ending the query
3.   while Not receiving feedback information from the querier
4.   send query result p to the querier
5.   end while; //receive the feedback information
6.   calculate $\angle pq'M$;
7.   if $\angle pq'M \in$ the range of θ
8.   send query result p to the querier
9. end while.

5 Performance Analysis

This section analyses the privacy security, accuracy and query efficiency of FINN method. The query efficiency mainly considers the communication overhead and query client computing overhead.

5.1 Privacy Security Analysis

Given the execution process of FINN method, the attacker can filter the querier meeting the specific conditions. Without additional information, the number of the query can indicate the security of privacy.

Definition 3 Privacy Area Ψ [8]. The set of all possible users that an attacker can infer.

For a query, the angle information sent to the LBS server is recorded as θ. Assuming that the number of messages sent by the server is m and the capacity of the message is c, the range of Ψ can be limited by the following formula:

$$\begin{cases} dist(q',p) + \min_{1 \leq i \leq (m-1)c}^{k} dist(q,p_i) > dist(q',p_{(m-1)c}) \\ dist(q',p) + \min_{1 \leq i \leq mc}^{k} dist(q,p_i) \leq dist(q',pmc) \\ \forall p \in \Psi, \angle pq' \, M \in \theta \end{cases} \qquad (1)$$

Property 1. FINN method has the ability of regulating privacy security it can provide.

Proof. The angle the user sends to the server is marked as θ. The attacker infers that the real location of the user is located in the light gray sector S with the anchor as the origin and theta as the angle and the radius of the distance between the final query result of the server and the anchor as the radius, as shown in Fig. 5. Note that the area of the light grey ring is L. Assume that the user initiates two queries, and the angles are θ_1 and θ_2, respectively, and $\theta_1 > \theta_2$, hence $S_1 = L_1 \cdot \theta_1$, $S_2 = L_2 \cdot \theta_2$. Considering the arbitrariness of node distribution and the communication mechanism between querier and LBS server, $P(L_1 > L_2) = 1/2$, and L and θ are independent of each other. It can be deduced that:

$$P(S_1 > S_2) > P(L_1 > L_2 \wedge \theta_1 > \theta_2) = P(L_1 > L_2) \cdot P(\theta_1 > \theta_2) = P(L_1 > L_2) = 1/2$$

It demonstrates that the larger the feedback angle, the larger the privacy area and the stronger the privacy security. That is to say, FINN has a regulatory effect on privacy security.

Property 2. If POI and users in the query space are uniformly distributed, then the attacker guesses that the probability of the user's real location is $\dfrac{N}{M\left(c - \sqrt{\frac{kc_0}{2\pi m}}\right)}$. Among them, N is the number of POI, M denotes the number of users, k is the number of nearest neighbor POI that users need to get.

Proof. For $\frac{\pi\gamma^2}{S} = \frac{k}{N}$, radius γ of final demand space is $\sqrt{\frac{Sk}{\pi N}}$. Similarly, when the query node receives (m–1) and m messages, the corresponding supply space radius $\tau_{m-1} = \sqrt{\frac{(m-1)cS}{\pi N} \cdot \frac{2\pi}{\theta}}$, $\tau_m = \sqrt{\frac{mcS}{\pi N} \cdot \frac{2\pi}{\theta}}$. The area of the area is denoted as S. Number

of users in the privacy area is: $\pi\left[(\tau_m - \gamma)^2 - (\tau_{m-1} - \gamma)^2\right] \cdot \frac{M}{S} \cdot \frac{\theta}{2\pi} =$

$$\left[c + \theta\left(\sqrt{\frac{2(m-1)ck}{\pi\theta}} - \sqrt{\frac{2mck}{\pi\theta}}\right)\right] \cdot \frac{M}{N} = \frac{M}{N}\left(\beta - \frac{2ck}{\pi\left(\sqrt{\frac{2(m-1)ck}{\pi\theta}} + \sqrt{\frac{2mck}{\pi\theta}}\right)}\right) \approx$$

$\frac{M}{N}\left(\beta - \sqrt{\frac{kc\theta}{2\pi m}}\right)$, the probability that the real location can be inferred is $\dfrac{N}{M\left(\beta - \sqrt{\frac{kc\theta}{2\pi m}}\right)}$.

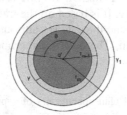

Fig. 5. Privacy area of the querier

5.2 Accuracy and Query Cost Analysis

FINN algorithm makes tangent line along the requirement space, and the effective query area of LBS server (the sector area with anchor q′ as the vertex and θ acting as circle angle) contains the requirement space, so the LBS server returns to the user POI containing the real k nearest neighbor of the querier, so FINN algorithm obtains accurate results.

The LBS server filters the query results according to the feedback angle and returns them to the user. So the communication cost of FINN is less than that of SpaceTwist, and the smaller the feedback angle is, the smaller the communication cost is.

Queriers usually acquire LBS services from mobile terminals such as smart phones. Although the performance of mobile terminals has been greatly improved, it is still the bottleneck of system computing performance compared with servers. Therefore, this paper mainly analyses the computational overhead of queriers. Compared with SpaceTwist method, FINN needs to calculate the location of anchor and requirement space in the early stage of query, but its computation amount is less than that of filtering POI in the later stage of query, so the computation cost of FINN is less than that of

SpaceTwist, and the smaller the feedback angle is, the smaller the computation cost of querier is.

6   Experimental Analysis

The experiment mainly investigates two aspects: one is to compare and analyze the differences between FINN method and SpaceTwist method in terms of query time and communication cost from the number of neighbors and POI nodes that queriers need to find; the other is to analyze the regulation of feedback angle theta on the privacy security and query performance of FINN. Response time refers to the time when the querier initiates the query to the LBS server to obtain the final query result. Traffic refers to the number of messages sent by the LBS server to the query client.

6.1   Experimental Environment

The algorithm is implemented in Java language. The experimental hardware environment is 2.9 GHz processor, 4G memory, and the operating system is Windows 7. The data set is generated by Thomas Brinkhoff Road Network Generator [15], which is widely recognized by the industry. It is based on the traffic network data of Aldenburg City, Germany. The user can define the data set attributes by himself. The area of the data set is 23.57 km × 26.92 km. The default experimental parameters are shown in Table 2.

Table 2.   Default parameters of experiments

Parameters	Default value
Number of queriers	10000
K(Number of target POIs)	10
Number or POIs	100000
Feedback angle enlarging range	10%

The maximum transmission unit of network between user and server is 576 bytes, the header of message is 40 bytes, the length of each communication message is 8 bytes, and the size of message capacity is (576–40)8 = 67.

6.2   Experimental Result

6.2.1   Comparison of FINN and SpaceTwist Algorithms

Figures 6 and 7 demonstrates variation trend of communication traffic and response time. It can be seen the communication overhead and response time increase with increasing K. This is because the demand space expands with the increase of query demand, and the server needs to query more POIs to make the supply space cover the demand space. In addition, the communication overhead and response time of FINN method are better than those of SpaceTwist method. The larger the number of nearest neighbors, the more obvious the advantage of FINN method shows. The reason lies in that only those POIs located within specific angle range need sending from server to query client in FINN, which reduces the communication overhead and the computing overhead at client side.

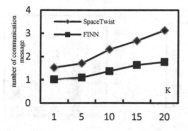

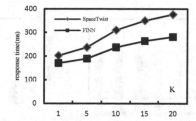

Fig. 6. Communication cost vs. varying K   **Fig. 7.** Response time vs. varying K

Figures 8 and 9 show communication cost and response time varying trends with increasing K. It is obviously that the communication overhead and response time increase with increasing K. The reason is with querier's demand space not changed, higher POI density leads to more POIs in the supply space. It is noted that the communication overhead and response time of FINN are better than that of SpaceTwist, and the advantage increases sharply with larger K.

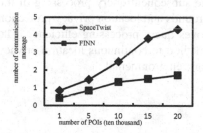

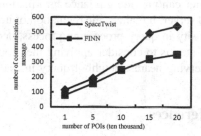

Fig. 8. Communication cost vs. number of POIs   **Fig. 9.** Response time vs. number of POIs

The experimental results show that FINN is suitable for different POI densities and different K, and has good scalability. In addition, compared with SpaceTwist, FINN significantly improves query performance, and the extent of improvement increases with K.

6.2.2 Controllability of Feedback Angle to FINN

As shown in Figs. 10 and 11 communication cost, as well as response time, increase with enlarging feedback angel. This is because with feedback angle increases, the LBS server needs to find and process more POIs, and the communication and computing overhead will increase accordingly.

Figure 12 represents the impact of feedback angle on privacy security, the privacy preserving effect is measured by the number of users in the privacy area Ψ. The larger the feedback angle is, the more users locate inside the privacy area, and the smaller the probability that the attacker infers that the real query users are, the stronger the privacy security is.

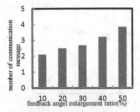

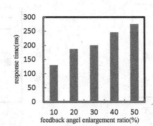

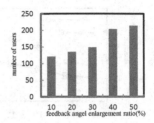

Fig. 10. Communication cost vs. feedback angel

Fig. 11. Response time vs. feedback angel

Fig. 12. Privacy protection vs. feedback angel

7 Conclusion

Aiming at the problem of high query cost, as well as uncontrollable query performance and privacy protection strength, in location perturbation-based privacy-preserving nearest neighbor queries, a feedback angle-based privacy-preserving nearest neighbor query method FINN is proposed. By sending feedback angle to the LBS server, the querier can provide guidance information for the subsequent query processing of the LBS server. It realizes the regulation of iteration rounds and location privacy protection intensity in query processing process, and improves query processing efficiency. The next step is to consider applying the proposed method to continuous location privacy preserving nearest neighbor query in road network environmental.

References

1. Kalnis, P., Ghinita, G., Mouratidis, K., Papadias, D.: Preventing location-based identity inference in anonymous spatial queries. IEEE Trans. Knowl. Data Eng. **19**(12), 1719–1733 (2007)
2. Um, J.-H., Kim, H.-D., Chang, J.-W.: An advanced cloaking algorithm using Hilbert curves for anonymous location based service. In: Proceedings of 2010 IEEE Second International Conference on Social Computing, pp. 1093–1098 (2010)
3. Hossain, A.-A., Hossain, A., Yoo, H.-K., Chang, J.-W.: H-star: Hilbert-order based star network expansion cloaking algorithm in road networks. In: Proceedings of IEEE 14th International Conference on Computational Science and Engineering (CSE), pp. 81–88, August 2011
4. Gruteser, M., Grunwald, D.: Anonymous usage of location-based services through spatial and temporal cloaking. In: Proceedings of 1st International Conference on Mobile Systems, Applications and Services, pp. 31–42 (2003)

5. Wu, J., Ni, W., Zhang, S.: Generalization based privacy-preserving provenance publishing. In: Meng, X., Li, R., Wang, K., Niu, B., Wang, X., Zhao, G. (eds.) WISA 2018. LNCS, vol. 11242, pp. 287–299. Springer, Cham (2018). https://doi.org/10.1007/978-3-030-02934-0_27
6. Hong, J.I., Landay, J.A.: An architecture for privacy-sensitive ubiquitous computing. In: Proceedings of the 2nd International Conference on Mobile Systems, Applications, and Services, pp. 177–189 (2004)
7. Kido, H., Yanagisawa, Y., Satoh, T.: An anonymous communication technique using dummies for location-based services. In: Pervasive Services, Proceedings of International Conference, pp. 88–97 (2005)
8. Yiu, M.L., Jensen, C.S., Huang, X.G., Lu, H.: SpaceTwist: managing the trade-offs among location privacy, query performance, and query accuracy in mobile services. In: IEEE 24th International Conference on Data Engineering, pp. 366–375 (2008)
9. Huang, Y., Huo, Z., Meng, X.: CoPrivacy: a collaborative location privacy-preserving method without cloaking region. Chin. J. Comput. **34**(10), 1975–1985 (2001). (in Chinese)
10. Gedik, B., Liu, L.: Protecting location privacy with personalized k-anonymity: architecture and algorithms. IEEE Trans. Mobile Comput. **7**(1), 1–18 (2008)
11. Chow, C.Y., Mokbel, M.F., Aref, W.G.: Casper*: query processing for location services without compromising privacy. ACM Trans. Database Syst. **34**(4), 1–45 (2009)
12. Khoshgozaran, A., Shahabi, C.: Blind evaluation of nearest neighbor queries using space transformation to preserve location privacy. In: Papadias, D., Zhang, D., Kollios, G. (eds.) SSTD 2007. LNCS, vol. 4605, pp. 239–257. Springer, Heidelberg (2007). https://doi.org/10.1007/978-3-540-73540-3_14
13. Papadopoulos, S., Bakiras, S., Papadias, D.: Nearest neighbor search with strong location privacy. Proc. VLDB Endow. **3**(1–2), 619–629 (2010)
14. Paulet, R., Kaosar, M.G., Yi, X., Bertino, E.: Privacy-preserving and content-protecting location based queries. In: Kementsietsidis, A., Salles, M.A.V. (eds.) Proceedings of the IEEE 28th International Conference on Data Engineering (ICDE 2012), pp. 44–53. IEEE Computer Society, Los Alamitos (2012)
15. Brinkhoff, T.: A framework for generating network based moving objects. GeoInformatica **6**(2), 153–180 (2000)

Semi-supervised Sentiment Classification Method Based on Weibo Social Relationship

Wei Liu and Mingxin Zhang[✉]

Department of Computer Science and Engineering,
Changshu Institute of Technology, Suhzou, China
mxzhang163@163.com

Abstract. Sentiment classification of microblog text is one of the hotspots and important research issues in text sentiment analysis. Aiming at the problem that the existing researches mostly assume that the micro-blog sentiments are independent of each other and have strong dependence on the training set, a semi-supervised sentiment classification method based on Weibo social relationship is proposed. The method utilizes the user's theme sentimental consistency and the approval of social relationships (like and repost) in Weibo to establish the sentimental relationship between microblogs to solve the problem that microblog sentiments are independent of each other. Semi-supervised sentimental classification model is constructed by establishing the sentimental relationship between labeled micro-blog and unlabeled micro-blog, which reduced the dependence on training set. Specifically, the semi-supervised sentiment classification method was constructed by constructing a microblog sentimental relationship matrix using the Laplacian matrix of the above microblog social relationship graph, and adding to the text content based classification model. Climbing the real dataset of Sina Weibo for experiment, the experimental results showed that the method is superior to other typical sentiment classification methods in terms of accuracy and recall rate. The validity of this method is verified and the dependence on training data set is reduced to a certain extent.

Keywords: Sentiment classification · Semi-supervision · Weibo · Social relationship

1 Introduction

With the development of Web2.0, online social networking sites such as Twitter and Sina Weibo have attracted a large number of users and have developed rapidly. Domestically, as of December 2018, Weibo's monthly active users reached 462 million, of which 93% were mobile users, and the average daily active users reached 200 million. Because of such a large number of audience users, the Weibo platform has penetrated into all aspects of life with amazing influence, and has become the source and source of hot events in various industries. Opinion mining has become one of the most vigorous research areas in NLP field [1–3], by analyzing the sentiments of these massive audience users, government departments can better understand and monitor the dissemination of public topic events, and timely guide the development of topic events. Enterprises can

© Springer Nature Switzerland AG 2019
W. Ni et al. (Eds.): WISA 2019, LNCS 11817, pp. 480–491, 2019.
https://doi.org/10.1007/978-3-030-30952-7_47

efficiently obtain customer product feedback [4, 5] to improve products quality and develop more efficient product promotion programs. Therefore, the analysis of Weibo user sentiments is of great significance to enterprise product promotion and marketing [6], public opinion monitoring and guidance [1, 7, 8].

Compared with traditional texts, microblog text has the characteristics of short, colloquial, and including online words, which makes the traditional text sentiment analysis method not effective for microblog text. Existing researches are mainly carried out from methods based on sentimental knowledge and methods based on feature classification. Mostly based on two theories of emotional consistency [9] and emotional infection [10]. Zhang [11] based on the sentimental dictionaries such as basic sentimental dictionary and network dictionary, through four operations of all sentimental words in sentences, he got the sentiments of sentences, and further used the relationship between sentences to calculate the sentiments of micro-blog text, the unsupervised method solves the dependence on sufficient training corpus and the problem of containing online words in Weibo, but did not consider the sentimental relationship between Weibo, so there is still much room for improvement in classification performance. Hai [12] introduced the sentiment recognition layer based on the LDA theme model, and proposed a supervised joint level sentiment model (SJASM), which solved the problem of microblog sentiment and topic association, however, the model has strong dependence on large-scale training corpus for classification performance, resulting in a large amount of manpower and material resources to construct large-scale sufficient training corpus. Some researchers have also conducted research on the sentimental relationship between Weibo, Hu [13] constructed the sentiment classifier by making the sentimental polarities of two microblogs published by the same two users or two users with friends as similar as possible, although considering the social relationship between microblogs, the two users of the friend relationship are not completely consistent with the same thing, so the accuracy of the sentimental classification needs to be improved, Zou [14] constructed the sentimental classifier by make microblog polarity as similar as possible between micro-blogs with three relationships: user context (two micro-blogs published by the same user), topic context (whether two micro-blogs belong to the same topic) and structural similarity context (structural similarity between the two micro-blog publishers), which solves the problem of sentimental relationship between microblogs, but does not effectively solve the problem that classifiers have strong dependence on large-scale training sets.

In summary, the existing research shortages mainly include the following two points: (1) the sentimental relationship between microblogs is not effectively established; (2) the dependence on the training data set is strong. This paper selects Sina Weibo data as the research object. For the first question, this paper uses the user's theme sentimental consistency and the social relationship (like and repost) in Weibo to establish the sentimental relationship between Weibo to solve the problem. For the second problem, the semi-supervised sentiment classification method based on Weibo social relationship (SASR, Sentiment Analysis using Social Relationships) is constructed by establishing the sentimental relationship between labeled and unlabeled microblogs, and introducing into the sentimental classification model based on text content to solve this problem. By crawling Sina Weibo's real topic dataset and experimenting with five typical Weibo sentiment classification methods, the SASR method is superior to the other five methods in terms of accuracy and recall rate.

2 Semi-supervised Sentiment Classification Method Based on Weibo Social Relationship

The microblog text has short features, colloquial, and includes network words, which leads to unclear sentimental expression. The effect of using traditional text sentiment analysis methods is not ideal. At the same time, Weibo is a social network with social relationship [15]. In order to make full use of this social feature, we propose a semi-supervised sentiment classification method based on Weibo social relationship, That is, on the basis of micro-blog text content, we use social relations to establish micro-blog sentimental relations, and construct a semi-supervised sentimental classification method, to some extent, it overcomes the short and colloquial text of Weibo text, and also reduces the cost of manual labeling.

2.1 Problem Definition

First, we define the symbols as follows.

Given a microblog corpus $P = [X,Y] \in R^{n \times (m+c)}$, where $X \in R^{n \times m}$ is the microblog content matrix, $Y \in R^{n \times c}$ is the sentimental tag matrix, n is the number of microblog messages, m is the number of features, c is the number of sentiments to be classified, and q is the number of topics. In this paper, we focus on the classification of sentiment polarity, i.e. $c = 2$. If the microblog sentimental polarity is positive, then $Y_{i*} = [+1,-1]$, if the microblog sentimental polarity is negative, $Y_{i*} = [-1, +1]$, it should be noted that it is also feasible to extend the number of classifications into multiple types of sentiment classification tasks. $u = \{u_1, u_2, ..., u_d\}$ is the user set, where d is the number of different users in the corpus, $U \in R^{d \times n}$ is the user-microblog matrix, $U_{ij} = 1$ means the message t_j is issued by the user u_i. $A \in R^{n \times n}$ is a microblog sentimental relationship matrix, $A_{ij} = 1$ indicates the social relationship described is exist between the user of microblog i and the user of microblog j. Since the social relationship between the two users in the text is not necessarily symmetrical, A is not symmetrical.

According to the above symbol definition, the microblog sentimental polarity classification problem in this article can be defined as: given microblog corpus P, message content matrix X, and corresponding sentiment label matrix Y, microblog sentimental relationship matrix A, the target is learning classifier W to facilitate the classification of Weibo of unknown sentiments

2.2 User Topic Context

It has been proven in the social sciences that the same user's opinion on the same subject are consistent over time [16], which we call the user topic context. Therefore, if two microblog are published by the same user and are related to the same topic, we believe that the sentiments of these two microblog are consistent in a short time.

We use $M \in R^{n \times q}$ to represent the microblog-topic matrix, $M_{ij} = 1$ means that the microblog i belongs to the topic j, otherwise, $M_{ij} = 0$. We denote the user topic context matrix as $A_{topic} \in R^{n \times n}$, indicating whether the two microblogs belong to the same user and belong to the same topic, the calculation formula is as follows:

$$A_{topic} = \left(U^T \times U\right) \circ \left(M \times M^T\right) \tag{1}$$

Where $U^T \times U$ indicates whether the two microblogs belong to the same user, $M \times M^T$ indicates whether the two microblogs belong to the same topic, and $\circ$ indicates the hadamard product.

2.3 Approval Relationship

Approval the relationship in this article refers to repost or like. If the Weibo user has a repost or like relationship on a certain topic, the two Weibo users have more consistent opinion on the topic. This assumption is closer to the actual situation. Weibo users generally only respost or praise this message when they agree with other users' opinion on a certain topic. This theory is also verified in other literature [17].

We use $F^k \in R^{d \times d}$ to indicate whether there are repost or like relationships between the two users on the topic k, where $k = 1, 2, ..., q$. Use A^k to represent the social relationship matrix in the topic k between the microblogs, that is, whether the users of the two microblogs have a repost or like relationship on the topic k, and the calculation is as follows:

$$A^k = U^T \times F^k \times U \tag{2}$$

Where U is a user-microblog matrix. If $A_{ij}^k = 1$, there is a forwarding or like relationship between the two users of microblog i and j on the topic k, otherwise, $A_{ij}^k = 0$. Therefore, we express the approval social relationship matrix as A_{ex}, and its calculation formula is as follows:

$$A_{ex} = \sum_{i=1}^{q} A^k \tag{3}$$

Among them, $A_{ex} \in R^{n \times n}$, A_{ex} indicates that two users belonging to two microblog have repost and like relationships on a specific topic.

2.4 Semi-supervised Sentiment Classification Method Based on Weibo Social Relationship

In order to better represent the text of the microblog, we use the best unigram model in the Pang [18] experiment to construct our feature space. We use the least squares method to learn the microblog content model, and the least squares method aims to learn the classifier by solving the following optimization problems:

$$\min_{W} \frac{1}{2} \left\| X^T W - Y \right\|_F^2 \tag{4}$$

Among them, W is a classifier for learning.

We further use the above two social relationship matrices to construct the micro-blog relationship matrix A:

$$A = \lambda_1 \times A_{topic} + \lambda_2 \times A_{ex} \tag{5}$$

Where λ_1 and λ_2 are parameters that combine the two social relationships.

The basic idea of sentimental classification in this paper is that if there are two kinds of social relations or one of them, that is to say, satisfying the sentimental relationship constructed by the above two kinds of social relations matrix, the sentimental polarity of the two microblogs is as similar as possible. We solve this problem by minimizing Eq. (6).

$$\min_{W} \frac{1}{2} \sum_{i=1}^{n} \sum_{j=1}^{n} F_{ij} \left\| \hat{Y}_{i*} - \hat{Y}_{j*} \right\| \tag{6}$$

$$= \min_{W} \sum_{l=1}^{c} \hat{Y}_{*k}^T (D - A) \hat{Y}_{*k} \tag{7}$$

$$= \min_{W} tr(X^T W L X^T W) \tag{8}$$

Where $tr(\bullet)$ is the trace of the matrix, $\hat{Y} = X^T W$ is the fitted value of the sentiment label Y, $L = D - A$ is the Laplacian matrix, A is the user social relationship matrix, and $D \in R^{n \times n}$ is the diagonal matrix.

For different social relationships existing between Weibo, different microblog sentimental relationships A can be constructed, so that different sentimental relationship matrices A can be used to obtain Laplacian matrix L, and microblog sentimental relationships can be integrated into message content sentiment classification. In the model, the resulting final sentiment classification model can be expressed using the following formula:

$$\min_{W} \frac{1}{2} \left\| X^T W - Y \right\|_F^2 + \frac{\alpha}{2} tr(X^T W L X^T W) \tag{9}$$

Among them, α is the weight of controlling the sentimental relationship of microblog in the sentiment classification model.

3 Algorithmic Solution

3.1 Sparse Processing

Compared with the text in the traditional sentiment classification, the microblog text has the characteristics of colloquial and short words, which leads to two problems. First, due to large-scale corpus and vocabulary, the text representation model (such as word bag or n-gram) usually leads to high dimensional feature space, and secondly, short and noisy microblog text makes the data representation very sparse. These problems create difficulties in building models with superior performance.

Therefore, we consider providing sparse reconstruction for the sentiment classification model. Sparse regularization has been widely applied to many data mining methods that have been obtained more stable and efficient. The linear reconstruction error based on L_1 norm can be minimized to obtain the sparse representation of text [18], which is robust to noise in features. We can optimize the following formula to learn the classification model.

$$\min_{W} \frac{1}{2}\left\|X^T W - Y\right\|_F^2 + \frac{\alpha}{2} tr(X^T WLX^T W) + \beta\|W\|_1 \tag{10}$$

Where β is the weight of regularization. In the objective function, the first term is the least squares loss, the second term is the microblog sentimental relationship, the third term is the L_1 norm regularization of the sentiment classification matrix W, and the third term causes some coefficients to be zero. Therefore, lasso [18] performs a continuous subset selection and can control the complexity of the model.

3.2 Optimization Solution

Since $\|W\|_1$ is non differentiable, the proposed objective function is non-smooth. We use the SANT optimization algorithm in [13] to solve this problem. It proposes to solve the non-smooth optimization problem in Eq. (10) by optimizing its equivalent smooth convex surface. Equation (10) can be re-expressed as constrained smooth convex optimization problem, which is specifically expressed as follows.

$$\min_{W \in Z} f(W) = \frac{1}{2}\left\|X^T W - Y\right\|_F^2 + \frac{\alpha}{2} tr(X^T WLX^T W) \tag{11}$$
$$where \ Z = \left\{W\|\|W\|_1 \le z\right\}$$

When considering the constraint Z and the given β in the Eq. (11), the $t + 1_{th}$ W can be calculated by the following equation.

$$W_{t+1}^j = \begin{cases} (1 - \frac{\beta}{\lambda_t\|u_t^j\|})u_t^j & if \|u_t^j\| \ge \frac{\beta}{\lambda_t} \\ 0 \ otherwise \end{cases} \tag{12}$$

The convergence ratio of the above method is $1/\varepsilon$. As discussed in [13], our constrained smooth convex optimization problem can be further accelerated to achieve optimal convergence $1/\sqrt{\varepsilon}$.

4   Experimental Results and Analysis

4.1   Experimental Data Set

We use the scrapy crawler framework to obtain related microblog users, select 1000 seed users, and use the depth-first traversal method to crawl the relevant microblog users to obtain the Sina Weibo real topic data set. We select six topics of topic time from June 2, 2017 to July 21, 2017, including "Xu Yuyu Case", "2017 NBA Final", "Li Wenxing Event", "Shunfeng Rookie's Data Disconnection", "Sichuan's Fighting Orphans", "Chinese-style Dating Disdain Chain" and so on, they are named Topic 1, Topic 2, Topic 3, Topic 4, Topic 5 and Topic 6 in turn. The labeled microblogs are positive, negative and unlabeled. The size of the labeled corpus is shown in Table 1.

Table 1. Experimental data.

Topic	Positive	Negative	Unlabeled
Total	42735	56854	24938
Sample ratio	38.20%	45.55%	16.25%

4.2   Experimental Verification Analysis

In the experiment, we compares the sentiment analysis using social relationships (SASR) with the following typical sentiment classification methods:

(1). LS: Least squares method [20] is a supervised classification method widely used for independent and identically distributed data.

(2). Support Vector Machine (SVM): SVM [18] is a classifier widely used in the fields of text and image classification.

(3). Naïve Bayes (NB): Like SVM, NB [20] is also a supervised classifier widely used in many fields.

(4). SANT: Hu [12] proposes a classification method that combines the theory of sentimental consistency and sentimental infection.

(5). SASS: A sentiment classification method proposed by Zou [14] that combines user context, topic context and structural similarity (friend relationship).

There are four important parameters in the experiment, α, β, λ_1 and λ_2, α is the parameter that controls the weight of the social relationship in the classification model, β is the sparse regularization parameter, and λ_1 and λ_2 are combined with three kinds of microblogs. The parameters of the sentimental relationship. α and β can be adjusted by cross-validation. In the experiment, we set $\alpha = 0.05$, $\beta = 0.1$, $\lambda_1 = 1$, $\lambda_2 = 1$.

In order to verify whether the Weibo social relationship used in the text can represent the consent relationship between Weibo users. We validated the correlation between Weibo social relationships and Weibo sentimental tags on the six Weibo topic datasets. Related research has verified the positive correlation between social relationships and tweet sentiment tags in Twitter through some analytical methods [13, 21]. Here, we use another data analysis method to evaluate the correlation between microblog social relationships and microblog sentimental tags.

Given the microblog relationship graph $G = (V,E)$ for establishing the social relationship in this article, we calculate the proportion of the corresponding nodes with the same sentiment label for all edges in E, which is expressed as

$$r = \frac{\sum_{i=1}^{n}\sum_{j=1}^{n} 1(Y_{i*} = Y_{j*}, e_{ij} \in E)F_{ij}}{\sum_{i=1}^{n}\sum_{j=1}^{n} 1(e_{ij} \in E)F_{ij}} \tag{13}$$

Among them, $1(\bullet)$ is an indicator function.
The analysis results of the six topic data sets are shown in Table 2.

Table 2. Correlation analysis of social relationships and sentimental tags.

Topic	Topic1	Topic2	Topic3	Topic4	Topic5	Topic6
A_{topic}	0.74	0.67	0.75	0.72	0.76	0.68
A_{ex}	0.72	0.64	0.77	0.75	0.74	0.64
$A = A_{topic+} A_{ex}$	0.84	0.73	0.87	0.85	0.86	0.74

As can be seen from Table 2, in the six different topics, there is a clear positive correlation between the two social relationships and the sentimental tags, and the correlation between the two social relationship combinations and the sentimental tags is stronger. However, for the topic "2017 NBA Finals" and "Chinese-style blind links", their relevance is relatively weak, because the textual diversity in these two topics weakens the role of the two social relationships. The diversity here is reflected in the different themes or description objects of the two microblogs, but express similar sentiments, such as the following two microblogs from the same user A, who also supports the Cavaliers, but the description objects are different.

A: "#2017NBA Finals #No matter what, Du soft eggs are always soft eggs, and the brave dogs are always brave dogs."

A: "#2017NBA Finals #The finals are averaging three pairs, which is awesome for my LeBron. Although I lost, we all know, don't blame you."

Throughout the experimental corpus, we used a 10-fold cross-validation to compare the accuracy, precision, and F1 values of the SASR method with the other five typical methods. The results are shown in Table 3.

It can be seen from the Table 3 that the SANT, SASS and SASR methods have improved the classification performance compared to the three benchmark methods, it shows that it is effective to consider the social characteristics of micro-blog when

Table 3. Comparison of classification performance of six methods.

Topic	Precision		Recall		F1	
	Positive	Negative	Positive	Negative	Positive	Negative
LS	0.741	0.669	0.726	0.705	0.733	0.687
NB	0.766	0.674	0.737	0.712	0.751	0.692
SVM	0.781	0.683	0.758	0.706	0.769	0.694
SANT	0.817	0.741	0.783	0.763	0.800	0.752
SASS	0.839	0.762	**0.821**	0.781	0.830	0.771
SASR	**0.863**	**0.790**	**0.821**	**0.811**	**0.842**	**0.801**

classifying micro-blog sentiments. Compared with SANT and SASS methods, SASR method has the same positive recall rate as SASS method, and improves the performance of other aspects by by 1–2% points, which proves that we consider the user topic context and the approval social relationship more reasonable.

One difficulty with sentiment analysis is the lack of manually labeled training data. In the supervised approach, public data sets are not always sufficient for better training purposes. In order to further verify the sensitivity of the model to the size of the training data, in Fig. 1, we select 5% to 30% training data from the whole microblog corpus and compare the SASR method with SANT and SASS methods which also consider the characteristics of microblog.

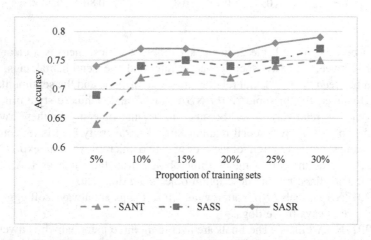

Fig. 1. Classification performance of small scale training sets.

In Fig. 1, the classification performance of SASR is always better than the other two methods in different size training sets. When the size of the training data set is reduced, the SASR method achieves a greater improvement than the other two methods, and when the size of the training data set changes, the SASR classification performance varies less than the other two methods. This indicates that the SASR method is less sensitive to the size of the training data set and can save the cost of constructing the training set to a certain extent.

Finally, we evaluated the effects of the parameters α and β on the classification performance of the SASR method, using 80% of the entire Weibo corpus for training and the rest for testing. When $\beta = 0.1$, the effect of the change of α on the classification performance of SASR is shown in Fig. 2. When $\alpha = 0.05$, the influence of the change of β on the performance of SASR is shown in Fig. 3.

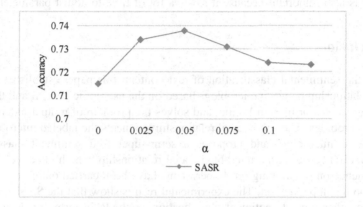

Fig. 2. Change of relative α performance of SASR.

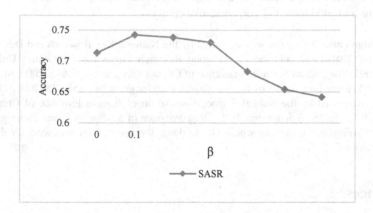

Fig. 3. Change of relative β performance of SASR.

As can be seen from Figs. 2 and 3, the performance of SASR is less sensitive to changes in α and β. When α is too small, SASR does not make full use of user social relationships. Therefore, performance increases as α increases. Then, when α is too large, the overall performance of the model mainly depends on the user's social relationship, and the classification performance of using the social relationship alone is not good. Therefore, the overall performance of the classification model begins to decline. Figure 2 shows that when $\alpha = 0.05$, the performance of SASR is optimal.

Figure 3 shows that the performance of SASR is optimal when $\beta = 0.1$. When β is too large, the performance of our model is degraded because it mainly depends on sparse regularization, and many features are filtered out by regularization; when β is too small, sparse regularization is not fully utilized, and there is still a lot of noise in the training set, so When β increases from 0 to 0.1, the classification performance of SASR also increases. Therefore, SASR is less sensitive to changes in α and β, and this advantage is also important because it saves a lot of time to adjust parameters.

5 Conclusion

Based on the sentimental classification of text content, this paper establishes the sentimental relationship between microblogs based on the user topic context and the social relationship (repost or like) in Weibo, and solves the problem of mutual independence between microblogs. The sentimental relationship connects the labeled microblog with the unlabeled microblog, and proposes a semi-supervised sentiment classification method (SASR) based on the microblog social relationship, which effectively reduces the dependence on the training set. Experiments have been carried out by crawling the Sina Weibo real topic dataset. The experimental results show that the SASR method is superior to other typical sentiment classification methods in terms of accuracy and recall rate, and has low sensitivity to the size of the training set. The microblog sentiment classification research in this paper is of great significance for public opinion monitoring and guidance and corporate marketing.

Acknowledgments. This work was supported by the National Key Research and Development Plan (2016YFC0101500) and the Fundamental Research Funds for the Central Universities (N161602002), the Natural Science Foundation of China under grant (No. 61532007, 61370076), the Natural Science Foundation of Jiangsu Province under grant No. 15KJB520001. This work was partly supported by the Natural Science Foundation of Jiangsu Province of China under grant NO. BK2012209, Science and Technology Program of Suzhou in China under grant NO. SYG201409. Finally, the authors would like to thank the anonymous reviewers for their constructive advices.

References

1. Zhao, X., Zhang, Y., Guo, W., Yuan, X.: Jointly trained convolutional neural networks for online news emotion analysis. In: Meng, X., Li, R., Wang, K., Niu, B., Wang, X., Zhao, G. (eds.) WISA 2018. LNCS, vol. 11242, pp. 170–181. Springer, Cham (2018). https://doi.org/10.1007/978-3-030-02934-0_16
2. Keshavarz, H., Abadeh, M.S.: ALGA: adaptive lexicon learning using genetic algorithm for sentiment analysis of microblogs. Knowl. Based Syst. **122**, 1–16 (2017)
3. Eliacik, A.B., Erdogan, N.: Influential user weighted sentiment analysis on topic based microblogging community. Expert Syst. Appl. **92**, 403–418 (2018)

4. Dave, K., Lawrence, S., Pennock, D.M.: Mining the peanut gallery: opinion extraction and semantic classification of product reviews. In: Proceedings of the 12th International Conference on World Wide Web, pp. 519–528. ACM (2003)

5. Yu, J., An, Y., Xu, T., Gao, J., Zhao, M., Yu, M.: Product recommendation method based on sentiment analysis. In: Meng, X., Li, R., Wang, K., Niu, B., Wang, X., Zhao, G. (eds.) WISA 2018. LNCS, vol. 11242, pp. 488–495. Springer, Cham (2018). https://doi.org/10.1007/978-3-030-02934-0_45

6. Hu, M., Liu, B.: Mining and summarizing customer review. In: Proceedings of the Tenth ACM SIGKDD International Conference on Knowledge Discovery and Data Mining, pp. 168–177. ACM (2004)

7. Han, Z., Jiang, X., Li, M., Zhang, M., Duan, D.: An integrated semantic-syntactic SBLSTM model for aspect specific opinion extraction. In: Meng, X., Li, R., Wang, K., Niu, B., Wang, X., Zhao, G. (eds.) WISA 2018. LNCS, vol. 11242, pp. 191–199. Springer, Cham (2018). https://doi.org/10.1007/978-3-030-02934-0_18

8. Wu, Y., Liu, S., Yan, K., et al.: OpinionFlow: visual analysis of opinion diffusion on social media. IEEE Trans. Vis. Comput. Graph. 20(12), 1763–1772 (2014)

9. Abelson, R.P.: Whatever became of consistency theory? Pers. Soc. Psychol. Bull. 9(1), 37–64 (1983)

10. Hatfield, E., Cacioppo, J.T., Rapson, R.L.: Emotional contagion. Curr. Dir. Psychol. Sci. 2(3), 96–100 (1993)

11. Zhang, S., Wei, Z., Wang, Y., et al.: Sentiment analysis of Chinese micro-blog text based on extended sentiment dictionar. Future Gen. Comput. Syst. 81, 395–403 (2018)

12. Hai, Z., Cong, G., Chang, K., et al.: Analyzing sentiments in one go: a supervised joint topic modeling approach. IEEE Trans. Knowl. Data Eng. 29(6), 1172–1185 (2017)

13. Hu, X., Tang, L., Tang, J., et al.: Exploiting social relations for sentiment analysis in microblogging. In: Proceedings of the sixth ACM International Conference on Web Search and Data Mining, pp. 537–546. ACM (2013)

14. Zou, X., Yang, J., Zhang, J.: Microblog sentiment analysis using social and topic context. PLoS ONE 13(2), 36–60 (2018)

15. Sluban, B., Smailovic, J., Battiston, S., et al.: Sentiment leaning of influential communities in social networks. Comput. Soc. Netw. 2(1), 1–21 (2015)

16. Wu, F., Huang, Y., Song, Y.: Structured microblog sentiment classification via social context regularization. Neurocomputing 175, 599–609 (2016)

17. West, R., Paskov, H.S., Leskovec, J., et al.: Exploiting social network structure for person-to-person sentiment analysis. Trans. Assoc. Comput. Linguist. 2(1), 297–310 (2014)

18. Pang, B., Lee, L., Vaithyanathan, S., et al.: Thumbs up? Sentiment classification using machine learning techniques. In: Empirical Methods in Natural Language Processing, pp. 79–86 (2002)

19. Hu, X., Sun, N., Zhang, C., et al.: Exploiting internal and external semantics for the clustering of short texts using world knowledge. In: Conference on Information and Knowledge Management, pp. 919–928 (2009)

20. Friedman, J., Hastie, T., Tibshirani, R.: The Elements of Statistical Learning. Springer, New York (2001). https://doi.org/10.1007/978-0-387-21606-5

21. Tan, C., Lee, L., Tang, J., et al.: User-level sentiment analysis incorporating social networks. In: Proceedings of the 17th ACM SIGKDD International Conference on Knowledge Discovery and Data Mining, pp. 1397–1405. ACM (2011)

Organization and Query Optimization of Large-Scale Product Knowledge

You Li[1], Taoyi Huang[2], Hao Song[2], and Yuming Lin[2(✉)]

[1] Guangxi Key Laboratory of Automatic Detecting Technology and Instruments,
Guilin University of Electronic Technology, 541004 Guilin, China
[2] Guangxi Key Laboratory of Trusted Software, Guilin University of Electronic
Technology, 541004 Guilin, China
ymlin@guet.edu.cn

Abstract. Knowledge graph is essential infrastructure of lots of intelligent Web applications. Recently, various types of knowledge graphs are designed and deployed to make the applications more smarter. However, the large amount and heterogeneity of product knowledge bring new challenges for managing such knowledge data. In this work, we propose a scalable framework for organizing large-scale product knowledge, which includes the objective product knowledge and the subject users' opinion knowledge. In order to improve the efficiency of knowledge query, we design a hybrid index structure with a learned model and several B-Tree indexes. Finally, a join strategy based on the variable combination of aspect and opinion is proposed to implement the query optimization. The experimental results show that the proposed method can improved the query efficiency significantly on a large-scale product knowledge compared with a states-of-the-art knowledge management system.

Keywords: Product knowledge · Organization · Query optimization

1 Introduction

With the rapid development and popularization of internet technology and E-Commerce, the count of data including product information increases dramatically on Web. However, it is till hard to meet the users' demand on acquiring accurate information on products. One of the fundamental reasons is that such massive information exists on Web in unstructured or semi-structured form, which limits severely them to be applied automatically and intelligently. On the other hand, the lack of effective management mechanism on such information makes users confront directly fragmented and redundant information, which exacerbates the problem of information overload.

Knowledge graph is an effective way to solve such problem above, which targets at extracting the knowledge from Web and managing these knowledge efficiently. Recently, various types of knowledge graph projects (such as YAGO [1], Freebase [2]) and knowledge management systems (such as RDF-3X [3],

© Springer Nature Switzerland AG 2019
W. Ni et al. (Eds.): WISA 2019, LNCS 11817, pp. 492–498, 2019.
https://doi.org/10.1007/978-3-030-30952-7_48

gStore [4]) have been proposed and developed. However, product knowledge includes the objective knowledge (such as product taxonomy) and the subject knowledge (such as user opinion) in general, both of them are important to many intelligent Web applications such as product recommendation [5]. This heterogeneity of knowledge brings new challenge for managing such knowledge. Moreover, existing systems often need to transform texts like URIs into ID values for query processing, which leads to extra cost for accessing the index frequently.

In this work, we propose a presentation framework for product knowledge to organize the objective product knowledge and the subject one uniformly at first. Then, we design a hybrid index structure to improve the retrieval efficiency based on the learned model and the traditional B-Tree index. Further, we propose join strategy based on the variable combination of aspect and opinion to accelerate the product knowledge retrieval. At last, a series of experiments carried on a large-scale dataset show that the proposed method can improve the query performance on product knowledge significantly compared with a state-of-the-art knowledge management system.

2   The Organization Framework of Product Knowledge

The product knowledge can be divided into two groups generally: the object knowledge and the subject one. The former describes the commonsense knowledge such as product taxonomy, product's aspects; the latter includes mainly users' opinion on a product or product's aspects (Fig. 1).

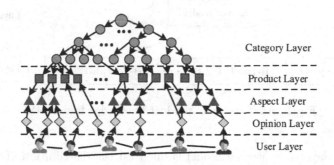

Fig. 1. An overview of product knowledge framework

Specifically, let product concept set $C = \{c_1, \cdots, c_{n1}\}$, product instance set $P = \{p_1, \cdots, p_{n2}\}$, and product aspect set $A = \{a_1, \cdots, a_{n3}\}$. Then, the entity set $E = C \cup P \cup A \cup T \cup F$, where $F = \{f_1, \cdots, f_{n5}\}$ is the set of facts, $f_i = <x, p_i, y>$ is a fact, $x \in E$, $y \in E$ and p_i is a predicate. We define some predicates used in this work in Table 1. Based on these symbols, we can present the product knowledge with the form of RDF (Resource Description Framework) triples.

Table 1. Some defined predicates and the corresponding descriptions

Predicate	Fact	Description
subCategory	<x, subCategory, y>	x is the sub category of y
productOf	<x, productOf y>	x is a product of y
aspectOf	<x, aspectOf, y>	x is an aspect of y
write	<x, write, y>	user x writes review y
reviewOn	<x, reviewOn, y>	review x is on product/aspect y

3   The Hybrid Mapping Index Structure

In existing knowledge management systems, lots of URI texts need to be transformed into ID values with mapping index for query processing. The B-Tree index is often used to speed up this process. However, as the size of index increasing, the query efficiency is gradually reduced. The learned index structure is a novel technique to build indexes by utilizing the distribution of data being indexed, which could provide benefits over state-of-the-art database indexes [6].

In order to applying machine learning model to index the URI texts, each URI is transformed into an one-dimensional array with ASCII code. Based on these sorted one-dimensional arrays, we design a two-layer mapping structure shown in Fig. 2, which combines a learned model and multiple B-Tree indexes.

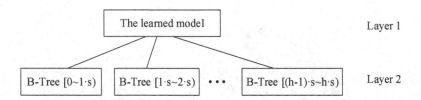

Fig. 2. An overview of product knowledge framework

The first layer is a learned model trained on the distribution of the sorted URI arrays, which tries to find the approximate location of each URI. Specially, we apply the neural network with one fully-connected hidden layer as the learned model, where the activation function is the *ReLu* function defined as follows.

$$RuLu(x) = \begin{cases} 0 & x \leq 0 \\ x & x > 0 \end{cases} \qquad (1)$$

In the training phase, we apply the variance of predicted value and true value of URI's location as the loss function. Since the product knowledge is coming from the Web, the URI of entity is relatively long. Then, it needs more nodes in input layer of neural network, which reduces the execution efficiency of model.

Considering the types of namespaces included in URIs is relatively few and fixed, we construct a prefix tree to compress the redundant namespace prefix. By this way, we can reduce the count of nodes in input layer greatly.

Ideally, the learned model would predict the exact value of each URI. However, there is a relatively large error between the predicted value and the true value of a URI's location because of the data distribution's complexity. Then, the second layer contains h B-Tree structures, which locate the URI accurately based on the approximate location predicted by the learned model. In order to improve the retrieval performance, we apply a threshold s to divide the URIs into multiple blocks for keeping the B-Trees even.

4   Query Optimization of Product Knowledge

The *join* operation is an important factor for influencing the query performance. When a SPARQL query is executed on product graph, the query is transformed into a query graph at first. Then, we will generate the candidates of each variable according to the adjacency of the variable node at first. At last, the *join* operation is executed to generate the result set by joining the candidate sets of variables based on the structure of query graph.

In the existing graph-based query systems, the query processes focus on the cases of the edge label being constant in query graph. When the query involves variables of edge label, the node variables are joined at first, then generate the values of edge variables. This strategy often leads to relatively low efficiency because of lots of intermediate results. However, user opinion is an important knowledge in product knowledge, on which users often query. There are lots of opinion expressions in product graph, which are in the forms shown in Fig. 3a and b. The former describes two reviews express different opinions on the same product aspect, the latter describes two reviews contain same opinion on different product aspects.

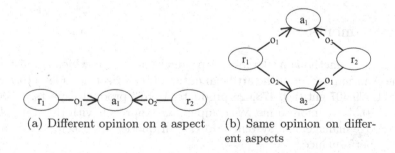

(a) Different opinion on a aspect   (b) Same opinion on different aspects

Fig. 3. Two forms of opinion data in product knowledge

When user try to find out all reviews including same opinion, the existing methods treat the product aspect and opinion separately, which would generate lots

---

Algorithm 1. Joining on a variable combination of product aspect and opinion

Input: Current table of intermediate results IRT,
           Variable combination (e, v)
Output: Final table of intermediate results $nIRT$

1:   **if** the candidate set of v is empty **then**
2:       **return** $nIRT$;
3:   **for each** intermediate result $r \in IRT$ **do**
4:       $tmp = \varnothing, tmpv = \varnothing$;
5:       **for each** $ele \in r$ **do**
6:           **if** there is not edge between ele and e **then**
7:               **continue**;
8:           $tmpv \leftarrow ele$;
9:           generating the other candidate set S of v by ele;
10:          **if** $tmp = \varnothing$ **then**
11:              $tmp \leftarrow S \cap C_v$;   /* C_v is the candidate set of v */
12:          **else** $tmp \leftarrow S \cap tmp$;
13:      **if** $tmp = \varnothing$ **then**
14:          **continue**;
15:      **for each** $ele \in tmp$ **do**
16:          generating the candidate set $tmpe$ of edge ele;
17:          **for each** $edge \in tmpe$ **do**
18:              generating the candidate set S by ele and $edge$;
19:          **if** $tmpv \subseteq S$ **then**
20:              creating r's duplicate r_c and $r_c \leftarrow (edge, ele)$;
21:              $nIRT \leftarrow r_c$;
22:      **return** $nIRT$;

---

of noneffective intermediate results and reduce the query process effectiveness. In reality applications, product aspect and opinion are often retrieved together. Then, we should treat the aspect variable of and the opinion variable as a combination for query processing. The detail of the *join* strategy based on the combination of aspect variable and opinion variable is shown in Algorithm 1.

5   Experiments

We evaluate our method on a large-scale product knowledge, which is constructed with the Amazon data and the artificial data. This dataset includes 116,174,460 triples, 11,979,407 entities, 478,626 products, 10,573 product aspects, 1,000,000 users and 10,566 opinion terms. We compare our approach with the state-of-the-art knowledge management system *gStore*. Four queries are executed to evaluate the query performances:

– Q1: Querying the products in the product category c_1, of which users hold the opinion term o_1 on product aspect a_1.
– Q2: Querying users' opinion and corresponding aspects of the product p_1.

- Q3: Querying the count of product aspects users are more focused on and the products belong to category c_1.
- Q4: Querying the users who have bought the same product(s) and have the same opinion on certain product aspects as those of user u_1.

In the first experiment, we compare the query respond time of the proposed method and the gStore for the four queries Q1–Q4 above. As shown in Table 2, our method outperforms the gStore significantly on query respond time for the Q2, Q3 and Q4. That is because the proposed method avoids lots of useless intermediate results are avoided in the query processes. For the query Q1, our method's query performance is similar to gStore's. The main reason is that the proposed optimization strategy does not work for Q1, because Q1 does not involve the aspect variable and opinion variable.

Table 2. Query respond time

	Q1	Q2	Q3	Q4
gStore	290 ms	913 ms	>30 min	>30 min
Our method	287 ms	271 ms	2285 ms	2389 ms

In the following experiment, we verify the performance of the hybrid mapping index structure (HMIS). Firstly, we analyze the influence of B-Tree's order on respond time, where the count of querying URI is set to 100,000. As shown in Fig. 4a, We can find that HMIS takes less average respond time than B-Tree. Moreover, the performance of HMIS improves more significantly with the increase in B-Tree's order. The Fig. 4b shows that the average respond time of HMIS is less than that of the B-Tree for different numbers of random access.

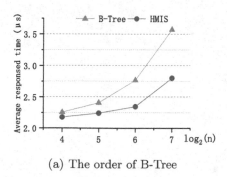

(a) The order of B-Tree

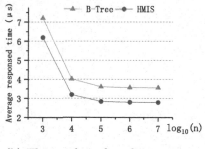

(b) The number of random access

Fig. 4. Performance comparisons of two index structures

Acknowledgements. This work is supported by National Natural Science Foundation of China (61562014), Guangxi Natural Science Foundation (2018GXNSFDA281049), the project of Guangxi Key Laboratory of Automatic Detecting Technology and Instruments (YQ17111), and the general scientific research project of Guangxi Provincial Department of Education (2017KY0195).

References

1. Hoffart, J., Suchanek, F.M., et al.: YAGO2: a spatially and temporally enhanced knowledge base from Wikipedia. Artif. Intell. **194**, 28–61 (2013)
2. Bollacker, K.D., Evans, C., et al.: Freebase: a collaboratively created graph database for structuring human knowledge. In: SIGMOD, pp. 1247–1250 (2008)
3. Neumann, T., Weikum, G.: The RDF-3X engine for scalable management of RDF data. VLDB J. **19**(1), 91–113 (2010)
4. Zou, L., Ozsu, M.T., Chen, L., et al.: gStore: a graph-based SPARQL query engine. VLDB J. **23**(4), 565–590 (2014)
5. Yu, J., An, Y., Xu, T., Gao, J., Zhao, M., Yu, M.: Product recommendation method based on sentiment analysis. In: Meng, X., Li, R., Wang, K., Niu, B., Wang, X., Zhao, G. (eds.) WISA 2018. LNCS, vol. 11242, pp. 488–495. Springer, Cham (2018). https://doi.org/10.1007/978-3-030-02934-0_45
6. Kraska, T., Beutel, A., Chi, E.H., et al.: The case for learned index structures. In: SIGMOD, pp. 489–504 (2018)

Drug Abuse Detection via Broad Learning

Chao Kong[1(✉)], Jianye Liu[1], Hao Li[1], Ying Liu[1], Haibei Zhu[2], and Tao Liu[1]

[1] School of Computer and Information, Anhui Polytechnic University, Wuhu, China
kongchao315@163.com, 17855368016@163.com, lhthomas@163.com,
liuyinglyj_011@163.com, liutao@ahpu.edu.cn
[2] School of Electrical Engineering, Anhui Polytechnic University, Wuhu, China
zhb877097717@163.com

Abstract. Prescription drug abuse is one of the fastest growing public health problems in the USA. This work develops a broad learning method for Drug Abuse Detection (DAD). In this paper, we propose a new broad learning-based method named ILSTM, short for *Improved Long Short-Term Memory*, to study the data fusion and prediction from heterogeneous data sources for DAD. The algorithm utilizes the broad learning framework to handle data fusion broadly and information mining deeply simultaneously. Moreover, the effectiveness and prevalence of Holt-Winter inspire our work in the temporal property for DAD.

Keywords: Broad learning · Data fusion · Drug Abuse Detection

1 Introduction

The United States is experiencing a national crisis regarding the use of synthetic and nonsynthetic opioids, either for the treatment and management of pain (legal, prescription use) or for a recreational purpose (illegal, non-prescription use). Federal organizations such as the Centers for Disease Control (CDC) are struggling to save lives and prevent negative health effects of this epidemic, such as opioid use disorder, hepatitis, HIV infections and neonatal abstinence syndrome."[1] To address this epidemic, we develop a broad learning method for Drug Abuse Detection (DAD). The broad learning method is an ubiquitous model to achieve better learning or mining performance on solving the real-world problem in the era of big data, which all kinds of data are available. It has been widely used in many applications such as POI recommendation [1], link prediction [2], fraud detection [3] and so on.

Generally speaking, DAD is the problem of monitoring the illegal drug or prescription medication abuse, predicting drug abuse trends, classifying people who may be caught in drug abuse or not. To perform predictive analytics on DAD, it is crucial to first fuse or integrate multiple available heterogeneous data

---

[1] http://www.cdc.gov/features/confronting-opioids/index.html.

© Springer Nature Switzerland AG 2019
W. Ni et al. (Eds.): WISA 2019, LNCS 11817, pp. 499–505, 2019.
https://doi.org/10.1007/978-3-030-30952-7_49

sources. However, existing works have primarily focused on the single data source such as Tweets [4]. Following the pioneering work of [4], these methods typically apply a three-step solution: first collecting drug abuse-related tweets at a large-scale, then designing an annotation strategy (drug abuse vs. non-drug abuse), and last developing a deep learning model that can accurately classify tweets into drug abuse risk behavior.

To the best of our knowledge, none of the existing works has paid special attention to DAD with broad learning framework. In this work, we focus on the problem of broad learning for DAD. We propose ILTSM (short for *Improved Long Short-Term Memory*), which addresses the aforementioned limitations of existing works. Below we highlight our major contributions in this work.

– To account for both the explicit information and implicit associations (spatio-temporal information and socio-economic data), we propose a new broad learning framework. In particular, the algorithm can handle data fusion broadly and information mining deeply simultaneously.
– To address the characteristic of time-lag, we set the dual gate to update new and effective data, forget old or invalid data in the pre-processing phase. Moreover, we employ Holt-Winter to predict the drug abuse trends in the time dimension, which can revise the prediction curve to denoise.

The rest of paper is organized as follows. We shortly discuss the related work in Sect. 2. We present the DAD method in Sect. 3 and report our empirical study in Sect. 4. Finally, we conclude this paper in Sect. 5.

2   Related Work

Drug abuse detection aims at predicting prevalence and patterns of abuse of both illegal drugs and prescription medications. The study of DAD problem has become a hot topic in recent years, and some earlier studies can go back to 2009s [5]. As an ubiquitous social media, Twitter is one of the most popular social networks, with more than 115 million monthly active users and over 58 million tweets per day.[2] Twitter is currently being used as a major resource in various detection tasks, including discrimination detection [6], influenza epidemic discovery [5], sentiment analysis [7], drug abuse detection, sexual health monitoring, and pharmacovigilance. To date, the existing works have primarily focused on the detection and analysis of illicit and prescription drug abuse using tweets. In general, the existing studies of DAD can be mainly divided into two categories: automatic monitoring and bag-of-words model. The former employs machine learning methods for an automatic classification that can identify tweets which are indicative of drug abuse [8]. Chary et al. [8] discussed how to use artificial intelligence techniques to extract content useful for purposes of toxicovigilance from social networks. However, the latter employs the bag-of-words model to build a dictionary, computes the similarity of the data items with a probabilistic manner and predicts the drug abuse trends based on a proposed decision or

---

[2] http://www.statisticbrain.com/twitter-statistics/.

score function [9]. Traditional DAD methods are developed on single social data without considering the data fusion of spatio-temporal information and socio-economic data. So it is difficult to guarantee better learning and optimization performance. These are the focuses of this work.

3 Drug Abuse Detection Approach

3.1 Overview of ILSTM

Our proposed approach consists of three components as following.

Step1. Feature selection via CNN. CNN can learn and provide the mapping between input and output, which does not require any precise mathematical expression between the input and output. The input features are converted to a two-dimensional matrix. The training phase compresses it to obtain the actual features, enabling the convolutional neural networks to map the input data into eigenvalue accurately.

Step2. Data fusion via ILSTM. Due to the characteristic of time-lag in datasets, we try to employ LSTM's [10] improved variant ILSTM to train. First, we set a dual gate to determine what information should be discarded from memory state: $f_t = \sigma(W_f \times [h_{t-1}, x_t] + b_f)$, where W_f and b_f represent weight and bias of **sigmoid** activation function respectively. Then, the sigmoid layer of the input gate layer determines which information needs to be updated by **tanh** layer. Finally, the ILSTM unit controls the output information: $o_t = \sigma(W_o \times [h_{t-1}, x_t] + b_o)$, $h_t = o_t * tanh(C_t)$. For ILSTM model, we treat the output layer of CNN as the input layer of bidirectional GRU to perform the feature extraction and data fusion.

Step3. Multi-class classification. We perform the information fusion and prediction by employing bidirectional ILSTM, and the predicted results are learned forward in full connection layer. The connection parameter batch is denoted as o^t and h^t respectively in the final fusion layer. Then, we make the derivation of weight W and biases coefficient b. Finally, we normalize the output layer by softmax function and give the probability distribution of all the attribute values: $P_r(a) = \frac{\exp(o_{t_i})}{\sum_{i=1}^{n} \exp(o_{t_i})}$.

3.2 Parameters Inference and Prediction

For each dataset, we define the attribute feature matrix as A_f. We put N attribute feature matrices $A_{f_i}(1 < i \leq N)$ into the input layer and extract the information through B-GRU [11]: $\overrightarrow{H}(A_{f_i}) = \overrightarrow{GRU}(A_{f_i})$, $\overleftarrow{H}(A_{f_i}) = \overleftarrow{GRU}(A_{f_i})$, $H(A_{f_i}) = \overrightarrow{H}(A_{f_i}) \times \overleftarrow{H}(A_{f_i})$, where $\overleftarrow{H}(A_{f_i})$ and $\overrightarrow{H}(A_{f_i})$ represent the information extracted by performing forward GRU and backward GRU respectively. When the last fusion layer extracted the information by ILSTM, $\overrightarrow{h_{i_l}}^t$ and $\overleftarrow{h_{i_l}}^t$

represent the information from ILSTM to full connection layer to perform a multi-class classification task.

First, we take the partial derivatives of output o^i : $\frac{\theta o^i}{\theta h_{ij}} = \frac{\sum_{j=1}^{length(input)} w_{ij} * h_{ij}}{\theta x_j} = w_{ij}$, where a_i represents the output of the i-th layer, and w_{ij} denotes the weight of j-th input in i-th layer. Then, we take the partial derivatives of loss:

$$\frac{\theta loss}{\theta h_{ij}} = \sum_{j}^{length(output)} \frac{\theta loss}{\theta o^i} \frac{\theta o^i}{\theta h_{ij}} = \sum_{j}^{length(output)} \frac{\theta loss}{\theta o^i} * w_{ij}, \tag{1}$$

which proves the backpropagation from $(i+1)$-th layer to i-th layer. Next, we can derive the weight w by

$$\frac{\theta loss}{\theta w_{h_{il}}} = \frac{\theta loss}{\theta o^i} \frac{\theta o^i}{\theta w_{h_{il}}} = \frac{\theta loss}{\theta o^i} * h_{ij} \tag{2}$$

The output O^t from full-connection layer is: $O^t = w_{\overrightarrow{h_{i_l}}}\overrightarrow{h_{i_l}} + w_{\overleftarrow{h_{i_l}}}\overleftarrow{h_{i_l}}$, where $w_{\overrightarrow{h_{i_l}}}$ and $w_{\overleftarrow{h_{i_l}}}$ denote the weight of forward pass and backward pass respectively. Once parameters O^t is estimated, ILSTM can classify each class V_{R_i} by computing its probability. Due to the page limitation, we omit the proofs and computations. To select the computation of probabilities significantly, here we employ the softmax function:

$$V_{R_i} = \frac{e^{\sum_{t=0}^{T_k-1} O^{t_k}}}{\sum_{i=0}^{C-1} e^{\sum_{t=0}^{T_i-1} O^{t_i}}}, \tag{3}$$

where C means the classes of each feature. Finally, we choose the *top*-k highest probability to predict the counties which most likely outbreak the drug abuse at a certain time.

4   Empirical Evaluation

We crawled three datasets from MCM/ICM contest 2019[3] and Twitter[4] for our experiments. The *ACS* covers socio-economic factor in each county. The *DEA* describes the number of opioid abuse reports of counties in each year. *Twitter* collects the comments from Twitter' users about drug abuse. The descriptive statistics about the datasets are shown in Table 1. We find a prevalent DAD's solution, called Support Vector Machine (shorted in SVM), to be the comparative baseline. As mentioned above, we evaluate our proposed method using *Precision* and *Recall*.

We extracted drug abuse reports, geographic location information, age, education and other socio-economic information from Twitter to integrate DEA

---

[3] https://www.comap.com/undergraduate/contests/.
[4] http://followthehashtag.com/datasets/free-twitter-dataset-usa-200000-free-usa-tweets/.

Table 1. Descriptive statistics of datasets

Data source	#records	#features
DEA	24063	10
ACS	3245	151
Twitter	204821	15

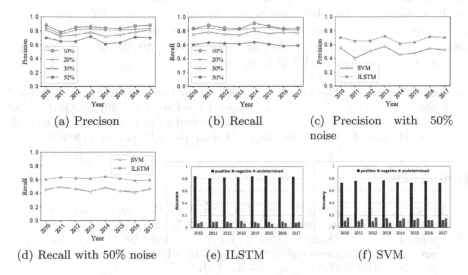

(a) Precison

(b) Recall

(c) Precision with 50% noise

(d) Recall with 50% noise

(e) ILSTM

(f) SVM

Fig. 1. Performance comparison of DAD on DEA, ACS and Twitter

and ACS to make the prediction. Moreover, Holt-Winter [12] was employed to smooth the output and remove noise. Figures 1(a) and (b) manifest the robustness of ILSTM by injecting noise whose ratio is varied from 10% to 50%. We observe that the accuracy of ILSTM is promised. If 10% noise is injected into data, almost 90% of drug abuse counties can be found by ILSTM in 2010. Even 50% noise is injected into the data, the precision in 2010 is almost 70%. Figures 1(c) and (d) illustrate that ILSTM outperforms the baseline significantly. We now turn to DAD on three heterogeneous data sources without noise namely clean DEA, ACS and Twitter. We manually annotated the accuracy from 2010 to 2017 labeled by ILSTM and SVM. As shown in Figs. 1(e) and (f), we find that the accuracy for the DAD of ILSTM from 2010 to 2017 is more than 80%, but it is about 65% for SVM. This illustrates that both ILSTM and SVM are quite good in returning the correct drug abuse counties from 2010 to 2017. ILSTM also returns fewer undetermined results than SVM. In summary, the result indicates that implicit associations are helpful to achieve better performance for DAD.

5   Conclusion

In this paper, we have studied the problem of Drug Abuse Detection via broad learning across two kinds of heterogeneous data sources. It is a challenging task due to the characteristic of time-lag, implicit associations and data fusion. We propose a supervised method to deal with the mentioned challenges. We have illustrated our proposed method on three real data sources. Experimental results demonstrate the effectiveness and rationality of our ILSTM method. In the future, we plan to expand the datasets to insert more explicit and implicit features, make the predictive effect of DAD closer to the truth and assess the impact of more drugs on DAD in target areas.

Acknowledgements. This work is supported by the Initial Scientific Research Fund of Introduced Talents in Anhui Polytechnic University (No. 2017YQQ015), Pre-research Project of National Natural Science Foundation of China (No. 2019yyzr03) and National Natural Science Foundation of China Youth Fund (No. 61300170).

References

1. Wang, F., Qu, Y., Zheng, L., Lu, C.-T., Yu, P.S.: Deep and broad learning on content-aware POI recommendation. In: CIC 2017, USA, pp. 369–378 (2017)
2. Zhang, J., Xia, C., Zhang, C., Cui, L., Fu, Y., Yu, P.S.: BL-MNE: emerging heterogeneous social network embedding through broad learning with aligned autoencoder. In: ICDM 2017, USA, pp. 605–614 (2017)
3. Cao, B., Mao, M., Viidu, S., Yu, P.S.: HitFraud: a broad learning approach for collective fraud detection in heterogeneous information networks. In: ICDM 2017, USA, pp. 769–774 (2017)
4. Hu, H., et al.: Deep learning model for classifying drug abuse risk behavior in tweets. In: ICHI 2018, USA, pp. 386–387 (2018)
5. Ginsberg, J., Mohebbi, M.H., Patel, R.S., Brammer, L., Smolinski, M.S., Brilliant, L.: Detecting influenza epidemics using search engine query data. Nature **457**(7232), 1012–1014 (2009)
6. Wu, Y., Wu, X.: Using loglinear model for discrimination discovery and prevention. In: DSAA 2016, Canada, pp. 110–119 (2016)
7. Keith Norambuena, B., Lettura, E.F., Villegas, C.M.: Sentiment analysis and opinion mining applied to scientific paper reviews. Intell. Data Anal. **23**(1), 191–214 (2019)
8. Chary, M., Genes, N., McKenzie, A., Manini, A.F.: Leveraging social networks for toxicovigilance. J. Med. Toxicol. **9**(2), 184–191 (2013)
9. Balsamo, D., Bajardi, P., et al.: Firsthand opiates abuse on social media: monitoring geospatial patterns of interest through a digital cohort. In: WWW 2019, USA, pp. 1–7 (2019)
10. Han, X., Xu, L., Qiao, F.: CNN-BiLSTM-CRF model for term extraction in Chinese corpus. In: Meng, X., Li, R., Wang, K., Niu, B., Wang, X., Zhao, G. (eds.) WISA 2018. LNCS, vol. 11242, pp. 267–274. Springer, Cham (2018). https://doi.org/10.1007/978-3-030-02934-0_25

11. Wang, S., Wu, B., Wang, B., Tong, X.: Complaint classification using hybrid-attention GRU neural network. In: Yang, Q., Zhou, Z.-H., Gong, Z., Zhang, M.-L., Huang, S.-J. (eds.) PAKDD 2019. LNCS (LNAI), vol. 11439, pp. 251–262. Springer, Cham (2019). https://doi.org/10.1007/978-3-030-16148-4_20
12. Raikwar, A.R., et al.: Long-Term and short-term traffic forecasting using holt-winters method: a comparability approach with comparable data in multiple seasons. IJSE **8**(2), 38–50 (2017)

Semi-supervised Meta-path-based Algorithm for Community Detection in Heterogeneous Information Networks

Limin Chen[(⊠)], Yan Zhang, and Liu Yang

School of Computer and Information Technology,
Mudanjiang Normal University, Mudanjiang 157012, China
chenlimin_clm@126.com

Abstract. Similarity between target objects is mostly computed based on meta-paths for semantic-based community detection algorithm in heterogeneous information networks. The meta-path-based similarity between target objects is very efficient, but the semantics of meta-path-based similarity is not integrity and can not truly express relations of target objects. And now there is lack of better expression for similarity semantics in heterogeneous information networks. The complex topology structure is usually neglected in semantic-based community detection algorithms for heterogeneous information networks. To effectively improve accuracy of community detection algorithm in heterogeneous information networks, a semi-supervised meta-path-based algorithm for community detection is proposed in this paper. First, spectral method is used to analyse the topology structure of heterogeneous information networks to select representative objects. Then the similarity of target objects is adjusted by representatives in every cluster. Last, NMF method is used to detect communities. Through experiments in simulation datasets and real datasets, the experimental results showed the proposed algorithm is effective.

Keywords: Heterogeneous information networks · Community detection · Semi-supervised · Meta-path · Topology structure

1 Introduction

Heterogeneous information networks are very common. The real world is reflected truly by heterogeneous information networks. Analyzing heterogeneous information networks can better understand the hidden structure of networks and the roles represented by each community data [1, 2].

Community detection algorithms in heterogeneous information networks are the research foundation, and community detection algorithms based on semantic similarity in heterogeneous information networks are mainstream methods [3, 4]. Among them, most of the semantic similarity measures are based on meta-path computation. The typical meta-path-based similarity measures are PathCount, PathSim [5] and JoinSim [6]. However, the similarity semantic of target objects based on meta-path is incomplete, and it can not truly reflect the relations of target objects. Now, there is no more accurate method to express the similarity semantics of target objects. Affected by it, the

© Springer Nature Switzerland AG 2019
W. Ni et al. (Eds.): WISA 2019, LNCS 11817, pp. 506–511, 2019.
https://doi.org/10.1007/978-3-030-30952-7_50

accuracy of the results of community detection for heterogeneous information networks is not high. Semantic-based community detection algorithms for heterogeneous information networks often neglect the complex topology of networks, Analyzing heterogeneous information networks from the view of topological structure can grasp the global distribution of data [7, 8]. Semi-supervised method can improve the accuracy of community detection [9], so the accuracy of community detection can be effectively improved in heterogeneous information networks while similarity adjusted semi-supervised from global structure.

Spectral clustering can capture the global distribution of datasets. In addition, NMF (Non-negative Matrix Factorization) method and spectral clustering method have supported by reliable theory [9]. Integrating the two methods to analyze communities of heterogeneous information networks, SMpC (Semi-supervised meta-path-based algorithm for community detection in heterogeneous information networks) is proposed in this paper. First, the SRC algorithm [8] is used to obtain the global distribution of the target dataset, and the representative objects in each cluster are selected to construct the prior information. Then, the meta-path-based similarity of target objects is computed, and the similarity of target objects is adjusted by prior information. And last, the target objects are partitioned by NMF algorithm to find reasonable communities.

2   Similarity Based on Meta-path

Definition 1: Given a dataset $X = \{X_t\}_{t=1}^{T}$ with T types, where $X_t = \{x_1^{(t)}, x_2^{(t)}, \ldots, x_{nt}^{(t)}\}$ is a dataset belonging to the t-th type, a weighted graph $G = <V, E, W>$ on X is called an information network; if $V = X$, the E is a binary relation on V and W: $E \to R^+$. Such an information network is called a heterogeneous information network when $T > 1$ [10].

An information network $G = <V, E, W>$ on $X = \{X_t\}_{t=1}^{T}$ with T types is given, X_1 is the target dataset and X_t $(t \neq 1)$ is the attribute dataset, $x_n^{(t)} \in X_t$ is the n-th object of X_t. The meta-path is a concatenation of multiple nodes or node types linked by edge types between object $x_u^{(1)}$ and $x_v^{(1)}$, denoted by $P^{<u,v>}$. Meta-path count S from object $x_u^{(1)}$ to $x_v^{(1)}$ is the total number of meta-paths between $x_u^{(1)}$ and $x_v^{(1)}$. Let $P_s^{<u,v>}$ is the s-th meta-path between $x_u^{(1)}$ and $x_v^{(1)}$, the weight of $P_s^{<u,v>}$ is α_s. The typical computing similarity method of target objects based on meta-path is as follows [5, 6]:

$$\textbf{PathCount}(u, v) = \sum_s^S \alpha_s P_s^{<u,v>}$$

$$\textbf{PathSim}(u, v) = \sum_s^S \alpha_s \left(2P_s^{<u,v>} / \left(P_s^{<u,u>} + P_s^{<v,v>}\right)\right)$$

$$\textbf{JoinSim}(u, v) = \sum_s^S \alpha_s \left(2P_s^{<u,v>} / \left(P_s^{<u,u>} \cdot P_s^{<v,v>}\right)^{1/2}\right)$$

3 Multi-type Data Partitioning Based on Topological Structure

From the view of topological structure, SRC algorithm solves the problem of collaborative clustering heterogeneous data well. Given $G = <V, E, W>$ on $X = \{X_t\}_{t=1}^T$ with T types and M relation matrices, where $\{W^{(pq)} \in R^{n_p \times n_q}\}_{1 \leq p,q \leq T}$ is the relation matrix between X_p and X_q. β_{pq} is the weight of $W^{(pq)}$, where $\sum \beta_{pq} = 1, \beta_{pq} > 0$. Let

$$L = \sum_{1 \leq p,q \leq T} \beta_{pq} \left\| W^{(pq)} - C^{(p)} A^{(pq)} \left(C^{(q)}\right)' \right\| \tag{1}$$

where $\|\cdot\|$ refers Frobenius norm, $C^{(p)} \in \{0,1\}^{n_p \times k_p}$, $A^{(pq)} \in R^{k_p \times k_q}$, $\left(C^{(p)}\right)' C^{(p)} = I$ and $\sum_{i=1}^{k_i} C_{ij}^{(p)} = 1$, I is unit matrix, $0 \leq i \leq n_i$. While L is the smallest, C is the best indicator matrix. Equation (1) can also be expressed as:

$$L = \sum_{1 \leq p,q \leq M} \beta_{pq} \mathrm{tr} \left\{ \left(W^{(pq)} - C^{(p)} A^{(pq)} \left(C^{(q)}\right)' \right)' \left(W^{(pq)} - C^{(p)} A^{(pq)} \left(C^{(q)}\right)' \right) \right\} \tag{2}$$

where, tr denotes the trace of matrix. While $\partial L / \partial A^{(pq)} = 0$, L is minimum. i.e.

$$\max_{(C(p))'C(p)=I} \mathrm{tr} \left(\left(C^{(p)}\right)' \Phi^{(p)} C^{(p)} \right) \tag{3}$$

where $\Phi^{(p)}$ is

$$\Phi^{(p)} = \beta_{pp} \left(A^{(pp)} \left(A^{(pp)}\right)' \right) + \sum_{p < i \leq M} \beta_{pi} \left(A^{(pi)} C^{(i)} \left(C^{(i)}\right)' \left(A^{(pi)}\right)' \right) + \sum_{1 < i \leq p} \beta_{pi} \left(\left(A^{(ip)}\right)' C^{(i)} \left(C^{(i)}\right)' A^{(ip)} \right) \tag{4}$$

The local maximum of Eq. (3) is easy obtained by iterative algorithm. First, given $M - 1$ indicator matrix $C^{(i)}$, then determining the optimal indicator matrix $C^{(p)}$, where $i \neq p, 0 \leq i, p \leq M$. The algorithm is as follows:

Algorithm 1: SRC for multi-type data
1) input matrices $\{W^{(pq)} \in R^{n_p \times n_q}\}_{1 \leq p,q \leq T}$, weight $\beta^{(pq)}$, cluster number $\{k_i\}_{1 \leq i \leq K}$;
2) initialize $\{C^{(p)}\}_{1 \leq p \leq T}$ with orthogonal normal matrices;
3) repeat
4)   for p=1 to T do
5)   { compute Eq.(4) $\Phi^{(p)}$; update $C^{(p)}$ with eigenvectors of $\Phi^{(p)}$;}
6)   until convergence
7)   partition indicator data $C^{(1)}$ of target objects by k-means algorithm.
8)   output cluster indicator matrices $\{C^{(p)}\}_{1 \leq p \leq T}$.

4 Semi-supervised Meta-path-based Algorithm

4.1 Constructing Prior Information

An information network with a star network schema is selected to analyse the global distribution of target objects in this paper in order to reduce computation complexity. There only exist relation matrices $\{W^{(1q)} \in R^{n_1 \times n_q}\}_{1 < q \leq T}$ in a star network schema. Then Eq. (1) is expressed as:

$$L = \sum_{1 < q \leq T} \beta_q \left\| W^{(1q)} - C^{(1)} A^{(1q)} \left(C^{(q)} \right)' \right\|$$

where β_q is the weight of $W^{(1q)}$, $\sum \beta_q = 1$ and $\beta_q > 0$.

Target indicator matrix $C^{(1)}$ is got by partitioning target dataset X_1 into K_1 clusters using Algorithm 1. To select representative objects in the k-th cluster, where $1 \leq k \leq K_1$, given threshold δ, first, select a seed $c_u^{(1)}$ in the k-th indicator cluster randomly, and $x_u^{(1)}$ is as the representative object corresponding to indicator object $c_u^{(1)}$. Then compute $dis = \left\| c_u^{(1)} - c_v^{(1)} \right\|$ between object $c_u^{(1)}$ and $c_v^{(1)}$ in the same indicator cluster. if $dis > \delta, c_v^{(1)}$ is selected as the next seed, and $x_v^{(1)}$ is as the representative object corresponding to $c_v^{(1)}$, repeat the step until no seed exists. Selecting the object whose distance from the seed is more than δ as the representative object, it can make representative objects associated in the topology structure to adjust the bias or incompleteness of the semantic similarity based on meta-path. Let $Z \in R^{n_1 \times n_1}$ is the priori information relation matrix of X_1. Given $z_{uv} \in Z, z_{uv} = 1$, if both $x_u^{(1)}$ and $x_v^{(1)}$ are represent objects and belong to the same cluster, otherwise $z_{uv} = 0$.

4.2 SMpC Algorithm

Semi-supervised meta-path-based algorithm for community detection proposed in this paper is as follow:

> **Algorithm 2: SMpC** Algorithm
> 1) input cluster number K_1, threshold δ, weight a, weight b;
> 2) compute the similarity matrix H of target objects based on meta-path and regularized H;
> 3) partition X_1 into K_1 clusters by algorithm 1;
> 4) select representative objects, construct priori information matrix $Z \in R^{n_1 \times n_1}$ and regularized Z;
> 5) construct relation matrix $aH+bZ$;
> 6) decompose $aH+bZ$ by NMF method;
> 7) allocate target objects into clusters $\{Y_k\}_{1 \leq k \leq K_1}$ by indicator matrix;
> 8) output clusters $\{Y_k\}_{1 \leq k \leq K_1}$ of target objects.

5 Experiment

5.1 Experiment Data and Parameter Analysis

S_s is the small test dataset as in the literature [10] and extracted from the DBLP dataset contains 4 areas related to data mining. S_l is the large test dataset and extracted from the Chinese DBLP dataset, which are sharing resources released by Institute of automation, Chinese Academy of Sciences. S_l includes 34 computer science journals, 2,671 papers, 4,576 authors and 4,962 terms.

The object similarity matrix H is computed respectively by PathCount, PathSim and JoinSim in this experiment. a is the parameter of similarity matrix H, and b is the parameter of the priori information matrix Z, and $a + b = 1$. The dataset S_s is used in this experiment to analyse parameters. b ranges from 0.1 to 1, While $0.4 \le b \le 0.6$, the accuracy of communities is higher as shown in Fig. 1. $b = 0.5$ is used in the next experiment.

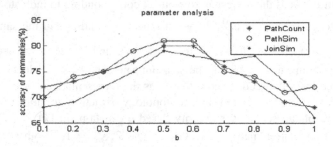

Fig. 1. Parameter analysis for papers in S_s

5.2 Accuracy Comparison of Community Detection

In this experiment, authors and papers as target objects are analyzed in datasets S_s and S_l. Target object similarity is computed by using PathCount, PathSim, JoinSim respectively and all the similarity matrices are adjusted by SMpC. Communities is obtained in S_s and S_l by NMF method. Accuracy of communities by different methods is compared. The experimental results are shown in Table 1. It shows that SMpC can effectively improve the accuracy of communities in heterogeneous information networks.

Table 1. Comparision of community accuracy (%)

Similarity	PathCount	SMpC	PathSim	SMpC	JoinSim	SMpC
Papers on S_s	73.91	80.87	71.54	81.72	72.83	79.65
Authors on S_s	74.41	82.33	69.13	75.13	67.91	81.54
Papers on S_l	70.84	76.36	68.28	72.89	72.93	79.92
Authors on S_l	71.02	78.94	68.29	73.01	70.01	75.32

6   Conclusion

The incompleteness of semantic similarity is effectively adjusted by using priori information matrix. Therefore, SMpC algorithm in this paper can effectively improve the accuracy of communities in heterogeneous information networks, and SMpC algorithm does not need manual intervention, so that it improves the self-adaptability of algorithm. However, because of analysis of the topology of heterogeneous information networks, the computational complexity of the algorithm is increased. Reducing the complexity of the algorithm on ensuring the accuracy of communities will be solved in the future.

Acknowledgments. This paper is supported by Heilongjiang Natural Science Foundation: LH2019F051, F2016039; Science Project of Heilongjiang Education Department: 12521578.

References

1. Sun, Y., Han, J.: Mining heterogeneous information networks: principles and methodologies. In: Proceedings of Mining Heterogeneous Information Networks: Principles and Methodologies, vol. 3, no. 2, pp. 1–159 (2012)
2. Li, Y., Li, C., Chen, W.: Research on influence ranking of chinese movie heterogeneous network based on PageRank algorithm. In: Meng, X., Li, R., Wang, K., Niu, B., Wang, X., Zhao, G. (eds.) WISA 2018. LNCS, vol. 11242, pp. 344–356. Springer, Cham (2018). https://doi.org/10.1007/978-3-030-02934-0_32
3. Shi, C., Li, Y., Zhang, J., et al.: A survey of heterogeneous information network analysis. IEEE Trans. Knowl. Data Eng. **29**(1), 17–37 (2016)
4. Sun, Y., Han, J., Yan, X., et al.: PathSim: meta path-based top-k similarity search in heterogeneous information networks. In: Proceedings of VLDB Endowment (2011)
5. Shi, Y., Chan, P.W., Zhuang, H., et al.: PReP: path-based relevance from a probabilistic perspective in heterogeneous information networks. In: KDD 2017, Canada, pp. 13–17 (2017)
6. Xiong, Y., Zhu, Y., Yu, P.S.: Top-k similarity join in heterogeneous information networks. IEEE Trans. Knowl. Data Eng. **27**(6), 1710–1723 (2015)
7. Yang, J., Chen, L., Zhang, J.: FctClus: a fast clustering algorithm for heterogeneous information networks. PLoS ONE **10**(6), e0130086 (2015)
8. Long, B., Zhang, Z.M., Wu, X., et al.: Spectral clustering for multi-type relational data. In: Proceedings of the 23rd International Conference on Machine learning, Pittsburgh, pp. 585–592 (2006)
9. Ma, X., Dong, D.: Evolutionary nonnegative matrix factorization algorithms for community detection in dynamic networks. IEEE Trans. Knowl. Data Eng. **29**(5), 1045–1058 (2017)
10. Sun, Y., Yu, Y., Han, J.: Ranking-based clustering of heterogeneous information networks with star network schema. In: Proceedings of the 15th ACM SIGKDD international conference on Knowledge discovery and data mining, Paris, pp. 797–806 (2009)

CLMed: A Cross-lingual Knowledge Graph Framework for Cardiovascular Diseases

Ming Sheng[1], Han Zhang[2], Yong Zhang[1(✉)], Chao Li[1],
Chunxiao Xing[1], Jingwen Wang[3], Yuyao Shao[3], and Fei Gao[4]

[1] RIIT&BNRCIST&DCST, Tsinghua University, Beijing 100084, China
{shengming, zhangyong05, li-chao,
xingcx}@tsinghua.edu.cn
[2] Beijing University of Posts and Telecommunications, Beijing 100876, China
zhanghan3281@bupt.edu.cn
[3] Beijing Foreign Studies University, Beijing 100089, China
{wjwen, shaoyuyao}@bfsu.edu.cn
[4] Henan Justice Police Vocational College, Zhengzhou 450046, China
68521617@qq.com

Abstract. Currently, knowledge graphs can be used to better query complex related information, understand user intentions from the semantic level, and improve search quality. With the increase in cross-lingual medical data recently, the construction of cross-lingual knowledge graph in the medical domain is urged. Meanwhile, there also exist many challenges due to the specific features in the medical area: (1) the scarcity of Chinese knowledge and the inner relations within the entities in existing open source cross-lingual knowledge bases; (2) the scarcity of prior knowledge defined by medical experts; (3) the availability of Electronic Medical Records (EMRs) and (4) the inconsistent semantic rules in different languages. To overcome these limitations, we propose CLMed platform which can be used in constructing the disease-specific, cross-lingual medical knowledge graphs from multiple sources. This platform provides several toolsets to process the cross-lingual data. Meanwhile, based on this platform, we generate a large scale, cross-lingual medical knowledge graph named CLKG.

Keywords: Cross-lingual · Knowledge graph · Domain-specific · EMR · Cardiovascular diseases

1 Introduction

Over the past years, the amount of knowledge grows rapidly due to the development of information technology. Knowledge graph [1] can describe the massive knowledge structurally upon the semantic knowledge base. As the Web is evolving to a highly globalized information space, knowledge sharing across different languages becomes very important [2]. Thus, how to create a cross-lingual knowledge graph is a very important issue.

In medical domain, there are multiple data sources of knowledge graph, which can be divided into two categories, one is the medical standard databases, the other one is

© Springer Nature Switzerland AG 2019
W. Ni et al. (Eds.): WISA 2019, LNCS 11817, pp. 512–517, 2019.
https://doi.org/10.1007/978-3-030-30952-7_51

the medical records databases collected from hospitals, physician institutes, etc. Also, there are two kinds of knowledge graphs in the medical domain: **Conceptual Knowledge Graph (CKG), Instance Knowledge Graph (IKG)**. One IKG with the corresponding CKG is called a **Factual Knowledge Graph (FKG)**. CKG contains only conceptual nodes and relations, while IKG contains instance nodes and relations.

The main challenge in constructing cross-lingual knowledge graph is the imbalance amount of knowledge in different languages in the existing knowledge bases and the inconsistence in the semantic rules. Moreover, for FKGs in medical-domain, the scarcity of EMRs and doctors' prior knowledge could lead low accuracy in construction process.

We propose a platform CLMed, which is used to build cross-lingual knowledge graphs for cardiovascular diseases. In this platform, we integrate both Chinese and English concepts and relations to construct cross-lingual CKG, meanwhile we construct cross-lingual IKG by processing instances from both Chinese and English EMRs. After the fusion of CKG and IKG, we construct cross-lingual FKG - CLKG.

This paper is organized as follows. Section 2 introduces the related work. Section 3 gives a brief description about the architecture of CLMed platform. Section 4 shows the process to construct cross-lingual CKG and FKG for cardiovascular diseases. At last, we conclude our current work and state our future work in Sect. 5.

2 Related Work

Cross-lingual knowledge bases such as DBpedia [3], YAGO and ConceptNet are becoming vital sources of knowledge for people and AI-related applications. These knowledge bases are normally modeled as knowledge graphs that store two kinds of knowledge: the monolingual knowledge and the cross-lingual knowledge that matches the monolingual knowledge among different languages. However, among these cross-lingual knowledge graphs, non-English knowledge is scarce. For instance, in Wikipedia-based knowledge bases: the English Wikipedia has 13 times more infoboxes than the Chinese Wikipedia and 3.5 times more infoboxes than the second largest Wikipedia of German language; DBpedia contains only 2% instances, properties and links in Chinese; YAGO collects more than 1200 million triples, and less than 5% of them are expressed in Chinese [4].

The imbalance situation in cross-lingual knowledge bases will cause scarcity in inner links between cross-lingual entities. The current solutions provided by cross-lingual knowledge graph such as XLORE, WiKiCiKE, ConceptNet5.5 is to provide classification-based method to correct the semantic rules, employ the cross-lingual knowledge linking via concept annotation [5], entity alignment method, etc. These methods work out in the general knowledge graphs but are not enough to construct a comprehensive cross-lingual knowledge graph in medical domain.

Integrating the methods investigated in cross-lingual general knowledge graph and medical knowledge graph, we realize that in order to fill the huge gap between the amount of Chinese and English knowledge, we should enrich the Chinese concepts as well as cross-lingual knowledge links by using techniques such as concept fusion, relation fusion, etc. In addition to the commonly used AI-related algorithm, we integrate doctor-involved toolsets to augment the efficiency in building cross-lingual FKG for cardiovascular diseases [6].

3   Architecture of CLMed

In this section, we introduce the architecture and data model of CLMed. The architecture presents the cross-lingual data sources, the output of these data, the toolsets and end users of the CLMed.

CLMed is a platform which can not only present large-scale, cross-lingual knowledge graphs for cardiovascular diseases, but also can provide doctor-involved toolsets to process the cross-lingual data into the nodes and edges in the knowledge graphs. Figure 1 shows the general architecture of the CLMed framework.

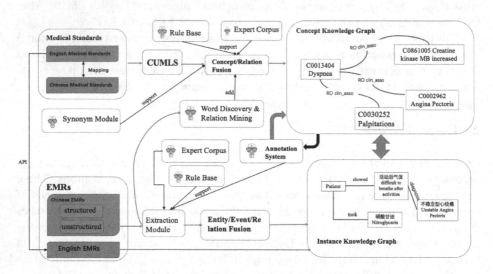

Fig. 1. General architecture of CLMed.

As shown in Fig. 1, the architecture of the CLMed can be divided into two parts: construction of CKG and construction of IKG. CLKG is the fusion of CKG and IKG, which is able to provide services such as similar-patient linking and similar-treatment linking. This service can compare different treatments about similar patient in Chinese and English, which will lead a better understanding of certain disease for users.

4   Construction of CLKG

Based on the doctor-involved toolsets, CLMed platform can be used to construct large-scale, cross-lingual FKG for cardiovascular diseases, CLKG. CLKG contains CKG and IKG from multiple, cross-lingual data sources.

4.1   Construction of CKG

Construction of CKG contains the construction of CUMLS, concept and relation fusion process.

4.1.1   Construction of CUMLS

In the construction of cross-lingual CKG, the biggest challenge is the imbalance of the amount of knowledge in different languages. To overcome this problem, we generate the cross-lingual knowledge base, CUMLS, based on integrating Chinese medical standards, Chinese health medical knowledge bases and UMLS.

Since UMLS contains large amounts of medical concepts and relations, we map the Chinese medical concepts with the existing English concepts in UMLS through the unique identification code, CUI. Meanwhile in UMLS, AUI can represent the source vocabulary of that concept. Therefore, in UMLS, a CUI may correspond to multiple AUIs. By analyzing the vocabulary features of UMLS, all strings in the thesaurus have a unique code. According to these features, we design to map the string's unique code to its source vocabulary, then get its AUI. After that, map the CUI through its AUI.

Based on this mapping procedure, we collect Chinese medical knowledge bases which contain medical thesaurus that UMLS stores. After merging all medical knowledge bases, we construct heterogeneous, multi-sources database, CUMLS, which stores 200,000 nodes contains Chinese expression.

4.1.2   Concept and Relation Fusion

The fusion of concepts and relations from multi-sources is necessary to construct an exhaustive cross-lingual CKG. We integrate the doctor-involved toolsets with ML algorithms to support the fusion process.

Doctor-involved toolsets in CLMed contain five modules as follow:

1. The rule base for the concept and relation fusion process concentrate on the data conversion rules from ER to RDF.
2. In expert corpus, doctors define and add the most common used Chinese medical concepts for cardiovascular diseases that not store in the CUMLS.
3. In synonym module, doctors define the synonym words in Chinese to avoid meaningless repetitive work in fusion process. Meanwhile, the fusion subsystem provides candidate synonym words based on the input words, doctors select the word with the closest meaning to join the thesaurus.
4. The annotation module focuses on discovery new cross-lingual links (CLs). Learned from the CLs from existed datasets and provided by doctors, the annotation module will provide several candidate CLs with the help of ML algorithms. Then, the new CLs with the highest weight in annotated features will be added into CKG.
5. Doctor-involved toolsets provide word discovery and relation mining module in order to integrate concepts and relations that occur frequently in medical instances into CKG.

Through these toolsets, cross-lingual concepts and relations from different sources can achieve effective and accurate fusion.

4.2 Construction of IKG

The IKG in CLKG includes concept nodes, entity nodes and event node, and relations between them. The data source of the IKG contains the EMRs from hospitals, open datasets and social media.

4.2.1 Extraction Module

The key construction process of IKG contains entity extraction, event extraction and relation extraction. When dealing with English EMRs, we use the API provided from UMLS, which was allowed to extract entities and relations based on clinical rules and users' need. For the Chineses EMRs, we use ML technologies and rule base defined by doctors to achieve accurate and effective extraction process [7].

In this process, we use expert corpus, rule base and annotation module to support the extraction. As described before, expert corpus totally based on doctors' prior medical knowledge. Rule base generates the extraction rules based on EMRs' semantic pattern and users' need and it help to transform unstructured EMRs into entities and relations. Doctor-involved annotation module in entity extraction provide sequence annotation, which is based on the entity corpus. In the relation extraction process, the doctor-involved toolsets generate pattern-based extraction model, which is based on the relation corpus. The extraction of events is also an important component of the IKG construction since timelines are an intuitive way to provide an overview of events over a certain period of time [8]. Rule base defined by doctors can support event extraction. Here is an example of event extraction rule: *"The patient was admitted to the hospital at the date of ****"*.

The extracted entities, events and relations will align with its unique code, which can achieve the cross-lingual entity mapping.

4.2.2 Entity, Event and Relation Fusion

The fusion of the nodes and edges in IKG depends on their features such as unique code, source CUI, time, etc. With the alignment of the unique code, instances from EMRs, open datasets and social media can be linked. Due to the massive data in IKG, we apply the same ML algorithm system as concept and relation fusion process, which will help automatically find more inner links between instances.

4.3 CLKG – A FKG for Cardiovascular Diseases

CLKG is the fusion of IKG and CKG, which contains over 200,000 concepts and a million relations from cross-lingual knowledge bases, and instances and relations from Chinese and English EMRs, social media, etc.

5   Conclusion and Future Work

In this paper, we propose an end-to-end platform named CLMed, which can be used construct the large-scale, cross-lingual knowledge graph. In CLMed platform, we integrate the doctor-involved toolsets to support the cross-lingual knowledge extraction and fusion process. Along with doctor-involved toolsets, ML algorithm can enhance the effectiveness in the cross-lingual knowledge fusion. Based on CLMed platform, we construct CLKG. CLKG is a fusion of both medical CKG and IKG. By cause of the integrity of source data and in CLKG, we are able to provide similar patient finding service to support cardiovascular knowledge sharing across different languages.

In the future, we intend to extend the scale of CLMed. Currently, CLMed works only for cardiovascular diseases. With integration of more medical databases and doctors, we hope CLMed could cover another field in medical domain.

Acknowledgments. This work was supported by NSFC (91646202), National Key R&D Program of China (2018YFB1404400, 2018YFB1402700).

References

1. Kejriwal, M.: Domain-Specific Knowledge Graph Construction. Springer, Heidelberg (2019). https://doi.org/10.1007/978-3-030-12375-8
2. Su, Y., Zhang, C., Li, J., Wang, C., Qian, W., Zhou, A.: Cross-lingual entity query from large-scale knowledge graphs. In: Cai, R., Chen, K., Hong, L., Yang, X., Zhang, R., Zou, L. (eds.) APWeb 2015. LNCS, vol. 9461, pp. 139–150. Springer, Cham (2015). https://doi.org/10.1007/978-3-319-28121-6_13
3. Li, W., Chai, L., Yang, C., Wang, X.: An evolutionary analysis of DBpedia datasets. In: Meng, X., Li, R., Wang, K., Niu, B., Wang, X., Zhao, G. (eds.) WISA 2018. LNCS, vol. 11242, pp. 317–329. Springer, Cham (2018). https://doi.org/10.1007/978-3-030-02934-0_30
4. Wang, Z., Li, J., Tang, J: Boosting cross-lingual knowledge linking via concept annotation. In: Rossi, F. (ed.) IJCAI, pp. 2733–2739 (2013)
5. Wang, Z., Li, Z., Li, J., Tang, J., Pan, J.Z.: Transfer learning based cross-lingual knowledge extraction for wikipedia, pp. 641–650. The Association for Computer Linguistics (2013)
6. Holzinger, A.: Interactive machine learning for health informatics: when do we need the human-in-the-loop? Brain Inf. **3**(2), 119–131 (2016)
7. Archer, V.: MuLLinG: multilevel linguistic graphs for knowledge extraction. In: Banea, C., et al. (eds.) TextGraphs@ACL, pp. 69–73. Association for Computational Linguistics (2010)
8. Gottschalk, S., Demidova, E.: EventKG+TL: creating cross-lingual timelines from an event-centric knowledge graph. CoRR. abs/1805.01359 (2018)

How to Empower Disease Diagnosis in a Medical Education System Using Knowledge Graph

Samuel Ansong$^{(\boxtimes)}$ [iD], Kalkidan F. Eteffa, Chao Li, Ming Sheng,
Yong Zhang, and Chunxiao Xing

Research Institute of Information Technology, Beijing National Research Centre
for Information Science and Technology, Department of Computer Science
and Technology, Institute of Internet Industry,
Tsinghua University, Beijing 100084, China
{ssm18,ajql8}@mails.tsinghua.edu.cn, {li-chao,
shengming,zhangyong05,xingcx}@tsinghua.edu.cn

Abstract. Disease diagnosis is an important function in a medical training system, an integrated system which is aimed at providing the necessary skills and know-how to health practitioners. As one of the most vital features of a medical training system, many researchers and industry alike have channelled time and resources to engage several techniques and practices in a bid to find a way to accurately predict diseases with minimal margin of error. This has motivated several variations in feature selection, data representations and techniques in machine learning. In this paper, we explore some of these variations with prime focus on how knowledge graphs have helped address issues like insufficient data and interpretation to help empower the construction of a disease diagnosis feature in a medical training systems.

Keywords: Knowledge graph · Neural networks · Disease diagnosis

1 Introduction

Today, the pathway to build robust educational technologies is to integrate several vital features tailored for its target group. Medical training systems just like any other training system will provide a way for medical practitioners and medical researchers to accomplish tasks ranging from predicting adverse drug reactions, viewing statistical health data and insights, calculating patient similarity and diagnosing patients. A medical training system will complement traditional medical training methodologies to provide a holistic approach to medical training. There are several features that make up a medical training system, however the disease diagnosis feature is the more relevant for junior medical doctors. According to research by Singh, 5% of patients, that is approximately 12 million patients are misdiagnosed a year in the united states of America alone. This contributes significantly to the overall percentage of medical mistakes resulting in death or serious complications for patients, making it the third leading cause of death in the USA according to a recent publication by CNBC.

© Springer Nature Switzerland AG 2019
W. Ni et al. (Eds.): WISA 2019, LNCS 11817, pp. 518–523, 2019.
https://doi.org/10.1007/978-3-030-30952-7_52

The disease diagnosis component of a medical training system demands a very high accuracy in its prediction task, an accuracy that cannot be guaranteed solely by the use of EHR (Electronic Health Record). It is therefore important that at the construction stage of such a feature the diagnosis of a medical practitioner is compared to that of a disease prediction algorithm and their differences reconciled by medical experts. However, building an accurate disease prediction algorithm does come with its own challenges. Competent and accurate implementation of such a system needs a comparative study of various techniques available.

This paper will explore how machine learning algorithms combined with knowledge graphs can improve the construction of a disease diagnosis feature in a medical training system. The paper will discuss how Knowledge graphs have contributed to resolving the challenges associated with EHR.

2 Methodology

The orientation of the paper follows the SLR guidelines for computer engineering. The key phases of this methodology include planning, selection and searching of primary studies which consists of formulating queries and keywords, performance of quality assessment and a selection procedure. The studies were limited to recent publications in the field of computer science and medicine which addresses the disease diagnosis task. Recent publications between the years 2013 and 2019 were considered.

3 Overview of Disease Diagnosis

Clinical decision making has evolved to become more complex and requires the evaluation of large volumes of data expressive of clinical information Liang et al. [1]. Artificial intelligence methods have emerged as possibly powerful tools to mine EHR data to aid in disease diagnosis and management. Data mining techniques have been adopted in variety of applications in the healthcare industry. For example, data mining proved essential in diagnosing diseases with good outcomes according to Zriqat et al. [2]. Due to the nature of the EHR, Hug et al. [3] suggests that comparable to spam filtering, sentiment analysis and language identification, disease prediction is an important medical text classification problem. Feature extraction from medical text has been a prevailing issue for most researchers over the years and several techniques have been deployed to address this issue, a study of eleven publications shows that most researchers favoured the bag of words and N-gram methodologies as seen in Fig. 1. However, BOW (Bag of Words) does not consider word order and grammar a worrying sign in medical text classification. To address this, several authors chose N-gram, although the approach takes into consideration word order, it does not solve the issue entirely since it does not account for word inversion, a persistent problem in healthcare data reports.

The general traditional approach process of diagnosing a patient can be seen as a classification task since it may be possible to achieve an acceptable level of conviction of a diagnosis with only a few features without having to process the whole feature set.

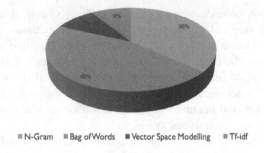

Fig. 1. Feature extraction techniques for medical text classification

This analogy has paved way for several implementations of classification algorithms for disease prediction. Researchers over the years have preferred neural networks over statistical approaches. These authors [4–6] used RNNs (Recurrent Neural Networks) in an attempt to model the sequential relations amongst medical codes. RNNs and all other deep learning algorithms however suffer from insufficient data and lack of interpretation. In the next section, we will focus on knowledge graphs and how they have helped in addressing these issues.

4   Knowledge Graph Methodology

Traditional machine learning algorithms take as input numeric or categorical qualities of an object known as a feature vector [7]. Knowledge graphs model information in the form of entities and relationships. Thus, a knowledge graph can contain an object together with its relationships to other objects in the form of a graph made up of nodes and labelled edges (which represents the relationships between the nodes). Knowledge Graphs (KGs) therefore offer semantically organized information that is interpretable by computers something that is needed in order to build more intelligent systems [7].

The speedy growth in volume and multiplicity of health care data from EHR and other sources [8] further necessitates the use of KGs. Training deep learning models typically involves large amounts of data that often cannot be met by a single health system or provider organization. Using knowledge graphs, distributed medical data sources can be aggregated into one meaningful data source. Several authors [8–14] have used knowledge graphs to solve the data inefficiency and interpretation drawbacks of neural networks. Choi et al. [8] applied knowledge graph in supplementing the EHR with hierarchical information from medical ontologies to improve interpretation whereas Ma et al. [9] used it to improve consistency by learning medical code representations. Authors [10–12] used knowledge graph to aggregate domain related data sources to solve the data insufficiency problem and achieved admirable results. Deploying knowledge graphs in the healthcare services space has proven to be an effective method to map relationships between the enormous variety and structure of healthcare data. Graphs provide an uncanny ability to model concealed relationships between information sources and capture linked information that other data models fail to capture. This enables researchers to more easily embed the information they need among a wide array of variables and data sources.

4.1 Knowledge Graph Representation

Machine learning on graphs is an important and pervasive task. The principal challenge in this area is finding an efficient method for the representation and encoding of graph structures in a way suitable for machine learning models [15]. Learning the representations of graphs is a hot research topic which has driven the proposal of various methods. Unlike traditional hand-engineered approaches, representation learning approaches treat this problem as machine learning task itself, using a data-driven approach to learn embeddings that encode graph structure [15]. Node embedding which is an approach use by several authors, is a process where nodes are encoded into low-dimensional vectors keeping intact their structure and relationships. DeepWalk, Node2Vec, LSHM, LINE, Metapath2Vec and Struc2Vec are some approaches proposed by authors.

All the listed methods focus on learning good representations for graph data, the majority of these methods belong to the direct encoding class under node embedding. Direct encoding however, fails to leverage node attributes which are sometimes highly helpful with regard to the node's position and role in the graph and are sometimes also computationally inefficient. To address these issues, methods like Deep Neural Graph Representations and SNDE (Structural Deep Network Embeddings) [16] have been proposed where graphs are directly incorporated into the encoder algorithm.

4.2 Discussion on Knowledge Graph Methodology

The task of disease prediction is an interesting field yet challenging one which has motivated several research and optimizations in pre-processing techniques, feature extraction techniques, feature selection techniques, algorithm design techniques and evaluation techniques. Knowledge graphs present us with the opportunity to incorporate large domain related datasets from arbitrary sources into one meaningful data source as input for neural networks to improve accuracy as demonstrated by authors [8, 10]. EHR are considered rich in data and can be greatly utilized [17]. In a sector like health where interpretation and accuracy is on a higher demand, knowledge graphs offer a way for understandable readings into the task of diagnosing a disease since it provides concise and meaningfully accurate relationships between features of a dataset. The integration of knowledge graphs into E-health systems will go a long way to drastically improve healthcare by offering an appreciable level of support to health professionals. In Liu et al. [12], the extraction of prediction rules generated from observing both classically professional paediatrics textbooks and clinical experiences of paediatric doctors respectively to form a knowledge graph indicates that junior doctors can benefit widely from existing knowledge graphs generated from more experience doctors.

The support from models powered by knowledge graphs could help junior doctors perform both differential diagnosis which requires an iterative step likened to the work of a classifier more efficiently and also allow junior doctors perform pattern recognition diagnosis which requires the use experience to recognize a pattern of clinical characteristics. Pattern recognition diagnosis can only be achieved if a practitioner has enough experience. Representing the knowledge of experience doctors and clinical books as a knowledge graph as demonstrated in Liu et al. [12] will result in a more efficient disease prediction algorithm which can predict diseases with minimal error.

5 Conclusion

In this paper, we establish the fact that, in order to build an effective disease diagnosis, feature for a medical training system, a hybrid approach must be utilised where results from a disease prediction algorithm are compared to that obtained from querying an electronic health record and reconciled by experts. The paper then systematically review the various techniques and methodologies deployed by researchers in building accurate disease prediction algorithms whose results can be used to validate query results from an EHR in constructing an efficient disease diagnosis feature for a medical training system. The paper has shown that researchers over the years deployed knowledge graphs as a tool to remove ambiguity in medical concepts to help improve feature extraction. Also, researchers have applied knowledge graph as a tool to aggregate the various data sources of EHR and also provide sufficient input for deep learning models to help improve their accuracy. An accurate disease diagnosis algorithm will not only provide a means for validating diagnosis records in an EHR but can also become central in the construction of a diagnosis feature in a medical training system.

Acknowledgements. This work was supported by NSFC (91646202), National Key R&D Program of China (2018YFB1404400, 2018YFB1402700).

References

1. Liang, H., et al.: Evaluation and accurate diagnoses of paediatric diseases using artificial intelligence. Nat. Med. **25** (2019). https://doi.org/10.1038/s41591-018-0335-9
2. Zriqat, E., Altamimi, A., Azzeh, M.: A comparative study for   predicting heart diseases using data mining classification methods (2017)
3. Huq, K.T., Mollah, A.S., Sajal, Md.S.H.: Comparative study of feature engineering techniques for disease prediction. In: Tabii, Y., Lazaar, M., Al Achhab, M., Enneya, N. (eds.) BDCA 2018. CCIS, vol. 872, pp. 105–117. Springer, Cham (2018). https://doi.org/10.1007/978-3-319-96292-4_9
4. Choi, E., et al.: Multi-layer representation learning for medical concepts. In: Proceedings of the 22nd ACM SIGKDD International Conference on Knowledge Discovery and Data Mining (KDD 2016) (2016)
5. Choi, E., et al.: RETAIN: an interpretable predictive model for healthcare using reverse time attention mechanism. In: Advances in Neural Information Processing Systems (NIPS 2016) (2016)

6. Ma, F., Chitta, R., Zhou, J., You, Q., Sun, T., Gao, J.: Dipole: diagnosis prediction in healthcare via attention-based bidirectional recurrent neural networks. In: Proceedings of the 23rd ACM SIGKDD International Conference on Knowledge Discovery and Data Mining (KDD 2017), pp. 1903–1911 (2017)
7. Nickel, M., Murphy, K., Tresp, V., Gabrilovich, E.: A review of relational machine learning for knowledge graphs. Proc. IEEE **104**(1), 11–33 (2016)
8. Choi, E., Bahadori, M.T., Song, L., Stewart, W.F., Sun, J.: GRAM: graph-based attention model for healthcare representation learning. In: KDD (2017)
9. Ma, F., You, Q., Xiao, H., Chitta, R., Zhou, J., Gao, J.: KAME: knowledge-based attention model for diagnosis prediction in healthcare, pp. 743–752 (2018). https://doi.org/10.1145/3269206.3271701
10. Jha, A., et al.: Deep Convolution Neural Network Model to Predict Relapse in Breast Cancer (2018)
11. Jiang, J., et al.: Medical knowledge embedding based on recursive neural network   for multi-disease diagnosis (2018)
12. Liu, P., et al.: HKDP: a hybrid knowledge graph based pediatric disease prediction system. In: Xing, C., Zhang, Y., Liang, Y. (eds.) ICSH 2016. LNCS, vol. 10219, pp. 78–90. Springer, Cham (2017). https://doi.org/10.1007/978-3-319-59858-1_8
13. Yang, et al.: GrEDeL: a knowledge graph embedding based method for drug discovery from biomedical literatures. IEEE Access, p. 1 (2018)
14. Bean, D.M., et al.: Knowledge graph prediction of unknown adverse drug reactions and validation in electronic health records. Sci. Rep. **7**, 16416 (2017)
15. Hamilton, W.L., Ying, R., Leskovec, J.: Representation Learning on Graphs: Methods and Applications (2017)
16. Cao, S., et al.: Deep neural networks for learning graph representations. In: AAAI Conference on Artificial Intelligence (2016). pag.web. 14 April 2019
17. Tian, B., Xing, C.: Deep learning based temporal information extraction framework on Chinese electronic health records. In: Meng, X., Li, R., Wang, K., Niu, B., Wang, X., Zhao, G. (eds.) WISA 2018. LNCS, vol. 11242, pp. 203–214. Springer, Cham (2018). https://doi.org/10.1007/978-3-030-02934-0_19

Blockchain

Blockchain Retrieval Model Based on Elastic Bloom Filter

Xuan Ma, Li Xu, and Lizhen Xu[✉]

Department of Computer Science and Engineering, Southeast University,
Nanjing 211189, China
jiujiuqiniao@163.com, lzxu@seu.edu.cn

Abstract. Blockchain as emerging technology is revolutionizing several industries, especially the education industry, which has high requirements for the
authenticity of data. The proposed blockchain technology realizes decentralization and time-sequence chain storage of data blocks, ensuring that the stored data
blocks are not tamperable and unforgeable, and satisfy the high trust of data
authenticity. However, current League Chains (such as Hyperledger Fabric)
generally have problems such as low throughput and lack of indexing technology,
which leads to inefficient data retrieval problems. To this end, this paper proposes
a new elastic Bloom filter model that combines smart contracts. This model
provides an adaptive adjustment method for Bloom filters, it can effectively reduce
the false positive probability under the condition of low memory consumption and
improve the efficiency of data retrieval. The experimental results based on
Hyperledger Fabric show that compared with the standard Bloom filter model, the
proposed model guarantees a lower false positive probability and verifies its high
efficiency under data retrieval.

Keywords: Blockchain · Bloom filter · Smart contract

1 Introduction

Every year, there are endless incidents of fabricate academic credentials [1], which has
a very negative impact on the recruitment of talents, national education work and social
development. The most authoritative solution to this problem is the Xuexin.com, which
is directly endorsed by the Ministry of Education. However, Xuexin.com is a centralized data management system with the drawbacks of data leakage and loss. In terms
of this issue, this paper applies blockchain technology to solve the problem of fabricate
academic credentials.

Students' learning experience data (e.g., Class records (Class videotaping);
Schedule of work and rest; Homework; Experiments; Awards of competitions; Grad-
uation projects; Diploma, etc.) are stored on the blockchain. These multi-dimensional
data can more accurately describe the unique value of each student, so as to trans-
parently and safely evaluate the ability of students. Blockchain, as a distributed database storage technology which is untampered, can record students' learning experience
data completely as historical data (these data are traceable and multi-party trusted). In
addition, based on the reliability of cryptography, it can also guarantee the privacy of

© Springer Nature Switzerland AG 2019
W. Ni et al. (Eds.): WISA 2019, LNCS 11817, pp. 527–538, 2019.
https://doi.org/10.1007/978-3-030-30952-7_53

students' data. However, these data are diverse and will grow rapidly over time. In such a large amount of data, data retrieval speed becomes a big problem. However, the low throughput of current League Chains (such as Hyperledger Fabric [2]) cannot satisfy the requirements of fast data retrieval.

In the field of big data, Bloom filter is a widely-used data structure. It is a very simple yet powerful data structure to handle large-scale data by sacrificing a small amount of memory space. Unlike a conventional hash data structure, Bloom Filter is more space efficient and Bloom Filter is able to reduce memory space consumption by order of magnitude [3].

In this paper, a Blockchain retrieval model based on Elastic Bloom filter will be proposed, which can effectively abbreviate the delay time of data retrieval. This model provides an adaptive method to adjust the memory consumption by Bloom filter under the condition of low false positive probability. In order to evaluate the performance of this solution, this solution was experimentally verified and the results were compared with the original solution.

2   Related Work

The collection, storage and analysis of big data all require distributed application scenarios. In other words, blockchain technology is highly consistent with the storage requirements of big data. The symbiotic development of big data and blockchain has become a trend, and the combination of them greatly enhance the ability of information acquisition and data anti-counterfeiting. Wang et al. recommended the precision analysis technology based on the Big Data and Blockchain to establish an intelligent contract system, use the weighted decision model to implement the decision-making process effectively [4].

However, miners and validators in the existing blockchains execute scripts of smart contracts in turn, which do not support concurrency, so the execution efficiency of smart contracts is not high. The serial execution of smart contracts largely limits the throughput of the blockchain system [5]. Dickerson et al. proposes a method that allows parallel execution of smart contracts by discovering a serializable concurrent schedule [6]. In addition, improving the speed of contracts' data retrieval is also expected to improve this issue.

Siddiqi et al. employed a scheme to securely log the data from wearable body sensing devices by storing the fingerprints of data in Bloom filters [7]. Guan et al. divided users into different groups and each group has a private blockchain to record its members' data. In addition, the Bloom filter is adopted for fast authentication [8]. As the running environment of smart contract, the Ethereum virtual machine (EVM) stores some log data in Bloom filter, so that developers can search logs efficiently and safely [9]. Liu et al. proposed a strategy for realizing access control of big data by using blockchain's transactions, and using the policy management method based on Bloom filter to realize fast retrieval of access control strategies [10]. However, the above schemes only use the standard Bloom filters, which have a relatively high probability of query failure.

The work in this paper is related to improvement of standard Bloom filter. The specific model will be shown in Sect. 3.

3   Model

3.1   Blockchain

Blockchain is derived from the underlying technology of bitcoin [11]. It is a specific data structure that combines data blocks in a chain according to the time sequence based on transparent and trusted consensus rules in the peer-to-peer network environment [12].

Blockchain is not a single technological innovation, but the result of deep integration of distributed technology, cryptography, consensus mechanism, smart contract and other technologies. Cryptography guarantees that the data on the blockchain has the characteristics of not tampering, not forging and traceable. The "distribution" of blockchain is not only reflected in the backup storage of data, but also in the data record, that is, all nodes participate in the data maintenance. The tampered or destroyed data of a single node will not affect the data stored in the blockchain, so it can effectively avoid the occurrence of a single point of failure.

Encrypted structures of chains are used to validate and store data, distributed technologies and consensus mechanisms are used to validate and communicate between nodes, and smart contracts enable complex business logic and automate data operations.

Smart Contract. Smart contract is a digital protocol that uses algorithms and programs to compile contract terms, deploy them on a blockchain, and automatically execute them according to rules [13]. Smart contract has access to block and state data. It can trigger the contents of contracts sequentially and complete a series of secure automation operations in a de-trusted environment. Meanwhile, it expands the functionality of blockchain, making it possible to build blockchain-based applications.

Figure 1 depicts the smart contract system. Users can easily send a transaction to the contract's address to run it. This transaction will then be executed by miners (i.e. peers) in the network to reach a consensus on its output. Then, the contract's status stored on the blockchain is updated accordingly.

Nevertheless, there is a performance issue in smart contracts. In blockchain systems, smart contracts are executed sequentially (e.g., one contract at a time). This would affect the performance of the blockchain negatively as the number of smart contracts that can be executed per second will be limited. This process also results in the blockchain being unable to scale [14].

Smart contracts can be developed and deployed in different blockchain platforms (e.g., Ethereum, Bitcoin and Hyperledger Fabric). Different platforms offer distinctive features and programming languages for developing smart contracts. We will only focus on Hyperledger Fabric in this paper. Hyperledger Fabric uses the Docker container as a sandbox environment for smart contracts. Docker containers provide isolation and security for smart contracts.

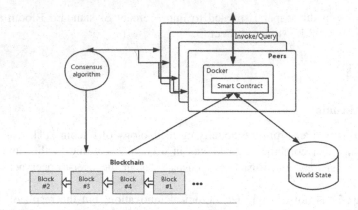

Fig. 1. Smart Contract system

3.2 Bloom Filter

Bloom filter is an algorithm proposed by Bloom in 1970 to detect whether a data element exists in a set [15], it is a space-efficient probabilistic data structure. The main idea of Bloom filter is to use binary array to describe a set, and use multiple different hash functions to judge whether data elements exist in this set. Its query efficiency and memory utilization are far more than other algorithms. However, the Bloom filter offers a compact probabilistic way to represent a set that can result in false positives (claiming an element to be part of the set when it was not inserted). Bloom filters include:

(1) An m-bit binary array M, and each bit of the array is initialized to 0.
(2) n elements in the set D.
(3) k independent hash functions ($hash_1$, $hash_2$... $hash_k$).

Figure 2 presents an overview of a Bloom filter. The Bloom filter consists of a binary array M of length m. Three elements have been inserted, namely key_1, key_2 and key_3. Each of the elements have been hashed using three hash functions to bit positions in the array M. The corresponding bits have been set to 1, otherwise 0. When verifying whether x exists in the set, it will be hashed using the same three hash functions into bit positions. If the corresponding position in the array M is 1, the element x is considered to be an element in the set, otherwise it is not.

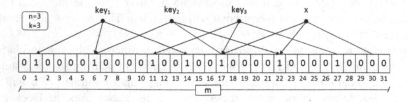

Fig. 2. Overview of the Standard Bloom Filter

3.3 Elastic Bloom Filter

In this paper, the standard Bloom filter has been enhanced, and the enhanced Bloom filter is called Elastic Bloom filter (EBF). EBF provides an adaptive adjustment method for Bloom filters, which is based on a Bloom filter using Partitioned Hashing.

Bloom Filters Using Partitioned Hashing. Partitioned Hashing is done in two steps, as illustrated in Fig. 3. In the first step, a hash function h is used to divide a key set into groups $g_1, g_2, \cdots, g_t$. Each group g_i is associated with k independent and distinctive hash functions $h_1, h_2, \cdots, h_k$ (Any k of the H distinctive hash functions). In the second step, each key is hashed k times to the Bloom filter by using the k hash functions of its group, and the corresponding bits in the filter are set to be 1 [16].

Since the hash function in each group of the partition bloom filter is not completely the same, the repetition rate of converting the data element to the binary array by the hash function is considerably reduced.

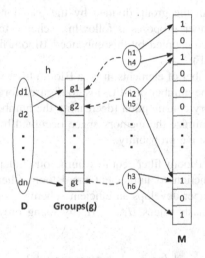

Fig. 3. Partitioned Hashing

Elastic Bloom Filter. In order to reduce the false positive probability of Bloom filter, this paper proposes an improvement based on multiple groups of hash functions for the partitioned Bloom filter. First, the hash function used by a group g_i in the partitioned Bloom filter is converted into several groups $hg_1, hg_2, \ldots, hg_l$, hg_i corresponding to one Bloom filter unit BFU_i. Each key in g_i is then mapped to the corresponding bit in each Bloom filter unit BFU_i by all hash functions in hg_i. Finally, the union of multiple BFU_i is used to indicate whether the element x exists. If x exists in g_i, then the corresponding bit of all BFU_i hash function mapping values must be 1, otherwise the element is considered to be absent. The model is described in Fig. 4.

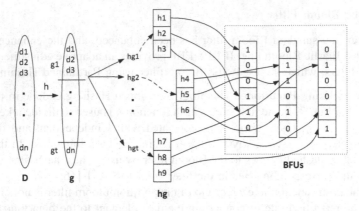

Fig. 4. Overview of the Elastic Bloom Filter

Since the elements in the group divided by the partitioned hash function h are different in size, this paper proposes a following scheme to solve the problem of excessive memory space occupied by this enhanced Bloom filter, which is called an Elastic Bloom filter (EBF):

According to the number of elements in g_i, a Bloom filter with a low false positive probability (increasing the number of BFUs) is configured for a large g_i, and a Bloom filter with a small memory consumption (decreasing the number of BFUs) is allocated for a small g_i. It can minimize the memory space occupied by the Bloom filter while ensuring a low false positive probability.

Hash Function. Elastic Bloom filter requires numerous independent hash functions. Using existing hash function to implement the Bloom filters consumes too much resources. Hence, this paper develops an efficient scheme.

We can generate c hash values $H_1,\ldots,H_c$ by using only k seed universal hash values $h_1,\ldots,h_k$, where

$$k = \begin{cases} log_2 c + 1, & c\,is\,power\,of\,2 \\ \lceil log_2 c \rceil, & c\,is\,not\,power\,of\,2 \end{cases} \tag{1}$$

H_i is addressed in the following equation:

$$H_i = (p_k h_k(x)) \oplus (p_{k-1} h_{k-1}(x)) \oplus \cdots \oplus (p_1 h_1(x)) \tag{2}$$

$$i = p_k 2^{k-1} + p_{k-1} 2^{k-2} + \cdots + p_j 2^{j-1} \cdots + p_2 2^1 + p_1 2^0,\ p_j = 0\,or\,1 \tag{3}$$

where x is the object key of g_i, $\oplus$ is bit-wise XOR operation.

3.4   EBF-Based Smart Contract Model

This paper proposes a new Elastic Bloom filter model that combines smart contracts, it can improve the speed of data retrieval in the blockchain as well as ensure simple and accurate verification.

EBF-Based Smart Contract Model. This paper designs a smart contract model based on EBF. Compared with the standard smart contract in blockchain, this model improves the efficiency of data retrieval and query by storing data in Elastic Bloom filter. Algorithm 1 gives the pseudocode that shows how the data set D is grouped and how the size of the BFUs adaptively changes. Algorithm 2 presents the pseudocode for the insertion operation of EBF-based smart contract model. Algorithm 3 gives the pseudocode for the membership test of a given element x in the EBF-based smart contract mode.

Algorithm 1 Partitioned Hashing

Input: $Set\ D$
Output: $Name_{sc}$

  1: **function** PARTITIONEDHASHING(D)
  2:     create new Smart Contract sc.
  3:     **for** each key in set D **do**
  4:         $V_{key} \leftarrow h(key)$
  5:         $s \leftarrow size_g$
  6:         $i \leftarrow V_{key}\ mod\ s + 1$
  7:     **end for**
  8:     Set a threshold θ
  9:     **for** each $group_i$, where $i = 1...s$ **do**
10:         $n_i \leftarrow size\ of\ group_i$
11:         $m \leftarrow size\ of\ bits\ in\ M$
12:         $length_i \leftarrow \lceil \frac{n_i * \theta}{m} \rceil$,/* $length_i$ changes adaptively according to $\frac{n_i}{m}$ */
13:         put$(group_i, length_i)$
14:     **end for**
15:     return $Name_{sc}$.
16: **end function**

Algorithm 2 The EBF-based smart contract insertion algorithm

Input: $Set\ D, Name_{sc}$
Output: $true\ or\ false$
1: create new Smart Contract SC_{Insert}.
2: $SC_{PH} \leftarrow PartitionedHashing(D)$
3: **for** each $group_i$ of SC_{PH} **do**
4:      **for** each key_i in $group_i$ of SC_{PH} **do**
5:           **for** each hash value H_t of hg_z, where $z = 1...length_i$ **do**
6:                $p_j \leftarrow 0\ or\ 1$
7:                $t \leftarrow p_k 2^{k-1} + p_{k-1} 2^{k-2} + ... + p_j 2^{j-1} + ... + p_2 2^1 + p_1 2^0$
8:                $H_t \leftarrow (p_k h_k(key_i)) \oplus (p_{k-1} h_{k-1}(key_i)) \oplus ... \oplus (p_1 h_1(key_i))$
9:                /* h_j is the hash function of $group_i$, where $j = 1...k$. */
10:               **if** B_{h_t} of $BFU_z == 0$ **then**
11:                    $B_{h_t} \leftarrow 1$
12:               **end if**
13:          **end for**
14:     **end for**
15:     SetBloomFilter(i,z)
16: **end for**
17: SC_{Insert} returns true or false depending on weather the insert operation is in error.

Algorithm 3 The EBF-based smart contract membership test algorithm

---

Input: x is the object key for which membership is test
Output: $true\ or\ false$
1: create new Smart Contract SC_{test}.
2: $V \leftarrow h(x)$
3: $i \leftarrow V\ mod\ s + 1$
4: $length \leftarrow$ get$(group_i)$
5: $flag \leftarrow 1$
6: **for** z from 1 to $length$ **do**
7:      $BFU_z \leftarrow$ getBloomFilter(i, z)
8:      **for** each hash value H_t of hg_z, where $z = 1...length$ **do**
9:           $p_j \leftarrow 0\ or\ 1$
10:          $t \leftarrow p_k 2^{k-1} + p_{k-1} 2^{k-2} + ... + p_j 2^{j-1} + ... + p_2 2^1 + p_1 2^0$
11:          $H_t \leftarrow (p_k h_k(key_i)) \oplus (p_{k-1} h_{k-1}(key_i)) \oplus ... \oplus (p_1 h_1(key_i))$
12:          **if** B_{h_t} of $BFU_z == 0$ **then**
13:               $flag \leftarrow 0$
14:          **end if**
15:     **end for**
16: **end for**
17: SC_{test} returns true or false depending on weather $flag$ is 1.

The hash function group hg_i in this model adaptively varies its size according to the constraint condition $length_i$. It maps different situations to different BFU models and then runs the BFU model.

4 Experiments

4.1 Comparisons with Standard and Elastic Bloom Filter

This experiment tested the performance of Standard, Multidimensional and Elastic Bloom filter, and were conducted for test sets of different query sizes. Multidimensional Bloom filter (MDBF) is composed of several Standard Bloom Filters (SBFs) with the same length of binary array. MDBF decomposes the representation and query of multidimensional element x into the representation query of single attribute value subset, and each SBF represents a one-dimensional attribute of x.

As shown in Table 1, compared with Standard Bloom Filter, MDBF and EBF can guarantee lower false positive probability on the same data sets.

Table 1. False positive probability between Standard, Partitioned and Elastic Bloom filter

Order	n	Number of queries	SBF	MDBF	EBF
1	500	300	0.06%	0.012%	0.013%
2	1000	1000	0.23%	0.077%	0.080%
3	1500	2000	0.98%	0.201%	0.213%
4	2000	3000	5.7%	0.930%	0.950%
5	2500	2000	9.22%	1.66%	1.71%
6	3000	3000	12.69%	3.01%	3.12%
7	3500	3000	14.28%	4.78%	4.89%
8	4000	3000	17.98%	6.09%	6.32%
9	4500	3000	22.25%	8.23%	8.46%

As show in Fig. 5, the memory consumption of EBF is far less than that of MDBF under the condition of similar false positive probability. In a word, although Elastic Bloom filter has made some sacrifices in terms of memory consumption, it has higher accuracy than Standard Bloom filter. Elastic Bloom filter's false positive probability is within the acceptable range and meets the requirements of data retrieval.

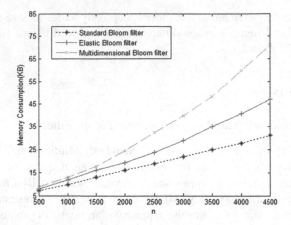

Fig. 5. Memory consumption of Standard and Elastic Bloom filter

4.2   Comparisons with Standard and EBF-Based Smart Contract

This experiment tested the performance of Standard and EBF-based Smart Contract on the optimization effect of data retrieval efficiency, and were conducted for test sets of different query sizes. As shown in Fig. 6, in terms of small-scale data sets, the advantages of EBF-based Smart contract are not obvious. However, Fig. 7 shows that EBF-based Smart contract significantly raises the efficiency of data retrieval in terms of large-scale data sets. EBF-based Smart contract meets the requirement of data retrieval efficiency.

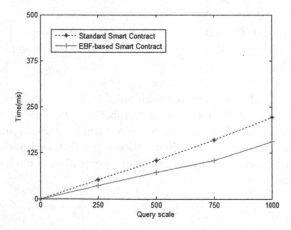

Fig. 6. The time about data retrieval in terms of small-scale data sets

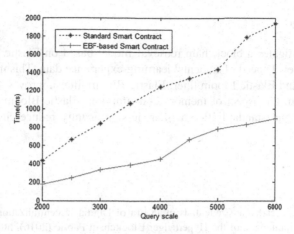

Fig. 7. The time about data retrieval in terms of large-scale data sets

4.3   Analysis

False Positive Probability. The equation of the probability of a false positive f in the standard Bloom filter is defined as follows:

$$f \approx \left(1 - e^{-kn/m}\right)^k \tag{4}$$

where m is the filter size in bits, n and k are the number of elements and hash functions respectively.

Since the Partitioned Hashing hashes the original data set D to $g_1, g_2, \cdots, g_t$, and each group g_i has an independent hash function group, the repetition rate of the bits in M is considerably reduced. In addition, Elastic Bloom filter expands each g_i's Bloom filter horizontally. Assuming that each group g_i is equipped with l_i BFUs, the formula for calculating the false positive probability of the g_i's BF is:

$$f_i = (f_u)^{l_i} \tag{5}$$

where f_u is the false positive probability of the one BFU, l_i changes adaptively according to size of g_i.

Since f_u is always less than 1, the false positive probability f_i will decrease as l increases. Therefore, the error probability f_E of EBF will be greatly reduced.

Retrieval Time. As the retrieval time of Bloom Filter is less affected by the data set size, compared with traversal retrieval of standard smart contract, EBF-based smart contract can achieve high performance. At the same time, EBF-based smart contract can effectively save cache space, reduce the number of cache requests and improve the efficiency of data query.

5 Conclusion

This paper investigates a blockchain retrieval model based on Elastic Bloom filter to improve the retrieval speed of students' learning experience data. This model combines smart contracts and Elastic Bloom filter, Elastic Bloom filter improves standard Bloom filter. Although in the aspect of memory consumption, Elastic Bloom filter has made more sacrifices than standard Bloom filter, it significantly reduces its false positive probability.

References

1. Smolenski, N.: Academic Credentials in an Era of Digital Decentralization (2016)
2. Cachin, C.: Architecture of the Hyperledger Blockchain Fabric (2016). https://www.zurich.ibm.com/dccl/papers/cachin_dccl.pdf
3. Patgiri, R., Nayak, S., Borgohain, S.K.: Role of bloom filter in big data research: a survey (2019)
4. Wang, X., Hu, Q., Zhang, Y., Zhang, G., Juan, W., Xing, C.: A kind of decision model research based on big data and blockchain in eHealth. In: Meng, X., Li, R., Wang, K., Niu, B., Wang, X., Zhao, G. (eds.) WISA 2018. LNCS, vol. 11242, pp. 300–306. Springer, Cham (2018). https://doi.org/10.1007/978-3-030-02934-0_28
5. Papadimitriou, C.H.: The Serializability of Concurrent Database Updates. ACM, New York (1979)
6. Dickerson, T., Gazzillo, P., Herlihy, M., et al.: Adding concurrency to smart contracts. In: ACM Symposium on Principles of Distributed Computing, pp. 303–312. ACM (2017)
7. Siddiqi, M., All, S.T., Sivaraman, V.: Secure lightweight context-driven data logging for bodyworn sensing devices. In: International Symposium on Digital Forensic and Security, pp. 1–6. IEEE (2017)
8. Guan, Z.: Privacy-preserving and efficient aggregation based on blockchain for power grid communications in smart communities, cryptography and security (cs.CR). https://arxiv.org/abs/1806.01056
9. Dannen, C.: Introducing ethereum and solidity: foundations of cryptocurrency and blockchain programming for beginners. In: Introducing Ethereum and Solidity. Apress (2017)
10. Liu, A., Du, X., Li, S.: Big data access control mechanism based on block chain. J. Softw. 1–18. https://doi.org/10.13328/j.cnki.jos.005771
11. Nakamoto, S.: Bitcoin: a peer-to-peer electronic cash system. White Paper (2008)
12. Shao, Q., Jin, C., Zhang, Z., et al.: Blockchain technology: architecture and progress. J. Comput. Sci. 6(5), 691–696 (2018)
13. Szabo, N.: Formalizing and securing relationships on public networks. First Monday 2(9) (1997)
14. Alharby, M., Van Moorsel, A.: Blockchain-based smart contracts: a systematic mapping study (2017)
15. Bloom, B.H.: Space/time trade-offs in hash coding with allowable errors. Commun. ACM 13(7), 422–426 (1970)
16. Hao, F., Kodialam, M., Lakshman, T.V.: Building high accuracy bloom filters using partitioned hashing. Perform. Eval. Rev. 35(1), 277–287 (2007)

An Empirical Analysis of Supply Chain BPM Model Based on Blockchain and IoT Integrated System

Ruixue Zhao

The Hong Kong University of Science and Technology,
Clear Water Bay, Hong Kong
rzhaoaf@connect.ust.hk

Abstract. Cross-organizational supply chain as a generic form of business process management has become a hot topic of research and application. Blockchain and IoT technology provide effective technical support for this business process model. However, the business process model design and analysis based on blockchain and IoT technology has the following problems and challenges: (1) Lack of targeted BPM model. (2) The analysis issue of the BPM model across organizational supply chains are not clear enough. This paper proposes a cross-organizational luxury supply chain BPM(Business Process Management) Model based on Internet of Things(IoT in short) and Blockchain integrated system, also gives the simulation design and simulation analyzing results of the BPM model. The simulation results show that the designed BPM model has no deadlock and is a model that can be used directly.

Keywords: BPM · COMT supply chain · Blockchain · IoT

1 Introduction

BPM is a specification that provides guidelines and methods for business process modeling, automation, management, and optimization [1]. Using BPM can achieve the overall goal of integrating various business segments of the enterprise. BPM includes an optimized compose of content for people, money, equipment, desktop applications, and enterprise back-office applications to enable business operations across applications, departments, partners, and customers. BPM not only covers the scope of process flow and process monitoring of traditional "workflow", but also breaks through the bottleneck of traditional "workflow" technology. The launch of BPM is an epoch-making leap in workflow technology and management concepts [1–3].

BlockChain [4–6] is a new application mode of computer technology such as distributed data storage, point-to-point transmission, consensus mechanism, and encryption algorithm. The effective integration of the IoT and the Blockchain has become one of the feasible technical frameworks for solving the security and distribution architecture. The combination of the both technology has become the current hot research field [7, 8].

© Springer Nature Switzerland AG 2019
W. Ni et al. (Eds.): WISA 2019, LNCS 11817, pp. 539–547, 2019.
https://doi.org/10.1007/978-3-030-30952-7_54

It is not difficult to analyze the characteristics of RFID IoT and Blockchain. The IoT and Blockchain provide effective information technology support for Cross-Organizational Multi-Transaction (COMT in short) business process management. The decentralization, openness, concealment and autonomy provided efficient support for COMT [4, 5] and the IoT technology ensures that COMT services can be effectively tracked and traced, providing information assurance for the production, transportation, sales and after-sales services of goods [9–11].

However, there are still many challenges in designing and analyzing a COMT BPM models based on IoT and Blockchain integrated systems. Expressed as follows:

- Lack of targeted BPM model. How to generate corresponding BMP models based on blockchain and IoT technology principles is a new requirement and challenge.
- The analysis issue of the BPM model across organizational supply chains are not clear enough. How to define analytical issue, how to simulate and analyze the feasibility of BPM models under given issue is the need and challenge of experimental analysis.

2   Business Process Model Based on Fabric Blockchain

Hyperledger Fabric [12] is a Blockchain framework implementation and one of the Hyperledger projects hosted by The Linux Foundation. There are three types of nodes:

- Client or submitting-client: a client that submits an actual transaction- invocation to the endorsers, and broadcasts transaction-proposals to the ordering service.
- Peer: a node that commits transactions and maintains the state and a copy of the ledger. Besides, peers can have a special endorser role.
- Ordering-service-node: a node running the communication service that implements a delivery guarantee, such as atomic or total order broadcast.

The workflow can be express as the Fig. 1.

Figure 2 is a BPM model corresponding to Fig. 1. The business process diagram consists of two parts (pools), one is the Luxury Customer and the other is the BlockChain Information System. The first part is a business process for the customers who purchase the luxury goods. The second part is a corresponding business process model for the message sequence diagram of Fig. 1.

An information interface Clerk that interacts with the consumer is added. It is responsible for submitting orders and user query services with users. The Blockchain business process model is described as follows:

- After receiving the message of prepayment it starts multiple Transaction Endorsed Peers to perform simulation operations; and
- feedback the simulation results to the customer; and
- Requesting a sorting service to provide an information channel for the proxy process;
- Submit settlement information, complete account modification and block winding operation.

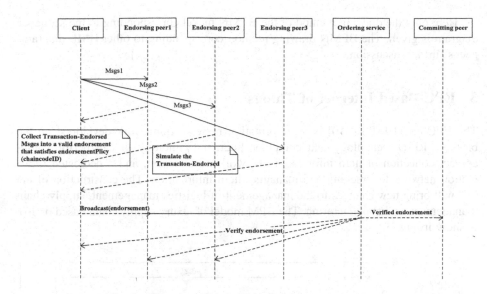

Fig. 1. Diagram of fabric blockchain message sequence

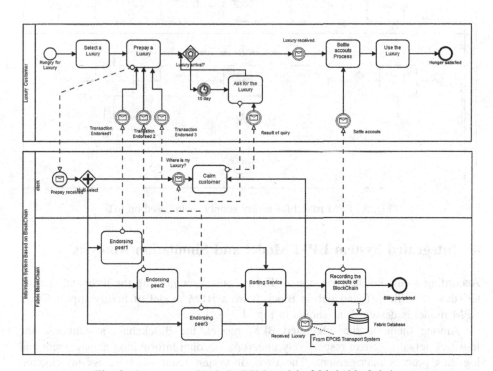

Fig. 2. Luxury supply chain BPM model of fabric blockchain

Here only the drawing method of the BPMN model corresponding to the sequence diagram is given. The BPMN drawing method corresponding to other UML diagrams needs further discussion.

3   EPC-Based Internet of Things

The EPC-based IoT [9, 10] is an information network framework developed on the basis of RFID technology and computer Internet technology. It can perform non-contact collection of item information, network transmission, and a dedicated information network for tracking and managing item information. The construction of the IoT will bring new changes to the management of logistics management, supply chain management, and manufacturing. The BPM model of luxury supply chain based on IoT is show in Fig. 3.

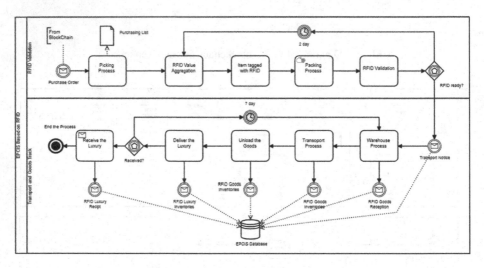

Fig. 3. BPM model for luxury supply chain based on IoT

4   Integrated System BPM Model and Simulation Analysis

According to the business process of Figs. 2 and 3, combining the technical characteristics of RFID IoT and Fabric Blockchain, a BPM model of luxury supply chain BPM model is designed, as shown in Fig. 4.

Among them, in the integrated BPM model, the Blockchain system and the logistics network system respectively undertake account information management and logistics process management. The two sub-system establish business information exchange through the information channel, the Blockchain provides ordering information for the IoT, and the IoT provides cargo tracking information for the Blockchain system. The both sub-systems are loosely coupled organizations, which have the advantages of clear system interface and convenient maintenance.

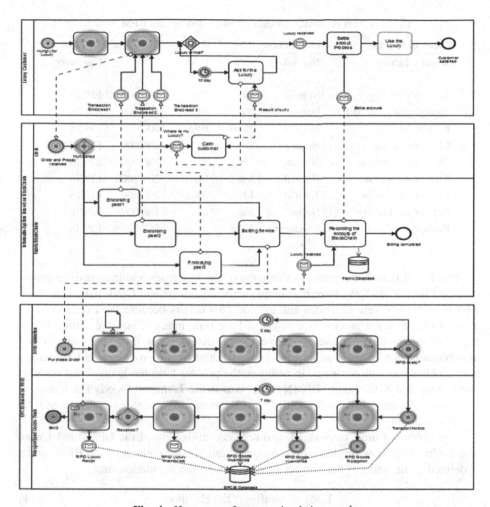

Fig. 4. Heat map of system simulation results

In order to evaluate the validity of the BPM model, an empirical analysis was conducted. It use the OMG recommended BPM simulator [13]. The data set for simulating is shown in Table 1. Meanwhile, the probability distribution of business occurrence is a key factor in system simulation. We analyzed and tested the key business processes, and by fitting the probability density curve and sensitivity analysis, we finally obtained the probability distribution of these business processes. The business process Select a Luxury to Deliver the Luxury shown in Table 1 is the internal business process of the system. These processes are distributed and measurable. The frequency of occurrence is in accordance with the normal distribution Normal, and the business process Receive the Luxury is the external process. It is impossible to control the frequency at which it occurs, which is an exponential distribution under a typical Poisson process.

Table 1. Part of simulation data of integrated system BPM model

Name of Process	Distribution	Interval	Offset/m	Cost/Total	Resource
Select a Luxury	Normal	30/m	10/m	0	Customer
...	...	...	...	...	...
RFID Value Aggregation	Normal	10/s	5/s	10/2000	EPCIS
Item tagged with RFID	Normal	20/s	10/s	10/2000	EPCIS
Packing process	Normal	30/m	10/m	10/2000	EPCIS
RFID validation	Normal	20/s	5/s	10/2000	EPCIS
Warehouse process	Normal	1/h	20/m	30/2000	EPCIS
Transport Process	Normal	3/d	1/d	200/2000	EPCIS
Unload the Goods	Normal	1/h	20/m	20/2000	EPCIS
Deliver the Luxury	Normal	3/h	1/h	30/2000	EPCIS
Receive the Luxury	Exponent	30/m		5/2000	EPCIS

The Fig. 4 is simulating results that most of the processes and intermediate events in the RFID IoT (EPCIS) have heat maps, while Customer and Clerk have heat maps with only the first few processes and events. This occurs because the execution time interval of the above process is assigned a long time interval, and the setting of the event occurrence time allocates a relatively small time interval, thereby forming a heat map phenomenon of the waiting time process. Other white processes and events are start-up elements with no waits. In addition, the process "use the luxury" did not start because the end of the luxury BPMN model was in the Transport Goods Track process pool, after "Receive Luxury", and the "Luxury received" event was fired to end the model execution.

The execution time, throughput, and resource utilization of the BPM model were also evaluated during the simulation. Among them, the throughput of the BPM model is defined as the ratio of the number of instances to the simulation time.

$$T(m) = num(ins)/\Sigma cycle\ time \tag{1}$$

Resource utilization is defined as the ratio of the number of processes to the time of one execution.

$$U(m) = num(proc)/cycle\ time(m) \tag{2}$$

The actual simulation results of throughput of Information system Based on Block-Chain, EPCIS Based on RFID, and System of Luxury Supply Chain respectively are 55/m, 33/m and 13/m; and utilization are 49%, 64% and 58%.

Table 2 is the design of the performance evaluation parameters of the simulation system. Here we only give the evaluation results of the throughput and utilization of the model. The utilization is the evaluation given automatically by the simulator, and the throughput is the author's proposed evaluation benchmark. The resource utilization of the model and the throughput of the system are defined by Eqs. (1) and (2).

Table 2. Performance of luxury supply chain based on COMT

Resource Pool	Num. of Instance	Time of Simulating	Num. of Process	Time of executing	Throughput	Utilization
Information system based on BlockChain	10	18/m	6	12.2/m	55/m	49%
EPCIS based on RFID	10	30/m	10	15.6/m	33/m	64%
System of luxury supply chain	10	72/m	20	34.2/m	13/m	58%

5 Related Works and Conclusion

The supply chain management model has been extensively studied, from the maintenance, reconstruction, performance evaluation and risk management of supply chain to the theory and practice [14, 15]. However, how the supply chain is related to the design and development of information systems, especially the relationship between the BPM model of the supply chain and the IS model is still rare.

Reyna et al. [8] emphasized that blockchain and IoT have many technical and system advantages, mainly reflected in the blockchain can assist IoT data management and improve the security, privacy and reliability of logistics data. While analyzing the advantages, it also proposes the challenges that the combination of the two may encounter. There are still many problems to be solved in the data storage capacity, processing speed and data consistency of the blockchain. Ju et al. [16] emphasized that Internet-based technology applications have been extensive, but the general business model based on the Internet of Things has not been thoroughly studied. The author uses literature review, market research and other methods to give an analysis of the business model research and practical application status based on the Internet of Things. The paper did not see an introduction to the research related to the Internet of Things and business process management model. Wang et al. [9] believe that blockchain can effectively solve the security problem in Internet of Things technology, but when the blockchain is combined with the Internet of Things, it will also encounter challenges such as low bandwidth, limited computing power and heterogeneous network structure. The article reviews the research status and future prospects of the two fusion technologies.

Giaglis et al. [17] and Rahimi et al. [18] considered that information system models and business analysis models have been well studied. Various models have been proposed, but the correlation analysis between business models and information system models is not enough. In-depth. For example, the use of the BPM model with the information system model has not been discussed in depth. The paper gives the correlation analysis and classification of BPM model and various information system models, and provides method support for constructing information system from business process.

Cai et al. [19] design a distributed rule engine based on Kafka and Structured Streaming (KSSRE), and propose a rule-fact matching strategy using the Spark SQL engine to support a large number of event stream inferences. KSSRE uses DataFrame to store data and inherits the load balancing, scalability and fault-tolerance mechanisms of Spark2.x.

In this paper, based on the cross-organizational business process modeling problem and the architecture modeling problem of COMT Internet of Things and blockchain fusion system, the BPM model of luxury supply chain based on blockchain system is given. Based on the above model, The BPM model based on RFID Internet of Things is designed, and the method of UML message graph to BPMN graph transformation is introduced. The BPM model of luxury goods supply chain integrating RFID Internet of Things and blockchain fusion system is proposed. The fusion system is given. The simulation design and simulation results of the BPM model of the luxury supply chain show that the designed BPM model has no deadlock. In actual use, the parameter settings can be adjusted according to the actual situation.

The cross-organizational business process management model based on the Internet of Things and blockchain is still a new research direction. Let's continue our research in the following directions:

- Optimization of the BPM model and detection of exchange protocols between organizations.
- Research on the BPM model of deep integration of the IoT and Blockchain
- Research on BPM simulation algorithm and simulation tool development technology.

References

1. Dumas, M., La Rosa, M., Mendling, J., Reijers, H.A.: Fundamentals of Business Process Management. Springer, Heidelberg (2013). https://doi.org/10.1007/978-3-662-56509-4
2. Ko, R.K.L., Lee, S.S.G., Lee, E.W.: Business process management standards: a survey. Bus. Process Manag. J. **15**(5), 744–791 (2009)
3. Weber, I., Xu, X., Riveret, R., Governatori, G., Ponomarev, A., Mendling, J.: Untrusted business process monitoring and execution using blockchain. In: La Rosa, M., Loos, P., Pastor, O. (eds.) BPM 2016. LNCS, vol. 9850, pp. 329–347. Springer, Cham (2016). https://doi.org/10.1007/978-3-319-45348-4_19
4. Blockchains: The great chain of being sure about things. The Economist, 31 October 2015
5. Morris, D.Z.: Leaderless, blockchain-based venture capital fund raises $100 million, and counting. Fortune. http://fortune.com/2016/05/15/leaderless-blockchain-vc-fund/. Accessed 23 May 2016
6. Popper, N.: A venture fund with plenty of virtual capital, but no capitalist. The New York Times. https://www.nytimes.com/2016/05/22/business/dealbook/crypto-ether-bitcoin-currency.html. Accessed 23 May 2016
7. Block Chain. https://en.wikipedia.org/wiki/Blockchain
8. Reyna, A., Martín, C., Chen, J., Soler, E., Diaz, M.: On blockchain and its integration with IoT. Challenges and opportunities. Future Gener. Comput. Syst. **88**, 173–190 (2018)

9. Wang, X., et al.: Survey on blockchain for Internet of Things. Computer Communications, 11 January 2019
10. Nakandala, D., Samaranayake, P., Lau, H., Ramanathan, K.: Modelling information flow and sharing matrix for fresh food supply chains. Bus. Process Manag. J. **23**(1), 108–129 (2017)
11. Research and Markets, RFID Industry-A Market Update, June 2005, http://www.researchandmarkets.com/reports/c20329
12. https://cn.hyperledger.org/projects/fabric
13. http://www.qbp-simulator.com/
14. Manataki, A., Chen-Burger, Y.-H., Rovatsos, M.: SCOlog: a logic-based approach to analyzing supply chain operation dynamics. Expert Syst. Appl. **41**, 23–38 (2014)
15. Bastas, A., Liyanage, K.: Integrated quality and supply chain management business diagnostics for organizational sustainability improvement. Sustain. Prod. Consumption **17**, 11–30 (2019)
16. Ju, Jaehyeon, Kim, M.-S., Ahn, J.-H.: Prototyping business models for IoT service. Procedia Comput. Sci. **91**, 882–890 (2016)
17. Giaglis, G.M.: A taxonomy of business process modeling and information system modeling techniques. Int. J. Flex. Manuf. Syst. **13**(2), 209–228 (2001)
18. Rahimi, F., Møller, C., Hvam, L.: Business process management and IT management: the missing integration. Int. J. Inf. Manag. **36**, 142–154 (2016)
19. Cai, D., Hou, D., Qi, Y., Yan, J., Lu, Yu.: A distributed rule engine for streaming big data. In: Meng, X., Li, R., Wang, K., Niu, B., Wang, X., Zhao, G. (eds.) WISA 2018. LNCS, vol. 11242, pp. 123–130. Springer, Cham (2018). https://doi.org/10.1007/978-3-030-02934-0_12

A Trusted System Framework for Electronic Records Management Based on Blockchain

Sixin Xue[1], Xu Zhao[4], Xin Li[2], Guigang Zhang[3(✉)],
and Chunxiao Xing[4]

[1] Archives of Tsinghua University, Tsinghua University, Beijing 100084, China
xue@tsinghua.edu.cn
[2] Beijing Tsinghua Changgung Hospital Medical Center, Tsinghua University,
Beijing 100084, China
[3] Institute of Automation, Chinese Academy of Sciences, Beijing 100190, China
guigang.zhang@ia.ac.cn
[4] Research Institute of Information Technology, Tsinghua University,
Beijing 100084, China
xingcx@tsinghua.edu.cn

Abstract. The credibility and authenticity of electronic records determine whether electronic documents have legal effect, preservation value, use effectiveness and reproduction history. Firstly, the existing research was analyzed and the consensus reached is that the reliable technical system is the key to ensure the dual requirements of evidence retention and information utilization during the activities of electronic records management. Secondly, based on the life cycle theory of electronic records and the methodology of whole process continuous management, an application framework of electronic records trusted management based on critical business process management is constructed. And then, according to the technical characteristics and application principle of block chain, the evidence chain model of trustworthy managing electronic records is proposed and the construction of evidence chain model based on type or time limitation with the help of electronic records classification management is studied. Finally, the related key technologies and their application scenarios are analyzed.

Keywords: Block chain · Electronic records management ·
Credibility and authenticity · Key business process · Evidence chain model

1 Introduction

With the rapid development and in-depth application of social informatization, various types of electronic records are making in large quantities, and gradually become an important part of institutional data assets, national information resources and social memory. For example, the electronic medical records such as doctor's diagnosis, medical prescription and patient's hospitalization in the hospital information system; the student's status card, transcript, degree examination and dissertation formed in the E-Campus system; and also the application form, approval transfer form and administrative approval conclusions formed in the E-Government system, and so on.

© Springer Nature Switzerland AG 2019
W. Ni et al. (Eds.): WISA 2019, LNCS 11817, pp. 548–559, 2019.
https://doi.org/10.1007/978-3-030-30952-7_55

How to manage these electronic records scientifically has become an interdisciplinary worldwide problem, in which how to ensure the authenticity and credentials of electronic records is the focus of research. The existing research mainly focuses on the legal status and evidential power of electronic records, the authenticity of electronic records and metadata control methods, based on data encapsulation, file format and data encryption technology, as well as the establishment of a secure, credible and fully functional electronic records management system.

In the research of the legal evidence and validity of electronic records, many countries, including China, have begun to adopt legislative forms to clarify the role of legal evidence of electronic records [1]. The Law of the People's Republic of China on Electronic Signatures clarifies the legal status of data messages, electronic signatures and their authentication services, that is, a true electronic records can be used as judicial audit evidence by "stamping" a compliance electronic signature. However, electronic evidence still has doubts about the qualification of evidence and disputes about the power of proof. Chang [2] suggested that the independent evidence status of electronic records should be more clearly defined in legislation. Dai [3] holds that specific rules should be formulated for the verification, collection and determination of electronic records in order to establish a unified standard of enforcement. Liu [4] believes that the real objectivity of electronic records is the key to being evidence. Zhang [5] thinks that electronic records' archiving and managing process needs to use electronic signature, third-party authentication, process control technology and methods to ensure its certification.

In the research of the basic theory of electronic records management, metadata integrity and content authenticity are considered as the important methods to guarantee the certification and evidence traceability of electronic records. Zhang [6] holds that metadata is an important tool to guarantee the authenticity of electronic records. He proposes that the business background, legal requirements, technical features and management process of electronic records should be described in detail in the form of metadata [7]. The principle and method of metadata control in electronic records should be clarified and the metadata phase of electronic records management should be worked out. The standards are implemented on the basis of intelligent system to protect the credentials of electronic records. Many scholars from at home and abroad believe that authenticity is the key element of the certification of electronic records. Inter-PARES [8], the International Research on Permanent Authentic Records in Electronic systems, initiated and chaired by Professor Luciana Duranti of Columbia University of Canada, proposed that an electronic record in the system can only be identified as authentic if it satisfies the characteristics of its complete background information, content solidification, fixed form and archival connection. The legal evidences of electronic records and the authenticity of electronic records are in essence the same. To ensure the authenticity of electronic records is the basic requirement to maintain their legal evidences. Zhang [9, 10] recognizes that authenticity should be guaranteed throughout the life cycle of electronic records, and its system framework should be constructed on multiple dimensions such as the environmental cognition, institutional norms, management activities and technical systems.

In the research of key technologies application, standards and system integration, it is believed that a compliance and reliable electronic records management system is the key to ensuring the authenticity of electronic records. Tao [11] puts forward a method to guarantee the evidence of electronic records based on ID card technology, and researches the application architecture of key technologies such as encapsulation technology, unstructured data storage and encryption, which aiming at solving the core problem of electronic records in heterogeneous system environment. Xue Sixin put forward to build a double-layer model in cloud computer system [12] to guarantee the dual demands of electronic documents management in the system, that is, evidence retention and social utilization.

Figure 1 illustrates the dual requirements of information utilization and voucher preservation in the whole process of continuous managing electronic records. The original records generated in the various key business processes, such as drafting, approval, issuance, distribution, archiving, preservation, identification and disposal, need to be fixed and kept to ensure the credibility of the electronic records from creation to long-term preservation or destruction. Previous systems commonly used technologies such as PDF format solidification, record encryption and other technologies for these specialized certificate locking.

2 The Ecological Environment of Continuous Managing Electronic Records

The electronic records produced in management information system should be managed on the basis of the system. Safe and reliable system is the key to guarantee the authenticity, integrity and validity of the electronic records. In 2007, the Australian National Archives proposed the Design and Implementation of Record Keeping Systems (DIRKS) [13] method based on the continuum model of electronic records, aiming to guide the planning, construction and implementation of electronic records management system. It provides a useful implementation methodology for the design and implementation of records keeping systems. Highly concerned in the field of management and used. In 2008, the model of Chain of Preservation (COP)and Business-Driven Recordkeeping (BDP) was created by InterPARES according to the IDEF0 modeling method. The model of COP was constructed on basis of the theory of records life cycle and the model of BDP was based on records continuum theory. The objective is to embed policies, systems, procedures, standards and rules for document management into electronic recording systems for automatic implementation. The objective is that the electronic record management policies, systems, procedures, standards and rules can be automatically implemented in the electronic record management system (ERMS) by embedding them in system solution. China has also begun to study and made a series of standards for constructing electronic records systems, such as guidelines, functional requirements and testing methods. Testing laboratories have been established and standards conformance testing has been carried out to improve the quality and reliability of electronic records systems. It is a consensus that a secure and credible electronic records management system is an important measure to ensure the certification of electronic records.

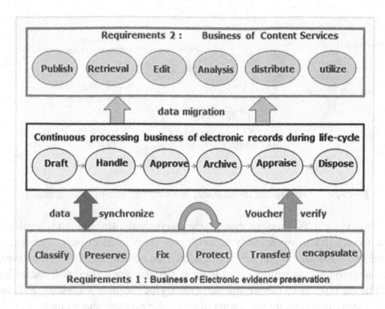

Fig. 1. The dual requirements of electronic records management

Different from the business such as ERP, OA, E-government, hospital informatization and E-campus systems, electronic records management system is a kind of special information system, which is used to keep records that have social preservation value and institutional assets attributes, voucher value. According to the division of functions in the current society, there are three kinds of information systems related to electronic documents, as shown in Fig. 2. Business information system (BIS), such as OA and ERP, aiming at supporting the units to carry out professional activities such as procurement, sales, inventory, finance and office automation business, is used to make the electronic documents. The electronic records management system (ERMS) consists of two parts: 2-1-ERMS and 2-2-ERMS. The main function of 2-1-ERMS is to keep the records form BIS by means of fixed format, make them archival process according the requirements and rules of archives management, as well as hand them over to the archives storage institutions. The main function of 2-2-ERMS is to manage the contents of electronic documents by means of data and information and provide information and content services. 2-1-ERMS integrates with BIS, captures the records and metadata from BIS at any time, solidifies them to retain evidence. 2-2-ERMS integrates with BIS and AMIS systems to provide current and historical document information services. Records preserving system such as TDR and AMIS is one kind of systems used by archival agencies for long-term preservation of electronic records.

BIS, ERMS and TDR are the information ecological environment on which electronic records exist. Evidence retention and information service are the basic functions of the system. In order to ensure the authenticity, integrity and validity of electronic records, third-party Certificate Authority (CA) solution is usually adopted. A typical application framework for trusted management of electronic student's records in Fig. 3. The CA technology is adopted in this system construction solution. The CA

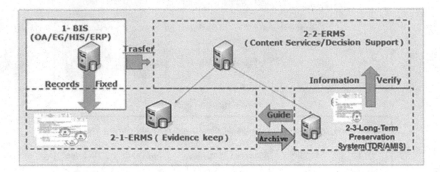

Fig. 2. The ecological environment of electronic records management

infrastructure system is used to ensure the credential attribute and legal effect of electronic records by using security technologies such as electronic signature, digital encryption and time stamp. In this application framework, the CA center issues certificates and provides authentication for each system, each user and each electronic records. The purpose is to ensure the authenticity of the user's identity and the reliability of the system, to ensure the authenticity of the process and transmission records, and to ensure the integrity of each electronic records, so as to form a credible ecological environment of electronic records. On the one hand, by issuing CA certificates to the key roles that affect the information of students school records, such as teaching teachers who grant student performance, administrators who manage student performance, teaching supervisors who audit academic degrees and archivists who keep student records, the safe and trustworthy operation of the key roles can be guaranteed. On the other hand, by adding electronic signatures to the electronic school records, such as electronic transcripts, electronic dissertations, electronic academic qualifications and degree certificates, the school records will have legal effect and can be recognized by the society just like paper certificates.

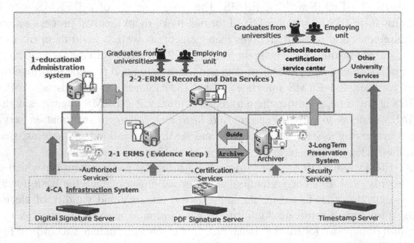

Fig. 3. An application framework of trusted management and certification service for student status electronic documents based on CA

3   Evidence Chain Model

The main content of this section is to design the blockachain evidence model of electronic documents with the characteristics of electronic document management business, aiming at the evidence retention and traceability requirements (process records) of key business parts of various electronic documents in their whole life cycle. Here, we present a basic evidence chain model. Based on this basic evidence chain model, richer functional modules can be realized and a more complete electronic document management system can be established.

(1) Basic Structure of evidence chain

The evidence chain model of electronic documents involves a large number of electronic file storage. According to the peer characteristics of block chain network nodes, each node needs to store all the files if using the public chain, which is unrealistic. Therefore, consortium chain is adopted here. The user nodes in the consortium chain are divided into two types. One is the bookkeeper nodes, which store all the data in the block chain and can mine to generate new data. The other is the common user nodes, they can not carry out mining to produce new blocks. Consortium chain features just meet the needs of the electronic document management system. Most business of electronic files management system generally involves several major institutions (For instance, the student status management system is made up of all the universities), the majority of users (ordinary student, recruitment companies) just do some general query or document authentication, such as operation, don't need to store all your data. As shown in Fig. 4, our consortium chain sets up a key bookkeeper node in each organization, and each key bookkeeper node will be accompanied by a database to store all documents in the whole chain. There is an important assumption of this evidence chain model: all the bookkeeper nodes in the consortium chain are trusted nodes, and the nodes authorized by the bookkeeper node are also trusted nodes within the scope of their authorization. The entire evidence chain model and the subsequent system frameworks are based on this assumption.

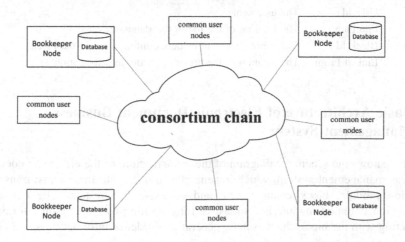

Fig. 4. Consortium chain structure

(2) Evidence chain transaction sheet design

This evidence chain model is based on blockchain systems with smart contracts, such as Ethereum and Hyperledger. Our evidence chain model needs to implement the functions including interaction with electronic document database, document operation legitimacy check, electronic document check, operator authority identification and so on using the smart contract.

Operations on documents include creation, review, issuance, circulation, distribution, archiving, identification, disposal, utilization, migration, conversion, destruction, etc. Every operation on the document will generate a certificate information to record the operation, and finally the certificate information will be saved on the block chain to realize the monitoring of the whole life cycle of the document. The evidence chain of this paper is based on the existing currency trading system block chain, such as Ethereum, Hyperledger, etc., which stores the evidence information in the transaction information of the transaction sheet, ensuring that the evidence information cannot be tampered with and prevented from being lost.

The main evidence information contained in each transaction sheet is given here. As shown in Table 1, only the evidence information of electronic documents is included in this table. For the time being, we do not pay attention to the rest of the transaction related information. The transaction related information depends on the selected blockchain system.

Table 1. Basic data structure of electronic document evidence information

Field Name	Description
File Name	Electronic document name
File Type	The document type
Time Stamp	Document generation timestamp
File Hash	Document fingerprint information
Operation	Operation on the document
Operation Info	Document operation details
User Id	The user name
File Path	The location of the file in the database
Block Id	Electronic document evidence information in the block
Linked From	The location of the previous evidence information

4 Basic Architecture of Electronic Document Business Management System

Figure 5 shows the schematic diagram of the infrastructure of the electronic document business management system, which is generally divided into three parts: transaction initiation module, block chain system and database file system. The transaction information initiation module is responsible for obtaining user operation information, interacting with the block chain system, generating evidence information and initiating a transaction in the block chain network so as to input evidence information into the block chain. Blockchain system is responsible for obtaining the operation evidence information and verifying the legitimacy and authenticity of the evidence information. If the operation is accompanied by database file reading and writing, it needs to interact with the database. Database file system is responsible for store all documents corresponding to the evidence information on the block chain. Note that this is different to user's own database file system. The electronic document management system can be used as a decentralized CA system (the CA system for document authentication is based on the entire network consensus, rather than through the center to issue a digital certificate to authenticate), so we can use the database file system as the CA database file system, it holds all the certification document backup in the process of the entire life cycle and business operations involved in the life cycle are saved through blockchain. At the same time, since each bookkeeper node on the consortium chain keeps a complete database backup, the system will not be affected by the outage of one bookkeeper node.

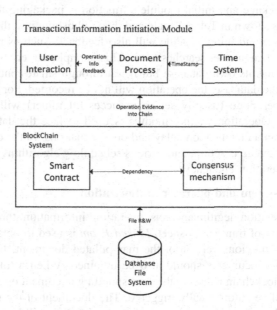

Fig. 5. Schematic diagram of the infrastructure

(1) Transaction Initiation Module

Transaction initiation module includes three sub-modules: user interaction, file processing and time system. The function of the user interaction module is to obtain the operation of the user and return the feedback result of the operation to the user. Document processing function is to package document information and operation information into a transaction sheet and broadcast to the whole network. The time system provides timestamp service for the document processing module. Generally, the time used by the system is UNIX system time.

Document fingerprint Generation Part is used to generate the fingerprint information of the document. First, the electronic document needs to be serialized to generate the binary document, hash the document with Sha256 to generate Hash1, and then Hash1+ timestamp is hashed together with Sha256 to generate Hash2 as the fingerprint information of the document. Each operation requires the manipulated document and timestamp to generate a fingerprint and store it on the block chain. When the miner verifies the transaction order, it needs to extract the manipulated document through File Path, and then connect it with the *Time Stamp* for hashing to verify with *File Hash* field. The specific consensus process will be described in detail later.

There are also operation analysis module and transaction packaging sending module. After receiving the user's operation, it is necessary to conduct atomic decomposition of the user's operation. Common atomic operation includes document creation, document review, document correction, document archiving, etc., and the user can provide complex operation. After conduction all atomic operation will be executed linearly. During the execution process a module called transaction package and initial module will be called.

Transaction package and initial module's function is packaging the atomic operation information as shown in Table 1 trade and then publishing it in the blockchain net. Transaction package and initial module will also feedback, such as returning a document in database when an user searches for it. This is implemented by smart contracts. Bookkeeper node has its own database but if not through smart contracts directly see the documents in the database, the operation will not be recorded. For the ordinary user node, a bookkeeper node (usually set as a successful miner) will send the target document to him. Operation waiting module is used to face the database read-write conflict, block chain authentication delay and other situations. In the operation waiting module, a waiting queue is set up, and various scheduling algorithms can be used here to improve the operation efficiency.

(2) **Block chain system and operator authorization**

Document operation legitimacy gets operation information through *Operation*, *Operation Info* field of transaction sheet. *Linked From* is used to obtain the Operation information of the previous version of the manipulated document. If new documents are generated, new documents should also be obtained. When a transaction sheet is submitted to the blockchain network, the corresponding document operation legitimacy smart contract will be automatically triggered. The document of the previous version, the current uploaded version, *Operation* and *Operation Info* will be input of the smart contract to check the legitimacy of the operation. If the transaction information fails to pass the check, it cannot be entered into the chain.

The consensus mechanism in this system is an improvement of the common electronic currency block chain system's consensus mechanism. When a trade sheet is published in the block chain network [14], all miners in the network will receive this trade order. If the miners think the miners' labor fee are reasonable, they will verify the transaction sheet. The verification process not only verifies the transaction amount and other information of the transaction sheet itself, but also needs to verify the legitimacy of document operation, operator identity and permission. Document operation legitimacy requires invoking smart contract execution as described above. Operator identity and permission verification requires that the operator permission information is obtained in order to verify the operator's permission. The following paragraph describes how to save and get operator permission information.

The last step is to verify the authenticity of the document. First, the manipulated document needs to be obtained. If the operation is not accompanied with the new document being entered into the library, the manipulated document can be obtained from the database according to FilePath. If this operation is accompanied with a new document being stored, the document uploaded by the user is used directly, because the smart contract code related to the database operation needs to be executed after the validation is passed and document has not been stored during the validation process. After obtaining the manipulated document, it needs to be linked with the TimeStamp field for Hash operation, and the result of calculation is compared with FileHash (document fingerprint) for verification.

After all the above verification, a document operation information will be accepted by the block chain system and packaged into the block chain by miners, and the smart contract code related to the database operation will be executed.

5 Application Framework of Electronic Document Management System Based on Blockchain Evidence Chain

Perform departmental division to confirm the bookkeeper node. The school roll electronic document trust management system within colleges and universities can be divided into departments according to the college, and each college sets up a bookkeeper node.

Determine the block formation mechanism and document classification and preservation structure. In the third chapter, the block formation mechanism and document classification and preservation structure based on time limit and type restriction are given. Here, we choose the block formation mechanism and document classification and preservation structure based on type restriction. According to the four departments shown in Fig. 3: Educational Administration System (EAS), Evidence Keeping System (EKS), Long-term Preservation System (LPS), and Records and Data Services (RDS). The first three systems involve the generation, solidification, and archiving of documents; the last system is responsible for verifying the information of the students. Here we divide document type according to the first three systems. RDS can be used as an ordinary user node to read the information of the EAS chain and verify the information of the LPS chain. The three chains based on the document types of the three systems are not completely independent. There will be a small number of mappings between the chains, such as document solidification from the EAS to the EKS and document archiving from EKS to LPS. When the document is solidified, the transaction sheet for creating the document on the EKS chain needs to be linked to the corresponding solidified document transaction sheet on the EAS chain. When the document is archived, the transaction sheet for creating a document on the LPS chain requires a Link to the corresponding archived document transaction sheet in the EKS chain. Therefore, when it comes to archiving, solidifying or other cross-departmental, cross-system operations, an operation requires adding a transaction sheet both on each of the two system chains.

Document operator node initialization. A normal document operator node needs to be authorized by the bookkeeper node. The document operator permission information, wallet nodes under his name, etc. need to be stored in a separate chain, called the operator information chain. User Id in each transaction on the other chains needs to point to a user on the operator information chain. During the system initialization process, each bookkeeper node can authorize some basic document management personnel, document auditors, etc. When the system is running the operator's authority can be changed and new operator can be added.

Choose Ethereum's architecture to build an evidence chain model. Ethereum DAPP blockchain application architecture supports Turing's complete smart contract language solidity. The system architecture needs the support of smart contracts in the process of transaction document verification and database operation, so choose Ethereum architecture to build the system. For estimation of miners' labor fee when initiating a transaction sheet Ethereum also has a mature interface that can be called for inquiry and estimation. Using node.js to build the application front-end interactive interface. Ethereum DAPP supports interaction with node.js. through smart contracts.

The database is built using MySQL database. MySQL is flexible and simple to use and can perform database operations through smart contract code.

6   Conclusion and Future Work

Block chains technology helps to establish a credible IT information ecological environment, which is undoubtedly of great significance to the authenticity of electronic records and the preservation of credentials. However, how to establish the evidence chain covering the whole process of electronic records and ensure the continuity and validity of the information in the evidence chain of electronic records with the current social function division system still needs further study. At the same time, after introducing the block chain, the cost of electronic records management, electronic records archiving and long-term preservation need to be studied.

Acknowledgement. Our work is supported by the National Social Science Fund of China (NSSFC) Project named "Principle and methods of protection of Electronic records evidence based on cloud computing environment" (No. 15BTQ079) and NSFC (61872443).

References

1. Zhenjie, X., et al.: The legal basis of electronic document voucher between China and foreign countries. Arch. Constr. **08**, 13–15 (2002)
2. Chang, Y., et al.: On the independent status of electronic evidence, 9 February 2015. http://www.procedurallaw.cn/zh/node/2926
3. Dai, Z.: Discussion on the rules of electronic evidence adoption. J. Hubei Police Officer Coll. **1**, 23–26 (2003)

4. Liu, J.: One of the voucher discussion of electronic documents and law-electronic documents. Arch. Constr. **1**, 16–18 (2000)
5. Zhang, W.: The Influence of the implementation of electronic signature law on the technical methods of electronic document filing management and its countermeasures. Arch. Res. **3**, 48–50 (2007)
6. Zhang, Z.: On the function research of electronic document management metadata. In: Chinese Archives Society 2008 Archivists Annual Conference Proceedings (Volume 1), 158–166 (2008)
7. Zhang, Z.: On the basic principles and methods of electronic document management metadata control. China Arch. **12**, 56–58 (2014)
8. Duranti, L.: The trustworthiness of digital records. In: International Symposium on Advanced Research Findings of Electronic Records Management, Beijing, China, 16 April 2010
9. Xie, L.: The concept of documents and their evolution in the digital environment: InterPARES perspective. Arch. Newsl. **3**, 46–50 (2012)
10. Zhang, N.: Re-recognition of the authenticity of electronic documents. Arch. Commun. **4**, 12–16 (2012)
11. Tao, S., et al.: Analysis of the status quo and countermeasures of electronic file voucher guarantee. Arch. Res. **1**, 57–60 (2012)
12. Xuc, S.: Implementation Mechanism of Electronic Document Management in Cloud Computing Environment, vol. 09, pp. 81-99. Shanghai World Publishing House, Beijing (2013)
13. http://www.naa.gov.au/records-management/publications/DIRKS-manual.aspx. 9 February 2015
14. Wang, X., Hu, Q., Zhang, Y., Zhang, G., Juan, W., Xing, C.: A kind of decision model research based on big data and blockchain in eHealth. In: Meng, X., Li, R., Wang, K., Niu, B., Wang, X., Zhao, G. (eds.) WISA 2018. LNCS, vol. 11242, pp. 300–306. Springer, Cham (2018). https://doi.org/10.1007/978-3-030-02934-0_28

A Blockchain Based Secure E-Commerce Transaction System

Yun Zhang, Xiaohua Li, Jili Fan, Tiezheng Nie, and Ge Yu$^{(\boxtimes)}$

School of Computer Science and Engineering, Northeastern University,
Shenyang 110819, China
1801762@stu.neu.edu.cn, yuge@cse.neu.edu.cn

Abstract. Traditional centralized e-commerce transaction systems have security problems such as untrustworthy third-party services, high cost of keeping funds, easy leakage of users' privacy. This paper designs and implements a decentralized secure e-commerce transaction system based on blockchain technology called "BSETS". The system adopts B/S and distributed architecture, and all the transactions are completed through blockchains to ensure security and convenience. The key techniques include user information encryption based on MD5 abstract algorithm to ensure anonymity; transaction process management based on smart contract to ensure security; transaction confirmation based on user's private key signature to ensure the non-repudiation of transactions. The blockchain can only be appended and read to protect the system from tampering. The use of smart contract realizes the decentralization of transactions and solves the problem of third-party untrustworthiness. Finally, the function and performance experiments verify that BSETS has completed transaction function and has good operation efficiency on the basis of ensuring information security.

Keywords: Blockchain · Transaction management · Ethereum · Smart contract · Distributed databases

1 Introduction

E-commerce is an important application with the development of Internet. In the existing payment methods, their transactions are based on the centralized platform. There are many drawbacks in it: the custody fee of transaction funds is required, and if the third party becomes untrustworthy, user's funds will face the risk of loss. To solve the above problems, the decentralized business mode has been proposed. Recently, with the development of blockchain, its distributed storage layer structure, consensus mechanism, encryption algorithm and other characteristics realize the reliability of transactions without third-party supervision, and promote the development of decentralized e-commerce transaction market [1]. OpenBazaar 1.0, a decentralized market for trading in bitcoin provides online p2p trading environment. It uses multiple signatures (multisig) [2], and the transaction on it does not charge fees and is not subject to platform review. The successful precedent of OpenBazaar provides a reference for the development of secure e-commerce transaction system which we called "BSETS".

© Springer Nature Switzerland AG 2019
W. Ni et al. (Eds.): WISA 2019, LNCS 11817, pp. 560–566, 2019.
https://doi.org/10.1007/978-3-030-30952-7_56

Ethereum [3] is the second generation blockchain platform after Bitcoin. Unlike Bitcoin, the Turing-complete smart contract makes Ethereum programmable. By replacing the third-party services with smart contracts, the custody cost of funds is reduced, the trustworthiness problem is also avoided. And whether using POW or POA [4], blocks are generated at a high rate on Ethereum. Transactions on it are also transparent, reliable, and verifiable. Sharding, the core technology of Ethereum 2.0 solve the problem of low throughput and poor scalability in Ethereum 1. X by partitioning large chains into smaller and faster ones, so BSETS can bear the huge transaction volume in practical e-commerce application. In 2017, Java launched the web3j class library, which implements the function of writing and deploying smart contracts and invoking contracts in Java language under the Maven framework [5]. And this makes java another language for writing smart contracts.

At present, there is no mature E-commerce transaction system based on Ethereum technology. The secure online system BSETS designed in this paper completes the blockchain-based transaction processing on the Ethereum, adopts SQL database to support additional transaction data management, and adopts web3j to support user interface management. Then construct a decentralized shopping website based on blockchain.

2   Framework Design

BSETS's structure is the combination of B/S and distributed system. The end user of browser can access Web server which communicates with blockchain network, and the miner digging and writing blocks while Web server records the detailed information on database.

2.1   Functional Design

Functions of system are divided into three modules. First one is account management. There should be only one user type to realize p2p trading mode [6], each user can be either a seller or a buyer. And users need two kind of account: an Ethereum account and a shopping website account. The first for payments in transaction and the second for login website. In addition, we add verification code mechanism to prevent DDoS attacks, and store the password salt into database to prevent storm.

The second module is transaction management. User's transaction includes creating goods, adding goods quantities, browsing, buying, signing and refunding. Deploy a contract for each newly created item, and then a series of operations can be done by invoking methods of the contract through its address we get after deploy. The ECDSA signature algorithm [7] is used in Ethereum, and when the status of contract changes, the transaction signature is verified which guarantees the msg.sender cannot be forged [8]. What's more, buyer and seller will both sign with their credentials during the transaction to ensure the transaction information is non-repudiation.

The third one is information query module. Users may query account information such as purchase record, sale record, balance and so on. When querying, the smart contract is called through eth_call in which the caller's signature is not required. At this time, caller needs to sign query with private key. Let the first 4 bytes of the function definition's hash value be F. (Account address on Ethereum is derived by hashing a public key with Keccak-256 [9] while we use SHA2-256 and RIPEMD 160 [10] on bitcoin.) Fill F to 32 bytes to get hash, set as Ft. The caller signs Ft with private key to get the (r, v, s) triple. And verifies the signature by ecrecover.

2.2 Architecture Design

The architecture of BSETS is shown in Fig. 1. The system is divided into three layers: data layer, middleware layer and application layer.

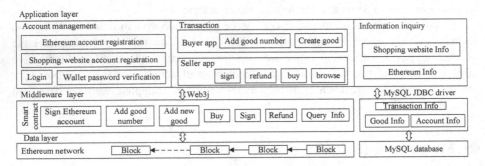

Fig. 1. The Architecture design of BSETS

In data layer, instead of relational database such as Oracle, OceanBase, smart contracts are deployed on p2p network built by blockchain system. Transactions will be broadcast to the Ethereum network and miners are responsible for mining and recording transactions. Considering that MyISAM engine of MySQL and distributed MySQL cluster are selected by main e-commerce platform Taobao, BSETS simplifies it and uses MySQL database to store additional data. Also, the password is salted and stored using the MD5 digest algorithm to secure the account. So data layer of BSETS includes the private Ethereum network built by geth and MySQL database. The middleware layer is implemented by java. It connects MySQL databases through MySQL JDBC driver and connects smart contracts through Web3j. The contract can't be modified once deployed. The application layer is implemented by Myeclipse developing web server, it provides a visual operation interface for the user.

3 Implementation of BSETS

3.1 Data Storage

MySQL database is used to store website data. There are three table including user, goods and transporting. However, MySQL databases don't hold user's assets information. So even if the data in MySQL is maliciously tampered with in the least probability, user's assets information on Ethereum won't be tampered with, not to mention Alipay's tampering with passwords security problems.

The purchase information stored in database is shown as Fig. 2, the goods can be indexed by contract address. An interface is provided for displaying user transaction information, including purchase information, sign-in status, and refund status.

good_id	good_name	seller_id	num	price	photo_src	good_kind	city	contract_address
5	bubble gum	manageruser	10000	100000000000000000	bubbleGum.jpg	food	Dalian	0xcbc8a25747cd702
6	noodles	userno6102	900	70000000000000000	noodles.jpg	food	Changsha	0x847e08ba4f994b6
7	Bjd doll	userno6103	30	230000000000000000	bjd.jpg	toy	Nanjign	0x5358382c5868c02

Fig. 2. Transaction Information

Other important trading information is stored on smart contract. The type of data on it include memory, storage, and calldata. Data that needs to be permanently saved in database LevelDB after transaction is set to the storage type. Every kind of good has its own contract. There are four main attributes in the contract: *name*, *owner*, *good_-trade_info*, and *buyer_record_info*. Where *owner* refers to the seller of goods represented by the contract. *good_trade_info* records the current status of each item including its owner and its buyer. *buyer_record_info* records the details of an order, giving contract address and user wallet address, the user's all purchase information can be track. *Buyer_record* has three main part, *balanceOf* record the purchases number, *buy_good_No* record good id in each order, *if_sign* is the index of *buy_good_No*. And Fig. 3 show the storage mechanism of *buy_good_No*.

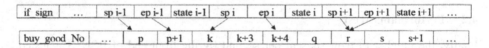

Fig. 3. *Buyer_record* storage mechanism

3.2 Function Module

Interaction. In the framework of Maven, the contract written by Solidarity is compiled into java, and the interaction with the contract is realized through web3j. In a transaction, the upper application sends a request to web3j, web3j calls the function of the deployed contract. Finally updates the SQL database after the execution.

Transaction function. Contract *owned* is used to record seller, and modifier is added to restrict the caller's identity. Contract *good* is used to record each item, and *good* inherits *owned* to match the item to its seller. Trading on a contract is a complete transaction, the results are output only if all conditions are met. And there is no rollback operation in the blockchain unless 51% of attacks occur.

The pseudo code of *buy* which is representative is shown as below. Then if the function *sign* is called, the good information in *good_trade_info* will be updated like in_whose_hand←msg. while the `balance` and `if_sign` will also be changed after verification. Another important function *refund* is similar to it.

Algorithm1 buy operation
Input: buy amount
Output: good_trade_info, buyer_record_info
1   check purchase amount and good balance
2   change the contract and seller's token number
3   add new purchase record of this buyer to if_sign
4   for i < amount-1
5   if good[i] have no buyer
6   Then update good[i]'s buyer and Current owner
7   buy_good_No←i

4 Performance Evaluation Experiments

4.1 System Settings

In the experiment, 2 geth nodes were started on one machine as miners. The code of smart contracts is written using Solidarity and is tested with mist client. The system is developed using MyEclipse. The Solidarity contracts are compiled using solc and the java contracts are generated by web3j command line. The testing tool for the Web site is Apache JMeter.

4.2 Performance Test

Firstly, the number of requests sent per second by the front end of BSETS is tested. Apache JMeter Settings are: Totally starts 20 threads and 1 thread is added every 5 s. After the maximum number of threads is reached, the execution proceeds for 30 s.

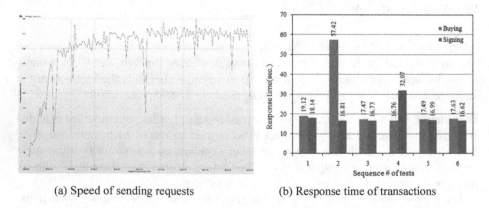

(a) Speed of sending requests   (b) Response time of transactions

Fig. 4. System performance

As Fig. 4a shows, the number of requests stabilize at 810 times per second when the threads increase to 6. Then the execution time of transactions on blockchain is tested. Figure 4b shows the response time of buying operations, which is basically stable at around 17 s. This is consistent with the nature of the Ethereum network confirming a block every 18 s.

5 Conclusion

The experiment verifies that the decentralized transaction system BSETS is realized, which is a new direction of e-commerce platform development. The transaction is recorded by miners into the blockchain. Records can only be appended, avoiding the risk of being tampered with. The additional data of BSETS except user assets is stored on the centralized database server, which guarantees the security of the user assets. What's more, with the Sharding technology in Ethereum 2.0, the throughput of Ethereum has been greatly improved, which makes it able to bear the huge transaction volume of e-commerce system. In the future, we need to add the logistics party as the participant, that is, the logistics party will also sign in a transaction as another type of users.

Acknowledgements. This work is supported in part by the Liaoning Science and Technology Foundation (20180550321), and the National Natural Science Foundation of China (61672142).

References

1. Gjermundrød, H., Chalkias, K., Dionysiou, I.: Going beyond the coinbase transaction fee: alternative reward schemes for miners in blockchain systems. In: Proceedings of the 20th Pan-Hellenic Conference on Informatics, PCI 2016, pp. 35:1–35:4. ACM, New York (2016). https://doi.org/10.1145/3003733.3003773
2. Wu, Y., Ivanov, A.: A multiple signature compaction scheme for bist. Microelectron. J. **23** (3), 205–214 (1992). https://doi.org/10.1016/00262692(92)90012-P
3. Iyer, K., Dannen, C.: The Ethereum Development Environment. In: Building Games with Ethereum Smart Contracts. Apress, Berkeley (2018)
4. Sukhwani, H., Martnez, J.M., Chang, X., Trivedi, K.S., Rindos, A.: Performance modeling of PBFT consensus process for permissioned blockchain network (hyperledger fabric). In: 2017 IEEE 36th Symposium on Reliable Distributed Systems (SRDS), pp. 253–255, September 2017. https://doi.org/10.1109/SRDS.2017.36
5. Varanasi, B., Belida, S.: Maven release. In: Introducing Maven. Apress, Berkeley (2014)
6. Gao, G., Li, R.: Collaborative Caching in P2P Streaming Systems. In: Meng, X., Li, R., Wang, K., Niu, B., Wang, X., Zhao, G. (eds.) WISA 2018. LNCS, vol. 11242, pp. 115–122. Springer, Cham (2018). https://doi.org/10.1007/978-3-030-02934-0_11
7. Doerner, J., Kondi, Y., Lee, E., Shelat, A.: Secure two-party threshold ecdsa from ecdsa assumptions. In: 2018 IEEE Symposium on Security and Privacy (SP), pp. 980–997, May 2018. https://doi.org/10.1109/SP.2018.00036

8. Weik, M.H.: Digital signature. In: Computer Science and Communications Dictionary. Springer, Boston (2000)
9. Dinur, I., Dunkelman, O., Shamir, A.: New Attacks on Keccak-224 and Keccak-256. In: Canteaut, A. (ed.) FSE 2012. LNCS, vol. 7549, pp. 442–461. Springer, Heidelberg (2012). https://doi.org/10.1007/978-3-642-34047-5_25
10. Dobbertin, H., Bosselaers, A., Preneel, B.: RIPEMD-160: a strengthened version of RIPEMD. In: Gollmann, D. (ed.) FSE 1996. LNCS, vol. 1039, pp. 71–82. Springer, Heidelberg (1996). https://doi.org/10.1007/3-540-60865-6_44

Query Processing

Data-Driven Power Quality Disturbance Sources Identification Method

Qi Li[1,2(✉)], Jun Fang[1,2], and Jia Sheng[1,2]

[1] School of Information, North China University of Technology,
Beijing 1000144, China
lq_941212@163.com

[2] Beijing Key Laboratory on Integration and Analysis of Large-scale Stream
Data, Beijing 100144, China

Abstract. Accurate identification of power quality (PQ) disturbance sources is the key to solve PQ problems. In order to improve the accuracy of classifying PQ disturbance sources, this paper proposes a data-driven PQ disturbance sources identification method. It takes some PQ data characteristics into account, such as diversity of indicators, non-linearity and time sequence characteristics. Firstly, the massive data are sifted and sampled. Then the feature subset is extracted by the sequence backward selection algorithm after evaluating the feature importance based on random forest (RF). Finally, the data are aggregated at fixed intervals, smoothed by the sliding average method, and put into Long Short-Term Memory (LSTM) network for model learning and prediction. Experimental results demonstrate that the proposed method is more effective than the traditional RF method.

Keywords: Power quality · Disturbance sources classification · Data sampling · Random forest · LSTM network

1 Introduction

With the construction and development of the ultra-high voltage (UHV) and smart grid, problems of power quality (PQ) caused by non-linear loads and equipment have been highlighted. Sometimes, the PQ indicators may fluctuate under the influence of disturbance sources, which can rise standard-exceeded problem and equipment damage [1]. PQ disturbance sources identification is to classify the PQ disturbance sources, for finding out the reason in time when the power grid breaks down. Disturbance events are divided into steady-state and transient disturbance events, electrified railways, photovoltaic power stations and wind farms are the main causes of the steady-state disturbance events, so the paper mainly studies this three PQ disturbance sources identification method.

In the current PQ analysis, the PQ disturbance sources identification is still in the preliminary phase. Traditional PQ analysis is generally based on a mechanism model and less practical [2, 3, 4, 5]. With the increasing importance of big data analysis, the State Grid has built the Power Quality Monitoring and Analysis System [6]. And a PQ disturbance sources recognition model [7] has been proposed based on this work,

© Springer Nature Switzerland AG 2019
W. Ni et al. (Eds.): WISA 2019, LNCS 11817, pp. 569–574, 2019.
https://doi.org/10.1007/978-3-030-30952-7_57

which used prior knowledge and statistical methods to extract features and then constructed a random forest (RF) [8] classifier to recognize the disturbance sources, but experiments show it has poor generalization performance on large samples.

Based on the above description, a power quality interference source identification method is proposed in this paper: Firstly, the massive data are sifted and sampled according to the data distribution characteristics. Then the method evaluates the feature importance based on the random forest and Out-Of-Bag (OOB) data [10], and extracts the feature subset with the sequence backward selection (SBS) [9] algorithm. Finally, the LSTM network model is built to learn and make predictions of the disturbance sources. In order to fix the issues about non-linearity and super-long sequence characteristics of the data and better fit the LSTM model, the data are aggregated at fixed intervals and then smoothed by the sliding average method before entering into the classifier.

2 Power Quality Disturbance Sources Identification

2.1 Disturbance Data Sampling

The PQ data mainly include the following characteristics: large volume, unbalanced sample distribution, lots of redundant indicators. Considering that the location and voltage level of monitoring points have great influence on the data range and trend, and the similarity of the monitoring point names is proportional to its geographical distance. In this study, stratified and systematic sampling is used to select samples. The sifting and sampling method is explained as below:

(1) Set the set of monitoring points to M. Take the daily active power from the monitoring points of line I and line II of each traction station, calculate the E_I and E_{II} by formula (1), if $E_I > E_{II}$, then put the point of line I into M.

$$E_i = \frac{1}{n}\sum_{j=1}^{j=n}|P_{ij}|, i = I, II; j = 1, \ldots, n \qquad (1)$$

(2) Calculate the data acquisition integrity rate r_i of each monitoring point in the experimental data period, ensure that the points in the set M satisfy $r_i > \varepsilon$.
(3) Stratify monitoring points of the three disturbance sources according to the voltage level, then take system sampling within the layer, and the sampling rate and interval of each group are determined by the distribution of the monitoring points.

2.2 RF-SBS Algorithm

For redundant indicators, the initial feature set is obtained according to business rules, and then the feature subset is extracted by RF-SBS algorithm. The idea of RF-SBS algorithm is to divide the data set into d equal parts and perform d-round iterations. In each iteration, one part is used as the test set to calculate the classification error rate, and the rest parts are used as the training set of the RF classifier. In training process, sort the feature importance [11] according to the OOB data of each decision tree, and use SBS algorithm to eliminate the feature with the smallest importance score, and then calculate the classification error rate again, and note that the feature is eliminated only when the error increment is less than threshold θ. Repeat d rounds, and select the feature subset with the smallest classification error in the d iterations. The procedure of RF-SBS algorithm is shown below.

Algorithm 1. RF-SBS algorithm

Input

$\quad D = \{(x_1, y_1), \ldots, (x_m, y_m)\}$, original data set.

$\quad F = \{f_1, f_2, \ldots, f_m\}$, original feature set.

$\quad d$, the number of parts of the original data set to be divided.

$\quad \theta$, threshold.

Output

$\quad F'$, final feature subset.

1.**begin**

2.Divide D into d equal parts randomly.

3.**for** i **in range**(d):

4. $\quad FI_i = \{f_1: 0, f_2: 0, \ldots, f_m: 0\}$, $Test_i := D_i$, $Train_i := D - D_i$;

5. $\quad OOB_i := Train_i - bootstrapSample(Train_i, N)$;

6. $\quad errOOB1_i := getErr(RF_i, OOB_i)$, $minErr_i := getErr(RF_i, Test_i)$;

7. $\quad\quad$ **for** t **in range**(m):

8. $\quad\quad errOOB2_{it} := getErr(DF_i, disturb(F_t, OOB_i))$;

9. $\quad\quad FI_i[f_s] := \frac{1}{k}\sum_1^k errOOB2_{is}\text{-}errOOB1_i$; //$k$ is the number of decision trees

10 $\quad\quad FI_i := sortByValue(FI_i)$; //sort by the feature importance scores

11. $\quad\quad\quad newErr := getErr(RF_i, Test_i, FI_i.pop())$

12. $\quad\quad$ **if** $newErr - minErr_i < \theta$, **then** $minErr_i := newErr$ **and** $FI_i.pop()$;

13. $\quad\quad$ **if** $minErr > minErr_i$, **then** $minErr := minErr_i$;

14. $\quad\quad F' := FI_i.keys()$;

15.**end for**

16.**return** F';

17.**end**

2.3 Data Aggregation and Smoothing

Considering that the LSTM step size should better not exceed 500 and the original data has 1,440 records per day, it is necessary to aggregate the data. Specifically, the initial aggregation period is set to 3 min. On this basis, try to increase the period and observe whether the data distribution will be affected to determine the final period. In addition, the original data jitter is serious, to enable the model learns the data time sequence characteristics better, the data should be smoothed. Specifically, the sliding window size is set to s, and replace the current data by the mean of each $\frac{s}{2}$ data before and after the current time as formula (2), when the number of samples before or after the current time is not to $\frac{s}{2}$, replace the current data with the mean of s data before and after the current time as formula (3). Figure 1 shows that there is little difference in data trend

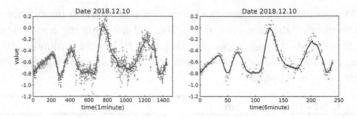

Fig. 1. Comparison of data trends before and after aggregation and smoothing processing.

before and after processing, which achieves the purpose of shortening the step size without affecting the overall trend.

$$D[t] = \frac{1}{s} \sum_{t=t-\frac{s}{2}}^{t=t+\frac{s}{2}} D[t] \tag{2}$$

$$D[t] = \frac{1}{s} \sum_{t=t-x}^{t=t+y} D[t], x + y = s \tag{3}$$

2.4 LSTM Model of Power Quality Disturbance Sources Identification

The LSTM network can fit the non-linearity and time sequence characteristics of the data well [12]. PQ data shows a certain regularity per day, so take the daily data of one monitoring point as one sample, and input the one-hot encoded label into the LSTM network simultaneously, the step size of LSTM takes the number of daily records of each monitoring point after data preprocessing. Finally, the softmax activation function is used to generate a vector containing the possibility of each label, and the label with the highest probability is selected as the final result. The model training process is shown in Fig. 2. In training process, the LSTM network structure and parameter are tuned by the control variable method.

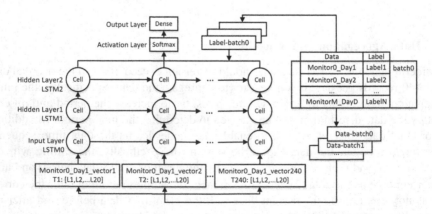

Fig. 2. Training process of power quality disturbance sources identification based on LSTM model.

3 Experiments and Analysis

In order to verify the feasibility of the power quality disturbance source identification method proposed in this paper, the data of all monitoring points during December 2018 were selected as the original data set, according to the above four steps, the data of 20 indicators of 300 monitoring points for 31 consecutive days were obtained. And the 20 indicators contain active power, harmonic voltage, voltage sequence component, three phase current imbalance etc. Table 1 shows the prediction results of the power quality disturbance sources identification model. It can be observed that the model can distinguish the positive and negative samples well.

Table 1. The prediction results of the power quality disturbance sources identification model

Label	Precision	Recall	F1-Score	Test Samples
Electrified Railways	0.88	0.90	0.89	1240
Wind Farms	0.90	0.91	0.90	1240
Photovoltaic Power Stations	0.92	0.93	0.92	1240

In addition, the paper compares with literature [5] in three aspects: (1) Sample selection. (2) Feature selection. (3) Data preprocessing. For comparison, data-set A is selected from adjacent monitoring points of the same voltage level, data-set B is randomly selected from all monitoring points, feature-set C is selected from the priori features in literature [5], feature-set D is selected according to RF-SBS algorithm, data preprocessing method in literature [5] is used as PE and data aggregation and smoothing method in this paper is used as PF. Figure 3 shows the comparison results: (1) The random forest model cannot extract the overall characteristics of random samples. (2) RF-SBS algorithm is more effective than the prior feature selection method. (3) It is necessary to consider the time sequence characteristics of data.

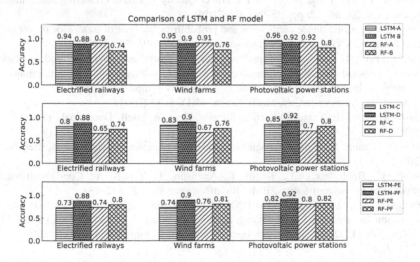

Fig. 3. Comparison of prediction accuracy of two models under different conditions.

4   Conclusion

In this paper, an effective data-driven power quality disturbance sources identification method is proposed. Its advantages are as follows: (1) Rational data sifting and sampling has reduced the problem scale, saved the computation cost, and improved the data skew problem. (2) RF-SBS algorithm can extract effective features, which is more universally applicable than the prior method. (3) After data aggregation and sliding average smoothing, the model can fit the samples better. The Experimental results have shown that the proposed method can effectively identify disturbance sources. Compared with traditional random forest method, it has better generalization performance and can be applied in practical engineering. The follow-up research will work on improving efficiency and building more lightweight model..

Acknowledgement. This work is supported by "The National Key Research and Development Plan (No. 2017YFC0804406), Public Safety Risk Prevention and Control and Emergency Technical equipment", and "Key Project of the National Natural Science Foundation of China, Research on Big Service Theory and Methods in Big Data Environment (No. 61832004)".

References

1. Xiao, X.: Analysis and Control of Power Quality. China Electric Power Press, Beijing (2010)
2. Lee, I.W.C., Dash, P.K.: S-transform-based intelligent system for classification of power quality disturbance signals. IEEE Trans. Ind. Electron. **50**(4), 800–805 (2003)
3. Qin, X., Gong, R.: Power quality disturbances classification based on generalized S-transform and PSO-PNN. Power Syst. Prot. Control **44**(15), 10–17 (2016)
4. Mercy, E.L., Arumugam, S., Chandrasekar, S.: Fuzzy recognition system for power quality events using spline wavelet. In: IEEE PES Power Systems Conference and Exposition, 2004 (2004)
5. Jiang, Z., Li, M., Yang, X.: About PV Power Quality Accurate Prediction and Simulation Studies, vol. 33, no. 12, pp. 95–99+304 (2016)
6. Liu, Y., et al.: Compliance verification and probabilistic analysis of state-wide power quality monitoring data. Global Energy Interconnection **1**(03), 391–395 (2018)
7. Liu, Y., Feng, D.: Identification of Power Quality Disturbance Events Based on Random Forest. National technical committee of standard voltages and current ratings and frequencies. In: 9th Proceedings of the power quality symposium (2018)
8. Breiman, L.: Random forests. Mach. Learn. **45**(1), 5–32 (2001)
9. Dash, M., Liu, H.: Feature selection for classification. Intell. Data Anal. **1**(3), 131–156 (1997)
10. Bian, Z., Zhang, X.: Pattern recognition, 2nd edn. Tsinghua University Publisher, Beijing (2000)
11. Strobl, C., Boulesteix, A.L., Kneib, T., Augustin, T., Zeileis, A.: Conditional variable importance for random forests. BMC Bioinformatics **9**(1), 307 (2008)
12. Zheng, H., Shi, D.: Using a LSTM-RNN based deep learning framework for ICU mortality prediction. In: Meng, X., Li, R., Wang, K., Niu, B., Wang, X., Zhao, G. (eds.) WISA 2018. LNCS, vol. 11242, pp. 60–67. Springer, Cham (2018). https://doi.org/10.1007/978-3-030-02934-0_6

Research and Implementation of a Skyline Query Method for Hidden Web Database

Zhengyu Li[1,2(✉)], Gui Li[1], Ziyang Han[1], Ping Sun[1], and Keyan Cao[1]

[1] Faculty of Information and Control Engineering,
Shenyang Jianzhu University, Shenyang 110168, China
1274850021@qq.com
[2] School of Information Science and Engineering,
Northeastern University, Shenyang 110004, China

Abstract. Skyline discovery from a hidden web database can enable a wide variety of innovative applications over one or multiple web databases in web information integration area. Due to the finite number of web accesses one can issue through per-IP-address or per-API-key, no more than k tuples of all matching tuples one top-k query can return, the restricted web interface types, etc., it is difficult to utilize traditional database Skyline computation methods for discovering Skyline of a hidden Web database. In the paper, we first analyze the key intersecting characters of skyline tuples by parallel coordinate system, and next define a search decomposition tree for searching skyline tuples of a hidden web database, and then prove that the tree is guaranteed to discover all skyline tuples. In the end, we put forward heuristic algorithms for getting skyline tuples of a hidden web database based on the mixture attributes. Theoretical analysis and real-world online and offline experiments demonstrate the effectiveness of our proposed methods and the superiority over baseline solutions.

Keywords: Database Skyline · Hidden web database · Intersecting tuples · Parallel coordinate system · Search decomposition tree

1 Introduction

Obtaining Skyline of web hidden database using the top-k query interface faces many challenges including: (1) each query result returns a maximum of k tuples; (2) the query conditions are limited by web interface types and attribute types; (3) the number of queries by the client is limited by the web server and so on. Therefore, how to reduce the query cost by reducing the number of remote queries becomes the key problem. Generally speaking, there are currently two implementation methods. One is to get all the tuples of the hidden web database through the web query interface, and then to generate Skyline of the database locally. Certainly, the query cost in this way is often high and limited by the number of queries on the web server. The other is to design a reasonable query decomposition algorithm and corresponding query conditions to obtain Skyline with less remote query times. In this paper, the second implementation method is studied, and the main contributions and organization are as follows.

© Springer Nature Switzerland AG 2019
W. Ni et al. (Eds.): WISA 2019, LNCS 11817, pp. 575–581, 2019.
https://doi.org/10.1007/978-3-030-30952-7_58

Section 2 gives an overview of Skyline query for hidden web database. Section 3 introduces concepts of Skyline and its characters in parallel coordinates. Section 4 is divided into two parts, the former describes how the intersecting tuples query tree is constructed and which crucial properties the tree possesses and then the baseline algorithm is presented to solve skyline for hidden web database, the latter discusses our novel heuristic query decomposition algorithm to reduce the baseline query cost, which is proved to satisfy completeness, and furthermore the upper and lower bounds of the heuristic query are obtained by theoretical analysis. We evaluate the baseline and heuristic algorithms together on the simulation and real data sets in 5 section.

2 Related Work

Literature [1, 2] defined respectively prior or frequent skyline points to reduce skyline points and used new metric called skyline priority or frequency to narrow candidate set to improve k-regret algorithms. Literature [3] proposed a mixed skyline query algorithm based on web query interface type and attribute category, but the query cost was high, which in some cases exceeded the cost of crawling the entire web database. Literature [4] proposed query result filtering method based on top-k and Skyline and progressive query strategy based on attribute importance and data source graph for Deep Web query failure. In literature [5], the calculation method of obtaining skyline from multiple web databases was studied under the condition of predicting the ranking function of web database query and obtaining all sorted tuples.

3 Concept and Properties of Skyline

3.1 Related Concept

Hidden web databases refer to the databases on the web server, which users can only access through the web query interface (e.g., top-k query).

Suppose a hidden web database D has n tuples, and each tuple t has m attributes, denoted as $A_1, \cdots, A_m$ respectively. $Dom(A_i)$ is the value domain of attribute A_i, while $t[A_i]$ is the value of attribute A_i and $t[A_i] \in Dom(A_i) \cup \{NULL\} (1 \leq i \leq m)$.

Definition 1 (Attribute Priority Relationship). The attributes of database tuples are sorted according to their importance.

Definition 2 (Tuple Domination Relationship). Given two tuples t_i and t_j, it exists domination relationship between them, or rather, the tuple t_i dominate t_j if $t_i[A_k] \leq t_j[A_k]$ for any attribute. Otherwise, the tuple t_i and t_j are nondominant.

Definition 3 (Tuple priority Relationship). Suppose that the tuple t_r and t_s are nondominant, then tuple t_r takes precedence over t_s, denoted as $t_r < t_s$, if some attribute $A_k(1 \leq k \leq m)$ exists such that $t_r[A_i] \leq t_s[A_i](1 \leq i < k)$ and $t_r[A_k] > t_s[A_k]$.

Definition 4 (Skyline for hidden web database). All nondominant tuples in the hidden web database D form Skyline of D, sometimes referred to as the outline of D.

Definition 5 (Dominant Consistency Constraint). The top-k query results $\{t_1, t_2, \cdots, t_k\}$, are the least dominated k ones among the result obtained by the complete query, and sorted according to '<' (the tuple priority relation determined by a specific ranking function). That is, t_i takes precedence over any one of $\{t_1, t_2, \cdots, t_k\}$.

3.2   The Properties of Skyline in Parallel Coordinate System

Parallel coordinates are a common way of mapping a n-dimensional data space to a 2-dimensional plane through n parallel axes, typically vertical and equally spaced. Each database tuple is represented as a polyline with vertices on the parallel axes.

Definition 6 (Intersecting and non-intersecting Relationship). Given the two tuples t_i and $t_j (1 \leq i, j \leq n)$, t_i and t_j have intersecting relationship if there are two attributes A_k and $A_r (1 \leq k, r \leq m, k \neq r)$ satisfying $t_i[A_k] < t_j[A_k]$ and $t_i[A_r] < t_j[A_r]$.

Definition 7 (completely intersecting relationship). For a group of k tuples, the k tuples have completely intersecting relationship if any tuple $t_i (1 \leq i \leq k)$ intersects the other $k-1$ tuples.

Lemma 1. Any tuple in a parallel coordinate system dominates all non-intersecting tuples above it.

Lemma 2. Tuples with intersecting relationships have non-dominant relationships.

Theorem 1. All of Skyline tuples in a database are completely intersecting.

Theorem 2. One tuple t added into a database, divides the database Skyline into two parts: one is tuple group S_1 intersecting t and the other is S_2 non-intersecting t. Then, if t is under S_2, the new Skyline is composed of S_1 and t. Otherwise, t must be above S_2, and Skyline remains unchanged.

Proof. Since the tuples of t and S_1 are intersected, we know from Lemma 2 that they are nondominant. For one thing, if t is located below S_2, t dominates S_2 because of Lemma 1, and the Skyline must be composed of S_1 and t by Definition 4. For the other thing, assume that t appears in the middle of S_2, such that S_2 will be divided into two disjoint parts by t, which contradicts the known fact that S_2 is part of Skyline and all tuples in S_2 are completely intersected by Theorem 1. So t can only be located above S_2. Again by Lemma 1, all of S_2 dominate t and obviously t cannot be part of the Skyline according to Definition 4, thus Skyline remains unchanged.

4   Skyline Query Methods for Hidden Web Database

4.1   Structure and Properties of Intersecting Tuple Query Tree

Suppose that T is the top-k result set obtained by query q_i which satisfies the dominant consistency constraint, and t is the first tuple of result set T.

    q_0: SELECT * FROM D;

According to the result T generated by query q_0, the first tuple t is utilized to recursively to construct m decomposed queries (i.e., $\{q_1, \cdots, q_m\}$) when $|T| \geq K$, while no more decomposition will be generated, that is, t acts as a leaf node, when $|T| < K$.

q_1: WHERE $A_1 < t[A_1]$ & $B\_Const(t)$; ...;

q_i: WHERE $A_1 \geq t[A_1] \& A_2 \geq t[A_2] \& ... \& A_{i-1} \geq t[A_{i-1}] \& A_i < t[A_i]$ & $B\_Const(t)$;...;

q_m: WHERE $A_1 \geq t[A_1] \& A_2 \geq t[A_2] \& ... \& A_{m-1} \geq t[A_{m-1}]$ & $A_m < t[A_m]$ & $B\_Const(t)$.

$B\_Const(t)$ represents the branch query condition of the parent node t, which is composed of the conjunction of conditions on the path from root node to the parent node.

$P\_Const(q_i) = B\_Const(t) \& A_1 \geq t[A_1] \& A_2 \geq t[A_2] \& ... \& A_{i-1} \geq t[A_{i-1}]$.

$P\_Const(q_i)$ represents the precondition of the query q_i.

From the above construction of the intersecting tuple query decomposition tree, we can know that it has the following properties.

The query decomposition conditions of intersecting tuples are mutually exclusive and fully covered in the range of attributes $A_1, A_2, \cdots, A_{m-1}$.

Completeness of Intersection Relation. The first tuple of the parent node generating m query decomposition, can find the highest priority tuples on m attributes that has intersecting relationship with the parent node, and as decomposition goes on, all tuples intersecting the parent node will be discovered in the end.

Completely Intersecting Relationship of the Same Branch. All node tuples on the same branch have completely intersecting relationship with each other.

Incompletely Intersection Relationship of Different Branches. Due to the limitation of query condition $P\_Const(q_i)$, node tuples on different branches may have dominant relations, that is, non-intersecting relations or non-completely intersecting relations.

Non-repeatability of the Same Layer Decomposition. Duplicate tuples aren't present in query decomposition of any node due to the mutual exclusion of query conditions.

Theorem 3 (Completeness). If a tuple t is one of Skyline tuples, namely $t \in Skyline$, there exists a query node q_i in the intersecting tuple query decomposition tree, such that the query result T of the node must contain tuple t, i.e., $t \in T$.

Proof. Due to $t \in Skyline$, there are at least two attributes $A_i, A_j (1 < i \leq m, 1 < j \leq m, i \neq j)$ so that $t[A_i] > t'[A_i]$ & $t[A_j] > t'[A_j]$ $(t' \in Skyline)$. It is known from Definition 8 that there will be a query node $t''(t'' \in Skyline)$ and a positive integer $k(1 < k \leq m)$ satisfying $t[A_1] \geq t[A_1] \& \cdots \& t[A_k] \geq t''[A_k]$, in the process of query decomposition of intersecting tuple. That is to say, tuple t is included in the query result T of a certain branch node of node t'', and when $|T| \geq k$, it appears as the first tuple.

4.2 Skyline Query Methods for Web Database

Basic Query Decomposition Method

The basic query decomposition method is proposed based on intersection tuple query decomposition tree and the properties of Skyline tuples. Its basic ideas are as followed.

(1) Establish the query decomposition tree of intersecting tuples through depth-first or breadth-first mode, and then we are able to obtain two intersecting tuples' sets S_1 and S_2 in hidden web database D, where S_1 is the collection composed of the first tuples of intermediate nodes in the query decomposition tree and S_2 is the collection composed of leaf nodes.

(2) For tuples in $S_1 \cup S_2$, the Skyline tuples' set of hidden web database D is generated according to the completely intersecting properties of Skyline tuples.

Heuristic Query Decomposition Method

To reduce query cost (mainly remote query times) and improve query efficiency, we present a heuristic query decomposition method. And its basic ideas are as follows.

(1) Different from the basic query decomposition, each query decomposition first tries to make a local query in the returned result set of the parent node. No remote query request need be issued if the query result is not empty. Otherwise, a remote query will be taken.

(2) if the first tuple t in the query result is dominated by another tuple t' in the query result of a certain node in the current query tree, t' will take place of t and then continue decomposition.

Theorem 4. Heuristic decomposition method satisfies completeness.

Proof. Consider the first case of heuristic algorithm, that is, a local query result is not empty, of which the first tuple is denoted as t_1. Without doubt, the result is also non-empty if our taking a remote query, where the first tuple is denoted by t_2. Therefore, we can get $t_1 = t_2$ because of dominant consistency constraints and then come to conclusion that the non-empty local queries instead of remote query is workable. (to ensure that the constraint holds, we just ensure that the order in which tuples appear in the parent node result remains the same in the local query result.)

Next, let's think about the second case of heuristic algorithm, i.e., t' take place of t due to t governed by t'. The new decomposition tree modified by t' will be shorter, because the upper bound of a certain subitem in t' conditional expression must be lower than that generated by t according to the way to generate conditional expression of the decomposition tree, thus it can filter non-skyline tuples more and faster. However, the remaining tuples in the decomposition tree (including intermediate nodes and leaf nodes) to constitute the Skyline is unchanged. In the end, combining the above two cases, the completeness of heuristic method can be guaranteed.

4.3 Query Cost Analysis

Considering the main factor of query cost is determined by the number of remote queries, we can figure it out that the number of remote queries equals to the number of remote query branches in the tree.

Theorem 5. The query cost C of heuristic query decomposition method, namely the number of remote queries, satisfies the following inequality when $K \leq m$.

$$m + 1 \leq C \leq (|S| + [n/k]) \times m, (k \leq n)$$

Proof. Suppose that $t_1(t_1 \in Skyline)$ is the first tuple of T_0 obtained by Q_0, and then ahead of each query Q_i to decompose t_1, a local query should be executed on the last query result T_0, and the result is denoted by T_i. When T_i is not empty, take the first tuple in T_i as the basis to proceed with m decomposition. Otherwise, remote queries are executed. When $|T_i| \geq K$, query decomposition is carried out m times using the first tuple t_1 of T_i. Otherwise, query decomposition is no longer performed. During query decomposition, the first tuple of query Q_i may be dominated by $S_1 \cup S_2$, so Skyline tuples will be repeatedly decomposed in the query decomposition tree. However, as the attribute ranges are mutually exclusive, the repeated decomposition of Skyline tuple must not be more than $[n/k]$ times. Therefore, the number of intermediate nodes will not exceed $|S| + [n/k]$, and the total number of branches will not exceed $(|S| + [n/k]) \times m$. As we know, the number of remote queries must be less than the total number of branches, that is, the remote query cost $C \leq (|S| + [n/k]) \times m$. Also, because there is at least one Skyline tuple in D, the remote query cost $C \geq m + 1$.

5   Evaluation

The student achievement database includes 10 attributes (5 numerical attributes and 5 classification attributes) and 100,000 tuples. The real estate database (www.house-book.com.cn) has 10 properties selected, containing more than 53 million records.

According to the theoretical query cost, $m + 1 \leq C \leq (|S| + [n/k]) \times m, (k \leq n)$, we analyzed the influence of parameter set $\{m, |S|, n, k\}$ on C between basic and heuristic decomposition algorithm. The experimental results show that the cost of heuristic decomposition algorithm is better than that of basic decomposition algorithm (Figs. 1, 2, 3, 4).

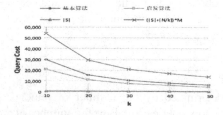

Fig. 1. Impact of Top - k on query cost

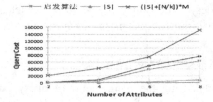

Fig. 2. Impact of the number of attributes

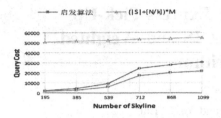

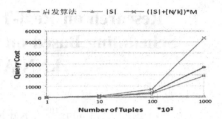

Fig. 3. Impact of the number of Skyline

Fig. 4. Impact of the number of tuples

References

1. Han, S., Zheng, J., Dong, Q.: Efficient processing of k-regret queries via skyline frequency. In: Meng, X., Li, R., Wang, K., Niu, B., Wang, X., Zhao, G. (eds.) WISA 2018. LNCS, vol. 11242, pp. 434–441. Springer, Cham (2018). https://doi.org/10.1007/978-3-030-02934-0_40
2. Han, S., Zheng, J., Dong, Q.: Efficient processing of k-regret queries via skyline priority. In: Meng, X., Li, R., Wang, K., Niu, B., Wang, X., Zhao, G. (eds.) WISA 2018. LNCS, vol. 11242, pp. 413–420. Springer, Cham (2018). https://doi.org/10.1007/978-3-030-02934-0_38
3. Asudehz, A., Thirumuruganathanz, S., Zhang, N., et al.: Discovering the skyline of web databases. PVLDB 9(7), 600–611 (2016)
4. Shen, D., Ma, Y., Nie, T.: A query relaxation strategy applied in a deep web data integration system. J. Comput. Res. Dev. 47(1), 88–95 (2010)
5. Lo, E., Yip, K., Lin, K., et al.: Progressive skylining over web-accessible databases. Data Knowl. Eng. 57(2), 122–147 (2006)

Research on Real-Time Express Pick-Up Scheduling Based on ArcGIS and Weighted kNN Algorithm

Yi Ying[1(✉)], Kai Ren[2], and Yajun Liu[3]

[1] College of Computer Science and Technology,
Sanjiang University, Nanjing 210012, China
907635255@qq.com
[2] Jinling College, Nanjing University, Nanjing 210089, China
[3] School of Computer Science and Engineering,
Southeast University, Nanjing 210096, China

Abstract. In the end delivery service of logistics, the research on real-time express pick-up scheduling is still in blank. Based on GIS technology, Web technology and mobile development technology, an intelligent logistics information system for "Last-Mile" distribution is constructed. In the framework of this system, weighted kNN classification algorithm is improved to make real-time express pick-up scheduling for delivery man in reality. Through the application of distribution activities on CaiNiao post station (an outlet to provide end delivery service), the intelligent logistics information system mentioned above can effectively improve the service quality of the logistics network. Meanwhile, the real-time express pick-up scheduling algorithm proposed in this paper has a significant effect on solving practical problems.

Keywords: Intelligent logistics information system ·
Real-time express pick-up scheduling · ArcGIS · Weighted kNN algorithm

1 Introduction

With the rapid development of information and communication technology and the popularity of various applications based on Internet/mobile Internet, e-commerce centered on commodity purchase and sales has gradually become a new bright spot in economic growth. Third-party logistics is the infrastructure for providing logistics services for e-commerce activities. Its basic process consists of three phases: warehousing system, transportation backbone network and the "Last-Mile" distribution [1].

Logistics management, especially logistics distribution process, is highly dependent on geospatial information. Geographic information systems (GIS) [2] have the ability to analyze and process geospatial information, and can support effective processing and help decision-making based on the acquisition and storage of relevant data. Literature [3] combined GIS means and SPSS analysis to improve the simulated annealing algorithm and thus solve the VRP. Literature [4] proposed a heuristic-simulated annealing hybrid search algorithm combing GIS model constraints to solve VRP. Literature [5]

© Springer Nature Switzerland AG 2019
W. Ni et al. (Eds.): WISA 2019, LNCS 11817, pp. 582–588, 2019.
https://doi.org/10.1007/978-3-030-30952-7_59

applied 3G technology to distribution management and designed a logistics distribution system model. Literature [6] combined network analysis function of MapGIS to explore location and path optimization of urban distribution centers.

However, under the framework of logistics information system, there has been no study on using GIS technology and data mining algorithm to solve the real-time express pick-up scheduling model at the outlet. Based on ArcGIS platform, this paper constructed an intelligent logistics information system and improved the weighted kNN classification algorithm to realize couriers' real-time express pick-up scheduling. The conclusions of this paper will provide theoretical support for the construction of smart logistics and the scientific scheduling of express vehicles.

2 Architecture of Intelligent Logistics Information System

Based on GIS technology, Web technology and mobile development technology, this paper constructed an intelligent logistics information system for "Last-Mile" distribution. The intelligent logistics information system adopts a three-tier architecture system, which consists of an application service layer, a business logic layer and a logistics data layer, which is shown in Fig. 1.

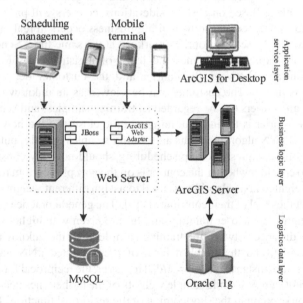

Fig. 1. Architecture of intelligent logistics information system

The business logic layer serves as the core of the system, including Web Server and ArcGIS Server. Web Server consists of the traditional Java EE server JBoss and ArcGIS Web Adaptor. The Web Adaptor component combines ArcGIS Server with a traditional Web server to offer richer functionality and greater security (such as user authentication). Web Adaptor is the single portal for GIS service consumers, which receives client-side requests, and then forwards requests to ArcGIS Server within the site. GIS-related services are provided by ArcGIS Server, for instance: spatial data management, map visualization and path analysis.

At the application service level, ArcGIS for Desktop [7] is the creator of GIS resources (such as map), which connects to ArcGIS Server via ArcMap to publish local resources as Web services. ArcMap is a major program in ArcGIS for Desktop, which boasts all map-based functions including map making, geographic data analysis and editing. Scheduling management and mobile terminal are users and consumers of GIS service, wherein mobile terminal is a handheld device of logistics distribution personnel.

The logistics data layer completes the data storage function. The geographic data objects are stored in the form of tables in the Oracle database; MySQL is the general business data (such as customer information) of the Java EE application storage system.

3   Real-Time Express Pick-Up Scheduling Algorithm

Currently, delivery and pick-up business are mixed together to be handled by the courier in most outlets. Based on cost considerations, couriers will pre-set the delivery path according to experience, but some mailing business occurs in real time. Moreover, in one area, there will be several couriers working at the same time. Scheduling which courier to pick up express determines the level of logistics costs to some extent. Intuitively, the distance between the courier and the customer determines whether express pick-up is timely. The customer can be viewed as an unknown sample in the kNN algorithm, all couriers can be regarded as training samples, and scheduling of the most appropriate courier is to find 1 training sample closest to the new sample.

The traditional kNN algorithm treats all training samples equally, but the classification [8] process of the express pick-up scheduling should not only consider the actual distance, but also consider whether the courier is on the way plus current road conditions. Therefore, the weighted kNN algorithm is used to assign different weights to the training samples to reflect the level of their contribution [9]. The general practice is to convert the distance into weights, and closer training samples are given with higher weights.

Let d_i be the distance between the training sample i and the unknown sample x, w_i represents its weight, and the simplest form of the weighted kNN algorithm is the reciprocal of the return distance, i.e. $w_i = 1/d_i$. However, the reciprocal form results in big weights of adjacent samples, while sample weights of the farther ones decay rapidly. The Gaussian function overcomes the shortcoming of the reciprocal function. The weight is 1 when the distance is 0. As the distance increases, the weight gradually decreases, but the attenuation is not too fast, and the attenuation is guaranteed not to be zero. Its form is:

$$w_i = ae^{-\frac{(d_i-b)^2}{2c^2}} \tag{1}$$

At the same time, the penalty factor δ is set, and the calculation of the sample weight is adjusted depending on whether the courier's driving is in the direction of the sender's address. The definition is as follows:

$$\delta = \begin{cases} 1; consistent\ direction \\ 0.5; opposite\ direction \\ 0.75; other\ cases \end{cases} \tag{2}$$

The weight calculation formula adjusted by the penalty factor δ is:

$$w'_i = \delta * w_i \tag{3}$$

Calculate w'_i of all training samples and training sample with highest w'_i is the most appropriate express pick-up person.

Based on the intelligent logistics information system, the real-time express pick-up scheduling process combining weighted kNN algorithm is as follows: ① The customer places an order and triggers express pick-up scheduling task, then the sender address is converted into location information through the place name and address retrieval technology; ② GPS/4G positioning technology is used to obtain the location information of each courier through the timed reporting function of the mobile APP; ③ NAServer module of ArcGIS GeoProcessing is invoked to calculate the actual shortest distance d_i between each courier and sender in the road network; ④ w'_i is calculated using weighted kNN algorithm based on d_i and direction information of the courier, the most appropriate express pick-up person is selected and scheduling instructions are given. The whole process is shown in Fig. 2:

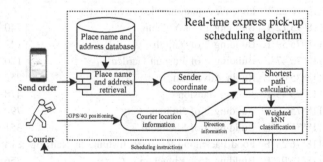

Fig. 2. Execution process of real-time express pick-up scheduling

4 Application Examples

4.1 Overview of the Experimental Area

The Mochou Xinyu outlet of CaiNiao post station provides customers with door-to-door service of self-service pick-up, delivery and receipt. The outlet distribution range covers a quadrilateral area consisting of Hanbei Street, Beiwei Road, Shuiximen Street, Mochouhu West Road and Hanzhongmen Street. The area is about 0.4 km^2 in size. Serving residential areas of Mochou Xinyu, Jinshun Garden, Jinmao New Apartment, Jinjitang City, it serves a population of about 32,000. The area has many residential communities with high population density, which poses a high requirement for logistics service quality. Through the investigation of the actual outlet distribution, application of the intelligent logistics information system designed in this paper achieved relatively successful results.

4.2 Analysis of Experimental Results

There are 4 courier service personnel in the outlet. Based on busyness of business, generally 2–3 people provide door-to-door service. Table 1 lists the actual express pick-up in the morning of February 25, 2019. The data indicates that the courier travels a distance of 300–800 m for each express pick-up, and the traveling speed of electric bicycles on urban roads is about 20 km/h, so the service response time of one express pick-up is less than 5 min.

Table 1. Actual pick-up information of a certain day

Geographical coordinates of courier	Pick-up address	Actual distance
32°04′01″N, 118°76′22″E	Phoenix Aquarium Flower Market	262 m
32°04′08″N, 118°75′72″E	Building 14 of Yulanli	740 m
32°03′93″N, 118°75′59″E	Building 7 of Yujinli	142 m
32°03′38″N, 118°75′47″E	Building 1 of Jinshun Garden	1202 m
32°03′88″N, 118°75′60″E	Ruxin Advertising	252 m
32°03′76″N, 118°75′45″E	Jinling Mansion	478 m
32°03′86″N, 118°75′59″E	Building 2 of Jinjitang City	790 m
32°03′96″N, 118°75′86″E	Building 3 of Shengqili	351 m
32°04′14″N, 118°76′28″E	Food Packaging Research Institute of NanJing	814 m
32°03′97″N, 118°76′02″E	Building 4 of Yingchunli	756 m

Table 2 is statistics of the overall door-to-door express pick-up in the outlet from the end of February to the beginning of March 2019. Since the service areas are mostly residential communities, express pick-up needs obviously shows characteristic of more needs in holidays. In 10 days, a total of 292 express were picked up, and the courier

Table 2. Whole pick-up information of outlets (10 days)

Date	Number of packages	Total distance of driving	Date	Number of packages	Total distance of driving
February 24th	37	24984 m	March 1st	30	20204 m
February 25th	23	12712 m	March 2nd	34	25033 m
February 26th	27	19293 m	March 3rd	37	24679 m
February 27th	29	19538 m	March 4th	25	17899 m
February 28th	28	18761 m	March 5th	22	15339 m

traveled a total of 198,442 m, averagely traveling nearly 680 m for each express pick-up. According to industry reports [10], a single picked-up package travels about 1 km. Compared with this, the real-time express pick-up scheduling method reduces the driving distance by 32%, and the path length is greatly shortened, which saves the time and cost of door-to-door service.

5   Conclusion

The modern logistics system has entered the development stage featuring informationization, networking and intelligence. Using key technologies of ArcGIS, Android, GPS, etc. this paper built an intelligent logistics information system applicable to logistics end distribution business. In this system, the improved weighted kNN classification algorithm is applied to courier's real-time express pick-up scheduling, which realizes visual push of the scheduling result on the mobile APP. The validity of the algorithm and the effectiveness of the application model of GIS technology in the logistics industry are verified by experiments, and the management efficiency of logistics information is improved. The parallel processing and personnel scheduling of the mailing order is the focus of further research. It is also a good research direction to make delivery path planning by improving the traditional VRP model under the intelligent system.

Acknowledgment. This research was supported by the Natural Science Foundation of the Jiangsu Higher Education Institutions of China (No. 17KJB520033), and QingLan Project of the Jiangsu Higher Education Institutions of China (No. Su Teacher [2018] 12).

References

1. Yang, J., Yang, C., Yao, X.: Research on the "Last-Mile" issue in the E-Commerce logistics system. J. Bus. Econ. **34**(04), 16–22+32 (2014)
2. Pan, K., Liu, Q.: Review of studies on application of GIS technology in logistics field. Logistics Technol. **33**(09), 26–27+55 (2014)
3. Pei, X., Jia, D.: Optimizing multi-objective vehicle routing problem in city logistics based on simulated annealing algorithm. Math. Pract. Theory **46**(02), 105–113 (2016)

4. Pan, G., Hu, J., Hong, M.: GIS-based logistics distribution area division and its VRP algorithm. J. Dalian Marit. Univ. **41**(01), 83–90 (2015)
5. Liu, Z., Zhou, F.: Design and realization of 3G-based logistic delivery system. Comput. Eng. Appl. **38**(18), 249–250+256 (2002)
6. Wang, L., Guo, J., Xuan, D., et al.: Optimization design of city distribution network based on MapGIS. J. Highway Transp. Res. Dev. **32**(08), 143–149 (2015)
7. Wu, J., Qui, Y.: ArcGIS Software and Application. Publishing House of Electronics Industry, Beijing (2017)
8. Zhao, D., Chang, Z., Du, N., Guo, S.: Classification for social media short text based on word distributed representation. In: Meng, X., Li, R., Wang, K., Niu, B., Wang, X., Zhao, G. (eds.) WISA 2018. LNCS, vol. 11242, pp. 259–266. Springer, Cham (2018). https://doi.org/10.1007/978-3-030-02934-0_24
9. Feng, G., Wu, J.: A literature review on the improvement of KNN algorithm. Libr. Inf. Serv. **56**(21), 97–100+118 (2012)
10. Insight Report on Express Courier Group in 2018, 19 June 2019. https://www.cbndata.com/report/983/detail

Similarity Histogram Estimation Based Top-k Similarity Join Algorithm on High-Dimensional Data

Youzhong Ma[1,2], Ruiling Zhang[1(✉)], and Yongxin Zhang[1]

[1] School of Information and Technology, Luoyang Normal University, Luoyang 471934, China
`ma_youzhong@163.com`, `ruilingzhanglynu@163.com`, `tabo126@126.com`
[2] Key Laboratory of E-commerce Big Data Processing and Analysis of Henan Province, Luoyang 471934, China

Abstract. Top-k similarity join on high-dimensional data plays an important role in many applications. The traditional tree-like index based approaches can't deal with large scale high-dimensional data efficiently because of "curse of dimensionality". So in this paper, we firstly propose an approach to construct the similarity distribution histogram using stratified sampling method, then to estimate the similarity threshold according to the number of the required returned results, finally we propose a novel Top-k similarity join algorithm based on similarity distribution histogram. We conduct comprehensive experiments and the experimental results show that our proposed approaches has good efficiency and scalability.

Keywords: Top-k similarity join · Similarity distribution histogram · Symbolic Aggregate Approximation · High-dimensional data

1 Introduction

Similarity join query is widely used in many applications, such as image clustering, duplicate web page detection, similar user recommendation and so on. Threshold based similarity join query always needs to know the similarity threshold in advance, while it is usually difficult or impossible for us to do this. In many cases, it is enough for the users to get the top-k similar pairs. If the distance of the kth similar pair(recorded as k-threshold) can be found previously, it can be

This research was partially supported by the grants from the National Natural Science Foundation of China (No. 61602231); Training plan for young backbone teachers of Colleges and universities in Henan (No. 2017GGJS134); Key Scientific Research Project of Higher Education of Henan Province (No. 16A520022); Outstanding talents of scientific and technological innovation in Henan (No. 184200510011); National key research and development program (No. 2016YFE0104600); Scientific and Technological Project of Henan Province (No. 192102210122).

W. Ni et al. (Eds.): WISA 2019, LNCS 11817, pp. 589–600, 2019.
https://doi.org/10.1007/978-3-030-30952-7_60

set as the threshold and the threshold based similarity join query algorithm can be used to get the top-k similar pairs. If the k-threshold is too big, there will exist many redundant comparisons; Otherwise, the number of the returned similar pairs will be less than k, it is necessary to conduct the second computation based on a bigger similarity threshold. So the selection of the k-threshold plays the key role on the performance of the top-k similarity join query.

A naive and feasible way is to select a number of distance thresholds in advance, $\epsilon_1, \epsilon_2, \ldots, \epsilon_n$, in which, $\forall i < j, \epsilon_i < \epsilon_j$. The similarity join query algorithm will be executed once for each distance threshold ϵ_i, if the number of the returned similar pairs is less than k for ϵ_i, the similarity join query algorithm needs to be executed again for ϵ_{i+1}, this procedure will end until the number of the satisfied similar pairs is equal or larger than k. The distance threshold can be selected according to the distance distribution of the vector pairs, e.g., $\epsilon_i = 0.05 * i, i = 1, 2, \ldots, n$. This method is easy to be implemented, but there exist many redundant, duplicated comparisons because the similarity join query algorithm needs to be executed again for each new distance threshold ϵ_i, so its performance will be bad.

In order to reduce the redundant comparisons, we propose a top-k similarity join algorithm based on similarity histogram estimation. The main contributions of the paper are as the follows:

- We propose a novel method to build up the similarity distribution histogram and to estimate the top-k threshold;
- A top-k similarity join query algorithm based on similarity threshold estimation is proposed;
- Comprehensive experiments are conducted and the experimental results show that our proposed methods has good performance and scalability.

The rest of the paper is organized as the following: The related research works are briefly introduced in Sect. 2; Sect. 3 gives the preliminaries and the problem definition; We describe the method of construction of the similarity distribution histogram, the method of the estimation of the top-k similarity threshold and the top-k similarity join algorithm based on estimated similarity threshold in Sect. 4; comprehensive experiments are conducted in Sect. 5; Sect. 6 concludes this paper.

2 Related Work

Many research works have been done to exploit similarity join problem [1–4]. Shim et al. [5] proposed a high-dimensional data similarity join algorithm based on ϵ-KDB-tree. Its main idea is to divide all the data into several partitions with equal width based on a selected dimension, each leaf node just needs to join with its two adjacent nodes. This approach requires enough memories to store the adjacent nodes. Zhu et al. [6] proposed a novel approach to return the top-k closest object pairs by using R-tree index. Other tree-like indexes have also been proposed to deal with similarity search or similarity join problem [7–9].

The distance between the objects become more smaller as the dimensionality increases, there will exist the "curse of the dimensionality" problem.

The space-partitioning based filtering scheme is also used to deal with the similarity join problem on large scale high-dimensional data [10–14]. In [10], the whole data space is divided into several equal-width grids, a novel grid coding method called Epsilon Grid Order(EGO) is used to code the grids, then the objects that need to be compared can be obtained based on the grid code number. An improved algorithm called Super-EGO is proposed in [11], Super-EGO designed a new EGO strategy by adopting the data-driven-based dimension reordering mechanism so that it can eliminate more unnecessary comparisons. The parallel Super-EGO is also developed and has high efficiency. An approximate similarity join algorithm on high-dimensional data by using space filling curve technique is proposed in [12]. MRHDJ [13] partitioned the objects using grid and designed a novel bit coding strategy to filter out the unnecessary comparisons, such that the network transmission cost can be reduced. MRHDJ [13] is easy to be parallelized, but it is suitable to deal with the low-dimensional data(<30), its efficiency decreases dramatically once the dimensionality is too high. In order to overcome this shortcoming, [14] expanded the solution of MRHDJ [13] and proposed a new approach called PHiDJ [14] based on MapReduce framework. PHiDJ [14] improved the similarity speed through dimension grouping and variable-length grid partitioning. However, the space-partitioning based filtering scheme also has the following problems: the grid granularity is related with the distance threshold, the data needs to be repartitioned for the different distance threshold; the grid number increases exponentially as the dimensionality increases.

Locality Sensitive Hashing(LSH) is widely used to deal with high-dimensional data similarity join [15]. Stupar et al. [16] adopted LSH to map the near points to the same hash bucket and proposed an efficient k nearest neighbor query algorithm based on MapReduce framework. Multiple hash tables and long hash codes are required to ensure the ideal query result, this will need too much storage space and the query time will increase. Multi-Probe-based LSH algorithm [17] can ensure the accuracy of the query results using less hash tables. Data Sensitive Hashing(DSH) [18] can solve the data skew problem effectively. DSH not only can make sure that the near objects are mapped in to the hash bucket, but also can ensure that the object number of each bucket is equal as far as possible. Fast CoveringLSH [19] can return all the required result. [20] proposed a distributed KNN query algorithm by combining P2P network. [21] proposed a personalized LSH based on a novel binding technique, it can reduce the false positives effectively and improve the query efficiency.

3 Preliminaries

3.1 Problem Definition

In this section, we mainly give the definition of Top-k Similarity Join (TSJ).

Definition 1. *(Top-k Similarity Join: TSJ) Given two d-dimensional data sets R and S, a distance function dist(.) and a positive integer k, then the Top-k Similarity Join on R and S aims to find out the Top-k object pairs whose distances are less than that of other object pairs. That is: $TSJ(R \bowtie S) = \{\langle r_1, s_1 \rangle, \langle r_2, s_2 \rangle, ..., \langle r_k, s_k \rangle\}$, in which, $r_i \in R(1 \le i \le k)$, $s_j \in S(1 \le j \le k)$, the distance function dist(.) refers to Euclidean distance. $TSJ(R \bowtie S)$ meets the following requirements:*

- *$r_i \in R$, $s_j \in S$*
- *$dist(r_1, s_1) \le dist(r_2, s_2) \le \cdots \le dist(r_k, s_k)$*
- *for each object pair $\langle r_m, s_m \rangle$, if $\langle r_m, s_m \rangle \notin TSJ(R \bowtie S)$, then $dist(r_k, s_k) \le dist(r_m, s_m)$*

For simplicity, we mainly focus on the Self-join in this paper, while our proposed solution can be easily extended to deal with two sets join.

3.2 Piecewise Aggregate Approximation and Symbolic Aggregate Approximation

Piecewise Aggregate Approximation(abbreviated as PAA) was firstly proposed as a new dimensionality reduction technique to deal with time series. For time series, the element is order dependent, actually PAA is not restricted with the order, so PAA can be used to reduce the dimensionality of the high-dimensional data. The basic idea of PAA is displayed through Definition 2.

Definition 2. *PAA vector X_P: Given a d-dimensional vector $X = \langle x_1, x_2, \ldots, x_d \rangle$, let d' be the dimensionality of the reduced space ($1 \le d' \le d$). For simplicity, we assume that d' is a factor of d, however this is not a requirement of PAA. The PAA representation of a vector X can be represented in d' space by a vector $X_P = \langle \overline{x}_1, \overline{x}_2, ..., \overline{x}_{d'} \rangle$. The element $\overline{x}_i$ is calculated as the following equation:*

$$\overline{x}_i = \frac{d'}{d} \sum_{j=\frac{d}{d'}(i-1)+1}^{\frac{d}{d'}i} x_j \tag{1}$$

It has been proven that the distance of PAA vectors is the lower bound of the original Euclidean distance.

Symbolic Aggregate Approximation for time series (abbreviated as SAX) is the first symbol representation of time series with dimensionality reduction and firstly proposed by Lin et al., it also works for high-diemnsional data. SAX is based on the PAA, it can map each element of the PAA vector to an appropriate symbol according to specific rules, then a d-dimensional vector X can be represented as a SAX string $X_S = (\tilde{x}_1...\tilde{x}_{d'})$.

Given two vectors X, Y and their SAX representations X_S and Y_S, the SAX distance function MINDIST is defined as the following:

$$MINDIST(X_S, Y_S) = \sqrt{\lambda} \sqrt{\sum_{j=1}^{d'} dist(\tilde{x}_j, \tilde{y}_j)^2} \tag{2}$$

It has been proved that the SAX distance MINDIST is the lower bound of the PAA distance D_P [14], and D_P is also the lower bound of the Euclidean distance D_E, so MINDIST is the lower bound of the Euclidean distance, that is:

$$MINDIST(X_S, Y_S) \leq D_E(X, Y) \tag{3}$$

Based on the above descriptions, we can get the following conclusion:

- The distance of PAA vectors and that of SAX strings is the lower bound of the original Euclidean distance, so they can be used to filter out the object pairs which are impossible to be similar at relative low cost.
- Different high-dimensional data may be mapped to the same SAX string, so SAX strings can be used to group the original high-dimensional data. The objects in the same SAX group have high possibility to be similar, while the possibility that the objects from different SAX groups to be similar may be low.

4 Top-k Similarity Join Algorithm Based on Similarity Histogram Estimation

A novel Top-k similarity join algorithm based on similarity histogram estimation is proposed in this section, it mainly contains three tasks: construction of similarity distribution histogram, estimation of Top-k similarity threshold and Top-k similarity join algorithm based on estimated similarity threshold.

4.1 Construction of Similarity Distribution Histogram

The construction of similarity distribution histogram aims to find out the number of the similar object pairs given different distance threshold, it is the foundation of estimation of Top-k similarity threshold.

Supposing that the cardinality of the high-dimensional data set is N, the distance threshold is ϵ, the straight-forward and native method is: selecting M samples randomly from the data set, then computing the distance of each object pair in the M samples and obtaining the number of the similar pairs whose distance is less than or equal with ϵ, recorded as R; finally, the total number of the similar object pairs in the whole data set can be estimated as $(\frac{N}{M})^2 \cdot R$. The previous method is easy to be implemented, but the number of the similar object pairs in the M samples may be very small, even be zero for a given relative small distance threshold ϵ, so it is likely to be inaccurate to estimate the final result through simple scaling up method. In order to ensure that the estimated result is as close as possible to the real value, we proposed a stratified sampling method based on SAX, the specific process is as the following:

(1) Vector grouping: reducing the dimensionality of the original high-dimensional vectors using PAA, then mapping the PAA vectors in to SAX strings. According to the previous conclusion, we know that different high-dimensional

vectors may be mapped to the same SAX string, so the original high-dimensional vectors can be divided into different groups based on their SAX strings.

(2) Supposing that the cardinality of the original vector set is N, divided into m groups totally, and the vector number of each group is $N_i, i = 1, \cdots, m$, then randomly selecting samples from each group, the number of the sampled vectors from ith group is S_i. Because the objects in the same SAX group have high possibility to be similar, we firstly figure out the number of the similar vector pairs of the samples from the same group. The number of the similar vector pairs of the samples from the ith group is R_i, then the number of the result of the ith group can be estimated as: $\frac{N_i(N_i-1)}{S_i(S_i-1)} \cdot R_i$, so the total number of the similar vector pairs from all the groups can be estimated as: $\sum_{i=1}^{m} \frac{N_i(N_i-1)}{S_i(S_i-1)} \cdot R_i$.

Although the possibility that the objects from different SAX groups to be similar may be low, they still may be similar, so we also need to estimate the number of the similar vector pairs form different group. Supposing that the number of the similar vector pairs of the samples from the ith group and the jth group is R_{ij}, then the estimated result value is: $\frac{N_i \cdot N_j}{S_i \cdot S_j} \cdot R_{ij}$, so the total number of the similar vector pairs between all the different groups can be estimated as: $\sum_{i=1}^{m-1} \sum_{j=i+1}^{m} \frac{N_i \cdot N_j}{S_i \cdot S_j} \cdot R_{ij}$. Finally the number of the whole similar vector pairs can be represented as: $\sum_{i=1}^{m} \frac{N_i(N_i-1)}{S_i(S_i-1)} \cdot R_i + \sum_{i=1}^{m-1} \sum_{j=i+1}^{m} \frac{N_i \cdot N_j}{S_i \cdot S_j} \cdot R_{ij}$.

(3) Based on the above estimation approach, for different distance threshold, we can figure out the according estimated result values. Finally we can construct the similarity distribution histogram based on those estimated values.

4.2   Estimation of Top-k Similarity Threshold

After the similarity distribution histogram has been constructed, the according distance threshold for a given k value can be estimated based on the similarity distribution histogram. The main procedure is as the following:

Supposing that the estimated result values of distance threshold $x_0, x_1, \cdots, x_m$ are $y_0, y_1, \cdots, y_m$, the according histogram is displayed in Fig. 1. We also suppose that the results number between two adjacent distance threshold increase linearly, then the linear equation between x_{i-1} and x_i can be represented as the following:

$$y = \frac{y_i - y_{i-1}}{x_i - x_{i-1}} \cdot x + \frac{x_i \cdot y_{i-1} - x_{i-1} \cdot y_i}{x_i - x_{i-1}}$$

When the required number of the Top-k query is k, we can obtain the according Top-k distance threshold through the linear equation, that is :

$$x_k = \left(k - \frac{x_i \cdot y_{i-1} - x_{i-1} \cdot y_i}{x_i - x_{i-1}} \right) \cdot \frac{x_i - x_{i-1}}{y_i - y_{i-1}}$$

4.3   Top-k Similarity Join Algorithm based on Estimated Similarity Threshold

Algorithm 1 describes the Top-k similarity join algorithm based on estimated similarity threshold, the main idea is as the following:

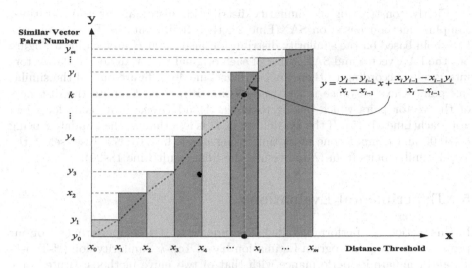

$$y = \frac{y_i - y_{i-1}}{x_i - x_{i-1}} x + \frac{x_i y_{i-1} - x_{i-1} y_i}{x_i - x_{i-1}}$$

Fig. 1. Estimation of Top-k Similarity Threshold

Algorithm 1. Top-k Similarity Join Algorithm based on Estimated Similarity Threshold

---

Input: d-dimensional data set: R, the number of the Top-k similar pairs: k.
Output: the Top-k similar pairs: Ω.

1   $\Omega \leftarrow \emptyset$
2   $H \leftarrow$ construct the similarity distribution histogram;
3   $\epsilon \leftarrow$ figure out the Top-k similarity threshold based on H;
4   **for** *each vector $v_i \in R$* **do**
5     $paa_{v_i} \leftarrow$ get the paa vector of v_i;
6     $sax_{v_i} \leftarrow$ get the sax string of v_i;

7   $\Lambda \leftarrow$ get all the unique sax strings;
8   grouping the original vectors according to their sax string;
9   **while** *the number of the similar pairs whose distance is less than k* **do**
10     $\Theta \leftarrow$ figure out all the similar sax pairs whose distance is less than ϵ;
11   **for** *each candidate sax pair $\langle sax_i, sax_j \rangle$* **do**
12   **for** *each $v_m \in G(sax_i)$* **do**
13   **for** *each $v_n \in G(sax_j)$* **do**
14   **if** $dist(v_m, v_n) \leq \epsilon$ **then**
15             $\Omega \leftarrow \Omega \cup \{\langle v_m, v_n \rangle\}$;

16   **if** $|\Omega| < k$ **then**
17       $\epsilon \leftarrow \epsilon + 0.05$;

18   **else**
19       $\Omega \leftarrow select\ Top - k\ similar\ pairs\ from\ \Omega$;
20   return the Top-k similar pairs: Ω;

Firstly, constructing the similarity distribution histogram by using stratified sampling method based on SAX(Line 2); then figure out the Top-k similarity threshold based on the similarity distribution histogram H and k(Line 3); figure out the PAA vector and SAX string of each original vector, divide all the vectors into groups according to their SAX strings(Line 4–8); figure out all the similar sax pairs whose distance is less than ϵ(Line 10); then compute the distances of the vector pairs which belong to $G(sax_i)$ and $G(sax_j)$ by using loop join approach(Line 11–15). If the cardinality of Ω is less than k, then update ϵ using $\epsilon + 0.05$ and compute one more time using new ϵ(line 16–17). Else, select the Top-k similar pairs from Ω and return the final result(Line 18–20).

5   Experimental Evaluations

In this section, we perform detailed experiments to test the performance of our proposed method: Histogram Estimation based Top-k Similarity Join(H-Top-k-J), and compare its performance with that of two naive methods: Brute Force Top-k Similarity Join(B-Top-k-J), Threshold based Top-k Similarity Join(T-Top-k-J). B-Top-k-J figures out the distance of all pairs of the high-dimensional data and select the Top-k closet pairs as the results; T-Top-k-J previously sets a threshold sequence incrementally, $\epsilon_1, \epsilon_2, ..., \epsilon_m$, and then conducts the threshold based similarity join algorithm using ϵ_1, if the number of the returned results is less than k, perform this algorithm again using ϵ_2 until the number of the returned results is over k.

Experimental Setup. We conduct the experiments on a laptop, the configuration for the laptop is as the follows: CPU: Intel Core i5-4210H 2.90 GHz, memory: 8 GB, disk: 1 TB, OS: 64bit windows 7. The main parameters and values used in the experiments are described in Table 1. The default values are presented in bold.

Table 1. Settings used in the experiments

Parameter	Values
k	500, 1000, 2000,**4000**, 8000
Distance threshold: ϵ	0.05, 0.1, 0.15, 0.2, 0.25, 0.3, 0.35, 0.4
Dimensionality: d	128, 256, **512**, 960
Data size: N	10000, 20000, **30000**, 40000, 50000

Datasets. The datasets used in our experiments are downloaded from the internet[1] that is used in [22]. We select part of the original dataset to do the tests, the details of the datasets are described in Table 2.

---

[1] http://corpus-texmex.irisa.fr/.

Table 2. Overview of Datasets

Dataset	Type	Num.	Dim.	DataSize
data-128-10000	Real	10,000	128	11.29M
data-128-20000	Real	20,000	128	22.51M
data-128-30000	Real	30,000	128	33.84M
data-128-40000	Real	40,000	128	44.98M
data-128-50000	Real	50,000	128	56.33M
data-256-30000	Real	30,000	256	67.62M
data-512-10000	Real	10,000	512	45.31M
data-512-20000	Real	20,000	512	90.26M
data-512-30000	Real	30,000	512	135.12M
data-512-40000	Real	40,000	512	180.11M
data-512-50000	Real	50,000	512	225.13M
data-960-30000	Real	30,000	960	253.42M

5.1 Performance vs. k

Figures 2 and 3 display the performance with different k. Figure 2 shows that the run time of B-Top-k-J is constant, because B-Top-k-J needs to compute the distance of all pairs of the high-dimensional data for different k. The run time of T-Top-k-J increases as k becomes larger, when k is over 2000, the run time of T-Top-k-J becomes larger than that of B-Top-k-J, the reason is that T-Top-k-J needs to execute more times for the bigger k, so its total run time may be larger than that of B-Top-k-J. The performance of H-Top-k-J is the best, because the Top-k threshold can be estimated through the similarity histogram, the desirable results can be obtained through one time execution of H-Top-k-J. The general change trend of Fig. 3 is almost the same with that of Fig. 2, but the run time of Fig. 3 is bigger than that of Fig. 2, because the former dimensionality is larger than the latter.

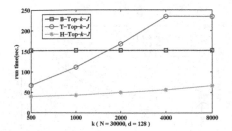 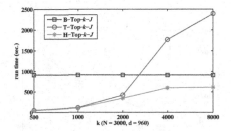

Fig. 2. Performance vs. k (N = 30000, d = 128)

Fig. 3. Performance vs. k (N = 30000, d = 960)

5.2 Performance vs. Data Size

Figures 4 and 5 display the run time with different data size. The experimental results show that the run time of the three methods increases as the data size becomes larger. When the data size is less than 20000, the run time of T-Top-k-J is bigger than that of B-Top-k-J, the reason is that the estimated threshold may not be correct, T-Top-k-J perhaps needs to be executed more than one time, so the total run time of T-Top-k-J may be greater than that of B-Top-k-J. The performance of H-Top-k-J is the best of the three methods.

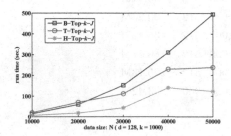

Fig. 4. Performance vs. Data Size N(d = 128,k = 10^3)

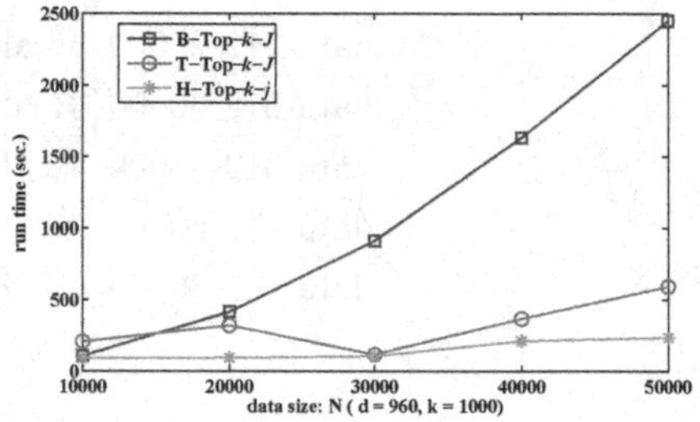

Fig. 5. Performance vs. Data Size N(d = 960,k = 10^3)

5.3 Performance vs. Dimension

In this subsection we conduct the experiments to test the run time with varying data dimensionality. Figures 6 and 7 show that the run time of B-Top-k-J increases almost exponentially with the dimensionality. In Fig. 6, the run time of T-Top-k-J and H-Top-k-J when dimensionality is 960 becomes smaller than that of the two methods when dimensionality is 512, this case is related with the data distribution, the estimated threshold of data-960-30000 may be more accurate and smaller than that of data-512-30000. In Fig. 7, the run time of T-Top-k-J is almost up to the run time of B-Top-k-J when dimensionality is 512, this is

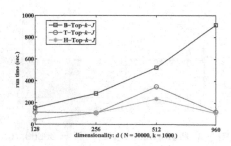

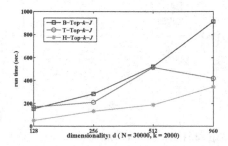

Fig. 6. Performance vs. Dimensionality d(k = 1000)

Fig. 7. Performance vs. Dimensionality d(k = 2000)

because T-Top-k-J may need to be executed many times, the total run time of T-Top-k-J may become larger. The performance of H-Top-k-J is still the best for all the different dimensionality.

6 Conclusions

In this paper an effective similarity distribution histogram estimation method is proposed based on Piecewise Aggregate Approximation(PAA) and Symbolic Aggregate Approximation(SAX) techniques, and we propose the Top-k similarity join algorithm on high-dimensional data by using the estimated similarity distribution histogram. The extensive experimental results prove that the efficiency of our proposed method. In the future, we plan to further study the efficiency and the accuracy of the similarity distribution histogram estimation, and try to exploit the Top-k similarity join problem on other data types, e.g., set data, strings or spatial data.

References

1. Pang, J., Gu, Y., Xu, J., Yu, G.: Research advance on similarity join queries. J. Front. Comput. Technol. **7**(1), 1–13 (2013)
2. Pang, J., Yu, G., Xu, J., Gu, Y.: Similarity joins on massive data based on mapreduce. Framework **42**(1), 1–5 (2015)
3. Yu, M., Li, G., Deng, D., Feng, J.: String similarity search and join: a survey. Front. Comput. Sci. **10**(3), 399–417 (2015)
4. Xu, W., Xu, Z., Ye, L.: Computing user similarity by combining item ratings and background knowledge from linked open data. In: Meng, X., Li, R., Wang, K., Niu, B., Wang, X., Zhao, G. (eds.) WISA 2018. LNCS, vol. 11242, pp. 467–478. Springer, Cham (2018). https://doi.org/10.1007/978-3-030-02934-0_43
5. Shim, K., Srikant, R., Agrawal, R.: High-dimensional similarity joins. In: Proceedings of ICDE, pp. 301–311 (1997)
6. Zhu, M., Papadias, D., Zhang, J., Lee, D.: Top-k spatial joins. IEEE Trans. Knowl. Data Eng. **17**(4), 567–579 (2005)
7. Yu, C., Cui, B., Wang, S., Su, J.: Efficient index-based KNN join processing for high-dimensional data. Inf. Software Technol. **49**(4), 32–344 (2007)
8. Sakurai, Y., Yoshikawa, M., Uemura, S., Kojima, H.: The A-tree: an index structure for high-dimensional spaces using relative approximation. In: Proceedings of VLDB, pp. 516–526 (2000)
9. Yu, X., Dong, J.: Indexing high-dimensional data for main-memory similarity search. Inf. Syst. **35**(7), 825–843 (2010)
10. Böhm, C., Braunmüller, B., Krebs, F., Kriegel, H.: Epsilon grid order: an algorithm for the similarity join on massive high-dimensional data. In: Proceedings of SIGMOD, pp. 379–388 (2001)
11. Dmitri, V.: Kalashnikov, Super-EGO: fast multi-dimensional similarity join. VLDB J. **22**(4), 56–85 (2013)
12. Lopez, M., Liao, S.: Finding k-closest-pairs efficiently for high dimensional data. In: Proceedings of CCCG, pp. 197–204 (2000)

13. Seidl, T., Fries, S., Boden, B.: MR-DSJ: distance-based self-join for large-scale vector data analysis with mapreduce. In: Proceedings of BTW, pp. 37–56 (2013)
14. Fries, S., Boden, B., Stepien, G., Seidl, T.: PHiDJ: parallel similarity self-join for high-dimensional vector data with mapreduce. In: Proceedings of ICDE, pp. 796–807 (2014)
15. Wang, J., Shen, H., Song, J., Ji, J.: Hashing for similarity search: a survey, pp. 1–29. arXiv:1408.2927 (2014)
16. Stupar, A., Michel, S., Schenkel, R.: Rankreduce-processing K-nearest neighbor queries on top of mapreduce. In: Proceedings of LSDS-IR, pp. 13–18 (2010)
17. Lv, Q., Josephson, W., Wang, Z., Charikar, M., Li, K.: MultiProbe LSH: efficient indexing for high-dimensional similarity search. In: Proceedings of VLDB, pp. 950–961 (2007)
18. Gao, J., Jagadish, H., Lu, W., Ooi, B.: DSH: data sensitive hashing for high-dimensional k-NN search. In: Proceedings of SIGMOD, pp. 1127–1138 (2015)
19. Pham, N., Pagh, R.: Scalability and Total Recall with Fast CoveringLSH, pp. 1–13. arXiv:1602.02620v1 (2016)
20. Haghani, P., Michel, S., CudreMauroux, P., Aberer, K.: LSH at large - distributed KNN search in high dimensions. In: Proceedings of WebDB, pp. 1–6 (2008)
21. Wang, J., Lin, C.: Mapreduce based personalized locality sensitive hashing for similarity joins on large scale data. Comput. Intell. Neurosci. **2015**, 1–13 (2015). Article ID 217216
22. Luo, W., Tan, H., Mao, H., Ni, L.: Efficient similarity joins on massive high-dimensional datasets using mapreduce. In: Proceedings of MDM, pp. 1–10 (2012)

Hybrid Indexes by Exploring Traditional B-Tree and Linear Regression

Wenwen Qu[✉], Xiaoling Wang, Jingdong Li, and Xin Li

Shanghai Key Laboratory of Trustworthy Computing, International Research Center of Trustworthy Software, East China Normal University, Shanghai, China
vinny_qu@163.com

Abstract. Recently, people begin to think that database can be augmented with machine learning. A recent study showed that deep learning could be used to model index structures. Such learning approach assumes that there is some particular data distribution in the database. However, we argue that the data distribution in the database may not follow a specific pattern in the real world and the learning models are usually too complicated, which makes the training process expensive. In this paper, we show that linear models can achieve the same precision as models trained by deep learning using a hybrid method and are easier to maintain. Based on this, we propose a hybrid method by exploring traditional b-tree and linear regression. The hybrid method retrieves data and checks whether the data can benefit from learning approach. We have implemented a prototype hybrid indexes in Postgres. By comparing with b-tree, we show that our method is more efficient on index construction, insertion, and query execution.

Keywords: Linear regression · Learned index · Hybrid indexes

1 Introduction

The index is an old but powerful technology which has widely used in industry and also commonly studied in the research area. Generally speaking, indexes are techniques that build extra structures for fast data accessing. B-tree and hash are the two most used methods in the index. It is suggested to build the b-tree index for range query and choose the hash index for point query. Both the b-tree index and the hash index are designed for the general purpose of data access, which means that the distribution of data has little effect on indexing.

Kraska et al. [12] have pointed index can benefit from data distribution by applying machine learning. The tuples are usually correlated with the location of the data in the OLAP database, e.g., the ordered data are organized in the disk sequentially. For example, Postgres, a well-known scientific database, will organize data in disk with their inserted orders. In an OLAP application such as signal system, the relationship between index keys and the addresses of the tuples can model as the CDF (Cumulative Distribution Function) of the index keys. Kraska et al. [12] reported that such data pattern could be learned by machine learning to construct a learned index.

© Springer Nature Switzerland AG 2019
W. Ni et al. (Eds.): WISA 2019, LNCS 11817, pp. 601–613, 2019.
https://doi.org/10.1007/978-3-030-30952-7_61

The learned index seems to be a choice to traditional indexes due to the low space usage. However, in the real world, the data pattern is not so well that the index can't learn it perfectly. Especially when insertion, deletion, and update happen, data pattern could be damaged. For example, Postgres will look up disk to find the block that can adequately store the data. It could cause a non-sequential data distribution. In such a case, the models in the learned index become no longer accurate. Of course, we can adjust the data to restore the data pattern or retrain models. Based on Kraska' approach, neither solution has a low update overhead. What's more, those abnormal points make the CDF model hard to learn. Besides, the cost of constructing and updating complex models should be taken notice. Some learning methods, like deep learning, could learn more complex models. However, the overhead of training or retraining such models can't be ignored. If we turn to simple models, the cost can be significantly reduced. However, we also have to guarantee the accuracy of prediction since the data pattern may be random. Besides, traditional indexes still have advantages to deal with random data patterns (no data patterns). However, the difficulty to use traditional indexes is massive space usage. Especially when the data file is large or users build many indexes on raw tables. And that's exactly the advantage that the learned index has.

In this paper, we propose a hybrid method which combines the advantages of traditional indexes and learning approach. The hybrid method can achieve lower space usage and more robust to random data patterns compared to Kraska' approach. It is based on the idea that the data are divided into two disjoint parts and use different methods to index them. The outliers are eliminated to make the linear models have a high accuracy of prediction.

To summarize, we have made the following contributions:

- We propose an approach that models index by linear regression called LR models. LR models can achieve good accuracy and low training latency. LR models can be quickly retrained. These are presented in Sect. 3.
- We introduce a hybrid approach to index the data in the database. We use three ways to distinguish whether the data can be beneficial from the learning approach. Also, traditional indexes are used for outliers in data. These are described in Sect. 4.
- We introduce how to maintain hybrid indexes and how to apply hybrid indexes to query execution. These are described in Sect. 5.
- We have implemented our hybrid indexes in Postgres, and it outperforms the traditional indexes such as b-tree and hash indexes based on the TPC-H benchmark and a real-world dataset. These are described in Sect. 6.

Furthermore, we mention some related works in Sect. 2, and the paper is concluded in Sect. 7.

2 Related Work

Traditional indexes such as b-tree, and hash have a long research history, and many works have been proposed to extend those. The storage media such as SSD and in-memory databases pose new challenges to traditional indexes. T-tree [16] mainly solves the problem that the index has significant overhead in memory. In addition to the high node space occupancy of T-tree, the search algorithm for traversing the tree also has an advantage in terms of complexity and execution time than b-tree. CSS-tree [18] can provide faster query speed than binary search in b-tree and take up less space. CSB+-Tree [19] is a variant of B+Tree that stores only the address of the first node to achieve the higher utilization of the cache and the addresses of other nodes are calculated by the offset. Besides, there are some works based on the SIMD system such as FAST [10].

Traditional indexes suffer from high space usage, which may consume the performance of the in-memory database. Several index compression techniques have been proposed. Prefix compression and dictionary compression are used in index compression [6, 7]. Also, a hybrid method [24] is used to distinguish between cold data and hot data. Cold data are indexed with compression because they are less to use.

Some other works try to reduce space usage using approximate index. BF-tree [1] uses Bloom Filters to index each region of datasets and organizes them with a tree structure. Same as our methods, A-tree [5] also uses linear models to index the datasets approximately. However, it may fail when the datasets are random. The learned index carried by Kraska et al. [12], which is mostly similar to our work, uses deep learning to learn the relationship between index keys and addresses. But the models are complicated and time-consuming. What's more, it can't handle random data, too.

3 LR Models

The critical idea to learning approach for the index is that there exist functions (think of the CDF in Kraska' work) to map the keys to the addresses of data. More specifically, considering the case we use in Fig. 1(a), there exists a one-to-one mapping among all the pairs (*key, position*). We first assume the mapping relation satisfies a regular pattern here and analysis the case of the irregular pattern in the next section.

To get a better understanding, we use $\omega(x)$ to present the real address of key x (the red line in Fig. 1) and $\varphi(x)$ to present the predicted address of the key x (the blue line in Fig. 1). We define the loss function as formulation (1). The training process is to minimize the loss function of L to make $\omega(x)$ and $\varphi(x)$ as close as possible. Then we can use $\varphi(x)$ to predict the addresses of tuples in the database.

$$L(\theta, \omega) = \sum_{i=1}^{n} (\omega(x) - \varphi(x))^2 \tag{1}$$

The question is how to describe and train the models (e.g., $\varphi(x)$). The following goals need to be satisfied in the database if we apply machine learning to indexes:

1. The speed of training models should be fast since the database concerns the latency of index maintenance and query execution.
2. The precise address needs to be returned. In other words, if the query falls into the database, the models should tell the database where to find the entry.

The learning approach proposed by Kraska et al. [12] uses deep learning to train models. The advantage of deep learning is that it can train complex models in complex

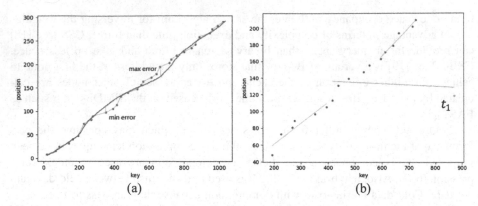

Fig. 1. (a) The red line represents the real addresses, and the blue line represents the predicted addresses given by models. (b) the outliers destroy the data distribution. If we use linear models to predict the address, the errors would be significant. (Color figure online)

situations. However, it is hard to design a proper network for users. Besides, the cost of retraining models is expensive, which means updating the index is time-consuming.

In this work, we improve Kraska's method. We observe that linear models can also guarantee the correctness of predicted address if the models are trained upon small data. The advantage of linear models is that they can be trained in $O(n)$ complexity and easier to retrain. The latter guarantees the efficiency of the learned index. To distinguish with the models proposed by Kraska et al. [12], we give the following definition:

Definition 1 (LR Models). Given a query key x defined by user's SQL, LR Models are a series of linear models constructed on each partitioned data, which is defined as formulation (2), and LR models return the predicted address to database and executor search the query entry near the predicted address.

$$\varphi(x) = \theta x + b \tag{2}$$

where θ and b are the weights generated during the training process.

To answer the second requirement, the models remember the worst case in the training process and store the predicted errors in the worst cases (e.g., max error ε_{min} and min error ε_{max}). The actual address always falls in the interval of $[\hat{p} + \varepsilon_{min}, \hat{p} + \varepsilon_{max}]$. For above, ε_{min} is the smallest error compared to the actual address (it may be negative) while ε_{max} is the biggest error compared to the actual address in Fig. 1(a). It guarantees that if the query key appears in the database, the tuples could always be found. Otherwise, *NULL* is returned.

4 Hybrid Indexes

In this section, we first analyze why learned index can't wholly replace traditional indexes and how to combine traditional indexes with a learning approach. Intuitively, we think the learned index performs well on data distribution with a specific pattern. However, learned models would suffer from irregular data distribution in practice. Besides, the data update can disrupt the distribution of data. Therefore, it is impractical to assume that the learning approach is a good solution to index.

Think about the example in Fig. 1(b). There is enough space to place the tuple t_1 to block 2. Therefore, the database inserts t_1 into block 2 instead of putting it into block 4 sequentially. It is the behavior happening in Postgres. But this makes t_1 an outlier. The outliers refer to the data that don't match an existing data pattern. It is obvious that the outliers increase the error boundaries since the least-squares method, which is used to train linear models, is very sensitive to outliers. More sophisticated models are demanded to fit the data distribution better. An alternative solution is to detect the outliers in the raw table and keep the remaining data good enough that can be well trained with linear models. It further reduces the complexity of datasets.

4.1 Approach

Considering a database, there are a set of tuples $T = \{t_1, t_2, t_3, \ldots, t_n\}$, where the tuples correspond to a set of addresses $A = \{a_1, a_2, a_3, \ldots, a_n\}$ and the pair $p_k(t_k, a_k)$ indicates the relationship between them. We divide all pairs p_k into S_1, S_2, where S_1 represents the set of all pairs that make a good data distribution while S_2 represents the set of all outliers.

There are three methods we consider in our hybrid indexes to distinguish S_1 and S_2. We call them Key-based Evaluation (KE), Cosine Similarity-based Evaluation (CSE) and Error-based Evaluation (EE):

KE: Here, we assume that the outliers are caused by the keys. For example, the spikes in a signal system or the change of distribution of signal data.

CSE: Since we use linear models to learn the distribution of data, the outliers could be those tuples that are far away from the center point based cosine distance.

EE: The outliers increases the errors of linear models. Therefore, we think outliers can be defined as the tuples with significant predicted errors. This method is costly because it needs to rescan the table to retrain the models after we remove the outliers.

Given a model m with weights θ and the pair $p_k(t_k, a_k)$, the index key for t_k is x. We use $\mathcal{A}$ to donate the variables set in aforesaid evaluations. For example, if we use EE to evaluate the outliers, $\mathcal{A}$ contains the predicted errors of all tuple addresses. For each $\alpha \in \mathcal{A}$, we use $\bar{\alpha}$ to indicate the mean of α and $\sigma_{\bar{\alpha}}$ to denote the standard deviation of α. We consider the parameters are subject to the normal distribution. By using these symbols, we define the outliers as any pair outside the range below, and γ is hyper-parameter, which can be set from 1 to 3:

$$[\bar{\alpha} - \gamma\sigma_{\bar{\alpha}}, \bar{\alpha} + \gamma\sigma_{\bar{\alpha}}] \tag{4}$$

By defining the outliers, we can eliminate the outliers from training data. Note the process is not convergence, and we only do it once. Here we consider using b-tree to index outliers. The mixture of using both b-tree and learning method to index data takes several advantages. First, we assume that the database is model-free. It means we don't know the data distribution in the database. The worst case is that a very complex model is demanded to fit the data. If we want to keep the models efficient and easy to maintain, it is a good choice to put hard-to-learn data into traditional indexes. Second, the LR models are very susceptible to outliers. If we don't eliminate the outliers, the errors of models will be much significant due to outliers. It further leads to a large interval when using the index to seek position, which reduces the performance of query execution.

Algorithm 1 Hybrid Train

Input: *(key, address)* pairs P, γ

1. $m = \text{LRTrain}(P)$;
2. compute $\bar{\varepsilon}$ and σ_{ε} for m
3. **FOR** each t_k *in P* **DO**
4.   **IF** ε_k **NOT IN** $[\bar{\varepsilon} - \gamma\sigma_{\varepsilon}, \bar{\varepsilon} + \gamma\sigma_{\varepsilon}]$
5. $btree.\text{insert}(t_k)$;
6.   remove t_k from m
7. retrain m

The algorithm we use in the hybrid learner is shown as Algorithm 1. First, the least-squares method is used to train LR models (line 1). The algorithm then checks all tuples in the raw table and finds out outliers (lines 3–5). If an outlier is found, the algorithm throws it to b-tree and indexes it (line 6). The retraining process (line 7) is incremental since we have kept some statistics during the first training process.

5   Index Maintenance and Query Execution

Given the above hybrid learner, we have implemented the hybrid indexes in Postgres. The source code can be found in [21]. In this section, we focus on the process of how to construct, update hybrid indexes, and how to help query execution with hybrid indexes.

5.1   Index Construction

Here we focus on the case that there are some data in the raw table and how to build an index on it. We use model capacity to define how much tuples are used to train models and bigger τ means we can get less models, but each model will have a bigger predicted error. We build hybrid indexes using Algorithm 2. First, the algorithm initializes a buffer named *tups* to store all *(key, address)*. The tuples are retrieved block by block, recording their keys and addresses into *tups*. If the number of tuples meets the model's capacity τ, the algorithm calls the hybrid learner in Algorithm 1 and saves models (lines 3–5). Outliers are detected and indexed by b-tree in *HybridTrain()*. Also, we construct the meta page (first page) to help looking up the target model for inserting and query.

Algorithm 2 index Construction

Input: the raw tables T, model capacity τ

1.　initialize *tups* and *metapage*
2.　**FOR** each heap tuple *ht* **IN** *T* **DO**
3.　　**IF** *tups.size() > τ*
4.　　　Model *m* = HybridTrain(*tmp_spool*);
5.　　　WriteModel(*m*)
6.　　**ELSE**
7.　　　*tups*.push(*ht.key*, *ht.addr*);
8.　　FinishConstruction ();

5.2　Index Insertion

In the database, the tuple t_k may be inserted into a new block, which means there is no corresponding model in LR models. In this case, a new model is built for tuple t_k. Then flush it into the disk (lines 1–2). The other case is a bit complex, that is, t_k belongs to an existing model. Therefore, the algorithm first locates the model corresponding to the block of the newly inserted tuple through the meta page (lines 4). Before updating, it judges whether the pair (t_k, a_k) is an outlier to the model and inserts it into b-tree if it's an outlier (lines 5–6).

It should be noted that we need to rescan the blocks' data in order to recalculate the error boundary if we retrain the model. It brings high IO cost. In order to reduce the frequency of model retraining, the algorithm calculates the predicted error of new tuples using the method *Cal_error()* by previous model, and the model remains unchanged unless the error boundary has changed (lines 7–8). Another way to avoid model retraining is to zoom in on the boundary of the model. That is, each time when we need to recalculate the boundary, we expand the boundary with several times the standard deviation of errors (line 9). It reduces the chance of models retraining but also enlarges error boundaries, which increases the cost of query. It should note that it is a trade-off to balance the cost of index updating and query execution.

5.3　Query Execution

The hybrid indexes scan process is shown in Algorithm 4. Like insertion, the algorithm first finds the target model *m* where the query key falls. Then the model will tell an interval (line 2) determined by the predicted position $\hat{p}$ and the error boundary of the model (e.g., min error and max error). The algorithm will search the interval along with b-tree associated with LR models until the exact record t_k is found (lines 3–5). If the scan fails, *NULL* is returned to indicate there is no record in the database.

Cost Estimation. The database adds both table scan and index scan into scan paths for query execution. The optimizer will then estimate the cost of each scan path and choose the cheapest. Our index scan strategy is first to get the prediction address using LR models and then get the exact address by doing a binary search near the prediction address in the raw table. If the above process fails, the query key may be an outlier indexed by b-tree. Then the algorithm retrieves the key in b-tree. The total cost contains the LR models and traditional b-tree. For the latter, we use the estimation model in Postgres. And for LR models, it only retrieves the meta page, the models' page, and raw table pages. Therefore, the total cost is the cost of LR modes plus $p_\tau \times$ the cost of b-tree, where p_τ is the probability that the key is an outlier with normal distribution.

Algorithm 3 Index Insertion

Input: inserted tuple t_k with its address α_k

1. **IF** t_k in a *fresh* block
2. └build new model m
3. **ELSE**
4. $m = $ binary_search(α_k, *metapage*);
5.   **IF** α_k **NOT IN** $[\bar{\alpha} - \gamma\sigma_\alpha, \bar{\alpha} + \gamma\sigma_\alpha]$
6.   └ *btree*.insert($t_k.key, \alpha_k$);
7.   **ELSE IF** m.Cal_error($t_k.key$) not in $m.boundary$
8. m.retrainModel($t_k.key, \alpha_k$)
9.   └ └$m.boundary = [lower\text{-}\sigma_\varepsilon, upper +\sigma_\varepsilon]$
10. WriteModel(m);

Algorithm 4 Query Execution

Input: query key *key*, the raw tables T
1. $m = $ binary_search(*key*, *metapage*);
2. $[\hat{p}, lower, upper] = m$.predict(*key*)
3. $t_k = $ do binary search in $T[\hat{p} + lower, \hat{p}+upper]$;
4. **IF** t_k is **NULL**
5. └*btree*.scan(*key*);
6. return t_k

---

6 Experiments

Experimental Setup. Our experiments ran on Postgres system with a computer of Lenovo of an Intel(R) Core (TM) i5-4460 M CPU at 3.20 Hz and 16 GB 1600 MHz DDR3 of RAM.

Datasets. We used two different data sets in our experiments. The first dataset, IMDB, has a random data distribution. Besides, we use the TPC-H DBGEN tool to generate the simulated table "orders", which has a good distribution. There are about three million rows in the IMDB dataset and fifteen million rows in the simulated table.

Comparison Methods

- B-tree: the implementation of the b-tree index in Postgres.
- Hash: the implementation of the hash index in Postgres.
- NN: the neural network used in Kraska' work to train the learned index.
- LR: the learned index without eliminating the outliers mentioned in Sect. 3.
- KE/CSE/EE: the learned indexes by removing the outliers estimated by KE/CSE/EE mentioned in Sect. 4.

6.1 Prediction Errors

First, we compared the predicted errors of models built by different methods. We compared the five aforesaid methods of training learned index. There are three layers in the neural network, each layer has eight nodes, which achieves the best performance. We set the total training epochs as 100, and most models are leverage in the early 20 epochs. The results are shown in Fig. 2. The x-axis represents the first 10 models while the y-axis represents the error boundaries compared to the real address in different methods. The positive values mean the predicted address is after the real address, while the negative values mean the predicted address is before the real address. The results show that in most case, LR can achieve performance as well as NN. The predicted errors of NN in Fig. 2(b) can be further reduced if we increase the training epochs. The error boundaries of NN in Fig. 2(b) are the largest in most case, which shows poor efficiency of NN in the real situation. Besides, we can see that KE, CSE and EE have lower error intervals since they remove the outliers. We can also observe that CSE and EE perform better than KE. And in most case, EE performs best.

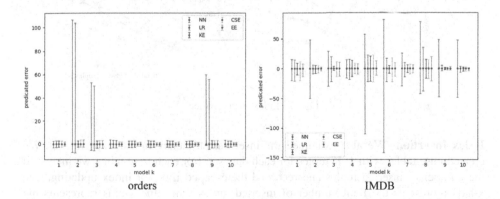

orders   IMDB

Fig. 2. Models precision

6.2   Efficiency

Index Construction. The following experiments are to examine the efficiency of our hybrid indexes compared to the b-tree index and hash index in Postgres. We mainly implement the hybrid indexes by eliminating the outliers using EE. We performed these experiments on two data sets. First, by increasing the size of raw tables, we recorded the time of building different indexes on those tables. The results are shown in Fig. 3. The x-axis represents the total number of tuples, while the y-axis represents the elapsed time during the indexes are built. It's clear that hybrid indexes have a better performance than the b-tree and hash indexes. In almost all configurations, hybrid indexes can be up to 2× faster. LR models only save the weights and boundaries for each model while b-tree holds information for each tuple. So LR models gain better performance on index construction due to lower IO. It should also be noted that b-tree not only saves information in the leaf nodes but also creates inner nodes. The IO cost of the b-tree index is expensive, which is more severe for the hash index.

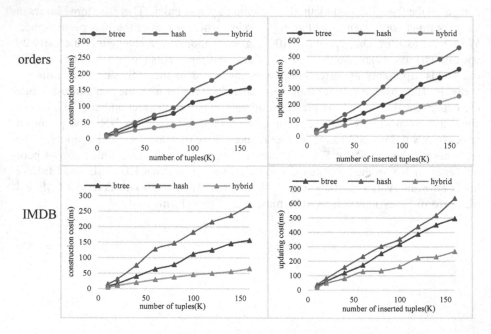

Fig. 3. Index maintenance

Index Insertion. We also examined the insertion efficiency in our experiments. We created base tables with 10K tuples for each data set. For each base table, we increased the number of inserted tuples and recorded the escaped time for index updating. The x-axis represents the total number of inserted tuples while the y-axis represents the updating cost. It shows that hybrid indexes also have a better performance when inserting with 1.5×–2× faster than b-tree and hash indexes. The main cost of LR model insertion is the high cost of retraining, but we have already thrown bad data to b-tree, which reduces the possibility of retraining (the bad data always lead to retraining the model). Besides, we zoom in on the boundaries of models, which also reduces the possibility of orders retraining. So in most case, there would be no retraining happens. Those make hybrid indexes faster to update compared to b-tree with a high cost of splitting and hash index which suffers from hash conflict.

Query Execution. We constructed tables ranging from 10 tuples to 1M tuples and did point queries on these tables. The results are shown in Fig. 4. The x-axis represents the total number of tuples with different table sizes while the y-axis represents the elapsed time for index scanning (without other overhead in query execution and without buffer). The result shows hybrid indexes almost have the same performance compared to b-tree and hash. Hybrid indexes gain performance improvement by exploring the LR models, which only have $O(1)$ lookup complexity. EE also helps to reduce the search time in error interval, and the size of b-tree is small compared with using b-tree only. Those make hybrid indexes have a better performance. Besides, we have also compared the query latency with buffer, and the results of the comparison are not significant.

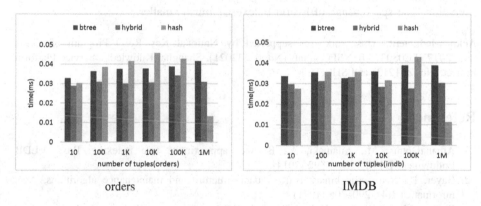

orders   IMDB

Fig. 4. Query latency

6.3   Space Usage

In the last experiment, we compared the space usage of hybrid indexes, b-tree and hash. The results are shown in Table 1. Overall, the size of hybrid indexes (the sum of LR models and b-tree) is much smaller than pure b-tree and hash indexes. It gains about $10\times$ space reduction in hybrid indexes. There are more outliers in the real word (e.g., IMDB dataset). So it shows different growth trends in the two datasets.

Table 1. Space usage (KB)

	Amount	10K	20K	40K	60K	80K	100K	120K	140K	160K
Orders	Hybrid	48	64	96	144	192	232	280	320	376
	B-tree	240	456	896	1328	1768	2208	2656	3096	3536
	Hash	528	1040	2064	2064	4664	4376	5216	4112	8208
IMDB	Hybrid	32	56	72	80	144	176	216	224	184
	B-tree	240	456	896	1328	1768	2208	2656	3096	3536
	Hash	528	1176	2352	2624	4664	4688	5248	5208	9288

7   Conclusion and Future Work

Traditional indexes suffer from high space usage, while learning approach does not handle outliers well and is hard to maintain. In this paper, we propose a hybrid indexing method by dividing the data into two parts and indexing them respectively. The hybrid indexes first partition dataset and distinguish the outliers from dataset to reduce the complexity of data, and then use linear regression and b-tree to index them. We have shown that the hybrid indexes perform well on index construction, insertion and query execution. The space usage of hybrid indexes is much smaller.

Acknowledgment. This work was supported by National Key R&D Program of China (No. 2017YFC0803700), NSFC grants (No. 61532021), Shanghai Knowledge Service Platform Project (No. ZF1213) and SHEITC.

References

1. Athanassoulis, M., Ailamaki, A.: BF-tree: approximate tree indexing. Proc. VLDB Endowment **7**(14), 1881–1892 (2014)
2. Bayer, R.: Symmetric binary B-trees: data structure and maintenance algorithms. Acta informatica **1**(4), 290–306 (1971)
3. Boehm, M., Schlegel, B., Volk, P.B., et al.: Efficient in-memory indexing with generalized prefix trees. Datenbanksysteme für Business, Technologie und Web (BTW) (2011)
4. Boyar, J., Larsen, K.S.: Efficient rebalancing of chromatic search trees. J. Comput. Syst. Sci. **49**(3), 667–682 (1992)
5. Galakatos, A., Markovitch, M., Binnig, C., et al.: A-tree: a bounded approximate index structure. CoRR, abs/1801.10207 (2018)
6. Goldstein, J., Ramakrishnan, R., Shaft, U.: Compressing relations and indexes. In: Proceedings of the 14th International Conference on Data Engineering, pp. 370–379. IEEE (1998)
7. Graefe, G., Larson, P.A.: B-tree indexes and CPU caches. In: Proceedings of the 17th International Conference on Data Engineering, pp. 349–358. IEEE (2001)
8. Graefe, G.: B-tree indexes, interpolation search, and skew. In: Proceedings of the 2nd International Workshop on Data Management on New Hardware, p. 5. ACM (2006)
9. Kang, D., Jung, D., Kang, J.U., et al.: μ-tree: an ordered index structure for NAND flash memory. In: Proceedings of the 7th ACM & IEEE International Conference on Embedded Software, pp. 144–153. ACM (2007)

10. Kim, C., Chhugani, J., Satish, N., et al.: FAST: fast architecture sensitive tree search on modern CPUs and GPUs. In: Proceedings of the 2010 ACM SIGMOD International Conference on Management of Data, pp. 339–350. ACM (2010)
11. Kissinger, T., Schlegel, B., Habich, D., et al.: KISS-tree: smart latch-free in-memory indexing on modern architectures. In: Proceedings of the Eighth International Workshop on Data Management on New Hardware, pp. 16–23. ACM (2012)
12. Kraska, T., Beutel, A., Chi, E.H., et al.: The case for learned index structures. In: Proceedings of the 2018 International Conference on Management of Data, pp. 489–504. ACM (2018)
13. Lehman, T.J., Carey, M.J.: A study of index structures for main memory database management systems. University of Wisconsin-Madison Department of Computer Sciences (1986)
14. Leis, V., Kemper, A., Neumann, T.: The adaptive radix tree: ARTful indexing for main-memory databases. In: 2013 IEEE 29th International Conference on Data Engineering (ICDE), pp. 38–49. IEEE (2013)
15. Li, Y., He, B., Yang, R.J., et al.: Tree indexing on solid state drives. Proc. VLDB Endowment 3(1–2), 1195–1206 (2010)
16. Lu, H., Ng, Y.Y., Tian, Z.: T-tree or b-tree: main memory database index structure revisited. In: Proceedings 11th Australasian Database Conference, ADC 2000 (Cat. No. PR00528), pp. 65–73. IEEE (2000)
17. Postgres database. http://www.postgresql.org/. Accessed 8 Apr 2019
18. Rao, J., Ross, K.A.: Making B+-trees cache conscious in main memory. ACM Sigmod Rec. 29(2), 475–486 (2000)
19. Rao, J., Ross, K.A.: Cache conscious indexing for decision-support in main memory. In: International Conference on Very Large Data Bases, pp. 78–89. Morgan Kaufmann Publishers Inc. (1999)
20. The datasets of IMDB. https://datasets.imdbws.com/. Accessed 8 Apr 2019
21. The hybrid indexes implementation. https://github.com/blankde/Learning-Postgres
22. The TPC-H Benchmark, http://www.tpc.org/tpch/. Accessed 8 Apr 2019
23. Yu, J., Sarwat, M.: Two birds, one stone: a fast, yet lightweight, indexing scheme for modern database systems. Proc. VLDB Endowment 10(4), 385–396 (2016)
24. Zhang, H., Andersen, D.G., Pavlo, A., et al.: Reducing the storage overhead of main-memory OLTP databases with hybrid indexes. In: Proceedings of the 2016 International Conference on Management of Data, pp. 1567–1581. ACM (2016)
25. Li, Y., Wen, Y., Yuan, X.: Online aggregation: a review. In: Meng, X., Li, R., Wang, K., Niu, B., Wang, X., Zhao, G. (eds.) WISA 2018. LNCS, vol. 11242, pp. 103–114. Springer, Cham (2018). https://doi.org/10.1007/978-3-030-02934-0_10

Crowdsourced Indoor Localization for Diverse Devices with RSSI Sequences

Jing Sun[✉], Xiaochun Yang, and Bin Wang

School of Computer Science and Engineering, Northeastern University,
Shenyang 110819, Liaoning, China
sunjing@stumail.neu.edu.cn, {yangxc,binwang}@mail.neu.edu.cn

Abstract. Indoor localization is of great importance for a range of pervasive applications, attracting many research efforts in the past decades. Received Signal Strength Indication (RSSI) has received much attention due to its simplicity and compatibility with existing hardware, which has been widely used for indoor localization. However, traditional fingerprint-based localization solutions require a process of site survey, in which radio signatures are collected and stored for further comparison and matching. It's a labor-intensive work to acquire the fingerprint database which costs much time and human resources. Meanwhile, users' heterogeneous devices may receive different RSSIs even at the same location. To alleviate these problems, we present an efficient localization method with crowdsourced users' RSSI sequences. We first cluster multiple users' RSSIs to construct a logical floor plan graph, and construct a physical floor plan graph from the real floor plan. Then we map the logical floor plan with physical floor plan with a path-based method to construct the radiomap. To solve the device diversity, we propose a novel RSSI distance metric. When localizing an query user, we propose a bit encoder method to prune the RSSIs that cannot be the result. We demonstrate the efficiency and effectiveness of the proposed solution through extensive experiments with two real data sets.

1 Introduction

In daily life, people often spend a considerable portion of their time in indoor space, such as office building, shopping mall, museum, airport, subway and other transportation facilities. With the prevalence of indoor location-based services, an increasingly important requirement for many novel applications need indoor localization.

There are two prevalent indoor localization technologies which are RFID localization [4,9] and the RSSI-based localization [1,11,13,16]. However, the RFID localization has to rely on lots of RFID sensors, which is costly buying and maintaining special hardware. Received Signal Strength Indication (RSSI) has received much attention due to its simplicity and compatibility with existing hardware, which has been widely used for indoor localization. However, traditional fingerprint-based localization solutions have three main shortcomings.

© Springer Nature Switzerland AG 2019
W. Ni et al. (Eds.): WISA 2019, LNCS 11817, pp. 614–625, 2019.
https://doi.org/10.1007/978-3-030-30952-7_62

The first shortcoming is the radiomap construction. It requires a process of site survey, in which radio signatures are collected and stored for further comparison and matching. Manually constructing an indoor map through site surveys is both expensive and time-consuming. For instance, the radiomap generation survey upon the installation of the *Ekahau* commercial localization system can cost $10,000 for a large office building with no maintenance included, as reported in [6]. The second shortcoming is the device diversity. As users typically carry diverse mobile devices, the RSSI values may be reported quite differently, even if they at the same location. This is because that different devices may equip with different WLAN cards. Moreover, two different devices may not detect the same number of APs at a specific location, due to variable receiver sensitivity. The problem of signal multi-path, fading and reflection make it worse. So the radiomap needs to be modified periodically with the environmental change, which significantly increases the maintenance cost.

To overcome these shortcomings, we present an efficient localization method with crowdsourced users' RSSI sequences. Crowdsourcing is a low-cost and efficient way to collect the RSSI of indoor space without expert surveyors and designated fingerprint collection points. Crowdsourced users carry with mobile devices, such as smartphones and tablets, can collect sequences of RSSI. By utilizing the RSSI data contributed by users, we can construct a logical floor plan graph. With the real indoor floor plan, we can construct a physical floor plan graph from the real floor plan. We map the logical floor plan with physical floor plan to construct the radiomap. To solve the device diversity, we propose a novel RSSI distance metric. When localizing an user, we not only match RSSI of single location, but also match the RSSIs sequence. Besides the challenge of no expert surveyors and device diversity proposed above, it also has the following challenges. The first challenge is how to map the logical floor plan with the physical floor plan. We cluster the RSSI sequences and construct a logical floor plan graph which the RSSIs in the same cluster is a vertex and the adjacent clusters have an edge. We partition the physical floor plan and construct a physical floor plan graph that each partition is a vertex and the adjacent partitions have an edge. We propose a graph mapping method to map the logical floor plan with physical floor plan. The second challenge is the multiple measurements for one location and the measurements are from heterogeneous devices. We propose a novel RSSI distance metric to measure the relationship of users' RSSI sequences. We use the following example to introduce our indoor localization method. The third challenge is the performance of localization. We propose a bit encoder method to prune the RSSIs that cannot be the result.

Example 1. As shown in Fig. 1, there is a set of RSSI sequences which is collected by k crowdsourced users walk in the indoor space. The k crowdsourced users may take diverse mobile devices and the RSSI measurements may be recorded at different time. Different users may take different devices. Each sequence which consists of a set of RSSIs is collected by one user. Each RSSI is a vector of signal strength values received from m APs at one position, and the first RSSI and last RSSI of each RSSI sequence have room-level position labels. Assuming

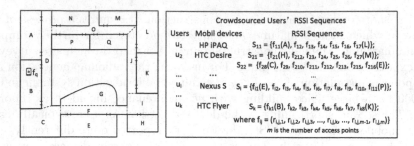

Fig. 1. Example of crowdsourced users' RSSI sequences.

that we know the physical floor plan, and the RSSI sequences almost cover the whole indoor space. In the localization, a query user u_q walks in the building, we can get a short sequence of RSSI $s_q = \{f_{q1}, f_{q2}\}$. The goal is to determine the RSSI-level position of u_q.

The main contributions of this paper are summarized as follows:

- We propose a RSSI-based indoor localization method without the expert surveyors and designated fingerprint collection points.
- We propose a novel RSSI distance metric to solve the problem of device diversity.
- We propose a path-based graph mapping method to map the logical floor plan with physical floor plan.
- we propose a bit encoder method to prune the RSSIs when locating.

The reminder of this paper is organized as follows: Sect. 2 proposes the preliminaries and problem statement. Section 3 proposes the radiomap construction. Section 4 introduces our localization method with RSSI Sequence. Finally, we report the experimental study and previous work.

2   Preliminaries and Problem Statement

In this section, we firstly introduce the preliminaries for well understanding indoor RSSI, and then we define the crowdedsourced indoor localization with RSSIs. We utilize the RSSIs to represent the indoor location, where an RSSI is a vector of received signal strength from different APs.

Definition 1 (Indoor Location RSSI). *Given an indoor space with m APs(access points), the indoor RSSI of a location is the Received Signal Strength (RSS) values from these m APs at that location, we denote the indoor location RSSI of location p_i as $f_{p_i} = (r_{i1}, r_{i2}, ..., r_{ij}, ..., r_{im})$, where the r_{ij} is the received signal strength value from the j-th AP.*

The RSSI-based indoor localization is one of the most popular indoor localization methods. The Wi-Fi fingerprint localization consists of two main phrases:

off-line training phrase and online positioning phrase. In the offline training phrase, some location points are determined, and the Wi-Fi signal strength of different locations are surveyed. The crowdsourced users take diverse mobile devices and the RSSI measurements may be recorded at different time. Each location point can receive m signal strength values from these m APs. This can get a radio frequency map. In the online positioning phrase, given a RSSI, the location is estimated by matching the similarity to the RSSIs in the off-line training phase. We formalize crowdedsourced indoor RSSI localization problem.

Definition 2 (Crowdsourced Indoor Localization with RSSIs). *Given a set $S = \{s_1, s_2, ..., s_n\}$ of n RSSI sequences which are collected from k crowdsourced users, each RSSI sequence $s_i = \{f_{i1}(p_i), f_{i2}, f_{i3}, ..., f_{it}(p_j)\}$ is a sequence of RSSIs from a crowdsourced user. The first RSSI f_{i1} and last RSSI f_{it} of s_i are labeled with a room-level location p_i and p_j. During the localization, given the query RSSI sequence $s_q = \{f_{q1}, f_{q2}\}$ of user u_q with two RSSIs, we want to find the RSSI in S whose location is nearest to the location of u_q.*

3   Indoor Radiomap Construction

In this section, we will introduce how to construct the radiomap without the expert surveyors and designated fingerprint collection points. Crowdsourced users with mobile devices walk in the indoor. Their trajectories are RSSI sequences. We first cluster these RSSI sequences to construct a logical floor plan which can represent the topology of RSSI space. Then we map the logical floor plan with the physical floor plan to construct the radiomap.

3.1   RSSI Distance Metric

To solve the device diversity, we propose a novel RSSI distance metric. Although heterogeneous mobile devices report the RSSI values quite differently, several studies report a linear relation between the RSSI values measured by heterogeneous devices [5]. Given two users' RSSI sequences, $S_1 = \{f_{11}, f_{12}, f_{13}, ..., f_{1p}\}$ and $S_2 = \{f_{21}, f_{22}, f_{23}, ..., f_{2p}\}$, where $f_{1i} = (r_{1i,1}, r_{1i,2}, .., r_{1i,m})$ and $f_{2j} = (r_{2j,1}, r_{2j,2}, .., r_{2j,m})$. We want to compute the similarity of $f_{1,i}$ and $f_{2,j}$. Instead of using the single RSSI, we utilize the sequence information to compute the RSSI distance.

Firstly, given two RSSIs $f_i = (r_{i1}, r_{i2}, .., r_{im})$ and $f_j = (r_{j1}, r_{j2}, .., r_{jm})$, we construct the following signature vectors which can measure the signal difference within RSSI sequence.

$$diff f_{f_i} = (\widehat{r_{i12}}, \widehat{r_{i23}}, ..., r_{i\widehat{(m-1)m}}) \quad diff f_{f_j} = (\widehat{r_{j12}}, \widehat{r_{j23}}, ..., r_{j\widehat{(m-1)m}})$$

where $\widetilde{r_{ist}} = r_{is} - r_{it}$ and $\widetilde{r_{jst}} = r_{js} - r_{jt}$. we also construct the following signature vectors which can measure the signal difference between different RSSI sequences.

$$diff f_{f_i, f_j} = (\widetilde{r_{ij1}}, \widetilde{r_{ij2}}, ..., \widetilde{r_{ijm}})$$

where $\widetilde{r_{ijk}} = r_{ik} - r_{jk}$ is the signal difference between RSSIs in two different sequences, where the dis' is the L2 distance. The values of parameters α, β and γ are set by experience.

$$dis(f_{1,i}, f_{2,j}) = \alpha dis'(f_{1,i}, f_{2,j}) + \beta dis'(diff_{f_i}, diff_{f_j})$$
$$+ \gamma dis'(diff_{f_i, f_j}, diff_{f_{i+1}, f_{j+1}})$$

3.2 Logical Floor Plan Generation

Given a set of users' RSSI sequences, we propose a RSSI sequence-based clustering method. With the crowdsourced users' RSSI sequences, the RSSIs in the same sequences are connected together. To align the different RSSI sequences, we compute the RSSI distance between the RSSIs of different sequences. If the RSSI distance of one RSSI to RSSI in other sequence is less the RSSI distance to its adjacent RSSI in its own sequence, we connect this two RSSIs. After aligning all the RSSI sequences, in the same sequence, we compute the distance of adjacent RSSIs, in the different sequences, we compute the distance of aligned RSSIs. Then we connect the aligned RSSIs together. Based on the observation that the distance of RSSIs in same room and across a wall are quite different. Given a RSSI sequence, we compute the *average distance* of adjacent RSSIs. We compare the distance of adjacent RSSIs with the *average distance*, and delete the connections which are large then the *average distance*, until we get the clusters. After clustering the RSSI sequences, each cluster is represented as a node, the adjacent clusters have an edge. Then we can construct a logical floor plan graph as shown in Fig. 2.

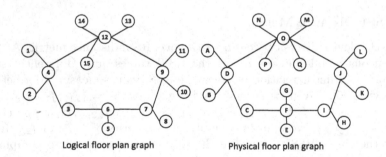

Logical floor plan graph   Physical floor plan graph

Fig. 2. Example of floor plan graph mapping.

3.3 Physical Floor Plan Generation

Given the physical floor plan, we can divide the indoor space into partitions with the room, the staircase, and walls. Each partition is a small piece of independent space, and two adjacent partitions are connected by a door. After partitioning the indoor space, each partition is represented as a node, and there is an edge between the adjacent partitions. Then we can construct a physical floor plan graph as shown in Fig. 2.

3.4 Path-Based Floor Plan Mapping

To match the logical floor plan and the physical floor plan, we propose a graph matching method which is based on the position label and the RSSI sequences. With the logical floor plan graph, each RSSI sequence can map to a connection path on the logical floor plan graph. Given two partitions, we can also find the physical paths between them on the physical floor plan graph. By comparing the connection paths on these two graphs, we can match the logical floor plan graph and the physical floor plan graph.

3.4.1 Connection Path and Physical Path Generation

As the crowdsourced user's RSSIs have been clustered into different clusters in Sect. 3.1, the logical floor plan graph is constructed with these clusters. Given an RSSI sequence, we can find the cluster that each RSSI belongs to, and we map this RSSI sequence to the logical floor plan graph. Then we can find the connection path of this RSSI sequence.

As the crowdsourced users may walk in the building with arbitrary paths, the connection path may have duplicate segment. So we remove the duplicate segments of connection path. For example in Fig. 2, the crowdsourced user's RSSI sequence is $S_{22} = \{f_{2,8}(C), f_{2,9}, , ..., f_{2,15}, f_{3,16}(E)\}$. The connection path is $CP_3 =< 3, 6, 7, 8, 6, 5 >$, and the duplicate segment is $< 6, 7, 8, 6 >$. After removing the duplicate segment, we can get the non-redundant connection path $CP_{22} =< 3, 6, 5 >$.

In $S_{22} = \{f_{2,8}(C), f_{2,9}, , ..., f_{2,15}, f_{3,16}(E)\}$, the first and last partition are C and E. We can find the path from A to D on the physical floor plan graph. The physical path without duplicate segments is $PP_{22,1} =< C, F, E >$ and $PP_{22,2} =< C, D, O, J, I, F, E >$.

3.4.2 Connection Path and Physical Path Matching

After finding the connection path and the physical paths, we can compare and match the connection path and the physical paths. Then we can match the RSSIs and the physical floor plan together. We use the path length and multiple paths to match the connection path and the physical paths. The *path length* is define as the number of partitions of the path. We first compare the length of connection path with the physical paths. Given $PP_{22,1}$ and $PP_{22,2}$, we can match $PP_{22,1}$ with the CP_{22}. The cluster 3, 6 and 5 can match the partition C, F and E. If they have the same path length, we compare these paths by other RSSI sequences' paths. For example in Fig. 2, an RSSI sequence is $S_i = \{f_{i,1}(E), f_{i,2}, ..., f_{i,10}, f_{i,11}(P)\}$. The RSSI connection path of a crowdsourced user is $CP_i =< 5, 6, 3, 4, 12, 15 >$. Given partition E and P, the physical path without duplicate segments is $PP_{i,1} =< E, F, C, D, O, P >$ and

---

Algorithm 1. Floor Plan Mapping

    Input: a set of RSSI sequences S, physical floor plan graph G_p

    Output: the set of mapping relationship MR

1 Initialize the set PM for path matching candidates;

2 Cluster the RSSI sequence s_i in S;

3 Construct the logical floor plan graph G_l;

4 **for** *each RSSI sequence s_i in S* **do**

5 | Find the labeled position p_i and p_j of s_i;

6 | Find all the non-redundant physical path PP_i from p_i to p_j on the G_p;

7 | Map each RSSI f_{ij} in s_i to the cluster c_i;

8 | Construct the non-redundant connection path CP_i;

9 | Put all the $< CP_i, PP_i >$ into PM;

10 Rank $< CP_i, PP_i >$ in PM with non-decreasing order of path length of CP_i;

11 **while** *the PM is not empty* **do**

12   **for** *each $< CP_i, PP_i >$ in PM* **do**

13   **if** *$PathLength(CP_i)==PathLength(PP_i)$ and there is only one PP_i match with CP_i* **then**

14   Match the RSSI f_i in CP_i with the partition p_k;

15   Put $< f_i, p_k >$ into MR;

16   **else**

17   Matching the $< CP_i, PP_i >$ with path length and the existing matching result $< f_i, p_k >$ in MR;

18 **return** MR;

$PP_{i,2} =< E, F, I, J, O, P >$. The $PP_{i,1}$ and $PP_{i,2}$ have the same path length with the CP_i. As the cluster 6 is match to partition F and 3 is match to partition C, we can determine that the CP_i is match to $PP_{i,1}$. So the cluster 5, 6, 3, 4, 12 and 15 can match the partition E, F, C, D, O and P, and the RSSIs in these clusters can match with the partitions of physical floor plan. Algorithm 1 describes the procedure of floor plan mapping.

With this path-based floor plan matching method, we can match the RSSIs to the physical floor plan. After floor plan matching, we label the RSSIs in the RSSI sequences with the partitions in the physical floor plan to construct the radiomap. Then we can get the result as shown in Fig. 3(b).

4 Localization with RSSI Sequence

As the query user walks in the building, we can get a short RSSI sequence of this user. When the query user wants to determine his location, we match the RSSI in the query user's RSSI sequence with the RSSIs in the radiomap.

4.1 Coarse Localization with Bit-Encoder Method

We utilize a bit-encoder method to encode the RSSIs with $0 - 1$ bits. As each RSSI $f_i = (r_{i1}, r_{i2}, ..., r_{im})$ is a vector of m dimensions where m is the number

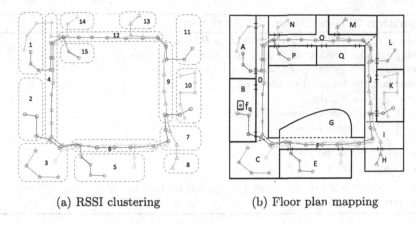

(a) RSSI clustering   (b) Floor plan mapping

Fig. 3. Example of floor plan mapping.

of APs. We find the top-3 entries with the strongest signal strength from the m entries. We label top-1 strongest signal strength entry with 11, the top-2 entry with 10, and the top-3 entry with 01. The other entries are labeled with 00. Then for each RSSI $f_i = (r_{i1}, r_{i2}, ..., r_{im})$, we construct a bit vector of $2m$ dimensions. For example, if the RSSI is $f_i = (-64.85, -51.45, -87.56, -99, -99)$, the bit label vector of f_i is $l_i = (1011010000)$.

Given query RSSI f_q, we encode the RSSI to the bit label vector l_q. We compare the bit label vector of f_q with that of RSSIs in the radiomap. If there are more than 6-bits differences between them, this RSSI can be pruned out. We put the RSSIs have less than 6-bits differences with f_q in the candidate set.

4.2 Fine Localization with RSSI Distance Metric

We match the query RSSI f_q with the RSSIs in the candidate set by the RSSI distance metric. The RSSI which has the minimum distance with the query RSSI is the location of the query user. The localization can return the $< partition, RSSI >$ pair as the result. Algorithm 2 describes the procedure of indoor localization with RSSIs.

5   Related Work

There are several approaches use RFID to locate the indoor location, such as the [2,9]. There are two kinds of RFID technologies, passive RFID and active RFID. RFID-based indoor localization has to equip with active transceivers or passive tags. RSSI-based indoor localization uses the pre-measured received signal strength of locations. RADAR [1] positioning system was considered as the first fingerprint-based system, which employs signal strength and signal-to-noise ratio with the triangulation location technique. The COMPASS system [11] takes advantages of WLAN infrastructures and digital compasses to provide low cost

---

Algorithm 2. Localization with RSSI Sequence

Input: A query RSSI f_q, the radiomap RM, bit label vector list BL
Output: The RSSI $< p_i, f_i >$ in RM that match with f_q
1 Initialize the candidate set $Cand$ for RSSI matching;
2 Construct the bit label vector l_q of f_q;
3 **for** *each l_i in BL* **do**
4 　| 　if *l_i match with l_q more than 6 bits* **then**
5 　| 　└ Put f_i into $Cand$;

6 **for** *each f_i in $Cand$* **do**
7 　└ Compute the $dis(f_i, f_q)$;

8 Find the f_i has the minimum $dis(f_i, f_q)$ and its partition p_i;
9 **return** $< p_i, f_i >$;

---

and relative high accurate positioning services to locate a user carrying a WLAN-enabled device. To reduce process of site survey, there are some localization methods which use the crowdsourcing by mapping the RSSI space with the floor plan [8,12,15]. [10] implement a WiFi fingerprinting and crowdsourcing-based indoor localization system called WicLoc which collect accelerometer and gyroscope readings and WiFi fingerprints in a crowdsourced manner. It needs to obtain the absolute coordinates of users. To solve the problem of diversity devices, some methods propose a new location signature to reduce the signal difference [3,7,14].

6 Experimentation and Evaluation

6.1 Experimental Setup

We use the following two real indoor RSSI data sets.

KIOS Dataset: The KIOS dataset contains RSSI data collected at KIOS Research Center. This is a $560\,\mathrm{m}^2$ typical office environment. There are 9 WiFi APs installed locally that provide full coverage throughout the floor. Moreover, there are 60 APs can be sensed in different parts of the floor and in some locations. Five different mobile devices were used for collecting RSSI data. It contains 2100 location-tagged fingerprints for each device collected at 105 reference locations corresponding to 20 fingerprints per reference location. We extract 300 location-tagged RSSIs totally by different devices.

Wi-Fi Received Signal Strength Measurements: We used the real indoor RSSIs that survey in our building. The area of the testbed is approximately $860\,\mathrm{m}^2$, and includes eight classrooms, four offices, and the main hallway. The area has 5 APs installed which have been deployed uniformly in the areas. There are 82 calibration points. We collect 400 location-tagged RSSIs with four different mobile devices(Thinkpad R400, Huawei Mate 10, Google Nexus and Samsung Galaxy S9). We extract 300 location-tagged RSSIs by different devices.

The data collection for the mobile devices was implemented using Android. All the algorithms were implemented using GNU C++. The experiments were run on a PC with an Intel 2.93 GHz Quad Core CPU i7 and 4 GB memory with a 500 GB disk, running a Windows 7 operating system.

6.2   Accuracy of Localization

We compute the average accuracy as the following function:

$$Average\ accuracy = \frac{number\ of\ positions\ with\ right\ results}{the\ total\ number\ of\ localization\ queries}$$

We compute the *Average accuracy* with 100 randomly generated localization queries. Figures 4(a) and 5(a) show the average accuracy when varying the number of labeled RSSIs with two real date sets. The more number label RSSIs, the higher average accuracy. The method *L2Dist* locates user's position with the L2 distance and finds the nearest neighbour as result.

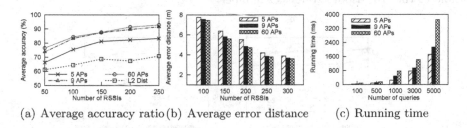

(a) Average accuracy ratio (b) Average error distance   (c) Running time

Fig. 4. Result on Dataset 1.

6.3   Average Error Distance

We also test the average error distance of localization. We compute the distance between the localization position with the ground truth position. Figure 5(b)

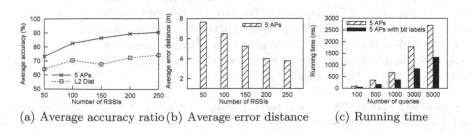

(a) Average accuracy ratio (b) Average error distance   (c) Running time

Fig. 5. Result on Dataset 2.

shows the average error distance when varying the number of RSSIs. We can find that the average error distance is less than 5 m when the number of labeled RSSIs is large than 250. The more number of labeled RSSIs, the less the average error distance is.

6.4   Performance of Localization Queries

To demonstrate the performance of Localization queries, we vary the number of queries. We compare the query time by the methods with and without bit label vectors. Then we randomly generate 100, 500, 1000, 5000 and 10000 queries. Figures 4(c) and 5(c) shows the total running time when varying the number of queries.

6.5   Effect of Diverse Devices

We also examine the effect of diverse devices. Figure 6 shows the average accuracy when varying the number of crowdsourcing devices. The localization accuracy is stable to the crowdsourced devices number.

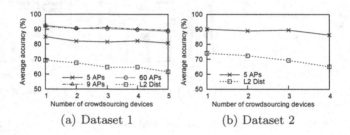

(a) Dataset 1   (b) Dataset 2

Fig. 6. Average accuracy varying crowdsourcing devices number.

7   Conclusion

In this paper, we propose a crowdsourced indoor localization method for diverse devices with users' RSSI sequences. To solve the device diversity, we propose a novel RSSI distance metric. We propose a path-based graph matching method to map the logical floor plan with physical floor plan and construct the radiomap. When localizing an query user, we propose a bit encoder method to prune the RSSIs that cannot be the result.

Acknowledgement. The work is partially supported by the National Key Research and Development Program of China (2018YFB1700404), National Natural Science Foundation of China (Nos. 61532021, 61572122, U1736104), and the Fundamental Research Funds for the Central Universities (No. N171602003).

References

1. Bahl, P., Padmanabhan, V.N.: RADAR: an in-building RF based user location and tracking system. Inst. Electr. Electron. Eng. **2**, 775–784 (2000)
2. Fernando, S., Christian, P., Jiménez, A.R., Wolfram, B.: Improving RFID-based indoor positioning accuracy using Gaussian processes. In: International Conference on Indoor Positioning and Indoor Navigation (IPIN), pp. 1–8. IEEE (2010)
3. Hossain, M., Jin, Y., Soh, W.-S., Van, H.N.: SSD: a robust RF location fingerprint addressing mobile devices' heterogeneity. IEEE Trans. Mob. Comput. **12**(1), 65–77 (2013)
4. Kiers, M., Krajnc, E., Dornhofer, M., Bischof, W.: Evaluation and improvements of an RFID based indoor navigation system for visually impaired and blind people. In: International Conference on Indoor Positioning and Indoor Navigation (2011)
5. Laoudias, C., Piche, R., Panayiotou, C.G.: Device self-calibration in location systems using signal strength histograms. J. Location Based Serv. **7**(3), 165–181 (2013)
6. Ledlie, J., Park, J.-g., Curtis, D., Cavalcante, A.: Mole: a scalable, user-generated WiFi positioning engine. J. Location Based Serv. **6**(2), 55–80 (2012)
7. Liu, X., Li, M., Xia, X., Li, J., Zong, C., Zhu, R.: Spatio-temporal features based sensitive relationship protection in social networks. In: Meng, X., Li, R., Wang, K., Niu, B., Wang, X., Zhao, G. (eds.) WISA 2018. LNCS, vol. 11242, pp. 330–343. Springer, Cham (2018). https://doi.org/10.1007/978-3-030-02934-0_31
8. Luo, C., Hong, H., Chan, M.C.: PiLoc: a self-calibrating participatory indoor localization system. In: IPSN, pp. 143–153. IEEE (2014)
9. Ni, L.M., Liu, Y., Lau, Y.C., Patil, A.P.: LANDMARC: indoor location sensing using active RFID. Wireless Netw. **10**(6), 701–710 (2003)
10. Niu, J., Wang, B., Cheng, L., Rodrigues, J.J.: WicLoc: an indoor localization system based on WiFi fingerprints and crowdsourcing. In: 2015 IEEE International Conference on Communications (ICC), pp. 3008–3013. IEEE (2015)
11. Thomas, K., Stephan, K., Thomas, H., Christian, L., Wolfgang, E.: COMPASS: a probabilistic indoor positioning system based on 802.11 and digital compasses. In: Proceedings of the 1st International Workshop on Wireless Network Testbeds, Experimental Evaluation & Characterization, pp. 34–40. ACM (2006)
12. Wu, C., Yang, Z., Liu, Y.: Smartphones based crowdsourcing for indoor localization. J. Location Based Serv. **14**(2), 444–457 (2015)
13. Xue, W., Qiu, W., Hua, X., Yu, K.: Improved Wi-Fi RSSI measurement for indoor localization. IEEE Sens. J. **17**(7), 2224–2230 (2017)
14. Yang, S., Dessai, P., Verma, M., Gerla, M.: FreeLoc: calibration-free crowdsourced indoor localization. In: INFOCOM, pp. 2481–2489. IEEE (2013)
15. Yang, Z., Wu, C., Liu, Y.: Locating in fingerprint space: wireless indoor localization with little human intervention. In: Proceedings of the 18th Annual International Conference on Mobile Computing and Networking, pp. 269–280. ACM (2012)
16. Yiu, S., Dashti, M., Claussen, H., Perez-Cruz, F.: Wireless RSSI fingerprinting localization. Sig. Process. **131**, 235–244 (2017)

Grading Programs Based on Hybrid Analysis

Zhikai Wang[1,2] and Lei Xu[1(✉)]

[1] Department of Computer Science and Technology, Nanjing University,
Nanjing 210023, China
`xlei@nju.edu.cn`
[2] Department of Computer Science and Technology, Sichuan University,
Chengdu 610211, China

Abstract. Grading programming assignments often take a lot of time and effort, so there is a growing need for automatic scoring systems. The existing program scoring system gives scores mainly by executing test cases. It cannot identify homework plagiarism or give grades according to steps. Therefore, this paper proposes an automatic scoring technique for Python programming assignments based on hybrid program analysis. Our scoring tool Paprika combines dynamic analysis and static analysis methods. Based on the Python abstract tree, it uses static analysis to realize the detection of clones and code style assessment and uses dynamic analysis to score with running test cases. It has a good tolerance for various formats of output of the code. The experiment shows that on the one hand, Paprika can reduce the burden of teachers and teaching assistants; on the other hand, it can accurately and objectively give grades and detailed feedback, which is helpful for students to improve learning efficiency and programming level.

Keywords: Automatic grading system · Dynamic analysis · Static analysis

1 Introduction

With the rapidly growing enrollment rates of computer science majors, large classes have gradually become normal, and teaching tasks have become much heavy too. To understand the students' learning state and measure their programming skills, teachers have to assign programming tasks and grade them. The advantage of manual evaluation is its subjective judgment and effective feedback. However, manual evaluation costs too much time and effort.

CACM Magazine published the first report on automatic grading systems [1] in 1960. Since then, researchers have conducted a series of researches in automatic grading [2–8]. They are based on dynamic analysis, static analysis or a combination of both.

In the field of dynamic analysis, there are some tools like CourseMarker[1], WebCat[2], JavaBrat[3], etc. They evaluate a program through test cases. If the matching result is True, then the application is considered correct. On this basis, some of them also take

---

[1] https://coursemarker.software.informer.com.

[2] https://web-cat.cs.vt.edu/Web-CAT/WebObjects/Web-CAT.woa.

[3] https://scholarworks.sjsu.edu/etd_projects/51/.

© Springer Nature Switzerland AG 2019
W. Ni et al. (Eds.): WISA 2019, LNCS 11817, pp. 626–637, 2019.
https://doi.org/10.1007/978-3-030-30952-7_63

factors such as time complexity and space complexity into consideration. The advantage of dynamic analysis is its objectivity and ease of use. The disadvantage is that it relies heavily on test cases. Inexperienced beginners sometimes make syntax mistakes which cause programs unable to be executed and result in low grades. Such circumstances are not conducive to the cultivation of students' programming ability, and teaching efficient cannot be improved either.

In the field of static analysis, there are tools like Findbugs, Checkstyle[4], Classifying Python Code Comments tool [16], etc. The static analysis evaluates programs through analyzing their structure, lexical grammar, semantics. Then, based on the result of the initial evaluation, it will check the programs' specification, plagiarism, semantical error, algorithm similarity, etc. Static analysis still works well even if the programs cannot be compiled in dynamic analysis, and it can provide more specific and diverse feedback than dynamic analysis. However, static analysis cannot analyze the correctness of programs and sometimes generates invalid feedback.

The combination of static and dynamic analysis is a popular solution in the field of automatic grading nowadays, because it combines the advantages of both approaches of analysis and circumvents their shortcomings. Therefore, this paper proposes an automated grading approach based on static analysis and dynamic analysis and designs a primary tool named Paprika. Section 2 lists some motivating examples. Section 3 introduces the program structure, scoring rules, and algorithms. Section 4 discusses our experiment and evaluation of Paprika, Sect. 5 is about related work, and Sect. 6 is about the summary.

2 Motivating Example

The manual grading of programming is usually time-consuming and inaccurate, and dynamic analysis cannot be used to verify code specification and code plagiarism. The sample program includes one question and two corresponding student codes.

Q: Given an array of integers, return indices of the two numbers such that they add up to a specific target.

```
1: nums = eval(input('Please input a set of num-
      bers:'))
2: target = eval(input('Please input your sum
      target:'))
3: for i in range(len(nums)):
4:    for j in range(i + 1, len(nums)):
5:       if nums[i] + nums[j] == target:
6:          print([i, j])
7:          break
```

```
1: def twoSum(nums, target) :
2:    for i in range(len(nums)):
3:       for j in range(i+1, len(nums)):
4:          if nums[i]+nums[j]==target:
5:             ans = [i, j]
6:             return ans
7: nums = eval(input())
8: target = eval(input())
9: print(twoSum(nums, target))
```

Fig. 1. Student assignments

---

[4] http://checkstyle.sourceforge.net.

When using the manual grading method, assuming that the TA spends an average of 20 s grading a programming assignment, there will be three questions to grade and three test cases to prepare for each question. There will be a total of 30 students' assignments, and the TA will spend at least 1 h and 30 min. This does not consider things like giving feedback on defects in each job and checking for plagiarism, so it will undoubtedly take longer than expected. Also, the manual audit process will inevitably produce fatigue, which will affect the objectivity of the scoring results. Therefore, we believe that the development of the automatic scoring system can save a lot of time for teaching staff and improve the quality of teaching.

When carrying out automatic grading based on dynamic analysis, all programming assignments submitted by the above two students can pass the test cases. However, there are problems in the two tasks that cannot be identified by dynamic analysis: assignment 1 has a high structural similarity with other student assignments, which may be suspected of plagiarism; assignment 2 has two code irregularities: the first line adds Spaces before the colon, and the fourth and fifth lines do not have Spaces around the binary operator. Plagiarism is not conducive to the cultivation of students' programming skills and violates academic ethics. Non-standard coding will reduce the readability of the program, hinder the maintenance of the program and drag down the development efficiency of the whole team. Therefore, we believe that the above issues should be considered.

3   Design

This Section mainly describes the overall structure (Sect. 3.1) and the grading functions (Sect. 3.2) of our method. Detailed feedback is generated automatically based on the result of grading functions (Sect. 3.3).

3.1   Architecture

Figure 2 shows the overall architecture of the automatic grading system:

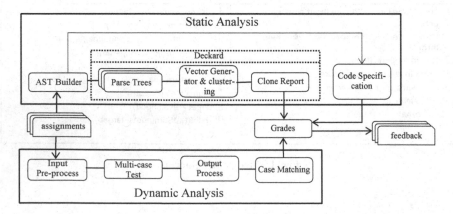

Fig. 2. System architecture

In terms of static analysis, first, the programming job is transformed into a Python abstract syntax tree; Then, Deckard[5] generates cloning report according to the AST; Finally, it conducts the code specification.

In terms of dynamic analysis, it preprocesses the input data first, then obtains the output results through test cases, and then processes the output results, and finally, it evaluates the results.

Finally, the output results of static analysis and dynamic analysis are combined to generate feedback information and job scores.

The instructor can assign the grading weight in this method. The main attributes that participated in the scoring included the accuracy rate of the program, whether there was plagiarism, and the code specification.

Correctness. The accuracy rate is determined by the number of tests the program passes and the proportion of correct results in each test to the sample results.

Plagiarism. We use the clone code detection tool to check. The selection of the clone code tool is not fixed. We choose Deckard - the AST-based tool.

Code Specification. The degree of coding specification is evaluated by using regular expressions and by analyzing the corresponding AST nodes.

3.2   Functions

Our scoring method includes two functions: dynamic scoring and static scoring.

Dynamic Scoring. The dynamic scoring uses test cases to assess accuracy. However, in daily homework, problems not specifying the input/output format and students not complying with the format requirements may lead to weak input (Definition 3.1) or weak output (Definition 3.2).

Definition 3.1 (weak input). A program submitted is said to have weak input if the input sequence or data type is different from the input required by the test case.

Definition 3.2 (weak output). A program submitted by a student is said to have a weak output if the output, output order, output number or output format of the program is different from the output required by the test case.

Considering the example in Fig. 1, a set of sample output is preset in this example, but there are three different sets of actual output in student homework, as shown in Fig. 3:

Sample output	Problematic outputs
1: [1, 3]	1: (1, 3)
	2: the result is [1, 3]
	3: 1 3

Fig. 3. Sample and problematic outputs

---

[5] https://github.com/skyhover/Deckard.

In traditional dynamic analysis, such as Online Judge, although the three output results in Fig. 3 are correct, their weak output will lead to the three student programs being considered wrong, resulting in low scores. Therefore, this paper proposes four states to process the output text based on regular matching to solve this problem.

Given the Boolean variables x and y. x = false indicating that the type of output variable is not specified explicitly by the question. So, all numeric variables will be typecasted to integers. y = false, indicating that the question does not require the order of the output variables, and then the output will be sorted.

According to the different values of x and y, the program will score in the following four different states:

State 1.

$$
\frac{\&[x]=true \ \&[y]=true}{<if(x\&y)s1 \ else \ if(x)s2 \ else \ if(y) \ s3 \ else \ s4;s,\&> \to <s1;s,\&>} \tag{1}
$$

State 2.

$$
\frac{\&[x]=true \ \&[y]=false}{<if(x\&y)s1 \ else \ if(x)s2 \ else \ if(y) \ s3 \ else \ s4;s,\&> \to <s2;s,\&>} \tag{2}
$$

State 3.

$$
\frac{\&[x]=false \ \&[y]=true}{<if(x\&y)s1 \ else \ if(x)s2 \ else \ if(y) \ s3 \ else \ s4;s,\&> \to <s3;s,\&>} \tag{3}
$$

State 4

$$
\frac{\&[x]=false \ \&[y]=false}{<if(x\&y)s1 \ else \ if(x)s2 \ else \ if(y) \ s3 \ else \ s4;s,\&> \to <s4;s,\&>} \tag{4}
$$

Static Scoring. The static scoring part includes two functions: clone detection and code specification checking.

Clone Detection. This project uses Deckard [9] to detect clone code fragments. Deckard is a clone detection tool based on AST and written in C++ language. At present, Deckard has supported C/C++, Java, Php5, Solidity. This project expands Deckard to enable the cloning detection of Python code.

The main structure of Deckard includes three parts: generating parse tree, generating vector and vector clustering, and it uses Yacc/Lex to create abstract syntax tree corresponding to different programming languages.

The work of extending the Python language focuses on how to generate a parse tree of Python code. There are two alternatives to generate a Python parse tree that meets the requirements of Deckard. One is to use the Yacc/Lex tool for syntax analysis, and the other is to use the ast module. Since Yacc uses BNF for context-free grammar [10], and Python's INDENT and DEDENT mechanisms determine that Python is context-related-grammar [11–14] language, we decided to use scheme 2 to generate the abstract syntax tree.

To convert a Python program to a parse tree, first, we read the source file and call the ast.parse function to generate the parse tree. Then pre-order traverses the parse tree and output to the temporary file. Finally, the prepared C code is used to read the intermediate file and convert it into a parse tree meeting the requirements of Deckard.

Code specification. We have selected ten indicators according to the Google Python Style Guide[6] and listed them in Table 1.

Table 1. Code specification

Type	Description
Semicolons	Do not terminate your lines with semicolons, and do not use semicolons to put two statements on the same line
Line length	Maximum line length is 80 characters
Parentheses	Do not use parentheses in return or conditional statements unless for implied line continuation or to indicate a tuple
Indentation	Indent your code blocks with 4 spaces. Never use tabs or mix tabs and spaces
Blank lines	Two blank lines between top-level definitions. One blank line between method definitions and between the class line and the first method. No blank line following a def line
Space 1	No white space inside parentheses, brackets or braces No white space before a comma, semicolon, or colon
Space 2	Surround binary operators with a single space on either side
Imports	Imports should be on separate lines
Statements	Generally, only one statement per line
backslash	Do not use backslash line continuation except for *with* statements

In the implementation, Line Length only needs to calculate the length of each line in programs; Semicolons, Indentation, White space 1&2, backslash, and Imports formatting require to match strings with regular expressions. Parentheses, Blank line, statements need to find the position of suspect points and then using these positions to search for node information in abstract trees. Finally, use the information to determine whether there is a problem of non-standard code style.

3.3 Feedback

Feedback consists of five parts: default input, student output, detailed grades, clone report and code specification.

Default Input. The default input section displays the sample input that was set before the assignment was marked. The dynamic analysis uses default input to generate output.

---

[6] https://google.github.io/styleguide/pyguide.html.

Student Output. The student output section shows the output of the student program when the input variable is the default input. The assignment is considered correct if the output of the student matches the preset output after weak output processing.

Detailed Grades. Final grades are influenced by grading functions and instructors can assign scoring weights according to their requirements. For example, if the teaching assistant sets the proportion of correct rate of students' homework to 60%, the proportion of plagiarism to 20%, and the score of code style to 20%, then this student's score is: $1.0 * 0.6 + 1 * 0.2 + 0.7 * 0.2 = 0.94$.

Clone Report. When there is no plagiarism, the plagiarism part of the student is empty. In the case of plagiarism, the student's plagiarism section shows the code clone report, as shown in the Fig. 4. The first column represents other assignments with high similarity to the student code, and the second column represents the starting and ending lines of similar code segments.

```
Clone report of 171850532_2.py:

171098064_2.py LINE:3:5
171840659_文件_3.8 practice2 .py LINE:3:5
171850532_2.py LINE:11:6
171870632_2.py LINE:1:5
```

Fig. 4. Clone report

Code Specification. The code specification section shows the code style problems of student assignments detected by static analysis.

4   Implementation and Evaluation

Paprika supports the case-by-case analysis of single file and simultaneous analysis of multiple files. Most functions and parameters are customized through configuration files. In this chapter, we discuss the practical implementation of Paprika, and compare it with the results of manual scoring and traditional dynamic scoring methods with three research questions:

RQ1: Can the dynamic scoring solve the weak output problems in programs?
RQ2: Can case-by-case scoring analyze code problems correctly and provide accurate feedback?
RQ3: Is Paprika more effective than manual marking and Online Judge marking?

The data set[7] is selected from three algorithm problems submitted by 27 students, and the three exercises are selected from LeetCode, respectively Two Sum[8], Move Zeroes[9] and Maximum Subarray[10].

Consider a sample program in Two Sum (Fig. 5):

Q: Given an array of integers, return indices of the two numbers such that they add up to a specific target.
Sample Input: [2,7,11,15]   22   Sample Output: [1, 3]
``` 1: numbers =eval(input("Please input a list:")) 2: target =eval(input("Please input the target:")) 3: for x in range(len(numbers)): 4:   for y in range(x + 1 , len (numbers)): 5:   if numbers[x] + numbers[y] == target: 6:   print((x, y)) ```

Fig. 5. Sample assignment

4.1 RQ1: Can the Dynamic Scoring Solve the Weak Output Problems in Programs?

To solve RQ1, in the dynamic scoring, considering that the output result specified in example 1 is two array indices, the type of output variables is not specified explicitly, and the sequence of output variables is not required, so the scoring state is state 4. The result of dynamic scoring shows that although the preset sample output is [1, 3] and the actual product is (1, 3) (weak output problem), the sample program is judged to be correct in our project.

```
Output of 171840012_1.py:

Please input a list:Please input the target:(1, 3)

Please input a list:Please input the target:(1, 6)

-------------------------------------------
score of 171840012_1.py:

score = 1.0 * 0.6 + 0 * 0.2 + 0.8 * 0.2 = 0.76

-------------------------------------------
Clone report of 171840012_1.py:

171840012_1.py LINE:1:5
171840013_1.py LINE:1:5
```

Fig. 6. Dynamic grading result

[7] https://github.com/YiBinRanMian/assignments.

[8] https://leetcode.com/problems/two-sum/.

[9] https://leetcode.com/problems/move-zeroes/.

[10] https://leetcode.com/problems/maximum-subarray/.

4.2   RQ2: Can Case-by-Case Scoring Analyze Code Problems Correctly and Provide Accurate Feedback?

To solve RQ2, in the feedback of the previous dynamic scoring, the grades and the actual running input and output results can be observed clearly. In the static scoring, for the clone detection, the feedback in the last four lines of the clone report in Fig. 6 shows that the sample program is similar to another student's program, and from Fig. 7 we can see that although the two code snippets differ in variable naming, the program structure of them are the same. Therefore, we can conclude that the feedback of clone detection is correct.

```
1: numbers =eval(input("Please input    a        1: nums = eval(input('Please input a set    of num-
list:"))                                          bers:'))
2: target =eval(input("Please input   the         2: target = eval(input('Please input your      sum
target:"))                                        target:'))
3: for x in range(len(numbers)):                  3: for i in range(len(nums)):
4: for y in range(x + 1 , len   (numbers)):       4: for j in range(i + 1, len(nums)):
5:   if numbers[x] + numbers[y] ==   target:      5:   if nums[i] + nums[j] == target:
```

Fig. 7. Clone code segments

As for the code specification, the feedback in this project (Fig. 8) suggests that there are two problems in the sample program, and in fact, the fourth line and the first two lines are consistent with the feedback.

```
---------------code specification---------------
No white space inside parentheses, brackets or braces.
Surround binary operators with a single space on either side.
```

Fig. 8. Feedback of code specification

4.3   RQ3: Is Paprika More Effective Than Manual Marking and Online Judge Marking?

To solve RQ3, we compared the results of Paprika, Online Judge and manual scoring. Figure 9 shows the line chart of the scores of three exercises under the Online Judge dynamic scoring and the Paprika scoring. It can be seen from the chart that OJ's score is lower than Paprika. After analysis the students' work, we find that the low marks from OJ are mainly distributed in weak input programs, weak output programs, and error programs, and as for Paprika, they are primarily distributed in weak input programs and error programs. Compared with the performance of the Online Judge, we can conclude that Paprika is inclusive to weak output problems. Besides, the static analysis part is not affected by weak input problems.

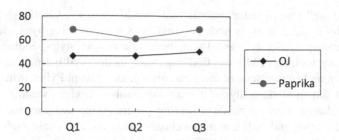

Fig. 9. Average scores of three questions

Figure 10 shows the line chart of the scores of three exercises respectively under the manual scoring and Paprika. Here, the scoring rules of Paprika need to be adjusted according to the manual scoring criteria. That is, assuming the final grades are 100, for each question, if the score above 80, then final grades minus zero, if the score above 60, then final grades minus five, and if the score below 60, then final grades minus ten.

From Fig. 10 and the feedback of the programs, we can conclude that on the on hand, the weak input problem that our project has not yet solved leads to the lower score of some questions (1, 4, 6, 7, 8, 13, 15, 16, 18, 26, 27). On the other hand, the project's ability to detect plagiarism and poor code style, which is difficult for manual scoring to identify, leads to lower scores.

Even with manual inspections, Paprika can save more than half of the manual scoring time. What is more, our project has the function of static analysis and generating feedback. These attributes make Paprika more effective than manual scoring.

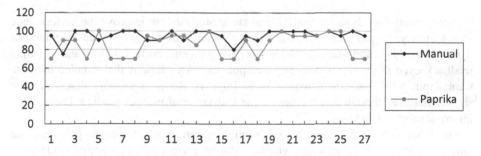

Fig. 10. Grades from manual grading and Paprika

5 Related Work

At present, there are a lot of researches conducted in the field of automatic scoring system, which can be divided into three categories: static analysis, dynamic analysis and hybrid analysis.

The most well-known static analysis tools are FindBugs[11], Checkstyle, Slithice[12], PMD[13], Pylint[14], etc. FindBugs finds possible bugs by comparing bytecode with a set of defect patterns, Checkstyle and Pylint both rate coding styles, Slithice is a Java program slicing tool, and PMD can find bugs such as declared but not used variables, empty pointer, and so on. None of them can give grades except Pylint. Although Pylint does a great job in static analysis, it must be combined with dynamic analysis, or otherwise, it cannot meet the needs of scoring programming assignments. In addition, clone code detection and code comments classification are also static analysis, and the well-known tools mainly include CCFinder [15], CCLearner [18], cp-Miner [19].

There are dynamic tools such as WebCat [3], JavaBrat [5], Petcha [8], Marmoset [2], etc. Most of them are based on test cases and cannot perform static analysis. JavaBrat does not support static analysis and does not give procedural scores. The uniqueness of Marmoset lies in its ability to save the first line of code saved by students to the database. Petcha is a test case based scoring system that does not include static analysis.

There are also some automatic grading systems combining dynamic analysis and static analysis, such as CourseMarker [6] and AutoLEP [7]. The predecessor of CourseMarker is Ceilidh [17], which was developed in the university of Nottingham in 1988, and improved Ceilidh in aspects of ease of use, maintainability, extensibility and feedback. However, CourseMarker does not support grading Python code. Nor can it deal with the input-output format irregularities that are common among beginners. AutoLEP is functionally similar to CourseMarker, but does not support code plagiarism detection.

6 Conclusions

The combination of dynamic analysis and static analysis can improve the performance and accuracy of program analysis. Paprika's main advantage is that it combines dynamic analysis and static analysis and can generate correct scores and accurate feedback even in programs with weak output. Paprika's flaw is that it failed to solve weak input problems, which require us to improve the next version. Practices have shown that even though it still relies on the assistance of manual grading, Paprika can improve scoring efficiency.

Also, we believe that based on the existing system, it is necessary to provide some other functions such as automatic guidance during programming, warning students of irregularities, and indicating the location of possible errors.

Acknowledgment. The work is supported by National Key R&D Program of China (2018YFB1003901), National Natural Science Foundation of China (61832009, 61728203).

---

[11] http://findbugs.sourceforge.net.

[12] https://github.com/juqian/Slithice.

[13] http://pmd.github.io/.

[14] https://www.pylint.org/.

References

1. Hollingsworth, J.K.: Automatic graders for programming classes. Commun. ACM **3**(10), 528–529 (1960)
2. Spacco, J., Hovemeyer, D., Pugh, W., Emad, F., Hollingsworth, J.K., Padua-Perez, N.: Experiences with marmoset: designing and using an advanced submission and testing system for programming courses. ACM SIGCSE Bull. **38**, 13–17 (2006)
3. Edwards, S.H., Perezquinones, M.A.: Web-CAT: automatically grading programming assignments. In: Technical Symposium on Computer Science Education, vol. 40, no. 3, p. 328 (2008)
4. Rodriguez-del-Pino, J.: A Virtual Programming Lab for Moodle with automatic assessment and anti-plagiarism features (2012)
5. Patil, A.: Automatic Grading of Programming Assignments. Master's Projects. 51 (2010). https://scholarworks.sjsu.edu/etd_projects/51
6. Higgins, C., Hegazy, T., Symeonidis, P., Tsintsifas, A.: The CourseMarker CBA system: improvements over Ceilidh. Educ. Inf. Technol. **8**(3), 287–304 (2003)
7. Wang, T., Su, X., Ma, P., et al.: Ability-training-oriented automated assessment in introductory programming course. Comput. Educ. **56**(1), 220–226 (2011)
8. Queirós, R.A.P., Leal, J.P.: PETCHA: a programming exercises teaching assistant. In: Proceedings of the 17th ACM Annual Conference on Innovation and Technology in Computer Science Education, pp. 192–197 (2012)
9. Jiang, L., Misherghi, G., Su, Z., Glondu, S.: DECKARD: scalable and accurate tree-based detection of code clones. In: ICSE, pp. 96–105 (2007)
10. Hopcroft, J.E., Ullman, J.D.: Chapter 4: Context-free grammars, pp. 77–106; Chapter 6: Properties of context-free languages. In: Introduction to Automata Theory, Languages, and Computation, pp. 125–137. Addison-Wesley (1979)
11. Linz, P.: An Introduction to Formal Languages and Automata, p. 291. Jones & Bartlett Publishers, Sudbury (2011). ISBN 978-1-4496-1552-9
12. Meduna, A.: Automata and Languages: Theory and Applications, p. 730. Springer, London (2000). ISBN 978-1-85233-074-3
13. Davis, M., Sigal, R., Weyuker, E.J.: Computability, Complexity, and Languages: Fundamentals of Theoretical Computer Science, 2nd ed., p. 189. Morgan Kaufmann (1994). ISBN 978-0-08-050246-5
14. Martin, J.C.: Introduction to Languages and the Theory of Computation, 4th edn, p. 277. McGraw-Hill, New York (2010). ISBN 9780073191461
15. Kamiya, T., Kusumoto, S., Inoue, K.: CCFinder: a multilinguistic token-based code clone detection system for large scale source code. TSE **28**(7), 654–670 (2002)
16. Zhang, J., Xu, L., Li, Y.: Classifying Python code comments based on supervised learning. In: Meng, X., Li, R., Wang, K., Niu, B., Wang, X., Zhao, G. (eds.) WISA 2018. LNCS, vol. 11242, pp. 39–47. Springer, Cham (2018). https://doi.org/10.1007/978-3-030-02934-0_4
17. Ihantola, P., Ahoniemi, T., Karavirta, V., Seppälä, O.: Review of recent systems for automatic assessment of programming assignments. In: Proceedings of the 10th Koli Calling International Conference on Computing Education Research, pp. 86–93. ACM (2010)
18. Li, L., Feng, H., Zhuang, W., Meng, N., Ryder, B.: CCLearner: a deep learning-based clone detection approach (2017). https://doi.org/10.1109/icsme.2017.46
19. Li, Z., Lu, S., Myagmar, S., Zhou, Y.: CP-Miner: a tool for finding copy-paste and related bugs in operating system code. In: OSDI, pp. 289–302 (2004)

Transferring Java Comments Based on Program Static Analysis

Binger Li[1], Feifei Li[1], Xinlin Huang[1], Xincheng He[1,2], and Lei Xu[1,2(✉)]

[1] Department of Computer Science and Technology, Nanjing University,
Nanjing, Jiangsu, China
{1124122506,1132702169,1515202411,1183413300}@qq.com, xlei@nju.edu.cn
[2] State Key Laboratory for Novel Software Technology, Nanjing University,
Nanjing, Jiangsu, China

Abstract. In the process of software development and maintenance, code comments can help developers reduce the time of reading source code, and thus improve their work efficiency. For large Java software projects, comments tend to appear in front of the main program entry or method definition. Readers need to search and read other source files containing the comments, which is very inconvenient. For this purpose, based on program static analysis technique, this paper has realized the process of automatically transferring comments in source code to their corresponding methods' callers. This method first processes the source code using the program static analysis tool, which identifies the method call and the variable Define-Use and other dependency relations and confirms the comment-transferring paths; then extracts the text information in the comments that are in front of method definitions. Finally, it completes the extraction and transferring of comments in Java projects. The average precision of the experiment with 14 open source Java projects is 82.76%.

Keywords: Comment transferring · Program static analysis · Software maintenance

1 Introduction

Existing research results show that software staff spend much time on searching, browsing and comprehending source codes of software projects, instead of editing them [1]. Software personnels read source codes to have enough understanding of them, in order to detect potential defects and repair them [2].

Currently there are two major approaches to generate source code comments: automatically generating comments with templates formulated by programming language grammar rules; automatically generating comments based on templates

The work is supported by National Key R&D Program of China (2018YFB1003901), National Natural Science Foundation of China (61832009, 61728203), 2019 Nanjing University College Student's Innovation Training Program (X201910284053).

© Springer Nature Switzerland AG 2019
W. Ni et al. (Eds.): WISA 2019, LNCS 11817, pp. 638–643, 2019.
https://doi.org/10.1007/978-3-030-30952-7_64

that are obtained by data mining and rule extraction [3–6]. However, these approaches tend to have the following drawbacks: based on a set of fixed templates, they are not able to analyze source code flexibly and comprehensively and will easily ignore the structural information of source code [7].

In this passage, we apply program static analysis techniques to analyze the call-relation of program entities, so as to obtain efficient comment-transferring paths. Then a certain approach is used to extract effective comment information of the source code. Finally, the picked comments are passed to the location where their related methods are called. Experiments show that this method possesses good accuracy.

2 Related Work

2.1 Automatic Generation of Program API Documentation

Javadoc, Doxygen, KDoc, LDoc and other tools have been applied in automatic generation of program API documentation.

For example, Javadoc passes the document comments with a specific set of rules to a specific program document. This kind of document comments coexist with the source code in the same files, which reduces the redundancy, reflects the association between API and source code [5] and helps the software personnel to understand software projects to a certain extent.

To improve the accuracy of the generated API documentation, Rahman et al. [8] proposed an approach using machine learning method RevHelper to predict the usefulness of code comments based on text characteristics and developer experience.

2.2 Program Static Analysis

The static analysis of the program can get the program's structural information without actually executing the program. And the process of program static analysis is fast, in contrast to program dynamic analysis. Sridhara et al. [9] performed program static analysis to preprocess the source code to facilitate the generation of annotations.

2.3 Comment Classification

Previous study in Python comments provides an idea of classifying comments according to their intention [14]. Using supervised learning method, the precision of classification can arrive at 75%.

3 Passing Comments Based on Static Analysis

3.1 Method Overview

In this passage, we present a method to transfer comments using program static analysis techniques. The skeletal implementation of the process of transferring comments is shown by Fig. 1.

Fig. 1. Process of transferring comments.

First, we analyze source code of all the files in the whole project to extract the dependencies of the various layers of source code (Sect. 3.2), including methods, classes and files so that we can get the paths of transferring comments (Sect. 3.3). Then we transfer the description text we extract from their defined location to their called location.

3.2 Analyzing Source Code

We use Understand, an existing code static analysis tool, to parse Java code into a UDB file. With code dependencies presented in UDB file, we can extract the call relationships between entities. Since code comments usually exist in three kinds of form: "//", "/*...*/" and "/**...*/", we can use Understand to extract the comments in the source code.

In general, there is a strong semantic affinity between a fragment of source code and its corresponding comment [10]. Studies show that abstract methods, class declarations, interface declarations, and package declarations are almost always commented before code snippets [11]. And comments that are associated with an entity's functionality in source code are also usually near the entity's declaration and are not so close to other entities [12]. So when we analyze the source code statically, we mainly extract the comments in front of the definition or declaration of the method.

3.3 Generating Call Relationships

Using the Perl interface provided by Understand, we extract all the definition information of the methods. Then we generate the call relations, class dependencies and file dependencies, and store them in a text file in the form of dependency.

3.4 Transferring Comments

Extracting Call Information. In order to quickly and effectively organize inter-entity call relations, we need to preprocess the text files. First, we filter the methods provided by the environment out of all the extracted methods and only reserve the user-defined methods. Second, we create two arrays to store all the definition information and call-relation information.

What code written by us does is to read the "called relationship" file generated by Understand, get the call relationship between x and y, and then link x to y in the "function call information array".

Transferring Comments. After the preparations above, we need to insert comments into where their corresponding methods are called. When inserting comments, we use the way of reading and writing concurrently. The transferring process of the comments in the source code is shown in Algorithm 1.

Algorithm 1. Transferring(Inserting) comments into source code file

---

Input: $MethodDefinition$Array, $MethodCall$Array
Output: $Processed.java$Source code file with comments inserted
 1: **for** $i = 0 \rightarrow MCIndex - 1$ **do**
 2:     $rfile \leftarrow$ FOPEN($MethodCall[i]- > CallFileName$)
 3:     $wfile \leftarrow$ FOPEN($TempDir$)
 4:     $pp \leftarrow MethodCall[i]- > CallList$
 5:     $cntLine \leftarrow 0$
 6:     $temp \leftarrow$ GETLINE(temp, rfile)   ▷ $temp$ is a $string$
 7:     **while** Read file successfully **and** Not in file end **do**
 8:         **if** $cntLine == pp- > callLine$ **and** $pp- > calledComment! = NULL$ **then**
 9:             FWRITE(pp -¿ callComment, wfile)
10:         **end if**
11:         FWRITE($temp, wfile$)
12:         $temp \leftarrow$ GETLINE(temp, rfile)
13:         $++cntLine$
14:     **end while**
15: **end for**

---

4   Experiments

Considering that Java is a static programming language, the static analysis results of Java source code can be viewed as basically exact. So the evaluation of the effectiveness of our method is mainly focused on the following research question: RQ: Are the comments extracted by this approach qualified to describe their corresponding methods correctly and provide useful information of the methods?

The main factor that may contribute to the wrong transferring of comment information is that, the features of comments on which the judgement of whether to extract and transfer a certain comment is based can be inaccurate.

To assess the precision of the comment-transferring method, we select 14 projects from Github that have relatively more stars and use the approach mentioned above to extract and transfer comments for the methods in source code. Then evaluation of the transferred comments is done manually to see if they are capable of correctly describing the function or intention of their corresponding methods.

Previous research [13] has proposed a way to evaluate the accuracy of comments for methods, the formula is:

$$precision = \frac{correct\ comments}{total\ comments} \times 100\% \tag{1}$$

In our experiment, we apply a similar standard and precision-calculating formula to our evaluation process. We classify the comments transferred into 2 categories: accurate and inaccurate, 'accurate' means the comments transferred can describe the function of the method properly, while 'inaccurate' means the comments transferred cannot. The formula of precision is:

$$precision = \frac{accurate\ comments}{total\ comments} \times 100\% \tag{2}$$

The result is listed below (Table 1).

Table 1. Accuracy of comments extracted and passed

Project	Accurate	Inaccurate	Total	Precision
BaseRecyclerViewAdapterHelper	100	28	128	78.13%
butterknife	90	18	108	83.33%
dubbo	97	16	113	85.84%
netty-4.1	103	26	129	79.84%
redisson	22	3	25	88.00%
SlidingMenu	90	14	104	86.54%
zipkin	104	29	133	78.20%
retrofit	77	17	94	81.92%
seata	96	20	116	82.76%
okhttp	99	17	116	85.35%
MSPaintIDE	36	5	41	87.80%
Duke	107	25	132	81.06%
mobius	54	16	70	77.14%
grant	24	5	29	82.76%

The average precision of the comments extracted and transferred in these projects is 82.76%, which shows that our approach of extracting method-related comments from the location in front of method definition is relatively reliable.

5   Conclusion

In this paper, we preprocess the source code by the program static analysis, and then extract the effective information of the comments in front of method definition. We finally transfer Java comments automatically. Experimental results show that the method has good accuracy.

In the future, we will expand the scope of the experiment and continuously improve the accuracy of the experiment. In addition, we plan to get more accurate analysis of the call relationship on the Android system intent mechanism to achieve annotation delivery.

References

1. Moreno, L., Aponte, J., Sridhara, G., et al.: Automatic generation of natural language summaries for Java classes. In: IEEE International Conference on Program Comprehension. IEEE (2013)
2. Murphy, G.C., Kersten, M., Robillard, M.P., Čubranić, D.: The emergent structure of development tasks. In: Black, A.P. (ed.) ECOOP 2005. LNCS, vol. 3586, pp. 33–48. Springer, Heidelberg (2005). https://doi.org/10.1007/11531142_2
3. Wong, E., Liu, T., Tan, L.: Mining existing source code for automatic comment generation. In: IEEE 22nd International Conference on Software Analysis, Evolution and Reengineering (SANER), pp. 380–389. IEEE (2015)
4. Raychev, V., Vechev, M., Krause, A.: Predicting program properties from "Big Code". ACM SIGPLAN Not. **50**(1), 111–124 (2015)
5. Kramer, D.: API documentation from source code comments: a case study of Javadoc. In: International Conference on Documentation, SIGDOC 1999, New Orleans, Louisiana, USA, pp. 147–153. DELP, September 1999
6. Panichella, S., Aponte, J., Penta, M.D., et al.: Mining source code descriptions from developer communications. In: IEEE, International Conference on Program Comprehension, pp. 63–72. IEEE (2012)
7. Wong, E., Liu, T., Tan, L., et al.: Convolutional neural networks over tree structures for programming language processing. In: Thirtieth AAAI Conference on Artificial Intelligence, pp. 1287–1293. AAAI Press (2016)
8. Rahman, M.M., Roy, C.K., Kula, R.G.: Predicting usefulness of code review comments using textual features and developer experience. In: IEEE/ACM. MSR (2017)
9. Sridhara, G., Hill, E., Muppaneni, D., et al.: Towards automatically generating comments for Java methods. IEEE/ACM (2010)
10. McBurney, P.W., McMillan, C.: An empirical study of the textual similarity between source code and source code summaries. In: 2016 Empirical Software Engineering and Measurement, ESEM 2016, pp. 17–42 (2016)
11. Haouari, D., Sahraoui, H., Langlais, P.: How good is your comment? A study of comments in Java programs. In: 2011 International Symposium on Empirical Software Engineering and Measurement (ESEM 2011), pp. 137–146 (2011)
12. Fluri, B., Wursch, M., Gall, H.C.: Do code and comments co-evolve? On the relation between source code and comment changes. In: 14th Working Conference on Reverse Engineering 2007, WCRE 2007, pp. 70–79 (2017)
13. Chen, H., Liu, Z., Chen, X., et al.: Automatically detecting the scopes of source code comments. In: IEEE Computer Software & Application Conference. IEEE Computer Society (2018)
14. Zhang, J., Xu, L., Li, Y.: Classifying Python code comments based on supervised learning. In: Meng, X., Li, R., Wang, K., Niu, B., Wang, X., Zhao, G. (eds.) WISA 2018. LNCS, vol. 11242, pp. 39–47. Springer, Cham (2018). https://doi.org/10.1007/978-3-030-02934-0_4

A Monitoring Mechanism for Electric Heaters Based on Edge Computing

Jing Wang[1,2(✉)], Zihao Wang[1,2], and Ling Zhao[1,2]

[1] Institute of Data Engineering, North China University of Technology,
Beijing 100144, China
wangjing@ict.ac.cn, 13681272894@163.com,
416141690@qq.com
[2] Beijing Key Laboratory on Integration and Analysis of Large-Scale Stream
Data, North China University of Technology, Beijing 100144, China

Abstract. With the rapid proliferation of IOT sensors, the traditional cloud computing paradigm confronts several challenges such as bandwidth limitation and high latency. Therefore, edge computing paradigm has been paid more attention. In order to guarantee the real-time monitoring of electric heaters, this paper proposes a monitoring architecture combining cloud and edge nodes, and an anomaly detection method and a heating prediction method based on the architecture. Experiments show that the presented monitoring mechanism can reduce response time, improve data transmission efficiency and realize real-time monitoring and management.

Keywords: Edge computing · Monitoring mechanism · Anomaly detection

1 Introduction

Heavy air pollutions occur mainly in the heating season in most parts of the northern area, and pollutions caused by heating has become one of the most important causes of winter haze. Electric thermal storage heaters can provide all-day heating by making use of low-priced electricity during nighttime, which benefit peak shaving and valley filling for grid, and heating expenses saving for users.

The monitoring system for electric heaters encounters following requirements: firstly, as the accuracy of monitoring is required, real-time monitoring data needs to be collected. Secondly, accurately anomaly detection is needed when there is high heating consumption occurs. Thirdly, the relationship between the thermal storage and the thermal demand of users need to be discovered, to avoid the energy waste caused by reserving thermal energy that exceeds the user's demand, or energy shortage caused by insufficient reservation.

Traditional smart home monitoring systems mainly use LAN-based monitoring technology [1]. Many electric enterprises have developed smart home systems, such as Hisense Smart Home Control System and Tsinghua e-Home Digital Home. However, the LAN-based monitoring method lacks flexibility and is limited by distance. Cloud-based monitoring systems [2] use heterogeneous sensors for monitoring, transform sensor data to cloud, and perform data analysis in the cloud, which will decrease the

© Springer Nature Switzerland AG 2019
W. Ni et al. (Eds.): WISA 2019, LNCS 11817, pp. 644–649, 2019.
https://doi.org/10.1007/978-3-030-30952-7_65

real-time performance of the monitoring system. In the aspect of monitoring based on edge computing, anomaly detection technology are presented in [3], which uses edge-cloud collaboration to detect data anomalies of real-time sensor data, which can reduce the load of the cloud center

This paper proposes a monitoring architecture combining cloud and edge nodes. An anomaly detection method and a heating prediction method were proposed based on this architecture. Through the mechanism, the effect of reducing response time, improving data transmission and real-time monitoring can be achieved.

2 System Architecture

Aiming at the demand of real-time monitoring for electric heaters, this paper proposes the monitoring system architecture based on edge-cloud collaboration. The system is divided into three parts: the edge node, the cloud center, and the mobile terminal, as Fig. 1 shows. The electric heaters send sensing data to the edge node, and real-time calculation is performed on the edge node according to the monitoring requirements. Then, the calculation results are sent to the cloud to be stored. The cloud is also responsible for computational tasks with large computational complexity and low real-time requirements, such as neural network model training tasks, and training results are transmitted to the edge. The mobile terminal can connect with the edge or the cloud center based on the network environment to receive monitoring data.

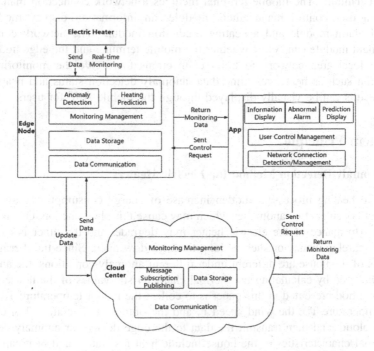

Fig. 1. The monitoring system architecture for electric heaters

Edge Node. The edge node contains a data communication module, a data storage module and a monitoring management module.

The data communication module is responsible for data transmission among the edge node, the cloud center and the mobile device. It provides multiple ports that can be configured to provide different sampling frequencies and data contents, thereby avoiding sending all sensor data to reduce network bandwidth consumption.

The data storage module store sensor data received from the heaters and external data acquired from the cloud. The data storage strategy is to store the latest data in the edge node and send the historical data to the cloud center for persistent storage.

The monitoring management module is divided into two parts: anomaly detection and heating prediction. The anomaly detection can find abnormal faults of electrical heaters and abnormal house ventilation. The heating prediction can predict the user's heating requirements for the next day according to the historical data under different weather conditions, and plan the heating amount to achieve energy saving.

Cloud Center. The cloud center includes a data communication module, a message publish-subscribe module, a data storage module, and a monitoring management module. The data communication module is responsible for data transmission between the cloud and the edge. After receiving the data, the message publish-subscribe module sends the data to the data storage module. The data sent from the edge device is combined with the stored historical data to form a training set. The monitoring management module trains the set and sends the trained model to the edge.

Mobile Terminal. The mobile terminal includes a network connection management module, a user control management module, an information display module, an abnormal alarm module and a heating prediction module. The network connection management module can detect whether the mobile terminal and the edge node are in the same local area network to build connections to receive the monitoring data. Information such as heater real-time data, anomaly detection status and heating prediction results can be visually displayed through the mobile APP for users.

3 System Principle

3.1 Anomaly Detection Method for Electric Heaters

During the heating process, a sudden increase of energy consumption may happen. This paper considers two anomalies which may cause this phenomenon. One is that the heating performance of the electric heater may degrade, and the other is caused by residents' careless behavior such as long-time window open. Since the thermal characteristics of the house are different under different anomaly conditions, the anomalies can be analyzed by calculating the thermal characteristic values of the house.

The method presented in this paper is to collect the indoor temperature T_{in}, the air outlet temperature T_w, the wind speed F, and the outdoor temperature T_{out} collected from the cloud, and then transmit the data to the edge device for anomaly detection. The thermal characteristics of the house include heat loss rate ε and heat capacity C.

In thermodynamics, energy analysis is often based on the degree days [4]:

$$DD_{Period A} = \sum\nolimits_{Period A} (T_{in} - T_{out})\Delta t \tag{1}$$

The energy provided by the heater is used both to compensate for energy losses and to increase the indoor temperature:

$$Provided\ Energy_{Period A} = \sum\nolimits_{Period A} (T_w \cdot F)\Delta t \tag{2}$$

Combined with heating and non-heating periods, the heat loss rate ε and the heat capacity C can be obtained:

$$\varepsilon = \frac{Provided\ Energy}{DD + \frac{\Delta T_{in}}{\mu}} \tag{3}$$

$$C = \frac{\varepsilon}{-\Delta T_{in}} \times DD \tag{4}$$

This paper designs the algorithm based on the conclusions in [4]. Specifically, when the heating performance of the electric heater degrades, both C and ε increase; when there is an anomaly due to the long ventilation, C does not change significantly but ε increases remarkably. By comparing the values of C and ε, the reason of anomalies can be detected. The pseudo code of the algorithm is described as follows:

Input: Sensing data collected by the heater and weather data collected by the cloud
Output: anomaly alarm

```
Start
1.   ProvidedEnergy = PE(Tw, F) //Calculate ProvidedEnergy by equation (2)
2.   DD = Degree Day(Tin, Tout) //Calculate the value of DD by equation (1)
3.   μ = ΔTin/DD
4.   ε = Loss(ProvidedEnergy, DD, ΔTin, μ)
     //Calculate the value of the heat loss rate ε by equation (3)
5.   C = Capacity(ε, −ΔTin, DD)
     //Calculate the value of the heat capacity C by equation (4)
6.   If (ε>ε0 && C>C0)
     //ε0, C0 is the thermal characteristic value under normal conditions
7.      Send the heating anomaly alarm
8.   Else If (ε>ε0 && C=C0)
9.      Send the house ventilation anomaly alarm
10.  End If
End
```

3.2 Prediction Method for Electric Heaters

Since the sensor data of the electric heater is mostly time series data, the Recurrent Neural Network (RNN) can achieve high precision. LSTM RNN [5, 6] is a special kind of RNN, which are characterized by the addition of valve nodes of each layer outside the RNN structure and long-term dependency problems can be avoided and long-term storage memory information can be supported.

In this paper, the training task is placed in the cloud and the prediction task is placed on the edge. Then the training model is sent to the edge and the prediction of the electric thermal usage is performed. Based on the above principles, this paper presents the algorithm for heating prediction described as follows:

Input: Sensing data collected by the heater and weather data collected by the cloud
Output: Prediction model

```
Start
1.  scaled = fit_transform(sensedata.values)
2.  reframed = series_to_supervised(scaled,1,1)
3.  DEFINE train_sets
4.  train_X=train_X.reshape(sensedata.sample, timesteps, sensedata. features)
    /*Standardized Data*/
5.  construct Model = LSTM(input1_dim, output1_dim, return_sequence=True))
6.  add_Layer=LSTM(output1_dim,output2_dim,return_sequence=True)
    /*Design the LSTM layer */
7.  add_Dropout()
8.  add Layer = Dense(output_dim=1)
9.  compile loss = mae
10. compile optimizer=adam
11. construct fit(train_X, epoch_num, batch_size)
    /*Define fit function*/
12. produce model
13. send to the edge
End
```

4 Results and Analysis

The Raspberry Pi development board is used to simulate the edge node and the data is collected from a temperature sensor and processed in real time. The basic parameters of the Raspberry Pi are as follows: BCM2837BO 64-bit, 1.4 GHz quad-core, 802.11AC wireless network card, memory 1 GB, Bluetooth 4.2. The cloud cluster environment consists of six servers, which includes one communication server, three HBase servers, one proxy server and one web console server.

The anomaly detection algorithm was tested in two cases. In experiment 1 the detection is conducted at the edge node, and in experiment 2 the detection is performed in the cloud. The response time is recorded and shown in Fig. 2. The results show that in the cloud mode, as the amount of data increases, the efficiency of data processing decreases due to the influence of network transmission and the response time increases significantly. In the edge mode, the data is processed on the edge node which is closer to the data source, and it improves the overall efficiency of the data processing and the response time is significantly reduced.

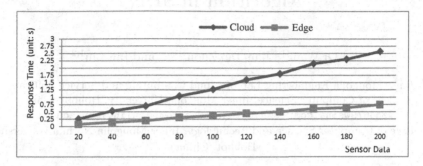

Fig. 2. Response time comparison

5   Conclusion

This paper proposes a monitoring mechanism for electric heaters based on edge computing, which combines the cloud and the edge to monitor electric heaters, and can reduce the response time and improve the data transmission efficiency. Based on edge-cloud collaboration, real-time monitoring and management is guaranteed. We will continue to study on anomaly analysis and pre-warning maintenance for IOT devices in the future.

Acknowledgement. This work is supported by the Key projects of the National Natural Science Foundation of China (No. 61832004).

References

1. Kusriyanto, M., Putra, B.D.: Smart home using local area network (LAN) based arduino mega 2560. In: 2nd International Conference on Wireless and Telematics (ICWT 2016), pp. 127–131. IEEE (2016)
2. Khattak, A.M., Ho, T., Hung, D., et al.: Towards smart homes using low level sensory data. Sensors **11**(03), 11581–11604 (2011)
3. Zhang, Q., Hu, Y., Ji, C., et al.: Edge computing application: real-time anomaly detection algorithm for sensing data. J. Comput. Res. Dev. **55**(03), 524–536 (2018)
4. Tabatabaei, S.A., Ham, W.V.D., Klein, M.C.A., et al.: A data analysis technique to estimate the thermal characteristics of a house. Energies **10**(9), 1358 (2017)
5. Gers, F.A., Schmidhuber, J., Cummins, F.: Learning to forget: continual prediction with LSTM. Neural Comput. **12**(10), 2451–2471 (2000)
6. Zheng, H., Shi, D.: Using a LSTM-RNN based deep learning framework for ICU mortality prediction. In: Meng, X., Li, R., Wang, K., Niu, B., Wang, X., Zhao, G. (eds.) WISA 2018. LNCS, vol. 11242, pp. 60–67. Springer, Cham (2018). https://doi.org/10.1007/978-3-030-02934-0_6

GRAMI-Based Multi-target Delay Monitoring Node Selection Algorithm in SDN

Zhi-Qi Wang[1], Jian-Tao Zhou[1(✉)], and Lin Liu[1,2]

[1] Inner Mongolia Engineering Lab of Cloud Computing and Service Software,
College of Computer Science, Inner Mongolia University, Hohhot, China
wzq4679@sina.com, cszhoujiantao@qq.com, liulin@imnu.edu.cn
[2] College of Computer Science and Technology, Inner Mongolia Normal University,
Hohhot, China

Abstract. Delay is a representation of the link state information in the network. Therefore, the real-time acquisition of the link delay data is the basis for effective network activity. Using the characteristics of SDN centralized control, the GRAMI-based multi-target delay monitoring node selection algorithm is proposed. Based on the GRAMI architecture, the algorithm minimizes the number of monitoring nodes and the total length of the detection path as the offline phase target, and combines NSGA-II to design the algorithm. The algorithm also satisfies the constraints on the measurement devices. The experimental results show that the algorithm can obtain satisfactory suboptimal solutions or even optimal solutions in a short time.

Keywords: Software-defined networking · GRAMI · NSGA-II · Link delay measurement

1 Introduction

In recent years, with the rapid development and popularization of the Internet, real-time monitoring and understanding of the current performance status of the network, and accordingly making a fast and effective management solution has become very important. Link delay, bandwidth and packet loss rate are important parameters of network performance [1–3]. Among them, link delay is the representation of link state information in the network, so real-time acquisition of link delay data is the basis for effective network activities.

Traditional network architecture (usually refers to the IP architecture) is becoming more and more complex and rigid in the process of development, the network protocols at each layer are bloated and complex, and the network scale is huge, which brings serious challenges to the network measurement task [4].

SDN (Software-defined networking) is a new type of network architecture, which uses hierarchical thinking to separate data from control [5]. In the control

© Springer Nature Switzerland AG 2019
W. Ni et al. (Eds.): WISA 2019, LNCS 11817, pp. 650–661, 2019.
https://doi.org/10.1007/978-3-030-30952-7_66

layer, including the logic centered and programmable controller, it can grasp the global network information, facilitate the management of the configuration network, and make the network control function highly centralized [6,7]. The emergence of SDN, effectively solved the shortcomings of network measurement in the traditional network architecture. By using the characteristic of SDN centralized control, the logic implementation of delay measurement can be completed by the controller.

In SDN delay measurement technology, the controller part plays a leading role, especially the selection of the measurement node and the planning of the measurement path affect the real-time and accuracy of the measurement. At present, the existing paper only aim at choosing measurement nodes, without balancing the relationship between measurement nodes and their corresponding paths. In order to solve this problem, this paper abstracts the problem into a multi-objective problem and proposes GRAMI-based multi-target delay monitoring node selection algorithm combined with NSGA-II.

The rest of this paper is organized as follows. Section 2 introduces the division of delay measurement technology in SDN and related paper in detail and puts forward the problem in active-switch measurement. Section 3 introduces GRAMI-based multi-target delay monitoring node selection algorithm to solve this problem. The experiment is presented in Sect. 4 and the conclusion is in Sect. 5.

2   Related Work

At present, SDN link delay measurement technology can be mainly divided into two categories, passive measurement and active measurement [8].

2.1   Passive Measurement

Passive measurement means that it does not send probe packets to the network, it uses real-time traffic in the network for acquisition and analysis, and uses statistical methods to estimate the time. [9–11] is based on this idea. [9] uses the mathematical relationship between expectation and variance, through rigorous mathematical derivation, the true end-to-end delay is estimated by the values of variance and covariance. And designs an end-to-end delay measurement method. [10,11] infer the link delay cumulants from end-to-end measurements without the cooperation between the network internal nodes. [10] uses the relationship between link delay cumulants and path delay cumulants to construct and solve linear equations. [11] presents a new method for estimation of internal link delay distributions using the end-to-end packet pair delay statistics gathered by back-to-back packet-pair unicast probes.

Passive measurements do not generate any detection messages and therefore do not incur additional overhead. However, when the network is complex, the design of sampling method is more important, and the analysis mechanism of data is more complex. The unreasonable of the two may lead to inaccurate measurement. At the same time, compared with the traditional network, one of the main features of SDN is centralized control, which is not fully utilized by passive measurement.

2.2 Active Measurement

Active measurement, as the name suggests, means that the measuring device sends probe packets to the network actively for delay measurement tasks. Active measurement is divided into controller measurement and switch measurement according to the different devices that send probe packets.

Based on the idea of controller measurement, SLAM and OpenNetMon schemes are proposed in [12,13]. The main idea of the scheme is to use the controller to dominate all activities, including calculating the route, sending the flow entries, sending the probe packet, and calculating the delay.

Based on the idea of switch measurement, which is GRAMI [14–16], the measurement task is divided into two parts, and the controller distributes part of the power to the switch, the controller is only responsible for the selection of the measurement node, the calculation of the route, and the delivery of the flow entries. Sending the probe packet and the collection and calculation of delay data are performed by the switch. GRAMI divides delay measurement into two phases: the offline phase and current phase, which corresponds to the controller part and switch part of the active-switch measurement. [14] proposed a far-distance priority greedy algorithm. The algorithm gives the number K of MPs (representing measurement nodes), and then uses a given set of nodes, or randomly chooses a node as the initial set of nodes, and then repeatedly tries to add the farthest nodes to the set until all links are covered. In [16], using the GRAMI framework, the topology is abstracted by constructing an auxiliary bipartite graph in the offline phase, and the problem is modeled with the minimum number of measurement nodes. The algorithm of the maximum flow minimum cost model is designed to solve the problem. Based on the initial solution, the algorithm continuously tries to delete some measurement nodes to minimize the number of measurement nodes. This method greatly reduces the work of the controller and effectively avoids the bottleneck of the controller caused by the busy controller.

2.3 Related to the Paper

Combining the background of SDN centralized control, the current mainstream delay measurement technology in SDN is switch-active measurement. In the offline phase of the GRAMI architecture, [14] specifies the number K of MPs, and selects the MPs to minimize the total length of the detection link. In [16], the minimum number of MPs is targeted, and the factors of the detection path are not considered. Neither of the two papers considers the constraint relationship between the monitoring node and the detection path. However, the choice of monitoring nodes and planning of the detection path greatly affects the real-time and accuracy of the delay measurement task. This problem is an NP-hard problem. It is well known that the NP-hard problem cannot find the optimal solution in a finite time, and generally replaces the optimal solution with a satisfactory suboptimal solution.

In order to solve the above problems, this paper proposes GRAMI-based multi-target delay monitoring node selection algorithm. The algorithm aims at minimizing the number of monitoring nodes and the total length of detection path. The NSGA-II algorithm is used to find a balance point between two goals [17,18].

3 Link Delay Measurement

3.1 Problem Description

As mentioned earlier, the main work of the active-switch measurement technique focuses on the selection of measurement nodes and the planning of measurement paths. Two problems need to be considered here: the cost of measurement hardware and the computing and storage capacity of network nodes as measurement nodes.

(1) For the cost problem, this paper abstracts it to solve the problem of minimizing the number of monitoring nodes and the total length of detection path, covering all links;
(2) For the calculation and storage capacity of the network nodes, manually set the measurement depth and measurement capacity of the measurement nodes according to the current network conditions.

We formulate the above questions. Like GRAMI, we called the calculated network node as a measurement node as MP. The link contains weights and is assigned to 1.

Definition 1. X_i *is the variable to be solved, with values of 0, 1. When = 1, the network node is selected as the measurement node.*

$$X_i = \begin{cases} 0 & X_i \text{ is the MP} \\ 1 & X_i \text{ isn't the MP} \end{cases} i = (1...N)$$

Definition 2. Y_{iy} *is the variable to be solved, and the value is 0, 1. When = 1, the measurement node i measures the delay of link y.*

$$Y_{iy} = \begin{cases} 0 & i \text{ measurement the delay from MP } i \text{ to link } y \\ 1 & i \text{ doesn't measurement the delay from MP } i \text{ to link } y \end{cases} i = (1...N), y = (1...M)$$

Formula (1) describes the goal of minimizing the number of monitoring nodes, which is the sum of all is the smallest. Formula (2) minimizes the total length of detection path. The values of and are both 0, 1. When $X_i = 0$, it means that i is not MP, and the sum of nodes i is 0. When $Y_{iy} = 0$, it means that i does not measure y. Only when X_i and Y_{iy} are both 1, $X_iY_{iy}Z_{iy}$ represents the distance from i to y. Two goals are described in mathematical form as follow:

$$\min \sum_{i=1}^{N} X_i \tag{1}$$

$$\min \sum_{i=1}^{N} X_iY_{iy}Z_{iy} \tag{2}$$

The problem constraint consists of two parts. Formula (3) describes that the measurement depth of a single measurement node does not exceed MPd. Wherein, the left side of the expression represents the measurement path length of the node, and the right side represents the product of the measured depth MPd and the sum of the number of measured links of the node. Formula (4) describes that the maximum number of links measured by node i does not exceed MPc. On the left side of the expression, X_iY_{iy} represents whether node i measures link y or not. After summing up, the total number of links measured by the node is obtained. According to the constraint description, this value does not exceed MPc. In this paper, the constraints are used to select the final satisfactory solution in the pareto solution. For each node i:

$$\sum_{i=1}^{N} X_iY_{iy}Z_{iy} \leq MPd \sum_{i=1}^{N} Y_{iy} \tag{3}$$

$$\sum_{i=1}^{N} X_iY_{iy} \leq MPc \tag{4}$$

Parameter settings are shown in the following Table 1.

Table 1. Parameter setting.

Symbol	Description
N	Constant, node number
M	Constant, link number
MPd	Measurement depth
MPc	Measurement capacity
X_i (i = 1...N)	Variables to be solved, represents he measurement node
Y_{iy} (i = 1...N, y = 1...M)	Variables to be solved, represents that i measurements the delay from measurement node i to link y
Z_{iy} (i = 1...N, y = 1...M)	Variables to be solved, constant, depth of node i to links

3.2   Algorithm Design

Based on the problem description, this paper proposed GRAMI-based multi-target delay monitoring node selection algorithm to solve the problem. The algorithm is divided into two parts, establish fitness function and use NSGA-II algorithm to solve the multi-objective problem.

Algorithm 1. Monitoring node selection algorithm

    Input: Topo, N, maxGen, MPc, MPd;

    Output: Pareto //Pareto optimal set;

               MPs //measurement nodes;

               Edges //corresponding link;

1 $topo \leftarrow toTopo(Topo);$// transform topo into an undirected graph;

2 $pop \leftarrow initialPopulation(topo);$

3 $fastNonDominationSort(pop);$

4 $plist \leftarrow choose(pop);$

5 $crossover(plist);$

6 $mutate(plist);$

7 $gen \leftarrow gen + 1;$

8 **while** $gen < maxGen$ **do**

9 | $mergePop()$ // merge parent and children;

10 | $fastNonDominationSort(pop);$

11 | $congDegree()$ // compute congestion degree;

12 | $plist \leftarrow choose(pop);$

13 | $crossover(plist);$

14 | $mutate(plist);$

15 | $gen \leftarrow gen + 1;$

16 **end**

17 $selected(Pareto, MPc, MPd);$

18 $print(selectedMPs)$ //print MPs;

19 $print(selectedEdges)$ //print corresponding links;

In first part, the fitness function to be established corresponds to two objective functions respectively. The objective function (1) is to accumulate all the values of X_i. In order to satisfy the objective function (2), the idea used here is the process of overlay network in GRAMI offline stage [19]. First calculate the shortest distance from each link to each network node, then, for each individual of the initial population, compare the distance within each individual, each MP distance from the link, select and record the distance of each link from the nearest MP, and sum the minimum distances of all links to get the objective function of the individual.

In second part, NSGA-II is a multi-objective evolutionary algorithm [17]. The algorithm gets the non-dominated solution set sorting by genetic algorithm, and obtains the multi-objective optimal solution set, which effectively solves the multi-objective optimization problem [17,18]. This paper chooses NSGA-II algorithm to solve the problem, which can obtain multiple Pareto optimal solutions [20]. According to the actual requirements of the network (depth and capacity of MPs), the final satisfactory solution can be selected. Algorithm 1 describes the main idea of the algorithm.

4   Experimental Results and Analysis

The experiment is based on Spyder development environment. Topology Zoo is chosen as the topology set [21]. Table 2 shows main parameters in Topology Zoo.

Table 2. Main parameters in Topology Zoo.

No	Name	Links	Switches	Max depth
1	GetNet	8	7	3
2	Peer1	20	16	4
3	Airtel	37	16	3
4	BICS	48	33	6
5	ATT	57	25	4
6	GEANT	61	40	5

In order to show the experimental results in a clear and concise manner, topo 2 is chosen for experiment. The topology is shown in Fig. 1.

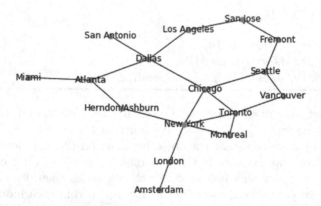

Fig. 1. Topo2.

Fitness function 1 represents the goal of minimizing the number of nodes, and 2 minimizes the total length of detection path. Let the initial population contain 20 individuals. For example, individual [0, 1, 1, 1, 1, 1, 1, 0, 0, 1, 1, 0, 0, 0, 0, 1, 1, 0] is randomly generated, each representing a network node, and value 1 representing the network node being selected as an MP. The result of fitness function 1 is the number of 1 in the individual, which is 9. After running the program, the result of fitness function 2 is 24, which represents the length of the shortest path obtained by the node as MP in the individual.

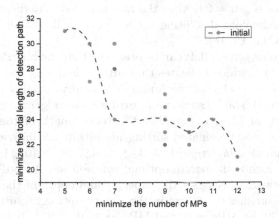

Fig. 2. Result of initial population.

Figure 2 shows the results of the initial population, which has a wide choice of individuals. The different lengths of the detection paths are different for different components under the same MP number. From the trend of the points in the figure, the horizontal and vertical coordinates show a negative correlation. After the fast non-dominated sorting algorithm, the resulting non-dominated solution set is: [[4, 6, 8, 14, 16, 19], [7, 9, 11, 12, 13, 15], [10, 0, 3, 17], [1, 18], [5, 2]].

Figure 3 shows the results of Gen = 10, 20 and 30.

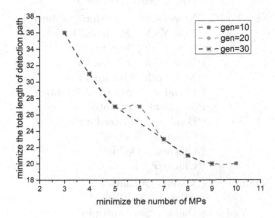

Fig. 3. Result of gen = 10, 20, 30. (Color figure online)

Green dotted line is gen = 10. After 10 generations, the overall trend remains unchanged, but the length of the detection path corresponding to the number of different MPs begins to converge. After the fast-non-dominated sorting algorithm, the resulting non-dominated solution set is: [[0, 1, 2, 3, 4, 5, 6, 7, 8, 9, 10, 11, 12, 13, 14, 15, 16, 17], [19, 18]].

Red dotted line is gen = 20. After the fast-non-dominated sorting algorithm, the resulting non-dominated solution set is: [[0, 1, 2, 3, 4, 5, 6, 7, 8, 9, 10, 11, 12, 13, 14, 15, 16, 17, 19, 18].

Blue dotted line is gen = 30. It can be observed that the blue dot line coincides with the red dot line, which indicates that after calculating the fitness function for the individual, the value of the function basically remains unchanged, and the whole tends to be stable, so no more evolutionary algebras are needed. The larger the number of MPs, the smaller the total length of the corresponding link. After the fast-non-dominated sorting algorithm, all individuals belong to the first non-dominated layer: [[0, 1, 2, 3, 4, 5, 6, 7, 8, 9, 10, 11, 12, 13, 14, 15, 16, 17, 18, 19]], which is the pareto optimal solution set. According to the state of the current network device, set the capacity and depth of the MP, and select the solution that satisfies the target in the pareto optimal solution set.

When gen = 30, let $MPc = 16$ and $MPd = 3$, [0, 0, 0, 1, 0, 0, 1, 0, 0, 0, 0, 1, 0, 0, 1, 0] meets the constraint. Four MPs are obtained. Table 3 shows the MPs selected by the algorithm and its corresponding measurement links. The number of MPs is 4, and the total length of the measurement path is 31.

Table 3. MPs and corresponding measurement links.

MP	Edges
'San Jose'	('San Jose', 'Fremont')
	('San Jose', 'Los Angeles')
	('Fremont', 'Seattle')
	('Los Angeles', 'Dallas')
'New York'	('New York', 'Herndon/Ashburn')
	('New York', 'Montreal')
	('New York', 'London')
	('New York', 'Chicago')
	('New York', 'Toronto')
	('Herndon/Ashburn', 'Atlanta')
	('Montreal', 'Toronto')
	('London', 'Amsterdam')
	('Chicago', 'Seattle')
	('Chicago', 'Dallas')
	('Chicago', 'Toronto')
'Miami'	('Miami', 'Atlanta')
	('Atlanta', 'Dallas')
	('Dallas', 'San Antonio')
'Toronto'	('Toronto', 'Vancouver')
	('Vancouver', 'Seattle')

Greedy approximation algorithm is often used to solve the minimum problem. The algorithm constraints are consistent with this paper, but the goal is only to minimize the number of MPs. It first creates a routing tree for each node, and then iteratively selects the node with the largest intersecting of the routing tree and the uncovered links through iteration until all links are covered. Let $MPc = 16$ and $MPd = 3$, the result of the algorithm is that the number of MPs is 4, and the total length of the measurement path is 35. Table 4 shows the result. Comparing the results with the algorithm of this paper, it can be clearly seen that the results obtained by this paper are better.

Table 4. MPs and corresponding measurement links from greedy approximation algorithm.

MP	Edges
'Chicago'	('Chicago', 'New York')
	('Chicago', 'Seattle')
	('Chicago', 'Dallas')
	('Chicago', 'Toronto')
	('New York', 'Herndon/Ashburn')
	('New York', 'Montreal')
	('New York', 'London')
	('New York', 'Toronto')
	('Seattle', 'Vancouver')
	('Seattle', 'Fremont')
	('Dallas', 'Los Angeles')
	('Dallas', 'San Antonio')
	('Dallas', 'Atlanta')
	('Toronto', 'Montreal')
	('Toronto', 'Vancouver')
'Herndon/Ashburn'	('Herndon/Ashburn', 'Atlanta')
	('Atlanta', 'Miami')
'Fremont'	('Fremont', 'San Jose')
	('San Jose', 'Los Angeles')
'London'	('London', 'Amsterdam')

5 Conclusion

In this paper, the problem of monitoring node selection in link delay measurement under SDN is studied and analyzed. Based on GRAMI framework, combined with NSGA-II algorithm, GRAMI-based multi-target delay monitoring node selection algorithm is proposed. According to the experimental results, the algorithm can effectively balance the constraints between monitoring nodes and detection paths. At the same time, it satisfies the storage and computation constraints of monitoring nodes. When compared with the greedy approximation algorithm, the algorithm proposed in this paper gets better results. Therefore, the algorithm is a good choice for selecting the network link delay monitoring node problem. In addition, for the preset MP measurement capacity and depth in the problem description, there is no clear method to put forward on how to set it up. In the following work, it can be analyzed according to other specific network parameters to obtain suitable values.

Acknowledgement. The research is supported by Natural Science Foundation of China under Grant No. 61662054, 61262082, Inner Mongolia Science and Technology Innovation Team of Cloud Computing and Software Engineering and Inner Mongolia Application Technology Research and Development Funding Project "Mutual Creation Service Platform Research and Development Based on Service Optimizing and Operation Integrating" under Grant 201702168, Inner Mongolia Engineering Lab of Cloud Computing and Service Software and Inner Mongolia Engineering Lab of Big Data Analysis Technology.

References

1. Sinha, D., Haribabu, K., Balasubramaniam, S.: Real-time monitoring of network latency in software defined networks. In: IEEE International Conference on Advanced Networks and Telecommuncations Systems (ANTS) (2015)
2. Cai, Z., Yin, J., Liu, F., Liu, X.: Efficiently monitoring link bandwidth in IP networks. In: IEEE Global Telecommunications Conference, GLOBECOM 2005, pp. 354–358 (2005)
3. Panwaree, P., Kim, J., Aswakul, C.: Packet delay and loss performance of streaming video over emulated and real OpenFlow networks, July 2014
4. Yan, X.-c., Yin, J.-p., Cai, Z.-p.: A survey on network topology discovery algorithm. Comput. Eng. Appl. **43**(14), 131 (2007)
5. Zhang, C.-K., Cui, Y., Tang, H.-Y., Wu, J.-P.: State-of-the-art survey on software-defined networking (SDN). Ruan Jian Xue Bao/J. Softw. **26**, 62–81 (2015)
6. Mckeown, N.: Keynote talk. Software-defined networking. IEEE Infocom **51**(2), 1–2 (2009)
7. Kreutz, D., Ramos, F., Verssimo, P., Rothenberg, C.E., Azodolmolky, S., Uhlig, S.: Software-defined networking: a comprehensive survey. ArXiv e-prints, 103, June 2014
8. Zhang, H., Cai, Z., Li, Y.: An overview of software-defined network measurement technologies. Scientia Sinica (Informationis) (2018)
9. Nguyen, H.N., Begin, T., Busson, A., Lassous, I.G.: Evaluation of an end-to-end delay estimation in the case of multiple flows in SDN networks. In: International Conference on Network and Service Management (2017)
10. Fei, G., Hu, G., Jiang, X.: Unicast-based inference of network link delay statistics. In: IEEE International Conference on Communication Technology (2011)
11. Hero, A.O., Shih, M.F.: Unicast-based inference of network link delay distributions with finite mixture models. IEEE Trans. Signal Process. **51**(8), 2219–2228 (2003)
12. Yu, C., Lumezanu, C., Sharma, A., et al.: Software-defined latency monitoring in data center networks. In: Software-Defined Latency Monitoring in Data Center Networks, pp. 360–372 (2015)

13. Van Adrichem, N.L.M., Doerr, C., Kuipers, F.A.: OpenNetMon: network monitoring in OpenFlow software-defined networks. In: Network Operations and Management Symposium (2014)
14. Atary, A., Bremler-Barr, A.: Efficient round-trip time monitoring in OpenFlow networks. In: IEEE Infocom -the IEEE International Conference on Computer Communications (2016)
15. GRAMI Software Enhancement on SDN Networks. Ph.D. thesis
16. Wang, X.: Research on network link delay measurement and flow table management method in SDN data center. Ph.D. thesis, University of Electronic Science and Technology of China (2018)
17. Deb, K., Pratap, A., Agarwal, S., Meyarivan, T.: A fast and elitist multiobjective genetic algorithm: NSGA-II. IEEE Trans. Evol. Comput. **6**(2), 182–197 (2002)
18. Li, H., Zhang, Q.: Multiobjective optimization problems with complicated pareto sets, MOEA/D and NSGA-II. IEEE Trans. Evol. Comput. **13**(2), 284–302 (2009)
19. Bremler-Barr, A., Atary, A.: Grami source code. https://github.com/alonatari1/GRAMI
20. Meng, X., Li, R., Wang, K., Niu, B., Wang, X., Zhao, G. (eds.): WISA 2018. LNCS, vol. 11242. Springer, Cham (2018). https://doi.org/10.1007/978-3-030-02934-0
21. The Internet Topology Zoo. http://www.topology-zoo.org/

Research on Fuzzy Adaptive PID Fuzzy Rule Optimization Based on Improved Discrete Bat Algorithm

Xuewu Du[1,2], Mingxin Zhang[1(✉)], and Guangtao Sha[1,3]

[1] School of Computer Science and Engineering, Changshu Institute
of Technology, Changshu 215500, China
mxzhang163@163.com
[2] College of Computer Science and Technology, Soochow University,
Suzhou 215006, China
[3] School of Computer Science and Technology, China University of Mining
and Technology, Xuzhou 221116, China

Abstract. The intelligent optimization algorithm has obvious advantages in solving the nonlinear and complexity optimization problems, and is widely used in the optimization of fuzzy control rules. However, it also has the disadvantages of low optimization precision and easy to fall into local optimum. The discrete bat algorithm (DBA) optimizes the fuzzy rules in the fuzzy adaptive PID controller. Firstly, the neighborhood search operator is designed according to the correlation between adaptive fuzzy rules to improve the search accuracy. Secondly, the chaotic mutation operator is used to avoid the algorithm falling into local optimum. Finally, the ITAE integral performance index is used as the optimization objective function and the particle swarm. The algorithm and genetic algorithm optimize the control effect for comparison analysis. The results show that the controller optimized by discrete bat algorithm has advantages in adjusting time and overshoot, and has strong adaptability and robustness.

Keywords: Fuzzy rule optimization · Fuzzy adaptive PID control ·
Discrete bat algorithm · Fuzzy control

1 Introduction

Fuzzy adaptive PID control uses fuzzy mathematics as the theory to express the rule conditions and operations with fuzzy sets, and stores the fuzzy control rules into the knowledge base. The control system uses fuzzy reasoning to realize the optimal adjustment of PID parameters according to the actual response, and improves PID control Performance [1]. Due to its simple structure, good stability and high reliability, it has been widely used [2–4]. However, in the system design, fuzzy rules, as the core of fuzzy control, still rely mainly on artificial prior knowledge in the determination. When the system scale becomes complex, the selection of fuzzy rules is greatly increased, and it is impossible to obtain better by artificial experience alone. Control rules. The intelligent optimization algorithm has obvious advantages in solving

© Springer Nature Switzerland AG 2019
W. Ni et al. (Eds.): WISA 2019, LNCS 11817, pp. 662–674, 2019.
https://doi.org/10.1007/978-3-030-30952-7_67

nonlinear and complex optimization problems, and is widely used in the optimization of fuzzy rules. The literature [5] used adaptive genetic algorithm to optimize the fuzzy rules of DFB laser fuzzy PID temperature control system. Improve the accuracy of the PID control algorithm. In [6], particle swarm optimization was used to optimize the underwater submersible depth adaptive fuzzy control, so that the overshoot and steady state error of the control variables are better than the traditional fuzzy adaptive control. In [7], the genetic algorithm was used to optimize the adaptive fuzzy PID parameters to obtain the optimal controller parameters, which improves the control effect of the shearer drum heightening system and meets the application requirements. However, these methods are too much guided by the objective function in the face of large-scale optimization problems. The lack of optimization precision and easy to fall into local optimum limits the accuracy of fuzzy rules selection.

Bat algorithm is a heuristic group intelligence algorithm proposed by Professor Yang [8]. It has been shown that the bat algorithm has better search accuracy and computational efficiency than the particle swarm algorithm and genetic algorithm [9], and has been successfully applied to continuous domain function optimization. In the discrete domain, the literature [10] proposed that the discrete bat algorithm solves the combinatorial optimization problem with good results, but it is not suitable for discrete non-combination optimization problems. In order to solve the fuzzy adaptive PID control rule discrete domain optimization problem, this paper discusses the discrete bat algorithm. The neighborhood search operator is introduced to improve the optimization precision, and the chaotic mutation operator avoids the algorithm to get the local optimal IDBA (Improved Discrete Bat Algorithm).

In this paper, the improved discrete bat algorithm is used to optimize the fuzzy control rules of fuzzy adaptive PID controller, and the experimental results are compared with the optimization results of particle swarm optimization algorithm and genetic algorithm. Experiments show that the fuzzy adaptive PID controller optimized by IDBA has better control results.

2 Fuzzy Adaptive PID Control

The fuzzy adaptive PID controller obtains the modified output variable to control the output of the controlled object according to the deviation e between the actual and expected control values and the deviation change rate e_c after the fuzzy controller performs fuzzy, fuzzy inference and defuzzification.

The PID parameter correction method is as shown in formula (1):

$$\begin{cases} K_p = K_{p0} + \Delta K_p \\ K_i = K_{i0} + \Delta K_i \\ K_d = K_{d0} + \Delta K_d \end{cases} \tag{1}$$

Among them, K_{p0}, K_{i0} and K_{d0} are the initial PID controller settings, ΔK_p, ΔK_i, ΔK_d is the amendment, and K_p, K_i, K_d are the adjusted final values.

The schematic diagram is shown in Fig. 1.

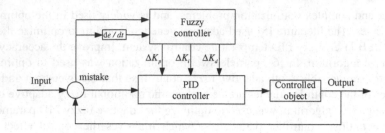

Fig. 1. Fuzzy adaptive PID control principle.

3   Bat Algorithm

Bat algorithm is an intelligent optimization algorithm proposed by Professor Yang Xinshe based on the principle of bat foraging. In the D-dimensional space, n bats search for prey through ultrasonic waves with frequency f_i, pulse emissivity r_i, and loudness A_i. The flight velocity v_i and position x_i are changed according to the returned information, and the pulse is gradually adjusted when approaching the prey process. Emissivity and loudness.

The process of updating the frequency, position and speed of the bat during flight is as follows:

$$f_i = f_{\min} + (f_{\max} - f_{\min})\beta \tag{2}$$

$$v_i^t = v_i^{t-1} + (x_i^t - x_*)f_i \tag{3}$$

$$x_i^t = x_i^{t-1} + v_i^t \tag{4}$$

Where f_i is the frequency of the i-th bat, randomly assigned at initialization and $f_i \in [f_{\min}, f_{\max}]$, β is a uniformly distributed random variable in the range [0, 1], and v_i^t represents the speed of the i-th bat at the t-th iteration, x_i^t represents the position of the i-th bat at the t-th iteration, and x_* is the current global optimal position, $i = 1, 2, \cdots, n$.

When performing a local search, the location of the current optimal solution is selected for random search, and the update manner of the bat position is:

$$x_{new} = x_{old} + \varepsilon A^t \tag{5}$$

Where x_{old} is the current optimal solution, x_{new} is the new solution produced, ε is a random number belonging to [–1, 1], and A^t is the average of all bat loudness at time t.

As the iteration increases, in order to better locate the target bat, the loudness A_i is gradually reduced, and the pulse transmission frequency r_i is increased. The update method is:

$$A_i^{t+1} = \alpha A_i^t \tag{6}$$

$$r_i^{t+1} = r_i^0[1 - e^{-\gamma t}] \tag{7}$$

Where α is the loudness attenuation coefficient, γ is the pulse transmission frequency increase coefficient, r_i^0 is the maximum pulse emission frequency of the i-th bat, and $0 < \alpha < 1$, $\gamma > 0$, and

$$A_i^{t+1} \to 0, r_i^t \to r_i^0 \quad as \quad t \to \infty \tag{8}$$

Usually, $\alpha = \gamma$, the specific parameters need to be determined according to the experiment.

4 Discrete Bat Optimization Algorithm Based on Fuzzy Adaptive PID Control

4.1 Coding Strategy

Since it is necessary to optimize the three fuzzy rules of ΔK_p, ΔK_i and ΔK_d at the same time, the three correction values of the two-input and three-output fuzzy controller outputs ΔK_p, ΔK_i and ΔK_d as shown in Fig. 1 are modified to correct the PID controller to the error e. And the error rate of change e_c is used as the two-dimensional input of the fuzzy controller. For the two-dimensional fuzzy control system, each input and output is usually represented by 7 language values {NB, NM, NS, ZO, PS, PM, PB}, in order to ensure the integrity and consistency of the rules, the output variables in the fuzzy rules table are arranged in order, each rule is represented by a decimal number, and the seven language values are respectively used {1, 2, 3, 4, 5, 6, 7} Code representation, the number of rules in the fuzzy rule table is 49, and the number of rules for each individual of the initial population is 147, of which 1–49 means ΔK_p fuzzy rule sequence, 50–98 is ΔK_i fuzzy rule sequence and Articles 99–147 are ΔK_d fuzzy rule sequences. In order to simplify the calculation, the membership function uniformly adopts the triangular membership function, and each individual randomly assigns a fuzzy rule sequence during initialization.

4.2 Relevant Operator

Neighborhood Search Operator. The fuzzy adaptive PID controller first acquires the error e and the error rate of change e_c of the actual and expected control values in the sampling period. The correction amount ΔK_p, ΔK_i and ΔK_d of the PID output by the fuzzy controller are optimized, and three fuzzy control needs to be optimized simultaneously in the optimization. In the rule table, the same position in the three fuzzy control rules table acts together on a PID control adjustment, that is, the fuzzy rule in the same position is relatively independent from other positional fuzzy rules. Combine the language values of the same position in the three fuzzy control tables into coordinate points in the three-dimensional space, as shown in Fig. 2:

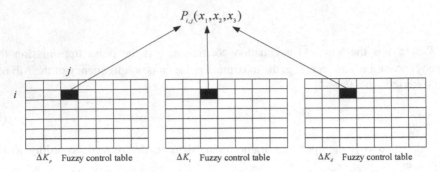

Fig. 2. Fuzzy rule table same position combination map.

The optimization of the same position control rule is equivalent to searching in the three-dimensional space composed of ΔK_p, ΔK_i and ΔK_d to the current optimal solution direction. In the solution space, there are often multiple local best advantages in the vicinity of the current optimal solution, and the true optimal. The solution may be located in it, and the neighborhood search operator is introduced according to this feature to solve the problem of low precision.

When the bat is updated, the random search is performed in the neighborhood with radius r as the center and the update of the neighborhood search operator is shown in Fig. 3.

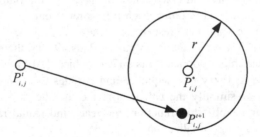

Fig. 3. Neighborhood search operator location update diagram.

Where $P_{i,j}^*$ is the current optimal fuzzy control table i row j column optimal solution, $P_{i,j}^t$ is the coordinate point located in the t-th iteration of the fuzzy control table i row j column language value, $P_{i,j}^{t+1}$ is located in the fuzzy control table i row j column language value The coordinate point at t+1 iterations, where r is the search radius.

The t-th iteration speed change rate $dv_{i,j}$ is:

$$dv_{i,j} = P_{i,j}^{t+1} - P_{i,j}^{t} \tag{9}$$

The way $P_{i,j}^{t+1}$ is selected is:

$$P_{i,j}^{t+1}(x_k) = P_{i,j}^{*}(x_k) + r * rand \quad k = 1, 2, 3 \tag{10}$$

Where $P_{i,j}^{t+1}(x_k)$ is the kth dimension of the t+1th iterative neighborhood search point, and $P_{i,j}^{*}(x_k)$ is the current best advantage kth dimension, $rand \in (0, 1)$.

Chaotic Mutation Operator. Chaos is a kind of non-periodic motion phenomenon unique to nonlinear systems. It has the characteristics of randomness, ergodicity and regularity. Using this feature to introduce the chaotic mutation operator to improve the search efficiency and improve the diversity of the population. Locally optimal.

Tent chaotic map has better ergodic uniformity and fast iteration, and has extremely complex dynamic behavior. This paper uses Tent chaotic mapping function, and its mathematical expression formula is:

$$y_{n+1} = \begin{cases} 2y_n, & y_n < 0.5 \\ 2(1 - y_n), & y_n \geq 0.5 \end{cases} \tag{11}$$

The interval mapping function and the inverse mapping function are as shown in Eqs. (12) and (13):

$$y_{best}^{i} = \frac{2(X_{best}^{i} - X_{min})}{X_{max} - X_{min}} - 1 \tag{12}$$

$$X_{best}^{'i} = \frac{1}{2}(X_{max} - X_{min}) \cdot y_{best}^{i} + \frac{1}{2}(X_{max} - X_{min}) \tag{13}$$

Where X_{min} and X_{max} are the upper and lower bounds of the search.

In order to make the current search optimal individual jump out of the local optimum, the current optimal solution $X_{best} = (X_{best}^{1}, X_{best}^{2}, \cdots, X_{best}^{d},)$ chaotic mutation. Firstly, according to formula (12), it maps to the chaotic mapping function domain $(-1, 1)$, then iteratively generates chaotic sequences by Eq. (11). Finally, the generated chaotic sequence is inversely mapped to the original search region according to Eq. (13) to get new. The optimal individual updates the current optimal individual position according to the fitness function until the maximum number of iteration searches is reached. The chaotic mutation operator steps are as follows:

Step1 Map X_{best} code sequence to Tent function domain.
Step2 Substituting chaotic variables into mapping functions to generate chaotic variable sequences.
Step3 Inverse mapping of chaotic variable sequences to the original search interval.
Step4 Adapting the new solution to the fitness and updating the optimal individual position.
Step5 Determine whether the maximum number of searches has been reached, if it is over, if not to step 2.

Discrete Correction Operator. The discrete correction operator is used to discretize and correct the variables in the fuzzy rule language value to the correct and reasonable range. The discretization method rounds to the nearest integer by simple and efficient rounding method; if the maximum language is out of range Value correction. For example, $P_{i,j}^t = (2.4, 8.9, 4.7)$ after the t-th iteration and $P_{i,j}^t = (2, 7, 5)$ after the $P_{i,j}^t$ correction.

4.3   Performance Indicator Function

In order to shorten the adjustment time, this paper uses the *ITAE* index [11] with good practicability and selectivity in engineering to evaluate the fuzzy adaptive PID control system as an index function for parameter evaluation.

Its discrete formula is:

$$ITAE = \sum_{k=0}^{n} |e(kT)| \cdot kT \cdot T \tag{14}$$

Where T is the sampling period and k is the sampling time.

4.4 Improved Algorithm Flow

The improved discrete bat algorithm is as shown in Algorithm 1:

Algorithm 1. improved discrete bat algorithm

Input: Population number n 、 iterations N ;

Output: optimum solution best_P;

begin

    initialize loudness A ;

    for i=1:n

        initialize P(i); //initialize the position of the bat i

        fitness(i) ← fit(P(i)); //calculation fitness

    end for

    best_P ← min_fit(fitness,P); //record the current optimal location

    for i_iter=1:N

        for i=1:n

            dv(i) ← calculateDv(Best_Sol,dv(i)); //calculate velocity variations

            P(i)=P(i)+dv(i); //update location

            if rand>r //satisfies the mutation condition

                best_P ← chaos(best_P); //chaotic variation

            end if

        P(i)=correct(P(i));

          if rand > A_i and fitness(i-1)<fitness(i)

             Increasing Pulse Frequency r_i ,Reduce loudness A_i ;

          end if

          if fitness(best_P)<fitness(P(i))

            best_P ← P(i);

          end if

        end for

    end for

    return best_P;

end

---

5 Experiment and Evaluation

Usually the industrial process can be equivalent to a second-order system plus typical nonlinear links, such as dead zone, saturation, pure delay, etc. The system is assumed here:

$$H(s) = \frac{20e^{0.02s}}{1.6s^2 + 4.4s + 1} \tag{15}$$

The control execution structure has a dead zone of 0.07 and a saturation zone of 0.7, and the sampling time interval is $T = 0.01$.

The membership function adopts the isosceles triangle membership function, and the deviation e of the control value and the e_c range of the deviation change rate are set to $[-6, 6]$, and $k_{p0} = 5$, $k_{i0} = 0.1$, $k_{d0} = 0.01$ in the fuzzy control.

In order to verify the feasibility and effectiveness of the algorithm, performance analysis will be performed from convergence speed and optimization accuracy, fixed deviation e range, deviation change rate e_c range, ΔK_p output range, ΔK_i output range, ΔK_d output range, iteration number, etc. Compare with Genetic algorithm (GA) and Particle swarm optimization (PSO) optimization results.

The parameters in the experiment are set as follows: iteration number $N_{iter} = 100$, code length $d = 147$, IDBA: population number $n = 20$, loudness $A = 0.25$, pulse emissivity $r = 0.5$, frequency lower limit $Q_{min} = 0$, frequency upper limit $Q_{max} = 1$, loudness attenuation coefficient $\alpha = 0.9$, pulse transmission frequency increase coefficient $\gamma = 0.9$; GA: population size $M = 50$, crossover rate $\zeta = 0.5$, mutation rate $\psi = 0.1$; PSO: particle number $n = 50$, speed lower limit $V_{min} = -2$, speed upper limit $V_{max} = 2$.

The experimental environment is MATLAB R2012a simulation software, Intel 2.8 GHz processor, 8G memory Windows 10 operating system. The fuzzy algorithm of fuzzy adaptive PID controller is optimized by IDBA algorithm, PSO algorithm and GA algorithm respectively, so that the index is close to the decreasing direction. The convergence curve of each algorithm is shown in Fig. 4.

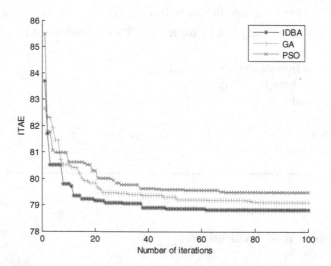

Fig. 4. Three algorithm convergence curves.

The horizontal axis represents the number of iterations, and the vertical axis represents the objective function *ITAE*. At the beginning of the iteration, all three algorithms can quickly approach the direction of *ITAE* reduction. With the iteration, IDBA is more likely to jump out of local optimum and the convergence speed is fast. Accuracy is improved compared to GA and PSO algorithms.

The improved discrete bat algorithm is used to optimize the three fuzzy control rule tables of ΔK_p, ΔK_i and ΔK_d. When the algorithm converges, the objective function *ITAE* approaches the minimum that the algorithm can optimize, and its corresponding coding strategy is the optimized fuzzy rule. This is the optimized fuzzy rule. The improved discrete bat algorithm obtains the fuzzy control rule table shown in Tables 1, 2 and 3 after 100 iterations, and loads the knowledge base as the optimized fuzzy control rule.

Table 1. Improved Δk_p fuzzy control rule table optimized by discrete bat algorithm.

E/EC	NB	NM	NS	ZO	PS	PM	PB
NB	PB	PB	PM	PM	NS	ZO	NS
NM	PB	PB	PM	PS	ZO	ZO	NS
NS	PB	PM	PM	PS	PM	NB	NM
ZO	PM	PM	PS	PS	NS	PB	PB
PS	PS	PS	PM	PS	ZO	NS	NM
PM	ZO	ZO	NS	NS	ZO	ZO	NS
PB	ZO	ZO	NS	NS	NM	NM	NM

Table 2. Improved Δk_i fuzzy control rule table optimized by discrete bat algorithm.

E/EC	NB	NM	NS	ZO	PS	PM	PB
NB	NB	NB	NM	NM	NS	ZO	ZO
NM	NB	NB	NM	NS	ZO	ZO	PM
NS	NB	NM	NS	PB	PB	NS	PB
ZO	NM	NM	NS	PB	PB	PS	PS
PS	NS	ZO	PS	PM	PM	PM	PS
PM	ZO	ZO	PS	PS	PM	PB	PB
PB	ZO	ZO	PS	PM	PB	PB	PB

Table 3. Improved Δk_d fuzzy control rule table optimized by discrete bat algorithm.

E/EC	NB	NM	NS	ZO	PS	PM	PB
NB	PB	PS	PM	NB	NS	NB	PM
NM	PM	ZO	ZO	PM	PS	ZO	PM
NS	PS	PM	PB	PM	PB	ZO	ZO
ZO	ZO	PB	PM	PB	PM	PB	NM
PS	ZO	NM	PB	PM	PM	ZO	ZO
PM	PS	PS	PM	NS	NS	NS	PM
PB	PM	ZO	PM	ZO	ZO	PB	PM

The optimal control rules obtained by PSO algorithm, GA algorithm and IDBA algorithm are used in fuzzy adaptive PID controller. The fuzzy is based on Mamdani reasoning method. The area center method is used clearly, and then the optimized fuzzy control rule is used. Simulation in the second-order closed-loop control system of Eq. (15). And when the control system is stable, it will be disturbed. The output response of the control system is shown in Fig. 5:

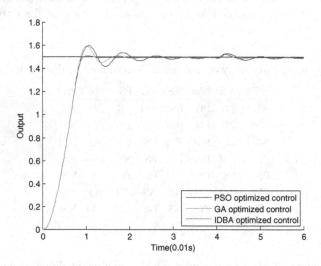

Fig. 5. Three algorithms optimize the output response of fuzzy rule control system.

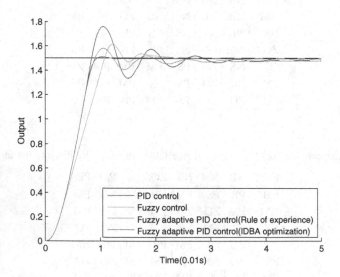

Fig. 6. Four control schemes control system output response.

It can be seen from the comparison of the three algorithms that the over-modulation of the fuzzy adaptive PID controller optimized by IDBA algorithm and PSO algorithm and GA algorithm is relatively small, the adjustment time is greatly improved, and the steady state can be quickly reached during the disturbance.

In order to fully verify the effect of improving the fuzzy rules by the improved discrete bat algorithm, PID control, fuzzy control, fuzzy PID control of empirical fuzzy rules in literature [12] and fuzzy adaptive PID control of fuzzy rules optimized by discrete bat algorithm are used. In the second-order system control in (15), the steady state is set to 1.5, and the control output is shown in Fig. 6:

By comparison, the improved discrete bat algorithm has obvious advantages in overshoot and adjustment time compared with the other three control schemes, and the steady state error is small. The effectiveness of the improved discrete bat algorithm for fuzzy rule optimization in fuzzy adaptive PID controllers is proved.

6 Conclusion

This paper introduces the neighborhood search operator and chaotic mutation operator to improve the discrete bat algorithm, improve the optimization precision of the discrete bat algorithm and expand the search range, and apply it to the fuzzy rule optimization of fuzzy adaptive PID controller. The fuzzy adaptive PID controller is applied to the typical second-order system control. The simulation results show that the improved control system of the discrete bat algorithm is compared with the particle swarm optimization algorithm and the genetic algorithm optimized control system in overshoot, adjustment time. There is a clear advantage in steady state error. It can quickly recover stability when external disturbance is applied, and has strong adaptability and robustness.

Acknowledgments. This work was supported by the Natural Science Foundation of China under grant (No. 61532007, 61370076), the Natural Science Foundation of Jiangsu Province under grant No. 15KJB520001. This work was partly supported by the Natural Science Foundation of Jiangsu Province of China under grant NO. BK2012209, Science and Technology Program of Suzhou in China under grant NO. SYG201409. Finally, the authors would like to thank the anonymous reviewers for their constructive advices.

References

1. Xiong, Z.G., Liu, X.Y., Jin, X., et al.: The intelligent control system of parameter self-adaptive PID based on fuzzy theory. J. Agric. Mechanization Res. **41**(05), 33–38 (2019)
2. Dettori, S., Iannino, V., Colla, V., Signorini, A.: An adaptive Fuzzy logic-based approach to PID control of steam turbines in solar applications. Appl. Energy **227**, 655–664 (2018)
3. Bouchebbat, R., Gherbi, S.: Design and application of fuzzy immune PID adaptive control based on particle swarm optimization in thermal power plants. In: 6th International Conference on Systems and Control, ICSC 2017, pp. 33–38 (2017)

4. Xi, Y., Zhuang, X., Wang, X., Nie, R., Zhao, G.: A research and application based on gradient boosting decision tree. In: Meng, X., Li, R., Wang, K., Niu, B., Wang, X., Zhao, G. (eds.) WISA 2018. LNCS, vol. 11242, pp. 15–26. Springer, Cham (2018). https://doi.org/10.1007/978-3-030-02934-0_2
5. Ma, X.Y.: Fuzzy PID temperature control system for DFB laser based on adaptive genetic algorithm. J. Shenyang Univ. Technol. **39**(04), 454–458 (2017)
6. Peng, P.F., Jiang, J., Huang, L.: The adaptive fuzzy control method for steady depth of underwater vehicle based on particles swarm optimization. Control Eng. China **24**(02), 441–445 (2017)
7. Wang, Z.C., Liu, J.W., Zhang, C., Cheng, J.: Study on drum lifting system of shearer based on genetic algorithm and fuzzy PID. China Min. Mag. **23**(01), 132–136 (2014)
8. Yang, X.S.: A New Metaheuristic Bat-Inspired Algorithm. In: González, J.R., Pelta, D.A., Cruz, C., Terrazas, G., Krasnogor, N. (eds.) Nature Inspired Cooperative Strategies for Optimization (NICSO 2010), pp. 65–74. Springer, Heidelberg (2010). https://doi.org/10.1007/978-3-642-12538-6_6
9. Yang, X.S., Gandomi, A.H.: Bat algorithm: a novel approach for global engineering optimization. Eng. Comput. **29**(5), 464–483 (2012)
10. Li, Z.Y., Ma, L., Zhang, H.Z.: Discrete bat algorithm for solving minimum ratio traveling salesman problem. Appl. Res. Comput. **32**(02), 356–359 (2015)
11. Xing, Y.L., He, X., Sun, S.Y.: Optimization design of fuzzy controller based on improved ant colony algorithm. Comput. Simul. **12**(01), 131–134 (2012)
12. Liu, H.B., Wang, J., Wu, Y.H.: Study and simulation of fuzzy adaptive PID control of brushless DC motor. Control Eng. China **21**(04), 583–587 (2014)

Recommendations

A Tag-Based Group Recommendation Algorithm

Wenkai Ma$^{(\boxtimes)}$, Gui Li, Zhengyu Li, Ziyang Han, and Keyan Cao

Faculty of Information and Control Engineering, Shenyang Jianzhu University,
Shenyang 110168, China
syjzmwk@163.com, ligui21c@sina.com

Abstract. Existing algorithms in group recommendation usually focus on studying the rating data instead of the social tag data. Finding the group based on tags is a novel problem for group recommendation. In this paper, we first employ locality-sensitive hashing to quickly find potential user groups of similar types. Next, we adopt higher-order singular value decomposition to explore the potential semantic relevance among the ternary relationship data of <user-item-tag> and fill the sparse data of users for items. Finally, we apply a fairness aggregation strategy based on proportionality and envy-freeness to generate the fairest group recommendation list that is simulated to address the problem of finding the maximum coverage. We use the MovieLens dataset to demonstrate that the fairness strategy proposed here has a better fairness. What's more, under the nDCG indicator, the algorithms based on the three strategies all generated a fairly good recommendation quality.

Keywords: Group recommendation · Social tag ·
Locality-sensitive-hashing (LSH) ·
Higher Order Singular Value Decomposition (HOSVD) · Fairness

1 Introduction

Recommender systems have been widely used in the mobile recommender system, context-aware recommender system, social network-based recommender system and various fields recently [1–6]. It is usually about personalized recommendation for a single user [4,5]. However, in the daily life, multiple users in group participate many activities. This kind of recommended objects varying from a single user to a group is called group recommender.

Based on the given group, most existing group recommenders are proposed. In fact, finding groups is a dynamic process in many cases [2].

In this paper, we adopt tag data marked by users and provide a group recommendation algorithm based on tags to quickly find group, improve the satisfaction of group members to group recommendation and optimize the group recommendation. We make the following contributions to deal with group discovery, users' preferences, preference aggregation method, preference aggregation strategy and other problems in the group recommendation:

1. We use LSH to quickly find potential user groups. (Sect. 2)

© Springer Nature Switzerland AG 2019
W. Ni et al. (Eds.): WISA 2019, LNCS 11817, pp. 677–683, 2019.
https://doi.org/10.1007/978-3-030-30952-7_68

2. In order to generate the personalized recommendation list of the single user in group, we use HOSVD to process the <user, item, tag> ternary relationship data. (Sect. 3)
3. In order to aggregate the personalized recommendation list in groups, we use a fairness aggregation strategy based on proportionality and envy-freeness, and propose a corresponding approximate greedy algorithm. (Sect. 4)

2   Quickly Find Similar Groups Based on the LSH

Two important elements can be used to find the potential groups usually including the size of groups and the similarity of users in groups. The reference [7] shows that the similarity of users in group is proportional to the recommendation effect. This section adopts the LSH to find the potential similar groups.

Data models in a group recommendation system are defined as follows. If user set $U = \{u_1, ..., u_m\}$, then $|U| = m$. If item set $I = \{i_1, ..., i_n\}$, then $|I| = n$. If tag set $T = \{t_1, ..., t_l\}$, then $|T| = l$. Every user can mark many tags for the same item. The <user, item, tag> ternary relationship is $A \subseteq \{(u, i, t) : u \in U, i \in I, t \in T\}$. If group $G = \{u_1, ..., u_g\}, G \subset U$, then $|G| = g$. If the recommendation package $P = \{i_1, ..., i_k\}, P \subset I$, then $|P| = K$. $S_G(P)$ is this group's satisfaction of group recommender.

In this section, we use user-tag data that is processed as a boolean type. It is recorded as the initial feature matrix $BM_{l \times m}$. bm_{ij} represents the element in the i row and j column. $bm_{ij} = 1$ means that user u_j prefers tag t_i. $bm_{ij} = 0$ means that user u_j dislikes tag t_i or the unknown situation.

Algorithm 1: MH-LSH
Input: The user-tag bool matrix $BM_{l \times m}$
Output: Group hash table HT
1. Initialising the signature matrix $SIG_{q \times m}$ and making all values be inf.
2. Calculating the minhash function $h_{mh_1}(r), h_{mh_2}(r), ..., h_{mh_q}(r), r \in [0, l-1]$.
3. Through the matrix $BM_{l \times m}$, if the element $bm_{r,c}$ at the c column and r row is 1, and the element $sig_{h,c}$ at the c column and j row is $\min(sig_{j,c}, h_{mh_j}(c))$ through the q minhash value.
4. The signature matrix is horizontally divided into b band, and the row number of every band is row, that is $b * row = q$.
5. First, traversing each band and performing a LSH calculation, and then dividing users into buckets, finally, returning the group hash table HT.

3   Recommendation of Groups Based on HOSVD

The reason why this paper adopts the ranking aggregation in the recommended aggregation is that [2]: (1) Group preference modeling is easily affected by sparse data, and the ternary relational data used in this paper: <user, tag, item> co-occurrence frequency is regarded as tensor element, and there are many sparse data. (2) When implicit processing of user preferences (such as SVD, etc.), the ranking aggregation effect is higher.

This paper uses HOSVD in the ternary tensor of users, tags, and items, and the physical meaning of every element represents whether user marks the tag t of the item i or not and shows the preference item i of user u according to tag t.

The relationship between multivariate data can usually be represented by tensors. In this paper, $(A, B, ...)$ represents tensors. $(\boldsymbol{A}, \boldsymbol{B}, ...)$ represents matrices. $(\boldsymbol{a}, \boldsymbol{b}, ...)$ represents vectors. $(a, b, ...)$ represents scalars. This paper focuses on the third-order tensor $A \in R^{U_m \times I_n \times T_l}$ of <user, item, tag>, and the element is represented as $a_{u,i,t}$.

In this paper, we use HOSVD to calculate the third-order tensor A, and the tensor A should be expanded into three matrices $\boldsymbol{A}_1$, $\boldsymbol{A}_2$, $\boldsymbol{A}_3$ that are decomposed to get the left-singular vectors $\boldsymbol{U}^{(1)}$, $\boldsymbol{U}^{(2)}$, $\boldsymbol{U}^{(3)}$. Through the extension of SVD, the HOSVD of three-dimensional tensor is defined as [8]

$$A = S \times_1 \boldsymbol{U}^{(1)} \times_2 \boldsymbol{U}^{(2)} \times_3 \boldsymbol{U}^{(3)} \tag{1}$$

Tensor A is calculated through HOSVD to get an approximate tensor $\hat{A}$ of low rank. Its formula is as

$$\hat{A} = \hat{S} \times_1 \boldsymbol{U}_{c_1}{}^{(1)} \times_2 \boldsymbol{U}_{c_2}{}^{(2)} \times_3 \boldsymbol{U}_{c_3}{}^{(3)} \tag{2}$$

$\hat{S} \in R^{c_1 \times c_2 \times c_3}$ is the approximate tensor of the core tensor S.

Algorithm 2: HOSVD algorithm based on LSH
Input: Tensor $A(u_g)$, composed of user u_g's neighboring users in group G.
Output: Personalized recommendation list R_g of u_g.
1. Initialising $R_g=[]$ and making the initial tensor $A(u_g)$ according to ternary data $\{u, t, i\}$.
2. HOSVD is used in tensor $A(u_g)$ to calculate related u_g's weights $W(u_g)$ towards all items based on tags.
3. Return R_g, recommending Top-N to u_g based on $W(u_g)$.

4   Aggregation Strategy of Fairness

In this section, we adopt the fairness of the proportionality and envy-freeness [9], and compare with two traditional aggregation strategies of average and least misery [10].

Proportionality. Given package P and parameter Δ, if the item $i, i \in P$ ranks $\Delta\%$ of the user u's preference towards I, user u prefers item i.

Definition 1. *Given user u, package P and parameter $m (m \geq 1)$, if there is user u of P likes m items at least, P meets the m-proportionality of u.*

If $m = 1$, the package of m-proportion is $single - proportional$. If $m > 1$, it is $multi - proportional$.

Definition 2. *m-proportionality: Given group G and Package P, the m-proportional metrics's definition of package P from group G is as*

$$F_{prop}(G, P) = \frac{|G_P|}{|G|} \tag{3}$$

Envy-Freeness. Given package P and parameter Δ and group G, in the set $\{w(v,i) : v \in G, i \in P\}$ centered by the item, if the association weight score $w(v,i)$ ranks $\Delta\%$ of the users' preference towards item i, and envy-freeness can be used to describe user $u \in G$ of any item i in P.

Definition 3. *Given user u, package P and parameter $m(m \geq 1)$, if there is user u of P likes m items at least, P meets the m-proportionality of u.*

Definition 4. *m-envy-freeness: Given group G and package P, the formula of package P's m-envy-freeness in group G is that $G_{ef} \subseteq G$ represents the group meeting of package P in group G.*

$$F_{ef}(G, P) = \frac{|G_{ef}|}{|G|} \tag{4}$$

This section changes the fairness maximization problem into a maximum set coverage problem. That is the maximization $F(G, P)$, where F represents fairness metric, m-proportionality or m-envy-freeness. Particularly, if $m = 1$, the fairness maximization problem is equal to the maximum set coverage problem. $SAT_G(i)$ is the number that satisfied users of item $i \in I$ in G. $SAT_G(P)$ represents all users of package P. $f_G(P, i) = |SAT_G(P \cup \{i\}) \backslash SAT_G(P)|$ is the utility function of item i added to P. The utility is the largest when new item i is added into package P. Maximizing the single proportionality metric is named as SPG_{REEDY}. Maximizing the envy-freeness metric is EFG_{REEDY}.

Algorithm 3: the fairness maximization greedy algorithm
Input: Group G, items I, value K.
Output: Group recommendation package P.
1. Initialising $P=[]$, $Candidates = I$.
2. If item i meets $max_{i \in Candidates} f_G(P, i)$ after K times and $P = P \cup \{i\}$, $Candidates = Candidates \backslash \{i\}$.
3. Return P.

5   Experiment

5.1   Experimental Designs

The ml-$20m$ in the public data set MovieLens is used in this experiment, including 465564 tag records for 19545 movies made by 7801 users. Besides, the number of tags per user is not less than 10 in this experiment. Without any relevant group records of data set, Algorithm 1 is used to find the potential and similar user groups. If it is not clearly stated, $|G| = 8$, $|P| = 4$, $\Delta = 5\%$.

In this section, the running efficiency of Algorithm 1 and sequential search are compared, and Algorithm 2 is used to fill 3D tensor data. The four algorithms are compared with the two fair measures defined in Sect. 4. The four algorithms are defined as follows:
SPG_{REEDY}: Using proportionality strategy to maximize $F_{prop}(G, P)$ based on the Algorithm 3.

EFG_{REEDY}: Using Envy-freeness strategy to maximize $F_{ef}(G, P)$ based on the Algorithm 3.

$AVRG_{REEDY}$: Choosing items to maximize average weight score of the group recommendation based on the greedy algorithm of average strategy.

$$AVR(G, P) = \frac{1}{|G||P|} \sum_{u \in G} \sum_{i \in P} w(u, i) \tag{5}$$

LMG_{REEDY}: Choosing items to maximize the least misery of the group recommendation based on the greedy algorithm of the least misery strategy.

$$LM(G, P) = \min_{u \in G} \min_{i \in P} w(u, i) \tag{6}$$

With the order of the group recommendation, this paper uses the evaluation measures for an information retrieval system nDCG to evaluate the group members' satisfaction with the group recommendation list (Top-k recommendation), which is defined as follows [2,3,6]:

$$DCG_k^u = r_{uc_1} + \sum_{i=2}^{k} \frac{r_{uc_i}}{\log_2(i)} \tag{7}$$

$$nDCG_k^u = \frac{DCG_k^u}{IDCG_k^u} \tag{8}$$

5.2   Experiment Results

Figure 1 shows that Algorithm 1 has obvious advantages if the number of tagging records is more than 80,000. Therefore, the larger the data, the more obvious the effect of Algorithm 1.

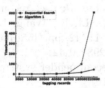

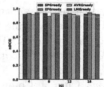

Fig. 1. Time cost   **Fig. 2.** nDCG

The Fig. 3 shows results of the four algorithms. This section uses the two fairness metrics defined in Sect. 4 as indicators. The SPG_{REEDY} algorithm performs more excellent both in proportional metrics and envy-freeness metrics. At the same time, the $AVRG_{REEDY}$ algorithm has a better performance in the two fairness indicators compared with the LMG_{REEDY} algorithm.

In this section, four ranking aggregation strategies are used to evaluate the groups with highly similar users. As shown in Fig. 2, different ranking strategies have little effect on the quality of group recommendation if the group size is 4, 8, 12, 16 respectively. The four algorithms have good results under the nDCG indicator.

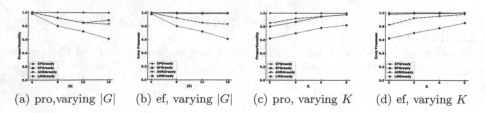

(a) pro,varying $|G|$ (b) ef, varying $|G|$ (c) pro, varying K (d) ef, varying K

Fig. 3. Performance comparison, m = 1

6   Conclusion

In this paper, we propose an LSH-based algorithm to quickly find potential and similar user groups. We propose a HOSVD-based algorithm to generate a personalized recommendation list for users in groups. We introduce two definitions of fairness based on proportionality and envy-freeness, and propose corresponding approximate greedy algorithm for group recommendations. Our experimental results show that the efficiency of finding groups are improved greatly in Algorithm 1. The group recommendation generated by the algorithm based on fairness strategy is superior in fairness. The four algorithms all have good recommendation results with the guide of nDCG indicator. Using the group recommendation results to alleviate the problems such as cold start problems, will be researched in the further study.

References

1. Xiong, X., Zhang, M., Zheng, J., Liu, Y.: Social network user recommendation method based on dynamic influence. In: Meng, X., Li, R., Wang, K., Niu, B., Wang, X., Zhao, G. (eds.) WISA 2018. LNCS, vol. 11242, pp. 455–466. Springer, Cham (2018). https://doi.org/10.1007/978-3-030-02934-0_42
2. Zhang, Y., Du, Y., Meng, X.: Research on group recommender systems and their applications. Chin. J. Comput. **39**(4), 745–764 (2016)
3. Jin, T., Xie, J., Yang, Z.: Research on fast build group and group recommendation in social networks. J. Chin. Comput. Syst. **38**(3), 483–488 (2017)
4. Symeonidis, P., Nanopoulos, A., Manolopoulos, Y.: Tag recommendations based on tensor dimensionality reduction. In: Proceedings of the 2008 ACM Conference on Recommender Systems, pp. 43–50. ACM (2008)
5. Li, G., Wang, S., Li, Z.: Personalized tag recommendation algorithm based on tensor decomposition. Comput. Sci. **42**(2), 267–273 (2015)
6. Wang, J., Zhang, C.: Research and analysis of group recommendation system. Comput. Technol. Dev. **28**(5), 164–169 (2018)
7. Ye, B., Xu, J., Yan, S.: Group recommendation method based on ratings and item features. Appl. Res. Comput. **34**(4), 1032–1035, 1046 (2017)
8. Lathauwer, L.D., Moor, B.D., Vandewalle, J.: A multilinear singular value decomposition. SIAM J. Matrix Anal. Appl **21**(4), 1253–1278 (2000)

9. Serbos, D., Qi, S., Mamoulis, N.: Fairness in package-to-group recommendations. In: Proceedings of the 26th International Conference on World Wide Web, pp. 371–379. ACM (2017)
10. Masthoff, J.: Group recommender systems: combining individual models. In: Ricci, F., Rokach, L., Shapira, B., Kantor, P.B. (eds.) Recommender Systems Handbook, pp. 677–702. Springer, Boston (2011). https://doi.org/10.1007/978-0-387-85820-3_21

A Graph Kernel Based Item Similarity
Measure for Top-N Recommendation

Wei Xu, Zhuoming Xu$^{(\boxtimes)}$, and Bo Zhao

College of Computer and Information, Hohai University, Nanjing 210098, China
{xwmr, zmxu, bzhao}@hhu.edu.cn

Abstract. The item neighborhood-based graph kernel (INGK) has recently been proposed to compute item similarity on the Linked Open Data (LOD) graph and then produce top-N recommendations (the similarity measure is abbreviated as INGK-LOD). This paper explores how to use the graph kernel to compute item similarity on the basis of user-item ratings. We transform the user-item ratings matrix into an undirected graph called a user-item ratings graph, and define the graph kernel based on the graph, which can be used to compute item similarity (the similarity measure is abbreviated as INGK-UIR). We applied INGK-UIR, INGK-LOD and two baseline similarity measures, Cosine and Pearson correlation coefficient, to top-N recommendation, and conducted experimental evaluation of recommendation accuracy using the MovieLens 1M benchmark dataset. The results show that our INGK-UIR significantly outperforms INGK-LOD and the two baseline measures in terms of precision and recall.

Keywords: Item similarity · Graph kernel · Top-N recommendation ·
User-item ratings · Linked Open Data

1 Introduction

The LOD [1] based recommender systems [2] exploit background knowledge from LOD to compute item similarity or user similarity [3]. Graph kernel [4] is a kernel function that computes inner products between two vectors in the Hilbert space to measure the similarity between a pair of graphs. Kernel methods can be used to solve recommendation problem, forming the kernel-based recommendation method [5].

Ostuni et al. [6] recently proposed the item neighborhood-based graph kernel (INGK) that is used to compute item similarity on the LOD graph and then produce top-N recommendations (the similarity measure is abbreviated as INGK-LOD). Their experimental evaluation results on the benchmark dataset MovieLens lM [7] show that INGK-LOD outperforms vector space model, naive bayes classifier and walk-based kernel in terms of accuracy and novelty. We argue that the user-item ratings in the recommender system are a more important source of knowledge which should be used as the (primary) basis for the computation of item similarity. This paper explores how to use the graph kernel to compute item similarity on the basis of user-item ratings (the similarity measure is abbreviated as INGK-UIR). The purposes of this work include: (i) proposing the method for computing item similarity using INGK based on the

© Springer Nature Switzerland AG 2019
W. Ni et al. (Eds.): WISA 2019, LNCS 11817, pp. 684–689, 2019.
https://doi.org/10.1007/978-3-030-30952-7_69

ratings data; (ii) comparing the performance of the two graph kernel-based similarity measures on the task of top-N recommendation; (iii) comparing the performance of graph kernel-based similarity measures with that of two baseline similarity measures, Cosine and Pearson correlation coefficient (PCC) [8].

We applied INGK-UIR, INGK-LOD and the two baseline similarity measures to the top-N recommendation task and conducted experimental evaluation of recommendation accuracy using the MovieLens 1M benchmark dataset [7]. The results show that our INGK-UIR significantly outperforms INGK-LOD and the two baseline measures in terms of precision and recall.

2 INGK-UIR: Graph Kernel Based Similarity Measure

The basic ideas of INGK-UIR are as follows: First, the user-item ratings matrix is transformed into an undirected graph. Second, for each item involved in the recommender system, an item neighborhood graph is extracted from the undirected graph. Finally, a kernel function is defined based on the neighborhood graphs, and item similarities are computed using the kernel function.

2.1 Construction of User-Item Ratings Graph

Given a user-item ratings matrix $R_{m \times n}$, the (u, i) th entry r_{ui} of the matrix indicates the rating of user u for item i. A user-item ratings graph as described in Definition 1 can be constructed based on the ratings matrix.

Definition 1 (user-item ratings graph). A user-item ratings graph is an undirected graph $G = (V, E)$, where $V = U \cup I$ is the set of nodes, U the set of user nodes, and I the set of item nodes, whereas $E \subseteq U \times I$ is the set of edges. There is an undirected edge $p_k(u, i) \in E$ between the node of user u and the node of item i if and only if the rating r_{ui} exists in $R_{m \times n}$, where $p_k = r_{ul}$ is the label on the edge and k is the serial number of discrete, ordered numbers in the interval-based ratings [9]. For example, the rating in 5-point ratings usually takes the value of ordered set $\{1, 2, 3, 4, 5\}$ or $\{-2, -1, 0, 1, 2\}$, in this case, $k = 1, \ldots, 5$.

2.2 Extraction of Item Neighborhood Graph

Similar to the extraction method of the h-hop item neighborhood graph proposed by Ostuni et al. [6], we extract the h-hop item neighborhood graph from the constructed user-item ratings graph. The h-hop neighborhood graph is described in Definition 2.

Definition 2 (h-hop item neighborhood graph). For item i, its h-hop item neighborhood graph $G^h(i) = (V^h(i), E^h(i))$ is a subgraph of G induced by the set of nodes that are reachable in at most h hops from i along the shortest paths.

Further, $\overline{V}^h(i) = V^h(i) \backslash V^{h-1}(i)$ is defined to represent the set of nodes exactly h hops far from i along the shortest paths. Likewise, $\overline{E}^h(i) = E^h(i) \backslash E^{h-1}(i)$ is defined to represent the set of the last edges on the shortest paths starting from i with their length being exactly h. The two sets are the basis for computing the item similarity using the graph kernel.

2.3 Computation of Item Similarity via Kernel Function

We can then use the same method in [6] to define the kernel function based on the h-hop item neighborhood graph. The kernel function is represented as the inner product of two vectors, as defined by Eq. (1).

$$k_{G^h}(i,j) = \langle \phi_{G^h}(i), \phi_{G^h}(j) \rangle \tag{1}$$

where $\phi_{G^h}(i)$ and $\phi_{G^h}(j)$ is a mapping function that transforms the h-hop item neighborhood graph $G^h(i)$ and $G^h(j)$ into a feature vector, respectively. Similar to the method in [6], the mapping functions are defined as Eqs. (2) and (3).

$$\phi_{G^h}(i) = \left(w_{i,u_1}, w_{i,u_2}, \cdots, w_{i,u_s}, \cdots, w_{i,u_m}, w_{i,i_1}, w_{i,i_2}, \cdots, w_{i,i_t}, \cdots, w_{i,i_n} \right) \tag{2}$$

$$\phi_{G^h}(j) = \left(w_{j,u_1}, w_{j,u_2}, \cdots, w_{j,u_s}, \cdots, w_{j,u_m}, w_{j,i_1}, w_{j,i_2}, \cdots, w_{j,i_t}, \cdots, w_{j,i_n} \right) \tag{3}$$

where w_{i,u_s} and w_{i,i_t} represents the weight associated with user node $u_s \in U$ and item node $i_t \in I$ in $G^h(i)$, respectively. The meaning of the symbols in Eq. (3) is similar to that in Eq. (2). The weight w_{i,i_t} is calculated using Eq. (4).

$$w_{i,i_t} = \sum_{l=1}^{h} \alpha_l \cdot c_{\overline{E}^l(i),i_t} \tag{4}$$

where the coefficient α_l is a non-negative real number, which can be regarded as an decay factor [10], denoted $\alpha_l = \frac{1}{1+\log(l)}$, for the distance l. $c_{\overline{E}^l(i),i_t}$ is the number of edges in $\overline{E}^l(i)$ that contains item node i_t, meaning the occurrence of i_t in the item neighborhood at distance l, and can be calculated using the similar method in [6], as defined by Eq. (5).

$$c_{\overline{E}^l(i),i_t} = \left| \left\{ p_k(u_s,i_t) | p_k(u_s,i_t) \in \overline{E}^l(i) \wedge i_t \in \overline{V}^l(i) \right\} \right| \tag{5}$$

Each feature vector $\phi_{G^h}(i)$ or $\phi_{G^h}(j)$ is normalized to unit length using the L2 norm, as the neighborhood sizes for each item is different. Finally, the similarity between items i and j can then be obtained by calculating the inner product of the two vectors with Eq. (1).

2.4 Top-N Recommendation Using Item Similarity

On the basis of the obtained item similarities, top-N recommendations can be produced using support vector regression (SVR) [11]. The recommendation process consists of two steps [6]: (i) for each user, a user model is learned using SVR based on the known relevance scores, which assigns a relevance score to each item; (ii) using the learned model to predict the score associated to all the unknown items and generate a list of top-N items for the user.

3 Experimental Evaluation

3.1 Experimental Design

To validate the effectiveness of our INGK-UIR, we applied INGK-UIR, INGK-LOD [6], and two baseline similarity measures, Cosine [8] and Pearson Correlation Coefficient (PCC) [8], to top-N recommendation and used the benchmark dataset to conduct experimental evaluation of recommendation accuracy. Two popular accuracy metrics, Average Precision@N and Average Recall@N [12], were used in the evaluation. As in [6], the calculation of accuracy values only considered such items that are included in the test dataset and used to generate the recommendation list.

Implementation. The four similarity measures were implemented through Java programming. The experimental results in [6, 10] show that $h = 2$ is a good compromise between time complexity and recommendation accuracy. Hence, our experiment set $h = 2$ for INGK-UIR and INGK-LOD. We used support vector machine library LIBSVM [13] (cf. https://www.csie.ntu.edu.tw/~cjlin/libsvm/) to implement support vector regression (SVR) and set the penalty factor C of SVR with the range being $\{0.1, 1, 10, 100, 1000\}$ [10]. We found in our experiment that the recommendation accuracy is highest when C = 1,000. Hence, our results were generated at C = 1,000.

Experimental Datasets. Our experiment used exactly the same datasets as in [6]: the MovieLens 1M dataset, DBpedia 2016-04 release, and DBpedia-MovieLens 1M dataset. Only the INGK-LOD method requires the use of the last two datasets.

- The MovieLens 1M dataset. This benchmark dataset [7] contains 1,000,209 ratings (5-point ratings $\{1, 2, 3, 4, 5\}$) from 6,040 users on 3,883 movies. The dataset was divided into two parts: training data (80%) and testing data (20%).
- The DBpedia 2016-04 release. The English version of DBpedia dataset [14] provides most of the movies in the MovieLens 1M dataset with a wealth of background knowledge in the form of RDF triples described with DBpedia's ontological properties and rdf:type, dcterms:subject and skos:broader.
- The DBpedia-MovieLens 1M dataset. This dataset [15] contains the mappings from MovieLens movie identifiers to DBpedia entity URIs.

3.2 Experimental Results and Discussion

Figure 1 shows the accuracy comparison between the four top-N recommendation methods that use Cosine, PCC, INGK-LOD and INGK-UIR, respectively, in terms of the two accuracy metrics, with N being 5, 10, 15, 20, and 25.

As shown in the figure, our INGK-UIR outperforms INGK-LOD and the two baseline measures in terms of all the two accuracy metrics in all the cases of five different values of N, and the most improved performance occurs when $N = 5$. More specifically, (i) when $N = 5$, INGK-UIR achieves the maximum increases of 9.11%,

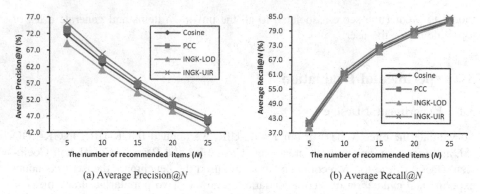

(a) Average Precision@*N*   (b) Average Recall@*N*

Fig. 1. Accuracy comparison of the four similarity measures.

5.01%, and 2.91% in Average Precision@N than INGK-LOD, Cosine, and PCC, respectively, and the average increases reach 7.6%, 4.12%, and 2.48%, respectively; (ii) when $N = 5$, INGK-UIR achieves the maximum increases of 6.78%, 3.82%, and 2.37% in Average Recall@N than INGK-LOD, Cosine, and PCC, respectively, and the average increases reach 3.4%, 2.44%, and 1.67%, respectively. The results show that our INGK-UIR significantly outperforms INGK-LOD and the two baseline measures, Cosine and PCC, in terms of precision and recall.

4 Conclusions

This paper proposes a graph kernel based item similarity measure (INGK-UIR) that can be used for top-N recommendation. Unlike INGK-LOD which uses LOD as the source of knowledge, INGK-UIR uses the user-item ratings as the source of knowledge. The top-N recommendation experiment on the MovieLens 1M benchmark dataset shows that INGK-UIR significantly outperforms INGK-LOD and the two baseline measures in terms of precision and recall.

References

1. Heath, T.: Linked data - welcome to the data network. IEEE Internet Comput. **15**(6), 70–73 (2011)
2. Di Noia, T., Tomeo, P.: Recommender systems based on linked open data. In: Alhajj, R., Rokne, J. (eds.) Encyclopedia of Social Network Analysis and Mining, 2nd edn, pp. 2064–2080. Springer, New York (2018). https://doi.org/10.1007/978-1-4939-7131-2_110165
3. Xu, W., Xu, Z., Ye, L.: Computing user similarity by combining item ratings and background knowledge from linked open data. In: Meng, X., Li, R., Wang, K., Niu, B., Wang, X., Zhao, G. (eds.) WISA 2018. LNCS, vol. 11242, pp. 467–478. Springer, Cham (2018). https://doi.org/10.1007/978-3-030-02934-0_43
4. Kriege, N.M., Johansson, F.D., Morris, C.: A survey on graph kernels. CoRR abs/1903.11835 (2019). https://arxiv.org/pdf/1903.11835

5. Fouss, F., Françoisse, K., Yen, L., Pirotte, A., Saerens, M.: An experimental investigation of kernels on graphs for collaborative recommendation and semisupervised classification. Neural Netw. **31**, 53–72 (2012)
6. Ostuni, V.C., Di Noia, T., Mirizzi, R., Di Sciascio, E.: A linked data recommender system using a neighborhood-based graph kernel. In: Hepp, M., Hoffner, Y. (eds.) EC-Web 2014. LNBIP, vol. 188, pp. 89–100. Springer, Cham (2014). https://doi.org/10.1007/978-3-319-10491-1_10
7. Harper, F.M., Konstan, J.A.: The MovieLens datasets: history and context. ACM Trans. Interact. Intell. Syst. (TiiS) **5**(4), 19:1–19:19 (2016)
8. Aggarwal, C.C.: Neighborhood-based collaborative filtering. In: Recommender Systems: The Textbook, pp. 29–70. Springer, Cham (2016). https://doi.org/10.1007/978-3-319-29659-3_2
9. Aggarwal, C.C.: An introduction to recommender systems. In: Recommender Systems: The Textbook, pp. 1–28. Springer, Cham (2016). https://doi.org/10.1007/978-3-319-29659-3_1
10. Oramas, S., Ostuni, V.C., Di Noia, T., Serra, X., Di Sciascio, E.: Sound and music recommendation with knowledge graphs. ACM Trans. Intell. Syst. Technol. (TIST) **8**(2) (2017). Article no. 21
11. Ho, C.H., Lin, C.J.: Large-scale linear support vector regression. J. Mach. Learn. Res. **13**, 3323–3348 (2012)
12. Aggarwal, C.C.: Evaluating recommender systems. In: Recommender Systems: The Textbook, pp. 225–254. Springer, Cham (2016). https://doi.org/10.1007/978-3-319-29659-3_7
13. Chang, C.C., Lin, C.J.: LIBSVM: a library for support vector machines. ACM Trans. Intell. Syst. Technol. (TIST) **2**(3) (2011). Article no. 27
14. Lehmann, J., Isele, R., Jakob, M., et al.: DBpedia - a large-scale, multilingual knowledge base extracted from Wikipedia. Semant. Web **6**(2), 167–195 (2015)
15. Di Noia, T., Ostuni, V.C., Tomeo, P., Di Sciascio, E.: SPRank: semantic path-based ranking for top-n recommendations using linked open data. ACM Trans. Intell. Syst. Technol. (TIST) **8**(1), 9:1–9:34 (2016)

Mining Core Contributors in Open-Source Projects

Xiaojin Liu[1], Jiayang Bai[1], Lanfeng Liu[1], Hongrong Ouyang[1], Hang Zhou[1,2], and Lei Xu[1,2(✉)]

[1] Department of Computer Science and Technology, Nanjing University,
Nanjing, China
xlei@nju.edu.cn
[2] State Key Laboratory for Novel Software Technology, Nanjing University,
Nanjing, China

Abstract. Open-source projects are becoming more and more prevalent, and the number of contributors involved in open-source project development is increasing. The evolution of a project is driven by all contributors. However, the role of core contributors is more important, especially in evaluating the progress and discovering bottlenecks of the project, etc. Visualization technology can show distribution and intrinsic relationship of data visually and reduce the complexity of information analysis. And it's appropriate to use different visualization methods in different research topics. The paper is based on common contribution behavioral indicators, such as adding code lines, submitting new files and solving a bug by committing, etc. We analyze the relationship between multiple indicators, measure the cooperation activities of the contributor, mine and analyze open-source projects on GitHub combining with a variety of visualization techniques, which reveals the core contributors in projects.

Keywords: Open source project · Data mining · Core contributor · Visualization

1 Introduction

With the rise of the open project, which breaks the monopoly position of software giant, improve the software development efficiency and shorten the cycle of the software evolution, more and more developer can participate in large project development. There are more than 2000 contributors have taken part in the Tensorflow, which is a well-known machine learning framework. Moreover, the similar neural framework Pytorch has attracted more than 1000 contributors.

The lack of advanced talents is usually the key which slows down the development of both open-source project and business software. With the increase in the number of open-source project contributors, but the core members of the open-source project are still a small number of people, usually no more than 3% of the total number. Therefore, discovering the core contributors in the project and their behavioral characteristics is of great significance for understanding the composition and evolution of the project.

W. Ni et al. (Eds.): WISA 2019, LNCS 11817, pp. 690–703, 2019.
https://doi.org/10.1007/978-3-030-30952-7_70

This paper is based on the Git API and GitHub API to obtain open-source project data, with the contribution as the basic indicator, select a variety of methods to measure contribution, use data mining and data statistics to process data, and use diversified visualization means to display the results. The results show the synergy among contributors of the project and reveal the behavioral characteristics of the core contributors.

In Sect. 2, we introduce the related work, including visualization means used by some popular software hosting platforms for stored data, researches about how to measure the contribution as well as paper about synergy among contributors. The Sect. 3 shows the source of data and the method of acquisition. The visualization results and analysis are presented in Sect. 4. The conclusion and outlook of our research are put in the Sect. 5.

2   Related Work

In the field of Mining Software Repository (MSR), current research hotspots focus on the field of software evolution. There are two main research directions in developer behavior characteristics: (1) research on the measurement of contribution of project participants; (2) research on the way of project participants' division of labor.

Meanwhile, the visualization means used by popular software hosting platforms mainly include line charts and histograms, which only show the simple contribution of developers but can't display the project's personnel contribution and personnel collaboration in depth and intuitively.

2.1   The Measurement and Calculation of Contribution

Early researchers' measurement of contribution is based on the amount of words in the work unit. With the development of programming language, lines of code replace the amount of words as the measurement, such as the classic LOC (Lines of Code). However, the development and management of modern software projects is not limited to personal development and submission. Simply measuring the contribution in units of code, functions, etc. is not enough.

Therefore, Gousios summarize a series of behaviors of participants in the open-source software project (Table 1). In the table, there are two kinds of values in the column named type: P (positive) denotes the behavior has a positive impact on the project, while N (negative) means a negative impact.

Recently, research on the contribution measurement of contributors has gradually turned to analysis of developers' commit and the commit standard of the project. Besides, research on the measurement of contribution in the dimension of time is gradually emerging (a project is usually composed by the beginning, the development and maintenance phases), trying to find developers' contribution behavior characteristics over different time periods.

Table 1. The actions of participants.

Asset	Action	ID	Type
Code and Documentation Repository	Add lines of code	CADD	P
	Remove code	CREM	P
	Update code	CCGN	P
	Commit new source file	CNS	P
	Commit new directory	CND	P
	Commit code that generates a bug	CGB	N
WIKI	Start a new wiki page	WSP	P
	Update a wiki page	WUP	P
	Link a wiki page from documentation file	WLP	P

2.2 Visualization of Mainstream Software Hosting Platform

GitHub is one of the most popular software hosting platforms at present. In order to show the contribution data, there are some visualization of the data presenting on the GitHub webpages. Including the following:

① Demonstrate the amount of contribution received by the entire project over time (Fig. 1)

Fig. 1. The amount of contribution received by the entire project

② Demonstrate the contributor's contribution code volume changes over time (Fig. 2)

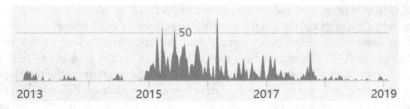

Fig. 2. Contributors' contribution code volume changed over time

③ Demonstrate the trend of commits over time in project (Fig. 3)

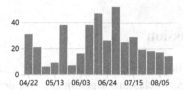

Fig. 3. Contributors' contribution code volume changed over time

④ Demonstrate the number of additions and deletions each week (Fig. 4)

Fig. 4. The number of additions and deletions each week

As shown above, GitHub's visualization method is simply based on number. The project manager can see the trend of the overall contribution of the project and the trend of each developer's contribution to the project, but cannot directly see the relationship between the developer's contribution and the development of the project, cannot known the contrast neither the collaboration among developers.

This research aims to more intuitively demonstrate the quantitative results of the developers' contribution behavior, as well as compare developers' contribution to the project. At the same time, it tries to depict the collaboration among developers, finding the way and scale of collaboration among them.

3 Data Source and Acquisition

3.1 Data Acquisition

JGit is a pure Java library that implements the Git version control system by the Eclipse Foundation. JGit can be used to get a local repository object reference through the underlying serialization, which contains all the information about the repository. This provides a stable and reliable data source for the research.

3.2 Data Source: Alluxio, the Open-Source Project

In order to study the behavioral characteristics of core contributors in open source projects, it is necessary to select a project that is large in scale, which has a large number of contributions and is still active. The Alluxio is a virtual distributed storage system project hosted on GitHub. The Alluxio project has evolved by following the rapid development of big data technology. By April 2019, the project contains 952 contributors, 29,267 commits and 48 releases, which is suitable for research.

4 Results and Discussion

This section will show the details of data acquisition and visualization results and then give corresponding analysis conclusions. Specifically, we demonstrate the evolution of overall contribution of the project, the relationship between the individual contribution and the overall contribution, the similarity of different contributors' contributions, the preferences of the contributors, and the collaborative relationship between the contributors to study the behavioral characteristics and reveal the core contributors in the project.

4.1 The Life Cycle of the Project

Commit Sequence Evolution of the Project

For an open source repository, a commit can be thought of as a code update. Therefore, the user's modification of the code can also be reflected by the number of commits. The number of commits can also reflect the activity of the open-source repository. We count the number of the commits from the starting time of Alluxio to March 2019 by months, and draw a line chart as shown in Fig. 5.

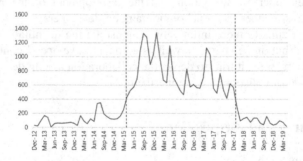

Fig. 5. The number of commits per month

The entire life cycle of the Alluxio project is divided into three phases: pre-, mid- and post-phase based on the number of commits per month that if it reaches 400.

Especially in the mid-phase, from April 2015 to December 2017, the number of commits accounted for 83.69% of the total commits. And the average monthly number of commits is 761. The mid-phase is a period of rapid development and stability of the project. The number of commits per month has been significantly improved compared with the pre-phase, peeking in September 2015, January 2016, May 2016 and April 2017.

Information Mining Based on the Classified Commit

In many open-source projects, commit message needs to briefly describe the purpose and modification of the code. Commitizen is a tool designed for teams. Its main purpose is to define a standard way of committing rules and communicating it. Focusing on the contribution of the code, the above types are summarized into five types as shown in Table 2. And we extract the keywords commonly used in the corresponding types.

Table 2. Commit type and keywords

Type	Keywords
Add	add, increas(e), creat(e), imple(ment)
Mod	modif(y), merg(e), refactor, improv(e), format(e), organi(ze), updat(e)
Fix	fix, bug, chang(e)
Del	delete(e), deprecat(e), clean, remov(e)

We collect commit message from the Alluxio and classify commits by keywords. The number of commits for each type is counted by month. The line graph is drawn as shown in Fig. 6.

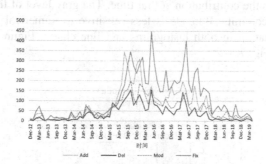

Fig. 6. Time variation graph of the number of commits in the four types

We find that the four types of commits have similar trends over time. Among them, fix accounts for the largest proportion of contribution, which is 39.83%. In the post-phase, the proportion of fix type increases significantly. From January 2018 to April 2019. And the proportion of fix type reached 52.27% of the total number of commits during the period. The peak period of fix type always appears with the peak period of add type, as adding new features has a risk of having a new bug.

4.2 The Time Sequence Relationship Between Individual Contributions and Overall Contribution

We measure the contribution with the number of commits per week as evaluation indexes and analyze the time sequence relationship between individual contributions and overall contribution. Figure 7 shows Bin Fan's contribution, Haoyuan Li's contributions and the overall contribution over time. It can be seen that these visualization methods are sensitive to numerical values, but it is difficult to distinguish different stages of the project. And when the number of individual commits is large, it is difficult to distinguish the relative relationship between the individual and the overall.

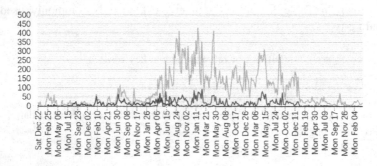

Fig. 7. Time variation graph of contributions by Bin Fan, Haoyuan Li and overall project

Barcode is chosen to visualize developers' contribution here. The barcode consists of vertical parallel lines. Lines re arranged in chronological order from left to right. Each line represents the contribution at that time. The gray level of lines is determined by the number of commits. Barcode is less sensitive to numerical values and retain most of the information in both histograms and line charts. Figure 8 is the barcode transformed from Fig. 7.

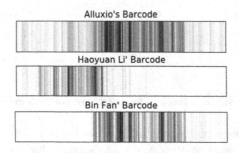

Fig. 8. Barcode of Alluxio, Haoyuan Li and Bin Fan

With the barcode in Fig. 8, we can divide the project into three phases, which is same as the life cycle of the project. The peak of Haoyuan Li's contribution is at the beginning of the project so Haoyuan Li can be inferred to be one of the founders of the project. The trend and gray level of Bin Fan's barcode is almost consistent with those of the overall project's barcode, and we can infer that Bin Fan is the core contributor who have make great contribution during the rapid development period of the project.

4.3   Contribution Similarity Analysis

The process of contributor barcode abstraction eliminated the sensitivity of methods such as line graphs to numerical values so the barcode can visually show the contributor's contribution in the whole project life cycle. To solve this problem, we introduce an indicator to measure the similarity of curves: Fréchet Distance (Fig. 9).

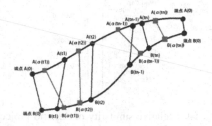

Fig. 9. Fréchet distance

The Fréchet Distance is a path space similarity indicator proposed by the French mathematician Maurice René Fréchet in 1906. It focuses on the distance in path space, making it more efficient in evaluating the similarity of curves with a certain spatial time series.

We select contributors with more than 100 commits Alluxio project for comparative study, and sorts them in descending order according to the times of commits. Here, we defined a new indicator: Relative Similarity.

Relative Similarity
Relative similarity is a measure of the similarity between individual contribution and total contribution, which is defined as the reciprocal of Fréchet Distance. In order to enhance the visualization, we normalize the data before calculation, so that the area between contribution curve and transverse axis (time axis) is 1 in the whole life cycle.

Due to the normalization of data, the relative similarity measures the contributors whose individual contributions are most similar to the overall contribution of the project throughout the life cycle. As shown in Fig. 10, although the number of commits is more than ten, the contributions of Yanqin Jin, David Capwell and Shaoshan Liu are similar to overall curve, so it can be inferred that they participate in the project more frequently throughout the life cycle and follow the development process of the project.

Combining the relative similarities of the 30 contributors with highest LOC contribution in the Alluxio project, we can get a relative similarity matrix. As shown in Fig. 11, the darker the color of the pixel in the figure, the more similar the contribution behavior between the corresponding two contributors, which can reflect the synergy between the contributors to a certain extent. As we can see from the graph, although the data is normalized, the synergies among the contributors with more commit times can still be reflected from the graph.

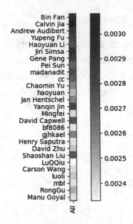

Fig. 10. Relative similarity of Alluxio project

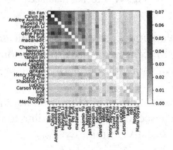

Fig. 11. Relative similarity matrix of Alluxio

4.4 Analysis of Contribution Heat Map

The contribution heat map shows contributors' contribution on project files. Here we select 20 contributors with the most LOC contribution and 50 files with the largest cumulative number of modified rows in project Alluxio to form the "contributor-file-LOC" matrix, and the contribution heat map is drawn according the matrix, as shown in Fig. 12. The shade of the gray square represents the LOC contribution of the corresponding contributor on the project file, where the darker color indicates a higher contribution, and vice versa.

To better analyze contribution behavior, we define two new indicators: contribution depth and contribution breadth. Contribution depth is the average contribution LOC on the file that the contributor has contributed:

$$Depth_{Contri} = \frac{LOC_{total}}{Files_{Contri}}$$

Contribution breadth is the percentage of the total number of contributions contributed by the contributor to the total number of files:

$$Width_{Contri} = \frac{Files_{Contri}}{Files_{total}}$$

Fig. 12. Contribution heat map of Alluxio

By observing and analyzing Fig. 13, the following phenomena and conclusions can be drawn:

① Most contributors who contribute a lot have high LOC contribution.

In order to analyze the relationship between contribution breadth and contributor LOC contribution, a scatter plot of contribution breadth and LOC contribution is drawn according to the contribution heat map, as shown in Fig. 13.

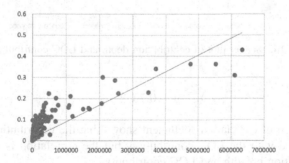

Fig. 13. Scatter plot of contribution breadth and LOC contribution

In can be seen from Fig. 13, the contributor who contributed the most, the breadth of their contributions reached nearly 0.45. In other words, the contributors with the most contribution breadth have contributed to nearly half of the code files throughout the whole project.

It also can be seen from Fig. 13 that the contribution LOC is basically positively correlated with the contribution breadth, and the correlation coefficient is calculated as below:

$$\rho = 0.8122$$

This shows that there is a strong correlation between contribution breadth and LOC contribution. At the same time, if can be seen in Fig. 13 that the vast majority of contributors (95%) have a LOC of less than 100,000 and a contribution breadth of less than 0.15, which indicated that the core contributors in the project are less than 5% of minority, and most of them are peripheral contributors to the project.

② The contribution of some contributors to certain files (modules) is very large.

Figure 14 shows a scatter plot of contribution depth and LOC contribution in the Alluxio project. Intuitively, the two are basically positive related, which is consistent with the contribution breadth and LOC contribution. Calculate the correlation coefficient:

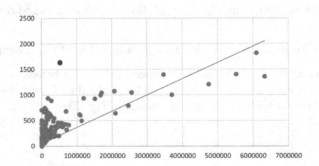

Fig. 14. Scatter plot of contribution depth and LOC contribution

$$\rho = 0.7632$$

The calculation of correlation coefficient shows that the contribution depth is still strongly correlated with the LOC contribution, but the correlation is slightly weaker than the contribution breadth and LOC contribution.

4.5   Contributor Network and Contribution Synergy Analysis

Contributor Network

The contributor network is used to visualize the synergies between contributors in the project. The contributor network is an undirected graph where each node represents a contributor in the project. Whether there is an edge between the two contributors. When

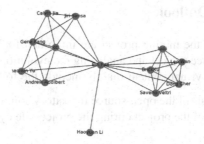

Fig. 15. Contributor network when K is 10

this number exceeds a certain threshold K, an edge is drawn between the two contributor nodes. Figures 15, 16 show the contributor network for the alluxio project when K is 10 and 15, respectively.

When K is taken as 10, the number of nodes in the contributor network is 13, which is less than when K is 5. Besides the synergy between contributors, a valuable phenomenon can be found in Fig. 15: The network forms three branched centered on Bin Fan. For the two branched located at the left and right, the number of contributor nodes is 6 and 5 respectively, and the internal network of the branch is dense. This shows that there is still a clear division of labor among the core contributors of this project. Some contributors focus on contributing code on the modules they are responsible for, and the collaborative behavior of the internal contributors of the modules is very frequent.

When K is increased to 15, the number of nodes in the contributor network is reduced to 4 (Fig. 16). It can be observed that Bin Fan is still the center of contributor network for the four contributors with the largest number of co-contributing files in the project, indicating that its position in the project is very important. These four contributors contribute to almost all modules of the project. Comparing the results of the previous barcode, it can be concluded that the four contributors have a high contribution in all periods of the Alluxio project life cycle, and their contribution behavior throughout the whole life cycle of the project.

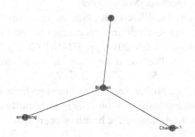

Fig. 16. Contributor network when K is 15

5 Summary and Outlook

The core contributors in the mining project are important for understanding the composition of the open-source project and analyzing the evolution process. The content of mining analysis and the visual content mainly include the following points:

① Mining analysis of the open-source repository commit time series, analyzing the evolution of the project during the project life cycle according to the line chart;

② Use keyword matching to classify the commits of open-source repositories, and analyze the differences in the proportion of commits in different categories and their evolution over time;

③ Based on the Fréchet distance, the relative similarity and the absolute similarity are used to measure the contribution of the contribution over time, and the collaborative relationship between the contributors is analyzed;

④ Draw and analyze the contribution hotspot map, find the division of labor in the project and reveal the contributors in the project that are in charge of the project.

In the future work, we will continue to conduct in-depth research on the measurement of code quality, quantitative analysis of large-scale project architecture and side-by-side data analysis of open source projects, so as to further reveal the contribution composition of the project and discover the core contributors of the project from different views.

References

1. German, D.M.: The GNOME project: a case study of open source, global software development. Softw. Process Improv. Pract. 8(4) (2004)
2. Godfrey, W., Tu, Q.: Evolution in open source software: a case study. In: Proceedings 2000 International Conference on Software Maintenance. IEEE (2000)
3. Gousios, G., Kalliamvakou, E., Spinellis, D.: Measuring developer contribution from software repository data. Electron. Notes (2009)
4. Kalliamvakou, E., Gousios, G., Blincoe, K., Singer, L., German, D.M., Damian, D.: The promises and perils of mining GitHub. In: Proceedings of the 11th Working Conference on Mining Software Repositories, MSR 2014, pp. 92–101 (2014)
5. Crowston, K., Howison, J.: The social structure of open source software development. J. Internet 10(2), 132–141 (2005)
6. Nakakoji, K., Yamamoto, Y., Nishinaka, Y.: Evolution patterns of open source systems and communities. In: International Workshop on Principles of Software Evolution. ACM (2003)
7. Crownston, K., Howison, J.: Assessing the health of open source communities. Computer 39(5), 89–91 (2006)
8. John, W.M.: Fencing a bar chart. J. Constr. Div. 107(3), p497–p507 (1981)
9. Asplund, E.: Frechet differentiability of convex functions. Acta Mathematica 121(1), 31–47 (1968)

10. Carliss, Y.B., Kim, B.C.: The architecture of cooperation: how code architecture mitigates free riding in the open source development model. Harvard Business School (2003)
11. Zhang, J., Xu, L., Li, Y.: Classifying Python code comments based on supervised learning. In: Meng, X., Li, R., Wang, K., Niu, B., Wang, X., Zhao, G. (eds.) WISA 2018. LNCS, vol. 11242, pp. 39–47. Springer, Cham (2018). https://doi.org/10.1007/978-3-030-02934-0_4

A QueryRating-Based Statistical Model for Predicting Concurrent Query Response Time

Zefeng Pei, Baoning Niu$^{(\boxtimes)}$, Jinwen Zhang, and Muhammad Amjad

Taiyuan University of Technology, 209 University Street, Yuci,
Jinzhong 030600, Shanxi, China
1194089714@qq.com, niubaoning@tyut.edu.cn,
259313322@qq.com, amjadsadiq786@yahoo.com

Abstract. Predicting query response time plays an important role in managing database systems. It can be used for tasks such as query scheduling, resource allocation, and system capacity planning. Due to the uncertainty of database systems, especially query interactions between concurrent queries, query response time varies greatly from one run to another. At present, there are two types of models for predicting query response time, namely, analytical models and statistical models. Since these models do not quantify query interactions and select the modeling metrics properly, they are only suitable for predicting the response time of the medium size queries. To address the issues, this paper proposes to quantify query interactions using QueryRating, and introduce a new statistical model TMT&TP to predict the response time of queries without the size constraint. Experiments show that the average prediction accuracy of response time for any size of queries is as high as 83%.

Keywords: Query response time · Query interaction · QueryRating · Query size

1 Introduction

Predicting the query response time is the key to managing database systems, such as scheduling queries to meet service level agreements. The predicted query response time must take into account the interactions between the queries executed in parallel [1]. In database systems, due to various uncertainties, especially query interactions between queries, query response time varies greatly when it runs in different query mixes. Therefore, it is challenging to accurately predict query response time.

The prediction of query response time can be used for query scheduling, progress indication [2], load balancing, cloud computing, etc., to reduce query cost, improve query efficiency and optimize user experience. Therefore, it is desirable to accurately predict query response time.

At present, the prediction method for query response time is based on the establishment of analytical models [3] and statistical models [4]. Analytical models are significant to predict the response time of a single query. Due to the complexity of analytical models and the implementation details of databases, when queries are running in parallel, analytical models have low prediction accuracy. At present, response time prediction of concurrent queries is primarily based on the establishment of statistical models. The state of the art statistical models do not quantify query interactions and improperly select the modeling metrics, so they are only applicable to the medium size queries [4, 5]. That is to say, statistical models are within a certain range of query modeling, and the scope of use is limited. In the real world, we cannot estimate the size of a query without certain auxiliary means before the query is executed. Whenever a light size or heavy size query appears in a query mix, the existing statistical models cannot function properly. It is desired to have a statistical model without the restriction on the query size. This paper proposes a TMT&TP model by removing the size restriction of BAL&B2L model. The innovations of this paper follow.

- QueryRating is used to quantify query interactions and applied to the model;
- A mapping TMT (Two-Many-Time) from two concurrent queries to multiple concurrent queries response time is proposed;
- A new response time prediction model TP (Time Prediction) is proposed without the size restriction on queries.

The rest of the paper is organized as follows. Section 2 discusses the related work. Section 3 introduces a BAL&B2L model. Section 4 introduces the TMT&TP and explains its modeling process. Section 5 reports the experimental studies. Section 6 concludes the paper.

2   Related Work

The existing prediction methods for query response time focus on establishing analytical models or statistical models. The analytical model obtains query response time by performing a detailed analysis of query process performed on the database system. The main research contents are: the speed at which the system executes query, and the amount of data that query needs to process. For the use of analytical models to predict the response time of a single query, if we estimate the above two factors, the query response time can be obtained. Chaudhuri et al. [3] and Luo et al. [6] published a paper respectively on query progress indicators in 2004, and studied progress indicators and query response time prediction models based on analytical methods. They refer to segmenting the query plan, computing the workload of the segment and the execution rate of the system, and then obtaining the execution time of the segment, and finally sum up the execution time of all segments. For the prediction of concurrent query response time, it is also necessary to consider the resource contention between queries, that is, query interactions. Because of the contention for computer resources (mainly CPU and IO) between concurrent queries, simply establishing a classical analytical model [7] for predicting query response time will cause a large error. The concurrent query analytical model first divides the query into pipes, and then refines the query's parallel resource contention to the query pipeline's parallel resource contention [8].

It is of significance to use analytical models to predict the response time of a single query. Because the analysis of the query for this type of model focuses on the operator level and is independent of the specific query plan, analytical models can model various queries. However, because the analytical method under concurrent query is also very complicated, and its final model is often related to the database system platform, it is very difficult for the analytical model to predict the query response time of concurrent query. Therefore, many scholars use statistical model to overcome the above shortcomings.

Statistical model [5, 8–12] refers to selecting and running a certain amount of samples in advance, giving the input and output according to the results of the sample, using statistical methods to train a model according to the training set. The model is then judged by applying the test set, and finally gives the prediction. The prediction accuracy of concurrent queries using statistical models is better than using analytical models. The current statistical models are mainly divided into linear query interaction model and its improvement, B2L model and its improvement. The linear query interaction model [1, 13] refers to the interactions between row queries as query interactions, and maps the number of each query in the query mix to the query response time. The B2L model [5] estimates the interactions by two concurrent queries, to measure the degree of the query interactions for a query in a query mix. Because the existing statistical models do not quantify the query interaction and select the modeling metrics properly, the model can only be used to predict the response time of medium size queries, and the scope of the model is limited.

In recent years, with the increment of data, the unstructured configuration of database parameter, and the increasing enrichment of query types, the existing statistical models cannot meet the needs for accurately predict the response time of concurrent queries of different query sizes. This paper proposes a statistical model TMT&TP based on the BAL&B2L statistical model. The model can predict the response time of different sizes of concurrent queries, which expands the scope of use of the model to meet actual needs while maintaining a certain degree of precision.

3   THE B2L (Buffer 2 Latency) Model

This section briefly introduces the B2L model, from which our proposed model is derived. The terms used are giving first.

3.1   Terms

See Table 1.

Table 1. The definition of related terms.

Terms	Description
MPL	Multi programming level, it represents the number of concurrent queries in the database
Query mix	Query collection when two or more queries run in parallel
Query interaction	The influence between queries when a query mix runs
Query size	Using TPC-H benchmark with a 10G database, queries are divided into light size, medium size and heavy size queries according to their average response time less than 50 s, 50 to 300 s, greater than 300 s

3.2 The Introduction to BAL&B2L Model

BAL&B2L [5] is a statistical model for predicting the response time of concurrent queries. This model has low cost and high prediction accuracy than other models. The model first calculates the average block fetch time of the query as its BAL (Buffer Access Latency) value, and then predicts the performance B2L (Buffer 2 Latency) of the query according to its BAL value.

BAL (Buffer Access Latency) Model. The BAL value of the query q is equal to the total time spent by the query to read the blocks divided by the total number of blocks read by this query. The BAL model for a query is as follows:

$$BAL_q = \alpha T_q + \beta \sum_{i=1}^{n} T_{c_i} + \gamma_1 \sum_{i=1}^{n} \Delta T_{q/c_i} + \gamma_2 \sum_{i=1}^{n} \sum_{j=1}^{n,i \neq j} \Delta T_{c_i/c_j} \tag{1}$$

In the above formula, T_q represents the BAL value of the query q when it is executed alone. $\Delta T_{q/c_i}$ represents the difference between the BAL value of q when the query q and c_i are executed in parallel and the BAL value when the query q is executed alone. $\Delta T_{c_i/c_j}$ represents the difference between the BAL value of c_i when the query c_i and c_j are executed in parallel and the BAL value when the query c_i is executed alone.

B2L (Buffer 2 Latency) Model. After obtaining the BAL value of the query q, B2L model predicts the response time of the query q according to its BAL. The B2L model is established as follows.

$$B2L_q = O_q + p_q \times BAL_q \tag{2}$$

In the above formula, O_q represents the fixed CPU overhead of query q, and P_q represents the total number of logical reads of query q.

4 The New Model TP

4.1 QueryRating to Quantify Query Interaction

From the above B2L model, we can see that the BAL value of the query is the key to establishing the B2L model. The BAL value of a query is equal to the total time spent reading the blocks by the query divided by the total number of blocks read by this query. In the actual database system, as the needs of enterprises and users increase, the types of queries are becoming more and more abundant, and the configuration of the database is often unable to change at will, if a light size or heavy size query appears in a query mix, the model cannot be modeled using the B2L. Therefore, it is necessary to use a new query interaction quantification method, which can complete the modeling of concurrent query response time without considering the query size before modeling.

We know that in the database system, CPU usage, buffer pool hit rate, I/O physical read and write are the main factors affecting the performance of a query. These factors are microscopic and sequential. They are often considered when modeling with analytical models, but statistical models can avoid the complexity. The response time is the most direct and macroscopic performance of query interactions. Therefore, QueryRating [17] is used to measure the degree of query interactions between two queries in parallel. The expression for establishing QueryRating is as follows:

$$\gamma_{p/q} = \frac{t_{p/q}}{t_p} \tag{3}$$

In the above formula, $t_{p/q}$ represents the response time of query p when query q and p are executed in parallel, and t_p represents the response time of query p when p is executed alone.

When $\gamma_{p/q} < 1$, it means that the response time of p, when executed in parallel with q, is shorter than the response time when it is executed alone. Due to the addition of q, the response time of the p is reduced. We can see that resources are shared by p and q, and there is a cooperative relationship between the two queries. The smaller the $\gamma_{p/q}$ value, the greater the response time promotion of q to p, and the smaller the degree of interactions between the two queries.

When $\gamma_{p/q} = 1$, it means that the addition of q has no effect on the execution of p query, and the response time of p does not change.

When $\gamma_{p/q} > 1$, it means that the response time of p when executed in parallel with q is longer than the response time when it is executed alone. Due to the addition of q, the response time of p increases, and it can be seen that there is a resource competition relationship between p and q. The larger the $\gamma_{p/q}$ value, the greater the response time hindrance of q to p, and the greater the degree of interactions between the two queries.

4.2 A Mapping $TMT_{p/q}$ (Two-Many-Time)

As we can see from Sect. 4.1, QueryRating can quantify the degree of interaction between two queries in parallel. If there are more than two queries in parallel in the database, we might model the query response time when multiple concurrent queries are in parallel via two queries interactions. We need to find a mapping from two querie interactions to multiple query interactions. We propose a mapping method TMT (Two query interactions to Many query interactions Time) to do so. The expression for establishing TMT is as follows:

$$\text{TMT}_{p/q} = \rho \times \gamma_{p/q} \times t_{p/q} \tag{4}$$

In the above formula, $\gamma_{p/q}$ represents the degree of the interaction of q on p when p and q are executed in parallel, $t_{p/q}$ represents the response time of p when p and q are executed in parallel, and ρ is referred to as the weighting factor, used to indicate the size of the $\gamma_{p/q}$.

We explain the $TMT_{p/q}$ as follows: When the query p and q are executed in parallel, the response time of p is $t_{p/q}$. Due to the addition of other queries, multiple queries (MPL $\geq$ 3) in the database are in parallel. For the effect of q on p, based on the original response time $t_{p/q}$ of p, q will block or promote the execution of p with the degree of $\gamma_{p/q}$. Considering the difference in the type of queries added to the database, we set the weighting factor ρ to indicate the size of the impact of q on p.

4.3   The Prediction Model of Response Time TP

From Sect. 4.2, we know that $TMT_{p/q}$ means that when there are multiple concurrent queries in the database, not only the effect of q on p makes the p response time change, but also other queries will also affect the execution of p. Based on this, the query response time prediction is augmented as follows:

$$\text{TP}_p = \alpha t_p + \beta \sum_{i=1}^{n} t_{q_i} + \gamma_1 \sum_{i=1}^{n} \text{TMT}_{p/q_i} + \gamma_2 \sum_{i=1}^{n} \sum_{j=1}^{n,i \neq j} \text{TMT}_{q_i/q_j} \tag{5}$$

In the above formula, t_p represents the response time of p executed alone, t_{q_i} represents the response time of q_i executed alone, TMT_{p/q_i} represents that when there are multiple queries in parallel in the database, the change of the response time of p contributed by q_i, and TMT_{q_i/q_j} represents that when there are multiple queries in parallel in the database, the change of the response time of q_i caused only by q_j.

We explain the model TP_p as follows: When there are multiple queries in parallel in the database, the response time of query p will not only be affected by other queries, but also the interaction between other queries. This causes changes in the allocation of database resources, so it also affects the query p. It has been experimentally verified that the lack of any of the above four modeling indicators will directly reduce modeling accuracy.

5 Experiments

5.1 Experimental Settings

The test benchmark used in the experiment is TPC-H. The query templates are 2, 4, 5, 6, 10, 11, 12, 13, 15, 18, 19, 21, covering various types such as light size, medium size and heavy size queries. The database is PostgreSQL 9.5.5 and runs on a Centos 7 server. The size of the database table is 1G and 10G, and the corresponding buffer pool size is 300M and 2000M. The rest is the initial configuration of the database. The programming environment is Eclipse, running on Windows 2008 Server operating system, the processor is Inter(R) Xeon(R) CPU E3-1225 V2 @3.2 GHz, and the memory is 8G. In this paper, the response time of the medium size queries is modeled and predicted by the TP model, compared with the previous B2L model. The effectiveness of the proposed method is proved when predicting the response time of the same type of medium size query. If other types of queries appear in multiple queries in parallel, this paper uses TP model to model and predict the response time of the query, which proves that the proposed method can maintain high prediction accuracy when different types of queries are in parallel. This paper uses the ARE (average relative error) to judge the prediction performance. The expression for ARE is as follows:

$$ARE = \frac{1}{n} \sum_{i=1}^{n} \frac{|PredictedValue_i - MeasuredValue_i|}{MeasuredValue_i} \tag{6}$$

In the above formula, $MeasuredValue_i$ represents the true value of the query i response time, and $PredictedValue_i$ represents the predicted value of the query i response time. From the above formula, we can see that the smaller the ARE value, the higher the prediction accuracy.

5.2 Performance Evaluation

Predicting the Response Time for Medium Size Queries. This experiment models and predicts the response time of queries 5, 10, 11, 12, 13, and 19 under different parallelisms MPL = 2, 3, 4 when table size is 1G, 10G. As the establishment of the

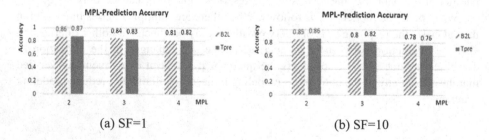

(a) SF=1   (b) SF=10

Fig. 1. The prediction precision for medium size queries when SF = 1 and 10.

B2L model, the sample selection method is the Latin hypercube sampling method to control the uniqueness of the variables. The experimental results are shown in the following figure:

It can be seen from Fig. 1(a) that when SF (scale factor) = 1, that is the data table capacity is 1G, the average prediction accuracy of the response time of the medium size queries predicted by the TP model can reach 84%, which is equivalent to the average prediction accuracy of the previous B2L model of 83.6%.

It can be seen from Fig. 1(b) that when SF = 10, that is the data table capacity is 10G, the average prediction accuracy of the response time of the medium size queries predicted by the TP model can reach 81.3%, which is equivalent to the average prediction accuracy of the previous B2L model of 81%.

Prediction of Response Time for Light Size and Heavy Size Queries. Assuming that the query templates are 2, 4, 6, 11, 12, 18, 19, 21, this experiment models and predicts the response time of light size queries 2, 4, 6 and heavy size queries 18, 21 under different parallelisms MPL = 2, 3, 4 when data table size is 1G, 10G. The sample selection method also uses the Latin hypercube sampling method. The experimental results are shown in Fig. 2.

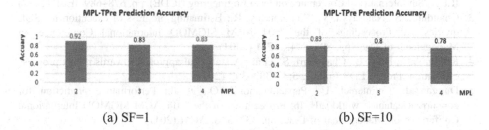

(a) SF=1   (b) SF=10

Fig. 2. The prediction precision for light and heavy size queries.

It can be seen from Fig. 2(a) that when SF = 1, that is the data table capacity size is 1G, the average prediction accuracy of the response time of the light size and heavy size queries predicted by the TP model is as high as 86%. This is because in TP model, we not only use QueryRating to quantify the query interactions, but also map the response time from two concurrent queries to multiple concurrent queries through TMT.

It can be seen from Fig. 2(b) that when SF = 10, that is the data table capacity size is 10G, the average prediction accuracy of the response time of the light size and heavy size query predicted by the TP model is 80.3%. The reason why the accuracy rate has dropped is that as the amount of data increases, the interactions between queries becomes more and more complicated.

6 Conclusions

This paper systematically quantifies query interactions based on QueryRating. Through the response time mapping $TMT_{p/q}$ from two queries to multiple queries, this paper proposes a modeling method TP, which is suitable for response time prediction of various types of queries. There is no need to consider the query size before modeling, and it is more in line with real needs. The future direction can be the performance model for distributed databases.

Acknowledgement. This work is supported by the National Natural Science Foundation of China under Grant 61572345.

References

1. Ahmad, M., Aboulnaga, A., Babu, S., et al.: Modeling and exploiting query interactions in database systems. In: Proceedings of the 17th ACM Conference on Information and Knowledge Management, pp. 183–192. ACM (2008)
2. Li, J., Nehme, R.V., Naughton, J.: GSLPI: a cost-based query progress indicator. In: 2012 IEEE 28th International Conference on Data Engineering (ICDE), pp. 678–689. IEEE (2012)
3. Chaudhuri, S., Narasayya, V., Ramamurthy, R.: Estimating progress of execution for SQL queries. In: Proceedings of the 2004 ACM SIGMOD International Conference on Management of Data, pp. 803–814. ACM (2004)
4. König, A.C., Ding, B., Chaudhuri, S., et al.: A statistical approach towards robust progress estimation. Proc. VLDB Endow. **5**(4), 382–393 (2011)
5. Duggan, J., Cetintemel, U., Papaemmanouil, O., et al.: Performance prediction for concurrent database workloads. In: Proceedings of the 2011 ACM SIGMOD International Conference on Management of Data, pp. 337–348. ACM (2011)
6. Luo, G., Naughton, J.F., Ellmann, C.J., et al.: Toward a progress indicator for database queries. In: Proceedings of the ACM SIGMOD International Conference on Management of Data, Paris, France, pp. 791–802. ACM (2004)
7. Hacigumus, H., Chi, Y., Wu, W., et al.: Predicting query execution time: are optimizer cost models really unusable? In: Proceedings of the 29th International Conference on Data Engineering, Brisbane, Australia, pp. 1081–1092. IEEE (2013)
8. Wu, W., Chi, Y., Hacígümüş, H., et al.: Towards predicting query execution time for concurrent and dynamic database workloads. Proc. VLDB Endow. **6**(10), 925–936 (2013)
9. Ahmad, M., Duan, S., Aboulnaga, A., et al.: Predicting completion times of batch query workloads using interaction-aware models and simulation. In: Proceedings of the 14th International Conference on Extending Database Technology, pp. 449–460. ACM (2011)
10. Sheikh, M.B., Minhas, U.F., Khan, O.Z., et al.: A Bayesian approach to online performance modeling for database appliances using gaussian models. In: Proceedings of the 8th ACM International Conference on Autonomic Computing, pp. 121–130. ACM (2011)
11. Wu, W., Wu, X., Hacigümüş, H., et al.: Uncertainty aware query execution time prediction. Proc. VLDB Endow. **7**(14), 1857–1868 (2014)
12. Amjad, M., Zhang, J.: Gscheduler: a query scheduler based on query interactions. In: Meng, X., Li, R., Wang, K., Niu, B., Wang, X., Zhao, G. (eds.) WISA 2018. LNCS, vol. 11242, pp. 395–403. Springer, Cham (2018). https://doi.org/10.1007/978-3-030-02934-0_36

13. Ahmad, M., Aboulnaga, A., Babu, S., et al.: Interaction-aware scheduling of report-generation workloads. VLDB J. **20**(4), 589–615 (2011)
14. Luo, G., Naughton, J.F., Yu, P.S.: Multi-query SQL Progress indicators. In: Ioannidis, Y., et al. (eds.) EDBT 2006. LNCS, vol. 3896, pp. 921–941. Springer, Heidelberg (2006). https://doi.org/10.1007/11687238_54
15. Ganapathi, A., Kuno, H., Dayal, U., et al.: Predicting multiple metrics for queries: better decisions enabled by machine learning. In: IEEE 25th International Conference on Data Engineering, ICDE 2009, pp. 592–603. IEEE (2009)
16. Markl, V.G., Haas, P.J., Aboulnaga, A.I., et al.: System and method for updating database statistics according to query feedback: U.S. Patent 7,831,592, 9 November 2010
17. Zhang, J., Niu, B.: A clustering-based sampling method for building query response time models. Comput. Syst. Sci. Eng. **32**(4), 319–331 (2017)
18. Zhang, J., Niu, B., Liu, A.: An optimized sampling method for query interaction aware respond time modeling. J. Chin. Mini-Micro Comput. Syst. **36**(10), 2240–2244 (2015)

Application of Patient Similarity in Smart Health: A Case Study in Medical Education

Kalkidan Fekadu Eteffa$^{(\boxtimes)}$ [ID], Samuel Ansong, Chao Li, Ming Sheng, Yong Zhang, and Chunxiao Xing

Research Institute of Information Technology, Beijing National Research Centre for Information Science and Technology, Department of Computer Science and Technology, Institute of Internet Industry, Tsinghua University, Beijing 100084, China
{ajql8, ssml8}@mails.tsinghua.edu.cn, {li-chao, shengming, zhangyong05, xingcx}@tsinghua.edu.cn

Abstract. Patient similarity relies on computations that synthesize EHRs (Electronic Health Records) to give personalized predictions, which inform diagnoses and treatments. Given the complexities in pre-processing EHRs, representing patient data and utilizing the most suitable similarity metrics and evaluation methods, patient similarity computations are far from the era of regular use in hospitals. This paper aims to both support further patient similarity research and to inform the importance of its application in medical education. It accomplishes this by examining relevant literature that offer techniques to tackle the computational challenges and by presenting their various applications in the healthcare industry.

Keywords: Patient similarity · EHR (Electronic Health Record) · Medical education

1 Introduction

A patient similarity computation uses collected medical data to find a cohort of patients with similar attributes to any given patient. The application of patient similarity computation in medical education has the potential to help medical students deeply understand complex medical concepts. For instance, Using patient similarity, students can easily compare and analyze all the similar patients encountered in the past and the treatments doctors used for each case. In lectures, professors could provide a plethora of examples for particular illnesses or display trends among patients afflicted by the same disease. In hospitals, physicians would have a reliable archive of past medical cases to arrive at informed treatment decisions.

Over several years, researchers have applied complex algorithms to EHRs to produce meaningful information from the raw data - i.e., generating predictions that guide diagnoses, treatment methods, and so on. However, often obtaining results is difficult because EHRs are large, complex, and differently formatted. Moreover, much of the data is redundant or irrelevant. This paper examines current research into patient similarity computation by presenting challenges, their solutions and various applications in the healthcare system.

W. Ni et al. (Eds.): WISA 2019, LNCS 11817, pp. 714–719, 2019.
https://doi.org/10.1007/978-3-030-30952-7_72

In this paper, Sect. 2 presents the methodology adopted for literature selection. Section 3 addresses the techniques that researchers apply to overcome the many computational challenges aforementioned. Section 4 points to the various ways in which the healthcare industry already uses patient similarity computation. Section 5 identifies a direction for future work, while Sect. 6 concludes the case study.

2 Methodology

The search for patient similarity literature was conducted through DBLP and Google Scholar using the keyword "patient similarity". The selected papers are from top journals and conferences with publication dates between January 2014 and March 2019.

3 Techniques Used in Patient Similarity Computations

Figure 1 displays the main elements of building a patient similarity computation, beginning by collecting EHRs from several hospitals. The data then enters the pre-processing stage, which is responsible for setting a unified format, removing redundancy, and selecting useful features. Researchers then use different techniques to represent a single patient; they will then utilize the patient representation as an input for the chosen similarity metric. After applying the similarity metric, researchers evaluate whether or not they have met their objectives. Lastly, the designed patient similarity computations are applied in different fields.

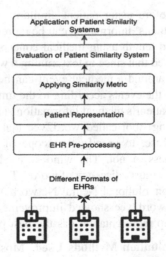

Fig. 1. Layout of patient similarity computations

The techniques used in the main elements of patient similarity computation can be classified into functional and non-functional components. The functional components include techniques used in data pre-processing, patient representation and similarity metric and evaluation. The non-functional components include scalability and other important decisions made, such as whether to consider time as a variable or to address the possibility of encountering patients afflicted with multiple diseases.

3.1   Functional Components of Patient Similarity

Electronic Health Record Pre-processing. EHRs contain patients' information including demographic information, laboratory test results, diagnoses, and billing information. Working with EHRs, therefore, presents a unique difficulty. Initially, EHRs collected from various hospitals do not have the same format. Because of this, researchers who want to use records from multiple hospitals have to develop a formatting standard. Barkhordari et al. [1] collected information from multiple sources that use different formats and converted data to the unified standard that fits their MapReduce architecture. In order to then clean the EHRs with the same format, sparse feature selections have been used to choose the features that are related to the task at hand. It is important to note that there is a dependency among features, and in order to avoid redundancy, we need to make sure that the uncorrelated features are selected. Huai et al. [2] suggested a way researchers can clean the data to remove redundant and irrelevant information, while still maintaining the information that is relevant and has lower correlations with other information. The authors did this by calculating the correlation coefficients between features and placing rules that restrict those features with higher correlations from being chosen in the same selection.

Patient Representation. Before using patient similarity metrics, it is important to address the way in which patients' information can be represented. Zhu et al. [3] gave a vector representation of fixed length using medical concept embedding that considers time. The authors hypothesized that since in text mining a word can be predicted by the context, events that happen around a medical event can be used as medical contexts. Zhu et al. [3] summed up all the events according to the time progression and obtained a paragraph that explains a patient's medical information. The authors then used a NLP (Natural Language Processing) tool called word2vec that uses a sliding window of adjustable length, in their case, to specify the scope of the context depending on whether it is a chronic illness or not. The window used in this method considers timestamps themselves and not just adjacency of words. They also combined word2vec embeddings with CNNs (Convolutional Neural Networks) and derived a deep representation. They found that word2vec showed improvement as compared to one-hot representations, whereas deep representation has the best performance.

Common Metrics and Evaluation Methods Used. Most similarity methods apply similarity calculations on the patient vectors. Cosine similarity has been widely used as a patient similarity metric. Lee [4] used the representation vector of two patients and found the cosine angle while some, like [5], used Euclidean Distance, Pearson Correlation, and Tanimoto Similarity. Zhu et al. [3] implemented an unsupervised method that utilized RV and dCov coefficient and supervised model using CNN.

For evaluation, researchers define methods based on the main objective of their research. Lee [4] measured prediction performance using the area under the precision-recall curve (AUPRC) and the area under the receiver operating characteristic curve (AUROC), while [5] used precision, recall, and F1 to measure the relevancy and completeness of their recommendation.

3.2 Non-functional Components of Patient Similarity

Temporal vs. Non-temporal Patient Similarity Calculations. Researchers are increasingly considering time as part of their patient similarity computation to follow the progression of diseases in similar patients. As we saw in the patient representation section, [3] arranged medical events using timestamps and used time to determine the scope of their sliding window. Other researchers have used time in their calculations as another variable to determine if patients are indeed similar; most patient similarity metrics consider similarity at one point in time, which may not provide the full picture. Unfortunately, there are no uniform sampling schedules followed by doctors. For instance, Sha et al. [6] pointed out that hospitals more closely observe patients in critical conditions than those that are not. Sha et al. [6] assumed that similar patients' laboratory tests follow similar procedures with comparable time-intervals and display similar results. They implemented a modified version of the Smith-Waterman algorithm and evaluated whether their temporal method has better predictive power than non-temporal methods. Their method achieved higher sensitivity and F-measure scores in predicting mortality in septic patients. It did not achieve high specificity, but they argued that higher sensitivity is more useful because it can help us avoid adverse consequences. Other researchers bring forth a mtTSML (multi-task triplet constrained sparse metric learning method) [7]. In this method, patient similarity is learned based on different timestamps, simultaneously. The results showed that multi-task performance is more efficient than single-task and that mtTSML is the best method for multi-task methods.

Scalability. EHRs contain large amounts of information that continues to grow exponentially. This makes patient similarity calculations a big data problem. Some researchers suggested a scalable solution that uses MapReduce and collects data from different sources with varying formats [1]; others carried out the data pre-processing and patient similarity analysis within the Database Management System (DBMS) [8]. The application of big data is very promising in the health field [9], therefore, researchers can make their computations more scalable by using these techniques.

Considering Multiple Diseases. Often patients do not have one disease, they have many. Zhao et al. [10] accounted for many diseases in patients by creating a similarity label representation that takes into account 197 diseases and diagnoses of a pair of patients. If a patient has the disease, then it is labeled as 1 and 0 otherwise.

4 Applications of Patient Similarity Systems

Researchers utilize patient similarity computations in various ways: Chomutare [5] recognized social media as a self-managing tool for chronic diseases and assigned top-N threads by applying patient similarity computation on community structures; others used this computation to identify cohorts of patients that share the same diseases. Among patients afflicted with AF (Atrial Fibrillation), [11] identified and determined the characteristics of a patient group with a low risk of experiencing an IS (Ischemic Stroke). Li et al. [12] considered a complex disease like Type 2 Diabetes and found subgroups and the characterization of these subgroups using patient-patient similarity network.

In regards to effectively designing treatments and appropriately allocating resources in ICUs (Intensive Care Units), [4] researchers have also created an ICU mortality prediction. In the pharmaceutical field, Keshava [13] identified patients who experience adverse drug events after taking Nifedipine. Wang et al. [14] designed a prototype that provides a treatment plan that derives its recommendations from similar patients' data. All in all, these applications decrease medical errors and, in turn, prevent adverse care.

5 Future Work

Though patient similarity shows promising applications in healthcare, its widespread use requires clear interpretations of similarity results. Knowledge Graphs may improve clarity as it not only displays the similarity score among patients but also explains the reasons behind the similarities [15]. For further research, Knowledge Graphs may provide better performance if designed to take into account the progression of diseases.

6 Conclusion

This paper introduces challenges associated with patient similarity computation and showcases the techniques that researchers use to overcome them. In doing so, researchers are advancing a system that can transform healthcare and medical education as the results of this analysis could be a valuable resource to medical students, medical doctors as well as medical researchers. Given the practical benefits of patient similarity computations, overcoming the challenges, furthering techniques, and improving interpretation of results with the use of Knowledge Graphs are essential for its widespread use in hospitals and research.

Acknowledgement. This work was supported by NSFC (91646202), National Key R&D program of China (2018YFB1404400, 2018YFB1402700).

References

1. Barkhordari, M., Niamanesh, M.: ScaDiPaSi: an effective scalable and distributable mapreduce-based method to find patient similarity on huge healthcare networks. Big Data Res. **2**(1), 19–27 (2015)
2. Huai, M., Mino, C., Suo, Q., Li, Y., Gao, J., Zhang, A.: Uncorrelated patient similarity learning. In: SIAM International Conference on Data Mining, pp. 270–278 (2018)
3. Zhu, Z., Yin, C., Qian, B., Cheng, Y., Wei, J., Wang, F.: Measuring patient similarities via a deep architecture with medical concept embedding. In: IEEE 16th International Conference on Data Mining (ICDM), pp. 749–758 (2016)
4. Lee, J.: Personalized mortality prediction for the critically ill using a patient similarity metric and bagging. In: IEEE-EMBS International Conference on Biomedical and Health Informatics (BHI), pp. 332–335 (2016)
5. Chomutare, T.: Patient similarity using network structure properties in online communities. In: IEEE-EMBS International Conference on Biomedical and Health Informatics (BHI), pp. 809–812 (2014)
6. Sha, Y., Venugopalan, J., Wang, M.: A novel temporal similarity measure for patients based on irregularly measured data in electronic health records. In: 7th ACM International Conference on Bioinformatics, Computational Biology, and Health Informatics, pp. 337–244 (2016)
7. Suo, Q., Zhong, W., Ma, F., Yuan, Y., Huai, M., Zhong, A.: Multi-task sparse metric learning for monitoring patient similarity progression. In: 2018 IEEE International Conference on Data Mining (ICDM), pp. 477–486 (2018)
8. Tashkandi, A., Wiese, I., Wiese, L.: Efficient in-database patient similarity analysis for personalized medical decision supports systems. Big Data Res. **13**, 52–64 (2018)
9. Wang, X., Hu, Q., Zhang, Y., Zhang, G., Juan, W., Xing, C.: A kind of decision model research based on big data and blockchain in eHealth. In: Meng, X., Li, R., Wang, K., Niu, B., Wang, X., Zhao, G. (eds.) WISA 2018. LNCS, vol. 11242, pp. 300–306. Springer, Cham (2018). https://doi.org/10.1007/978-3-030-02934-0_28
10. Zhao, F., Xu, J., Lin, Y.: Similarity measure for patients via a siamese CNN network. In: Vaidya, J., Li, J. (eds.) ICA3PP 2018. LNCS, vol. 11335, pp. 319–328. Springer, Cham (2018). https://doi.org/10.1007/978-3-030-05054-2_25
11. Liu, H., Li, X., Xie, G., Du, X., Zhong, P., Gu, C.; Precision cohort finding with outcome-driven similarity analytics: a case study of patients with atrial fibrillation. In: 16th World Congress on Medical and Health Informatics, Hangzhou, China (2017)
12. Li, L., Cheng, W., Glickberg, B., Gottesman, O., Tamler, R., Chen, R.: Identification of type 2 diabetes subgroups through topological analysis of patient similarity. Sci. Transl. Med. **7** (311ra174) (2015)
13. Keshava, N.: Measuring the performance of an integrative patient similarity measure in the context of adverse drug events. In: IEEE EMBS International Conference on Biomedical & Health Informatics (BHI), pp. 93–96 (2017)
14. Wang, Y., Tian, Y., Tian, L., Qian, Y., Li, J.: An electronic medical record system with treatment recommendation based on patient similarity. J. Med. Syst. **39**, 55 (2015)
15. Wang, X., Wang, D., Xu, C., He, X., Cao, Y., Chua, T.: Explainable reasoning over knowledge graphs for recommendation. CoRRabs (2018)

Author Index

Amjad, Muhammad 704
An, Chengzhi 3
Ansong, Samuel 518, 714

Bai, Jiayang 690

Cao, Bin 79
Cao, Keyan 113, 575, 677
Cao, Weiwei 257
Chang, Xingong 40
Chang, Yaomin 79
Chen, Chi-Hua 98
Chen, Chuan 79
Chen, Jing 239
Chen, Jinmei 368
Chen, Limin 506
Cheng, Zhi 28

Dong, Yongquan 85
Du, Xuewu 662
Duan, Qiang 3

Eteffa, Kalkidan Fekadu 518, 714

Fan, Jili 560
Fan, Ying 201
Fang, Jun 569
Feng, Li 449
Feng, Shuo 397
Feng, Yixuan 169
Feng, Zhigang 467

Gao, Fei 512
Gao, Jianye 226
Gao, Ming 3
Gao, Yanfang 176
Ge, Yu 449
Gu, Qing 85
Gu, Yu 213
Guan, Heng 72
Guo, Canyang 98
Guo, Wenzhong 98
Guo, Xiefan 418

Han, Daoqi 15
Han, Yanbo 257
Han, Ziyang 113, 575, 677
He, Tieke 92, 309
He, Xiaomei 340
He, Xincheng 638
He, Ying 430
Hu, Yang 85
Huang, Mengxing 47
Huang, Shaoqiong 47
Huang, Taoyi 492
Huang, Tianhao 298
Huang, Xinlin 638

Ji, Cong 467
Ji, Yuhang 113
Jing, Lin 443
Jingbin, Wang 443

Kang, Kai 201
Kening, Gao 449
Kong, Chao 499
Kou, Yue 397, 404

Lee, Wang-Chien 347
Li, Binger 638
Li, Chao 169, 512, 518, 714
Li, Cheng 189
Li, Feifei 638
Li, Gui 113, 575, 677
Li, Han 270
Li, Hao 499
Li, Jiajia 368
Li, Jian 455
Li, Jingdong 601
Li, Menglong 47
Li, Qi 569
Li, Rui 3
Li, Sizhuo 418
Li, Xiang 397
Li, Xiaohua 560
Li, Xin 151, 548, 601
Li, Yanhui 138

Li, You   492
Li, Yuzheng   79
Li, Zhengyu   113, 575, 677
Li, Zhenxing   79, 169
Lian, Hao   92
Lian, Zengshen   169
Liang, Ye   298
Lin, Li   430
Lin, Nanzhou   361
Lin, Yuming   492
Liu, Chen   257
Liu, ChengHou   28
Liu, Genggeng   98
Liu, Hai   285
Liu, Huimin   72
Liu, Jianye   499
Liu, Jiaqiang   382
Liu, Lanfeng   690
Liu, Lin   650
Liu, Pengkai   418
Liu, Qing   15
Liu, Tao   499
Liu, Wei   480
Liu, Xiangyu   368
Liu, Xiaojin   690
Liu, Xinyue   138
Liu, Xueyan   340
Liu, Yajun   582
Liu, Yali   85
Liu, Ying   499
Liu, Yuxin   404
Lu, Tingting   340
Luo, Haiqi   327
Lv, Yan   455

Ma, Jingjiao   226
Ma, Wenkai   677
Ma, Xiao   126
Ma, Xuan   119, 527
Ma, Youzhong   589

Ni, Weiwei   382
Nie, Tiezheng   397, 404, 560
Niu, Baoning   704

Ouyang, Hongrong   690

Pan, Heng-Xi   285
Pan, Zhenkuan   430
Pang, Jun   213
Pei, Zefeng   704
Peng, Wei   151

Qu, Wenwen   601

Ren, Kai   582
Rong-yi, Cui   319

Sha, Guangtao   662
Shan, Xiaohuan   226
Shao, Hua   189
Shao, Yuyao   512
Shen, Derong   397, 404
Shen, Yuan   361
Sheng, Jia   569
Sheng, Ming   512, 518, 714
Shi, Wenqiang   40
Shi, Yilin   201
Song, Baoyan   163, 226
Song, Hao   492
Song, Shiyuan   327
Song, Wei   361
Sun, Jing   614
Sun, Ping   575
Sun, Yan   361

Tang, Yong   285

Wan, Jiabing   92, 309
Wang, Bin   347, 614
Wang, Hui   92, 309
Wang, Jian   169
Wang, Jing   644
Wang, Jingwen   512
Wang, Leixia   347
Wang, Ranran   126
Wang, Xiangcheng   3
Wang, Xiaoling   601
Wang, Xin   98, 418
Wang, Yan   163
Wang, Yun   169
Wang, Zhikai   626
Wang, Zhi-Qi   650

Wang, Zihao   644
Weng, Lifen   59
Wu, Ning   382
Wu, Yongqi   251
Wu, Yu   361

Xia, Chunyu   309
Xia, Xiufeng   368
Xian-yan, Meng   319
Xie, Linfeng   467
Xie, Xiaoqi   59
Xing, Chunxiao   169, 512, 518, 548, 714
Xing, Qingbin   201
Xu, Bin   22
Xu, Chao   327, 382
Xu, Huarong   59
Xu, Lei   239, 626, 638, 690
Xu, Li   119, 527
Xu, Lili   298
Xu, Lizhen   119, 527
Xu, Wei   684
Xu, Zhuoming   455, 684
Xu, Zihuan   92
Xu, Zixuan   226
Xue, Sixin   548

Ya-hui, Zhao   319
Yan, Sheng   22
Yan, Yongming   251
Yang, Bingqing   163
Yang, Dan   22
Yang, Liu   506
Yang, Nan   15
Yang, Xiandi   361
Yang, Xiaochun   347, 614
Yang, Xiaotao   340
Yao, Lan   176, 251
Yin, Qingshan   3
Ying, Yi   582
Yu, Bin   270
Yu, Ge   189, 560
Yu, Jinyue   176
Yu, Minghe   72, 213, 251
Yu, Yaowei   455

Yu, Zhang   449
Yun, Hongyan   430

Zeng, Hui   163
Zhang, Chen   28
Zhang, Fei   40
Zhang, Guigang   169, 548
Zhang, Han   512
Zhang, Jinwen   704
Zhang, Juntao   361
Zhang, Junyao   15
Zhang, Mingxin   480, 662
Zhang, Ran   418
Zhang, Ruiling   589
Zhang, Tiancheng   189
Zhang, Xin   28
Zhang, Xinlei   151
Zhang, Xiuhua   430
Zhang, Yan   506
Zhang, Yin   126
Zhang, Yong   512, 518, 714
Zhang, Yongxin   589
Zhang, Yu   47
Zhang, Yun   560
Zhao, Bo   684
Zhao, Heng   239
Zhao, Ling   644
Zhao, Ruixue   539
Zhao, Ting   270
Zhao, Xu   548
Zhao, Zhibin   72
Zhao, ZhongQi   28
Zhenguo, Zhang   319
Zhou, Hang   690
Zhou, Jian-Tao   650
Zhou, Yuhang   15
Zhou, Zhichao   251
Zhu, Haibei   499
Zhu, Huaijie   347
Zhu, Rui   368
Zhu, Yongjin   467
Zhu, Zhu   163
Zong, Chuanyu   368
Zong, Yunbing   3
Zou, Yunfeng   327, 382

Printed in the United States
By Bookmasters